Aerodynamic Theory

A General Review of Progress

Under a Grant of the Guggenheim Fund for the Promotion of Aeronautics

William Frederick Durand

Editor-in-Chief

Volume VI

Div. P · Airplane as a Whole · W. F. Durand
Div. Q · Aerodynamics of Airships · Max M. Munk
Div. R · Performance of Airships · K. Arnstein and W. Klemperer
Div. S · Hydrodynamics of Boats and Floats · E. G. Barrillon
Div. T · Aerodynamics of Cooling · H. L. Dryden

With 127 Figures and 2 Plates

Berlin · Julius Springer · 1936

ISBN-13: 978-3-642-89627-9 e-ISBN-13: 978-3-642-91484-3
DOI: 10.1007/978-3-642-91484-3

GENERAL PREFACE

During the active life of the Guggenheim Fund for the Promotion of Aeronautics, provision was made for the preparation of a series of monographs on the general subject of Aerodynamic Theory. It was recognized that in its highly specialized form, as developed during the past twenty-five years, there was nowhere to be found a fairly comprehensive exposition of this theory, both general and in its more important applications to the problems of aeronautic design. The preparation and publication of a series of monographs on the various phases of this subject seemed, therefore, a timely undertaking, representing, as it is intended to do, a general review of progress during the past quarter century, and thus covering substantially the period since flight in heavier than air machines became an assured fact.

Such a present taking of stock should also be of value and of interest as furnishing a point of departure from which progress during coming decades may be measured.

But the chief purpose held in view in this project has been to provide for the student and for the aeronautic designer a reasonably adequate presentation of background theory. No attempt has been made to cover the domains of design itself or of construction. Important as these are, they lie quite aside from the purpose of the present work.

In order the better to suit the work to this main purpose, the first volume is largely taken up with material dealing with special mathematical topics and with fluid mechanics. The purpose of this material is to furnish, close at hand, brief treatments of special mathematical topics which, as a rule, are not usually included in the curricula of engineering and technical courses and thus to furnish to the reader, at least some elementary notions of various mathematical methods and resources, of which much use is made in the development of aerodynamic theory. The same material should also be acceptable to many who from long disuse may have lost facility in such methods and who may thus, close at hand, find the means of refreshing the memory regarding these various matters.

The treatment of the subject of Fluid Mechanics has been developed in relatively extended form since the texts usually available to the technical student are lacking in the developments more especially of interest to the student of aerodynamic theory. The more elementary treatment by the General Editor is intended to be read easily by the average technical graduate with some help from the topics comprised in Division A. The more advanced treatment by Dr. Munk will call for some familiarity with space vector analysis and with more advanced mathematical methods, but will commend itself to more advanced students by the elegance of such methods and by the generality and importance of the results reached through this generalized three-dimensional treatment.

In order to place in its proper setting this entire development during the past quarter century, a historical sketch has been prepared by Professor Giacomelli whose careful and extended researches have resulted in a historical document which will especially interest and commend itself to the study of all those who are interested in the story of the gradual evolution of the ideas which have finally culminated in the developments which furnish the main material for the present work.

The remaining volumes of the work are intended to include the general subjects of: The aerodynamics of perfect fluids; The modifications due to viscosity and compressibility; Experiment and research, equipment and methods; Applied airfoil theory with analysis and discussion of the most important experimental results; The non-lifting system of the airplane; The air propeller; Influence of the propeller on the remainder of the structure; The dynamics of the airplane; Performance, prediction and analysis; General view of airplane as comprising four interacting and related systems; Airships, aerodynamics and performance; Hydrodynamics of boats and floats; and the Aerodynamics of cooling.

Individual reference will be made to these various divisions of the work, each in its place, and they need not, therefore, be referred to in detail at this point.

Certain general features of the work editorially may be noted as follows:

1. Symbols. No attempt has been made to maintain, in the treatment of the various Divisions and topics, an absolutely uniform system of notation. This was found to be quite impracticable.

Notation, to a large extent, is peculiar to the special subject under treatment and must be adjusted thereto. Furthermore, beyond a few symbols, there is no generally accepted system of notation even in any one country. For the few important items covered by the recommendations of the National Advisory Committee for Aeronautics, symbols have been employed accordingly. Otherwise, each author has developed his system of symbols in accordance with his peculiar needs.

At the head of each Division, however, will be found a table giving the most frequently employed symbols with their meaning. Symbols in general are explained or defined when first introduced.

2. General Plan of Construction. The work as a whole is made up of *Divisions*, each one dealing with a special topic or phase of the general subject. These are designated by letters of the alphabet in accordance with the table on a following page.

The Divisions are then divided into chapters and the chapters into sections and occasionally subsections. The Chapters are designated by Roman numerals and the Sections by numbers in bold face.

The Chapter is made the unit for the numbering of sections and the section for the numbering of equations. The latter are given a double number in parenthesis, thus (13.6) of which the number at the left of the point designates the section and that on the right the serial number of the equation in that section.

Each page carries at the top, the chapter and section numbers.

W. F. Durand

Stanford University, California
January, 1934.

GENERAL LIST OF DIVISIONS WITH AUTHORS

Volume I.

A. Mathematical Aids

W. F. DURAND — Professor (Emeritus) of Mechanical Engineering, Stanford University, Calif., Member of the National Advisory Committee for Aeronautics.

B. Fluid Mechanics, Part I

W. F. DURAND

C. Fluid Mechanics, Part II

MAX M. MUNK — Lecturer in Aerodynamics at the Catholic University of America, Washington, D. C., and Technical Editor of the "Aero Digest".

D. Historical Sketch

R. GIACOMELLI — Lecturer in History of Mechanics at the University of Rome, Italy, and Editor of "L'Aerotecnica".

with the collaboration of

E. PISTOLESI — Professor of Mechanics at the Royal School of Engineering at Pisa, Italy, and Editor-in-Chief of "L'Aerotecnica".

Volume II.

E. General Aerodynamic Theory—Perfect Fluids

TH. VON KÁRMÁN — Director of the Guggenheim Aeronautics Laboratory, California Institute of Technology, Pasadena, Calif., and formerly Director of the Aerodynamic Institute. Aachen, Germany.

J. M. BURGERS — Professor of Aero- and Hydrodynamics at the Technische Hoogeschool at Delft, Holland.

Volume III.

F. The Theory of Single Burbling

C. WITOSZYŃSKI — Professor of Aerodynamics at the Warsaw Polytechnical School and Director of the Warsaw Aerodynamic Institute, Poland.

M. J. THOMPSON — Assistant Professor of Aeronautical Engineering at the University of Michigan, Ann Arbor, Mich.

G. The Mechanics of Viscous Fluids

L. PRANDTL — Professor in Applied Mechanics at the University of Göttingen, Germany, and Director of the Kaiser Wilhelm Institute for Fluid Research.

H. The Mechanics of Compressible Fluids

G. I. TAYLOR — Yarrow Research Professor of the Royal Society, Fellow of Trinity College, Cambridge, England.

J. W. MACCOLL — Research Officer, Department of External Ballistics, Ordnance Committee, Woolwich, England.

I. Experimental Methods—Wind Tunnels

A. TOUSSAINT — Director of the Aerodynamic Laboratory, Saint-Cyr-l'École, France.

E. JACOBS — Associate Aeronautical Engineer, in charge of the National Advisory Committee for Aeronautics' variable-density wind tunnel, Langley Field, Virginia.

Volume IV.

J. Applied Airfoil Theory
A. BETZ — Professor at the University and Director of the Aerodynamic Research Institute at Göttingen, Germany.

K. Airplane Body (Non-Lifting System) Drag and Influence on Lifting System
C. WIESELSBERGER — Professor of Aerodynamics and Director of the Aerodynamic Institute, Technische Hochschule, Aachen, Germany.

L. Airplane Propellers
H. GLAUERT[1] — Past Fellow of Trinity College, Cambridge, England; Principal Scientific Officer at the Royal Aircraft Establishment, Farnborough.

M. Influence of the Propeller on other Parts of the Airplane Structure
C. KONING — Rijks-Studiedienst voor de Luchtvaart, Amsterdam, Holland.

Volume V.

N. Dynamics of the Airplane
B. MELVILL JONES — Professor of Aeronautical Engineering in the University of Cambridge, England, Member of the Aeronautical Research Committee of Great Britain.

O. Airplane Performance
L. V. KERBER — Former Chief Aerodynamics Branch Materiel Division, U. S. Army Air Corps, and former Chief, Engineering Section Aeronautics Branch, Department of Commerce.

Volume VI.

P. Airplane as a Whole—General View of Mutual Interactions Among Constituent Systems
W. F. DURAND[2] — Professor (Emeritus) of Mechanical Engineering, Stanford University, Calif., Member of the National Advisory Committee for Aeronautics.

Q. Aerodynamics of Airships
MAX M. MUNK — Lecturer in Aerodynamics at the Catholic University of America, Washington, D. C., and Technical Editor of the "Aero Digest".

R. Performance of Airships
K. ARNSTEIN — Chief Engineer of the Goodyear Zeppelin Company, Akron, Ohio.

W. KLEMPERER — Research Engineer of the Goodyear Zeppelin Company, Akron, Ohio.

S. Hydrodynamics of Boats and Floats
E. G. BARRILLON — Director of the Naval Experimental Tank, Paris, France.

T. Aerodynamics of Cooling
H. L. DRYDEN — Physicist in the United States Bureau of Standards, Chief of the Aerodynamics Section, Washington, D. C.

---

[1] Deceased August 4, 1934.

[2] In the original plan, it was expected that this Division would be prepared by Professor M. Panetti of the R. Scuola di Ingegneria of Turin. Unfortunately, at the last, Professor Panetti found himself unable to give the needed time for this work, and in order not to delay publication, the General Editor has undertaken to prepare a brief treatment of the subject.

CONTENTS

DIVISION P

AIRPLANE AS A WHOLE — GENERAL VIEW OF MUTUAL INTERACTIONS AMONG CONSTITUENT SYSTEMS

By **W. F. Durand,**

Professor (Emeritus) of Mechanical Engineering, Stanford University, Calif., Member of the National Advisory Committee for Aeronautics

DIVISION Q

AERODYNAMICS OF AIRSHIPS

By **Max M. Munk,**

Lecturer in Aerodynamics at the Catholic University of America, Washington, D. C., and Technical Editor of the "Aero Digest"

DIVISION R

PERFORMANCE OF AIRSHIPS

By **K. Arnstein,**

Chief Engineer of the Goodyear Zeppelin Company, Akron, Ohio

and **W. Klemperer,**

Research Engineer of the Goodyear Zeppelin Company, Akron, Ohio

DIVISION S

HYDRODYNAMICS OF BOATS AND FLOATS

By **E. G. BARRILLON,**

Director of the Naval Experimental Tank, Paris, France

DIVISION T

AERODYNAMICS OF COOLING

By **H. L. Dryden,**

Physicist in the United States Bureau of Standards, Chief of the Aerodynamics Section, Washington, D.C.

NOTATION

The following table comprises a list of the principal notations employed in the present Volume. Notations not listed are either so well understood as to render mention unnecessary, or are only rarely employed and are explained as introduced. Where occasionally a symbol is employed with more than one meaning, the local context will make the significance clear.

DIVISION Q

a	Semi-longitudinal axis	
b	Semi-transverse axis	
γ	Radius of curvature	
x	Length along longitudinal axis	
S	Area or surface, usually area of section	
α	Angle between axis and direction of motion—angle of attack	
φ	Angle of yaw,	**9**
U	Axial velocity	
V	Velocity in general	
v	Lateral component velocity	
W	Resultant velocity,	**6**
ω	Angular velocity,	**10**
F	Force	
L	Lift	
M	Moment	
P	Pressure on surface	
c_1, c_2, c_3	Special correction factors,	**7**
k_1	Inertia factor for axial motion	
k_2	Inertia factor for lateral motion	
k'	Inertia factor for rotation	
η	Special inertia factor,	**2**, **8**

DIVISION R

x	Distance along longitudinal axis	
b	Transverse force breadth	III **2**
h	Change in metacentric height	
R	Radius of circular path	
A_D	Drag area	II **5**
S	Area, usually of cross section	
Q	Volume	
α	Angle of attack	
ε	Slope of ships path	III **1**
V	Speed	
L	Lift	
D	Drag	
F	Lateral air force on ship	
M	Mass	
P	Power	
C_L	Coefficient of lift	

C_D Drag coefficient
k_1 Longitudinal inertia factor II 4
k_2 Transverse inertia factor III 2
η Propeller efficiency
E Propulsive efficiency II 5
f Fuel consumption per unit of power
R Reynolds number
ϱ Density of air
T Temperature, absolute
t Time

DIVISION S

A Half wave height, III 2
b Width, usually of float
D Diameter of propeller
h Pitch of propeller
L Length
A, S Area or surface
σ Area or surface, III 7
i Angle of incidence or slope, or inclination between planes representing wings and hull bottom, III 7
θ Angle of inclination, usually to water surface
V Speed
B Lift due to buoyancy of water
H Hydrodynamic and resultant force, II 7
Sometimes also hydrodynamic force coefficient, III 5
M Mass
N, T Normal and tangential components of force
Q Torque of propeller
R Resistance, also aerodynamic coefficient in equation Force $= R V^2$
T Thrust of propeller
W Weight
α, β Special coefficients III 6, 8, 11 also sometimes special forms of expression, III 8
K Coefficient in general
K_1 Coefficient of aerodynamic lift, III 8
K_2 Coefficient of aerodynamic drag, III 8
K' Coefficient of hydrodynamic lift, III 8
K'' Coefficient of hydrodynamic drag, III 8
δ Weight of unit volume of water, III 8
ϱ Density of air
G Center of gravity
ר Hebrew *resch*, used for Reech-Froude ratio $V/\sqrt{g\,A}$

DIVISION T

x, y, z Coordinates along axes of X, Y, Z
x, r, φ Cylindrical coordinates (Fig. 3)
δ Thickness of boundary layer VIII 1
ε Apparent or eddy viscosity III 2
Also thickness of thin layer adjacent to wall V 2
l Mixing Length III 4
L Length
R Radius of cylinder or pipe

S	Area or surface
u, v, w	Component velocities along X, Y, Z
U or V	Velocity of flow in general
G	Average mass velocity over a cross-section
M	Momentum
p	Pressure, or stress
F_x	Skin friction force per unit area
τ	Frictional shearing stress III 3
c	Specific heat
k	Conductivity for heat flow
q	Heat transfer per unit area
h	Heat transfer divided by temperature difference
Q	Quantity of heat
β	Eddy conductivity III 2
β	Coefficient of thermal expansion IV 4
S	Rate of heat generation or disappearance V 3
ψ	Stream function
φ, ψ	Functional symbols VII 3
μ	Coefficient of viscosity
ν	Kinematic viscosity
g	Acceleration due to gravity
ϱ	Density of air
σ	Put for ratio $\mu c/k$
T	Absolute temperature
	Temperature
θ	Time

DIVISION P

DIVISION P

AIRPLANE AS A WHOLE
GENERAL VIEW OF MUTUAL INTERACTIONS[1] AMONG CONSTITUENT SYSTEMS

By

W. F. Durand,

Stanford University, California

PREFACE

The Editor regrets most sincerely to announce that the other engagements of Professor Panetti of the R. Scuola die Ingegneria di Torino, Turin, Italy, have not permitted him the time to prepare the manuscript for the present Division, as was originally planned. Pressed near the close of the period of publication by the need of some treatment of this topic, and in order to fill out the original schedule of Divisions and of subject matter to be treated, the Editor has undertaken to give a brief treatment of the subject which, it is hoped, however, may aid the reader in gaining a more comprehensive view of the airplane as a whole and of its performance as the resultant of a very considerable manifold of actions, interactions and reactions.

In the present Division, the airplane is viewed as a complex of four interacting systems—the lifting system, the non-lifting system, the propulsive system and the control system. These four systems may and do interact mutually, thus modifying in various ways the basic effect or purpose of each of these systems considered as isolated from the others. In all there are twelve such interactions, some of which, however, are of small or negligible importance.

Several of these interactions have been discussed in mathematical terms in other Divisions of this Series and it is the purpose in the present Division to give rather in a descriptive and non-mathematical way, a general view of this entire manifold of interactions, with suitable reference to other Divisions for more complete discussion of the more important in mathematical terms.

Such a bird's eye view, as it were, of these various actions and reactions seems desirable in a series of monographs of the character of the present series, and the Editor only regrets the need of any departure from the original plan regarding the authorship.

---

[1] See Journées Scientifiques et Techniques de Mécanique des Fluides, Vol. 1, p. 189, Paris 1935.

Also with special reference to sections 4, 5, 9, 11 see Division O, pp. 37—41.

Introductory. We may view the airplane as a complex of four systems as follows:

1) The lifting system.
2) The non-lifting system.
3) The propulsive system.
4) The control system.

The basic aerodynamic effect or purpose of each one of these systems will be in some measure influenced or modified by each one of the other systems. There will be, therefore, twelve such effects or influences as we may choose to call them. Not all of these effects are of equal importance, but it will be of interest to list them all, in order that they may be viewed as a whole and examined, each with reference to its place in the picture of an airplane as performing under the aggregate of this complex of disturbing actions and reactions.

These disturbing inflences are, therefore, as follows:

1) Influence of the lifting system on the non-lifting system.
2) Influence of the non-lifting system on the lifting system.
3) Influence of the lifting system on the propulsive system.
4) Influence of the propulsive system on the lifting system.
5) Influence of the lifting system on the control system.
6) Influence of the control system on the lifting system.
7 Influence of the non-lifting system on the propulsive system.
8) Influence of the propulsive system on the non-lifting system.
9) Influence of the non-lifting system on the control system.
10) Influence of the control system on the non-lifting system.
11) Influence of the propulsive system on the control system.
12) Influence of the control system on the propulsive system.

We proceed, then, with the examination of these in order.

1. Influence of the Lifting System on the Non-Lifting System. The lifting system is represented by the wings and the non-lifting system primarily by the fuselage, to which may be added such items as the landing gear (when not retractable), engine nacelles in multi-engined planes, struts, guy wires, etc. The fuselage also may and usually does contribute something to the lift and the stabilizer likewise contributes vertical forces, though often opposed to wing lift for purposes of longitudinal stability. However, for our present purpose it will be sufficient to consider the influence of the wings on the fuselage.

The purpose of the fuselage is, of course, to house a power plant, operating personnel, passengers and useful load, and to make connection between the wings and the empennage or otherwise to furnish a mounting for the latter. None of these is directly aerodynamic in character and the chief aerodynamic result of the presence of the fuselage is the production of so-called parasitic resistance or drag. At the same time, as noted above, the fuselage may, and at large angles of attack will, give an element of lift, functioning as a poorly formed airfoil of aspect ratio much less than unity. However our major interest in the present inquiry will relate to the influence of the wings on the production of fuselage drag.

The lift produced by the wings is associated with a circulation about them, with the trailing vortices streaming down the wake and producing the well known induced "downwash" or downward component in the resultant airflow over the plane and changing the geometrical angle of attack α by the so-called induced angle φ or α_i (see Division E I 12 and III, Part C).

The circulation about the wing itself will furthermore introduce another factor, especially influential at the leading edge where it tends to produce an upward component in the line of the airflow to the wing. The general character of the lines of airflow in approaching a wing as shown by photography and as indicated by diagram may be seen by reference to Division E I, Figs. 7, 10, 12.

The extent to which the airflow over the wings as influenced by circulation and downwash will influence the airflow to the fuselage, especially in respect to direction, will naturally depend on the location of the fuselage relative to the wings. In the general case, however, some degree of such influence may be expected and, to the extent to which it exists, it will enter as a more or less influential factor in modifying the effect of obliquity due to the attitude of the plane under the conditions of flight controlling at the moment.

Thus, in actual flight, at any one loading, there will be only one speed and one attitude of plane at which the axis of the fuselage will lie strictly in the line of flight. At all other speeds, the attitude will be such as to make some angle of obliquity between this axis and the line of flight.

The combination of these obliquities, of flow and of attitude, will in general, result in an obliquity of airflow relative to the direction for minimum fuselage drag, with increase in the size of the turbulent wake and with the result of an increase of drag, more or less pronounced as the resulting angle of obliquity is large or small. The same obliquity relative to the airflow will likewise influence such lift as the fuselage may give—in general, an increase of fuselage lift with increasing angle of wing attack, up to some angle presumably approaching the burble point for the wings.

In addition to these more or less obvious forms of reaction, there will result on the body of the airplane a positive lift resulting from the development of a positive circulation about the body due to the existence of the circulation about the wings. Where the wing extends continuously across from tip to tip over the airplane body, the wing circulation must in part extend around the body itself with the result of a positive lift. Where the wing is discontinuous, this action will be less pronounced, but even here there may be some lateral extension of the wing circulation in such a way as to at least partially inclose the body of the plane.

We have thus, in summary, first an attitude of the plane relative to the line of flight, dependent on the wings in that such attitude is necessary to enable the wings to realize the lift under the actual conditions of load; second, some modification of the resulting lines of airflow to the fuselage, due to the influence of the downwash and circulation about the wings; and third, some actual increase in the circulation about the fuselage, representing, in a sense, an extension of that about the wings. The combination of obliquities will, as noted, produce some increase in the fuselage drag while the increment of circulation will produce an increment of fuselage lift.

The amount of the increase in drag will be small unless the obliquity of the axis of the fuselage to the line of flight becomes large, in which case it may become serious. The increment of lift due to wing circulation will be relatively small, depending on the arrangement of wings and fuselage.

2. Influence of the Non-Lifting System on the Lifting System. For our present purposes, the non-lifting system may be represented by the fuselage, to which may be added such structures as engine nacelles in the case of multi-engined planes. On the other hand the lifting system is represented by the wings, so that the present question reduces primarily to the influence of the fuselage on the wings.

We may first refer to Division K III, wherein this subject is approached from the mathematical standpoint and results are given based on two methods of treatment as follows:

(1) Representation of the fuselage by an indefinitely long cylindrical body with wings attached.

(2) Treatment of the combination of wings and fuselage (or wings and nacelles) as a generalized form of wing with abrupt change of profile at the location of the fuselage or engine nacelles. The distribution of lift over the span is then developed by suitable mathematical procedures.

Method 1 admits of a certain simplification in the theoretical treatment and furnishes results of definite interest and significance. In general it is shown that there is produced what is called an "additional stream flow", conditioned by the body, and which will produce a downward component of velocity on the wing. It thus results that the body will cause a change in the effective angle of incidence of the wing and likewise a supplementary drag for the same, the character and value of which will depend on the relative locations of body and wing. Division K should be referred to for further details.

Method 2 permits of treatment by several different mathematical procedures of which three are given in brief abstract. This mode of approach to the problem shows, as might be anticipated, an abrupt decrease in lift over that part of the span represented by the body or by engine nacelles. It shows also what may be termed the spread of this influence over the rest of the wing, resulting in a general decrease of lift as compared with that for a wing alone of the same total span (see Fig. 1[1]). This departure of the lift distribution from that for the wing alone causes also an increase in the induced drag, the magnitude of which is, however, relatively small. See Division K, Fig. 45. Attention may be called to the considerable amount of experimental information on this subject, especially as regards the results of fillets between body and wings and the best relative location of engine nacelles and wings[2].

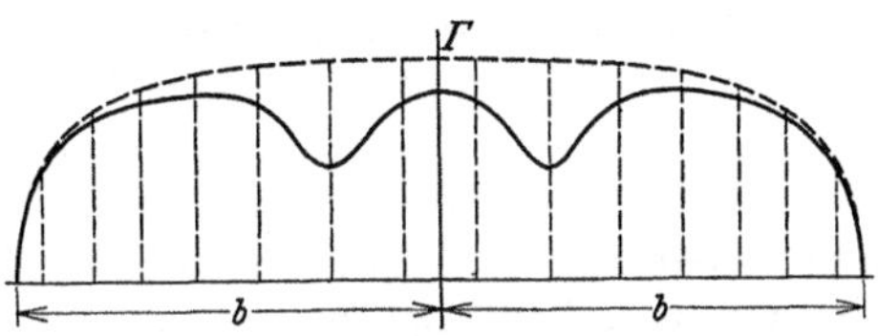

Fig. 1. Effect of nacelles on distribution of lift.

It may be desirable to give some further discussion of this general topic from a slightly different viewpoint and without direct reference to mathematical procedures. This will permit the development of a somewhat more complete picture of the subject as a whole, reaching however, the same general conclusions as above.

To this end we shall find it convenient to consider the question in two parts, depending on whether the wing is continuous across the span from tip to tip or is interrupted by the fuselage. In the latter case it may be considered either as a continuous wing structure with an extreme and abrupt change of form in the central portion, or as two half wings, with attachment at one end to the fuselage and with the other end free.

We take first the case of a continuous wing extending across the span above the fuselage, and recall that such a wing realizes lift by reason of its translation through the air combined with a circulation flow around the wing section. These two types of flow give a resultant

[1] Repeated from Division K, Fig. 44.

[2] Ergebnisse der Aerodynamischen Versuchsanstalt zu Göttingen, I. Lief., p. 118, 1925.

Muttray, H., Untersuchungen über die Beeinflussung des Tragflügels eines Tiefdeckers durch den Rumpf. Luftfahrtforschung, Bd. 2, Heft 2, 1928.

Gough, M. N., The Effect of Fillets Between Wings and Fuselage on the Drag and Propulsive Efficiency of an Airplane, U.S. N.A.C.A. Technical Note No. 299.

Klein, A. L., Effect of Fillets on Wing-Fuselage Interference, Transactions Am. Soc. Mech. Engrs., 1934.

Wood, Donald, H., Tests of Nacelle-Propeller Combinations I, II, III, U.S. N.A.C.A. Technical Reports Nos. 415, 436, 462; 1932, 1933.

air flow about the wing and we have now to inquire as to the extent to which such flow will be influenced or modified by the presence of the fuselage lying below its middle portion and of breadth some five to ten per cent of the wing span.

Stated in this way it is clear that this influence should not be of serious amount. Its major effect will be confined to a small part of the span near the center, and with usual dimensions, this effect will be relatively small in magnitude. There will be a tendency toward a slight compression of the flow between the fuselage and the central part of the wing, combined with a tendency for the flow to spread obliquely as the air travels along the under side of the wing from the leading to the trailing edge. Compression of the lines of flow will result in higher velocities, the result of which will depend on the form and attitude of the under surface of the wing. Insofar as the velocity is increased, the pressure will be reduced (Bernoulli's law) but to the extent the flow is deflected downward, the upward dynamic reaction will be increased. We have here two opposing effects, the resultant of which will depend on the special circumstances of the case and no general conclusion can be drawn. Obliquely spreading flow may likewise cause a disturbance over a width somewhat greater than that of the fuselage itself, and of a character to reduce the over pressure which would otherwise result from an uninterrupted flow. Viewed otherwise, it would appear that the effect of this crowding of the lines of flow may have the effect of decreasing the circulation velocity below the wing and to the extent to which this might result and for the parts so affected, the lift would be correspondingly reduced.

On the whole, however, with normal dimensions, it is clear that this influence, compared with the lift of the wing as a whole, will be small though perhaps not of vanishing importance.

With certain types of design however, notably with some of the high wing monoplane types, the wing is structurally continuous but with a much reduced if not vanishing clearance between it and the fuselage. In such cases these effects will be much more pronounced and may seriously affect the lift of the under surface of that part of the wing nearest the fuselage. On the other hand, the upper surface of the wing will still be effective, so that the result as a whole may not involve a serious loss in the total lift.

It is clear, therefore, that in any case, viewed in comparison with a continuous wing with unimpeded flow, the near presence of a body such as the fuselage will result in some distortion and change of flow, all of which will react unfavorably on the lift of the wing as a whole; and that the magnitude of such influence will depend in primary degree on the amount of clearance between the wing and the fuselage body.

At the same time such disturbances to the flow over the wing may likewise affect the drag, both form and induced; the first by an increase in the turbulent wake and the energy carried away in it, and the second as a result of relatively abrupt changes in lift distribution and in the resulting system of trailing vortices.

Turning now to the case of a wing interrupted by the fuselage and comparing it with a single wing of the extreme span from tip to tip, it is shown in Division K III 2, as indeed may naturally be expected, that at normal angles of attack the abrupt change in virtual wing form at the center will result in a loss in lift over this part of the span, and furthermore, that this general disturbance will cause a measurable loss over the remainder of the span. At the same time the abrupt change in lift distribution will also cause an increase in the induced drag as noted above. On the other hand it may be noted that at high angles of attack and with special fuselage forms, the combined lift of wings and fuselage may be more than for an uninterrupted wing of the same over all span. In this connection special attention should be given to the influence of the geometrical form at the abrupt transition from wing to fuselage. Such a transition of form, if relatively sharp angled, will result in the formation of turbulence in the angle, spreading outward on the wing and creating generally a disturbance in the normal flow over the wing in this vicinity. The consequences here depend further in marked degree on the character of the flow through the angle between the wing and the fuselage—whether divergent or convergent. In the former case there will be high probability of separation of flow with increase in drag; in the latter case this condition should not develop and this particular increment of drag will be avoided. Naturally the low wing monoplane will be more subject to the consequences of a divergent flow than will the high wing type. The results of such conditions will be, in general, a decrease of lift and an increase in drag.

However, generous and carefully designed fillet forms connecting wings and fuselage are found to reduce in marked degree the prejudicial results due to these conditions[1].

If now we view the lifting system as consisting of two half wings, one projecting on either side of the fuselage, we have a pair of wings, of which one end is free and the other shielded by the fuselage side. We shall therefore have, for each wing, the normal spilling of the air around the free end, while the fuselage form will act in some degree as a shield for the inner or root end of each wing.

In addition, there will be, as discussed above, turbulence and general disturbance to the flow about the root of each wing due to the more or less abrupt transition in form from wing to fuselage.

---

[1] See references on p. 5.

We now ask, with what form or porportion of wing shall we compare these two half wings in order to discuss the question of the influence of the fuselage on the lift of such a wing system. It will be of interest to consider two such ideal wing forms: (1) A continuous wing of span equal to the combined span of the two wings excluding the fuselage. (2) Two wings of span each equal to the span of the wing structure on one side of the fuselage, but with free ends. This will be, in effect, two wings each of half the span in (1) and hence with half the aspect ratio.

Taking these in order, we shall have, with the actual structure, and excluding the fuselage entirely, a close approach to (1) except for the turbulence and disturbance to the flow at and near the root of each half wing. The shielding due to the fuselage side should much reduce the loss due to spilling or in other words it should give, over the root of the wing, a close approach to two-dimensional flow. There remains, however, the loss due to turbulence and disturbance of flow as already noted. In addition we must remember that, as we have already seen, the disturbance to the flow caused by the presence of the fuselage will extend in some degree over the entire wing span and the present picture is, therefore, the same as that presented by Fig. 1, except that we now exclude from consideration the marked loss in the center over that part of the total span represented by the fuselage itself. There remains then the loss over the actual wings proper plus that due to turbulence caused at the roots of the wings. The latter, as noted, can be much reduced by careful filleting, but the former will persist, and over all, as compared with a single wing of combined wing span, there will result a definite loss in lift together with some increase in drag.

Taking now the second ideal, we have to compare on each side of the plane, two half wings, one of which has both ends free while the other has one end shielded by the fuselage side, with the resultant turbulence formation. The shielding considered by itself, compelling as it will a close approach to two-dimensional flow near the root of the wing, will be favorable to the lift. It will give, in effect, a lift distribution holding up to a considerable value at the wing root, instead of falling to zero as in the case of a free wing tip. There will be, however, some general decrease of lift over the wing due to what may be termed the "spread" in the effect of the fuselage; see Fig. 1.

There will be likewise some further loss in lift and increase in drag due to turbulence at the root of the wing as already discussed. Definite experimental measure seems to be lacking regarding the values of these effects, favorable and unfavorable, but with a well filleted wing junction with the fuselage, the beneficial effect should definitely outweigh the loss due to turbulence, and with a wing structure of this character, the lift should be definitely greater than for two separate wings, each of the span of the half-wing on one side of the fuselage, and with free ends. This would mean in effect, two wings each of approximately half the aspect ratio of the single wing as a whole.

Again, to the lift of the two half-wings as here considered, must be added the lift due to the fuselage, whatever it may be, usually small at low angles of wing incidence, but larger with increase of this angle.

On the whole, therefore, it appears that if the wing be considered as extending across the entire span from wing tip to wing tip, the presence of the fuselage, for small angles of attack, will result in a loss of lift, while for large angles, the combined lift of wings and fuselage may be greater than for a single wing of the same over all span. Likewise if we consider the wing as composed of the two half structures on either side of the fuselage, and compare this with a single wing of the same aggregate span, there will likewise result a loss in lift. But if we compare this latter structure with two wings with free ends, each of the span of the half-wing of the actual plane, the result will involve effects favorable and unfavorable, but with the balance, in any normal case of a well filleted junction, definitely on the favorable side.

In this connection attention may be called to the fact that other things the same, the two wings of half span would have double the induced drag of the single wing, thus giving a further advantage to the actual construction in comparison with this particular combination.

3. Influence of the Lifting System on the Propulsive System. The propulsive system in the simple case is represented by the propeller located at the nose of the fuselage. Propeller performance as based on the simpler theory and as estimated from the results of model or even full scale tests in aerodynamic laboratories, is considered relative to the case of motion in the direction of the shaft. That is, propeller traction (thrust) is considered as acting along the line of the shaft, and the relative motion of propeller and air is assumed to lie along the same line. It is shown, however, in Division L[1] that obliquity of flow of the air to the propeller has the effect of producing lateral forces on the propeller as well as other marked effects on its performance. The production of lateral forces on the propeller and hence on the shaft will have the effect of giving a resultant propeller force oblique to the line of the shaft and likewise in general, oblique to the direction of flight.

[1] See VIII 7 and XII 5; also:

Clarke, T. W. K., Effect of Side Wind on a Propeller, Br. A.R.C. R. and M. 80, 1913.

Bramwell, F. H., Relf, E. F., and Bryant, L. W., Experiments to Determine the Lateral Force on a Propeller in a Side Wind, Br. A.R.C. R. and M. 123, 1914.

Harris, R. G., Forces on a Propeller Due to Sideslip, Br. A.R.C. R. and M. 427, 1918.

Misztal, Franz, Zur Frage der schräg angeblasenen Propeller, Abhandlungen aus dem Aerodynamischen Institut an der Technischen Hochschule Aachen, Heft 11, p. 5, 1932.

In the case of the actual propeller in flight we have four directions to consider; (1) the direction of flight, (2) the direction of the shaft of the propeller, (3) the direction of the flow of air to the propeller, and (4) the direction of the resultant force on the propeller shaft and hence on the plane. In the general case, neither (2), (3) nor (4) will be the same as (1). The direction of the shaft will vary with every change in the angle of attack (quite aside from the rapid changes in maneuvers) and this will mean generally that with every change of speed the line of the shaft will change relative to the line of flight.

Regarding the line of airflow to the propeller, we remember that, due to the action of the wings on the direction of the airflow over the plane (see **1**) the lines of flow to the propeller will suffer more or less deflection (normally upward) as compared with the line of flight. This deflection will vary, furthermore, with the attitude of the plane, greater as the angle of attack is greater. We shall thus have, in general, a variable angle between the line of airflow to the propeller and the line of flight. This combined with the variation in the angle between the shaft and the line of flight will result in an angle between the shaft and the line of airflow to the propeller, subject to more or less irregular variation, and in the general case differing from zero.

This complex of obliquities may, therefore, be considered as due to the fact that the lifting system, in order to maintain a lift L equal to the weight W at varying speeds, must assume varying attitudes in flight; and in this sense these various consequences on the propulsive system may be considered as due to the characteristics of the lifting system.

It must, therefore, be accepted that the propeller normally does not find itself in the simple condition of operation usually assumed, and in particular that it must in general accept a flow of air with some degree of obliquity relative both to the line of its shaft and to the direction of flight. This will result, as we have seen, in the production of lateral forces on the shaft and in a line of action of the total force on the propeller oblique to the line of flight. The obliquity of the flow to the propeller will itself result in some loss of efficiency and the obliquity of the total resultant propeller force to the line of flight will cause a further loss in propulsive effect. Adequate experimental data bearing on these losses do not seem to be available, but it may be safely assumed that such loss will not be serious until the angles of obliquity approach values larger than those in usual practice. However in the case of a plane making long flights, in particular for those periods of time for which the angle of attack may be relatively high, these sources of propulsive loss will be operative in some degree and corresponding allowance should be made for them.

In addition to these influences traceable primarily to obliquity of airflow and shaft direction, there may be, especially with wing mounted engines, a further influence due to the wing itself by way of a slowing up of the airflow to the propeller, and similar to that of the fuselage as discussed in detail in 7, though smaller in amount. Reference to 7 may, therefore, be made for further details regarding this particular feature of the total reaction on the propeller.

In an extended examination of this same general problem by Betz[1] the conclusion is reached that the action of a propeller in the presence of a wing should be represented by a vector diagram including not only forces axially on the propeller but a vertical component also, equal in magnitude to the increment of lift on the wing (see 4). For some further details of this analysis, see Division L VIII 7.

Another possible source of influence may be found in the partial arrest by the wings of the rotary component in the wake flowing from the propeller. The effect of this rotary component on the wings is referred to in detail in 4. Inversely the presence of the wings in the wake may be reacted back through the rotary component on the propeller, affecting more directly the torque in some small degree.

4. Influence of Propulsive System on the Lifting System. The character and relative magnitude of this influence have been dealt with in mathematical detail in Division M and to which reference should be made.

The reaction between propeller and wing results primarily as a consequence of the wake which the propeller drives aft and in which the characteristics of the flow differ from those for the flow over the wings otherwise, in the following particulars. (1) It has an accelerated velocity aft. (2) It has a velocity of rotation—*i. e.* a rotary component of the total velocity. (3) The turbulence will be more pronounced than in the airflow generally over the wing. We have, however, to remember that this wake extends over only a part of the span of the wing—larger naturally in the case of a multi-engined plane than for a single propulsive unit.

We may first consider briefly and without mathematical detail the more obvious consequences of these three characteristics of the wake as noted above.

First with regard to the wake itself, it should be noted that as a result of the difference in the conditions within and without the wake, there will exist a surface or zone of discontinuity separating the cylindrical wake from the remainder of the airflow. The major influence of the propeller, or of the wake which it generates, will, therefore, be manifest on the parts of the plane lying within this wake. We cannot, however, assume that the wake is entirely without influence on the parts lying outside its boundaries. As shown in Division M the complete mathematical analysis of the problem requires the recognition of certain

[1] Betz, A., Der Wirkungsgradbegriff beim Propeller, Zeitschr. f. Flugtechnik u. Motorl. **19**, 171, 1928.

effects outside the wake boundary, even though their intensity may rapidly decrease outside the boundary limits.

Passing these over for the moment, however, we may restrict ourselves to the more obvious effects due to the three characteristics of the motion in the wake as noted above.

First with regard to the increased axial velocity in the wake. This will have the effect of producing an increased lift over the part of the wing lying directly in the wake, measured approximately by the square of the ratio of the two velocities, within and without the wake. At the same time this increased lift will be obtained at the price of an increased frictional drag over this part of the wing, practically in the same ratio. The increase of lift will also entail a corresponding increase of induced drag[1]. The net result is, therefore, an increase in lift and an increase in drag over this part of the wing, substantially as though it were moving with a speed of advance $(V + u)$ instead of V, where V is the speed of advance of the plane relative to the outside air and u is the speed of the wake to the rear.

Taking next the velocity of rotation, it is evident from elementary mechanics, that the torque which the engine exerts in turning the propeller must have its equal in the production of angular momentum in the medium acted on by propeller—*i. e.* in the air. The external evidence of this angular momentum is then found in the angular velocity imparted to the air in the wake in consequence of which the actual air particles stream backward down the wake in helical paths, forming ideally a sort of twisted rope of air constituting the wake proper and separated from the outlying body of air by the surface of discontinuity to which reference has previously been made.

In the usual case, however, the flow in the wake will not be permitted to form any such coherent ideal helical stream, but will be more or less broken up and diverted by parts of the structure lying directly in its path. Thus in the case of a single engined plane with propeller at the nose of the fuselage, the latter will lie directly in the propeller wake which must, in consequence, divide and pass around the fuselage with the assemblage of wings still farther to the rear. Similarly with multi-engined planes, the wake of the wing propellers will be obstructed by some form of engine nacelle and also by the leading edge of the wing itself, causing usually a complete separation of the wake into two parts, one passing above and the other below the wing.

Nevertheless, the air as it leaves the propeller will have this angular component of velocity and corresponding angular momentum and wherever it impinges on other parts of the structure will produce results accordingly. Thus on the wings, for the part directly within the wake, this angular component may be viewed either as a component velocity

[1] See Division E I (11.4).

impinging on the wings, above on one side and below on the other, or as a component of the total velocity increasing the angle of attack on one side and decreasing it on the other. In either case the result is the same, with an increase of lift on one side and a decrease on the other, thus producing a rolling moment about the longitudinal axis.

Obviously these two effects of increase and decrease will nearly balance, leaving a negligible effect on the lift as a whole. So far, therefore, as the rotary component of the velocity in the wake is concerned, we are left with the production of a rolling moment produced by the upflow and downflow components acting respectively on the under and upper sides of that part of the wing structure lying within the wake boundary, or otherwise, due to a difference in lift on the two sides caused by changes plus and minus in the effective angle of attack on that part of the wing lying within the slipstream, due to the opposite directions of the rotary component of flow on the two sides.

It may be noted in this connection that the engine torque transmitted through the engine frame to the structure of the plane will produce a reaction torque tending to rotate the plane as a whole in a sense opposite to that of the propeller. This is, in fact, the reactive torque, the equivalent of which is found in the production of angular momentum in the wake as noted previously. But the sense of rotation of the air in the wake is the same as that of the propeller and hence the sense of the rolling moment produced by the action of the wake on the wings will be in the same sense. Hence the two moments, one due to engine torque and the other to angular rotation in the wake will be opposite in sense and will thus tend to balance. Actually, however, the moment due to torque will greatly exceed in magnitude that due to angular rotation in the wake, leaving as a net result a distinct rolling moment in the sense opposite to that of the propeller, the existence of which must be recognized by suitable adjustments in the "rigging" of the plane. (Aileron tabs, differential aileron control, etc.)

Passing next to the effect of increased turbulence in the wake, it may be noted that the general subject of varying turbulence in an airstream and its effect on the force reactions on a body immersed in such a stream, is involved in a number of the interactions considered in the present Division, and for this reason certain general statements regarding the subject may be introduced at this point[1].

It has long been known that the behavior of a fluid flowing along the surfaces of a solid body is profoundly affected by the previous history of the fluid as regards the presence or absence of turbulence. Thus in the early experiments on the flow through pipes by Osborne Reynolds and others, it was found that the value of the parameter

[1] See Division G **19**; also Division I, Part 2, p. 329 *et seq.*

lv/ν (the so-called Reynolds number) at which the transition from laminar to turbulent flow occurs, was directly dependent on the condition of the fluid entering the pipe. With sharp edges at the pipe entrance, or in the usual case where some initial turbulence is present, the value of this parameter is known to be about 2,000; while with rounded entrance and by avoidance of initial turbulence, this value may rise to perhaps ten times this figure. This immediate effect is traceable to the more or less abrupt change from laminar to turbulent flow.

In the case of bodies placed in an indefinite fluid stream, in which the flow is always in some degree turbulent (as is the case for all conditions with which we are concerned) the results, first with respect to drag, are much more complex in character, depending on the form of the body and on the Reynolds number and with possibly opposite effects on the two components, "form" drag and "skin friction" drag.

For bodies of bluff form, for example a sphere or an airfoil or wing or body of similar form at a high angle of incidence, the form drag, represented by an eddying turbulent wake, will vary with the size of this wake and with the stream of energy which it represents. That is, the larger the wake the greater the form drag. But the formation of a wake of this character is due to the failure of the stream flow, and in particular the boundary layer, to follow around the form of the body and close in about its after part. Dependent on the form and surface texture of the body and on the Reynolds number there is a tendency of the flow to separate from the body thus forming a surface of separation or boundary, streaming to the rear, and within which the fluid is in a condition of mixed eddying turbulence forming a so-called "dead water" or "dead air" wake.

Now any condition which will tend to retard this separation, or enable the flow to more completely encompass the body will result in a decreased size of this eddying wake and a correspondingly reduced form drag.

But it is known that increase of turbulence in the boundary layer, or generally in the air flowing about the body has precisely this effect and hence we have the general result that with highly turbulent air the boundary layer more effectively resists separation and so in special cases, the resulting form drag is less than with air less turbulent.

A further effect, and of even greater inportance, results with forms intended to produce lift—airfoils, wings, symmetrical sections as used for control surfaces, etc. With such forms, as is known, the lift increases continuously with the angle of incidence, up to some value (the so-called burble point) at which separation develops and the lift suddenly decreases. However, the effect of increasing turbulence in the air flowing about the wing has the effect of postponing such flow separation to higher and higher angles and thus of enabling the wing to carry its normal law of lift increase to greater angles of attack and thus to higher values of the lift than with air of lesser turbulence.

Turning now to the effect of increasing turbulence on skin friction drag, it is known that, in general, greater turbulence in the boundary layer will increase this form of drag. This may be assumed to be due to a more energetic action between highly turbulent air and the surface texture of a body, than with air less turbulent. As a result, where the parasite drag is primarily "frictional" and in only small or negligible degree "form" (as with well streamlined bodies such as airship forms, airfoils and wings under small angles of attack, etc.) the drag will in general increase with increasing turbulence.

In attempting to assess the influence of turbulence, therefore, these two possibly opposite effects must be held in mind. For such forms as are used for wings and control surfaces, the result at small or moderate angles of attack will be an increase of drag with increasing turbulence—small in amount—with no sensible influence on the lift. And then as extreme angles of lift are reached, a marked increase in lift and decrease in drag as compared with values for less turbulent air.

It may be noted here that by the phrase "varying turbulence" we mean varying over the range readily producible by laboratory methods. The air in wind tunnels and as generally used in laboratory research is characterized by rather high turbulence, which may, however, be increased by suitable means (wire screens, etc.). Air in the open atmosphere, as met with in ordinary flying, is normally much less turbulent, but as it meets the various parts of an airplane its turbulence may be much increased by influences as noted, and with results generally as indicated by the preceding discussion[1].

In general, regarding the influence of turbulence, it may be said that its effects are produced primarily by way of its influence on the conditions (form, surface texture, Reynolds number) under which separation of the flow takes place. According as separation of the flow takes place earlier or later, so will the size of the turbulent wake vary and so will the relative areas of contact between the surface of the body and boundary layer flow on the one hand or dead air contact on the other, vary, all with resulting effects on lift and on form and skin friction drag.

We have therefore in summary for the effect of the propeller on the wings, a generally increased value of the force reactions over the

[1] Dryden, H. L., and Kuethe, A. M., The Effect of Turbulence in Wind Tunnel Measurements, U.S. N.A.C.A. Technical Report No. 342, 1930.

Millikan, C. B., and Klein, A. L., The Effect of Turbulence, Aircraft Engineering, August 1933.

Relf, E. F.: Results From the Compressed Air Tunnel, The Journal of the Royal Aeronautical Society, January 1935.

See also Division G **19**.

part of the wings lying within the wake, comprising an increase of both lift and drag, and the production of a rolling moment. The first of these is due primarily to the increase in axial velocity while the latter is due to the component of rotation which characterizes the motion of the air in the wake.

Regarding the effect of turbulence, it cannot be assumed that, under ordinary conditions of flight, increase of turbulence in the propeller wake will count as a factor of practical importance. It seems possible, however, that at extreme angles and when approaching the burble point, the influence of such turbulence, over that part of the wing on which it is effective, will be to delay the separation of the flow and aid the wing in carrying on to a somewhat greater angle of attack and higher lift than might be otherwise possible.

At this point also, mention may be made of a further influence on the phenomenon of separation[1] due to the possible influence of the propeller in modifying the angle of attack on the wing. Under given conditions otherwise, separation begins at a definite angle of attack which depends primarily on the geometrical form of the section. If then the effective angle of attack is increased, separation will be hastened; if the angle is decreased, it will be retarded. Depending on the combination of angles to which reference has been made above, the action of the propeller may, therefore, through the influence of the wake on the angle of attack, hasten or retard the beginning of separation of the flow from the wing. Normally, the result will lie on the side of retardation of separation, affecting primarily, of course, that part of the wing directly exposed to its influence.

A further influence on separation is to be found in the negative pressure gradient in the propeller wake. It is known that the increment of axial velocity in the wake is subject to a continued increase from its value just behind the propeller disc to twice this value at a great distance down the wake (∞ in strict theory). This will mean a continued acceleration in the wake with a negative pressure gradient along the line of the flow (Bernoulli's law). Now it is known that the phenomena of separation are intimately associated with the development of a positive gradient of pressure in the boundary layer, against which the flow cannot make headway and separation results[2]. Any influence, therefore, which will tend to relieve this condition by supplying a counter tendency, will, to that extent, tend to reduce the tendency to separation otherwise existing.

The results thus far stated are relatively obvious. The detailed analysis developed in Division M, and to which reference should be

[1] See Division M II **13.**

[2] See Division G **15;** Division I, Part 2, I **3.**

made, shows many modifications of this relatively simple picture. Thus by the vortex theory of the propeller[1], the propeller wake may be viewed as a stream of twisted vortex filaments, while again the surface of discontinuity bounding the wake may be viewed as a cylindrical vortex sheet. As may be readily seen, the possible interactions between the wing vortex system and such propeller systems would present a problem of very great complexity. It is also clear that in this broader view, the influence of the propeller wake will be no longer confined to the parts of the structure with which it comes into immediate contact, but will extend broadly to the whole flow system about the plane, such outside effects, however, being relatively small in comparison with those within the wake itself. Broadly this view relates the results of the interaction to changes in the circulation about the wing and to changes in the axial and vertical velocities of the airflow to the wing.

In the analysis of Division M the final results are expressed as due to the four major parts of the complex whole:

(1) The direct influence of the change in longitudinal velocity in the wake.

(2) The direct influence of the vertical component velocity in the wake.

(3) The indirect influence of the slipstream boundary.

(4) The direct influence of the slipstream boundary.

Again each of these is considered with reference to its effect both within and without the slipstream boundary, thus giving eight possible constituent elements of the whole, though in particular cases, certain of these may become zero.

Added to these influences, susceptible in some degree to theoretical treatment, there will be some further influence due to turbulence as referred to at an earlier point.

For further details, reference should be made to Division M. These brief references to the mathematical aspects of the problem have been made only to indicate how vastly more complicated the problem is than might at first sight appear. The more extended view furnished by such analysis shows that while the intensity of the effects of the propeller wake falls off rapidly outside the wake boundary, these effects as a whole, notably on the wings, are still of significance and must be considered in any measurably complete picture of the problem. These conclusions are also borne out by a limited field of experimental results to which reference is made in Division M.

5. Influence of the Lifting System on the Control System. As we have already seen, the action of the lifting system, significant for present purposes, is the production of lift through the combination of velocity and circulation, the latter for a given lift varying inversely as the speed.

[1] See Division L VI.

The effect of this circulation is to produce the familiar downflow velocity, the result of which is to modify by the induced angle α_i, the entire picture of the flow of air over the plane.

In addition to this general effect due to downwash, the direction of flow of the air leaving the following edge of the plane will be further modified in some degree, and we thus find the control system, working in a flow of air with generally a definite downward component as compared with the direction of airflow at a great distance. In general, therefore, the tail control surfaces must be so set for what may be called their zero position, as to take due account of these changes in the direction of airflow.

In addition to this change of direction, the presence of the wing will further induce some degree of turbulence in the air leaving it and meeting the tail surfaces. This effect will, therefore, conjoin with that due to the same general cause resulting from the presence of the fuselage and the action of the propeller, as referred to later in **9** and **11**.

However, as noted in **4**, we can hardly expect any pronounced effect on the control surfaces due to turbulence, except in positions of large or extreme movement from neutral. Under such conditions, as noted for the wings in **4**, it may be expected that some beneficial effect of the turbulence might result.

These remarks on the influence of turbulence refer more directly to what is sometimes called "fine grained" turbulence, such as is normally found within the boundary layer itself. On the other hand, the aerodynamic behavoir of a body of wing or control surface form, placed within a flow of relatively large sized eddies such as constitute in greater part a "dead air" or turbulent wake, is entirely unreliable, especially as regards the "lift" forces which it may be intended to develop. Such a condition may, to some extent develop as between wings and horizontal control surfaces depending on proportions and form and angle of attack. To the extent, therefore, to which such surfaces find themselves within a turbulent air wake (wake shadow) so will they become in relative degree ineffective and unreliable. In the extreme case, control may be lost and a spin may develop with resulting crash, unless suitable altitude is available for recovery[1].

A further influence on the control surfaces causing a loss of force reaction is due to the effect of skin frction on the wings. This slows up the velocity of the air in the boundary layer which then passes on forming the boundary layer wake and contacting the control surfaces. Since such force reactions vary with the square of the velocity, it is not difficult to see that this loss may be very considerable quantitatively and must be allowed for in design.

---

[1] See Division J III **13**; also Division O VIII.

On the whole, therefore, the normal effect of the wings on the tail control surfaces is threefold: first, by causing a change in the general direction of the airflow acting on these surfaces and to which they must be adjusted in the so-called rigging of the plane; second, an increase in the turbulence of the air flowing over the control surfaces, the effect of which may be, in positions of extreme throw (elevator or rudder), an increase of force reaction or perhaps rather, the saving or postponing of a loss of force reaction under possibly critical conditions of flight; and finally a loss in force reaction due to the effect of skin friction on the wings.

The above discussion refers primarily to the tail surfaces as representing that part of the control system located at the tail of the fuselage and hence entirely separated from the wings themselves. In addition to the tail structure, the ailerons, directly attached to the wings, form a second important feature of the control system. The general character of the influence of the wings on the ailerons may be inferred from the preceding remarks regarding the elevator or rudder. In the normal position they form simply a part of the wings themselves. In the operative position, they receive the air as it flows to them from the wing structure forward, modified as this may be by the circulation and downwash immediately at the rear of the wing. At low angles of attack or for such angles as permit a smooth continuous flow over both upper and lower wing surfaces to contact with the ailerons, the forces on the two ailerons, up and down, for equal angles of throw from mean position, will be practically the same. At large angles of wing attack, however, and especially near the burble point, the aileron turned upward may find itself in a turbulent eddying wake and for reasons noted above, subject to a much reduced force reaction per unit area than under normal conditions of flow. These conditions will result in a decrease of effectiveness of the aileron so situated, a decrease which must be recognized in connection with the use of the ailerons at or near the stalling attitude of the plane.

6. Influence of the Control System on the Lifting System. The control system is represented by the elevator, rudder and ailerons. In the neutral position the influence of these on the wings is small if not negligible. In such position the rudder forms simply an extension of the fin, the elevator an extension of the stabilizer, while the ailerons form extensions of the wings. Under these conditions the elevator together with the stabilizer will form a surface adapted to develop a vertical force, depending on the effective angle of incidence in which it finds itself relative to the air in the wake of the wings and fuselage. For purposes of stability the setting of the stabilizer will be such as to give, with the wings themselves, a combination stable longitudinally. The primary function of these parts of the control system is thus to provide a moment offsetting the naturally unstable moment of the wing by itself, rather than to develop lift as such.

However it is clear that, located as the stabilizer and elevator are, relatively far downstream from the wing, they can have no significant effect on the circulation about the wing, which, with the density and speed determine the amount of the lift on the wing itself. With the elevator thrown out of neutral position, as in the execution of a maneuver, the same general conditions prevail. The aerodynamic conditions about a body in relative motion with the air are highly sensitive to what takes place upstream from the body, but practically insensitive to conditions at any sensible distance downstream. We may conclude, therefore, that so far as the elevator and stabilizer are concerned, their influence on the wing lift is negligible.

Taking next the rudder, it will be clear that the same general conclusions hold here as for the elevator. Whether the rudder is in neutral or displaced position, it is evident that it can have no sensible influence on wing lift.

Coming now to the ailerons; these, when in the neutral position, form a part of the wing proper and thus constitute a part of the wing system. It is only when they are thrown out of neutral position that the question of their influence on the wing lift becomes of importance.

In such case, one aileron will be thrown up and the other down. The wing with the aileron thrown upward suffers immediately a loss of a part of its normal lifting surface. The oblique extension of the aileron upward, will furthermore, for that part of the wing over which it extends, disturb in more or less marked degree, the flow of air over the back of the wing and the circulation around it. In general, these results will all tend in the direction of a decrease in the lift over this part of the wing span, thus favoring the moment about the longitudinal axis which this movement of the ailerons is intended to produce. On the other hand, the oblique extension of the aileron downward, will, for that part of the wing over which it extends, produce a disturbance in the flow tending normally toward an increase in the lift on that part of the wing span and thus again aiding the moment which the aileron movement is intended to produce. This action of a depressed aileron is analogous but somewhat less in extent, than that due to the various forms of flaps which are now taking so important a place in connection with airplane control, especially in the way of a reduction of landing speeds through an increase in lift connected with a sharp increase in drag. While, therefore, the effect of displaced ailerons is to produce a change in the lift, the effects on the two sides of the wing will be opposite in direction and will usually nearly balance, so that the effect on the lift as a whole is not usually of importance.

We reach, therefore, the general conclusion that so far as the lift as a whole is concerned, the influence of the control system, whether in neutral or displaced position, is small or negligible.

However, it may not be out of place to note, with reference to the ailerons, that while their influence on the lift as a whole is small or negligible, the same is not the case with regard to the drag. Lift cannot be realized without drag, and while lift is the basic function of the wings, drag is an unavoidable attendant circumstance. In this respect, the ailerons, displaced from neutral position, will produce a definite influence on the drag. Thus the increase of lift associated with the "down" aileron will give an increment of induced drag, while the possible increase of turbulent wake following such aileron may increase also the form drag. On the other hand the reduced lift caused by the "up" aileron will cause a decrease of induced drag while the possible increase of turbulent wake will have the same effect as for the down aileron. The result will be a difference in drag giving a yawing moment opposing the turn for which the aileron shift is intended to provide the proper rolling moment, and thus requiring a further aileron movement than would otherwise be necessary.

The same increase of form drag for the plane as a whole may also be noted for the rudder or elevator displaced from neutral. Here, of course, the influence is not on the lifting system direct, but rather appears as a circumstance inseparable from the realization of the particular function required of these elements of the control system.

7. Influence of the Non-Lifting System on the Propulsive System[1]. The typical combination presenting this particular reaction is that of a propeller located at the nose of the fuselage, as in all single engined planes and in many multi-engined planes. This combination presents, in effect, the problem of an obstruction located just behind the propeller. To examine the effect of such an obstruction we have first to note that as the obstruction is more or less abrupt and large in transverse dimension, the lines of stream flow approaching the plane will spread, and the relative velocity will decrease in comparison with that of the air at outlying points. This will have the effect of slowing up the air passing through the propeller disk, and this, for a given value of N (the revolutions) will give an increased value of the thrust. In a free stream the form of the thrust coefficient curve corresponding to the equation

$$T = C_T \varrho n^2 D^4$$

shows increasing values of the coefficient for decreasing values of the

[1] For a general discussion of the subject matter of this and the following section, including also to some extent the subject matter of **3** and **4**, reference should be made to Division L VIII. The chapter includes mathematical analysis, formulae and methods for quantitative estimate of results and references to sources of experimental data.

parameter v/nD. But a decrease in the value of the velocity of flow of the air to the propeller will, for the same revolutions, decrease the value of v/nD and hence will give a higher value of C_T and with given n a higher value of T. This, of course, is fully borne out by experiment. In a wind tunnel, for example, with the revolutions held constant, there will be some speed of the air for which the value of T will be zero (speed for zero thrust). At higher speeds a negative value of T is developed, but we are not here concerned with this phase of operation. Starting then from the speed of zero thrust, the value of the thrust will continuously increase with decreasing wind speed, reaching its maximum value (aside from a reversal of the flow) when the tunnel fan is stopped and the air approaching the propeller has only the speed due to the propeller itself acting as a fan.

The simple and direct result of an obstruction such as the nose of a fuselage located directly in the rear of the propeller is, therefore, to give, for a fixed value of the revolutions, an increased value of the thrust. If now the efficiency of the propeller is measured by the product of thrust by speed through the air divided by power delivered by the engine to the propeller, there will be an apparent gain in efficiency since, with other things the same, bringing of the obstruction to a point just behind the propeller has resulted in an increase in the actual thrust developed by the propeller and hence apparently an increase in the numerator of the fraction defining the efficiency. This gain in efficiency however, is only apparent as will appear in **8** when we come to the reverse action of the propeller on the fuselage. These two reactions must in fact be considered together in order to obtain a correct view of the over-all result of this particular pair—the action of the fuselage on the propeller and that of the propeller on the fuselage. Without further discussion of the present phase of the matter, we pass, therefore, to the reverse action in the following section.

8. Influence of the Propulsive System on the Non-Lifting System[1]. As in the preceding section, the typical combination here is a propeller located at the nose of a fuselage. The immediate action of the propeller is to send to the rear a wake of velocity relatively greater than that of the outlying body of air, and such a stream of air parting and passing around the fuselage will give a definitely higher velocity of flow than for the case of like motion without the action of the propeller. If then we compare the case of a fuselage towed through the air with no propeller at the nose, with the actual case of a fuselage drawn through the air by the action of a propeller located just ahead of the forward end, we shall have a definitely higher velocity of flow over and about the fuselage in the latter than in the former case. This will result in an increase of

[1] See footnote to **7**.

resistance to motion, or in other words in an increase in the fuselage drag. It may be also noted that this increase in drag will be due both to the increase in the axial velocity of flow and also to the rotary component in the wake with the resultant disturbance to the lines of flow otherwise and the possible production of an increase in the turbulent wake.

If now we combine in one picture the results for the preceding and the present sections, we have a propeller in a location where it is able to develop an actual increase of thrust for a given number of revolutions, but at the expense of an increase of drag on the fuselage behind it. Comparing these results with those for the same propeller in a free stream and a fuselage in motion without the action of a propeller at its nose, we see that if we credit the propeller with the extra thrust developed, we must likewise charge against it the additional drag developed on the fuselage. The net result will be, therefore, the thrust which the propeller would develop in a free stream (taken as equal to the drag of the airplane as a whole at the given speed) plus the extra thrust developed by reaction with the fuselage and minus the additional drag developed on the latter. This result, as the net or useful thrust taken with the speed and with the shaft power to the propeller, will then give the over-all efficiency of the propeller including the mutual reactions between itself and the fuselage.

In such case, experiment shows that, at least for normal forms, the over-all result, over the operative range of values of v/nD, is a loss in efficiency. The combination of the propeller and the fuselage at close quarters is less efficient than if each could operate independent of all influence from the other.

In order to obtain a more detailed view of these mutual reactions over the whole range of values of v/nD, we may conveniently start again with the influence of the fuselage on the value of v/nD for zero thrust (see Fig. 2).

From what has been said above, and as shown quantitatively by experiment[1], the presence of an obstruction results in shifting to the right (to a larger value) on the axis of v/nD, the value of the latter for zero thrust. This condition is readily seen to follow as a result of slowing down the column of air actually operative on the propeller as compared with the air flowing at a distance. For any value of the wind velocity as based on the latter, the air column operative on the propeller will be slowed down, the thrust generally will be increased and the value of v/nD for zero thrust will be increased. It will result that the curve for thrust or thrust coefficient for the combination of propeller and fuselage, as compared with that for the propeller alone, will start at a larger value

[1] U.S. N.A.C.A. Technical Report No. 220, 1926.

of v/nD and, near the start must lie above that for the propeller alone (see Fig. 2).

However, as the slip becomes greater with decreasing values of v/nD, the gain in net thrust decreases and ultimately the two curves meet and cross. For the conditions covered by the investigation indicated by the reference, this point of crossing was not far from the value of v/nD for best efficiency. Beyond this value the curve for net thrust lies below that for the propeller alone, showing for this range of operative values of v/nD a loss in net thrust for the combination as compared with the propeller alone.

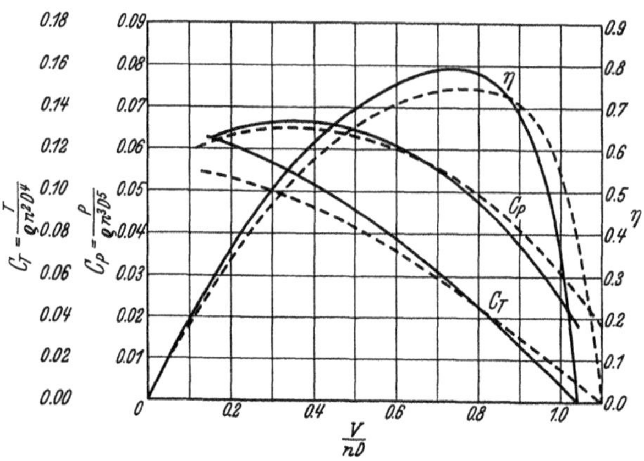

Fig. 2.

It will be noted, however, that the range of values of v/nD, for which there is a gain in net thrust, lies quite outside that for normal operation. The values of v/nD lying between that for best efficiency and that for zero thrust are values for small and decreasing values of the slip and, for a given thrust, would require a prohibitive diameter of propeller. The range of practically operative values of v/nD lies normally near and below that for best efficiency and for this range there is, for the combination, a loss in net thrust.

Similarly as for the thrust, the torque and hence the shaft power coefficient for the combination is increased for large values of v/nD and decreased for small values, with a crossing point apparently at a smaller value of v/nD than for the thrust (see Fig. 2).

In consequence of these relative changes in the values of the thrust and power coefficients, it results that, for very large values of v/nD, there is an actual gain in propulsive efficiency for the combination as compared with the propeller alone, while for moderate and smaller values (over the operative range) there is a loss in propulsive efficiency. This possible gain in propulsive efficiency is readily seen by taking the extreme case of the value for zero thrust with the propeller alone. Under these conditions, the thrust is zero while the torque will have a definite positive value and the efficiency is, therefore, zero. Now with the combination, for the same value of v/nD, the thrust is no longer zero but becomes positive and with a positive value of the torque gives a value of the efficiency greater than zero.

The two curves of efficiency will thus cross and the point of crossing, for the conditions of the reference, was at a value of v/nD somewhat greater than that for the best value on either curve.

Again the best value of the efficiency for the combination is less than that for the propeller alone and generally this relation will hold over the entire range of values of v/nD from that for the point of crossing of the two curves, indefinitely to the left, or toward smaller values of v/nD.

Thus, finally, over the range of practical values of v/nD, the net result for the combination of propeller and fuselage will be a loss in propulsive efficiency. The amount of such loss will obviously depend on many circumstances, but again for the conditions of the reference, the indicated values of the loss ranged from 2 to 5 per cent for normal operative values of v/nD.

The typical combination assumed for the preceding discussion has been taken as a propeller and a fuselage. Evidently the same general principles apply and the same general results will follow for the combination of a propeller and an engine nacelle in the case of multi-engined planes, or for any similar combination of a propeller with an obstruction placed near and in its direct wake.

In addition to the above noted primary sources of reaction between the propeller and the fuselage, mention may be made of a further small increment of drag which may arise as a result of the decreasing pressure gradient in the propeller wake, as referred to above in 4.

9. Influence of the Non-Lifting System on the Control System. For our present purposes, the control system may be considered as represented by (1) the tail surfaces forming the empennage, and (2) the ailerons. The first is located at the rear and in the direct wake of the fuselage; the second is located on the wings and at a distance on either side of the fuselage.

The chief effects of the presence of the fuselage, so far as we are here concerned, are in the disturbance to the direct flow of air to and past the plane, in the production of some degree of turbulence in the air approaching the tail surfaces, and in the effects due to skin friction.

In free air, some turbulence may in general be expected; but the presence of the fuselage will increase this in degree and we may, therefore, expect that the control surfaces forming the empennage must normally be met by air with a relatively high degree of turbulence. The actual condition in this respect, will, furthermore, be due, not only to the presence of the fuselage, but also in still greater degree to the action of the propeller located at the nose of the fuselage. The actual condition at the empennage is, therefore, due to these two agencies in combination, of which, however, at the present moment, we direct attention more especially to the part due to the fuselage.

The result which we are now considering is, therefore, simply that of turbulent air flowing over the control surfaces as compared with air relatively free from turbulence. The general effect of such condition in the air, as we have already seen in **4**, will be negligible for small or moderate angles of throw of the control surfaces. Under large angles of throw, however, the effect will tend to retard conditions of airflow separation, and as earlier noted, permit effective service of the control system at angles of throw perhaps somewhat greater than would otherwise be possible.

On the other hand the general disturbance of the flow to these surfaces due to the presence of the fuselage, will tend to produce prejudicial effects on the aerodynamic force reactions upon them. The skin friction will slow up the boundary layer flow, the same as for the flow over the wings referred to in **5**, and with a similar result in reducing the force reactions on the control surfaces over which such flow passes. The extreme of such disturbances may result in placing these surfaces in a dead air turbulent fuselage wake "shadow" with more or less complete loss of effective operation. The consequences of such partial or complete loss of control, as the leading factor in the development of spins, lends special importance to effects of this character[1].

On the ailerons, situated as they are, well outside the direct influence of the fuselage, the action here considered will obviously be negligible and need not be further considered.

10. Influence of the Control System on the Non-Lifting System. In more specific terms, this is a question of the influence of the elevator, rudder or ailerons on the fuselage. Put in these terms, it is clear that such influence must be small or negligible. The elevator and rudder are located at the tail end of the fuselage and the ailerons are on the wings usually far removed from the central body of the plane. Remembering then that the aerodynamic result of the fuselage, or of the non-lifting system generally, is the development of drag, it seems clear that the elements of the control system can have no sensible influence on the conditions which determine the amount of such drag and we may, therefore, dismiss this part of the subject without further consideration.

On the other hand reference has been already made to the addition to the total drag which may be due to the elements of the control system when displaced from neutral position. While, therefore, they can have no sensible influence on the fuselage drag as such, they will, when in operative position, cause a definite addition to the drag as a whole.

[1] See Division J III **13**; also Division O VIII.

11. Influence of the Propulsive System on the Control System[1]. In **5** and **9** reference was made to the influence of the wings and fuselage on the control surfaces by way of a disturbance to the airflow with generation of turbulence. The propeller, located at the nose of the fuselage, will augment in marked degree these same causes of disturbance at the tail surfaces, and in addition will give an increase in the axial velocity as well as a rotary component in the flow of air to and over the tail surfaces.

No further discussion is required regarding the influence due to turbulence other than to note that the actual final result at the tail surfaces will be the combination of the influences discussed in **5** and **9** and in the present section, and of which the part due to the propeller is presumably the greater of the two.

It may be noted, however, that cases have been known where the dead air wake caused by a propeller hub of unusually large size has had the effect of seriously reducing or even practically destroying the effective operation of the tail surfaces operative in such wake. See also similar condition noted in **9**.

With regard to the increase of the axial component in the propeller wake, this will, in itself, give rise to an increase in all force reactions on the surfaces over which it flows, while at the same time, in combination with the downwash velocity, it will have the effect of decreasing the angle of attack on the horizontal tail surfaces and likewise the rate of change of this angle, or otherwise, of varying this angle according as the increment of axial velocity varies with varying conditions of operation. Actually the condition is still more complex since we have here to consider three directions—the line of flight, the downwash direction, and the line of the shaft. These three directions with the actual values of the several velocities referred to above will determine the final angle of attack and velocity of flow to and over the horizontal tail surfaces.

Likewise for the fin and rudder, the increase in axial velocity will have the effect of modifying the angle of attack and the rate of change of this angle, with corresponding results on its effect in connection with maneuvers and flight stability.

Then we must further include the rotary component of the velocity in the propeller wake. It is clear that this will have the direct effect of modifying the angle of attack for the entire system of tail surfaces from the value which it would have under flow without rotation. For the horizontal surfaces which are located on either side of the center line, the result will be an increase on one side and a decrease on the other, thus introducing a condition of unbalance on the two sides with a resultant moment tending to rotate the plane about the longitudinal axis. This is the same action as that produced by the propeller on the wings and noted more especially in **3**.

---

[1] See Division M II **15**, **16**.

Likewise on the fin and rudder, usually located for the most part, or wholly, in the upper part of the propeller wake, the action will be a change in the angle of attack, to the right or left according to the direction of rotation of the propeller. This again will have a tendency toward rotation or at least lateral movement, similar to the action on the wings and horizontal surfaces, but naturally less in amount.

Attention may be directed to the remarks in **4** regarding the rolling moment due to the engine torque applied directly on the plane, and to the fact that the moment on the wings due to the rotation in the propeller wake is opposite to it in sense. The same observation applies here in regard to the rolling moment resulting from the action of the wake on the control surfaces. As a whole and as noted in **4** these moments due to the wake will only in part, and normally in small part, balance the moment due to the torque, and provision for such balance must be made otherwise as indicated in **4**.

Finally it may be noted that the increase of axial velocity in the propeller wake will give rise to two opposite effects on the force reactions on the horizontal tail surfaces. The increase in velocity will, in itself, tend to increase these forces and the effects on the angle of attack, as noted above, will usually tend to decrease them, with the final result a compromise between the two.

In order to simplify the discussion thus far, the presence of the wing and fuselage in the propeller wake, ahead of the tail surfaces has been ignored and the effects noted above are those which might be expected with a skeleton plane comprising only propeller and tail planes. Actually, the presence of the wings and fuselage will complicate these conditions in marked degree. Also the location of the tail surfaces relative to the wake, as in the case of multi-engined planes, will furnish controlling conditions with regard to interferences of the type here discussed. The discussions of the preceding sections will serve to indicate the general character of these intermediate influences, but the actual results will depend in primary degree on the details of proportion and arrangement in the individual case.

It is obvious, since the functions of the tail surfaces relate directly to problems of control and maneuverability, that changes in the forces on these surfaces due to changes in angle of attack or velocity of air-flow, or both in combination, will have a marked influence on all matters relating to stability and control. Here again, while the trend of individual causes can be clearly enough seen, in combination, the problem becomes very complex and will depend on the balance of interacting conditions resulting from the particular circumstances of the case.

Some discussion of these matters will be found in Divisions J III **9**, **13**; IV, Part B; M II **15**, **16** and N VIII and to which reference may be made.

12. Influence of the Control System on the Propulsive System. When in straight horizontal flight, the various items of the control system are normally in a mean or neutral position and have little influence on the flow of air over the plane and in particular little or no influence on the flow of air to the propeller or on the conditions in the wake. Furthermore, considering the main control items as located in the empennage, these are so far behind the propeller in all cases of normal design that even when thrown to an extreme angle (elevator or rudder) they can have no influence on the flow of air to the propeller itself; neither can there be any influence on the propeller wake which could reach forward to the vicinity of the propeller itself.

While in the extreme and exact sense it may be true that any disturbance from the normal neutral position of the controls will produce a change affecting the entire flow of air to and around the plane as a whole, it is very clear that, for practical purposes, such influence estimated with reference to the propeller must be entirely negligible and may, therefore, be dismissed without further consideration.

13. General Summary of Interferences. With a somewhat different manner of grouping, the more important features of the various interactions and interferences discussed in the preceding sections may be summarized as follows:

The wings will undergo interference: (1) By way of increase of lift and increase of drag, both friction and induced, due to the increase of velocity in the propeller wake. (2) Some small decrease in lift due to the reaction of the wing vortex system with the fuselage, resulting in a downward component of velocity in the air flow to the wing with decrease in the angle of attack[1]. (3) By way of a further decrease in lift and possible increase in drag, due to the production by the fuselage of a general disturbance in the airflow over the wing, such effect extending with decreasing amount from the fuselage outward to the tips of the wings. (4) Some increase in induced drag due to abrupt changes in lift distribution due to causes noted in the preceding number. (5) By way of local disturbance to the airflow and increased turblence generated about the junction of the wings and fuselage, resulting in some loss of lift and possibly a marked increase in drag. This entire effect may be much reduced by the fitting of generous fillets at the junction of wings and fuselage. (6) By way of a rolling moment due to partial arrest of the rotary component in the propeller wake, such moment balancing in some part the counter moment due to the engine torque. (7) At extreme angles of attack (at or near the burble point) a possible beneficial effect due to the action of increased turbulence in the propeller wake aiding in some degree to delay the separation of the flow and consequent sudden

[1] See **1, 2.**

loss of lift. (8) Possible delay of separation of flow due to negative pressure gradient in the propeller wake and to decrease of angle of attack due to increase of axial component of airflow in the wake. (9) From the ailerons considered as part of the control system, causing, when in operative position, a decrease of lift on one side and an increase on the other, thus producing a rolling moment on the plane in accordance with their intended purpose. (10) The increase of drag and the yawing moment due to the ailerons when in operative position may also be noted at this point:

The fuselage as representing the non-lifting system will be subject to interference: (1) By way of a change in the direction of flow to and across the fuselage due to the action of the wings and resulting usually in some increase of turbulent wake and hence increase in drag, and at the same time with some effect plus or minus on the lift which the fuselage may incidentally furnish. (2) By way of some small increment of lift due to reaction with the wings and representing a partial extension of the wing vortex system about the fuselage[1]. (3) Increase of skin drag and turbulence effects due to action of the propeller, both by way of the increased axial and the rotary components of velocity in the propeller wake. (4) Some small increment of drag due to negative pressure gradient in the propeller wake.

The propulsive system, or briefly the propeller will be subject to interference: (1) By way of a slowing down of the inflow air velocity due to the retarding action of the fuselage and wings (or nacelle and wings) immediately behind it, thus producing an increase of thrust, other conditions the same. This increase of thrust, we remember, is, in the general case, more than counterbalanced by an increase of fuselage and wing drag due to the influence of the increased axial velocity in the propeller wake. (2) By way of the influence of the wings in disturbing the direction of the airflow thus causing an inflow to the propeller in a direction oblique to the axis and resulting in the development of lateral forces and some loss in propeller efficiency. (3) By way of obliquity between the line of action of the total force acting on the propeller and the line of flight, resulting in a loss of propulsive effect. (4) By way of a reflection, so to speak, backward from the wings to the propeller, due to the partial arrest of the rotary component in the propeller wake velocity, producing some small increase in torque.

The control system will be subject to interference, **A**: As regards the stabilizer and elevator: (1) By way of a deflection of the air due to the form of the wing and to the downwash component of the airflow, requiring an initial setting of these surfaces in recognition of such effects. (2) By way of a decrease in the force reaction on the control surfaces due to the decrement of axial velocity in the friction wake of fuselage

[1] See **1**, **2**.

and wing. (3) By way of increased turbulence due to the combined influence of propeller, wings and fuselage—usually not of serious importance at small or moderate angles of incidence on these surfaces, and possibly of some help in delaying separation of the flow around the elevator when at extreme throw. (4) By way of a decrease in the effective angle of attack on the horizontal surfaces due to the increment in axial velocity in the propeller wake, varying with varying conditions of propeller operation, and with consequent effects on longitudinal stability and control. (5) By way of a further change in the effective angle of attack at different attitudes of the plane and consequent varying obliquity of the propeller shaft with the line of flight. (6) By way of an increase in the force reaction on the control surfaces due to the increment of the axial velocity factor in the propeller wake [(4), (5) and (6) must be considered together with reference to the final result on the force reactions]. (7) By way of a rolling moment due to the rotary component of the velocity in the propeller wake and to which the initial setting must be suitably adjusted. **B**: As regards the fin and rudder: (8) By way of increased velocity in the propeller wake with general disturbance of flow, influence on angle of attack and increased turbulence, all due to the same general causes as for the horizontal surfaces. (9) By way of a lateral force or rolling moment due to the rotary component in the propeller wake, the same as for the horizontal surfaces. (10) Note should also be made of the possibility of a serious or even almost total loss of useful effect on the tail control surfaces in case they are operative in a "dead air" wake due either to the fuselage or an oversized propeller hub or some combination of the two. The same condition may also sometimes develop in maneuvers where the elevator might find itself in the wake of the fin and rudder or *vice versa* the rudder in the wake of the stabilizer. **C**: As regards the ailerons: (11) Small influence due to turbulent flow over the wings at small or moderate angles of wing attack and small aileron displacement, and possibly some advantage in retarding separation of flow when at extreme angles of aileron throw. (12) Possible serious loss of useful effect for the aileron deflected upward when main wing is at extreme angle of attack at or near the burble point, thus placing aileron in the relatively dead or eddying air in the wake of the wing. (13) To the extent to which the ailerons may be subject to influence from the propeller wake, the rotary component in the latter will still further modify the line of airflow to the ailerons, with opposite effects according as the vertical component velocity in question is upward or downward. (14) Added to these effects plus and minus on the direction of flow, there will be a small rolling moment due to the same rotary component in the wake acting on the ailerons, less however, when displaced from neutral position than when in neutral and forming a part of the wing system proper.

DIVISION Q

AERODYNAMICS OF AIRSHIPS

By

Max M. Munk,

Washington, D.C.

PREFACE

Airship design leans more heavily on aerodynamic theory than does airplane design. Individual airships are much larger and more expensive than airplanes; the completed airship structure can much less readily be modified after its completion, so that the trial and error method is practically not as available for airships as for airplanes; furthermore, there is available comparatively much less experience from earlier airships because not many have been built, and even wind tunnel tests, although they have always been diligently undertaken, carry less persuasion in consequence of the larger scale effect and the larger sensitivity to such effect and to other doubt-inviting factors. All this is indicative of the need of a further development of airship aerodynamics as a foundation for further progress in the construction of large airships. Moreover, since airship design draws on the whole domain of aerodynamics and since special airship aerodynamics should contain as its most notable problem the full analysis of airship drag, it seems quite possible that from airship theory may some day come forward such fundamental progress in aerodynamics as shall revolutionize our technique of air travel.

Airship design involves, as a special field, the investigation of air forces brought into existence by the motion through the air of large, bulky, streamlined solids. The theory of the influence of air friction on these forces, in spite of strenuous efforts, has not yet been developed to a satisfactory status and has not been included in the treatment of the present Division. For this aspect of the general problem, the reader is referred to Division G. The present Division deals only with the theoretical motion of a perfect fluid, and constitutes an application of the principles and results developed in Division C.

The author presents herewith the results of an effort to organize airship aerodynamics along certain well defined logical steps, leading to a unitary, complete and convenient system of mathematical procedure. During the last decade this system has been received and used in the mathematical computations for the design of large airships built during that period. It is hoped that it may thus constitute a permanent nucleus for the development of applied airship aerodynamics.

The basic subdivisions for such foundation for an applied theory are as follows: (A) The resultant or integral aerodynamic effect of the entire airship structure is approximated by a superposition of the air forces on the bare hull, deduced from the laws of classical hydrodynamics, and of the air forces on the fin and control surfaces, assumed to follow the laws of modern airfoil theory. (B) The local distribution of the air forces along the axis and the pressure distribution is computed on the basis of a large elongation of the airship hull, thus reducing the actual three-dimensional flow around the hull to a superposition of two-dimensional flows. (C) The errors introduced by these assumptions are taken care of by the introduction of constant correction multipliers. (D) A three-dimensional flow for a mathematically simple shape is used for the computation of the pressure distribution over the bow region. (E) The general results of strict theory, valid for certain mathematically simple shapes, are generalized by means of engineering rules to cover practical shapes.

In studying the developments of the present Division, the reader will find it helpful to keep in mind these successive steps or stages, as guides or connecting links between the successive sections.

1. Introduction. The aerodynamic theory of airships deals with the loads imposed on the structural system of airships by the air forces, and with the problem of stability and the required fin areas. As a basic assumption the theory assumes the substitution of a perfect, non-viscous fluid for air, and for this reason fails to be of use for the computation of the performance, since solids moving in a perfect fluid experience no resistance. The actual resistance of airship hulls, while not indeed zero, is, however, surprisingly small relative to their bulk, and arises almost entirely from the direct action of viscous forces.

The present treatment is based chiefly on the theory and the solutions discussed in Division C.

The exact results comprised in that Division are confined to a very small number of shapes, all of great mathematical simplicity. It is the object of the present section to discuss the application of these results to airship shapes empirically given. This must be carried out through approximations in accordance with the usual procedure when applying the results of rigorous mathematical methods to the problems of nature. Indeed, improvements in the mathematical methods would be of little further use. The main source of the disagreement between the computed air forces and pressures and the actual ones is not the lack of better mathematical methods, but the incorrectness of the physical assumptions, especially the neglect of the viscosity. In view of the discrepancy between computation and observation caused by the viscosity of the air, the methods discussed in Division C are exact enough, and are furthermore sufficient for most practical purposes.

We proceed then with some discussion of approximate methods for the computation of the numerical values of such aerodynamic quantities as we shall need.

2. Area of Apparent Mass. As the first of these problems we shall take the question of the area of apparent additional mass in two-dimensional flow. In Division B VII 4 it is shown that the area of the apparent additional mass of a circle is equal to the area of this circle itself. The area of apparent additional mass of an ellipse moving in the direction of one of its principal axes is equal to the area of the circle on the diameter at right angles to the direction of motion. This rule includes the circle as a special case, and also the straight line as an ellipse of zero thickness. As a generalization of this theorem, we propose to apply this rule to any shape reasonably resembling an ellipse. The area of apparent additional mass of an airship cross section is accordingly equal to the circle over its height for horizontal transverse motion or components of motion, and to the circle over its width for vertical transverse motions or components of motion. The corresponding inertia factor will be denoted by η, and it must not be supposed that the error committed in adopting this rule is necessarily large for sections that are pear shaped or even with a sharp corner, since it is strictly true even for a straight line. For the common circular cross section

$$\eta = 1 \tag{2.1}$$

For other cross sections

$$\eta = \frac{\pi c^2}{4S} \tag{2.2}$$

where S denotes the area of the cross section and c its largest width or height at right angles to the motion considered, which motion is supposed to be in a principal direction, or in a direction of symmetry or at right angles to it. For lateral motion in other directions, the motion must be split up into components parallel to principal directions and the apparent masses computed separately for the components, as explained in Division C III. With a circle, every diameter is a principal direction.

3. Volume of Apparent Axial Mass. For an estimate of the volume of apparent additional axial mass of an airship hull, there are available in the exact theory the volume of apparent mass for two individual sources (Division C V 5), and the results obtained for the ellipsoid of revolution (Division C VII 5). Both results agree sufficiently well to indicate that with elongated airship hulls the volume of apparent additional mass for axial motion is only a fraction of the volume of the hull itself. This is not surprising as there are no large velocities produced by the axial motion of airship hulls except near the bow and stern. The magnitude of this volume depends much on the shape of the ends of the hull. Blunt ends have larger volumes of apparent mass than well tapered ones of the same volume. Two single sources give [see Division C V (5.5)]

$$K = \frac{2}{3}\pi b^3 \tag{3.1}$$

where b denotes the largest radius of the open body form produced by one source, which is a little larger than the largest radius of the shape equivalent to two sources with equal but opposite strength. The same value of the volume of the apparent additional mass is found for an ellipsoid of revolution with an elongation ratio a/b of about 4.5. In this case, however, b is actually the largest radius, and a the semi axis. With increasing elongation, the coefficient of additional apparent volume of the ellipsoid decreases towards zero, but only very slowly. The apparent volume for a shape equivalent to two concentrated sources approaches a fixed value [that given by (3.1)] if the maximum diameter is kept constant.

The elongation ratio of modern airship hulls is not very different from the ratio $a/b = 5$. The front end is a little blunter than the bow of the equivalent elliptical shape, the rear end somewhat more sharply tapered. Since the volume of apparent additional mass is only a fraction of the hull's own volume, there is no urgent necessity for its precise value. It is therefore exact enough for practical purposes either to apply (3.1) indiscriminately to all hull shapes or to use the apparent mass of the equivalent ellipse more exactly defined in the next section.

4. Lateral Motion. The inertia factor for the transverse motions of airship hulls depends to a large extent on the shape of their cross section. For an estimate, this influence can be taken into account by applying as a correction factor the inertia factor η of the cross section. This is given by (2.2) and with geometrically similar cross sections, η is constant along the entire axis. If the cross sections vary in shape as well as in size, the inertia factor of the hull for transverse motion can be computed by summing up the effects of all its portions by an integration along the axis. In most cases, the assumption of a constant inertia factor η, equal to its average value, is exact enough.

Aside from the influence of the cross section, the inertia factor for transverse motion depends on the elongation of the hull and the distribution of the cross sections along the axis, and is estimated exact enough by introducing the inertia factor $\bar{k}_2$ for an equivalent ellipsoid of revolution. This, however, can be chosen in different ways. We prefer the ellipsoid of revolution with the same area of meridian cross section S' as the hull, and the same length, L. The elongation ratio of the equivalent ellipsoid of revolution is accordingly

$$a/b = \pi L^2/4S' \tag{4.1}$$

Denoting the inertia factor for lateral motion of this ellipsoid by $\bar{k}_2$, the inertia factor of the hull for lateral motion will then be assumed to be:

$$k_2 = \eta\, \bar{k}_2 \tag{4.2}$$

See Fig. 1 for general diagram of ellipsoid coefficients.

5. Difference of the Inertia Factors. The resultant unstable moment of the hull is seen by Division C III (4.6) to be proportional to the difference $(k_2 - k_1)$ of the coefficients of the additional mass laterally and axially. For circular or nearly circular cross sections (the most common case) it seems most appropriate to take for this difference directly the value of $(k_2 - k_1)$ of the equivalent ellipsoid of revolution. For other cross sections, those, for instance, that are distinctly elliptical, it is necessary to compute k_2 and k_1 separately and then take the difference.

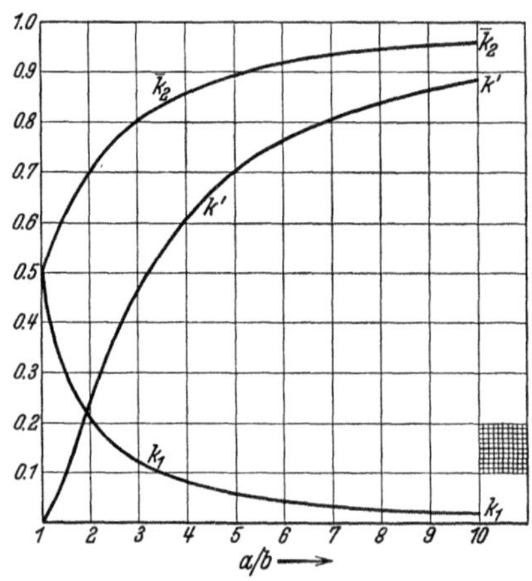

Fig. 1. Inertia coefficients of an ellipsoid moving in a fluid. k_1 axial; $\bar{k}_2$ transverse; k' rotation.

The same rules apply to the estimation of the inertia factor for rotation about an axis passing through the center of gravity of the volume of the hull, at right angles to the main axis. For circular cross sections, it will be exact enough to take the factor k' of the equivalent ellipsoid. For other cases, this factor, then denoted by $\bar{k}'$, must be multiplied by η, the inertia factor of the cross section, or by its average value over the length. Thus

$$k' = \eta\, \bar{k}' \tag{5.1}$$

where η refers to the direction in which the cross sections are moving when rotating.

6. Nose Pressure. The knowledge of the pressure distribution near the bow of airships is of particular importance, especially for ships of the semi-rigid or non-rigid type, because the pressure there reaches its maximum and minimum values, and thus determines the rigidity and strength of the bow stiffening to be provided.

Experience has shown that it is sufficiently exact for practical purposes to substitute, for the computation of this pressure distribution over the bow portion of the hull, an elliptical shell, and to compute the pressure distribution over this shell in accordance with Division C VII 6. Hulls with axial symmetry or nearly so are of chief importance, and we discuss therefore this case first. The generalization to the case of the ellipsoid with three unequal axes is not very different and offers no fundamental difficulties.

The equivalent ellipsoid of revolution used for the computation of the nose pressure is not identical with the equivalent ellipsoid of **3**, which represented the best approximation to the entire hull. It is rather chosen so as to represent the best approximation to the form of the bow only. The first requirement for this purpose is an agreement between the radius of curvature r at the front of the actual hull and of the equivalent ellipsoid. Should there be a small protruding cone or any other like formation at the front of the hull, as for mooring, it is necessary to determine the radius of curvature for the faired bow, treating the protrusion as a local attachment not virtually affecting the pressure distribution over any large area.

With this radius of curvature decided on, there remains only one more variable to be chosen. There are several possibilities, and we prefer to assume the minor axis of the ellipsoid to be equal to the maximum diameter of the hull. It is known that the radius of curvature of an ellipse at the extremity of its major axis $2a$ is b^2/a and at the extremity of its minor axis $2b$ is a^2/b. We have thus given or assumed, b and $r = b^2/a$. We have then $a = b^2/r$, the half length of the equivalent ellipsoid. Then the elongation ratio to be used in the computation of the inertia factors $= a/b = b/r$.

Since the point of maximum pressure occurs on the meridian we consider the motion symmetrical about the meridian plane. The tangent at the point of greatest pressure is at right angles to maximum velocity W of the flow, and the tangent at the point of greatest suction is parallel to the same direction. The magnitude and direction of the maximum velocity W is found by adding vectorially the axial component $u = U \cos \alpha$ multiplied by $(k_1 + 1)$ and the lateral component $v = U \sin \alpha$ multiplied by $(k_2 + 1)$. See Division C VII **12**. For very large values of the elongation ratio, the former factor, $k_1 + 1$ approaches 1 and the latter, 2. The angle between the maximum velocity and the axis is therefore larger than the angle of attack and for very elongated shapes approaches twice its value, provided the angle of attack is small. With a moderate elongation ratio of the bow, the small angle of attack is increased in the ratio $(1 + k_2)/(1 + k_1)$.

At the point where the normal to the surface is parallel to this direction, the relative velocity between air and hull is zero, and the air pressure exceeds the pressure of the undisturbed air at large distance from the airship by $\varrho U^2/2$ where U denotes the velocity of motion. See Division C II **4**. At the points where the tangent to the surface is parallel to the maximum velocity, the air velocity relative to the hull is equal to this maximum velocity

$$U \sqrt{\cos^2 \alpha \, (1 + k_1)^2 + \sin^2 \alpha \, (1 + k_2)^2}$$

and the pressure has its minimum, being smaller than the maximum pressure by the amount:

$$(1/2)\,\varrho\,U^2\,[\cos^2\alpha\,(1+k_1)^2+\sin^2\alpha\,(1+k_2)^2]$$

and hence smaller than the pressure in the undisturbed atmosphere by the amount

$$(1/2)\,\varrho\,U^2\,[\cos^2\alpha\,(1+k_1)^2+\sin^2\alpha\,(1+k_2)^2-1]$$

At all other points, the velocity is a component of the maximum velocity, and the pressure is computed from it in the same way as from the maximum velocity.

We see that the maximum pressure depends on the magnitude of the velocity only, in conjunction with the air density ϱ, and this theoretical value is realized by observation with surprising exactness. The minimum pressure, that is at the largest suction, depends besides on the shape. With bows that are ellipsoids of revolution, there exists relative to the maximum pressure, a theoretical limit to the largest sub-pressure $4\,U^2\,(\varrho/2)$, and obtained only if both the angle of attack and the elongation ratio are very large. If the hull moves strictly sideways, the theoretical greatest sub-pressure measured from the pressure at a great distance, is three times the dynamic pressure $U^2\,(\varrho/2)$, and measured from the maximum pressure, four times the dynamic pressure. In the extreme case, however, the sub-pressure actually realized would fall considerably short of the theoretical value. For small angles of attack, the theory gives quite good agreement with experience. Measured from the maximum pressure, the largest subpressure is then equal to the dynamic pressure, multiplied by $(\cos^2\alpha+4\sin^2\alpha)$ for very large elongation ratios and by $\left(\cos^2\alpha\,(1+k_1)^2+\sin^2\alpha\,(1+k_2)^2\right)$ for moderate ones.

On the other meridian sections, the points of maximum velocity and hence of minimum pressure are situated along an intersection of the ellipsoid with a plane, and hence, in a side view of the plane of symmetry, the points of minimum pressure occupy a straight line connecting the two points of minimum pressure on the meridian in the plane of symmetry. The minimum pressure is constant along this line. If the bow only is drawn, only one point of minimum pressure appears in the drawing of the meridian of symmetry. The line of minimum pressure in the side view can then be obtained approximately by connecting this point of minimum pressure with the center of curvature.

With hulls the cross sections of which are not circular, the method remains substantially the same for motions parallel to a plane of symmetry. The bow has then to be approximated by an ellipsoid with three unequal axes, and in most cases it is exact enough to compute the inertia factors by (4.2) rather than by evaluating the elliptical integrals of Division C VIII (3.9).

For unsymmetrical motion, the same procedure has to be applied to the two components of motion parallel to the principal planes through the axis, and the results superposed. The theorem referring to the curve of minimum pressure stands also for this most general case and likewise the mode of computing the minimum pressure from the maximum velocity. This maximum velocity is then, however, not necessarily parallel to the motion.

7. Stability. Spindle shaped solids moving in a perfect fluid parallel to their axis are in an unstable equilibrium, and tend to turn crosswise to the motion. This follows from the fact that the additional apparent mass for lateral motion of spindles is much larger than for axial motion. This theoretical result is borne out by experience for the motions of such solids through air. Bare airship hulls are unmaneuverable, and bare spindle shaped arrows have been known since time immemorial to fly unsatisfactorily. The remedy has likewise been known since before the dawn of history—the spindle is provided with fins near its rear end, flexible feathers for arrows, and more substantial ones for airship hulls.

Applying the theory of three-dimensional irrotational motion to spindles with fins does not give any explanation of the beneficial effect of the fins. The apparent mass is increased for lateral motion but not for axial motion, making things apparently worse. The effect of the fins can, however, be explained by going a step farther, and employing the wing theory. Solids with sharp trailing edges, such as fins, give rise to a motion of the air which is by no means irrotational throughout the entire space, but involves circulation and the development of forces perpendicular to the plane of the fin. With plane thin surfaces, in particular, there is in consequence produced an air force the magnitude of which is theoretically

$$L = \frac{\varrho V^2}{2} S \frac{\alpha}{1 + 2\, S/b^2} \tag{7.1}$$

where S denotes the area of the surface, b its lateral extension, α its small angle of attack, V the velocity of motion and ϱ the mass density of the air. The air force has the direction $\alpha \left[1 - \frac{1}{1 + 2\, S/b^2}\right]$ with respect to the motion. The center of action of the resultant air force is at 25 per cent of the chord from the leading edge. There is besides a resistance in consequence of the viscosity of the air.

The stabilizing action of the fins is the effect of this air force, the so-called lift. The fins tend to turn the hull into a direction parallel to the motion, counteracting thereby the natural instability.

The correct prediction of the fin area required for a hull is of high importance, as it is difficult to make any change after the ship has been completed. Fortunately, there seems to be a large margin of fin size, within which the stability is satisfactory, for nearly all airships built have been found to possess satisfactory stability on their first flight. Nevertheless is it very desirable to keep the fin size as close to the minimum as possible, for this not only diminishes the drag of the fins, but their weight and the weight of the structure necessary to support them reduce the useful load. Every source of information regarding the necessary fin size is therefore of definite value to the airship designer.

In addition to theory, the designer has the known experience with earlier ships as a guide in the selection of the fin dimensions, and it is of importance to interpret correctly this experience as between hulls of different shape. This interpretation is carried out by the use of some selected formula. A variety of such formulae are in use, leading to a corresponding variation in the results. To these, theory may add another at least as to its general form. The fin area is expressed by means of its ratio to some area computed from the dimensions of the hull, thus eliminating differences of hull shape, and obtaining results that can be directly compared with each other. Different standard areas are used for this purpose—usually the two-thirds power of the hull volume or the area of the meridian plane. These two areas by no means change in the same ratio if the hull shape changes, and hence the use of a fixed relation to the one or the other of two such areas will lead to different results for the fin area.

The theory of infinitely elongated hulls suggests a third standard area—the area of maximum cross section. It would even be better to employ the average cross section, that is the volume of the hull divided by its length. The ratio of these two areas is not very different for hull shapes actually in use, and therefore it would make little difference which of the two is used. Since the moment of instability of the hull with large elongation is proportional to its volume, and since (other things being equal) the stabilizing moment of the fins proportional to the area multiplied by their distance from the center of gravity, it would follow that the fin area should be proportional to the volume divided by a length—that is, to the mean cross section. But the ratio between the mean cross section and the maximum cross section is very nearly constant in modern airship construction and therefore so far as ratios are concerned, the maximum cross section may be properly and consistently used. Taking the maximum cross section then as a standard area, the next step would be the introduction of several correction factors to take care of the influences of the other elements. The rational formula suggested is therefore of the form

$$\text{Fin area} = c_1\, c_2\, c_3\, S_{max}\,. \tag{7.2}$$

where the different factors c are determined by the secondary dimensions.

There are unfortunately two effects, regarding which little is known, and this makes it useless to push the classical theory too far. In the first place, the instability of the hull is smaller than as indicated by theory, since its rear portions produce some stabilizing forces even when fins are absent. On the other hand, a portion of the fins is surrounded by air which has flown along the surface of the hull, and while doing so has, to a certain extent, equalized, under the influence of friction forces, its velocity with that of the hull, so that the relative velocity between the fin and the air is diminished. This effect diminishes the stabilizing moment and would of itself require larger fin areas. The two effects to a large extent seem to neutralize each other, but exact amounts are unknown and for this reason an exact prediction of the necessary fin area is impossible in the present state of our knowledge.

Of the factors c in (7.2), the correction factor for the elongation ratio of the hull,

$$c_1 = k_2 - k_1 \tag{7.3}$$

and the factor for the aspect ratio of the fins,

$$c_2 = \frac{1}{1 + 2\, S/b^2} \tag{7.4}$$

These are the most important ones. In the latter equation, (7.4), S is not the fin area proper but the entire area between the outer edges and including the projection of the portion of the hull between the fins.

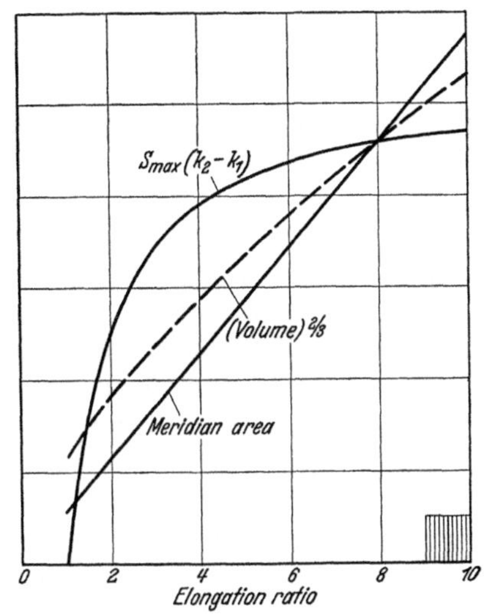

Fig. 2.

It is now interesting to compare the outcome of this method, referring the fin area to the cross section, multiplied by $(k_2 - k_1)$ with the other two methods mentioned. We illustrate the comparison by plotting against the elongation ratio of ellipsoids the three areas, meridian area, $(\text{Volume})^{2/3}$ and the proposed area $S_{max.}\ (k_2 - k_1)$. These three areas are multiplied by certain constant factors so chosen as to make them agree for the elongation ratio 8. The divergence for other values of the elongation ratio is shown in Fig. 2.

8. Lateral Forces in Straight Motion. The moment of instability

$$(k_2 - k_1)\ \text{Vol.}\ \frac{\varrho\, V^2}{2}\ \sin 2\,\alpha \tag{8.1}$$

of the air forces acting on the airship hull is produced by the variation of the pressure over the entire hull. As an intermediate step between dealing with the resultant moment only and with the entire pressure distribution, it is possible to make positive statements regarding the air forces which seem to act at different points of the axis, being the resultants of the pressure distribution around zones of the hull surfaces corresponding to the elements of the axis. These intermediate resultants lend themselves readily to the computations of the bending moments of the hull as a whole, and are therefore of direct and important use for the computation of the necessary structural strength.

According to Division C VI (4.5) very elongated surfaces of revolution when moving in a straight line with a velocity V at a small angle of attack relative to the long axis, experience lateral forces the magnitude of which, per unit length of the axis, is

$$P = \frac{\varrho V^2}{2} \frac{dS}{dx} \sin 2\alpha \tag{8.2}$$

where S denotes the circular cross section. If the cross section is not circular, the distribution of the lateral forces will be

$$P = \eta \frac{\varrho V^2}{2} \frac{dS}{dx} \sin 2\alpha \tag{8.3}$$

where η denotes the inertia factor of the cross section for a symmetrical inclination. A formula of the same form applies to the case of attack in the plane of an unsymmetrical inclination, but in the latter case the forces are not necessarily in the plane of attack.

Equations similar to (8.2) or (8,3) apply further, if the axis of the spindle shaped hull is slightly bent. In this case, the angle of attack α is considered variable and must be included under the differentiation, so that the equations become

$$P = \frac{\varrho V^2}{2} \frac{d}{dx} (S \sin 2\alpha) \tag{8.4}$$

and

$$P = \frac{\varrho V^2}{2} \frac{d}{dx} (\eta S \sin 2\alpha) \tag{8.5}$$

We proceed to the case where the elongation of the hull is not extremely large, but only moderate. The lateral forces along the axis are then modified, and there occur further couples distributed continuously along the axis. These couples arise from the conicity of the hull. If the meridian is inclined under a finite angle toward the axis, the pressure difference on opposite ends of a diameter not only gives rise to lateral components, but also to axial components of the air force, the former giving the lateral forces and the latter in the first place a couple and besides, axial force components. The couple can be computed from the assumption that the pressure around the zone is proportional to the distance from a diameter, in accordance with Division C VI (4.6). This computation gives

$$M = \frac{dp}{dy} \frac{dS}{dx} \frac{S}{2\pi}$$

where dp/dy denotes the apparent pressure gradient at the points of the surface. These couples diminish the resultant couple of the entire hull produced by the lateral forces.

In practice, these couples, continuously distributed along the axis, together with the axial forces, are often neglected, and the lateral forces are computed by a modification of (8.3), multiplying these lateral forces by a constant multiplier of such magnitude as to bring the added effect of all lateral forces in agreement with (8.1).

The constant multiplier is seen to be $(k_2 - k_1)$ and hence the usual formula for the computation of the lateral air forces is

$$P = \frac{\varrho V^2}{2} \frac{dS}{dx} (k_2 - k_1) \sin 2\alpha \tag{8.6}$$

for circular cross sections, and

$$P = \frac{\varrho V^2}{2} \frac{dS}{dx} \eta (k_2 - k_1) \sin 2\alpha \tag{8.7}$$

for other cross sections.

These lateral forces have a resulting couple, but their resultant force is zero.

9. Lateral Forces in Curved Motion. If the motion is not straight, the distribution of the lateral forces is different from that indicated by (8.6).

We begin with the hull moving uniformly along a circular path. The entire reultant air force is then given by Division C III (3.2) and (3.3). If the motion is steady, there is a resultant force passing through the center of rotation, located in general outside the hull. This gives rise to a moment of the air pressures with respect to the center of gravity of the volume of the hull, approximately equal to the couple occurring in straight motion under the same angle of attack at the center of gravity and with same velocity. The distribution of the lateral forces giving rise to the same moment is, however, entirely different in the two cases of straight and circular path. The computation of the distribution in both cases is, however, based on the same principle.

Suppose the hull to turn at an angle of yaw equal to φ, in a circular path of radius r. The momentum of each layer of air transverse to the axis is now

$$\eta v \varrho S\, dx \tag{9.1}$$

where ηS denotes the area of apparent mass of the cross section, v the lateral component of motion between hull and air, and dx the element of the axis. The transverse velocity is now variable, and is composed of the constant portion $V \sin \varphi$ produced by the yaw and the variable portion $V (x/r) \cos \varphi$ produced by the turning with radius r. The value $x = 0$ may represent the center of the hull. Hence the time rate of change of the momentum per unit length, or in other words the transverse force per unit length is

$$\frac{\varrho V^2}{2} \sin 2\varphi \frac{dS}{dx} + \frac{\varrho V^2}{r} \cos 2\varphi \left(S + x \frac{dS}{dx}\right) \tag{9.2}$$

or otherwise written

$$d F = d x \left[\frac{\varrho V^2}{2} \sin 2 \varphi \frac{d S}{d x} + \frac{\varrho V^2}{r} \cos^2 \varphi S + \frac{\varrho V^2}{r} \cos^2 \varphi \cdot x \frac{d S}{d x}\right] \qquad (9.3)$$

The first term agrees with the lateral forces on the hull flying straight at an angle of pitch φ. The direction of this lateral force is opposite at the two ends, and gives rise to an unstable moment. Ships in actual maneuver have the hull at its center portion turned inward when flying in a turn. Then the lateral force represented by the first term of (9.3) is directed inward near the bow and outward near the stern. The sum of the second and third terms in (9.3) gives neither a resultant force nor a resultant moment. The second term by itself gives a lateral force, being in magnitude and distribution almost equal to the lateral component of the centrifugal force of the displaced air, but reversed. This latter becomes distinctly apparent at the cylindrical portion of the ship, where the other two terms are zero. The front portion of the cylindrical part of the hull moves toward the center of the turn and the rear portion moves away from it. The inward momentum of the flow has to change into an outward momentum requiring an outward force acting on the air and giving rise to an inward force reacting on the hull.

The third term in (9.3) represents forces concentrated near the two ends, and their sum in magnitude and direction is equal to the lateral component of the centrifugal force of the displaced air. They are directed outward.

Hulls only moderately elongated have resultant forces the distribution and magnitude of which are different from those given by (8.2) and (9.3). The assumption of the layers remaining plane is more accurate near the middle of the hull than near the ends, and in consequence the transverse forces, diminished in some degree over the entire length, are diminished to a greater extent at the ends than near the center, when compared with the very elongated hull. In practice, it is usually exact enough to assume the same law of distribution for each single term of the equations, and to reduce the lateral forces by means of constant multipliers. These factors should be chosen different for the different terms of (9.3). The first term represents the forces giving the resultant moment proportional to $(k_2 - k_1)$ and hence it is logical to diminish this term by multiplying it by $(k_2 - k_1)$. The second and third term take care of the momenta of the air flowing transverse with a velocity proportional to the distance from the center. Moments of inertia enter in connection with the turning motion and hence it seems reasonable to affect the terms by the factor k', the ratio of the apparent additional moment of inertia to the moment of inertia of the displaced air.

The transverse component of the centrifugal force produced by the air taken along by the hull due to its longitudinal motion is neglected. It is concentrated near the ends, but its magnitude is small.

The distribution of the lateral force on an airship hull, turning at an angle of yaw φ with the velocity V and the radius r is, as the result of the preceding discussion (see Fig. 3)

$$d F = d x \Big[(k_2 - k_1) \frac{\varrho V^2}{2} \sin 2 \varphi \frac{d S}{d x} + \\ + k' \frac{\varrho V^2}{r} \cos^2 \varphi \, S + k' \frac{\varrho V^2}{r} \cos^2 \varphi \, x \frac{d S}{d x} \Big] \qquad (9.4)$$

This expression does not include the air forces on the fins, and refers to circular cross sections. With sections otherwise, the factor η, the inertia factor of the cross section, must be included in the formula.

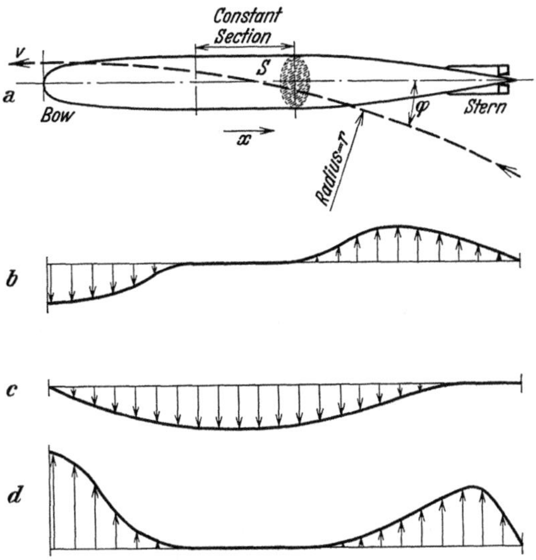

Fig. 3. Airship in curved flight and forces developed [see (9.4)]. a. Attitude of airship in relation to path. b. Curve showing 1st term of (9.4)—the same as in straight flight under pitch. c. Curve showing 2nd term of (9.4)—"negative centrifugal force". d. Curve showing 3rd term of (9.4).

The case of the air moving laterally with different velocities relative to the axis is equivalent to the motion of an elongated hull with bent axis, so that the angle of attack is in conformity with the relative motion between the air and the hull at all points. General formulae are not possible, but the principle used is the same as in the preceding case, and the computation depends on the assumed distribution of the air motion. It may be remarked that in the region of increasing cross section, that is in front, the reaction from the change of the momentum tends to move the hull into the direction of the moving air, so that after some time the relative velocity between hull and air can be expected to be decreased. However, this argument should be used with great caution, as the same reaction increases the angle of attack and may thereby neutralize to a great extent the beneficial effect just mentioned.

10. Lift of the Airship. The cross force on an airship, such as for instance the lift, can be computed to a first approximation from the preceding discussion, from the condition that the lift of an airship moving with constant translational velocity under a small angle of inclination of the axis is equal to the air force on its tail surface necessary to neutralize the unstable moment of the hull. This fin force is supposed to act at the fin, at the distance a from the center of the ship, and is supposed to counteract the moment exactly, although in practice a small "static" instability is admissible and, in consequence of "dynamic" effects, does not actually render the motion unstable.

The unstable moment was seen to be

$$M = (k_2 - k_1)\,\text{Vol.}\,\frac{\varrho V^2}{2}\,\sin 2\alpha \tag{10.1}$$

This gives the fin force

$$F = (k_2 - k_1)\,\frac{\text{Vol.}}{a}\,\frac{\varrho V^2}{2}\,\sin 2\alpha \tag{10.2}$$

where α denotes the angle of attack.

In practice, this equation is generally used the other way around. The resultant air force is then not computed from the angle of attack of the airship, but on the contrary this force is assumed in the form of excess or insufficient buoyancy, and the angle of attack necessary to create an equivalent air force is computed by means of (10.2).

This case of an airship with constant translational velocity with its nose turned up or down, producing thereby negative or positive lift, is closely allied to the case of an airship flying along a circular path while producing air forces substantially side-ways. In that case the resultant air force on the hull does not exactly reduce to zero, for even with constant tangential velocity there comes into effect a centrifugal force of the air mass virtually accompanying the ship, and this centrifugal force, together with the centrifugal force of the ship structure, including the air and gas enclosed therein, must be overcome and neutralized by the side force on the tail surfaces. Generally speaking, the strict definition of lift or side force and of drag breaks down as soon as there is no longer a translational velocity to the direction of which these components of the resultant air force can be referred. Different points of the airship move in different directions, and there is no longer a standard direction to which the drag and lift can be naturally referred. In its stead we have then a center of rotation about which the airship is turning at the instant considered. In case of a steady turn, it follows from considerations of energy, that the resultant moment with respect to that center must be zero, and hence the resultant air force must pass through that center. Such resultant air force passing through this center is therefore equivalent to the lift, and the resultant moment about the center represents the drag.

The resultant air force on a steadily turning hull is at right angles to the momentum of the air flow caused by the hull. Its magnitude is equal to the product of this momentum and of the angular velocity of the hull, or, more generally, to the vector cross product of the momentum and the angular velocity. In actual cases, the lateral component of the airship velocity is only small, and the apparent additional mass for axial motion is likewise small. From this it follows that the resultant air force is generally much smaller than the actual centrifugal force of the airship structure, and for many purposes can be neglected in comparison therewith.

If this is done, the centrifugal force of the airship just supported by its buoyancy assumes the magnitude

$$P = \varrho \text{ Vol. } \omega^2 r \tag{10.3}$$

where ω the angular velocity is equal to V/r, and hence the centrifugal force assumes the form,

$$P = \varrho \text{ Vol. } V^2/r \tag{10.4}$$

The "volume" in this expression should be increased by the apparent additional volume, if a more exact value is desired. Equating (10.2) and (10.4) gives the angle of yaw required to turn the ship at a large radius of turn,

$$\alpha = \varphi = \frac{a}{r\,(k_2 - k_1)} \tag{10.5}$$

This angle of attack is to be measured at the center of the hull. It is supposed to be small. Taking into account the apparent centrifugal force of the air leads to a slightly larger value.

These results permit of an interesting conclusion in the ideal case of a very elongated hull where, moreover, the distances from the center of the hull to the two ends are equal. We compute as follows the local angle of attack at the bow. Its lateral motion due to rotation is $a\,\omega$, directed toward the center of turn. This is equivalent to an angle of attack α such that $V \sin \alpha = V\alpha = a\omega$

and hence

$$\alpha = \frac{a\,\omega}{V} = \frac{a}{r} \tag{10.6}$$

This angle is directed opposite to the angle (10.5), for in that case the hull center was flying inward relative to the tangent of the circle of turn, so that the motion relative to the air is outward from the center of turn. Considering further that with a very elongated hull the expression $(k_2 - k_1)$ in (10.5) approaches the value one, it results that the angle (10.6) becomes sensibly equal in magnitude and opposite in direction to the angle (10.5). It is thus seen that the bow has the angle of attack zero; it has no lateral motion relative to the air, but heads directly into the wind in circular motion as well as along straight paths. This relation varies somewhat for ships with moderate elongation and different distances from the center to the two ends, but not in great degree. It can be generally stated that airships in circular flight head into the wind.

11. Conclusion. As the chief result of the foregoing investigation, it follows that in straight motion the lateral air forces acting on the airship hull are distributed proportional to the rate of change of the cross section. This gives a convenient and fundamentally sound specification for the loading assumptions for the structural computation.

The largest bending moment of the hull results proportional to its volume and to the square of the maximum velocity. All other factors can be included in a safety factor or load factor, obtained from experience and judgment. In the absence of any generally recognized method of computing the various air loads in detail from the motion of the ship, the specification of such a safety factor is preferable to the attempt to reach the same result through the specification of fictitious motions, either of the airship or of the air.

DIVISION R

PERFORMANCE OF AIRSHIPS

By

K. Arnstein and **W. Klemperer**
Akron, Ohio Akron, Ohio

PREFACE

In spite of the widespread interest in airships, actual experience in their construction and operation is accessible only to a rather limited number of engineers, so that it is not surprising that scientific literature regarding the aerodynamic performance characteristics of airships is scanty and mostly scattered through disconnected monographs, cumbersome to collect and not readily available to the student. It was therefore felt that the General Editor's desire to include a comprehensive survey of this knowledge in the present series of volumes is eagerly shared by all aeronautical students interested in lighter-than-air craft. The present authors accepted the invitation to contribute to this work with a full realization that they can do only partial justice to the presentation of such a vast and intricate subject.

They wish to acknowledge the valuable contributions to the present Division made by Mr. Herman Richard Liebert, Project Engineer and Mr. Thomas A. Knowles, Development Engineer, both of the Goodyear Zeppelin Corporation; the former who has contributed from his many years of experience in Project Work and Performance Prediction and from his familiarity with the associated problems; the latter to whom we are indebted for contributions involving navigational and operational problems, and for assistance in the systematic and didactic presentation of the subject matter and the compilation of bibliographical references.

If the experience accumulated by the Goodyear Zeppelin Corporation has been drawn upon profusely, it is only a natural result of the authors' association with this organization. However, a conscious effort has been made to base the treatment on as broad a foundation of international experience as was accessible. The frequent references to the literature are intended to help the student who is interested in more detailed information.

Much aerodynamic knowledge has been acquired through flight experience and research, and in this respect especial acknowledgment is due the "Luftschiffbau Zeppelin" of Friedrichshafen, Germany, whose pioneering of airships since 1900 forms a background for the bulk of our present operational knowledge. In America there have been three active agencies operating airships; the U.S. Navy making important contributions to the development of means for the handling of large rigid airships and smaller pressure airships, including the metal clad type; the U.S. Army developing and operating semi-rigid and non-rigid types; and Goodyear with its widespread commercial blimp operations providing opportunities for research and experiment.

Unfortunately there is no space here to give proper credit to all those who have made important contributions and advancements in the theory, construction, operation and testing of lighter-than-air craft. Many of their names will be found in the literature quoted especially in the beginning of the Division.

In attempting to make the present Division a self-contained unit, it has occasionally been necessary to briefly touch upon matters which have been more fully discussed in previous divisions, notably the one immediately preceding, but it is hoped that this will be pardoned in the interests of coherent presentation. The subject matter has been classified, as clearly as possible, into six chapters.

The first chapter is devoted to a brief discussion of Aerostatics, a subject which is not commonly thought of as a branch of Aerodynamic Theory. It was included here as far as it was thought helpful for a proper approach to many of the aerodynamic problems of airships. This chapter is not, however, a complete text on aerostatics. For instance free ballooning is not covered, even in spite of its interesting aerodynamic aspects.

The second chapter deals with the propulsion of the ship and with the accompanying aerodynamic performance in axial motion. Since the drag of airships is not embraced by the so-called classical aerodynamic theory, this chapter leans heavily on non-classical and empirical methods of approach and the subject is still largely dependent on experimental evidence and on its evaluation by scientific methods.

In the third chapter, the airship is studied in the condition of disturbed buoyancy equilibrium, where the ship flies with dynamic lift. In this condition there are many parallels with the wing lift of airplanes; the difference is mostly due to the lack of geometrical resemblance between an airship hull and an efficient wing.

The fourth chapter is devoted to departures from straight flight, such as turning, deliberate application of controls and disturbed flight through turbulent air. These conditions also have many aspects in common with airplanes, the main difference being the relatively large length of an airship.

The fifth chapter contains some of the aerodynamic problems arising with the mooring and ground handling of airships. These problems are peculiar to lighter-than-air craft.

The sixth and concluding chapter lists some of the outstanding problems and indicates the directions in which the expected future advancement of our knowledge appears to lie.

An effort has been made to deal only with the aerodynamic aspects of airship problems and in many instances the complications introduced by other vital considerations have been deliberately left out of the discussion. After all, no aircraft can be built to satisfy aerodynamic considerations alone, and while these considerations are of great importance, the practical engineer must also have in mind adequate strength and safety and the economy and practicability of construction and operation. For this reason the following pages do not, by any means, constitute a text on airship design.

CHAPTER I

BUOYANCY

1. Buoyancy Equilibrium and Its Maintenance. The flight of an airship differs from that of an airplane inasmuch as it floats in the air by virtue of the fact that the buoyant force due to the displaced air compensates the weight of the structure plus the light gas with which it is inflated. Buoyancy always acts as a vertical force and therefore the vertical equilibrium between lift and weight can be accomplished independently of the axial equilibrium between thrust and drag. Once an airship is in buoyancy equilibrium it can be nosed up and will climb without losing speed along its air path, or it can be nosed down without picking up speed. In fact, when the atmospheric temperature gradient is that actually accompanying the expansion or compression of the buoyant gas with altitude (where no radiation is present to offset the temperature balance between the ship's interior and the surrounding atmosphere), the ship can be made to rise and descend at will without incurring any change of buoyancy equilibrium since the gas and air inside will expand or contract and cool or warm up at the appropriate rate to offset the variation in the density of the air encountered. The ship will simply fly where pointed, and the purely aerodynamic problems of pitching and yawing are essentially the same. This condition will naturally prevail only while the gas containers (gas cells, balloons) are only partially filled, *i. e.*, below the pressure height[1]. In this condition the hull space not occupied by gas is filled with air either at the same pressure as the outside air, as in rigid airships having a load-carrying skeleton framework, or at a slight over-pressure maintained by blowers or by the velocity head of flight[2] and, possibly augmented

[1] UPSON and CHANDLER, Free and Captive Balloons, Ronald Press.

[2] FRITSCHE, C. B., The Metalclad Airship, Journal of the Royal Aeronautical Society, September 1931.

by the propeller slipstream, as in non-rigid or pressure airships. Upon ascending to pressure height (which can be increased by initially inflating to a lesser degree at the expense of useful load or *vice versa*) air is expelled and the gas containers become taut. Further climbing would result in expelling buoyant gas from the automatic safety valves and in making the ship "heavy", *i. e.*, heavier than the displaced air. Other possible causes of heaviness in flight are overloads from rain, snow or ice, and loads deliberately taken aboard, such as airplanes. Heaviness may also arise from cooling of the gas or flying into warmer air. On the other hand, the airship can become lighter than air by loss of load or ballast, by the combustion of fuel, or by heating. In either condition the unbalanced forces must be carried aerodynamically. Under the buoyancy equilibrium condition, however, the ship is "just as light as air".

When the atmospheric temperature gradient differs from that for actual gas expansion, the buoyancy equilibrium is disturbed by climbing or descending. When the temperature drops less with altitude (standard air temperature gradient is 3.566^0 F. per 1000 feet), or if the air is even warmer above than below, *i. e.*, if a temperature inversion exists, then the ship finds a definite zone of stability; descending lower it would become light, rising higher it would become heavy. Flying in such a zone is smooth, but landing through it requires pulling the ship down or releasing gas. On the other hand, where the air is colder above than according to the equilibrium gradient, both the air aggregate and the ship itself are statically unstable; altitude control, and especially landing, then require greater watchfulness. The allowances which the altitude navigator makes must be based upon a knowledge of the stationary temperature field through measurements obtained aboard—for instance by lowering a temperature indicator into the strata below—or received from ground stations by radio or signals. The theory involved is straightforward thermodynamics and need not be here reproduced[1]. Slide rules have been designed to facilitate routine computation of airship buoyancy problems[2].

The vertical temperature gradient in the atmosphere is intimately linked with the humidity of the air because moisture condensing acts as a powerful heat reservoir. The same is true of the humidity of the gas in the ship.

In a descent, when the gas volume shrinks, the gas becomes relatively drier and, as far as heat transfer through the cell walls may be neglected, an adiabatic exponent would govern the simultaneous changes of volume, pressure and temperature. This adiabatic exponent is larger

---

[1] Humphreys, W. J., Physics of the Air, Chap. II, 1929.

[2] Weaver, E. R. and Pickering, S. F., An Airship Slide Rule, U.S. N.A.C.A. Report No. 160, 1923.

for monatomic helium than the 1.4 valid for diatomic gases such as hydrogen, air, etc. Pure helium has $k = C_p/C_v = 1.667$. Actually, due to impurities (air and water vapor) it may vary from 1.65 to 1.58.

In a rapid ascent the same exponents hold only if the dew point is not reached in the gas. Unless super-saturation occurs, the moisture will then condense and fog forms in the gas. This moisture eventually drains off and drips to the bottom of the cells. Thence it will not immediately re-evaporate. Therefore in an ensuing ascent the gas may soon

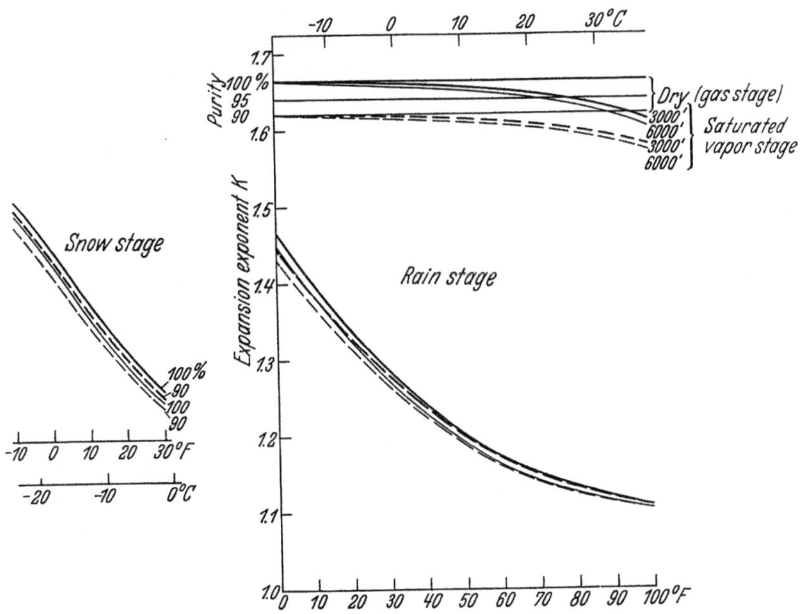

Fig. 1. Exponent for expansion of helium as influenced by temperature and purity.
Heavy full line . . . 100 % pure at 3000 ft.
Light full line. . . . 100 % pure at 6000 ft.
Heavy broken line 90 % pure at 3000 ft.
Light broken line . 90 % pure at 6000 ft.

reacquire a high exponent and retain it on subsequent descents. It is therefore possible to dry the gas by a deliberate ascent.

In such a first "wet" ascent the effective expansion exponent is lowered by condensation. This influence has been calculated for helium and found to depend largely on temperature and to some extent on gas purity and absolute pressure. The results of these calculations are reproduced in Fig. 1 for both the "rain" stage and the (problematic) "snow" stage.

In the preceding paragraphs only such rapid changes of altitude have been considered as preclude appreciable heat transfer between the gas and the air outside. However, in any slow change of altitude and under normal conditions of level flight, a continuous exchange of heat between the ship's interior and the exterior may occur due to four causes: 1) Ventilation with its bodily exchange of appreciable quantities of air drawn into and expelled from the space in the ship not filled with gas, especially emphasized when the ship is descending. 2) Conduction of heat from the gas and air through the outer cover into the boundary layer of the outside air flow, and *vice versa*. 3) Radiation or insolation. 4) Artificial super-heating.

Certain of these causes produce temperature differences between the inside and outside, while others tend to reduce any that may exist. These influences do not lend themselves readily to a generalized theoretical treatment because they depend largely on design details such as the provisions made for, and the effectiveness of, ventilation hoods, scoops and screens. As to the conduction of heat from the outer covering to the air, a theory, although not yet explicitly available, can probably be advanced. Indications are, however, that it makes little difference of what material the outer cover is made. In any case the value of such a theory for practical purposes would be quite limited because the interior phenomena, *viz.* the convection currents in the gas cells and the insulating action of the air space between outer cover and inner gas cells (where such provision is made) practically defy theoretical treatment. Yet it is these factors which, at high flying speeds, largely govern the rate of heat transfer. Therefore, calculations of thermodynamic buoyancy disturbances rely mainly on actual experience gained in the operation of similar ships. In sunny weather insolation may heat the ship to from 10° to 35° F. above outside temperature[1], causing a surplus lift of the order of ten tons with a large ship starting out without superheat and not quite fully inflated. Temperature lag, adiabatic cooling, evaporation of moisture and radiation from the ship may lower the temperature as much as 10° F. below the surrounding atmosphere. The superheat is measured by thermometers inside and outside, or by differential thermometers. However, since it is impracticable to cover the huge bulk of the ship with such instruments, it may well occur that the true average superheat of the ship as a whole will differ from that measured at particular thermometer locations[2]. It also often varies from bow to stern. Superheat in the dead air space in the ship gives a net gain in lift and may, on large ships under not uncommon circumstances, amount to a ton. Superheat of the buoyant gas is, of course, more effective in lift gain but is accompanied by unwelcome expansion of the gas cells and thus a reduction of the pressure height or service ceiling.

[1] Richmond and Scott, Effect of Meteorological Conditions on Airships, Journal R.A.S., March 1924.

[2] Bassus, K. and Schmauss, A., Zeitschr. f. Flugtechnik u. Motorl., December 16, 1911; Bassus, K., Zeitschr. f. Flugtechnik u. Motorl., March 30, 1912; Captain Gluud and von Soden, Zeitschr. f. Flugtechnik u. Motorl., Heft 7, 1912; Stern, J., Zeitschr. f. Flugtechnik u. Motorl., October 17, 1914 and October 30, 1915; Hovgaard, W., Journal of Mathematics and Physics II., December 4, 1923.

Artificial superheating in flight, using the exhaust and radiator heat of the engines, is probably of limited effectiveness but feasible to a moderate extent. Naturally the structural and mechanical complications of such a system must be weighed against the simplicity of dynamic lift for temporary periods. Artificial superheating before a take off, from energy sources ashore, is quite feasible[1].

A large rigid airship is seldom in exact buoyancy equilibrium. Its condition may be disturbed not only by the thermodynamic variations in air and gas density, but also by changes in the dead weight of the airship and the loads aboard. In the rigid airship these changes are often so small in comparison with the dynamic lift available under way that they do not hinder normal operation of the airship and pass unnoticed. However, under certain circumstances they may accumulate and their influence may gradually make itself felt in the elevator control of the ship.

Large departures from the equilibrium condition have decided disadvantages. They increase the drag (thereby penalizing either the flight speed or the fuel consumption) and the range of available control angles may be seriously reduced. The decrease in the airspeed, if serious, may so reduce the available dynamic lift that the airship "stalls" and rises or falls not unlike a free balloon[2].

The normal changes in loading due to rain, snow, ice, water absorbed by or evaporated from the fabric, flight superheat, and atmospheric changes, can be readily compensated for statically by jettisoning portions of the reserve ballast or by valving off lifting gas always carried aboard large rigid airships.

The most serious problem generally encountered is that connected with the loss of weight due to the consumption of liquid fuel in the airship engines[3]. The airship tends to become lighter by the weight of the fuel consumed. This condition, however, does not arise with a gaseous fuel of unit specific gravity or when burning the proper proportions of liquid fuels and fuel gases lighter than air[4].

The original method of intermittently restoring equilibrium on liquid-fuel airships was to climb to pressure height and to valve buoyant gas. With the advent of helium inflation this process became too expensive and alternative methods were found. The most commonly accepted method of keeping the loading sensibly constant and thus doing away

---

[1] Upson, Ralph, H., U.S. Patent 1,096,578, 1913; Blakemore, T. L., Artificial Payload for Coming Commercial Airships, Aeronautical Engineering (Am. Soc. Mech'l Engrs.), Apr.-June, 1933.

[2] Klemperer, W., The Stalling of Airships, Journal of the Institute of the Aeronautical Sciences, July 1934.

[3] Arnstein, Karl, Developments in Lighter-than-Aircraft, S.A.E. Journal, May 1929; Teed, Maj. P. L., Airship Propulsion Methods, Aircraft Engineering, December 1929.

[4] Lempertz, E., Flüssige und gasförmige Brennstoffe im Luftschiffbetrieb, Luftfahrt No. 6, pp. 82—83, Berlin 1927.

with the need for valving is one which owes much of its development to the U.S. Navy, *viz.*, the recovery of the water of combustion from the exhaust gases[1]. Other methods studied and tried with varying degrees of success are rain troughs, surface water pick-ups, and dessicating agents for atmospheric air and for exhaust gases.

Of recent years considerable thought has been given to a combination of the fireproof features of the helium inflated airship with the freedom of valving associated with the hydrogen inflated airship. Hydrogen amounting to a nominal percentage of the total inflation would be carried in special ballonets surrounded by the helium and would be valved freely for the maintenance of equilibrium. If one does not consider the careful use of small quantities of hydrogen as an increased fire hazard, this means of maintaining equilibrium appears attractive, particularly since the added hydrogen lift may compensate for the installation weight.

The large supply of liquid ballast always available in water recovery airships is a decided asset from the operational standpoint and it is very probable that the hydrogen ballast airships of the future will carry partial water recovery or ballast gaining equipment.

The normal method of deliberate valving is to climb slowly above the original pressure height so that gas escapes through all automatic valves without upsetting the trim of the airship.

The total volume of gas which has to be valved when a ship is driven deliberately, or accidentally above pressure height, is readily computed by combining the thermodynamic equation of the expanding gas with the barometric formula expressing the vertical gradient as equal to the ambient air weight density. Thus, for any rate of ascent a, there will be discharged, in unit time, a volume Q, such that

$$Q = \frac{a\,V}{R\,T\,k}$$

all in consistent units, where V is the volume of the cell, R and T the gas constant and ambient (abs.) air temperature at pressure height, and k is the effective adiabatic (or polytropic) expansion exponent already discussed[2].

[1] Barr, G., On Obtaining Ballast on an Airship During Flight by Means of Water from the Motor Exhaust, Br. A.R.C. R. and M. 234, p. 519, 1915—16; Crocco, G. A., Replacing the Weight of Materials Consumed on Airships, U.S. N.A.C.A. Technical Memorandum No. 211, 1923; Kohr, R. F., Condensation of Water from Engine Exhausts for Airship Ballasting, Bull. Bureau of Standards T-293, August 1925; Cave-Brown-Cave, T. R., Condensation of Exhaust Gases for Water Recovery, Journal of the Royal Aeronautical Society, January 1926; Burgess, C. P., Water Recovery Apparatus for Airships, Trans. Am. Soc. Mech'l Engrs. AER. 54, 11, 83, 1932.

[2] Bird, W. G., Atmospheric Humidity and the Static Lift of Airships, Jr. R. Aero Society, November, 1931.

If, in any design, the valve area A and contraction coefficient ψ are known and the permissible overpressure between the inside and outside of the valve is prescribed as p, then the "permissible rate of ascent" can be expressed as

$$a = \frac{\psi A}{V} \sqrt{p} \cdot k R T \sqrt{\frac{2g}{\gamma}}$$

where $\gamma =$ the weight density of the gas under the conditions prevailing at pressure height.

While the ship rises but stays below the pressure height, air must be vented out to make room for the gas to expand. In rigid ships not depending on internal pressure for form, this is done through ventilation orifices or screens. In pressure ships it is generally done through automatic over-pressure valves. Here again the amount to which the pressure will actually build up over the valve relief pressure setting in a rapid climb depends on details of the design and the aerodynamics of the outlets[1]. For instance, if the valves are exposed to the wash of the outside air flow, they may open under a different pressure in flight than at rest, because the air flow about the ship at the location of the valves or about the valve or parts of it may cause a "false" pressure or suction.

The flow of the escaping air or gas around the valve rim may also cause the opening characteristics (*i. e.*, the correlation between pressure and gap) to deviate from the equivalent characteristic under static loads. This influence is apt to become important since it is quite difficult to build a light valve of large diameter that will be quite tight up to a pressure of the order of an inch of water and open wide under a small increase of pressure without having much hysteresis in closing. Furthermore, chimney draft effects in vertical ducts filling themselves with the escaping gas may cause suction and the valve may fail to close at the correct pressure[2].

The impossibilitity of making the gas containers absolutely gas tight entails the necessity of measuring the permeability. Usually the rate of diffusion is slow so that very sensitive measuring methods must be resorted to[3]. The conventional theory of this diffusion is based

---

[1] KLEMPERER, W., Handbook of Experimental Physics, Vol. 4, Part 3, pp. 145—147.

[2] SCHERZ, W.: Luftfahrt, Vols. 22 and 23, 1927.

[3] FRENZEL, W., Die Gasdurchlässigkeit gummierter Ballonstoffe, Zeitschr. f. Flugtechnik u. Motorl., October 17, November 14, 1914; SHAKESPEAR, G. A., Br. A.R.C. R. and M. 317, 1917; R. and M. 447, 1918; ELWORTHY and MURRAY, Permeability of Balloon Fabrics to Hydrogen and Helium. Trans. Royal Soc. Canada 13, 37—45, 1919; EDWARDS, J. D., Determination of Permeability of Balloon Fabrics, Bureau of Standards, Tech. Paper No. 113; Journal Ind. and Eng'g Chemistry, 11, 966, 1919; NIEDNER, Zeitschr. f. Flugtechnik u. Motorl., p. 173, 1922; UPSON, R. H., Free and Captive Balloons, Ronald Press 1926; DAYNES, H. A., Gas Analysis by Measurements of Thermal Conductivity. Cambridge Univ. Press, 1933; also Br. A.R.C. R. and M. 360, 435, 504, 516, 614, 622, 640 by various investigators.

on the concept of the diffusion of gaseous media through pervious membranes and assumes the rate of diffusion of gas out and air in as inversely proportional to the square roots of their absolute densities and independent of the pressure drop. In reality pressure may have an indirect influence as the mechanical stress may cause minute seams to open in varying degrees. Comparative experiments with helium and hydrogen have not consistently confirmed the law that the diffusion of these two gases through fabric is always in inverse proportion to the square roots of their densities. There are marked differences which vary from material to material and with ageing, and therefore these observations lend weight to the theory that the diffusion process may also involve chemical or molecular interactions.

2. The Bulkhead Problem and Aerostatic Stability. Large airships of both the rigid and pressure types must be equipped with bulkheads in order that the buoyant gas may be confined to certain portions of the hull and not surge to the ends of the ship during inclinations. An airship without transverse partitions might be likened to a surface ship carrying a large liquid load unrestrained by bulkheads. It is a matter of practical experience that a surface vessel so loaded is unmanageable in a sea-way.

In general, airship bulkheads must possess a reasonable degree of tautness and stiffness and there are a number of bulkhead systems which have given acceptable solutions to the problem. The type most commonly and successfully used in rigid airships is a taut wire system in the plane of the main frames. Such bulkheads have been successfully built both with and without a central support. The central support helps to reduce the bulging of the bulkhead and its influence upon the static stability. An initially slack bulkhead can be stiffened by deliberately maintaining pressure differences between adjacent cells. Naatz[1] has proposed cylindrical slack bulkheads which may not need a differential pressure.

The stiffness and spacing of the transverse bulkheads may noticeably influence the metacentric height of an airship, and this influence has a bearing upon how low the center of gravity must be kept beneath the center of buoyancy in order to ensure sufficient static stability.

The elasticity of the bulkheads separating individual gas cells, and the flabby nature of the cell bottoms, allows some surging of the gas. A slack bulkhead would suffer under the disadvantage that its large bulge under inclination would reduce the metacentric height, especially at small angles, and might have an undesirable influence upon altitude steering.

The aerostatic stability of airships is governed by the locus of the metacenter and is measured by the metacentric height in a manner

[1] NAATZ, H., Neuere Forschungen im Luftschiffbau, Jahrbuch der Wissenschaftlichen Gesellschaft für Luftfahrt, 1923.

somewhat similar to its nautical parallel which is dealt with in textbooks on naval architecture[1]. The change of shape of the gasbags with shifting bottom level is, indeed, the inverted analogue to the behavior of liquid ballast in sea-going vessels.

The lateral and longitudinal metacentric heights may be different. Lateral stability against rolling is necessary to prevent heeling; longitudinal (trim) stability is desirable for airships, but need not be large.

The longitudinal shifting of the center of buoyancy of a partially full gas cell due to surging of the cell bottom can be computed from the geometry of the cell. In most naval vessels, where the side walls are nearly vertical, the reduction of metacentric height due to the presence of a free liquid ballast surface is simply the quotient of the equatorial moment of inertia of the mobile surface by the volume of the container. In an airship of circular cross-section the same relation would hold for the gas level only if the cell were but half inflated. For any reasonable degree of inflation the reduction of metacentric height is much smaller. The following table is an aid to such calculations and is based on the geometry of the cylinder intersected by a slant surface:

Fullness	93	94	95	96	97	98	99	100%
h	.040	.038	.036	.033	.030	.026	.018	0

where h is a non-dimensional quantity valid for unit radius and unit length of the cell. The actual reduction $-h_1$ of metacentric height due to the surging of the bottom of any cell of length L and radius R is then,

$$-h_1 = h\frac{L^2}{R}$$

For decidedly conical cells such as occur toward the extremities of a rigid ship, an exact solution is more complicated but the secondary effects from the bow and stern largely cancel each other.

The additional metacentric height reductions due to surging of bulkheads depend on the elastic mechanism of the latter. Plane, taut, radial wire net type bulkheads cause a change of metacentric height for a full cell of length L and bulkhead radius r of approximately,

$$-h_2 = \frac{5\pi}{24}\frac{r^3 L}{n T}\beta$$

where β is the buoyancy (weight density of air—weight density of gas),
n is the number of radial wires present, and
T their (average) initial tension.

---

[1] Johow-Foerster, Hilfsbuch für den Schiffbau, Vol. 1, Chap. 4, Berlin: Julius Springer, 1920; Hovgaard, William, General Design of Warships; Peabody, C. H., Naval Architecture; Hillhouse, Percy, A., Ship Stability and Trim; Niedermair, J. C., Stability of Ships after Damage, Soc. Nav. Arch. and Mar. Eng., New York, 1932.

(an initially slack diaphragm type of bulkhead would make $h_2 = \infty$ but if tension builds up under bulging, stability may be regained at small inclinations).

The metacentric height reduction of a whole ship is a weighted average of the reductions due to the individual cells, weighted in proportion to the cell volumes.

In actually computing the resultant metacentric height it must be borne in mind that not only the mentioned corrections, but also the actual loci of the centers of buoyancy and gravity, vary with fullness and altitude.

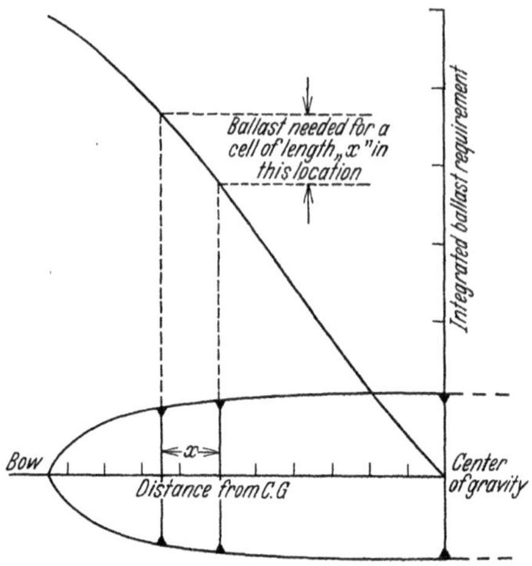

Fig. 2. Curve of integrated ballast requirement.

The inclination angle which gives equilibrium against pitching, or "trim angle" as it is called, can be effectively controlled by the shifting of loads such as fuel, ballast and, in an emergency, personnel and freight. Accidentally, the trim can be upset by such occurrences as the tearing of gas cells forward or aft, or the accidental loss of ballast or other loads. For such cases it is necessary to provide distant-controlled devices for the quick release of an equivalent amount of ballast or gas respectively, lest the ship get out of control with consequent danger of stranding if a low flying ship becomes suddenly heavy.

The hazard of a deflated cell must be provided for in the design. The supply of ballast aboard must be adequate for the restoration of trim and equilibrium in an emergency. The proper ballast requirements for any given airship shape may be obtained by the integration of the lift and deflation moments for axial slices lying between the limits of the given cell spacings[1]. Alternatively the spacing corresponding to any chosen amount of trim ballast may be determined by stepping off the allowable amount of ballast as an ordinate on a plot of integrated ballast requirement versus distance from the center of gravity (see Fig. 2).

3. Aerostatic Performance. The aerostatic performance is expressed in terms of attainable ceiling and intimately linked with the loading and gas capacity of the airship[2].

[1] Verduzio, Rodolfo, U.S. N.A.C.A. Technical Memorandum No. 285; Weiss, G., Schütte-Lanz Airship Projects After the War, U.S. N.A.C.A. Technical Memorandum No. 335.

[2] Burgess, C. P., Airship Design, Ronald Press, New York 1927.

The flight ceiling or service pressure height of an airship is determined by the volumetric percentage fullness to which the gas cells are inflated at the take-off with any given load. For "standard atmosphere" conditions[1] the ceiling for any sea level fullness at take-off can be read from the standard air density curve, the ratio of inflation in per cent of total inflation being inversely proportional to air densities at the respective heights. In case the take-off field is itself at considerable altitude, the maximum possible fullness at start is the ratio of the standard fullness for the desired ceiling to the standard fullness for the field altitude. Correction for variation of ceiling due to conditions varying in temperature or barometer from standard are made according to the well known laws of thermodynamics. For instance: A ship inflated with helium taking off from a field at 1100 feet elevation above sea level with a ground temperature of 50° F. and 28.90 inch local barometric pressure and no super-heat will have to cross a mountain range at 6,000 feet altitude under weather conditions expected to be equivalent to a rise of 0.6 in. Hg. in the sea level barometer and when the temperature is 25° F. at the altitude. What should be the initial inflation in order to prevent loss of gas enroute? Answer, 90 per cent.

The absolute ceiling or maximum pressure height at which the airship can still float is governed by the specific buoyancy of the gas and by the minimum inflation necessary to float the airship when stripped of all removable loads. The only weight then remaining to be lifted is the "dead weight". The dead weight referred to the total gas capacity (for instance, lbs. per cubic foot) is a structural constant. Its ratio to the specific lift will vary with the latter but will again become a constant if the specific lift[2] of a standard gas in standard air is used. This standardized value of the ratio is frequently used as a comparative weight criterion.

For high altitude performance it is desirable to reduce the dead weight to the lowest possible value. It is of interest to study the variations in dead weight for airships of a common type built to corresponding strength and speed requirements but differing in displacement. The methods of computing the dead weights for such a family of airships has been discussed elsewhere[3]. It may here suffice to note that modern practice still shows a trend toward decreasing values of the dead weight

[1] See Division B X, or, Diehl, W. S., Standard Atmosphere, U.S. N.A.C.A. Report No. 218, 1925.

[2] Havill, Lt. Comdr. Clinton, H., Helium Tables, U.S. N.A.C.A. Technical Note No. 276, p. 15, 1928.

[3] Hunsaker, J. C., Journal of the Royal Aeronautical Society, July 1920; Nobile, U., U.S. N.A.C.A. Technical Note No. 63; Crocco, G. A., U.S. N.A.C.A. Technical Notes Nos. 80 and 274; Lewitt, E. H., The Rigid Airship, Pitman and Sons, London 1925; Burgess, C. P., Airship Design, New York 1927.

per unit volume displacement if existing ships are to be further increased in size[1].

If for any given design specification[2] the dead weight of the airship parts performing individual functions can be expressed in terms proportional (factor k_n) to various powers n of linear ship dimensions l then the total dead weight W would be

$$W = \Sigma\, k_n\, l^n$$

The optimum size for least dead weight per unit volume is indicated by the condition $\Sigma\,(n - 3)\, k_n\, l^{n-4} = 0$

If the parts varying with the power n make up a portion w_n of the whole W, then the optimum condition is expressed by

$$\Sigma\,(n - 3)\, w_n = 0$$

If this critical sum is not 0, its sign indicates that the ship is above (+) or below (—) the optimum size in this respect.

The influence of size on the dead weight criterion is small beyond the capacities of the largest ships built to date. Variations in structural design or speed and strength requirements may have a greater influence. The dead weight of an airship is for instance modified by the choice of the frame spacing or of the number of main frames, bulkheads and cell ends required. An airship with a large frame spacing will have a lower dead weight and good altitude performance but for commercial operations may be under a serious handicap due to the large amounts of trim ballast which must be carried to meet the potential emergency of a deflated cell.

It is possible to compute and plot the variations in dead weight and ballast requirement for a series of alternative frame spacings. With existing airship types there is a definite spacing at which the sum of the dead weight and the required ballast is a minimum (and the remaining lift is a maximum). Such a spacing would give the most efficient airship for commercial purposes, although this optimum ship would not necessarily have the best altitude performance otherwise obtainable.

For airships equipped with ballast gaining equipment, a better criterion is the sum of the dead weight and the weight required for the apparatus (plus weight value of the drag) needed to recover or gain the desired amount of ballast. Since, on flights of normal duration, the conventional ballast gaining equipment gains many times its own weight in ballast, the ballast term has comparatively little influence and the larger frame spacings prove to be most efficient for airships of this type. Naturally also, they have good altitude performance.

---

[1] Arnstein, Karl, Some Design Aspects of the Rigid Airship, Trans. Am. Soc. Mech'l Engrs. p. 385, 1934; Ebner, Hans, Der heutige Stand des Luftschiffbaus, etc., Zeitschr. f. Flugtechnik u. Motorl. No. 12 (24. Jahrg., 1933).

[2] Fulton, Comdr. Garland, Some Features of a Modern Airship, Trans. Soc. Nav. Arch. and. Mar. Eng., Vol. 39, 1931.

In airships which normally maintain equilibrium by valving buoyant gas, the amount of ballast required is frequently reduced by flying the airship in a "light" condition; *i. e.*, with a reserve of buoyancy against the contingency of a deflated cell. The higher drag in the "light" (pitched) condition requires a larger fuel load for the accomplishment of a given mission. The criterion in this case is then the sum of the dead weight, the modified ballast load and the added fuel load.

In the measurement of aerostatic performance, chief concern is given to the determination of total lift in what is called the "100 per cent weigh-off". The airship is filled completely full, possibly to a slight over-pressure, and an inventory taken of the loading while the ship, kept almost in buoyancy equilibrium, rests on dynamometers. The dead weight of the airship is obtained by subtracting the known variable loads from the computed total lift. The total volume of the airship is computed from a consideration of the geometry of the cells and the mean specific lift of the buoyant gas. The determination of the mean specific lift involves the accurate measurement of the magnitude and distribution of the temperature, pressure, and humidity of the air and the lifting gas. The density, or chemical composition and purity of the lifting gas, can also be accurately measured by any of the following methods[1].

The density may be determined by measuring the gas pressure difference against air at different levels[2]. Another method consists in taking gas samples in small vessels of accurately known volume and weighing the latter on a sensitive analytical balance before and after the sample gas has been replaced by air of known pressure, or after rebalancing the same vessel with air of suitably reduced pressure[3]. Still another is the Bunsen method of timing the efflux of a gas sample and an air sample through a small orifice (Schillings, Edwards, Simmance)[4].

A portable interferometer has been developed by C. Zeiss to compare gas densities by the interference fringes of two branches of coherent light rays sent through the two samples. The velocity of sound has been proposed as means of indicating gas density[5]. Organ pipes and sonic resonator devices have been tried, also centrifugal compressors measuring speed and pressure generated.

The most common method of chemical analysis of pure helium is that of absorbing all impurities over refrigerated charcoal. The purity

---

[1] Austerweil, G., Die angewandte Chemie in der Luftfahrt; Oldenbourg 1914; Niedner, Zeitschr. f. Flugtechnik u. Motorl., p. 172, 1922; Bird, W. G., Journal of the Royal Aeronautical Society, November 1931.

[2] Barr, G., Br. A.R.C. R. and M. 88, 1912.

[3] Edwards, J. D., Bureau of Standards, T. 89; Bethuys, G., La Technique Aéronautique, March 15, 1911.

[4] Edwards, J. D., Bureau of Standards, T. 94, p. 359.

[5] Bird, W. G., Journal of the Royal Aeronautical Society, November, 1931.

of dried specimens can be accurately measured by hot wire instruments based on the high heat conductivity of helium and hydrogen (Engelhardt)[1]. Electrostatic measurements have also been tried on helium.

In warm climates the atmospheric humidity has a noticeable influence on the buoyancy of the airship. Not only has moist air less carrying capacity than the heavier dry air, but the moisture also penetrates through the fabric of gas cells and diaphragms into the gas and it may also be absorbed or adsorbed by the fabrics[2].

Many of the buoyancy problems apply to aerostats in general, since an airship, deprived of its motive power, becomes a free balloon. When an airship is in motion through the air, aerodynamic reactions arise which superimpose themselves on the independent aerostatic forces and in some respects are of predominant influence.

CHAPTER II

PROPULSION

1. Axial Motion. When the ship flies straight in buoyancy equilibrium, whether climbing, descending, or maintaining its altitude, the flow about the ship has axial symmetry with respect to the ship's axis, provided the ship's hull is built as a streamlined body of revolution. Rigid airships are usually polygonal in cross section, but the longitudinal or meridional edges follow the natural stream-lines and the number of polygon sides is usually sufficiently great to give a flow departing from axial symmetry only to a negligible degree. The cross sections of non-rigid airships are seldom perfect circles due to the loadings of gas pressure, fabric weight, and car suspensions[3]. When the latter consist of internal rigging or longitudinal curtains, the bag may have a decidedly notched appearance. It is also often the practice to build airship bags with lobed cross sections in order to reduce the maximum radius of curvature and hence the stresses in the fabric due to gas pressure. Although the pear-shaped and multilobed cross sections are not as favorable with regard to the ratio of surface area to volume, they have proved quite acceptable in service.

Airships having elliptical cross sections and therefore lacking flow symmetry have been frequently proposed. The elliptical shape, with

[1] DAYNES, H. A., Gas Analysis by Measurements of Thermal Conductivity, Cambridge, 1933.

[2] BIRD, *loc. cit.*; KLINE, G. M., Moisture Relations of Aircraft Fabrics, Bureau of Standards Jour. Research. 14, 67, 1935; HOUSTON, D. F., Effective of Protective Coatings on the absorption of Moisture by Gelatine Latex Gas-Cell Fabrics, Bureau of Stan'ards Jour Research 15, 163, 1935. .

[3] HAAS and DIETZIUS, The Stret hing of the Fabric and the Deformation of the Envelope in Non-rigid Balloons, U.S. N.A.C.A. Report No. 16; BLAKEMORE, T. L. and PAGON, W. W., Pressure Airships, Ronald Press, New York 1927; EVANS, F. G., The Cross Section of the Semi-rigid Airship, Journal of the Royal Aeronautical Society, August 1930.

the major axis vertical, offers a good approximation to the catenary needed for the transfer of the loads in the vertical plane. Placing the major axis horizontally has been advocated for many years[1], for the reasons that the aerodynamic lift would be considerably improved and the lateral forces incurred in ground handling somewhat reduced. Up to the present time considerations of simplicity and cost have so far precluded the construction of both elliptical types. If such cross sections are built in the future, attention must be given to differences in the aerodynamic force and stability characteristics arising through the lack of symmetry[2].

The surface of airship hulls is usually of fabric, either rubberized (non-rigid ships) or doped (rigid-ships). It has also been demonstrated that very thin sheet metal can be employed as an outer skin[3].

As the airship moves through the air with a flight velocity V, its hull surface is swept by an air stream the velocity of which varies along the longitudinal generatrices. Over the greater part of the ship's length this local air stream velocity is greater than V because the air displaced by the bow must flow around the hull to fill in the space at the stern, and this flow adds itself to the flight speed V. The energy vested in this velocity increment is intimately linked with the phenomenon of apparent additional longitudinal mass[4]. Toward the bow and tail where the surface elements form large angles with the direction of flight, the velocity of the airflow is reduced below flight speed. In between there are two zones, one forward and one aft, where the stream velocity is approximately the same as that of flight. Since the variations in surface velocities amount to many per cent[5] (Fig. 3), it may become of importance to account for them not only in determining the proper pitch of propellers[6] but also in locating airspeed meters[7]. Since the velocity disturbance decreases with distance from the ship, and since with various maneuvers the zone of zero speed increment at the hull proper travels sensitively fore and aft, and since there are marked changes in the magnitude of the disturbance at any fixed station, it is often preferred to suspend an airspeed head at least a ship's radius away from the hull[8].

---

[1] Burney, Comdr. Sir C. D. Bart, The World, the Air, and the Future, Knopf, London 1929.

[2] Jones, R., Williams, D. H. and Bell, A. H., Experiments on Model of a Rigid Airship of New Design, Br. A.R.C. R. and M. 802.

[3] Fritsche, C. B., *loc. cit.*

[4] See Division C III 2 and VII 7; Lamb, H., Hydrodynamics, 5th Edition, p. 116; Munk, M. M., The Computation of the Apparent Mass of Dirigibles, Journal of the Aeronautical Sciences, Vol. 2, No. 3, May 1935.

[5] Jones, R. and Bell, A., Br. A.R.C. R. and M. 1169, Table 13, 1929.

[6] Wieselsberger, C., Zeitschr. f. Flugtechnik u. Motorl., October 25, 1913.

[7] Stapfer, P., Bulletin Technique 57, March 1929, Ser. Tech. et. Indus. de L'Aéronautique.

[8] Eaton, H. N., Aircraft Instruments, p. 95, Ronald Press.

The velocity distribution about a given body of revolution and the corresponding pressure distribution can be computed by substituting a suitable system of sources and sinks or doublets[1], or in certain cases by the substitution of an equivalent ellipsoid[2]. It has been found that potential flow calculations and the corresponding pressure distributions agree with model[3] and full scale[4] experiments except for the extreme tail region. However if a model is not negligibly small as compared to the tunnel diameter, corrections for the wind restraint or expansion, in a closed or open tunnel respectively, become necessary. A knowledge of the pressure distribution is necessary for the design of the nose stiffening and the design and arrangement of valves, louvers, and other

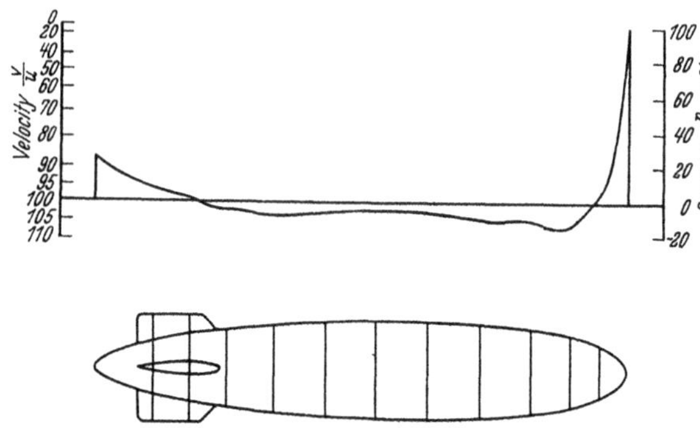

Fig. 3. Curve showing typical distribution of pressure and velocity.

ventilating equiment. The hull contour must be maintained against differences between inside and outside pressures. While the outside pressures vary decidedly over the ship's length, the interior air spaces are generally interconnected and the inside pressure remains substantially constant[5].

The potential flow from sources and sinks gives a faithful picture only at and beyond a certain distance from the hull skin. In the immediate vicinity of the skin, frictional forces act and form a boundary

[1] Fuhrmann, G., Theoretische und experimentelle Untersuchungen an Ballonmodellen, Jahrbuch V der Motorluftschiffahrts-Studiengesellschaft, 1911—12; Kármán, Th. v., Abhandlungen aus dem Aerodynamischen Institut an der Technischen Hochschule, Heft 6, pp. 3—17, Aachen, 1927 (U.S. N.A.C.A. Technical Memorandum No. 574); Smith, R. H., Longitudinal Potential Flow about an Arbitrary Body of Revolution with Application to the Airship *Akron*, Journal of the Aeronautical Sciences, Vol. 3, No. 1, September 1935; Kaplan, C., Potential Flow About Elongated Bodies of Revolution, U.S. N.A.C.A. Technical Report No. 516, 1935.

[2] See Division C VII 5.

[3] Jones, R., and Bell, A. H., Br. A.R.C. R. and M. 801.

[4] Richmond, V. C., Br. A.R.C. R. and M. 1044.

[5] Stapfer, P., *loc. cit.*

layer of air dragged along in greater or less degree. As the approximate dimensions of this boundary can be estimated, it is possible to correct for it in a second approximation by substituting such a system of sources and sinks as will produce, not the actual geometric shape of the hull, but that of the hull plus a "representative" boundary layer shell equivalent in momentum to that of the air dragged along. By selecting sinks slightly less powerful than the sources, the wake of the ship can be represented. The momentum of this additional flow can be interpreted in terms of another contribution to the longitudinal virtual mass which approximate calculations have indicated to be of the order of two per cent of the ship's mass[1].

The momentum imparted to the flow in the boundary layer is the reason for the major part of the drag of the airship and therefore the cause of the consumption of the power required to drive it forward.

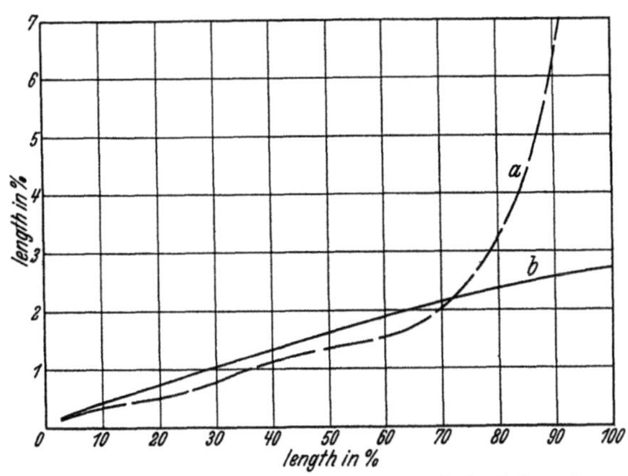

Fig. 4. Curves showing thickness of turbulent boundary layer. (a) Airship model. (b) Flat plate in turbulent flow at same Reynolds number.

The actual texture of the flow in the boundary layer shell and the velocity profile therein, is of much significance. It has a direct influence upon the parasite resistance of any small devices or accessories which may protrude from the hull, and also upon their operations, as in the case of radiators and other heat exchange apparatus, windmills for auxiliary power generation, ventilator hoods and aerodynamic measuring instruments.

The theory of the frictional boundary layer along curved surfaces has been greatly advanced recently[2]. Dr. C. B. Millikan[3] has combined the assumption of a law governing the friction velocity profile on a flat wall with the continuity requirement for elongated bodies of revolution such as airship hulls. He is able, in this manner, to account for the experimentally confirmed fact that in the bow region, where the

[1] SMITH, R. H., *loc. cit.*

[2] WILCKEN, H., Ing.-Arch., pp. 357—376, September 1930; FEDIAEVSKY, C. C., The Boundary Layer and the Drag of a Body of Revolution at Large Reynolds Number, Journal of the Aeronautical Sciences, Vol. 3, No. 1, September 1935; also see Division G.

[3] MILLIKAN, C. B., Transactions of the A.S.M.E., 1932. Also Phil. Mag. VII, p. 655, 1929.

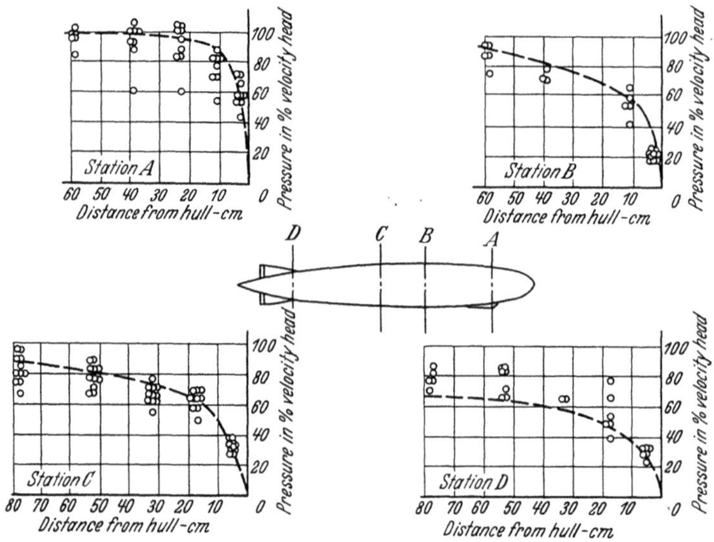

Fig. 5. Curves showing results of experiments on boundary layer velocity. Abscissae, distances outward from surface of hull. Ordinates, pressures in per cent of velocity head.

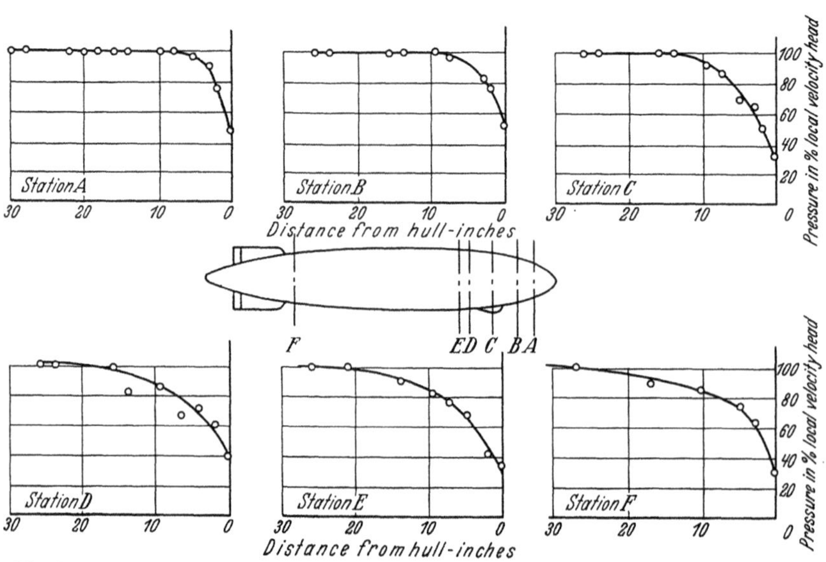

Fig. 6. Curves showing results of experiments on boundary layer velocity. Abscissae, distances outward from surface of hull. Ordinates, pressures in per cent of velocity head. Reynolds number $V\sqrt[3]{\text{Vol.}}/\nu = 635{,}000$.

streamlines locally converge outside the surface, the boundary layer grows much slower, and in the tail region where they locally diverge, it grows much faster than along a flat plate (Fig. 4).

The velocity profile has been experimentally determined in model size for the airship LZ-126 (U.S.S. Los Angeles)[1], for the U.S.S. Akron[2], for models of R-101[3], and for two stream-line bodies[4].

While in the middle body of the ship, the boundary layer follows quite well a fractional power law such as that given by the exponent 1/7, in the forebody there is a tendency for a "fuller" profile and in the rear for a "leaner" one.

In full size there have been a number of scattered experiments of limited scope made on German, English, and American ships and the results of some of these as yet unpublished experiments are here reproduced as Figs. 5 and 6. It must be kept in mind that full size experiments are much handicapped by the difficulties of averaging the fluctuating pressures measured by Pitot tubes in the turbulent zone, and last but not least, by the uncertainty regarding the ship's buoyancy equilbirium and maneuvers. In the presence of dynamic lift the boundary layer is compressed on the attacked side and expanded on the opposite side.

Although the experimental points shown were obtained under various conditions of dynamic lift, they may be considered as in general agreement with the quantitative theory for boundary layer thickness. In the middle body the boundary layer appears to have a thickness similar to that which would develop on a smooth flat plate at the equivalent Reynolds' number in terms of distance downstream from the leading edge or bow. The full scale boundary layer is definitely thinner than would correspond to geometric similarity with the boundary layers found on models. This reduction is reasonably in accord with recent assumptions of the influence of Reynolds' number on boundary layer thickness.

2. Resistance of Hull. The knowledge of the drag of an airship is of principal importance for the prediction in the design stage of its speed or power requirement. The most important single item of drag is that of the huge hull, although through careful streamlining it has become possible to reduce this drag to such a low figure that the sum of the drag contributions of the inevitable appendages and accessories of the ship, though themselves much smaller in size, may be of equal order of magnitude. The problem of aerodynamic improvement in hull shape would seem to resolve itself into finding the form of least drag for a given volume which latter dictates the gross lift obtainable.

Designers have not yet standardized on any particular shape or form of airship hull as "the best" for all purposes. It is commonly

[1] KLEMPERER, W., Windkanalversuche an einem Zeppelin-Luftschiff-Modell, Abhandlungen aus dem Aerodynamischen Institut an der Technischen Hochschule, Aachen 1932.

[2] FREEMAN, H. B., U.S. N.A.C.A. Report No. 430, 1932.

[3] JONES, R., and BELL, A., Br. A.R.C. R. and M. 1169, 1929; SIMMONS, L. F. G., Br. A.R.C. R. and M. 1268, 1929.

[4] OWER and HUTTON, Br. A.R.C. R. and M. 1271.

understood, however, that a smooth meridional curve preserving continuity up to derivatives of the second order, all along from stem to stern is desirable. Several simple mathematically defined curves, chosen for individual portions of the hull, are often combined for the sake of simplicity in mathematical calculations of such items as the buoyancy and pitching moments of various compartments when empty or partly deflated. If cleverly done, so as to match inclination and curvatures at the junction points, only negligible increase of drag may be incurred and the procedure justified. On the other hand, some investigators have tried to develop formulae from which an entire meridional curve of a good shape smooth in all derivatives can be developed[1]. Such formulae, especially when based on relatively simple source and sink concepts, may have practical advantages in the design office, but to what degree they can insure low drag for a given volume beyond securing smoothness, is problematical. However, there seems to be general agreement that the bow may, to good advantage, be somewhat blunter than an ellipsoid, although if mooring equipment requires a conical nose, no serious harm is done by such form. The insertion of a short cylindrical midship section does not seem to appreciably harm an otherwise good continuously curved shape. The curvature, usually decreasing from bow to master section is usually increased again toward the stern. This latter change, however, should be very easy and gentle. To what degree the tail end may be cut off more or less bluntly without serious harm is a matter of some uncertainty[2].

The question of the best fineness ratio (Diameter to Length) cannot be decided in a general way either. The history of airship design shows uncertain tendencies alternating between fuller and slenderer forms. Many ships, however, may have become more slender than their designers wished, either because they were to fit into available hangars, or because they were subsequently lengthened after some service in order to increase their useful load for more ambitious journeys. As the very slender form implies more surface per volume, it must cope with more frictional drag, whereas the more plump form introduces more severe curvature of the stream-lines, thus giving rise to more rapid growth of boundary layer in the rear and earlier separation of flow. In general there is but little to choose, as far as drag is concerned, from $L/D = 4$ to 8. The optimum is broad with reference to this ratio so that structural considerations which depend upon the details of the design may govern the choice. As a rule non-rigid airships are advantageously made more plump, rigid ships built of annular frames and longitudinals, more slender.

---

[1] MILARCH, Zeitschr. f. Flugtechnik u. Motorl., June 14, 1928 and August 28, 1929; COX, H. R., Journal of the Royal Aeronautical Society, p. 800, September 1929.

[2] ABBOTT, I., U.S. N.A.C.A. Report No. 451, 1932.

While with many other vehicles the ratio of resistance against motion to the gross weight carried represents the "frictional coefficient" or "gliding angle" which constitutes a measure of the degree of mechanical perfection, with airships this measure depends greatly upon both size and speed.

If the air resistance were always proportional to the exposed area and the velocity head, the refinement of any shape would be truly reflected by a drag coefficient C_D referred to unit velocity head and to the two thirds power of the volume according to the formula

$$D = C_D \frac{\varrho V^2}{2} Q^{2/3} \quad (2.1)$$ [1]

where D = drag, C_D = drag coefficient, ϱ = density of air, V = speed, Q = volume

The 2/3 power of the volume is preferable to the master section area commonly adopted in airplane fuselage aerodynamics, since the best shape for housing a given volume is not necessarily the same as that providing the best fairing for a given master section. The former is more slender than the latter. In some scientific publications a drag coefficient is determined by reference to the hull surface exposed.

The drag coefficient is, of course, not a true constant, but depends on the Reynolds number R for the ship's size and speed. Reynolds numbers are usually referred either to the cube root of the volume or to the length of the ship. Reynolds numbers of large rigid airships at commercial speeds are of the order of 10^8 to 10^9 and from ten to several hundred times larger than can at present be obtained in model experiments in wind tunnels. Insofar as the hull drag is essentially skin friction, its mechanism may vary sensitively with change of Reynolds number. Therefore the extrapolation from the value of the resistance for any known limited range of R to much higher ranges is quite uncertain, and even if data are available for one type of hull shape, it would be quite unsafe to presume similar relations for other shapes. For very large models tested in atmospheric tunnels giving a value of $L \times V$ greater than 100 m.^{2} per second, as well as in the moderate and high compression range of the N.A.C.A. variable density wind tunnel[2], and for full size airships of slender stream-line shape, a steady drop of the resistance coefficient varying at a rate somewhere between $R^{-0.17}$ and $R^{-0.08}$ has often been observed. A ship may thus have as little as half the drag coefficient shown by its model tests.

All this appears quite reasonable in the light of modern theories of the variations of turbulent boundary layer friction drag, postulating

[1] Burgess, C. P., Airship Design, New York 1927; also U.S. N.A.C.A. Technical Note No. 194.

[2] Abbott, I. H., U.S. N.A.C.A. Report No. 451.

either an exponent of -0.20[1] or more recently a logarithmic law[2] which expresses a lesser influence than an exponent of -0.20 with increasing Reynolds number.

Tests seem to indicate that the more slender the ship's form the more beneficial the "scale effect" to be expected. The more plump form apparently gives rise to an element of the drag due to actual flow detachment[3] at the tail, the magnitude of which would more nearly follow a velocity square law. Small scale model tests are severely handicapped and many show freak drag coefficients quite unsuitable for such extrapolation.

While there has been a great deal of airship model testing in various wind tunnels[4] many discrepancies were noted in the early data. In 1920 an international program[5] was instigated for the testing of two small airship models in many laboratories throughout the world. Even the results of these tests showed wide variations proving that there were obscuring influences due to the air flow in these laboratories or to the experimental technique employed.

Since that time knowledge has been greatly advanced and it appears that there are six major phenomena which are apt to obscure comparative model test results unless their influence is carefully determined and proper corrections made for them.

The first of these phenomena is the presence of a pressure gradient dp/dx in most closed wind tunnels. This causes an axial buoyancy of the order of $Q\,dp/dx$ or more accurately, $K_x\, Q\, dp/dx$[6], by which the measured drag appears too high. Where the pressure gradient is not constant along the region occupied by a long model, or in open jet tunnels where it is usually confined to a small region near the jet entrance nozzle, the product $Q\,dp/dx$ is more logically replaced by $\int_0^L S\,dp$ where S is the cross sectional area of the model at the station where, in its absence, the pressure p would prevail. This integration can be readily carried out as indicated in Fig. 7 especially if the gradient pressures vary in proportion with the tunnel velocity head without change in characteristics.

---

[1] Kármán, Th. v., Zeitschrift für angewandte Mathematik und Mechanik, Vol. 4, p. 1, 1921; Freeman, H. B., U.S. N.A.C.A. Report No. 430, 1932.

[2] Kármán, Th. v., Journal of the Aeronautical Sciences, Vol. 1, January 1934, and Proc. 3rd Int. Congress for Applied Mechanics, Stockholm 1930, and Werft, Reederei, Hafen, April 22, 1928.

[3] Gruschwitz, E., Über den Ablösungsvorgang in der turbulenten Reibungsschicht, Zeitschr. f. Flugtechnik u. Motorl. Vol. 23, No. 11, 1932. (U.S. N.A.C.A. Technical Memorandum No. 699.)

[4] See Bibliography.

[5] International Trials, Br. A.R.C. R. and M. 954, May 1925.

[6] Glauert, H., Br. A.R.C. R. and M. 1158 and 1159; Taylor, G. I., Br. A.R.C. R. and M. 1166; Zahm, A. F., U.S. N.A.C.A. Technical Note No. 23; Munk, M. M., U.S. N.A.C.A. Technical Report No. 114, 1921.

The second argument concerns the measurement of the effective velocity head of the test[1]. In an open jet tunnel obviously the difference between the total dynamic head and the static pressure prevailing in the experiment chamber surrounding the jet is a representative measure for the effective velocity head although not necessarily exactly identical with that of free flight conditions. In a closed tunnel, however, the velocity head varies over the entire space surrounding the model, so that a definition of speed measurement becomes necessary, or corrections may have to be computed if the tunnel speed is measured at an arbitrary place in the tunnel and referred to a different standard station. The different degree of flow constraint offered by the open and by the closed tunnel make it doubtful if the drag of any one model should be expected to appear the same in both tunnels. In the open tunnel a given model may act like a fatter one in free air; in the closed tunnel like a slimmer one in free air.

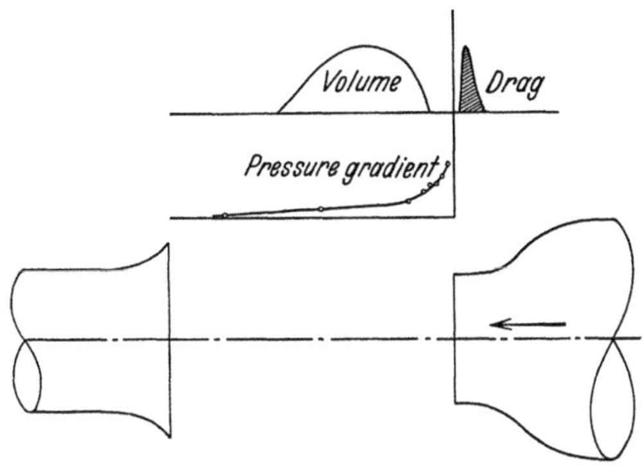

Fig. 7. Correction for pressure gradient.

The third phenomenon apt to obscure model drag test results is the tare drag and flow interference caused by the suspension system connecting the model to the balances. The determination of the net drag of airship models by wind tunnel experiment requires great experimental skill. The forces are small and it is often difficult to keep the suspension tare drag sufficiently low to prevent the final value resulting as the difference between two large quantities. Where the model is suspended by wires, it is necessary to exercise great care to avoid flow obstructions or disturbances at the attachment points without at the same time introducing mechanical constraint. In addition the tare drag area of the wires may vary with the tunnel speed.

The earlier method of determining the tare drag by means of a test with all suspension members doubled is usually less accurate than the dummy method in which the model is independently suspended in place while the original suspension system is alone acting on the balance without contact with the model. Spindle or jig suspensions are very treacherous unless the utmost care is taken to avoid mutual influence of flow. Even if only a tail spindle protrudes from the model, its anchorage on a strut or rig downstream may cause sufficient stagnation

[1] LAMB, H., Br. A.R.C. R. and M. 1010; LOCK, C. N., Br. A.R.C. R. and M. 1275.

of the flow upstream to obscure the delicate drag of the model, as was discovered at the Langley Memorial Laboratory[1].

The fourth obscuring influence is inherent in small Reynolds number experiments where a large part of the boundary layer[2] is laminar while with increasing Reynolds number the transition point from laminar to turbulent boundary layer creeps forward on the model until turbulent flow prevails throughout, as it undoubtedly does in full size. Jones[3] has demonstrated that the peculiar and seemingly erratic drop and rise of the drag coefficient observed in many small scale tests can be readily interpreted as due to such a travel of the inception of turbulence. He also demonstrated that the magnitude of the lowest observed drag values fits well a theoretical laminar friction of a flat plate of similar extension as the exposed surface and that for higher Reynolds number the theoretical turbulent friction of an equivalent flat plate is approached. Millikan[4] has refined this picture very much by reconciling these measured drags with those to be computed for a body of revolution retaining from the flat plate theory merely the $1/7^{th}$ power profile law. The next step along this line is the substitution of the Kármán-Prandtl[5] logarithmic profile law[6], with possibly the introduction of the influence of the surface taper and curvature upon this profile law.

The fifth obscuring influence is the turbulence inherent in the tunnel. It has a bearing on the prevalence of turbulence in the boundary layer. In more turbulent air, naturally more of the layer is turbulent than in smooth air at the same Reynolds number. This has been demonstrated by introducing artificial turbulence into the tunnel air either by annular protuberances placed on the bow of the model or by wire screens placed upstream[7]. It has therefore been suggested that airship model tests, if they must be done at Reynolds numbers insufficiently large to ensure

---

[1] Abbot, I., U.S. N.A.C.A. Report No. 394, 1931.

[2] Zahm, A. F., U.S. N.A.C.A. Report No. 139.

[3] Jones, B. M., Br. A.R.C. R. and M. 1199; Ower and Hutton, Br. A.R.C. R. and M. 1271 and 1409.

[4] Millikan, C. B., Transactions of the A.S.M.E., 1932; Freeman, H. B., U.S. N.A.C.A. Report No. 430.

[5] Prandtl, L., Neuere Ergebnisse der Turbulenzforschung, Z. V.D.I., Vol. 77, No. 5, April 28, 1933 (U.S. N.A.C.A. Technical Memorandum No. 720); Kármán, Th. v., Turbulence and Skin Friction, Journal of the Aeronautical Sciences, Vol. 1, January 1934.

[6] Moore, N. B., Application of Kármán's Logarithmic Law to Prediction of Airship Hull Drag, Journal of the Aeronautical Sciences, Vol. 2, No. 1, January 1935 and The Boundary Layer and Skin Friction for a Figure of Revolution at Large Reynolds Numbers, Daniel Guggenheim Airship Institute, Akron, Ohio, Publication No. 2, 1935.

[7] Relf, E. F., and Lavender, T., Br. A.R.C. R. and M. 597; Lyon, H. M., Br. A.R.C. R. and M. 1511; Prandtl, L., Der Luftwiderstand von Kugeln. K. Gesellschaften zu Göttingen, Math.-physikalische Klasse, 1914; Lyon, H. M., The Drag of Streamline Bodies, Aircraft Engineering, September 1934.

essentially turbulent boundary layer, be made with the air stream rendered artificially turbulent and this turbulence measured by the sphere drag[1] or other methods. However, Jones questions the adequacy of this artifice.

The last of the six factors to be considered is the smoothness of the surface. While some investigators have found large variations of drag with surface conditions, others have found practically none. These differences may be due to different turbulence regimes[2]. Smooth wax-polished model surfaces seem to give the lowest and most consistent drag results. In full size the skin of both metal clad and well doped or rubberized fabric covered airships can probably be considered as aerodynamically "smooth"[3].

For unusually rough hulls the theory of friction on rough surfaces would apply. Th v. Kármán[4] has shown that in order to be aerodynamically smooth the hull of airships should not have a roughness exceeding .03 to .04 mm over the greater part of their length, the very bow being the most sensitive. Well doped taut fabric and thin sheet metal under pressure are smooth within this specification. However, in actual service, fabric may flap when not taut and metal sheet may be wrinkled and studded with rivet heads. To what degree such surface irregularities may influence the mechanism of impulse transmission in the boundary layer is still problematical.

In summarizing it may be said that while an injudicious application of wind tunnel test drag measurements to a full size project can be quantitatively and qualitatively grossly misleading, the prediction of full size drag need not necessarily depend solely on a digest of actual flight experience and service performance. On the contrary careful model tests at suitable Reynolds number[5] under controlled turbulence and surface conditions are quite apt to reveal the degree of perfection of a proposed shape with respect to skin friction and pressure drag. For good shapes the full size drag can then be calculated with a satisfactory degree of reliability, confirmed by actual flight performances of ships built.

Theoretically it is interesting to compare the measured drag with the impulse left in the wake[6] which is a large portion of the whole; and

---

[1] Dryden, H. L., and Kuethe, A. M., U.S. N.A.C.A. Reports Nos. 342 and 392.

[2] Wieselsberger, C., Experiments on Model Balloons and the Resistance of Various Kinds of Surfaces, Zeitschr. f. Flugtechnik u. Motorl., September, 1915; Abbott, I. H., Airship Model Tests in the Variable Density Wind Tunnel, U.S. N.A.C.A. Report No. 394.

[3] Gehman, S. D., and Mallory, G. C., Skin Friction of Various Surfaces, Journal of the Franklin Institute, Vol. 216, No. 3, 1933.

[4] Kármán, Th. v., Turbulence and Skin Friction, Journal of the Aeronautical Sciences, Vol. 1, 1934.

[5] Ower, E., and Hutton, C. T., Br. A.R.C. R. and M. 1409.

[6] Betz, A., Zeitschr. f. Flugtechnik u. Motorl., p. 42, 1925; Schrenk, M., Luftfahrtforschung, Vol. 2, Heft 1, 1928.

with the pressure drag from normal pressure measurements by integration over the projected area elements which, with efficient shapes, is a very small portion of the whole[1].

3. Resistance of Accessories. It has not yet been found feasible to house all of the equipment of an airship within the streamlined hull, although there prevails a decided tendency to eliminate more and more of the outside appendages and protuberances. So long as individual propellers are disposed about the ship there are outriggers or complete power cars outside which require individual fairing[2]. The control car accommodating the navigating crew is usually located in the forward lower part of the hull. An empennage is carried at the tail. Furthermore, ground handling and mooring attachments usually protrude from the hull. Last but not least there are items of equipment such as radiators and other devices for the exchange of heat, and likewise hoods and vents for the intake or expulsion of ventilation air and again certain navigational instruments, all of which depend upon exposure to the outside air for their proper operation.

Evidently the drag of such protuberances and appendages can be calculated from model experiments made on much larger scale than for the ship as a whole. In all of this, allowance must of course be made for any local excess or deficiency of airspeed due to the potential flow or the boundary layer about the ship.

Occasionally, relatively small protuberances, especially on the forebody on a model tested at low Reynolds number, have shown an apparent influence by way of an increase of drag far beyond any normal expectation based on the drag of the protuberance itself. That similar freak influences would occur in full dimension seems rather doubtful in the light of experiments made with artificial "spoilers" on full size ships and in high Reynolds number model tests[3].

The drag of nacelles or power cars can be estimated from experiments on similar objects occuring in heavier than air design provided allowance is made, if necessary, for the proximity to the airship hull which acts as a mirror surface of the wash of the propeller slip stream, and if need be, of the flow through such a car containing a radiator vent, or the like.

Undoubtedly the most favorable location, from the viewpoint of drag, for a radiator, water condenser, or other heat exchange device,

---

[1] Fuhrmann, G., Jahrbuch V der Motorluftschiffahrts-Studiengesellschaft, 1911—12; Fage, A., and Stern, W. J., Br. A.R.C. R. and M. 107, 1914; Richmond, V. C., Airship Research and Experiment, Journal of the Royal Aeronautical Society, October 1926; Stapfer, P., Bulletin Technique 57, Ser. Tech. et Indus. de L'Aéronautique, March 1929; Freeman, H. B. U.S. N.A.C.A. Report No. 443, 1933.

[2] Arnstein, K., Some Design Aspects of the Rigid Airship, Trans. Am. Soc. Mech'l Engr's p. 385, 1934.

[3] Abbott, I. H., U.S. N.A.C.A. Report No. 451.

would be just inside the hull so that the skin of the ship, the friction of which is inevitable, could be utilized as a heat radiating surface without adding parasite resistance. The area required in such case is, of course, much larger, since the heat must traverse the entire boundary layer of the hull. However in view of the fact that the heat transfer varies with a fractional power of the velocity head, the handicap is not necessarily insurmountable. Just how far the finning of a heat exchange apparatus should protrude into the ship's boundary layer is a matter of design compromise into which considerations of space available, weight, complication and maintenance enter, aside from the mere question of drag.

The drag of the fins of an airship can be computed from model experimental data with a similar degree of accuracy as in the case of airplanes, excepting only that their huge size renders a beneficial scale effect of frictional resistance of importance. For conservative estimates, however, it is well to add a certain average of induced drag to the form drag of fins and control surfaces, because in flight they are, for reasons to be explained later, almost continuously under some attack—a condition which entails the development of induced drag along with the forces of control. The more stable the ship the less the allowance required in this respect.

A sample list of accessories drag is given in Table 1 for three typical rigid airships as estimated under certain experimental conditions, without water recovery apparatus. The accuracy of any analysis of the gross drag of an airship into various parts is naturally dependent upon the

Table 1.

Estimated Drag Area Breakdown Without Water Recovery	Bodensee		U.S.S. Los Angeles		U.S.S. Macon	
	Sq. M.	Sq. Ft.	Sq. M.	Sq. Ft.	Sq. M.	Sq. Ft.
A. Bare Hull	9.4	101	21.8	235	39.0	420
B. Fins and Rudders	2.5	27	4.9	53	14.0	151
C. Wing Power Cars or Outrigger Gears, Their Suspension, Ladders, Struts, Hoods, Radiators, Exhaust Mufflers	2.8	30	6.8	73	10.7	115
D. Rear Power Car with Handling Rails and Bumpers	2.4	26	2.2	23.5		
E. Control Car or Passenger Car with Handling Rails and Bumpers	2.4	26	4.5	48	2.8	30
F. Miscellaneous Protrusions-Mooring Mast Equipment, Hoods, etc.	0.5	5	0.8	8.5	1.8	19
	20.0	215	41.0	441	68.3	735
$(\text{Volume})^{2/3}$	790	8500	1845	19852	3528	37978
Resistance Coefficient C_D	.025		.022		.019	

availability of uniformly accurate data on the contributions of all these parts. By attributing different degrees of importance to the various component data derived from indirect evidence, conclusions can be shifted somewhat[1].

4. Experimental Determination of Drag. The combined drag of the hull and of the accessories—the gross drag—enters into speed and performance computations. Measurement of this gross drag may be attempted by direct experiment full size. It would be interesting to measure this drag directly by towing from another airship. So far this has not yet been accomplished, but undoubtedly will be some day. Measuring the thrust of the propellers would also furnish a measure of the drag. However, consideration must be given to the force reactions due to the presence of the propeller wake impinging on part of the structure. An analysis of the problem has been given by Durand[2]. Successful thrust dynamometers to be inserted between shaft and propeller hub have been constructed[3] and it would be only a matter of carrying out such a test program to obtain exhaustive data. However, the costs and elaborate preparations necessary have thus far prevented such a test. Thrust measurements on one of the five propellers of the U.S.S. Los Angeles were made by the Zeppelin Company and served to confirm the resistance estimates under various operating conditions and engine combinations. However, the experimental error multiplied by 5 and the uncertainty regarding the degree to which the five propellers could be considered identical and equally loaded, limit the accuracy of the conclusions.

There is, however, an indirect method available for the determination of the gross drag, the so-called "deceleration" or "coasting" test[4]. The ship is flown at its top speed and then suddenly on signal all engines are stopped. The ship gradually slows down and the deceleration process is recorded by suitable airspeed meters. The underlying theory of the evaluation of the deceleration records is based on the equilibrium between the aerodynamic drag and the inertia force.

$$D = -M\,(1 + k_1)\,\frac{dV}{dt} \tag{4.1}$$

where M is the ship's mass, k_1 the contribution of the virtual longitudinal mass due to potential flow and boundary layer[5], and V the velocity

[1] HAVILL, LT. C. H., The Drag of Airships, U.S. N.A.C.A. Technical Note No. 247, and 248.

[2] DURAND, W. F., U.S. N.A.C.A. Report No. 235, pp. 3—5, 1926.

[3] GOVE, W. D., and GREEN, M. E., U.S. N.A.C.A. Report No. 252, 1927; SEEWALD, F., Zeitschr. f. Flugtechnik u. Motorl. **13**, 1931.

[4] VON SODEN and DORNIER, Die Bestimmung des Schiffswiderstandes durch den Fahrtversuch, Zeitschr. f. Flugtechnik u. Motorl., No. 19, 1911; MUNK, M., The Drag of Zeppelin Airships, U.S. N.A.C.A. Technical Report No. 117; STAPFER, P.. *loc. cit.*; BURGESS, C. P., Airship Design, New York 1927.

[5] U.S. N.A.C.A. Technical Report No. 164.

at the time t. If the ship is in buoyancy equilibrium, $M = \varrho Q$ and if the drag is expressed as in (2.1) we have immediately

$$C_D = -2\,Q^{1/3}\,(1 + k_1)\,\frac{d\,V}{V^2\,d\,t}$$

However, since $-d\,V/V^2 = d\,(1/V)$ this becomes,

$$C_D = 2\,Q^{1/3}\,(1 + k_1)\,\frac{d}{d\,t}\left(\frac{1}{V}\right)$$

If then $1/V$ is plotted against time, the slope of the curve is indicative of the drag coefficient. Figure 8 shows a sample record of an original coasting test, and Fig. 9 its evaluation in terms of $1/V$. In some instances the curve of $1/V$ versus t appears quite straight, thus revealing no variation of the drag coefficient with speed. In other tests the curve appears concave as though indicating the regular "scale effect" of turbulent friction. Others invite interpretation as a broken line, as though two distinctly different slopes and drag coefficients prevail above and below a critical speed[1]. Again others show more erratic behavoir. However, all such detailed conclusions must be taken with due reserve considering the serious experimental difficulties which attend tests of this nature. The lag in the speed recorders, the influence of gusts and slight pitching or yawing of the ship, the drag due to rudder and especially to elevators, as well as the time required to bring the propellers to a stop often prevent experiments on even the same ship in the same flight repeated after only a short interval or with different measuring

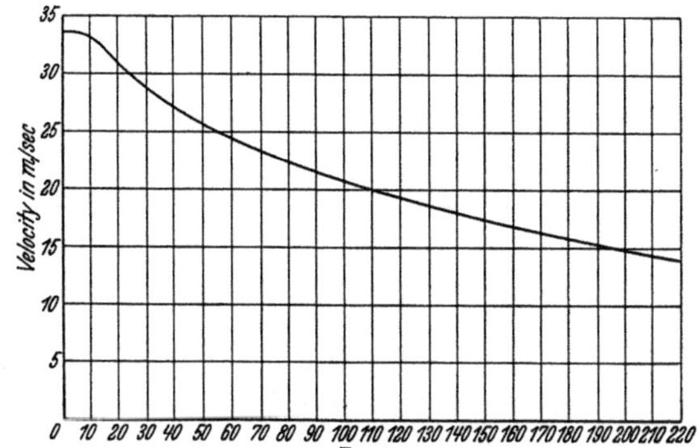

Fig. 8. Deceleration test, velocity on time.

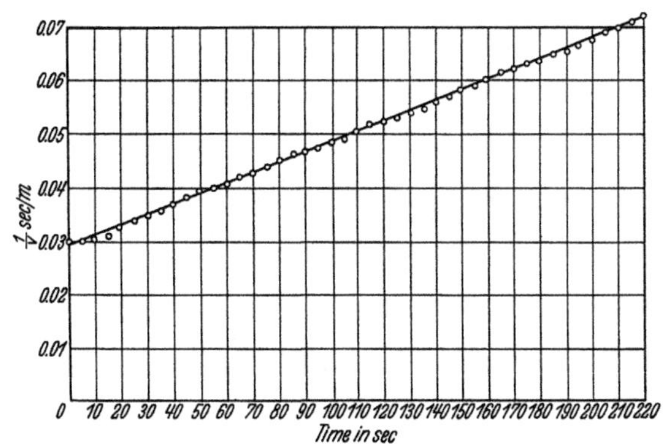

Fig. 9. Deceleration test $1/V$ on time.

[1] MUNK, M., U.S. N.A.C.A. Report No. 117, 1921; DEFRANCE, S. J., and BURGESS, C. P., U.S. N.A.C.A. Report No. 318, Figs. 4, 8, 12, 1929; THOMPSON, F. L., and KIRSCHBAUM, H. W., U.S. N.A.C.A. Report No. 397, Fig. 5, 1931.

instruments, from giving duplicate results[1]. For any accurate evaluation, automatic records of elevator angle, ship's inclination and altitude are indispensable. Proper correction for the drag due to pitch, elevator and propellers and the influence of possible lightness or heaviness of the ship have been found to straighten out the $1/V$ curves very remarkably in deceleration tests made in calm air, as shown in Figs. 10 and 11, taken from a typical experiment on a large rigid airship.

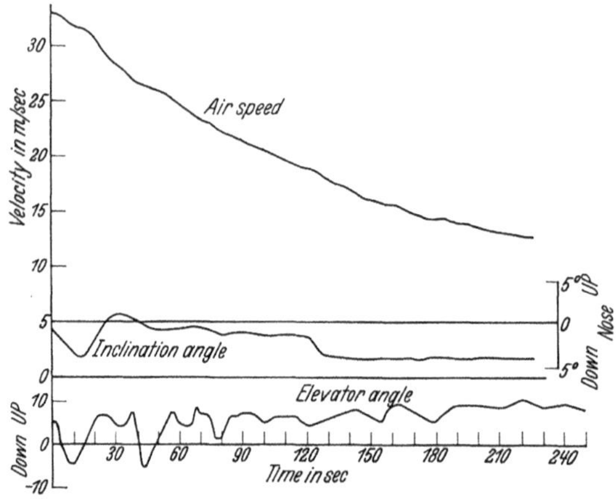

Fig. 10. Deceleration test, velocity on time with records of inclination (pitch) and elevator angles.

In evaluating coasting tests the drag of the dead (or idling) propellers must be separately determined and subtracted from the experimental result in order to obtain the ship's own drag[2].

If the ship was not in perfect buoyancy equilibrium at the time the coasting test was run, a correction for the induced drag of the dynamic lift (L) or dip ($-L$) and for the difference between the ship's actual mass and that of the air displaced, *viz.*, $\pm L/g$, is required.

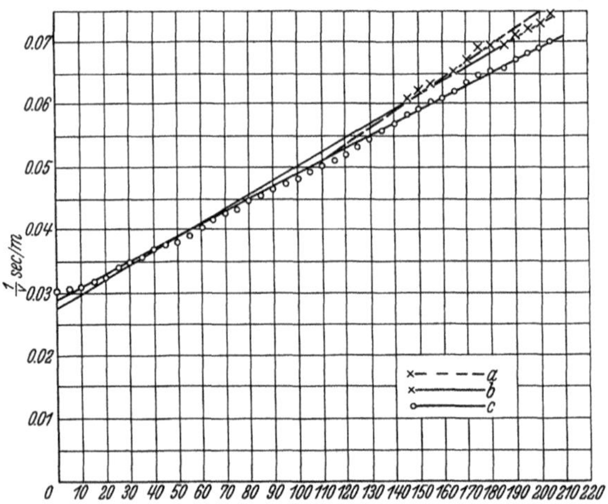

Fig. 11. Deceleration test, $1/V$ on time. Observations corrected for pitch and elevator angles. a. Interpreted as a critical region, uncorrected. b. Faired average, uncorrected. c. Corrected for pitch and elevator action.

5. Propulsive Efficiency. Speed Performance. The power required to drive a ship of known resistance characteristics at a given maximum air speed V is obviously[3],

[1] U.S. N.A.C.A. Report N. 397, Figs. 4 and 10.

[2] Hartman, E. P., U.S. N.A.C.A. Report No. 464.

[3] Jaray, P., Studien zur Entwicklung der Luftfahrzeuge, Zeitschr. f. Flugtechnik u. Motorl. Heft 11, 1920.

$$P = \frac{\varrho\, V^3\, Q^{2/3}}{2\,\eta}\, C_D$$

or

$$P = \frac{\varrho\, V^3 A_D}{2\,\eta} \qquad (5.1)$$

where the drag area $A_D = C_D Q^{2/3}$ and η = propeller efficiency. In horsepower and foot, pound, second units, this becomes

$$P = \frac{\varrho\, V^3 A_D}{1100\,\eta} \qquad (5.2)$$

In order to obtain maximum speed from a given engine power it is necessary to select a propeller type which will reach its highest efficiency and the maximum permissible engine speed at that top velocity.

If in cruising, all engines are throttled to some fraction of maximum power, the propeller efficiency changes only slightly because the engine speed automatically varies nearly proportional to the forward speed, thus leaving the value of (V/nD) (see Division L) nearly constant. This would be exactly so if the drag area A_D were independent of the Reynolds number or of the velocity V; but even scale effect alters this relation so slightly that one may speak of the "effective pitch" of the propeller or of its slip against the zero thrust pitch as a measure of the resistance condition of the ship and any change in this condition will manifest itself in a change of the effective pitch or slope of the V/rpm line. It is significant that the effective pitch or the airspeed made good for any given rpm is independent of the air density.

If in cruising some of the engines are shut down and the others made to do all the work, the disk loading of these propellers will be increased. This will result in an added drag of the dead propellers, an increase of slip, a decrease in the value of V/nD and a loss in efficiency. Adjustable pitch propellers have some advantages in such case as they will permit of better adaptation of the motor speed to the increased loading.

If the power absorbed by the propellers is known (it can be determined by measuring the torque and the shaft speed) then the observation of the air speed made good will give the "propulsive" efficiency E, which is defined as the quotient of the propeller efficiency by the gross drag coefficient.

$$E = \frac{\eta}{C_D} = \frac{\varrho\, V^3\, Q^{2/3}}{2\,P} \quad \text{(in consistent units).}$$

This is a non-dimensional number indicative of the degree of engineering success.

The efficiency of airship propellers is somewhat limited by the low pitch diameter ratio unavoidable for economic air speeds. Low speed motors and geared down propeller drives are therefore preferable from the viewpoint of efficiency alone, although the resulting large propeller diameters and increased weight and resistances of the rigging call for compromise values.

The conventional location of propellers is nearly amidships, somewhat distributed so as to avoid overlapping slipstreams. However, considerations of structural advantage may warrant a departure from this arrangement and invite tandem arrangements despite the contingent loss of propeller efficiency. The pitch of the rear propellers is, of course, made larger than that of the forward ones, and the sense of rotation is best alternated. Placing a propeller in the potential flow about the ship's hull should not cause a first order influence upon its efficiency inasmuch as the potential flow part of the inflow velocity is theoretically recovered from the slipstream in the form of pressure.

The measurement of the air speed of an airship is necessary, in service for navigation of the craft and in tests for the determination of the speed performance. An accurate measurement of the air speed is by no means an easy task. Either anemometric instruments such as cup-anemometers or spinning windmills which are directly responsive to air speed, or aerodynamic instruments such as Pitot tubes[1], Venturis, bridled windmills, pressure plates, etc., which really measure velocity head depending on air density, are commonly in use for this purpose. When the measuring elements are carried close to the ship due allowance must be made for the influence of potential flow (possibly modified by the presence of dynamic lift); when trailing on a long line payed out, they may swing and actually travel more air miles than the ship; also they may be trailing in a stratum of air where the density and the wind may appreciably differ from that at the ship's altitude. It has been a time-honored, though problematic, practice to check air speed measurements against ground speed measurements. While holding the engine speed and altitude constant the ship is made to fly three "legs" in different directions over a well surveyed ground area, crabbing if necessary, to follow straight landmarks between easily distinguished terminal marks. Thus, by clocking the time for each leg of known length and laying out the three ground speed vectors from an origin according to the known azimuths, a circumscribed speed circle is constructed whose center displacement from the origin is interpreted as the average wind vector and whose radius is the air speed. It is not uncommon to find discrepancies as high as 3%, and even 5%, between the vectorial average of the ground speed and of the speed indicated by calibrated air speed meters, even after allowances are made for the inherent instrumental errors already mentioned. Additional errors may occasionally creep in from the following sources: In trying to follow a marked ground track or, in more marked degree, in trying to approach a distant landmark without a guide track, the ship will occasionally swing off her course or pursue a more or less sinusoidal path. The same is true in the vertical plane for the ship will

[1] von Soden, A. F., and Dornier, C., Zeitschr. f. Flugtechnik u. Motorl., October 14, 1911.

occasionally gain or lose altitude. If compass course and inclination or altitude are carefully recorded, the distance flown can be corrected inversely according to the average cosine of the deviation of the actual from the average path. Furthermore, the wind in the trial region may change between runs. Pilot balloon ascensions made over the flight area and repeated throughout the test duration may reveal such changes and furnish clues for corrections. When the balloon runs are simultaneously observed from two theodolites so as to give accurate altitude references, the actual variation of the wind speed with altitude is determined. If the indicated average air speed, or the wind speed, vary slightly from leg to leg, a correction can be worked out by successive approximations.

Instead of the conventional ground speed vector triangle method, an abbreviated and quicker method of making a ground speed trial on two legs becomes feasible over a country where parallel landmarks are available; for instance in certain mid-western States of America where the roads are parallel and located at intervals of 1 or 5 miles. Here the ship simply flies over the area, steered on a compass course determined at right angles to the parallel landmarks. The pilot need not follow any definite track on the ground, but merely tries to maintain the average course by the compass. Since he does not have to mind his wind drift he can hold his course much better than if he had to also pass over definite points. By clocking the passage over the border lines the "crossing" speed is determined and if necessary corrected for mean course error cosine. The procedure is then repeated in the opposite direction. Provided that the wind did not change in the meantime, the algebraic average of the two opposite crossing speeds gives directly the air speed made good, as the wind speed components here cancel out. By running a third or a fourth leg and by optically measuring drift angles, ground speed, etc., additional checks are obtained.

The practical navigator often determines his ground speed and drift by means of optical drifts sights, by observation of the ship's shadow against landmarks, or, over water, against dropped "drift bombs".

6. Fuel Economy, Range. The range of action or distance which the craft can travel without refueling depends not only on its propulsive efficiency, but also on the amount of fuel carried, on the fuel economy, on the air speed made good, on the air density and on the wind. The fuel consumed not only per hour, but also per air mile travelled, decreases with reduced air speed in more marked degree than with airplanes. At modest speeds, large modern airships could stay out weeks and travel to the farthest point on the globe. The computation of the actual range for given initial conditions is a straight forward problem provided the propeller efficiency for various engine combinations and the fuel consumptions for various engine speeds and air densities are known. As with most motors the fuel consumed per hp hr. increases with throttling, investigation will be needed in each particular case to determine whether reduced speed cruising is more economical by throttling all motors, or by shutting some of them down, depending on whether the reduced fuel economy or the loss in propeller efficiency is the more important factor.

In the development of larger and faster airships it has been mandatory that parasite drag be reduced to a bare minimum in order that the desired speeds might be realized without the expenditure of prohibitive amounts of power. Certain matters, such as the removal of unnecessary protuberances and the cowling of those few which remained, were logical steps in the development of design technique and the modern airship now closely approaches a smooth streamlined body of revolution.

While some parts of an airship's propulsive system may be completely cowled, it is necessary that the propellers and the heat dissipating surfaces come into direct contact with the air. The decision as to how much of the auxiliary propulsive equipment should also be placed in an exposed position is one which can be decided from consideration of the relation of the differences in installed dead weight to the differences in drag, as reflected in the extra weight of fuel needed for a given cruising distance, or the extra power plant weight needed for a given design top speed.

Since the horsepower required for the propulsion of airships increases nearly with the cube of speed, the contribution of the propulsive system to the total drag is appreciable on high speed airships. Thus, changes in the weight and drag of the propulsive units become of great importance. As a matter of fact, for speeds now considered, the cleaning up of the design by the installation of inside power plants instead of outboard power cars, may so reduce the power requirement as to offset the higher specific weight per hp inherent with inside power plants[1]. The design top speed at which the dead weight of an airship equipped with inside power plants will be equivalent to that of a similar design incorporating outside power plants is given by

$$v = \sqrt[3]{\frac{1100}{\varrho} \frac{w_0 E_i - w_i E_0}{w_0 d_i - w_i d_0}}$$

where w_0 and w_i are the specific weights of the outside and inside power plants respectively (lbs/hp),

E_0 and E_i are the drive efficiencies of the outside and inside power plants respectively,

d_0 and d_i are the specific drags of the outside and inside power plants respectively (sq. ft./hp).

[1] Arnstein, Karl, Some Design Aspects of the Rigid Airship, Trans. Am. Soc. Mech'l Engr's p. 385, 1934.

For the case of comparable drive efficiencies, the airship having the smallest total drag area will have the lowest fuel consumption at any given air speed. Even when the design speed is not sufficiently high to warrant an inside installation purely from considerations of deadweight, the reduced fuel load required for the accomplishment of a given mission frequently makes such installations desirable from the standpoint of lift available for payload. The hours of flight needed to overcome an initial weight handicap are given by the quotient of the total excess power plant weight (inside-outside) by the hourly reduction in fuel consumption. Expressed with the foregoing symbols, the hours required to equalize the weights and loadings are given by

$$h = \frac{1}{f}\left(\frac{v_m}{v_c}\right)^3 \frac{\dfrac{w_i}{1100\,E_i - d_i\,\varrho\,v_m^3} - \dfrac{w_0}{1100\,E_0 - d_0\,\varrho\,v_m^3}}{\dfrac{1}{1100\,E_0 - d_0\,\varrho\,v_c^3} - \dfrac{1}{1100\,E_i - d_i\,\varrho\,v_c^3}}$$

where v_m and v_c are the maximum and cruising speeds respectively, and f is the specific fuel consumption of the engines (lbs/hp-hr) at the chosen cruising speed.

Such methods apply to comparisons of weight savings with the corresponding drag and power influences in passing upon the merits of many other auxiliaries such as radiators, water recovery apparatus, and propeller gearing.

The science of navigation of airships is, of course, replete with interesting theoretical problems such as the best course and airspeed to reach a certain destination with the largest margin of fuel reserve under given or predicted conditions of locally variable winds or cyclones along the route[1]; or again, how best to circumnavigate unfavorable weather conditions or cope with unforeseen wind changes en route. However, these problems are not strictly of an aerodynamic nature and can, therefore, not be pursued here in detail. It may only be mentioned that their solution sometimes differs for airships and airplanes due to the fact that

[1] Scott, G. H., Airship Piloting, Journal of the Royal Aeronautical Society, February 1921; Munk, M., U.S. N.A.C.A. Technical Note No. 89, 1922; Klemperer, W., Abhandlungen aus dem Aerodynamischen Institut an der Technischen Hochschule Aachen, Vol. II, p. 31, 1922; Scott, G. H., and Richmond, V. C., Effect of Meteorological Conditions on Airships, Journal of the Royal Aeronautical Society, March 1924; Bleistein, W., Effect of Speed on Economy of Airship Traffic, U.S. N.A.C.A. Technical Memorandum No. 302; Seilkopf, H., Die Wetterberatung der Amerikafahrt des LZ-126, Zeitschr. f. Flugtechnik u. Motorl., Heft 6, 1925; Silvester, N. L., The Use of Barometric Charts in the Navigation of Airships, Journal of the Royal Aeronautical Society, January 1927; Wagner, F., Der Einfluß des Windes auf die Reisegeschwindigkeit von Luftfahrzeugen, Annalen der Hydrographie und maritimen Meteorologie, Heft 12, 1927; Zermelo, E., Z.A.M.M., Vol. II, p. 124, 1931; von Mises, R., Fluglehre, 1933. See also Br. A.R.C. R. and M. 281, 389, 521 and 637.

the energy consumed per air mile depends mainly on the air speed with the one and mainly on the gradually diminishing fuel load with the other.

The fuel load that can be taken aboard depends, of course, on the useful lift and on the size of crew and payload to be carried. The former again depends on the buoyancy conditions of the air at the start (temperature and barometric pressure) and on the flight altitude to be reached. In the case of ships burning gaseous fuel the ceiling increases as fuel is consumed, so that high altitudes can be easily reached during the later stages of a flight, when the bulk of the gaseous fuel has been spent. The same is true of ships burning liquid fuel without ballast recovery, when much of the fuel load has been spent and the equivalent amount of the lifting gas has been valved.

The relations between speed, altitude and power requirement are different for the airship and the airplane since with the former the lift remains independent of density up to the ceiling and in addition there is no need to fly faster than at certain minimum air speeds or flight attitudes in order to remain aloft. As we have already seen, the horsepower required for any given airship speed varies directly with the air density. The air density decreases with altitude according to the formula

$$\varrho = \varrho_0 \frac{P - (3/8)\, e}{P_0} \frac{T_0}{T}$$

where ϱ is the density at an altitude where the pressure is P and the temperature T (abs.) and ϱ_0 is the density of dry air at a standard pressure P_0 and temperature T_0. The pressure of aqueous vapor e at the altitude is frequently neglected in which case the density, and hence the drag, is assumed to vary directly with the pressure and inversely with the absolute temperature.

The power output of the engines at any given altitude is given by the formula[1]

$$b\,h\,p = b\,h\,p_0 \left[\frac{P}{P_0}\left(\frac{T_0}{T}\right)^{1/2}\left(1 + \frac{\lambda - \lambda n}{n}\right) - \frac{\lambda - \lambda n}{n}\right]$$

where n is the mechanical efficiency at sea level and λ is the ratio of mechanical friction to friction horsepower at sea level.

If the mechanical efficiency variation is neglected, the power output would be assumed to vary directly with the pressure but inversely with the square root of the absolute temperature.

A comparison of these approximate expressions for power required and power available indicates that with unsupercharged engines the maximum speeds attainable are proportional to the sixth root of the absolute temperatures. Hence, an airship can attain its highest speed at high temperatures. With supercharged engines, the highest air speeds

[1] Gove, W. D., Variation in Engine Power with Altitude, etc., U.S. N.A.C.A. Technical Report No. 295. Also see Br. A.R.C. R. and M. 462, 960, 961 and 1099.

would be obtained at the maximum altitude at which the supercharger can deliver air to the engine at sea level pressure.

For any given cruising speed, however, the power required falls off directly with the density, and for an engine with a constant specific fuel consumption (lbs/hp-hr) the total fuel consumed per hour at a given air speed would also vary directly with the density. Actually the specific fuel consumption does not remain exactly constant with altitude but rather increases. However, this increase is small, particularly so in good airship engines (Fig. 12) and diminishes only slightly the potential saving in fuel inherent in cruising at higher altitudes or lower densities.

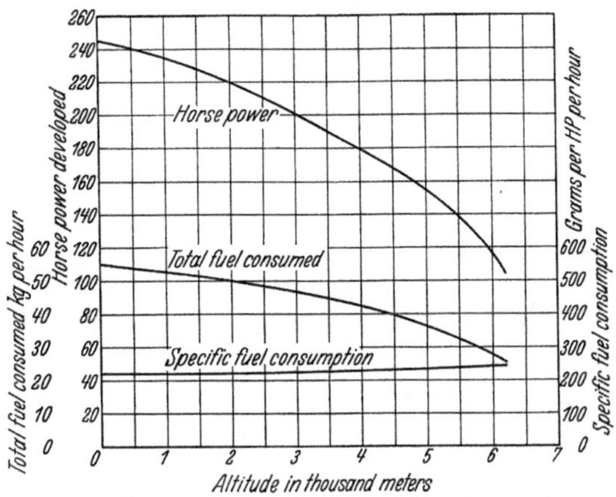

Fig. 12. Fuel characteristics of an airship engine.

The conclusion is that for economical operation, airships should be flown as near their ceiling as possible; *i. e.*, with due regard of course to the vertical structure of the winds en route and other operational considerations. In this respect the helium inflated airship, with no valving and a constant ceiling, may be at a disadvantage when compared with fuel gas and valving airships where the ceiling progressively increases as fuel is consumed and where the mean fuel consumption for the voyage may sometimes be appreciably reduced by flying at a higher average altitude. However, this comparison is tempered somewhat by considerations of meteorological conditions which may dictate the optimum flying altitude more predominantly than mechanical conditions.

It follows that there are differences between the fuel consumptions and operating economies of various types of airships. As a rule they all have a common merit however, in that their transport efficiency improves with increasing size. This transport efficiency is defined, for a given speed, by the quotient,

$$\frac{\text{Payload} \times \text{Distance Carried}}{\text{Energy Expended}}$$

and is expressed[1] by the formula

$$T = K\,(1 - d)\,\frac{\sqrt[3]{Q}\,(\alpha - \psi)\,E}{(V^2/2g)\,C_D} - f D$$

[1] Arnstein, Karl, The Development of Large Commercial Rigid Airships, A.S.M.E. Trans., January-April 1928.

where K = Ratio of gas capacity to gross volume,
d = Relative density of lifting gas with reference to air,
Q = Volume in cubic feet,
α = Ratio of loads other than deadweight to gross load,
ψ = Ratio of service loads to gross load,
$(\alpha - \psi)$ = Ratio of combined pay load and fuel load to gross load,
f = Specific fuel consumption (lbs per ft lb of work produced),
D = Flight distance in feet,
$V^2/2g$ = Velocity head in feet,
C_D = Drag coefficient (per Vol.$^{2/3}$),
E = Efficiency of propulsion.

The general statement of the improvement of the transport efficiency with size may be demonstrated by an analysis of the factors involved. Of these factors d, D, and $V^2/2g$ are inherently independent of size and f and E are essentially so. K tends to increase slightly with size because the waste space devoted to corridors, quarters and insulation space may not need to be increased in proportion to the gas volume.

The useful load ratio α tends to increase as the ratio $(1 - \alpha)$ (the deadweight ratio) tends to decrease with increasing size, at least for modern airships[1]. The service load ratio ψ tends to decrease with size, since the crew need not necessarily be increased in number proportional to size, if at all, once a sufficient number of pilots and mechanics are provided. The decrease of the drag coefficient with size due to the influence of Reynolds number has already been discussed[2].

In addition to the benefits derived from these variables the term $\sqrt[3]{Q}$ represents a factor directly proportional to size.

The range of an airship can be somewhat increased by carrying a part of the initial load dynamically[3].

CHAPTER III

DYNAMIC LIFT

1. Flight with Dynamic Lift. When an airship is propelled at an angle of attack, lift forces are created in a similar manner as by the wing of an airplane. It is true that the airship's shape as a wing is very poor and its aspect ratio extremely small; but the size of the exposed surfaces is so great that trememdous aerodynamic force components at right angles to the flight path can be evoked. A part of this "dynamic lift" is produced by the hull of the airship proper (not embraced by the classical treatment of the flow about the ship neglecting friction and

[1] See footnote reference 1, p. 62.
[2] See Division I, Part, II 4.
[3] BLAKEMORE, T. L., Artificial Payload for Coming Commercial Airships, A.S.M.E. Buffalo, June 1932.

circulation) and the rest by the fins and control surfaces which in appearance resemble stubby airplane wings and function to some degree as such.

Dynamic lift (upward) is resorted to whenever the ship becomes "heavier than air" or "heavy" as it is called in airship parlance. This may happen in various ways accidentally, or it may be brought about deliberately. Precipitation in the form of rain, snow or ice on the surface of a large ship may result in an added load of several tons. Running into a layer of warmer air will make the ship heavy due to the lag of the gas inside the ship in assuming temperature equilibrium. These are usually temporary conditions. Loss of buoyant gas through accidental injury of gas cells or in consequence of climbing above pressure height with resultant valving causes a permanent loss of buoyancy. On the other hand an overload may be taken aboard deliberately in the form of mail, passengers, or airplanes. In all these cases the ship flies "heavy", up by the nose at an angle of pitch which must be the larger the less the airspeed. In a similar manner a ship may become "light" and must be flown down by the nose (at a negative angle of pitch) when for instance load or ballast is dropped, or when radiation "superheats" the gas and air inside the ship, or again when liquid fuel is consumed.

It is the practice to avoid these conditions in any marked degree. Well planned navigation will usually succeed in anticipating their causes and in meeting them at least part way. However, they may occur on short notice or they may be accepted deliberately, and in consequence a study of their aerodynamic aspects assumes a definite importance. Transport economy is, of course, reduced by the induced drag accompanying the production of dynamic lift and it is easily seen that, if the voyage is long enough, the fuel consumed to overcome this induced drag might outweigh the increase of useful load so carried. However, for ships burning liquid fuel, which gradually become lighter as fuel is consumed, there would be a distinct advantage in taking off heavy and accepting the drawback of the induced drag for a short while, until equilibrium is regained. For instance if the "overload" at any time t is L, the time rate of change of L is given by

$$\frac{dL}{dt} = -fP \tag{1.1}$$

where f is the fuel consumption per unit of power and time and P is the instantaneous value of the motor horsepower. Assuming as a first approximation that the induced drag is proportional to the square of the lift and that the equivalent "wing aspect ratio" of the ship, so to speak, can be expressed by some "equivalent span", s, which may differ somewhat from the ship's maximum diameter, then the additional power which must be spent in excess of that due to the normal drag of the ship in equilibrium would be: $P_1 = \frac{L^2 V}{q \pi s^2 E}$

where L = lift produced,

q = velocity head = $(1/2)\,\varrho\,V^2$,

s = equivalent span, *i. e.*, the span of a wing (assuming elliptical distribution of the lift) which would give rise to the same drag increase and which must be known from experimental data,

E = propeller and drive efficiency.

Then referring to II (5.1), equation (1.1) takes the form:

$$\frac{dL}{dt} = -\frac{fV}{E}\left(q A_D + \frac{L^2}{q \pi s^2}\right)$$

This can be readily integrated and furnishes the gain in range $S = \int V dt$ for L from L_0 to 0 in the form

$$S = \frac{Es}{f}\sqrt{\frac{\pi}{A_D}}\, tan^{-1}\left(\frac{L_0}{q s \sqrt{\pi A_D}}\right)$$

or

$$S = \frac{E}{f}\sqrt{\frac{A'}{A_D}}\, tan^{-1}\left(\frac{L_0}{q \sqrt{A' A_D}}\right)$$

where $A' = \pi s^2$ the "influence" area. Without the induced drag the range would have been

$$S_0 = \frac{E}{f}\,\frac{L_0}{q A_D} \qquad (1.2)^1$$

The reduction of the range due to the induced drag can be expressed by the series $1 - \frac{1}{3}\,\frac{L^2}{q^2 A' A_D} + \frac{1}{5}\,\frac{L^4}{q^4 (A' A_D)^2} - \ldots$ The reduction becomes a noticeable percentage only for large overloads. The problem of carrying these during the take-off is a serious one. However, by taking off with artificial superheat secured from a heat source ashore it is possible to start with considerable overload provided the route does not require a high ceiling at the beginning of the flight and before the gas has cooled down. The possibility of airships with heavy overloads taking off like airplanes has been demonstrated with small ships. However, the idea of a combination airship-airplane, which so frequently fascinates inventors, would seem to have only very limited possibilities, unless means may be found for providing a very large wing spread which moreover must admit of folding in close to the body of the ship. Otherwise problems of housing would be complicated beyond any conceivable advantage to be gained.

The dynamic lift of airships is limited by the power available in a way similar to that of airplanes. There is an optimum combination of angle of attack and speed for which the maximum load can be carried

[1] This simple example omits many important considerations. In reality such factors as the increase of drag at a rate greater than as the square of the lift, the loss of propeller efficiency with increased thrust and with the angularity of the inflow, the arbitrariness of maintaining the same forward speed to be made good, and the possibility of cutting out part of the resistance of water recovery apparatus while not required, all have their practical influence.

with given power which is indicated by the maximum value of C_L^3/C_D^2 in a manner similar to that for the same problem with the airplane. This maximum carrying capacity would be attained at the angle of pitch for which the "induced" drag is three times the parasite drag[1], thus

$$C_L = \sqrt{3\pi\lambda C_D}$$

where λ = the equivalent "aspect ratio" which would have the same induced drag characteristic. Naturally C_L and C_D must be expressed with reference to the same area, for instance $Q^{2/3}$. In reality the maximum lift is much less than would appear from the λ valid for small angles of attack, because the validity of the parabolic induced drag law does not extend to sufficiently high angles, *i. e.*, λ is not constant.

Beyond the angle of pitch corresponding to the condition of maximum lift for given power looms the "stall"[2]. If the dynamic lift were proportional to the angle of attack up to the stalling angle, the latter would be

$$\alpha^* = \sqrt{\frac{C_D}{d\,C_L/d\,\alpha}}$$

and the stalling speed would be $\sqrt[3]{1/4}$ (63%) of the top speed attainable under the same power in the absence of dynamic lift.

In reality the power available drops with the reduction of speed. Rather, it is the engine torque which remains essentially unaltered. Insofar as the actual propeller thrust T available at any speed V is approached by a parabola, $T = T_0 - C_T\, v^2\, (\varrho/2)\, Q^{2/3}$, the power drop expresses itself in the form of an additional drag which makes the formula for the stalling angle α^* (in radians)

$$C_L = (C_D + C_T)\, d\, C_L\, d\, \alpha$$

or

$$\alpha^* = \sqrt{\frac{C_D + C_T}{d\,C_L/d\,\alpha}}$$

and the stalling speed in level flight would become $\sqrt{1/2} = 71\%$ of the top speed attained with the same engine throttle position in the absence of dynamic lift.

Similar to the airplane, the approach to the stall is essentially governed by the aerodynamic attack, and associated with a definite stalling angle of attack (pitch). However the overload that can be carried at this angle of attack depends on the slope ε of the ship's path. In a climb, less overload can be carried. The difference is in first approximation.

$$\frac{\triangle L}{L} = -\frac{\varepsilon}{(\alpha + \varepsilon) + (C_D + C_T)/C_L}$$

When a "heavy" ship, not yet heavy enough in level flight to approach a stall, is made to climb, it may prematurely stall; similarly a light ship,

---

[1] Klemperer, W., Zeitschr. f. Flugtechnik u. Motorl., p, 78, 1922.

[2] Klemperer, W., Stalling of Airships, Journal of the Aeronautical Sciences, p. 113, July 1934.

when it is made to descend. On the other hand when a heavy ship is allowed to descend or a light ship allowed to rise, an imminent stall is (temporarily) averted.

It is interesting to note that a heavy ship carrying its overload dynamically, when actually nosed down, ($-\varepsilon/\alpha > 1$) can "glide" and thereby pick up speed exactly as an airplane can. When light, and flying with dynamic down-dip it will "glide up" and pick up speed in so doing when permitted to nose up, unless the added drag of the elevator predominates.

On ships of conventional design, dynamic lift is associated with unwelcome stresses and demands upon controllability. This is due to the manner in which dynamic lift distributes itself unevenly over the length of the ship—a large part at the bow and a considerable amount at the stern, the two not necessarily in equilibrium about the center of buoyancy. In order to appreciate this it is convenient to consider the dynamic lifts of the hull and of the empennage separately as well as their mutual interference.

2. Dynamic Lift of the Hull. In a non-viscous fluid an elongated body (of volume Q) such as an airship hull moving at an acute angle of attack (α) between its longitudinal axis and its path would experience no force such as dynamic lift, but only an unstable deviating moment $(k_2 - k_1)\, Q \cdot q \sin 2\alpha$ [1] where $(k_2 - k_1)$ denotes the difference of the virtual mass coefficients for the transverse and axial flow components and q the velocity head. This moment tends to increase the angle of attack and is largely concentrated on the bow and stern parts of the ship, the components acting there in opposite directions. For the detailed distribution of these transverse forces along the axis of the ship a first approximation is given in Division Q [equations (8.6), (8.7)]

$$b = (k_2 - k_1) \frac{dA}{dx} \sin 2\alpha \qquad (2.1)^{2}$$

In wind tunnel tests the pitching moment weighed on the balances appears from 15 to 30 per cent smaller than this, and much less concentrated at the nose, especially for ships having a blunt bow. This is due to the fact that where the taper is pronounced, the equivalence between adjacent length elements and cylindrical slices acting upon the flow independent of each other is no longer valid. For an ellipsoid of

[1] See Division C III 4; and Q 8.
MUNK, M., U.S. N.A.C.A. Report No. 184, 1923.

[2] We may for present convenience call b the "transverse force breadth" because it would indicate the local breadth or width of a pail filled with water (or any other liquid) to the height equivalent to the velocity head q, in order to produce the same load distribution over the ship considered as a beam, as the actual air forces would. The total transverse force or the dynamic lift would then be $L = q \int_{\text{bow}}^{\text{stern}} b\, dx$.

revolution for which the exact pressure distribution is known[1] the integration around any conical slice at a station where the local taper angle between the tangent of the generatrix and the axis is τ and the local radius or ordinate to the generatrix is r, has the value

$$b = r \pi \sin 2 \alpha \sin 2 \tau$$

This is equivalent to the substitution of the variable $\cos^2 \tau$ for the constant $k_2 - k_1$ and even for ships whose bow is somewhat blunter than an ellipsoid gives a much better approximation, as pressure distribution measurements both on wind tunnel models and on ships in flight have shown[2].

A closer investigation and digest of wind tunnel results may require the introduction of corrections for the influences of the finite wind stream dimensions in the laboratory. In the open jet tunnel, at the jet boundary, the pressure influence due to the model is offset, so that there, actual velocity increments have faded out. In the closed tunnel there can be no radial velocity component at the tunnel walls in spite of the presence of the model, which at such distance in a free stream would, in most cases, still give rise to such a component.

For various reasons some designers prefer to choose a hull shape which is expressed by a relatively simple formula for the cross sectional area (S) in terms of the abscissa station (x) rather than for the ordinate r of the generatrix. For contours of this class it is sometimes convenient to express the transverse force breadth in terms of S and $S' = dS/dx$.

This is done by: $b = \dfrac{\sin 2 \alpha}{1/S' + S'/4 \pi S} = \dfrac{S' \sin 2 \alpha}{1 + S'^2/4 \pi S}$

In order to accurately determine the theoretical distribution of the transverse force breadth for any given shape of hull, recourse may be had to methods given by v. Kármán[3] or Kaplan[4] and by Lotz[5] of which the principal features are as follows:

A system of sources is determined and so distributed along the ship's axis or its hull surface as to represent the shape for the axial component

---

[1] See Division C VII 5.

LAMB, H., Hydrodynamics, 5th Edition, p. 132; UPSON, R. H., and KLIKOFF, W. A., U.S. N.A.C.A. Report No. 405.

[2] U.S. N.A.C.A. Technical Reports Nos. 223, 324, 405 and 443; JONES, R., Br. A.R.C. R. and M. 1061, 1927; BURGESS, C. P., Airship Design, New York 1927; KLEMPERER, W., Windkanalversuche an einem Zeppelin-Luftschiffmodell, Abhandlungen aus dem Aerodynamischen Institut an der Technischen Hochschule, Aachen 1932.

[3] KÁRMÁN, TH. v., Abhandlungen aus dem Aerodynamischen Institut an der Technischen Hochschule, Aachen, Heft 6, p. 1—17, 1927.

[4] KAPLAN, C., Potential Flow About Elongated Bodies of Revolution, U.S. N.A.C.A. Technical Report No. 516.

[5] LOTZ, I., U.S. N.A.C.A. Technical Memorandum No. 675, Calculation of Potential Flow Past Airship Bodies in Yaw. From Ingenieur-Archiv, Vol. II, 1931.

flow and upon these is superimposed a system of doublets in such manner as to maintain the hull form against the cross component flow.

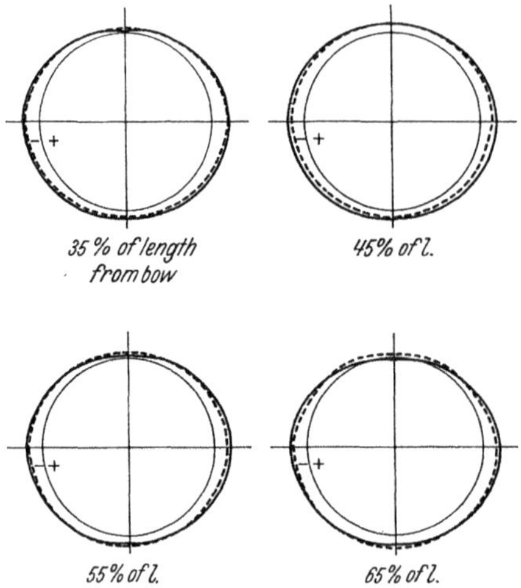

Fig. 13a. Pressure distribution (circumferential) on airship model at pitch angle of 12°. Full lines, calculated values. Dotted lines, measured values.

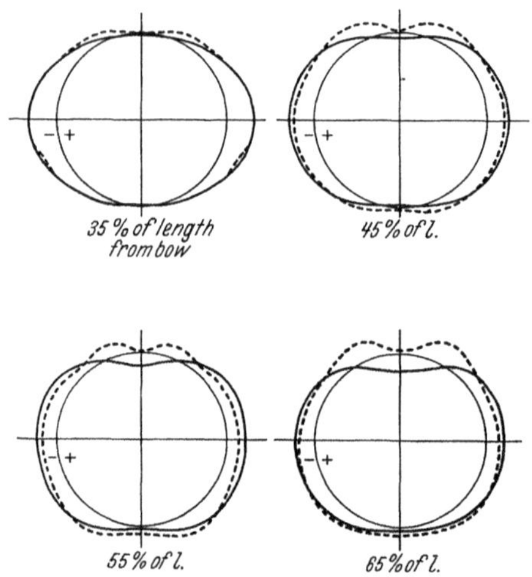

Fig. 13b. Pressure distribution (circumferential) on airship model at pitch angle of 18°. Full lines, calculated values. Dotted lines, measured values.

The pressures may then be computed and integrated around successive slices or zones from station to station, and the longitudinal distribution of transverse force thus determined.

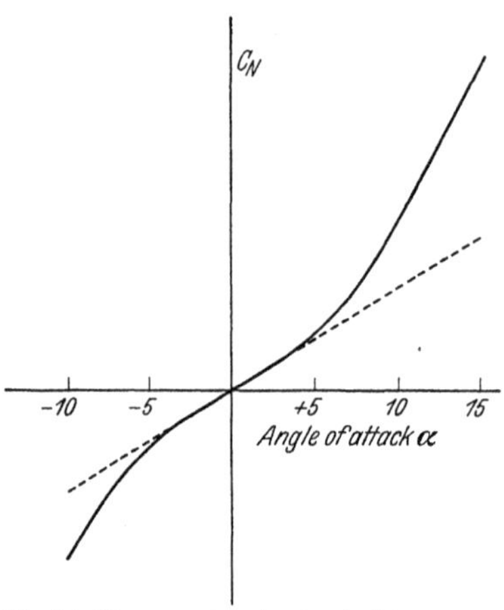

Fig. 14. Diagram showing typical increase of dynamic lift with angle of attack.

For ships that are not very slender, the pitching moment is slightly smaller than the integral of the moments of the transverse forces, *viz.*, $q\int b\,x\,dx$ because the longitudinal components of the pressure contribute a restoring moment. Theoretically this reduction should amount to $-2\pi q r^2 \sin 2\alpha \sin^2\tau$ so that the total moment at any station is reduced in the proportion $1:(1 - r/x \tan\tau)$. Again expressed in terms of cross section area only, this zonal moment correction is

$$d M = \left(\frac{2\pi x - S'}{2\pi/S' + S'/2S}\right) q \sin 2\alpha\, d x$$

However, in reality, in model size as well as full size, the superposition of axial and transverse potential flows gives a faithful picture only in the front windward and midship region as can be readily visualized.

When flying at an angle of pitch one longitudinal will be to the leeward; with a heavy ship the top one, with a light ship the bottom one. In this region and with actual fluids, the stream-lines will be unable to close in behind, and in consequence the pressures will depart from those for a purely potential flow. The skin friction imparts vorticity to the flow and the trailing vortices form the counter part of a circulation which builds up mainly aft of the master section. Pressure distribution experiments on models show that in the rear part of the hull the negative forces (due to defect of pressure) fall considerably short of theoretical values. Figures 13a and 13b show a comparison between calculated pressure distributions and those measured on a wind tunnel model[1]. This pressure deficiency is one of the causes of the difference between the theoretical moment of the hull and that weighed on the wind tunnel balances. It accompanies the development of a lift force. Th. v. Kármán[2] has begun a theoretical treatment of this hull lift adducing plausible assumptions concerning the shedding of circulation.

For higher angles, both pressure distribution and model balance measurements indicate a quicker increase than in the ratio of the sine of the angle of pitch (see Fig. 14). It would therefore appear that the phenomenon of the detachment of vortices on the lee side of an inclined streamlined body is controlled by a sensitive mechanism and that the area subject to it gradually expands upstream, both forward and circumferentially as the angle of attack is increased.

It is reasonable to expect that more insight into the mechanism of the lift of the hull or of the deviation of the pressure distribution from potential flow may be gained from a study of the vorticity in the wake of the acutely attacked hull. An elaborate study of this nature has been begun by Harrington[3]. A survey of the velocity vector field in the wake reveals the presence of two vortex systems trailing downstream through the wake and showing many traits in common with the tip vortices of wings. Figure 15 is a typical example of the results of Harrington's measurements. It is a picture of the transverse velocity components in a section of the wake 20 cm. behind the tail end of an ellipsoid of 99 cm. length and 16.5 cm. diameter attacked at an angle of 21.5° at an air speed of 22.3 m./sec.

To what degree the analogy of model and full size laws of hull lift are obscured by scale effect and turbulence is a question needing still further study.

---

[1] Freeman, H., Pressure Distribution Measurements on the Hull and Fins of a 1/40 Scale Model of the U.S.S. *Akron*, U.S. N.A.C.A. Technical Report No. 443.

[2] Kármán, Th. v., Abhandlungen aus dem Aerodynamischen Institut an der Technischen Hochschule, Aachen, Heft 6, p. 1—17, 1927.

[3] Harrington, R. P., An Attack on the Origin of Lift of an Elongated Body, the Daniel Guggenheim Airship Institute, Akron, Ohio, Publication No. 2, 1935.

At very large angles of attack, wind tunnel tests on conventional airship model sizes are likely to run into scale effect troubles as indicated by experiments on round and elliptical cylinders of such width and ellipticity as would correspond to the slant section of an airship parallel to the plane of flow at very high incidence.

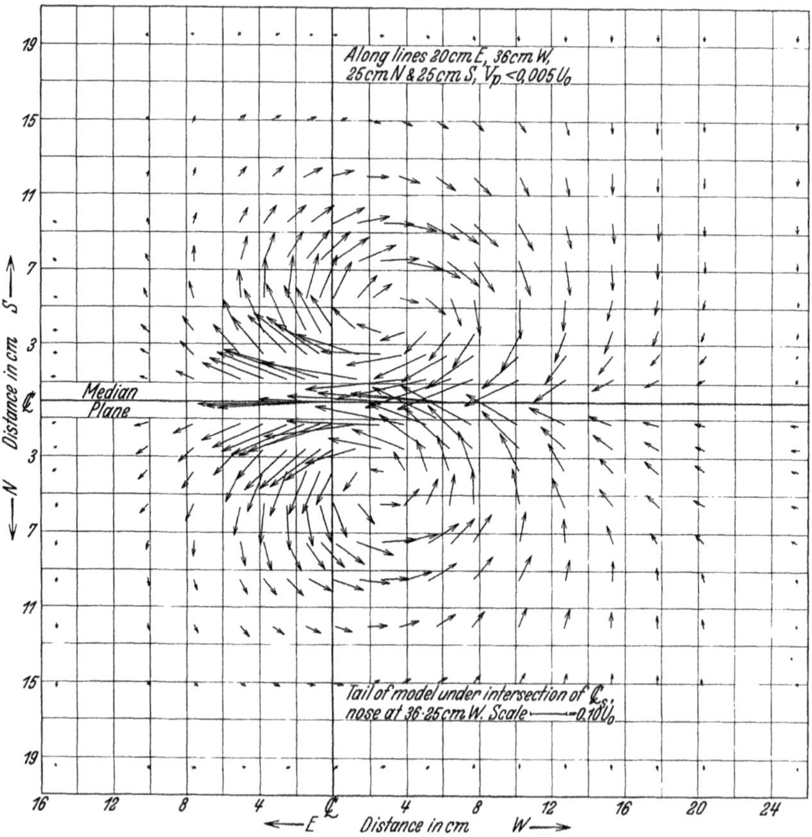

Fig. 15. Velocity field in the wake of an airship.
Induced velocity V_p in plane normal to U_0 located 20 cm. Behind ellipsoid at $\alpha = 21.5°$
$V_p^2 = v^2 + w^2$ Length of model = 99 cm.
$U_0 = 22.3$ m./s. Fineness Ratio = 6
Scale: |———| = 0.10 U_0.

The dynamic lift characteristics are also somewhat influenced by details of the form—whether round or polygonal or heart or pear shaped; likewise by unsymmetrical arrangements of form such as a pronounced keel structure or other features on the under side of the ship. In such cases the lift may not be zero for zero angle of attack.

The drag D of the bare hull (and indeed also of the ship with empennage) increases with the angle of pitch α, very approximately as the product of lift and $\tan \alpha$ or in other words the axial component T is, within wide angle limits, unaffected and the action of oblique attack is essentially the evocation of a force N normal to the ship's axis[1].

$$N = L \cos \alpha + D \sin \alpha$$
$$T = D \cos \alpha - L \sin \alpha$$

3. Lift Due to Fins. In order to neutralize the inherent directional instability of the elongated streamlined hull, airships are equipped with tail empennages in manner similar to an arrow. The action of these fins can, in first approximation, be approached by the airplane wing theory. They are airfoils, usually of either flat or biconvex symmetrical airfoil section, mostly tapered toward the rim. Their aerodynamic properties are somewhat difficult to compute and predict in terms of the classical wing theory because of five important secondary influences.

(1) Their shape is usually, for engineering reasons, long, rather than wide, so that in terms of wing theory their aspect ratio is extraordinarily low. Therefore the spill over the edge becomes an important rather than negligible factor. The whole fin is a wing tip rather than a wing.

(2) The part of the hull between opposite fins is usually so large that its size and shape have an important influence upon the flow about and the forces exerted upon the fins.

(3) The angle of attack of the fins is influenced in marked degree by the induced "downwash" which trails off the preceding parts of the ship's hull. The magnitude of this downwash will further vary over the span of the fin.

(4) The presence of fins when the ship is under an angle of attack influences again the pressures on the rear part of the hull, not only between and to the rear of the fins, but also considerably forward of them.

(5) The roots of the fins are in a region of diminished velocity within the boundary layer of the hull.

It is, of course, conceivable to develop a specific method for introducing all these influences properly into a fin theory. For instance the presence of the hull between the fins can be accounted for by the substitution of a fictitious system of sources and sinks or doublets in its place, as is done in v. Kármán's method for dealing with monoplane wings rooted on a fuselage[2]. However, if accurate representation of the actual facts is attempted, any such procedure suffers from the well known difficulties, attending the necessity of preserving the actual fore and aft distribution of lift, and, for the present, the problematic points regarding the generation of lift by the stern of the hull and its attendant downwash.

---

[1] NAATZ, H., Neuere Forschungen im Luftschiffbau, Jahrbuch der Wiss. Ges. f. Luftfahrt, 1923; ZAHM, A. F., SMITH, R. H., and LOUDEN, F. A., U.S. N.A.C.A. Technical Report No. 215, 1925. Also U.S. N.A.C.A. Technical Reports Nos. 394 and 432.

[2] LENNERTZ, J., Abhandlungen aus dem Aerodynamischen Institut an der Technischen Hochschule, Aachen, Vol. 8, pp. 1—30, 1928.

Many efforts have therefore been made to secure reliable experimental data. Full size experiments are very difficult and very expensive and have been limited largely to pressure distribution measurements. Many of the results are of limited value because of the extreme difficulties of measuring simultaneously the pressures at a sufficiently large number of distant orifice points while the aerodynamic condition of the ship is steady, though departing in marked degree from the simple condition of straight flight equilibrium. There is the further requirement that all parameters of the flight condition must be accurately determined. On the other hand most of the model tests suffer from uncertainty regarding the possible scale "effect". The larger the Reynolds number of the experiment the more valuable the results may appear. The least angular irregularity of flow in a wind tunnel when varying along the length of the experimental section may cause a first order error in the pitching moment measured on a long airship model, whereas with short airplane models, the corresponding error may appear negligibly small. In cases of large models, corrections for tunnel or jet wall influences upon induced drag and effective angle of attack as well as downwash may become in order, as with airplane models. In a closed tunnel the determination of the effective wind speed in the tunnel, as it is increasingly obstructed at larger angles of attack, deserves attention.

Practical experience has shown that a wide variety of fin forms and arrangements may be reasonably satisfactory and there are evidently a great number of variable parameters which may enter into any detailed appreciation of the actual aerodynamic characteristics of an airship empennage. A first approximation to the lift on a fin may be taken on the basis of the conventional airplane wing theory, [see Division E IV (2.15)].

$$L = \frac{2\,\pi\,q\,S\,\alpha}{1+2\,S/b^2}$$

where S is the fin area, b its (effective) span, q the velocity head and α the angle of attack. According to the more trustworthy among model tests in wind tunnels (and probably in a similar manner full scale), the actual stabilizing empennage force, as indicated by the difference of the lift with and without fins, is of the theoretical order of magnitude for very small angles of attack only, whereas for angles of practical interest and importance the force is much greater. Much of the surplus is of course borne by the part of the hull between the roots of the fins and even ahead of them. This share can be measured by pressure distribution experiments[1], an example of which is presented in Fig. 16. The

[1] FREEMAN, H., Pressure Distribution Measurements on the Hull and Fins of a 1/40 Scale Model of the U.S.S. *Akron*, U.S. N.A.C.A. Technical Report No. 443; KLEMPERER, W., Windkanalversuche an einem Zeppelin-Luftschiffmodell, Abhandlungen aus dem Aerodynamischen Institut an der Technischen Hochschule, Aachen, Vol. 12.

center of this stabilizing force is therefore not actually at the theoretical quarter chord point (a point not easily defined with fins whose leading edge gradually merges into the rim), but it may be farther forward when computed by dividing the difference in the stabilizing moments with and without fins by the difference of lift with and without fins. For the airship model of the U.S.S. Los Angeles to which Fig. 16 referred, this leverage is of the order of 78 m. from the center of buoyancy whereas the fins extend about 75 m. to 97-$^1/_2$ m. aft of this point. That neither the stabilizing force nor the stabilizing moment due to the fin appear to be even approximately proportional to the angle of attack may perhaps be regarded as an indication that this part of the hull which, between the fins, has the form of a well rounded body and thus does not offer a definite trailing edge, begins to build up its own contribution to the force only when higher angles of attack are reached.

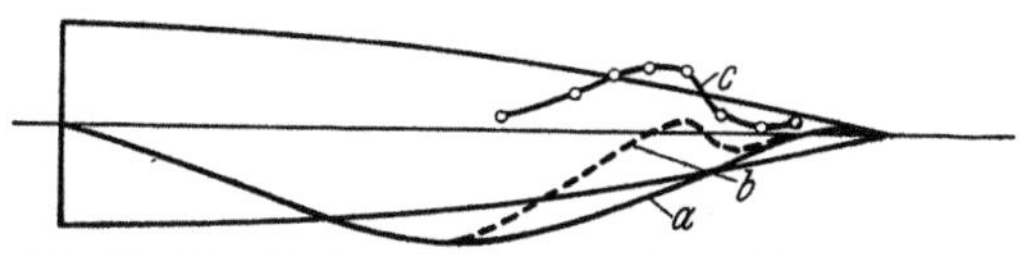

Fig. 16. Distribution of pressure about the empennage. (a) Zonal force integrated from typical pressure distribution without empennage. (b) Zonal force integrated from typical pressure distribution with empennage. (c) Difference due to empennage.

Of the innumerable varieties of fin forms proposed or used on airships, only the major features can be here indicated. For details, reference must be made to the general literature of this subject[1]. Flat fins produce slightly greater forces than fins built up of thick sections tapering from root to tip. The latter, however, offer structural and engineering advantages and are under certain circumstances preferred. Larger aspect ratio of a fin of otherwise fixed shape and location increases its action per unit area[2]. Changing the shape mainly influences the location of the center of action and the pressure distribution. More pronounced leading edge and receding rim moves the center of action forward; a more slanting leading edge gradually flaring into the rim moves it aft. The pressure distribution[3] is similar to that of wing tips. Most of the force is concentrated along the rim. Pressure distribution near the rim is influenced by an angle of yaw simultaneously present with an angle of pitch. Some

[1] Br. A.R.C. R. and M. 714, 799, 802, 1168; U.S. N.A.C.A. Technical Reports Nos. 215, 394, 432; Naatz, H., Neuere Forschungen im Luftschiffbau, Jahrbuch der Wiss. Ges. f. Luftfahrt 1923; Rizzo, Frank, A Study of Static Stability of Airships, U.S. N.A.C.A. Technical Note No. 204; Richmond, V. C., Airship Research and Experiment, Journal of the Royal Aeronautical Society, October 1926.

[2] This is not always an advantage. For slender ships it may be preferable, both structurally and from a weight saving standpoint, to accept a larger fin area if the forces are better distributed over a greater root length.

[3] Jones, R., and Bell, H., Br. A.R.C. R. and M. 1169, 1928; Br. A.R.C. R. and M. 808, 811; U.S. N.A.C.A. Technical Reports Nos. 223, 324, 443; Abhandlungen aus dem Aerodynamischen Institut an der Technischen Hochschule, Aachen, Heft 12, 1932.

sample pressure distributions are represented in Fig. 17. For further details reference may be made to the publications here noted. Moving a given set of fins farther forward or aft will increase or reduce respectively the fin forces but within certain limits may scarcely change the stabilizing moment. The most conventional form of empennage is an essentially symmetrical cruciform set of two pairs of fins, one upper and lower in the keel plane for directional stabilization and the other, port and starboard, for stabilization against pitching. Many other arrangements such as box frames, ring shapes[1] or more than four radial fins[2] have also been tried. Problems of slant attack and mutual shielding from tip-spilling as well as biplane influences come into question in connection with these arrangements.

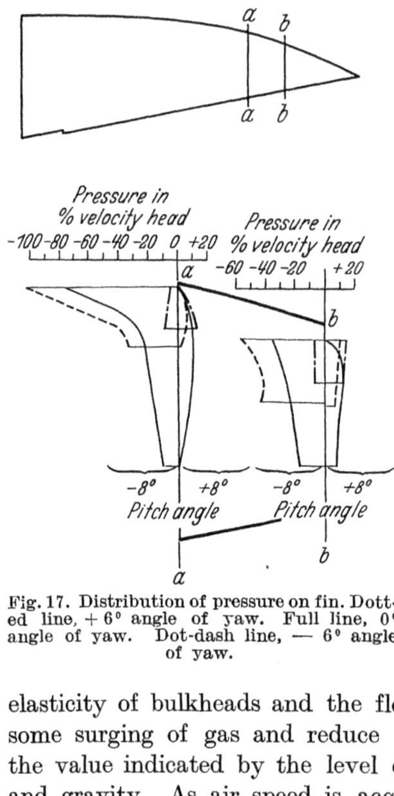

Fig. 17. Distribution of pressure on fin. Dotted line, + 6° angle of yaw. Full line, 0° angle of yaw. Dot-dash line, — 6° angle of yaw.

It is not necessary (nor held desirable by many) to make the fins so large that the moment about the C.B.[3] of the empennaged ship shall be stable for all angles or even for moderate angles of attack, without the aid of the movable control surfaces or even with their aid. Floating without air speed, there is usually a certain small aerostatic stability due to a positive metacentric height. However the elasticity of bulkheads and the floating of lower gas cell levels permit some surging of gas and reduce the metacentric height to less than the value indicated by the level difference of the centers of buoyancy and gravity. As air speed is acquired, an additional (dynamic) metacentric height comes into play, subtractive when the aerodynamical moment is unstable. This dynamic term can, as H. R. Liebert has proposed, be expressed in terms of the velocity height $h = v^2/2g$, *viz.*, $H^* = 2\,(k_2 - k_1)\; h\,(m/m_0)$ where m/m_0 is the ratio of the aerodynamic

[1] DÜRR, L., 25 Jahre Zeppelin-Luftschiffbau, V.D.I.-Verlag, 1924; KÁRMÁN, TH. v., Wind Tunnel Tests on a 1/75 Model of Goodyear Zeppelin Airship ZRS4 with Normal and Ring Tail Surfaces, Report No. 105 of the Guggenheim Aeronautics Laboratory, California Institute of Technology.

[2] UPSON, R., Metalclad Rigid Airship Development, Journal of the Society of Automotive Engineers, February 1926; ABBOTT, I., U.S. N.A.C.A. Technical Report No. 451.

[3] Center of Buoyancy.

moments of the actual empennaged ship to the theoretical moment of the bare hull.

Finally there is always the expedient of shifting ballast so that equilibrium can be established. Just how much fin area is desirable for flying with dynamic lift is therefore largely dependent on navigational problems and on the mechanical and control apparatus provided aboard. It may be mentioned as significant, however, that, with large rigid airships, the first sign of growing heavier usually appears as a tendency to become tail-heavy so that the ship must fly nose up in order to maintain altitude, but with the need of "down elevator" to hold the ship in this attitude. With the ship growing light, corresponding indications, reverse in character, appear. In small nonrigid airships this phenomenon is rarely observed.

4. Dynamic Lift Experiments. The experimental determination of the dynamic lift characteristics[1] of the complete ship, full size, is a very delicate problem. Aside from the difficulty of correctly measuring and averaging the observed angles of pitch and with airspeeds continuously fluctuating as they are, the exact amount of lightness or heaviness is a very elusive quantity. Theoretically the test program is simple, as follows.

(a) Weigh off to make sure that buoyancy equilibrium is established and then either valve a measured quantity of gas or better, drop a measured amount of ballast and determine a set of corresponding pairs of values of air speed and angle of pitch for which the ship will neither rise nor fall;

or otherwise:

(b) First valve a suitable amount of gas and then go through the above measurements and at last see how much ballast must be released in order to reestablish equilibrium. The latter method, especially when valving automatically by deliberately overclimbing the pressure height provides a check when the air density at the ceiling is observed.

An accurate record of elevator angles and of the ship's inclination oscillations must be kept during the experiments because the elevator contributes a considerable amount to the dynamic lift. Corrections required for variations of temperatures inside and outside, for fuel consumed and weights shifted during the time of the tests, render the procedure less simple. This and the reluctance of deliberately putting the ship through the ordeal are the main reasons for the scantiness of data available. Some are compiled in Table 2[2]. When high dynamic

---

[1] Regarding model experimentation:

Frazer, R. A., Bateman, H., Measurement of Normal Force and Pitching Moment of Rigid Airship R 33, Br. A.R.C. R. and M. 815; Zahm, A. F., Smith, R. M., and Louden, F. A., U.S. N.A.C.A. Technical Report No. 215, 1925; Jones, R., Br. A.R.C. R. and M. 1168, 1927.

[2] Burgess, C. P., Airship Design, New York 1927.

TABLE 2. *Some Dynamic Lift Experiments, Abstracted.*

Experiment	Dynamic Lift	Pitch Angle	Elevator Angle	Air-Speed	Temperature	Barometric Pressure	Altitude by aneroid
	kg.	Degrees	Degrees	m./sec.	^{0}C	mm. Hg	m.
British R-33, of 55000 m.^{3} nom. capacity Trial Flight on May 23, 1921, from Br. A.R.C. R. and M. 815	1100	— .8	8 up	23.	8.1		760
	1100	— .2	7 up	24.3	8.3		780
	1000	— 1.5	16 up	14.5	8.0		740
	1150	— 2.8	9 up	15.5	11.1		780
	1100	— 1.7	6 up	22.9	11.7		765
	2050	— 3.1	8 up	22.4	11.7		780
	1950	— 9.5	3 down	14.7	12.9		785
	1950	— 8.1	0	14.7	12.9		760
	1900	— 4.1	6 up	22.9	13.8		795
	2700	— 6.2	6 up	21.2	14.0		815
	2800	—10.0	3 down	16.2	14.0		790
U.S.S. Los-Angeles (LZ 126) of 70,000 m.^{3} nom. capacity, Trial Flight over Lake Constance, Germany, on Sept. 11, 1924	1700	∼ +2	2 up	25	11.5	685	895
	2600	+3 to 4	7 up	25	10.1	670	1030
	3300	+4 to 5.5	11 up	24	9.3	664	1150
	4000	+6 to 7	12 up	22	8.7	652.5	1300
	4800	+8 to 10	8 to 10 up	20	8.1	645.5	1400

loading occurs unexpectedly in practical navigation, the conditions are usually unfavorable for scientific investigations with neither time nor personnel available.

Attempts have been made to develop instruments to indicate currently the magnitude of the dynamic lift of a ship. Such instruments can be based upon the differences of pressures or airstream velocities prevailing on strategic stations above and below the ship's bow. A calibration must be obtained either from dynamic lift tests or from model tests.

Airships are known to have carried huge loads dynamically on various occasions. Thus the Graf Zeppelin was drenched by a torrential rain upon her start from Brazil in 1930. The rain-load thus carried was estimated to be of the order of five tons. The U.S.S. *Akron* once went through severe winter storms and collected 18,000 pounds of ice on her hull. She continued on her mission which lasted fifty-six more flight hours[1]. The U.S.S. *Macon* on part of a transcontinental trip carried 30,000 pounds by dynamic lift.

[1] ARNSTEIN, K., Über einige Luftschiffprobleme, Zeitschr. f. Flugtechnik u. Motorl., Heft 1, 1933.

It is not without interest that, at a given speed, small ships can carry a larger dynamic lift in proportion to their gross aerostatic lift because the former is proportional to the square of the linear dimensions and the latter to the cube. However, larger ships are usually faster, and loads due to rain and sleet are also proportional to the square of the linear dimensions, so that the proportion does not vary very widely.

CHAPTER IV

MANEUVERING

1. Curvilinear Flight. In the horizontal plane there never occurs a condition similar to that in the vertical plane in steady flight under an angle of pitch. An airship cannot proceed straight at an angle of yaw. It differs in this respect from the airplane which can sideslip. On the airship, whose lift is always vertical, there would be nothing to balance the lateral forces arising from flight at an angle of yaw; the ship would turn. Curved flight in a vertical plane is also theoretically possible, but in practice is confined to short arcs, as airships do not loop the loop.

In steady curved flight the centrifugal force must be balanced by aerodynamic forces acting inwards. Consequently, airships turn under an angle of yaw and are attacked from the outside of their turning circles. The condition of attack in a turn is as depicted in Fig. 18. Due to the considerable length of airships, even in comparison with the radius of curvature in a turn, the angle of yaw varies noticeably from stem to stern. In the region of the empennage it is almost twice as large as it is amidships. The large yaw angle at the fins is the essential factor making the empennage so powerful in curved flight.

The turning performance of airships has repeatedly been measured in flight trials[1]. In such tests the helm is put over to a measured angle and the ship allowed to turn. It will first react by swinging the tail outward to build up an angle of yaw and start along a spiral course which may, by the time the flight direction has changed something like 90°, attain approximately the terminal turning radius. The radius can be measured by any of the following methods: In the simplest method the airspeed is measured by a reliable instrument (itself unaffected by turning and yawing) and the angular velocity can be measured either by clocking the times when the ship in a sighting device appears parallel to

[1] Br. A.R.C. R. and M. 537, 631, 668, 675, 713, 716, 749, 779, 780, 782, 811, 812; U.S. N.A.C.A. Technical Reports Nos. 208, 333; Fairbanks, K. J., Pressure Distribution on the Nose of an Airship in Circling Flight, U.S. N.A.C.A. Technical Note No. 224, 1925; von Orel, E., Bildmessung und Luftbildmeßwesen, No. 2, 1929; Lacmann, O., and Block, W., D.V.L. Report, Photogrammetrische Lage- und Geschwindigkeitsbestimmung des Luftschiffs LZ 127, etc., Zeitschr. f. Flugtechnik u. Motorl., Vol. 11, 1930; Wien-Harms, Handbuch der Experimental-Physik, Vol. IV, 3, 1930.

identified ground objects of known azimuth, or by timing a compass swing (corrections should be made for vertical magnetic field component error and lag). In more elaborate methods, in which the yaw angles or the station of zero yaw are also obtained, the path of the ship and its relative attitude are photographed, either in a camera obscura or by stereophotography from two synchronized ground stations or by photographing the ground from aboard. In all these cases allowance must be made for wind drift which distorts the flight path into a cycloid. Gusts enhance the difficulties of measurements. The local yaw angle at any station on the ship has also been measured by a suspended yaw meter, which, however, unless far enough away from the ship's hull and fins, may be influenced by the flow around them. It was early discovered that the region of zero yaw is usually near the nose and that it does not travel perceptibly for different curvatures. Pressure distribution experiments in curved flight show fair symmetry about the nose, the presence of lateral forces amidships, and significant pressure differences on the fins.

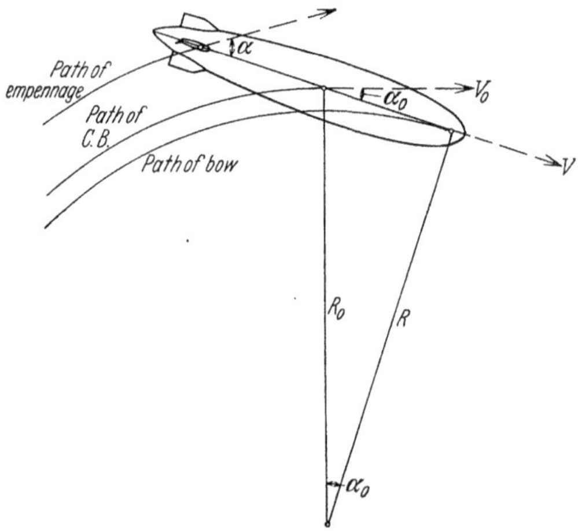

Fig. 18. Conditions of air attack on an airship in a turn.

As a rule an airship is considered satisfactorily maneuverable when its rudder will enable it to turn on a radius of approximately four times the ship's length.

The aerodynamic forces prevailing in curvilinear flight and the distribution of these forces are discussed in Division Q, where references are quoted. The centrifugal force arising in the turn is slightly larger than that of the mechanical mass of the ship. The increment is essentially due to the longitudinal virtual mass[1], and to the momentum of the boundary layer air. The distribution of forces can, in first approximation for the hull, be assumed as according to the potential flow theory of rotation and angle of attack combined[2]. However, for any more accurate appreciation of the part played by the generation of circulation along

[1] $(k_1 \cos^2 \alpha + k_2 \sin^2 \alpha)\, \varrho\, Q\, V^2/R$, according to MAX MUNK, U.S. N.A.C.A. Technical Report No. 184 and Technical Note No. 196; also P. FRANK and TH. v. KÁRMÁN in Riemann-Weber's Differential and Integralgleichungen der Physik, Vieweg, 1927; British writers in various A.R.C. R. and M. have proposed different expressions.

[2] Br. A.R.C. R. and M. 1061; U.S. N.A.C.A. Technical Report No. 405.

the ship's stern, its downwash upon the empennage and the actual empennage forces to be derived from any particular design, recourse must be had to experimental data.

Apparently the most direct imitation of curved flight on model scale consists of carrying a model around a circular path on a whirling arm. This experiment was performed in Italy in water (where however, questions of wave formation arise) and in England in 1925 on an ellipsoid (prolate spheroid) in air. The latter series of tests[1], although confined to a velocity of 12 m. per sec. and made on rather small scale (diameter of ellipsoid = 6″ and length = 24″) and not with the same coordination of yaw and curvature as that which characterizes a turning airship, are of fundamental interest. While it must be realized that these delicate measurements, perhaps obscured by the influence of the air dragged around in the room by the whirling arm, can claim only a limited accuracy, yet they seem to clearly indicate the presence of a resultant aerodynamic centripetal force due to yaw and another force due to curvature at zero yaw.

Fig. 19. Cross-sectional shape of U.S.S. Los Angeles.

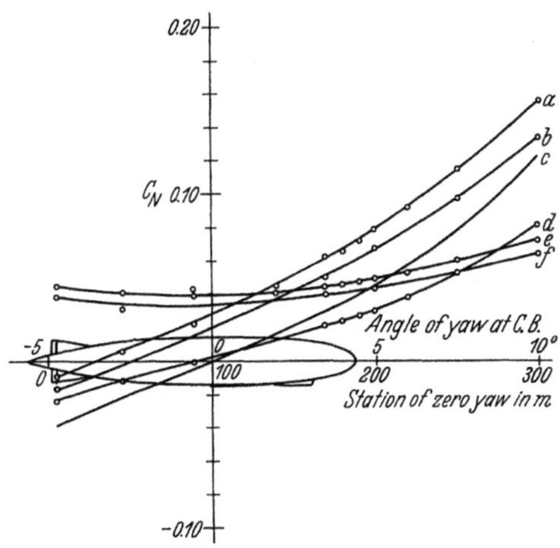

Fig. 20. Curves showing lateral force coefficients of curved model. a. Normal force with fins. b. Normal force with fins and screen. c. Normal force straight model with fins. d. Normal force without fins. e. Drag without screen. f. Drag with screen.

Th. Troller has built a new whirling arm laboratory at the Guggenheim Airship Institute at Akron, Ohio[2]. With this modern equipment it should be possible to attain Reynolds numbers much higher than were ever attained before. This laboratory has begun large scale experiments exploring into the conditions of curvilinear flight of airships.

In 1924, at the Luftschiffbau-Zeppelin Werft, W. Klemperer tested a model of the airship LZ-126 (the U.S.S. Los Angeles) constructed with a curved axis[3] in an attempt to thus reproduce steady turn condi-

[1] Jones, R., Br. A.R.C. R. and M. 1061, December 1925.

[2] Troller, T. H., The New Whirling Arm, Journal of Aeronautical Sciences, 1, 4, 1934; also Daniel Guggenheim Airship Institute, Publication No. 3, 1935.

[3] A British experiment on a curved model which, however, was made with a different purpose, is reported in Br. A.R.C. R. and M. 104, 1913.

tions in the wind tunnel. These experiments are quoted here in some detail because the results have not as yet been published elsewhere. The model was built to a scale 1:75 and the axis was curved to a radius of 15.6 m. representing a turning radius of 1168 m. in flight, which would correspond to a moderate rudder maneuver. The cross section is a regular polygon except at the bottom, where the 11th chords on both sides are extended until they meet, thus giving a form of keel, as shown in Fig. 19. The test results, as obtained in the open jet wind tunnel in Friedrichshafen at wind speeds ranging from 25 to 46 m. per sec., with and without fins are given in Figs. 20 to 23. Figure 20 shows the lateral force coefficients (forces per unit velocity head and volume $^{2/3}$) plotted against the angle of yaw at center of buoyancy and, by reference to the ship's contour, also against the position of the station of zero yaw to which each angle of model incidence corresponded. Figure 21 is a similar plot of the yawing moment coefficients (moments referred to unit velocity head and unit volume). Figure 22 depicts the apparent leverage of the fins expressed as the quotient of moment and force differences with and without fins. Figure 23 plots the "lateral force breadth" in terms of unit $\sqrt[3]{\text{volume}}$ as it is found distributed along the bare hull by zonal integration of many pressure measurements for a typical location of the station of zero yaw, *viz.*, at frame No. 180, *i. e.* 7.5 m. aft of the ship's nose. As will be seen, the lateral force and unstable moment of the bare hull due to curvature only, vary moderately from the corresponding influences due to straight yaw. Part of the variation may, of course, be ascribed to the keel, and the hump in the lateral force breadth to the control car. The aerodynamic influence of curvature upon the empennage is quite strong, even stronger than in proportion to the variation of local angle of yaw at the fins.

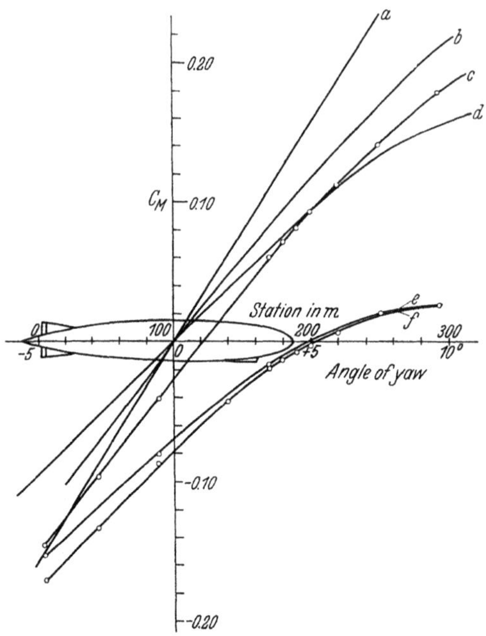

Fig. 21. Curves showing lateral moment coefficients of curved model. a. Theoretical. b. Straight model bare hull. c. Curved model with fins. d. Straight model with fins. e. Curved model with fins and with screen. f. Curved model without fins and without screen.

The analogy between the curved model and circular flight is, of course, somewhat strained. In the model the lengths of the surfaces exposed to the air stream are slightly different on the two sides. In full scale, the air in the actual boundary layer is subjected to centrifugal force and furthermore, the tail swings on a larger radius than the bow and is thus exposed to a higher air speed.

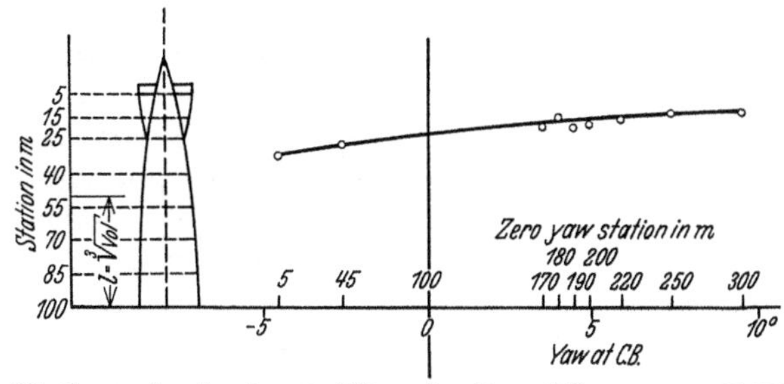

Fig. 22. Curve showing locus of line of action of fins on curved model.

The latter effect can be roughly allowed for by increasing and correcting the observed empennage forces in proportion to the calculated velocity heads. In the Luftschiffbau Zeppelin experiments, an attempt

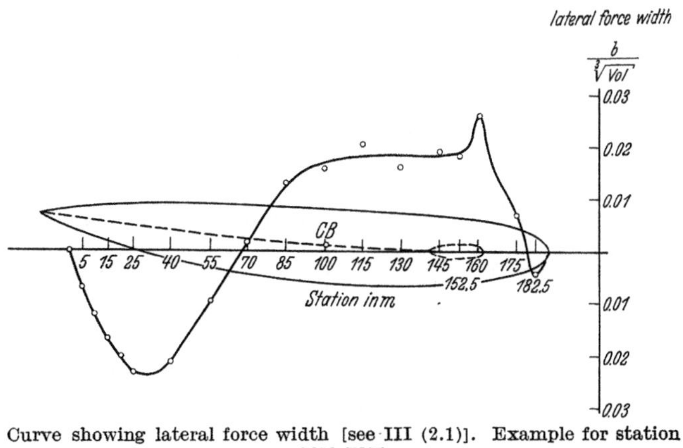

Fig. 23. Curve showing lateral force width [see III (2.1)]. Example for station 180 m at zero yaw.

was also made to imitate this condition by superimposing a velocity gradient upon the wind tunnel jet by means of a wire screen. The dotted curves in Fig. 22 and 23 refer to this condition. The velocity gradient produced in the experiment was, however, less than desired. The presence of the wire screen may also have changed the turbulence in the airstream[1].

[1] GOURJIENKO, Method of Curved Models and its Application to the Study of Curvilinear Flight of Airships, Trans. of the Central Aero-Hydrodynamic Institute, Moscow, 1934.

In 1930—1932, R. H. Smith independently studied the aerodynamics of curvilinear flight by means of curved models at the Massachusetts Institute of Technology, Cambridge, Mass. He tested several models of the U.S.S. Shenandoah with regular polygonal cross section and also a curved ellipsoid. His experiments indicate that the curved model method not only checks results by the older oscillation method (to be described later) but also confirms the usual theoretical assumption that the forces due to rotation are practically proportional to path curvature up to as sharp a curve as can be flown by airships under full rudder. The proportionality factors (rotary derivatives), however, appear to be appreciably influenced by, and not independent of, the simultaneously prevailing angle of yaw (see Fig. 24). This latter phenomenon appears more emphasized on Professor Smith's tests than in the Zeppelin tests. His researches have not yet been published but are quoted here by permission.

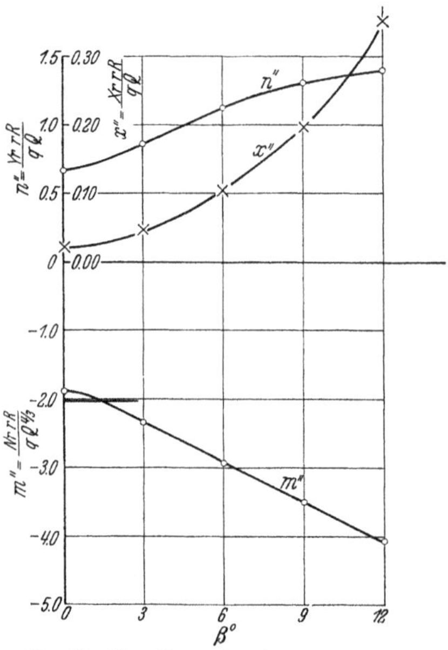

Fig. 24. Non-dimensional coefficients for Yr, Xr, Nr, *vs* yaw at zero rudder for U.S.S. Shenandoah from curved models.

Another less direct method for the determination by model experiments of one of the rotary derivatives, namely the "damping moment" due to rotation (*i. e.* curvature of path), was developed in England by way of the "oscillation" test[1]. The model was suspended elastically in the wind tunnel with its axis parallel to the wind tunnel axis in elastic equilibrium, but so that it could oscillate in yaw about its center of buoyancy. While the wind tunnel was in operation, the model was artificially deflected in yaw and left to oscillate, while observations of the rate of decay of the yaw amplitude were made. The theory underlying the evaluation of this experiment assumes that the aerodynamic moment has one component which is porportional to the angle of yaw (α) and the square of the speed, whereas the other is proportional to the product of the angular velocity of rotation and the air speed itself, while the whole must be equal to the product of the actual moment of inertia of the model, J, (including the virtual inertia) and angular acceleration, *viz.*

[1] Bairstow, L.,, and Maclachlan, L., Br. A.R.C. R. and M. 78, 1913; Simmons L. F. G., and Bateman, H., Br. A.R.C. R. and M. 665, 1920; Zahm, A. F., Smith, R. H., and Louden, F. A., U.S. N.A.C.A. Technical Report No. 215, 1925.

$$J\ddot{\alpha} = \varrho V^2 Q (k_2 - k_1)\alpha + V A \dot{\alpha}$$

by assuming damped harmonic oscillation $\alpha = \alpha_0 e^{-\mu t} \cos \omega t$ and substituting the derivatives it is readily seen from the $\sin \omega t$ terms that

$$A = -\frac{2 J \mu}{V}$$

Theoretically, it is easily shown that the constant A thus measured by observing the rate of decay μ, is nothing other than

$$A = \varrho a^2 S \frac{d C_L}{d\alpha}$$

where ϱ is the air density, a the lever of the fins, S their area and $dC_L/d\alpha$ their aerodynamic effectiveness including the share of the hull.

A forced oscillation method[1] has also been developed and applied for the study of airplane rotary derivatives. It should be applicable to airships in principle.

Oscillation experiments are naturally delicate and the results must be corrected for the inevitable damping decrement of the apparatus without wind. The objection that the model is exposed to the wake of an unsteady motion and to varying angles of yaw has also been raised. Yet the results obtained are of a similar order of magnitude to those given by other tests. The extension of the oscillation method to determine, for instance, the lift due to rotation by additional tests with the model suspended eccentrically would hardly seem to give promise of results with acceptable accuracy.

An additional drag arises in circular flight and the air speed decreases to a lower terminal velocity when turning at a set position of motor throttle. This drag is due partly to the tangential components of the aerodynamic forces acting on hull and empennage and it is further aggravated by the accompanying loss in propeller efficiency due to increased disk loading and yaw angle prevailing at the propellers.

The bending moment due to turning[2] which stresses the ship as a beam in bending can be computed for any station x_1 along the axis by integrating from either end (preferably from the nose) to that station, the moments due to the aerodynamic transverse or zonal forces and subtracting the integrated moment of the centrifugal forces of the part of the ship on the same side of the frame or station under examination. To a first approximation these centrifugal forces can usually be assumed distributed as the ship's cross sectional area S.

$$\text{Bending Moment} = \int_0^{x_1} (x_1 - x) \left(\frac{dF}{dx} - \frac{\varrho S V^2}{R} \right) dx$$

---

[1] Simmons, L. F. G., Br. A.R.C. R. and M. 711, January 1921; Relf, E. F., Lavender, T., and Ower, E., Br. A.R.C. R. and M. 809, September 1921.

[2] U.S. N.A.C.A. Technical Reports Nos. 115, 323, 325.

which expression can be simplified owing to the relations between dF/dx, the radius R, the angle of yaw, and the velocity V.

2. Dynamic Stability. Dynamic stability is defined as the quality under which the ship, after having accidentally deviated from an attitude of equilibrium of forces and moments (be this straight flight, axial or with an angle of pitch, or a steady turn) will return to this attitude by its own inherent aerodynamic and inertia reactions in such a manner that the disturbance eventually disappears, either in the form of diminishing oscillations or aperiodically[1]. If, on the contrary, the disturbance should grow by way of increasing oscillations or by way of a continuous movement toward a new attitude, the original attitude was dynamically unstable. It is, of course, conceivable that any particular situation may be dynamically stable for disturbances of a certain kind or magnitude and unstable for others. In fact such is actually the case with airships whose aerodynamic reactions for increasing angles of attack increase first with an increasing rate and later with a decreasing rate.

However, by confining the theory to very small deviations, neglect of the higher terms becomes legitimate and one can apply the well known doctrine of "small oscillations" in order to derive the "stability criteria". Since 1920, English writers on the subject have developed this idea by writing down the differential equations for the lateral force and yawing moment, assigning to each of the aerodynamical forces and moments a share or component "due to yaw"[2], and another "due to rotation"[2].

If the derivatives of these component forces F and moments M with yaw α and angular velocity ω are known, *viz.*,

$$F_\alpha = \frac{\partial F}{\partial \alpha}, \quad F_\omega = \frac{\partial F}{\partial \omega}, \quad M_\alpha = \frac{\partial M}{\partial \alpha}, \quad M_\omega = \frac{\partial M}{\partial \omega},$$

then evidently for a small displacement α from equilibrium and under a small angular velocity ω, the two simultaneous differential equations determining the d'Alembert reactions will read

$$\begin{aligned} m V [(1 + k_1)\,\omega - (1 + k_2)\,\dot{\alpha}] &= F_\alpha \alpha + F_\omega \omega \\ J (1 + k')\,\dot{\omega} &= M_\alpha \alpha + M_\omega \omega \end{aligned}$$

where m is the mass and J the moment of inertia, k_1, k_2, and k' the coefficients of longitudinal, transverse and polar additional apparent mass due to the flow about the ship[3] and V the velocity of flight. The lateral acceleration appears split up into two terms, one, $-m\,(1 + k_2)\,V\dot{\alpha}$ accounting for the actual reduction of the angle of yaw through lateral

[1] U.S. N.A.C.A. Technical Report No. 212 and Br. A.R.C. R. and M. 257, 602, 751; Burgess, C. P., Airship Design, New York 1927.

[2] The British A.R.C. R. and M. authors express them in terms of the transverse velocity component (v in yaw or w in pitch) and of the angular velocity (q in pitch or r in yaw).

[3] U.S. N.A.C.A. Technical Reports Nos. 164, 323.

yielding of the ship to the force, and the other, $m(1+k_1)\omega V$ the centrifugal force.

Since the derivatives F and M as defined are dimensional, depending on size and speed, one may prefer to transform the expressions with the help of non-dimensional coefficients depending on geometric properties only. Assuming that the aerodynamic properties of the ship under question had been determined by any one of the methods previously discussed, coefficients of force and moment could have been derived referring to unit $Q^{2/3}$ and unit Q, (volume of the ship) respectively, and to unit velocity head q in each case. The slope of the curve of these coefficients plotted against yaw α, near $\alpha =$ zero, would then constitute the non-dimensional derivative. Calling this n' for the normal force and m' for the moment, we have $F = n'ql^2$ and $M = m'ql^3$ where l stands for $\sqrt[3]{Q}$. Theoretically m' should be $2(k_2 - k_1)$ and $n' = 0$. In reality m' is a little less, and n' is of the order of 1/2.

In order to express rotation also by an angular measure of curvature, let us introduce the "curvature angle" ζ under which some unit length (for instance l) appears from the center of the turn: that is, $R\zeta = l$ where R is the radius of the turn. Then we shall have the relation $\omega/V = \zeta/l$. For instance, in the Zeppelin experiments with the curved model, ζ was $= 0.0367$ ($= 2^\circ 9'$). By virtue of the proportionality concept accepted for small curvatures, non-dimensional rotary derivatives n'' and m'' are found by dividing the coefficients measured for zero yaw (or the difference between the curved and the straight model coefficients at any particular angle of yaw) by the value of $\zeta = 0.0367$. They result for that particular model approximately

$$n'' = +\ .9$$
$$m'' = -2.5$$

If now we realize that for buoyancy equilibrium $m = \varrho l^3$ and since q stands for $(\varrho/2)V^2$, the density ϱ and most of the dimensions l cancel. To make everything non-dimensional, one may introduce the time unit $\tau = l/V$ during which the ship travels a distance equal to $\sqrt[3]{Q}$ and express the moment of inertia by means of a relative radius of gyration j in terms of l which is to include the hydrodynamic or virtual inertia so that $J(1+k') = \varrho l^5 j^2$. Then writing k_x for $1 + k_1$ and k_y for $1 + k_2$, the two equations are reduced to

$$-2\,k_y \tau \dot{\alpha} = n'\alpha + (n'' - 2\,k_x)\zeta \tag{2.1}$$

$$2\,j^2 \tau \dot{\zeta} = m'\alpha + m''\zeta \tag{2.2}$$

From (2.2) we have: $\alpha = \dfrac{2\,j^2\tau\dot{\zeta} - m''\zeta}{m'}$

whence $\dot{\alpha} = \dfrac{2\,j^2\tau\ddot{\zeta} - m''\dot{\zeta}}{m'}$

Substitution of this value in (2.1) then gives

$$4\,k_y j^2 \tau^2 \ddot{\zeta} + 2\,(j^2 n' - k_y m'')\,\tau\,\dot{\zeta} - (2\,k_x m' + n' m'' - m' n'')\,\zeta = 0 \qquad (2.3)$$

This is solved by assuming $\zeta = \zeta_1\, e^{\lambda_1 t} + \zeta_2\, e^{\lambda_2 t}$

where λ_1, λ_2 are the roots of the equation $A\lambda^2 + B\lambda + C = 0$ and the coefficients are taken from the above (2.3) interpreted as

$$A\ddot{\zeta} + B\dot{\zeta} + C = 0\,.$$

Since both $A = 4\,k_y\, j^2\,\tau^2$ and $B = 2\,(j^2\,n' - k_y\, m'')\,\tau$ are always positive, the criterion of stability, *viz.*, for the decrease of ζ, is simply that C must be positive, or, as it is most conveniently expressed

$$D = m' + \frac{n'\,m'' - m'\,n''}{2\,k_x} \leqq 0$$

In the case of the model of the U.S.S. Los Angeles this criterion is approximately zero, the ship appears practically indifferent, *viz.* $n' = 0.5$, $m' = 1.04$, $n'' = 0.9$, $m'' = -2.5$ and $D = 0$ within the accuracy of the determination. If it were not for the damping factor n''[1] and especially m'', the aerodynamic moment about the C. B. due to yaw, m', would indeed have to be zero or negative (restoring). Since with tail fins m'' is always negative, this is not necessary. The mechanism by which an airship with $m' > 0$ flies straight is somewhat equivalent to the phenomenon of a bicycle riding straight above a critical speed. It is as though the angle of yaw at the fins were mechanically compelled to grow (about twice the yaw of the ship) whenever the empennage is called upon, and as though curvature helped out with an invisible rudder always moved the right way. H. R. Liebert has proposed to visualize the influence of this type of dynamic stability as another dynamic metacentric height which can be expressed in our terminology as

$$H_d = -h\left(m' + m''\frac{l}{a}\right)$$

where h is the velocity height and a the distance from the C. B. to the station of zero yaw (the center of "swing").

3. Control Maneuvers. From this state of affairs it is evident that a ship which is even aerodynamically slightly unstable can be held on a practically straight course by a watchful helmsman although when left alone it will turn to port or starboard as the first incidental disturbance may dictate. As an example a diagram is shown in Fig. 25, of the path curvature as a function of rudder inclinations of an English ship as well as for the U.S.S. Los Angeles, the latter, however, while in a condition in which it had less dynamical stability than it later had under service conditions. It is seen that a definite curve is flown with

[1] R. Jones and other British authors having had no convenient method for determining n'', or the force due to rotation, substituted for it the empirical knowledge that for most ships the point of zero yaw is located ahead of the C. B. about 0.9 times the distance of the center of the fins aft of the same point.

the rudder neutral and there is a zone of angles in which the rudder seems to have a sort of hysteresis when reversing the rudder under a condition of turn. The reason for the steady rate of turn with rudder neutral being small, is that actually the aerodynamic forces increase more rapidly than in linear proportion to yaw. In terms of the theory of small deviations, this is equivalent to an increase of the derivatives and thence stability with increasing curvature. Thus, by taking the n', m', n'', m'' not at zero values of α and ζ but at definite small values, a theory of dynamical stability in the turn can be evolved. In order, however, to follow more in detail the consequences of any disturbance or, for that matter, of any definite sequence of control maneuvers, step by step methods may be resorted to[1].

The conventional, but not the only, way of controlling airships is by stern control surfaces hinged to the tail fins. Various unconventional types of controls have also been tried. Some early airships, like submarines, had bow control surfaces. Recently bow elevators were made the subject of extensive research both in flight and in wind tunnels by the Goodyear Zeppelin Corporation[2]. These experiments have demonstrated the following: It is possible, under certain circumstances, to steer an airship by the bow and it appears that this method of control has a tendency to involve less undulation of path than stern control, although when compared on terms of equal area and incidence angle, bow control surfaces appear less effective than stern control surfaces. Bow and stern controls, when applied simultaneously, afford a more powerful means to pitch or swing the ship than either alone. This would have certain advantages in clearing obstacles after the take-off and in counteracting gusts, but the additional drag, weight, and mechanism present difficulties which must be considered in connection with the question of supplementary bow control. In order to avoid their

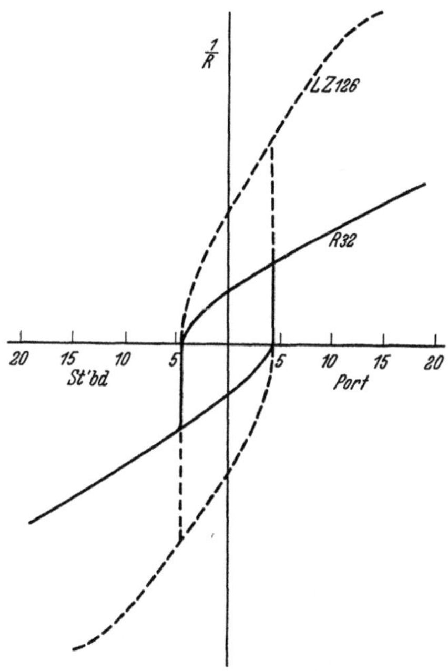

Fig. 25. Curves showing path curvature ($1/R$) as dependent on rudder angles.

[1] Changeux, P., Dynamique du Dirigeable, Librairie Aéronautique, Paris, Br. A.R.C. R. and M. 781, 1401.

[2] Unpublished as yet.

acting as fixed fins exposed to gusts ahead of the center of buoyancy, bow control surfaces have been successfully arranged "floating". In this condition, if alertly used to counteract gusts, they may tend to alleviate the bending moments imposed on the ship by the gusts whereas similar application of stern controls may enhance them. Bow controls cast a shadow or wake upon tail fins under certain circumstances. The aerodynamics of bow control surfaces are influenced by an induced angle of attack generated by the potential flow around the bow of the ship, especially when the ship flies at an angle of pitch.

The problem of balancing the control surfaces of an airship in order to keep the hinge moments small is of great technical importance. The aerodynamics involved, however, are closely related to the corresponding problems arising in heavier-than-air craft controls.

From the preceding, it will be appreciated why it requires skill and experience to steer an airship steadily, and also why the problem of automatic control sensitized for instance from compasses or gyrostatic apparatus is replete with difficulties. Such a device may indeed work quite satisfactorily in calm weather, once the desired flight attitude and course are attained and disturbances are checked within very small amplitudes. In gusty weather, however, or once a larger deviation is incurred "hunting" is likely to develop. Evidently, an aerodynamic element must be introduced into the control mechanism which, in effect, is equivalent to an increased response to yaw (or pitch) and curvature (*i. e.* increased dynamic stability) before a servo control sensitized from independent parameters can be successfully employed[1].

When the rudders or elevators are suddenly put over, equilibrium is disturbed and non-uniform motion ensues. The balancing force being usually applied at the stern, the ship pivots at a point well forward and the inertia forces are more concentrated near the source of the disturbance. If the unbalanced force F has a lever arm a aft of the C. B., then the pivot about which the ensuing motion can be viewed as an accelerated rotation is located at a distance

$$\xi = \frac{1+k'}{1+k_2} \cdot \frac{i^2}{a} = \frac{j^2}{k_y a}$$

ahead of the C. B., where i is the ship's radius of gyration for rotation about a transverse axis and k' and k_2 are again the virtual mass coefficients (including the shares of the fins). The angular acceleration is,

$$\dot{\omega} = \frac{F a}{Q i^2 (1+k')}$$

Likewise the bending moment[2] due to this motion, at any station x_1 is

$$M = \varrho \dot{\omega} \int_{x=0}^{x=x_1} (x - x_1)(x - \xi)(1+k) S \, d x$$

[1] CROCCO, G. A., I Timoni Automatici nei Dirigibili, Rendiconti delle Esperienze e degli Studi etc. Roma. Tip. Acc. dei Lincei 1912.

[2] BAIRSTOW, L., The Aerodynamic Loading of Airships, Br. A.R.C. R. and M. 794.

where k may be some constant (or better a variable) to allow for the (influence of the conicity on) virtual inertia. Integrating from the nose to the station of interest has the advantage of avoiding the uncertainties of the force distribution about the stern. Figure 26 shows typical bending moment and shear diagrams for sudden application of a rudder force.

The answer to the question of the magnitude of a suddenly applied rudder force can, of course, be sought from aerodynamic experiments with models having movable control surfaces. More or less sudden maneuvers are possible from various initial conditions of equilibrium, and not only from straight flight. The most violent maneuver appears to be a rapid reversal of the helm (rudder or elevator); because then the free force is that corresponding to the whole difference of the rudder angles before and after the maneuver, and furthermore, the new bending moments are added to the bending moments already attending the steady turn. However, as a rule, in a sharp steady turn of any duration, the additional drag has already slowed down the ship somewhat, so that the free rudder force will be less than if travelling at full speed. The apparently worst condition in calm air is, therefore, a sudden complete reversal of the controls immediately after starting a sharp turn or dive. Some experimental data of interest are given by Richmond and by Burgess[1].

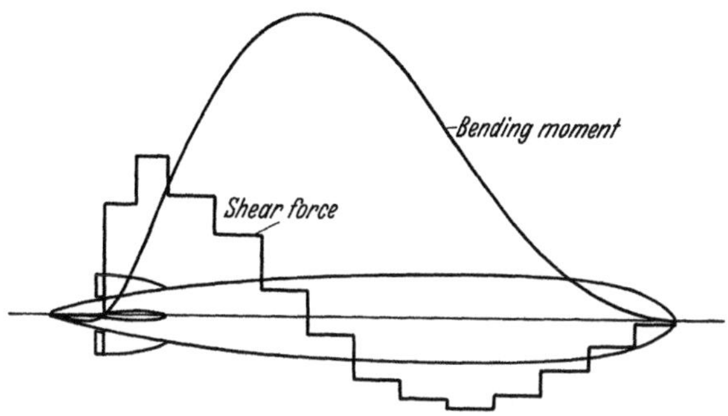

Fig. 26. Bending moment and shear diagram for sudden application of rudder force.

In a vertical reversal maneuver (checking a climb or dive) the stresses due to the more or less sudden application of the controls may occur in addition to the stresses due to carrying some overload (or excess buoyancy) by dynamic lift. In this case the ship's speed is, however, usually diminished owing to the additional drag accompanying the dynamic lift.

Combined elevator and rudder maneuvers, although hardly more severe for the ship than as corresponding to the vectorial resultant free force, may nevertheless cause increased concentrations of fin pressures along the windward fin edges.

4. Gusts. Whereas deliberate steering maneuvers normally produce unbalanced forces at the tail, gusts attack the ship first at the bow,

[1] Richmond, Lt. Col. V. C., Br. A.R.C. R. and M. 1044, 1926; Burgess, C. P., U.S. N.A.C.A. Technical Report No. 325.

when the ship runs into the gust zone, presumably with its own flying speed V. It is conceivable that the seat of the atmospheric irregularity may not always be travelling steadily with the average wind. If the disturbance is propagated through air[1] or if it does not partake of the atmospheric wind as for instance in a mountainous region, its front may be met slantwise or the speed of the gust relative to the ship may differ from the air speed of the latter. Refinements of the more simple theory, in order to take care of such influences will be obvious. An interesting study is afforded by the basic fictitious problem of the ship running with its air speed V into a "cross wind" zone. This picture has parallel applications on the one hand in yaw, as for example adjacent layers of wind of different direction or velocity, especially in mountain passes; and on the other hand, in pitch, for example, the entrance into a vertical stream of air such as may be caused by vertical deflection of wind through mountain ranges or by convection currents and thunderstorms. Vertical velocities of 3 to 6 m. per sec. are often encountered and velocities up to 10 m. per sec. and more have occasionally deliberately been hunted up, and utilized in the soaring flight of gliders and sailplanes[2].

This picture of the "cross wind"[3] zone introduces no difficulties of concept in regard to fluid continuity, but it leaves unexplained how the hydrodynamic equilibrium of the fluid shear force present is achieved, and leaves somewhat in doubt the character of the distribution of aerodynamic forces with the ship partly within and partly without the zone of disturbance. Evidently, if such conditions occur, the angle of attack will differ over the two parts of the ship and there is some question as to what degree the influence of the new angle of attack may precede the passing of the border line.

In a first approximation treatment, the zonal aerodynamic force "breadths" occurring at any station x along the ship may perhaps be assumed to be those which would obtain if the whole ship were flying at the angle of yaw (or pitch if a vertical gust is to be studied) actually

---

[1] ECKENER, H., Modern Zeppelin Airships, Journal of the Royal Aeronautical Society, June 1925; WATTENDORF, F. L., Preliminary Investigations of Atmospheric Turbulence; and KUETHE, A. M., Some Feature of Atmospheric Turbulence and the Passage of Fronts, both in Daniel Guggenheim Airship Institute, Publication No. 2, 1933; REICHELDERFER, F. W., Meteorological Aspects of Airship Operations, Daniel Guggenheim Airship Institute, Akron Ohio, Publication No. 3, 1935.

[2] KUSSNER, H. G., Zeitschr. f. Flugtechnik u. Motorl., Vol. 19, pp. 579—582, 1931; and LANGE, K. O., Zeitschr. f. Flugtechnik u. Motorl., Vol. 17, pp. 513—519, 1931.

[3] It may be noted that a gust of given velocity amplitude may be expected to exert the greatest disturbing force not when striking exactly at right angles to the ship's axis, but at an angle with a slight head-on component. However, even for the angle giving the maximum product of lateral and resultant axial relative wind component, the force increment is but small.

prevailing locally at that station[1]. The aerodynamic force due to rotation (if any) is assumed superimposed independently. Several alternative variations of the problem have been studied[1], for instance:

(a) The border of the "gust" can be assumed infinitely sharp, a jump from cross wind zero to v.

(b) A gradual increase from zero to v over a certain distance at a constant rate $dv/dx = b$ (constant vorticity).

(c) An exponential law of increase of the type $v = v_0\,(1 - e^{-rx})$.

(d) A sine wave $v = v_0\; sin\; x/a$.

In these expressions, v is the velocity of the gust and b or r indicate its degree of "sharpness". Various conditions may then be assumed regarding the course of the ship; for instance:

(1) The helmsman is able to hold the ship straight on its compass course. This eliminates rotation and all the complications associated therewith, but introduces unknown tail forces. It is true that they need not be known in order to compute the motion of the ship and the bending moments in the most important parts, but the question may still be asked: Is it possible to realize such a condition? Is the rudder system capable of producing the required forces?

(2) The helmsman holds the rudder straight and lets the ship deviate as conditions may determine. In some respects this is a simpler plan, but the computations are more complicated.

(3) The helmsman uses the controls either in an attempt to counteract the gust in some partial measure or he may also deliberately or accidentally enhance the deviation caused by the gust.

The procedure for any of these assumptions is to first write down the equations of motion *i. e.* for the lateral force (assuming a horizontal gust) and (in case 2) the yawing moment in terms of the aerodynamic and inertia forces for the ship as a whole, and (in case 3) including the control forces and moments as a function of time. As far as a first order of approximation is permissible, *i. e.*, as far as the aerodynamic reaction can be expressed as a sum of one component proportional to the angle of attack, and another independent component proportional to path curvature, the lateral and rotary accelerations can be brought into the form

$$\frac{d\eta}{d\xi} = a_1\eta + b_1\zeta + \frac{w}{u} f_1(\xi)$$

and

$$\frac{d\zeta}{d\xi} = a_2\eta + b_2\zeta + \frac{w}{u} f_2(\xi)$$

[1] Munk, M. M., The Aerodynamic Forces on Airship Hulls, U.S. N.A.C.A. Technical Report No. 184, 1924; Klemperer, W., Airships in Gusts, Daniel Guggenheim Airship Institute, Publication No. 3, 1935.

C. P. Burgess in U.S. N.A.C.A. Technical Report No. 204 and D. H. Williams and A. R. Collar in Br. A.R.C. R. and M. 1401 accepted a more drastic simplification, *viz.*, the substitution over the whole ship of the force distribution it would have under the angle of yaw or pitch which exists momentarily at the center of buoyancy.

where the variables are expressed by $\eta = v/u$ which is the ratio of lateral to forward velocity (tangent of the angle of yaw at the center of buoyancy before the passage of the gust border over this station) and $\zeta = \sqrt[3]{Q}/R$ the characteristic angle of path curvature introduced before. These variable flight parameters are expressed as functions of $\xi = x/L = ut/L$, *viz.*, the portion (x) of the ship's length (L) which has just penetrated into the gust, u being the flight velocity, t the time since the nose struck the gust border, and w the lateral velocity of the gust. The coefficients contain the inertia characteristics and aerodynamic derivatives. The functions $f(\xi)$ encompass the aerodynamic action of the lateral attack on the bow from the nose to the transient station to which the gust has just progressed. The above equations can be split into two explicit equations,

$$\frac{d^2 \eta}{d \xi^2} + A_1 \frac{d \eta}{d \xi} + B_1 \eta = F_1(\xi)$$

and

$$\frac{d^2 \zeta}{d \xi^2} + A_2 \frac{d \zeta}{d \xi} + B_2 \zeta = F_2(\xi)$$

In this form they can be solved for the angle of yaw at the C. B. (with respect to the undisturbed air) and for the curvature of path, as functions of time. Having these, the bending moments and if the need be the shear forces for any station can be computed as a function of time. Their maxima, when and wherever they may occur, are the data desired by the stress analyst.

The actual integration of the equations of motion can be performed by graphical or step-by-step methods. For ships whose forebody shape can be expressed by analytical formulae there is a chance of integrating by calculation.

Considerable simplification is afforded by an assumption of zero dynamic stability. In this case the factors B vanish and a solution can be obtained in the form.

$$\frac{d \eta}{d \xi} = - e^{-A_1 \xi} \int_0^{\xi} e^{A_1 \xi} F(\xi)\, d \xi$$

and similarly for $d\zeta/d\xi$.

Since the bending moments are the differences between moments of aerodynamic and inertia forces, all of which are crowded toward the ends, it is important that too drastic simplifications should not be applied in these regions. This and other refinements somewhat complicate the integration.

The following is an example calculated for a ship project equipped with a hypothetical empennage which would give it zero dynamical stability. This stability and the aerodynamical characteristics were arbitrarily assumed to remain constant while the ship would be exposed to a fictitious sharp-bordered gust of cross-path velocity w. Figure 27 shows how the lateral drift η and path curvature ζ build up as the ship penetrates into the gust zone and Fig. 28 depicts the deflected path the ship assumes while the helmsman does nothing to parry the gust. Figure 29 shows how, under certain assumptions concerning mass distribution, the bending moments at several reference stations ξ_f would build up and die down again as the gust border sweeps over the ship. The ordinates of the angles of Fig. 27 are here referred to unit ratio $\beta = w/u$ of assumed gust velocity to ship speed. For any particular gust or ship speed, they would have to be multiplied by β. Similarly the bending moment coefficient is referred to unit $\varrho u w Q$.

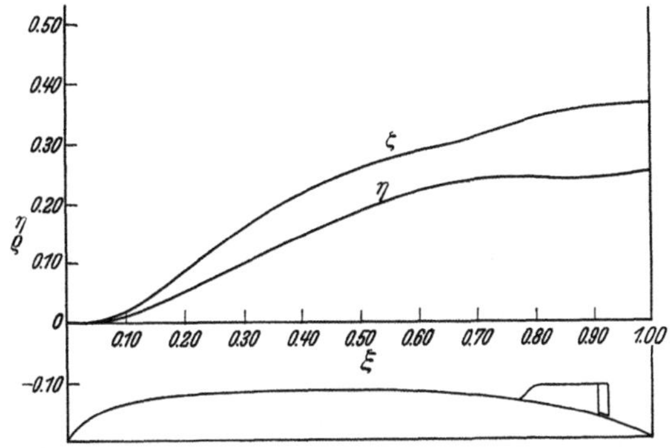

Fig. 27. Characteristics of motion in a sharp bordered gust.

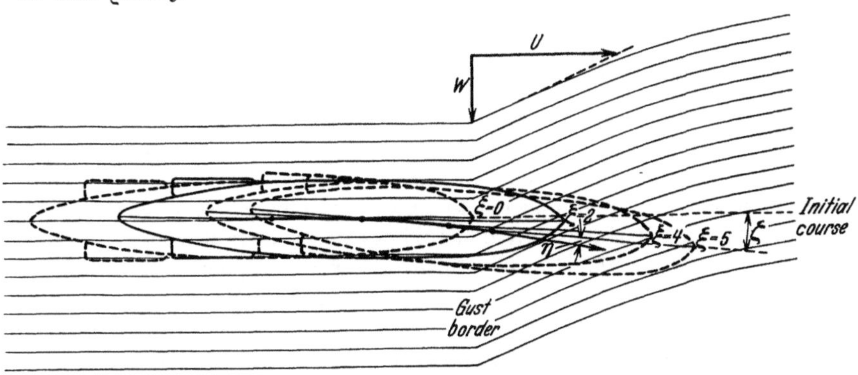

Fig. 28. Deflected path as airship enters a gust.

The examples here given were naturally based on some particular assumptions of the ship's aerodynamical and rotary characteristics. A ship of different empennage for instance would be expected to behave differently.

For any gradual increase of the cross wind to its full velocity w, the deviations and moments are obviously less than for a sharp-bordered gust.

It is also of interest to study the flight of an airship through a field of a moving vortex, which can be interpreted as a particularly simple and well defined form of a "gust". Investigations of this problem were undertaken by Poggi[1] and Oswald[2]. The forces and moments experienced by an ellipsoid exposed to a flow consisting of a straight component and a vortex located on the "port or starboard beam" or "dead ahead" were thus determined. These assumptions are equivalent to the conditions prevailing in the flight past the side of a tornado or toward one. As these calculations are based on classical fluid concepts, they still leave the lateral force on the hull and all actions of an empennage out of consideration. To take these and also deliberate rudder action into account, would constitute the next step toward a study of the actual dynamics of the ensuing motion.

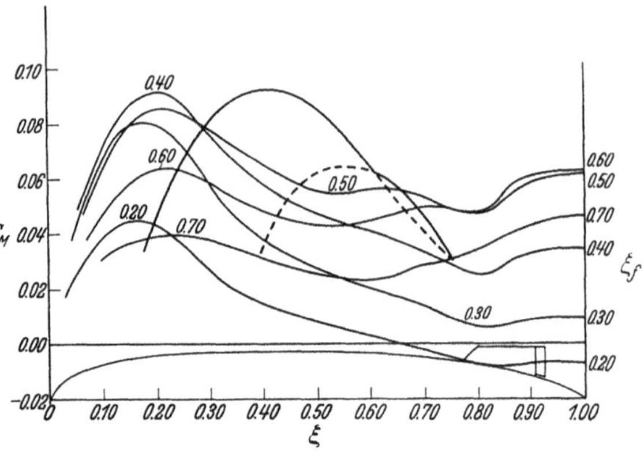

Fig. 29. Bending moments at several stations in relation to sweep of gust border along the ship.

CHAPTER V

MOORING AND GROUND HANDLING

1. Mast Mooring. Whenever it becomes necessary to anchor an airship in the open, it is desirable to hold it headed into the wind as this is the attitude imparting the lowest stresses. Provisions are therefore made to let the ship swing to follow any changes of wind direction in weathercock fashion. Evidently it is most advantageous for this purpose, to anchor the ship to a mast or tower at a point as far forward as possible, *viz.*, at the nose. This has the additional advantage of flow symmetry, thus minimizing any pitching and rolling tendencies in fluctuating winds. If the derivative of the aerodynamic moment about the mooring point due to yaw is zero or negative, the ship will ride stably at zero yaw. In the previous notation this condition is expressed by $(m' - n' a) = 0$ where $m' = \partial C_m / \partial \alpha$ and $n' = \partial C_n / \partial \alpha$, the moment

[1] Poggi, Azioni aerodinamiche su di una ellissoide di rotazione rivestito da un vortice, etc. Estratto d'Atti della Pont. Accademia delle scienze, nuovi Lincei Anno LXXXIV, December 21, 1930, Rome.

[2] Oswald, W. B., The Transverse Force Distribution on Ellipsoidal and Nearly Ellipsoidal Bodies Moving in an Arbitrary Potential Flow. Thesis, Calif. Institute of Technology, 1932; von Kármán, Th., Some Aerodynamic Problems of Airships, Daniel Guggenheim Airship Institute, Publication No. 1, 1933; Tollmien, W., The Motion of Ellipsoidal Bodies Through Curved Streams, Daniel Guggenheim Airship Institute, Publication No. 2, 1935.

and normal force derivatives due to yaw of the empennaged ship, expressed in terms of $\sqrt[3]{Q}$, with a the lever arm from the bow anchor point to the C. B. For small non-rigid ships with large empennages, this condition is usually amply fulfilled. In moderate winds they can even be moored satisfactorily at a "breast plate" some distance aft of the nose, thus permitting lower and simpler masts. Large rigid ships can be satisfactorily moored at the nose even though $(m' - n'a)$ may have a slight positive value at zero yaw. In this case they will ride at a small angle of yaw leaning on the wind to port or starboard as chance may determine and occasionally swaying over gently from one side to the other. These angles are, of course, the ones where the curve $(C_m - aC_n)$ plotted against the angle of yaw α crosses the zero lines under a negative slope as indicated in Fig. 30. Under suitable conditions of trim an angular deviation may also develop in the vertical plane, the position of equilibrium being determined by the balance between the out of trim moment and the aerodynamic forces due to the wind. These conditions can be duplicated and studied in wind tunnel experiments[1]. In such conditions of mooring, there is of course, a small transverse force exerted by the mooring device upon the nose of the ship.

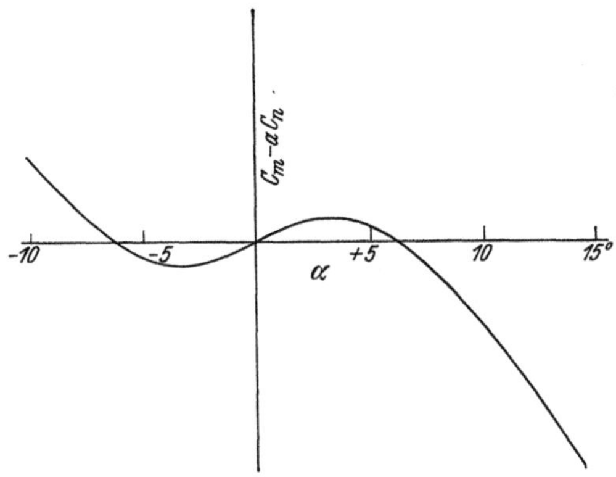

Fig. 30. Curve of $C_m - a\,C_n$ on angle of yaw.

The merits of various types of mooring and handling equipment have been frequently discussed in the technical literature[2]. Higher masts permit greater clearance between the ship and the ground or the water (in case of a floating mast such as the U.S.S. Patoka). However,

[1] Klemperer, W., Windkanalversuche an einem Zeppelin-Luftschiffmodell, Abhandlungen aus dem Aerodynamischen Institut an der Technischen Hochschule, Aachen, 12.

[2] Butcher, F. L. C., Airship Mooring and Handling, Journal of the Royal Aeronautical Society, February 1921; Richmond, V. C., Airship Research and Experiment, Journal of the Royal Aeronautical Society, October 1926; Scott, Maj. G. H., The Handling and Mooring of Airships, Journal of the Royal Aeronautical Society, November 1929; Bleistein, W., Zeitschr. f. Flugtechnik u. Motorl., pp. 362—365, 1929; Rosendahl, Lt. Comdr. C. E., Up Ship, p. 109, 1931 (Dodd Mead and Co.). Mooring Masts and Landing Trucks for Airships, Journal, Society of Automotive Engineers, July 1929; Bolster, Lt. C. M., Mechanical Equipment for Handling Large Rigid Airships, Aeronautical Engineering (A.S.M.E. New York) No. 3, 1933.

low masts are considerably less expensive than high masts, and have the further advantage that they permit of securing the tail to a circular track on the ground, so as to prevent any vertical movement of the ship. A special type of the low mooring mast is the mobile variety. It is ordinarily intended for ground transport, but may occasionally be called upon for temporary mooring services.

In gusty weather with capricious wind shifts, the ship may fail to follow quickly like a weather vane. Due to the great inertia of the ship, there may be a lag in turning and large angles of yaw and large lateral forces may thus occur temporarily. The dynamics of this motion can be dealt with in similar manner as in free flight, only that the motion is confined to rotation about the nose anchor point. Theoretical difficulties attending the arbitrary assumptions regarding the structure and the mechanism of the "gust" still remain.

2. Cable Mooring. There have been various proposals of methods for mooring airships in the air by means of cable systems and many of these methods have been tried[1]. In the cable type of mooring, the airship is attached to the apex of a three wire "pyramid" and is flown light or is pitched upward so that the airship lift keeps the cable system taut and creates the effect of a "virtual" mast. This type of mooring apparently originated with the Italians in 1908[2] but received more extended use in England. It was this type of mooring which was used for the R-34 at Long Island after her trans-Atlantic flight in 1919.

In addition to their full scale experimental work with this three wire system, the English undertook laboratory experiments with this and other forms of "free" wire systems[3]. The dynamics are essentially the same as with a solid mast but for the increased elasticity and the chance that the ship's bow may acquire excessive momentum due to the inevitable sag in the cables or to occasional slackness in one cable or another should the surplus lift be insufficient to keep them taut in some phase of oscillation.

A more complicated dynamical problem is offered when the ship is moored by one cable only, like a kite balloon[4]. This condition can occur while hauling the ship up to a mooring mast by means of a nose cable, should the side guy lines which are supposed to hold taut during

---

[1] KRELL, O., Zeitschr. f. Flugtechnik u. Motorl., pp. 401—438, 1928 (also U.S. N.A.C.A. Technical Memoranda 512 and 513, 1929); MASTERMAN, AIR COMMODORE E. A. D., The Evolution of Mooring and Handling Devices for Airships in Air Annual of the British Empire, 1930.

[2] CROCCO, G. A., Translations appearing in U.S. N.A.C.A. Technical Memorandum No. 283, 1923.

[3] FRAZER, R. A., Note on the Mooring of Airships by "Free" Wire Systems, Journal of the Royal Aeronautical Society, February 1921.

[4] BAIRSTOW, L., The Stability of Kite Balloons: Mathematical Investigation, Br. A.R.C. R. and M. 208, 1915.

the maneuver go slack. While the mooring cable is long or slack, the ship can take advantage of the dynamic stability in free flight without much influence upon the forces transmitted by the mooring cable until the latter snaps taut. If, however, the mooring cable has already been hauled in to a length comparable to the ship's length, any yawing of the ship and resultant lateral drift will produce large changes in the angle the cable makes with the wind. The multiple degree of freedom may give rise to complicated coupled oscillations. These, by the way, can be readily studied by model experiments in a water channel.

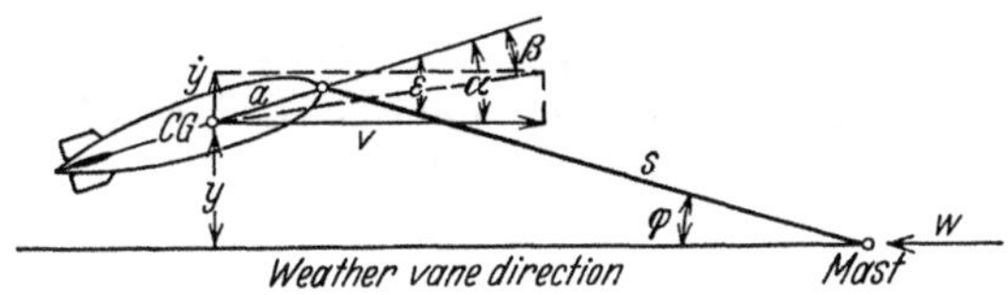

Fig. 31.

If the oscillations are small and confined to one plane and, if an infinitely thin, massless cable is assumed, they can be described by the following primitive theory. With notations and a general arrangement as given below and sketched in Fig. 31, the equations of motion, in terms of accelerations and d'Alembert reactions, are:

$$[k_y \ddot{y} - (k_y - k_x) U \dot{\alpha}] = \frac{U^2}{2l} [n'\beta + n'' \zeta - C_D \varepsilon] \tag{2.1}$$

$$j^2 l^2 \ddot{\alpha} = \frac{U^2}{2} [m' \beta + m'' \zeta - C_D \varepsilon a/l] \tag{2.2}$$

The notations aside from those already introduced are,

$l^3 = Q$ = volume of ship,

$j l = \sqrt{\frac{J k_p}{Q \varrho}}$ = radius of gyration of the ship about a master section diameter (including virtual mass),

y = lateral displacement of ship's center of gravity from the weather vane direction,

α = azimuth angle between ship's axis and absolute wind direction (weather vane direction),

β = angle of attack between ship's axis and relative wind,

ε = angle between cable and ship's axis,

U = wind velocity,

C_D = drag coefficient,

s = length of mooring cable,

a = distance from bow to center of gravity.

After elimination of ζ, ε, γ by recourse to the geometrical relations

$$\zeta = l \dot{\alpha}/U$$

$$\varepsilon = \frac{y + (a + s)\alpha}{s}$$

$$\beta = \alpha - \dot{y}/U$$

two simultaneous equations of the type

$$A_1 \ddot{y} + B_1 \dot{y} + C_1 y \qquad + E_1 \dot{\alpha} + F_1 \alpha = 0 \tag{2.3}$$

$$B_2 \dot{y} + C_2 y + D_2 \ddot{\alpha} + E_2 \dot{\alpha} + F_2 \alpha = 0 \tag{2.4}$$

are obtained.

The equations can be readily solved; the stability criterion for the gradual dying out of the oscillations is expressed by five coefficients, a_0 to a_4, which must all be positive and obey a relation

$$a_2 > \frac{a_1 a_4}{a_3} + \frac{a_3 a_0}{a_1} \tag{2.5}$$

These coefficients a_0, a_1, a_2, a_3, a_4 are derived from the coefficients A, B, C, D, E, F, of the equations (2.3) and (2.4). Inserting the respective original values they become

$$\left.\begin{aligned} a_4 &= 2\, k_y j^2 \\ a_3 &= j\, n' - k_y m'' \\ a_2 &= C_D \left[k_y \frac{a}{l}\left(1 + \frac{a}{s}\right) + \frac{j^2 l}{s}\right] - k_y D \\ a_1 &= -\, C_D \left[m_b' + \frac{m_b'' l - 2\,(k_y - k_x)\, a}{s}\right] \\ a_0 &= -\,\frac{C_D m_b' l}{s} \end{aligned}\right\} \tag{2.6}$$

For the sake of abbreviation and better visualization, we have introduced here the derivatives about the bow attachment point; *viz.*,

$$m_b' = m' \;-\, n'\, a/l$$

and

$$m_b'' = m'' - n''\, a/l$$

and the criterion of dynamic stability in free flight

$$D = m' + \frac{n' m'' - m' n''}{2\, k_x}$$

The most significant of the conditions for stability of the mooring lines is expressed by a_0 which dominates all others for short lengths of line. It postulates that for stable flight on a mooring line a ship would have to be so empennaged that even at small angles of yaw, the aerodynamic resultant force passes back of the bow mooring point (m_b' negative). This is easily fulfilled with small airships but has not always been completely fulfilled with large rigid airships. However, the undamped oscillation component which results from this slight deficiency is of very slow period and small. It is kept in bounds by the more rapid increase of the stabilizing forces with increasing angles of attack. The role which the dynamic stability criterion D plays (in a_2) becomes less and less important as the cable length is decreased and it vanishes for $s = 0$ when the ship is finally anchored by her nose. A small dynamic instability ($D \gtrsim 0$) is tolerable. Even with a very long cable a positive D need only meet the condition $D < C_D a/l$ in order to still be compatible with theoretic stability. In reality, of course, variations in cable pull and possible vertical forces due to lack of exact buoyancy equilibrium complicate the picture. The other requirements a_1, a_3, a_4 are always fulfilled, the two former due to the large negative values of the rotary moment derivatives.

The condition (2.5) is not easily visualized. It is usually also fulfilled for any cable length if conditions are such that a_0 and a_2 are positive. Artificial increase of drag (C_D) increases stability in mooring. The case of a streamlined, finned body, towed from aircraft is a related problem. It had been studied by Glauert[1].

3. Ground Transport. The ideal solution of housing or "docking" huge airships, from a mere aerodynamic standpoint, is the turntable hangar which can be turned into the wind direction so that the ship can, after landing, be walked into the shed by man or machine power without broadside exposure to the wind. Such hangars were built in Germany[2] during the war and were successfully operated. The first Zeppelin hangar was built as a floating structure on Lake Constance. The drawback of this type is the additional cost, the original investment required for any type of stationary airship hangar being already large. Where rotatable hangars are not available, stationary hangars or "docks" built with the axis parallel to the "prevailing" direction of moderate winds furnish the next best solution. The ship is then landed on the field into the hands of a landing party or by attachment to a stationary or mobile landing mast and then "walked" into the hangar. Occasionally it will occur that the "prevailing" wind does not prevail and the ship while being maneuvered into the dock has to pass through some phase of slant or broadside exposure to a cross hangar wind. The motion as a rule is slow so that the forces and stresses can be studied as though the ship was held stationary and without regard to inertia forces. There are three distinct aerodynamic problems arising in this connection[3].

(1) The aerodynamic forces due to very large angles of yaw of the airship, including attack broadside and from astern, and quartering.

(2) The influence of the close proximity of the ground.

(3) The wind obstruction formed by the hangar itself and its doors.

The aerodynamic force (and its distribution) normal to the ship's axis at a very large angle of yaw does not as yet lend itself well to a purely theoretical treatment. As a first approximation it may perhaps be assumed roughly proportional to the projected area, including fins and all appendages, with a drag coefficient in the vicinity of unity, both

[1] Glauert, H., Br. A.R.C. R. and M. 1312, 1930.

[2] Christians, Anlage und Betrieb von Luftschiffhäfen, Oldenbourg, 1914; Engberding, Luftschiff und Luftschiffahrt, V.D.I.-Verlag, 1928; Krell, O., Zeitschr. f. Flugtechnik u. Motorl., pp. 401—438, 1928; also U.S. N.A.C.A. Technical Memorandum Nos 512—513, 1929.

[3] Arnstein, K., and Klemperer, W., Aerodynamic Problems in Connection with Ground Handling and Docking of Airships, A.S.M.E., 1934.

for the near cylindrical sections and for the sharp edged empennage region. In view of the uncertainty regarding the other factors, including the wind speed to be assumed, this may suffice for many general purposes. Upon closer scrutiny it would appear that the results reached in this way may quite exaggerate the actual condition. This is for the reason that, for Reynolds numbers above about 300,000, the Kármán vortex street mechanism of the drag of cylindrical bodies breaks down and gives way to a vortex system of much smaller size which drops the drag coefficient to less than half its value for smaller Reynolds' number though not necessarily to a constant value. Model experiments of sufficient size or speed to reach into the Reynolds' range above 10^7, such as would correspond to ships of 30 to 40 m. diameter in wind speeds of the order of 5 to 10 m. per sec, have not yet been published. The unfortunate feature of most model tests thus far is the fact that it has been just barely possible[1] to reach into the above critical Reynolds' region, whereas at the bow and stern, where the diameter is smaller, conditions would remain in the sub-critical region so that the results obtained may be freakish and uncertain. The least little irregularity of the surface may cause the second type of flow to originate sooner on one side than the other, resulting in irregular pitching moments and lift even under symmetrical attack.

Even if the aerodynamic forces and their distribution about the airship under broadside attack in free air were well understood, there would remain the influence of the proximity of the ground. This makes itself felt in two ways. First, it constitutes an aerodynamic mirror surface forcing the near-by stream-lines to follow it, thereby tending to produce some acceleration below the ship. An attraction to ground might thus be expected. By integrating the pressures produced upon the ground surface by a series of repeatedly reflected two-dimensional doublets[2], the attraction between a cylinder and the mirror surfaces can be computed for hypothetical potential flow. In a first approximation by means of a single doublet and mirror image, representing a somewhat flattened hull cross-section, an attraction or negative lift can be calculated

$$\frac{dL}{dx} = -q\pi\frac{R^4}{h^3}$$

per unit length of the ship where q is the velocity head of the wind, h the height of the ship's axis above the ground, and R the radius of a cross-section equivalent to the doublet without its image. Furthermore the wake shed by a cylinder attacked broadside may be simulated by an unbalanced source-sink combination (with source stronger than sink)

[1] Even though the critical speed may be slightly lowered by making the model polygonal in cross section rather than round.

[2] Müller, W., Z.A.M.M., Vol. 9, No. 3, p. 200, 1929; Z.A.M.M., Vol. 11, p. 231, 1931; Lagally, Z.A.M.M., Vol. 9, No. 4, p. 299, 1929.

supplemented by its mirror images. In this manner one finds the attraction by the ground reduced and correlated with the drag.

An investigation into the forces on a large cylinder near a ground dummy was recently made at the Daniel Guggenheim Airship Institute, Akron, Ohio, by Th. Troller and F. D. Knoblock who measured the pressure distribution and the force reactions under special precautions, excluding disturbances from the ends of the cylinders[1]. Their results indicated for the section of such a cylinder (two-dimensional flow) that, during the transition from under-critical to above-critical flow regime, the stagnation point on the windward side travels from some 70^0 ahead of the point nearest the ground to 80^0, while the zero over-pressure point travels from about 30^0 to 44^0 ahead of the same location. The force between ground and cylinder is a repulsion in the completely under-critical flow as well as in the completely over-critical flow, whereas during some phase of the transition, when the suction peak is more pronounced in the interspace, the force is a pronounced attraction.

In reality it is possible that circulation builds up around an airship held close to the ground against a side wind. In fact such circulation is favored by the presence of wind friction on the ground. It is conducive to a repulsion or positive lift which is likely to outweigh the attraction postulated in its absence from potential flow. H. v. Sanden[2] has calculated this lift for a symmetrical profile exposed to a wind, the velocity U, of which, increases with height h, at a rate $U' = dU/dh$ and has shown that the lift is proportional to $Q \varrho U U'$, *viz.*; the volume, density, velocity and velocity gradient for any given profile.

Troller and Knoblock also measured the forces on their cylinder above a dummy ground board in the presence of an artificially produced velocity gradient simulating that present in a natural wind under certain conditions. In this particular case they found only repulsion; *i. e.*, positive lift forces.

In the practice of docking airships, especially when maneuvering to the leeward of the dock, it is also important to be on the watch for the presence of huge eddies trailing off from the building and its open doors. The "shelter" or wake region of these doors, and of artificially erected screens, has often been regarded as beneficial and may be so for small ships under certain circumstances. However, actual ground handling experience and extensive wind tunnel experimentation with model ships, docks and screens[3] demonstrate that there may be treacherous whirls

---

[1] Knoblock, F. D., and Troller, Th., Tests on the Effect of Sidewind on the Ground-Handling of Airships, Daniel Guggenheim Airship Institute, Publication No. 2, 1935.

[2] Sanden, H. v., Über den Auftrieb im natürlichen Winde, Zeitschrift für Mathematik und Physik, No. 3, pp. 225—245, 1913.

[3] Jones, R., and Levy, H., Br. A.R.C. R. and M. 338, 1917; Nayler, J. L., and Woodford, F. G., Br. A.R.C. R. and M. 428, 1917.

with vertical as well as horizontal and slant axes and high velocities originating in the spillover regions. Experiments in depicting the flow with streamers[1] indicate quite clearly the difficulties of dealing with this problem by mathematical formulae. On the other hand, there is a trend toward minimizing these field disturbances by a smooth design of the dock building and especially by the use of doors which exert a minimum of aerodynamic influence when opened. These thoughts were controlling in the design of the airship dock at Akron, Ohio (Fig. 32). When the ground handling of an airship into and out of the leeward door of a dock is interfered with by aerodynamic effects in the wake of the building the windward doors are sometimes partly opened, thus permitting some direct passage of air through the dock.

CHAPTER VI

OUTSTANDING PROBLEMS

It would be going far beyond the intended scope of the present work if any attempt were made to expose all of the respects in which aerodynamic (and certain aerostatic) phenomena have a bearing on the design of airships. They are so intimately interwoven with structural, mechanical, and navigational considerations that a design handbook would be required to do justice to the outstanding problems in this domain. If, nevertheless, a few such problems are mentioned here, it is done with a full appreciation of the arbitrary character of the selection, and in the thought that they may furnish some suggestions to students and investigators interested in carrying forward the boundaries of our knowledge on this subject.

Man's knowledge of an art can never be complete and there is no exception in the case of Airship Design. The airship designer and the airship operator are continually facing problems[2] which must be solved if airships are to be continually improved. As in all arts the major difficulties must either be solved or skillfully circumvented and the frequency of recurrence of a problem, or the difficulty of dismissing it from practical design considerations determines its lasting importance.

---

[1] Krell, O., Zeitschr. f. Flugtechnik u. Motorl., pp. 401—438, 1928 (also U.S. N.A.C.A. Technical Memoranda Nos. 512 and 513, 1929).

[2] Richmond, V. C., Airship Research and Experiment, Journal of the Royal Aeronautical Society, October 1926; Scott, G. H., Research Problem in Airship Development, Journal of the Royal Aeronautical Society, April 1926; Fulton, G., Airship Progress and Airship Problems, Journal of Americal Society of Naval Engineers, February 1929; Arnstein, K., Fulton, G., Hunsaker, J., and Kármán, Th. v., Daniel Guggenheim Airship Institute, Publication No. 1, 1933; Munk, M. M., On the Problems of Progressive Airship Research, Daniel Guggenheim Airship Institute, Publication No. 3, 1935.

Fig. 32. Modern Airship Dock.

In this connection the acceptance of a *routine method* for the application of aerodynamic data to design procedure and especially to the stressing schedule or assignment of safety factors for various loading assumptions, such as is already the practice with heavier than air craft, must still be regarded as an outstanding problem with lighter than air craft. In fact the appreciative judgment and balanced valuation of partly conflicting consequences of theoretical loading assumptions constitutes one of the most difficult problems confronting airship designers. If, therefore, thumb rules—rules expressing in general terms of a few parameters, such as size and speed and experience coefficients, the "practical" loading assumptions for such problems as the required nose stiffening, an equivalent bow force representing gusts likely to be encountered, a standard bow force to represent average mooring stresses, or even a good "all around" maximum bending moment to cover all possible events—if such rules are sometimes found in the literature of the subject they must be taken for what they were intended, *viz.*, an attempt to evaluate the order of magnitude found in successful designs. Any more searching or responsible procedure will inquire into the manifold aerostatic and aerodynamic variables and their individual as well as combined influences upon the stability, controllability, strength and safety of the ship as a whole and in all its vital and essential parts.

The large variety of possibilities has been a barrier to airship research. For instance in spite of the amount of theoretical and experimental research work which has been done on the aerodynamics of airship hulls, there is still a need for dependable information on the distribution of the air forces over the hull and empennage during certain conceivable flight conditions. Up to date it is a fact that designers have not always obtained as much full flight experimental data as was to be desired.

Difficulties lie in the need for a full *synchronization of all flight data* from all observation points, and for the record, along with the observed data, of such flight characteristics as airspeed, rudder and elevator positions, together with the ship's instantaneous position, attitude, rotation and acceleration. All of these flight observations and the desired test maneuvers must be executed with due regard for the safety of the ship and generally with the stipulation that they shall place no restrictions upon normal operation.

One of the basic problems, therefore, is the continued development of equipment and of experimental technique which will aid in obtaining this greatly needed flight information without undue interference with service operation, unless airships can be made especially available for an extensive program of flight testing.

The research required runs through the whole range of experimental technique from the accurate weighing of a complete 100 ton airship to the delicate measurement of structural stresses in an enormous number of members under conditions of flight through turbulent air of measured characteristics. The importance of the various design factors in determining the overall strength and the efficiency of the airship is known to the designer and he uses this knowledge to guide his research, keeping the goal of a well balanced airship design always foremost in his mind. In many cases over-caution regarding safety against one contingency may invite increased liability with regard to another[1].

The development of a complete *theory of dynamic lift* on airship hulls and fins under small, moderate, and large angles of attack with and without proximity to the ground, would appear welcome in order to give theoretical confirmation and interpretation to the results of delicate experiments.

The same is true regarding the theory of *air forces produced by gusts* although here progress will equally depend on the furtherance of elaborate experimental research into the nature and texture of gusts such as they actually occur.

Step-by-step investigations into what happens when a ship whether in flight, moored or towed, is *struck by a gust* of any assumed type, but with due regard to the variability of the dynamic stability characteristics with the varying phases of motion, should give valuable results. Such studies offer many interesting problems although their value may seem somewhat limited to the specific case and ship under investigation.

An extension of the *theory of airship drag* and of the velocity field in the boundary layer around the airship hull, perhaps including considerations of propulsion and the most advantageous placement from a drag standpoint of radiators and heat transfer apparatus, is a promising field of research especially in the light of the advances of knowledge recently made regarding the nature and laws of the impulse transmission in the boundary layer in general.

The problem of *retaining buoyancy equilibrium* despite the consumption of large quantities of fuel has been attacked in many ways. The recovery of ballast in flight has been tried by the chemical and hygroscopic collection of water from the air, the collection of materials and substances in the exhaust gases, the catching of rain water and the lifting of ballast from the surface. In ships built to date the most successful solution has been the recovery of water from the engine exhausts but it has been replete with problems since the structural and thermal efficiency requirements are antagonistic to weight and drag economy.

The "Graf Zeppelin" has avoided this problem, eliminating it at its source by the consumption of a *gaseous fuel* having a density about equal to that of air. The "space efficiency" of fuel gas ships is excellent but

[1] Arnstein, Karl, Über einige Luftschiffprobleme, Zeitschr. f. Flugtechnik u. Motorl., January 14, 1933.

there may be offsetting disadvantages in the added fire hazard; also the weight of the suspensions and fuel ballonets and the surging must be allowed for. Ballast is not likely to be available in as generous a quantity as on a ship burning liquid fuel and equipped with exhaust water recovery apparatus.

In assisting the ship to rise when heavy or to land when light, *tiltable* propeller arrangements have shown considerable merit. Since some forward speed may exist in combination with vertical thrust, a suitable theory for the determination of the action and stresses on propellers under large angles of slant inflow becomes of interest, as with travelling helicopters. Similarly studies of the practice of decelerating from flight speed when making an approach for landing by *reversing the propellers*, requires a knowledge of the theory of a propeller travelling at negative velocity and (unless reversible blades are resorted to) with attack on the back of the blade.

The problem of the best shape of the ship for lowest drag per volume housed may appear to be fairly well in hand. However, it is not possible to predict the extent to which the art of streamlining may be changed by some success in the *removal or repulsion of the boundary layer*. The problems of disposing of the dead air and of handling the appreciable volumes involved without incurring more loss in ducts and fans than the gain may be worth, constitute, however, possible stumbling blocks.

Similarly both theory and experiment may find an interesting field in an investigation directed toward the realization of *steering control* through means other than the conventional movable surfaces[1].

Technical improvement and simplification of the *technique for landing and housing* large airships will always be appreciated and may undoubtedly be helped by improvements in existing aerodynamic theories. Advanced knowledge along these lines cannot help but result in safer and more economical operations.

The original method of landing and housing rigid airships was on the water. Since then the trend of the development has been toward landing and housing them on land. However, *water landings* have occasionally again come to the fore. The merits of various suggestions made in connection with water landing depend upon the proper appraisal of the additional hydrodynamical problems. These arise with the necessity of holding and hauling the ship down and horizontally in winds and waves.

Gas for Airships. The inexplosive inert gas helium offers such evident advantages of safety over the inflammable gas hydrogen, that some means to offset its higher cost and lesser buoyancy would be greeted as most welcome. Whether, in some degree, this can eventually be achieved by mixtures of gases or otherwise is not directly a matter of

[1] MUNK, M. M., On Problems of Progressive Airship Research, Daniel Guggenheim Airship Institute, Publication No. 3, 1935.

aerodynamics, but inasmuch as the retention of the gases introduces aerostatic problems, it may properly receive present mention.

The enumeration of such problems could be continued at great length, and as rapidly as the problems are properly stated, airship inventors will conceive new means and devices for overcoming them. If the historical record of the past is an index to the future, the application of these new inventions will introduce further problems. What is really needed is less invention and more application of the time proven principles which have already been successfully demonstrated. After all, the airship is not an abstract thing; it is a man made vehicle operating in an earthly medium and its functioning is governed by nature's physical laws. There is probably no more fitting thought with which to close this brief discussion of airship problems than Count Ferdinand von Zeppelin's motto,

> "The forces of nature cannot be eliminated
> but they may be balanced one against the other."

Bibliography.

In addition to the many references supplied directly in connection with the text, the following more general references are given.

The U.S. N.A.C.A. Bibliographies of Aeronautics, and the indices of the Reports, Notes and Memoranda of the N.A.C.A.

The Index of the Reports and Memoranda (R. and M.'s) of the British Aeronautical Research Comittee.

The Indices to the publications of the Deutsche Versuchsanstalt für Luftfahrt.

The Indices of the Journal of the Royal Aeronautical Society.

The Indices of the Zeitschrift für Flugtechnik und Motorluftschiffahrt.

The Bibliography of Rigid Airships forming Appendix I of "The Strength of Rigid Airships", by Hunsaker, Burgess and Truscott, published in the Journal of the Royal Aeronautical Society, Vol. XXVIII, No. 162, June 1924.

The list of references forming Appendix III of "The Rigid Airship and Full-Scale Experiment", by R. A. Frazer, published in the Journal of the Royal Aeronautical Society, Vol. XXIX, No. 177, September 1925.

The Bibliography forming Appendix II to R. and M. 800. — "Report of the Airship Stressing Panel."

The List of Papers relating to Airships forming Appendix III to R. and M. 970. — "Report of the Airworthiness of Airships Panel."

The List of Reports given in Appendix IV to R. and M. 775. — "Report on the Accident to H. M. Airship R-38."

The Bibliography on p. 21 of R. and M. 1170. — "Report of the Airworthiness of Semi-Rigid Airships Sub-Committee."

The List of R. and M. Reports appended to "The Aerodynamical Characteristics of the Airship", by R. Jones, published in the Journal of the Royal Aeronautical Society, Vol. XXVIII, No. 158, February 1924.

The Bibliography in "Luftschiff-Messtechnik", by W. Klemperer, in Handbuch der Experimentalphysik, Band 4, 3. Teil, Leipzig 1930.

The Indices of the Publications of the Central Aero-Hydrodynamical Institute, Moscow.

American books on "lighter-than-air" published in the Ronald Press Series, as follows:

WARNER, E. P., Aerostatics.

CHANDLER and DIEHL, Balloon and Airship Gases.

UPSON and CHANDLER, Free and Captive Ballons.

BLAKEMORE and PAGON, Pressure Airships.

BURGESS, C. P., Airship Design.

The two principal British design texts are:

LEWITT, E. H., The Rigid Airship, Sir Isaac Pitman, London 1925.

COX, H. R., and COLLINS, T. S. D., Airship Design, in Handbook of Aeronautics, London 1931.

Following is a series of references to Wind Tunnel Tests on Airship Models, covering various phases and problems connected with pressure distribution, drag, control, etc.

American.

N.A.C.A. Technical Reports Nos. 138, 215, 291, 342, 394, 405, 432, 451 and Technical Note No. 264. Pubs. 1, 2 and 3 Daniel Guggenheim Airship Institute, 1933 and 1935 respectively.

SEIFERTH, R., Paper No. 24 of "Unpublished Papers Presented at A.S.M.E. Meetings in 1932" on file at the Engineering Societies Library, 29 West 39th Street, New York.

British.

R. and M.'s 17, 30, 43, 55, 104, 107, 151, 193, 245, 246, 302, 311, 361, 456, 457, 482, 541, 597, 607, 619, 646, 698, 714, 779, 802, 827, 1160, 1168, 1234, 1271, 1379, 1400, 1409, 1511.

German.

FUHRMANN, G., Jahrbuch der Motorluftschiff-Studiengesellschaft. Also Z.F.M., June 11; July 16, 1910, Band V, 1911—12.

WIESELSBERGER, C., Z.F.M., September 1915.

PRANDTL, L., Z.V.D.I., October 16, 1909. Also Jahrbuch der Mot. Stud., Vol. IV, 1910—11.

NAATZ, H., Neuere Forschungen im Luftschiffbau, Z.F.M., Beiheft 11, 1923 (U.S. N.A.C.A. Technical Memorandum No. 275).

KLEMPERER, W., Abhandlungen aus dem Aerodynamischen Institut an der Technischen Hochschule, Aachen, 1932.

Italian.

COSTANZI, CAPT. G., Costruzioni Aeronautiche del Genio, Anno II, No. 4, 1912 and Anno III, No. 9, 1913.

French.

EIFFEL, G., Nouvelles recherches sur la resistance de l'air et l'aviation faites an laboratoire d'Auteil (Dunod et Pinat, 1914).

Japanese.

The Resistance of the Airship Models Measured in the Wind Tunnels of Japan. Aero. Res. Inst. Tokyo Imp. Univ., No. 15, March 1926.

DIVISION S

HYDRODYNAMICS OF BOATS AND FLOATS

By

E. G. Barrillon,

Paris

EDITOR'S PREFACE

Any record of progress in aerodynamic theory would be quite incomplete without some reference to marine aircraft—the seaplane and the flying boat. The performance of these craft involves two distinct phases, on the surface of the water and in the air, the one merging into the other at the moment of take-off from the water, or of returning to the same. When flying, the basic problems of marine aircraft are similar to those of land craft except as they may be modified in detail by necessary changes in form and proportion resulting in relative differences in the location of the center of gravity and in the thrust line of the propeller.

For these reasons, the treatment of the present Division is limited to what be termed the "marine" phases of performance. In Chapter I a general view is given of the special conditions imposed on a seaplane, both in repose on the water and during the periods of take-off from the water and return to repose on the same. Special note is made of the significance of the *step* and of the part which it plays in facilitating take-off from the surface of the water.

Then follows Chapter II dealing in further detail with the varying phases during take-off from and return to the water, and under varying conditions of wind and sea. This is followed in Chapter III by a study of theory exemplified by diagrams and dealing analytically and geometrically with this same general range of phenomena. Then follow chapters dealing with certain differences between airplanes and seaplanes with reference to the aerial portions; differences and analogies between forms for hydroplanes and for seaplanes; calculations of displacement and of stability, and varying conditions affecting the latter; rules of extrapolation; tests in a marine tank on reduced scale models; strength, and the present gaps between theory and practice.

The diagrams, in addition to furnishing illustrations of the text, give many forms of graphical construction adapted to the treatment of special problems arising in connection with the study of seaplanes or flying boats and their design.

Tabular matter is also furnished as an aid in following some of the forms of computation, such as those relating to displacement and the location of the metacenter. An extended table is also given based on the Dornier designs and proportions and intended to indicate the possibilities and limitations in the direction of increased size.

With the material presented in this Division supplemented according to wish or need by reference to the literature of the subject as given in the text and Bibliography, the reader should be able to obtain a clear view of the major problems presented by the hydrodynamic phases of marine aircraft and of their relation to the field of aerodynamic theory in its broader aspects.

W. F. Durand.

CHAPTER I

HISTORICAL, DIFFERENT TYPES OF SEAPLANES PRINCIPLE OF THE RAMUS STEP

1. Introductory. A *seaplane* is a mechanical construction, capable as is the airplane, of maintaining itself in flight with a weight greater than that of the air displaced and furthermore capable of resting on the water, of rising from the water, and of alighting upon the water. The history of the seaplane is relatively brief, dating back only to 1910 or 1912. Following the pioneer achievements, however, progress was rapid, and since 1918 seaplanes have been looped and otherwise stunted.

From the first there have been seaplanes with floats and seaplanes with boats (flying boats); in the first the floats contain neither mechanism nor personnel; in the second, the boat may contain both mechanism and personnel.

Intermediate types have furthermore been realized or projected, forming almost a continuous series of forms in which are found successively, the seaplane with boat hull of considerable length carrying the ensemble of tail surfaces, the seaplane with one or two relatively short boat hulls in which the tail surfaces are at some distance from the boat, the seaplane with relatively short boat hull or hulls directly carrying an ensemble of fuselage and tail surfaces, and finally seaplanes with one or several floats separate from the ensemble of fuselage and tail surfaces.

To this classification based on the longitudinal aspect, there might be added another based on the transverse distribution and showing various forms intermediate between a central float associated with two lateral auxiliary floats, three equal floats or again two lateral floats only.

In a seaplane, there is an aerial sustaining structure, a part which may be immersed in the water and a propulsive unit. In the development of the seaplane, progress thus far has depended primarily upon the advance in the aerial structure and in the propulsive unit. These in turn have been made possible through the great development in land planes and through the many "sporting" competitions permitting the progressive improvement of this type of construction. From this condition, it has resulted that the part immersed in the water has, for a considerable time, suffered relative neglect and has been considered as a technical detail of secondary importance. The result of this has been the development of a series of seaplane constructions not having altogether satisfactory nautical qualities.

It is only during the past ten years that adequate recognition has been given to the need of a study of seaplane floats and boats carried out with the same care as in the case of ship design on a large scale. It is now, however, well recognized that the study of the nautical parts of a seaplane should be carried out with regard to both the hydrodynamic and the aerodynamic problems involved; the former in order to satisfy the many conditions imposed from this point of view and the latter in order that when in the passive state, in the air, the floats or the boat shall prejudice in minimum degree the aerodynamic qualities of the construction as a whole, either from the point of view of sustentation or propulsion.

2. Conditions Imposed on Seaplanes. Beyond the conditions imposed by the act of breaking clear of the water, flight and return to the surface of the water, a seaplane must meet other conditions regarding the state of repose on the water, either smooth or rough, conditions relative to movement on the surface of the water either smooth or rough, and either by its own power or by towing.

The conditions imposed by the state of rest on the water are those of any floating body. With smooth water, the examination of these conditions requires only an acquaintance with general laws of the equilibrium and the stability of floating bodies, and of the resistance offered to the wind by the superstructure.

The conditions imposed in connection with movement on the surface of the water, either under power or by towing, permit of study by methods already well known for boats of dimensions similar to the boats or floats forming the nautical part of seaplane construction.

The conditions imposed by flight are the same as those for an airplane carrying a heavy load, at the same time bulky and placed well below the axes of the propellers.

In a summary examination of the general subject, the features characteristic of the seaplane will then relate to the special phenomena which present themselves during the periods of transition from water to air or inversely—periods of takeoff from the surface of the water and of return to the same.

Special Conditions During Take Off From and Return to the Water. In the period of take-off, the seaplane passes progressively from a condition in which the buoyant force of the water just equals the weight, to a condition in which all hydrodynamic forces vanish; in other words, the static and dynamic forces due to the water are progressively replaced by forces due to air in relative motion acting on the surfaces of the structure. The seaplane cannot rise unless this relative motion is sufficiently rapid to furnish complete sustentation. It must therefore acquire, while still in contact with the water, a speed sufficient for sustentation and this speed is high in relation to the dimensions of the boat hull or floats; that is, the coefficient ר[1] $= V/\sqrt{g\Lambda}$ is relatively high.

The known facts regarding the resistance of floats of normal ship or boat form show immediately that, in order to realize such speeds with such forms, far greater power would be required than could be installed in a structure of this character. It has, therefore, been possible to meet the necessary conditions only through the use of a special underwater form characterized by the presence of the *step*.

Step. This step is formed as an abrupt discontinuity or jog in the vertical direction on the lower part of the hull form, such that the part forward is more deeply immersed than the part aft. The step is of advantage only in case the boat is moving at a speed characterized by a high value of the coefficient ר. For such speeds it has been known for nearly a century that a general decrease in the displacement of the boat form will result. The movement resulting in this decrease of displacement comprises a change in the angle of inclination of the hull to the horizon together with an elevation of the center of gravity with reference to the water level.

The search for forms with maximum lift has led to the application of the step to racing boats, hence called "hydroplanes" and which have been, in fact, the forerunners of the hull form for seaplane structures. In a hydroplane in motion, the water does not act by static pressure alone, but also by a vertical component dynamic force analogous to that of the air on the wings of an airplane.

In the period of gradually increasing speed on the water, the seaplane moves first as an ordinary boat and then as a hydroplane. There is, therefore, a period of decreasing displacement and a period of hydroplaning. During the period of decreasing displacement the wings of a seaplane provide a relatively small sustaining force.

The *period of hydroplaning* is that during which the seaplane moves over the surface of the water almost without displacement. The speed increases continuously, the wings provide in consequence a sustaining force continuously greater, the plane lifts further and the resistance due to the water decreases accordingly.

[1] The Hebrew letter *resch.*

The instant of take-off corresponds to the condition in which the air sustaining forces have become sufficient to render further support from the water unnecessary. The plane, from this moment, behaves as an ordinary airplane. During the period of return to the water, analogous phases, though not exactly inverse in character and order, are found.

During these two phases characteristic of the seaplane, numerous problems present themselves: first the study of the forces both aerodynamic and hydrodynamic to which the structure is subjected, the study of the changes which may be effected in these forces through various maneuvers, the study of the regime of the engine, and finally the study of the local forces between the water and the surface of the hull. The problem as a whole requires then studies of very diverse character, among which we shall here chiefly consider those of a hydrodynamic nature concerned with the take-off.

The first question relates to the sustentation which may be expected from diverse forms of underwater body for floats or boats. The order of magnitude of this sustentation can be estimated, as well as the resistance to motion in the case of a hydroplane. This will serve to show the difference between the propulsive resistance of a hydroplane and that of the usual boat or ship form.

3. Principle of Ramus. Let us consider, with Froude, a plane moving with a speed V and resting on a free water surface, making with the latter a small angle θ. This plane is subjected to a system of forces of which the components normal and tangential may be expressed (as to order of quantity) as follows:

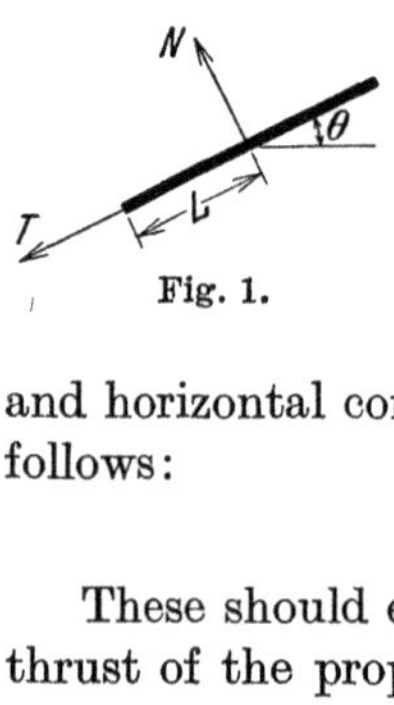

Fig. 1.

$$N = 55.35 \quad L\, b\, V^2\, \theta$$

$$T = \quad 0.186\, L\, b\, V^2.$$

L being the length in contact with the water and b the breadth of the surface, Fig. 1. The vertical and horizontal components are then, taking $\cos\theta = 1$ and $\sin\theta = \theta$, as follows:

$$F_v = N \quad - T\,\theta$$

$$F_h = N\theta + T.$$

These should equal, respectively, the weight of the structure and the thrust of the propeller, when the regime of hydroplaning is established.

The most favorable conditions will then be found when the ratio of thrust to weight is minimum. This ratio is

$$\frac{N\,\theta + T}{N - T\,\theta} = \frac{55.35\,\theta^2 + 0.186}{55.35\,\theta - 0.186\,\theta} \tag{3.1}$$

This varies as $55.35\ \theta + 0.186/\theta$, and is minimum for $\theta = \sqrt{.00336} = .058 = 3^0.32$. For this value of θ the ratio of thrust to weight, by substitution in (3.1), is found to be 0.120. We must therefore anticipate a resistance of the order of 120 kg. per ton of displacement.

With normal ship forms a resistance of the order of 50 kg. per ton of displacement is only encountered with the very highest nautical speeds—speeds at the limit of possibility for modern marine construction. From this it is easily seen that lifting planes would be of small value in connection with ordinary ship propulsion. Attempts in this direction have been entirely without useful result. But if very high values of the speed ratio ℸ must be realized, the seaplane here has the advantage that for its wings the ratio of thrust to weight does not increase with the speed as in the case of marine craft.

Furthermore, the figure of 120 kg. per ton found above by a very rough approximation is less than that actually obtained for hydroplanes (138 kg.) while for seaplanes the figure rises to the vicinity of 250 kg. per ton. The principal cause of this larger figure for seaplanes than for hydroplanes is found in the larger angle of attack for the former. In the seaplane the angle of attack is approximately three times the angle for optimum conditions.

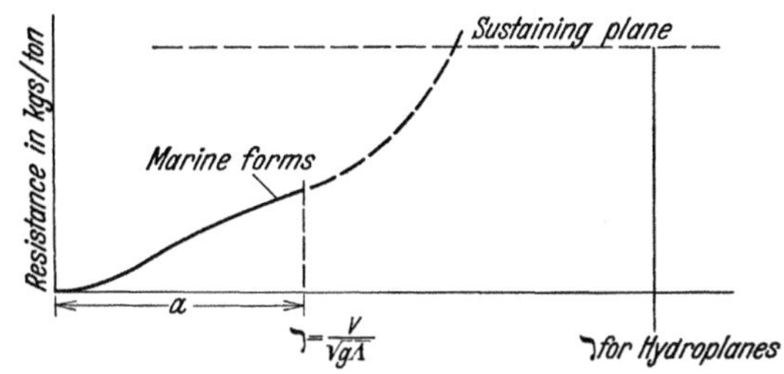

Fig. 2. Range of values for speed ratio $V/\sqrt{g\Lambda}$. *a* denotes approximate limit for Destroyer forms.

Figure 2 shows schematically the difference between the problem of a theoretical hydroplane and an actual ship and shows why the high speed required for the hydroplane phase of the seaplane cannot be realized through the use of normal marine forms.

It is to be here noted that in examining the conditions for hydroplaning, steady conditions have been assumed, that is to say, the plane is supposed to be in motion at uniform velocity. Between the period of displacement reduction and of flight, is found this phase of hydroplaning which, *a priori*, seems difficult to bring into harmony with the conditions of uniform motion.

Under steady motion conditions, the system of waves formed by the boat, and of which the continued development forms an important element of the resistance, accompanies the boat with the same speed as that of the boat itself. In a fast take-off with a seaplane, however, the plane must overrun the system of waves. There must, therefore, be expended not alone the energy required to extend a wave system previously formed, but also the energy required for the continued generation of a new system.

The procedure to be followed in the study of these problems may be found, as in most of the problems of practical hydrodynamics, in three different directions—pure theory, experiments with full scale structures, experiments with reduced scale models. Each of these methods has its limitations. Pure theory is thus far unable to give numerical values with the degree of precision needful for actual construction; experiments with full scale structures is time consuming, costly and sometimes dangerous, experiments with models may lie under the suspicion of not correctly representing all the conditions of the full scale structure.

4. Floatability and Stability at Rest. In a seaplane the necessity of floating on the water at rest imposes the provision of a certain water-tight volume and the condition of stability imposes the realization of a certain moment of inertia of the waterplane area.

If this moment is insufficient it can be augmented by the use of outrigger floats, thus preventing a plane unstable in the upright position from exceeding an inclination of a few degrees.

In flight the water-tight hull required for flotation augments the resistance and should therefore be given a form suitable for small resistance in the air. If the necessary volume is formed by a single hull its moment of inertia with reference to the longitudinal axis will be less than that with reference to a transverse axis. To meet this condition the single hull may be replaced by an ensemble of several hulls.

In a case of two hulls the arrangement is called a twin float or catamaran. A still larger number of hulls may furthermore be considered. Mixed arrangements are found where use is made of a boat hull with outrigger floats or boat hull with fin structures.

As a general rule for the same useful volume and moment of inertia of waterplane, the resistance in flight is diminished by the multiplication of units; on the other hand the weight of the structure is increased for the two reasons that the total exterior surface increases with their number and the weight of the connections with the wings is also greater.

This last consideration is, however, not of great importance under smooth water conditions, but becomes important if one considers either a seaplane resting on rough water or the forces which develop in connection with non-symmetrical take-off or return. The need of protecting the propellers and carburetors from the effects of spray, furthermore, necessitates placing the wing structure and the engines at a considerable height above the hull, and this results in a considerable increase in the weight of the necessary connections. Furthermore, in the case of a seaplane on rough water it is necessary to take account of the inertia of the entire mass of the structure with relation to the longitudinal axis passing through the center of gravity. It is then seen that the forces developed in rolling are strongly augmented by the multiplication of hull elements.

A second effect of the increase in total weight of the hull structure is the lowering of the center of gravity of the structure as a whole. In flight this effect would be unfavorable for the two reasons that if the engine slows down the plane will be in danger of losing speed rather than of beginning a glide and because in descending to alight on the water the pilot may be compelled to keep his engine in operation and to reach the condition of hydroplaning with a speed endangering the strength of the hull. Final mention may be made of secondary effects due to the fact that in a seaplane the general center of gravity is lower than the center of drift in flight. With the present limitation of the study of the seaplane to conditions other than those of flight, it will suffice to have mentioned these different points in order to show that the study of a seaplane cannot really be limited to a study of the phase of take-off from and return to the surface of the water. Rather the nature of the problem imposes the need of a study of all conditions through which the seaplane may pass; in such study contradictory requirements are met with, since in flight it would be desirable to raise the center of gravity, while on the water it would be of interest to lower it.

5. History of the Step. The effect of a step on the action of the hull of a boat is so fundamental in relation to the performance of a seaplane, that it may be of interest to recall the fact that the first observations dealing with the phenomena of the decrease of immersion through dynamic means go back to John Russell in 1834 and that in 1852 a patent relating to this subject was taken out by Apsey. When Ramus proposed the use of the step with ordinary ship forms, Froude was able to show that no advantage would result for the types of construction practicable at that time. In 1872 the ratio of horsepower to weight of machinery was about 7 horsepower per ton and it was impossible to foresee the tremendous advance which, in less than sixty years, has raised this figure to 3,000. With the decreased weight of machinery, the speed ratio $\urcorner$ has risen continuously. In 1881 Raoul Pictet and Ader made certain tests and in 1894, the Forban, under test, showed clearly the phenomena of decrease of immersion through the action of dynamic forces. Further steps before the realization of the seaplane itself, were taken by de Bonnemaison, by de Lambert and by Tellier. In 1905 Forlanini succeeded in lifting a boat completely from the water at a speed of 70 km/hr by the use of immersed sustaining planes and in 1906 Crocco and Ricaldoni realized absolute hydroplaning. Six years later, the seaplane had achieved take-off from the water and flight.

CHAPTER II

DESCRIPTION OF THE DIFFERENT PHASES OF TAKE-OFF FROM AND RETURN TO THE WATER

1. Normal Take-Off. The simplest case is that in which the water is absolutely calm and the air without wind. We assume first in these conditions: the engine in operation, the first effect of the thrust of the propeller is to cause the plane to nose down. Progressively the plane gathers speed and then begins of itself to rise forward. In planes with a single step and chiefly according to the relative positions of the center of gravity and of the step, the amplitude of the tilt nose-up may reach a limit of equilibrium or on the other hand pass through a maximum. For planes with several steps it may happen that the upward and backward tilting will increase continuously up to the take-off. In case the tilting is limited, the plane would continue to move on the water as an ordinary boat without mounting on the step. It will be necessary then for the pilot to use his controls to ease the plane and bring it upon the step. When the tilting passes a maximum and then descends to a lower value it is said that the plane passes of itself onto the step. In this condition the plane experiences, so far as the water is concerned, a relatively low resistance. It continues to hydroplane with a speed continuously increasing, and receives from the air a sustentation larger and larger, resulting in a decrease of the disturbance on the free surface of the water and in consequence a diminution of hydrodynamic resistance.

During this period the attitude varies very little. The speed increases continuously up to the point where the speed being sufficient, the air sustentation equals the weight, following which the plane rises. The duration of the period of hydroplaning may be diminished by increasing the upward tilting of the plane by means of the elevator in order to augment the lift. This manoeuver, however, must be carried out with caution in order to avoid the loss of speed which would cause the plane to fall back immediately upon the water.

2. Varying Conditions. The take-off which has just been described is of normal type such as would obtain in calm air with a plane of known type with motor power generously proportioned and under normal loading conditions. Several particular cases may be noted due to the non-fulfillment of one or another of these conditions.

Supposing always the sea calm, the succession of phenomena which has been given as normal may be interrupted either because some one of the various phases does not properly develop or because it may continue for too long a period of time or because secondary difficulties may present themselves.

If the entrance is too short the plane may tilt downward excessively. If on the other hand the entrance is too blunt the period of mounting upon the step may become too long and the same conditions may result with insufficient wing surface.

Certain planes before mounting on the step undergo a certain pitching movement. It has been assumed that this movement is caused by too flat a bottom or a step too far aft. If the under surface is too much developed forward, the float may leave the water at a sharp angle of upward tilt and at a speed insufficient to secure aerodynamic sustentation.

The plane then falls back upon the water, loses speed, may dip a wing-float, or turn on the side.

It may happen during the period of hydroplaning that the plane may develop a series of jumps more and more marked. The plane is then said to "porpoise". This defect is attributed to a position of the step too near the front. The jumps in this case are in the attitude nose high. Another type of jumping without marked inclination is attributed to too large an under surface area in relation to the surface of the wings or to an angle of incidence of the bottom of the boat too large in relation to the incidence of the wing.

Again it may happen that the waves produced by the boat driven back by the wind may by their blows damage the propeller or drown the carburetors. These defects are still further augmented when the sea is not entirely calm or in the case of any considerable wind or when these two conditions are united.

3. Take-Off With Bad Weather. The circumstances which may develop with bad weather are extremely variable and upon them must depend the particular maneuvers to follow. Here still more than in take-off under conditions of calm, the tactics to be followed depend largely on the skill of the pilot and no general rule can be formulated. For planes of moderate dimensions when the wave system is well formed and the wind is not strong, pilots usually endeavor to take off across the wind, maintaining the plane on the same wave crest, thus avoiding shock from the succession of waves. For larger planes this maneuver may result in the immersion of one of the wing floats, thus preventing the necessary acceleration. Take-off in a direction perpendicular to the wave crests is then necessary. In actual practice there can be no standard maneuver and it must be left to the quick judgment of the pilot to determine in each case, in accordance with the state of the sea, the best plan to follow. In general, with a strong choppy sea with wind approximating strength 4, a seaplane should take off with the least practicable throwing of spray. High waves with little wind may prevent a plane from taking off with full load, but a direct head wind alone is on the whole favorable. On the whole it is desirable that planes should take off across the waves. Planes with skimmers or with large floats near the boat hull have an advantage from this viewpoint. On the contrary, planes with two floats widely separated, cannot, in general take off across the waves.

4. Normal Alighting. With the plane descending normally, the pilot may propose to alight on the water tangentially, as in a normal landing on the ground. Phases the inverse of those in take-off will then present themselves. The pilot uses his elevator, the plane comes in contact with the water, first toward the stern, either extreme aft or on the step. The latter is preferable because it avoids a severe pivoting with the possibility of a rebound due to the resistance of the water. The speed decreases, the plane continues to skim with decreasing speed until, no longer supported by the dynamic reaction of the water, it settles rather abruptly into the water and returns to the conditions of an ordinary high speed boat. In general it is desirable to alight in an attitude nose high, rather marked, in order to touch the water at low speed and seaplanes should be so designed as to avoid any severe return to the attitude tail high.

5. Diverse Conditions. During the period of first contact with the water, violent shocks may result with changes of attitude on the water. These variations of attitude are rapid when the center of gravity is too far longitudinally from the step, either forward or aft. If the step is too far forward, the plane rises forward as soon as it touches the water and then may ricochet, perhaps several times, with possible loss of speed. If, on the contrary, the step is too far aft, the plane, on contact with the water, rises at the tail; this motion is followed by fore and aft rocking movements, very trying to the structure of the hull. Rebounds from the water on alighting are of the same two kinds previously considered and arise from the same causes.

6. Alighting on Rough Water. The difficulties here are similar to those indicated for take-off. The greatest danger is in an attitude insufficiently nose high. In such case the hull of the boat may receive from the water a blow of force sufficient to throw it again high in the air with danger of fall with loss of control. The danger in the inverse case is that, meeting with insufficient repulse from the water, the plane may plunge under with fatal consequences.

7. Graphical Representation. A somewhat greater precision can be introduced into the description of these phenomena by the use of a graphical representation in which it is assumed that for each speed, the plane is under steady conditions. A horizontal speed axis is then laid off and for each speed it is assumed that the plane is in equilibrium under the effect of its weight, aerodynamic forces, hydrodynamic forces, and the trust T of the propeller. Inertia forces are neglected. The weight W of the plane is known and applied at the center of gravity. The remaining forces depend on the speed, both with regard to their magnitude and their location.

For the aerodynamic forces, the influence of change of inclination is neglected and these forces, considered as depending on speed alone, are taken as equal to:

$R_y V^2$ in the vertical direction

$R_x V^2$ in the horizontal direction

For the hydrodynamic forces, account must be taken of the fact that these depend not only on the speed but on the attitude of the plane relative to the indefinite plane of the water surface. It is assumed, then, that these forces may be decomposed into a vertical component corresponding to the displacement of the immersed part and an inclined force due to the form of the immersed surfaces, the latter having again two components

H_y in the vertical direction

H_x in the horizontal direction

For the equilibrium of the plane, it is necessary that the general resultant of this system of forces and the general resultant moment shall both be zero. With the two components of the general resultant, we thus obtain three relations, Fig. 3,

$$R_y V^2 + B + H_y - W = 0$$
$$R_x V^2 + H_x - T = 0$$
$$(b R_x V^2 - a R_y V^2) + B u - (\beta H_x + \alpha H_y) - T \delta = 0$$

In these equations and with reference to G, the center of gravity, α and β are the coordinates of the step, a and b those of the point of application of the aerodynamic forces, δ the distance of the propeller shaft from G, and u the horizontal distance of the force of buoyancy B from G.

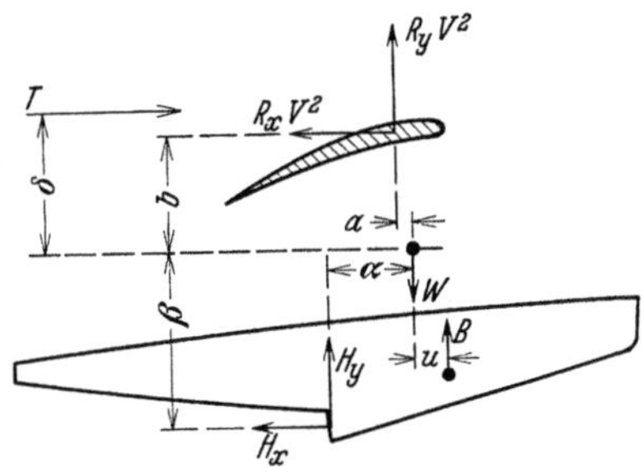

Fig. 3.

To represent graphically these three equations we may employ three diagrams in which may be plotted as ordinates respectively, the vertical components, the horizontal components, and the moments relative to the center of gravity. The axis of abscissae in these three diagrams is an axis of speed V. In order to establish the diagram of horizontal forces, we must have the law of variation of the thrust of the propeller as a function of the speed. We shall therefore first examine this law.

8. The Thrust of a Propeller Under Constant Torque With Variable Speed. The problem of a propeller under constant torque Q presents itself during the period of take-off of a seaplane. The reason for this condition is as follows:

For a seaplane engine the curve of power as a function of revolutions with full throttle and given altitude is nearly a straight line passing through the origin and in order to pass through the critical period of take-off, the pilot runs with full throttle. For a theoretical examination it is then sufficient to assume that the thrust given by the propeller of a seaplane will be determined by the condition necessary to give at any speed a constant torque. The problem is thus reduced to the determination of the law of thrust as a function of speed with constant torque.

To solve this problem use may be made of the results given by systematic experiments on propellers, experiments which give as a function of nD/V the values of T/n^2D^4 and Q/n^2D^5, D being the diameter of the propeller.

For a given propeller a series of values of nD/V (or of V/nh) is taken, h being the pitch of the propeller. To each of these values corresponds a value of T/n^2 and a value of Q/n^2, given by the experimental results, and since Q is known, or assumed, there results the value of n corresponding to each value of V/nh and since T/n^2 is known, the value of T follows.

The computation is developed as shown in the table, the first three columns being given by experimental results *(proportional figures)*.

$\frac{V}{nh}$	$\frac{T}{n^2}$	$\frac{Q}{n^2}$	n^2	n	V	T
0.9	11	26	386	621	559	42
0.8	33	40	250	500	400	82
0.7	54	50	200	447	312	108
0.6	73	56	179	423	254	131

It remains to trace the curve connecting T and V.

When the pitch ratio is moderate or low (for example, 0.7) the result approaches an inclined straight line as in Fig. 4. When the pitch ratio is relatively high (for example, 1.3) the result shows a curve descending at first slowly and with a horizontal tangent at the start, as in Fig. 5. In general the thrust decreases with increase of speed and the values

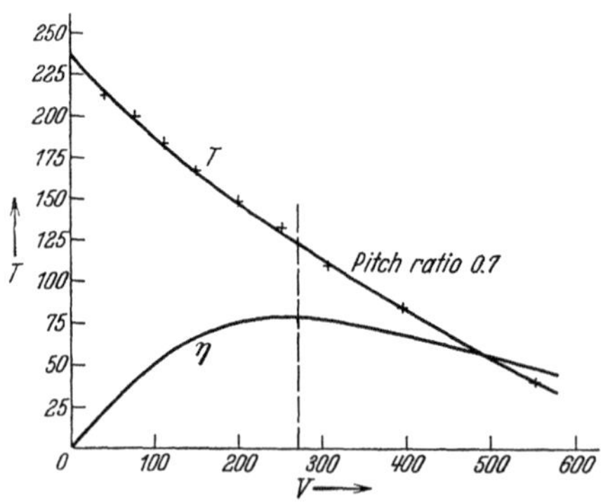

Fig. 4. Thrust and efficiency of propeller under constant torque: pitch ratio = 0.7.

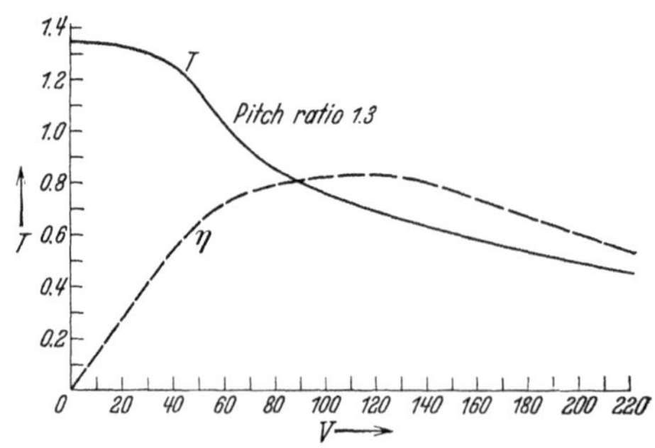

Fig. 5. Thrust and efficiency of propeller under constant torque: pitch ratio = 1.3. Ordinate scale in terms of $V = 60$ as unit.

of T for $V = 0$ and of V for $T = 0$ can be readily determined, at least to the general order of quantity.

The diagrams show two such curves for pitch ratios 0.7 and 1.3. For pitch ratios still higher it may result that the curves for T will rise at the start. This case, however, is only found with racing machines. The diagrams also give in addition the values of the efficiency η.

9. Determination of a Curve of Take-Off. Having in hand now the values of the thrust T as a function of the speed, three graphs may be established. For the first, Fig. 6, we trace a horizontal WW, giving the value of the ordinate W equal to the weight of the plane. This ordinate must be equal to the sum of the vertical forces due to the air and the water. The curve representing the change in the force due to the air will be a parabola if the coefficient R_y is constant, that is to say, if the wings have a constant angle of attack, if they are always in the same position with regard to the surface of the water, and if they are non-deformable as a whole and not influenced by the propeller.

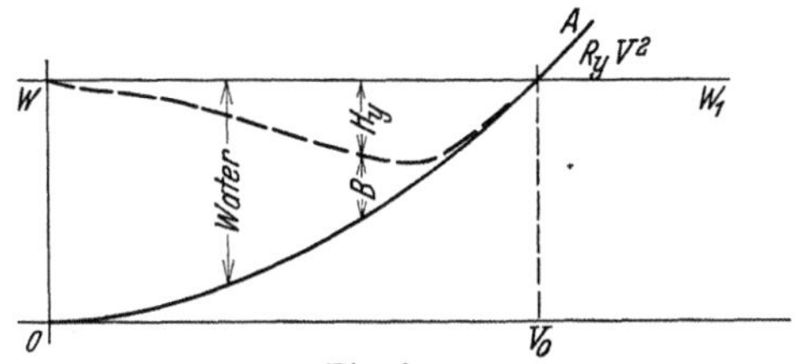

Fig. 6.

In reality the curve of air sustentation will have a form nearly parabolic which we may trace as $R_y V^2$. This parabola and the straight line WW intersect at a point of which the abscissa V_0 is the speed for which the air sustentation will equal the weight. This is the minimum speed for flight at sea level.

The vertical acceleration being neglected for speeds between 0 an V_0, the difference between the ordinates of the straight line and of the parabola will represent the sustentation due to the water. This may be divided into the two parts, one due to static upward thrust and the other due to dynamic reactions. We know only that for $V = 0$ the static upward thrust is exactly W and that for $V = V_0$ this force vanishes.

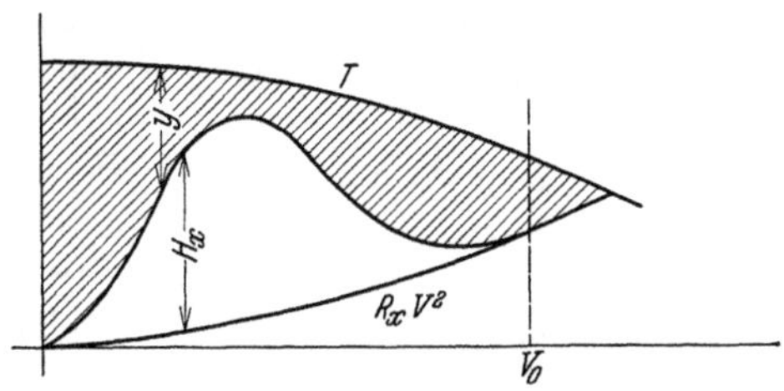

Fig. 7. Curves of propeller thrust and air resistance. The hatched area gives ordinates of water resistance plus the thrust available for acceleration.

On the second diagram (Fig. 7) let us trace the curve T representing the horizontal thrust of the propeller and the parabola $R_x V^2$ representing the air resistance of the plane. The difference of the ordinates of these two curves will represent for each speed the sum of the water resistance of the hull and of the thrust available to overcome horizontal inertia. The water resistance is, of course, 0 for $V = 0$ and for $V = V_0$.

We can form an idea of the character of these relations by experiments upon models. The resistance passes through a maximum R_m for a certain value V_m of the speed. We shall suppose here that the curve of total ordinates $H_x + R_x V^2$ does not intersect the curve T. There remains therefore for each speed an overbalance of thrust and in consequence the plane undergoes a horizontal acceleration.

On the third graphic, Fig. 8, we represent the moments with relation to the center of gravity, laying off above the axis moments tending to lift the bow and below the axis moments tending to lift the tail. The propeller gives rise to a moment tending to lift the tail, varying with the thrust of the propeller and represented by the curve $T\delta$. The air resistances give rise to a moment tending to lift the head, varying with the speed, with the distance between the center of gravity and the center of air resistance on the wings and with the increase of resistance of the parts emerging from the water. This moment will vary over all with the square of the speed and for $V = V_0$ will become equal and opposite to the moment due to the propeller. The moment of the resistance due to the water varies likewise in consequence of changes of speed and changes of position of the resultant lift force due to the water. It is an easy matter to determine the location of this force in the two conditions of rest and take-off.

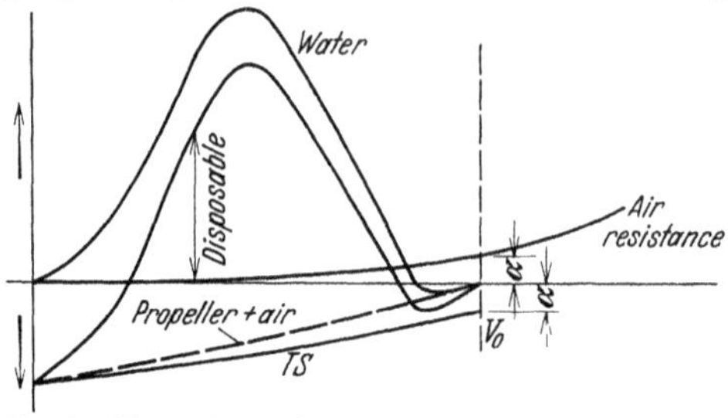

Fig. 8. Moments of forces and resistances during take off.

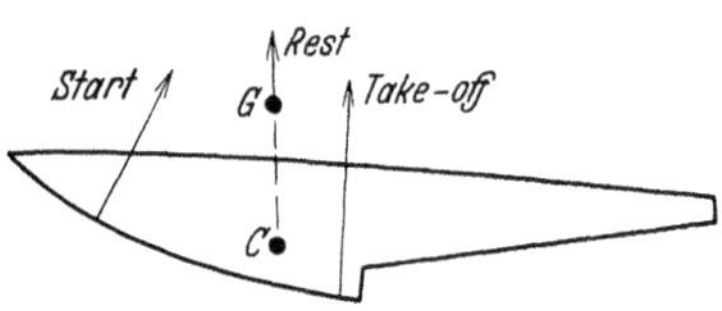

Fig. 9.

The moment, at first zero, begins with a direction tending to raise the bow (see Fig. 9) and with a considerable magnitude because both the lever arm and the resistance are large (although the speed is relatively low). At the end of the take-off the moment is of the reverse sign and tends to depress the bow, at the same time small in value because both the lever arm and the resistance are small (although the speed is large). These conditions develop as a consequence of the progressive reduction of the immersed part of the boat and of the consequent diminution of the sustentation required from the water.

In adding together with their signs the three moments considered, a curve is obtained giving the total acting moments. These moments must be balanced either by the action of the controls or by an inertia couple or by a moment of hydrostatic stability—that is to say, by a variation of the inclination of the plane as a function of the time.

It may be remarked that the curve of available moment is completed by a portion of the vertical axis passing through the origin representing the progressive growth of the moment tending to depress the bow between the two conditions of the plane—on the water with the propeller at rest and the propeller in motion at the instant of departure.

The second graphic indicates the facility of take-off and flight in relation to the power of the motor-propeller unit.

The third graphic indicates the difficulty which may be experienced in relation to the effort demanded of the pilot in his use of the controls.

10. Duration and Distance of Take-Off. The second chart provides for the determination of the duration of the take-off period. For this purpose, it is sufficient to evaluate the cross-hatched area, see Fig. 10, between the curves of water resistance and propeller thrust less air resistance. Hermann has indicated an elegant method for effecting this integration as follows:

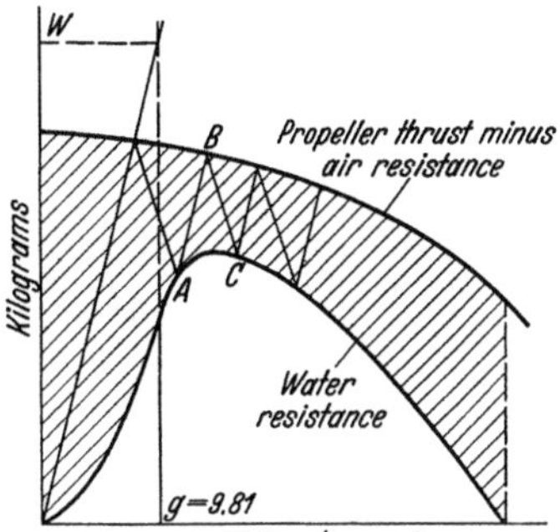

Fig. 10.

On the axis of W lay off the point W representing the weight of the plane. On the axis of speed in meters per second (or feet) lay down the point 9.81 (or 32.2). From the point determined by these two coordinates, draw a line to the origin. Then from the point of intersection with the upper curve, continue to draw between the two curves a series of intercepts having the same slope as the original line. The number of these intercepts will then equal the number of seconds required for the take-off.

Let us consider the course of events for the period during which the speed changes from V_1 to V_2, Fig. 11.

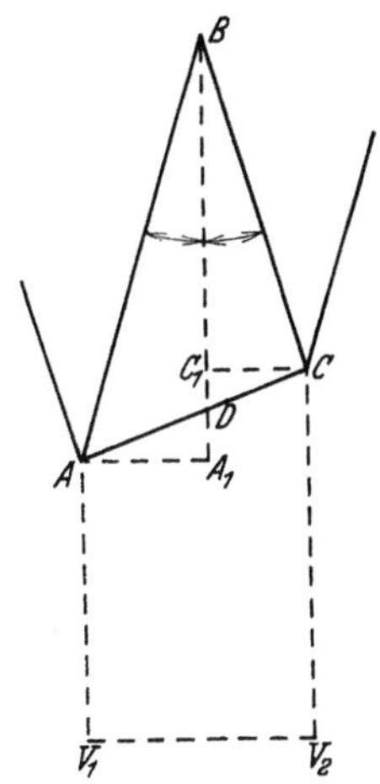

Fig. 11.

We may assume that in the change from speed V_1 to speed V_2 the average accelerating force is BD. Then the construction gives

$$\frac{BD - DC_1}{CC_1} = \frac{W}{g}$$

$$\frac{BD + DA_1}{AA_1} = \frac{W}{g}$$

Adding,
$$\frac{BD}{CC_1} + \frac{BD}{AA_1} = \frac{2W}{g}$$

Again since CC_1 and AA_1 will, on the average differ but little from $(V_2 - V_1)/2$, we have

$$\frac{4BD}{V_2 - V_1} = \frac{2W}{g}$$

Furthermore, the mass of the plane, W/g, will move according to the law

$$L = V_1 t + \frac{1}{2}\gamma t^2$$

where $$\gamma = g\frac{BD}{W} = \frac{V_2 - V_1}{2}$$

whence $$L = V_1 t + \frac{1}{4}(V_2 - V_1)\, t^2$$

$$V = V_1 + \frac{1}{2}(V_2 - V_1)\, t$$

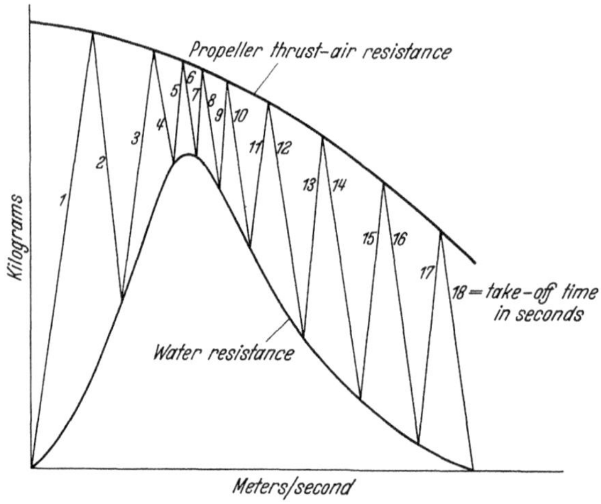

Fig. 12.

For $t = 2$, this gives $V = V_2$ showing that on this assumption, the time required to pass from the abscissa of A to the abscissa of C is two seconds.

On the scale of speeds, the abscissae of the summits ABC will then give the speeds at the end of one second, two seconds, etc. Each of these speeds continues for one second. By adding these abscissae, then, the sum will give approximately the distance run during the period of take-off. In the example given, Fig. 12, the total distance run was **143.3** m. and the duration was eighteen seconds.

CHAPTER III

DISCUSSION OF THE GENERAL PHENOMENA ARISING DURING THE PERIOD OF TAKE-OFF

1. Introductory. We have seen that the seaplane, during its period of contact with the water, passes through two phases of navigation, very different in character. At low speeds, it behaves like an ordinary boat. At high speeds it skims the surface. The instant which separates the first mode from the second is that when the water ceases to follow the hull clear to the after part; that is, the instant when there is suddenly formed a space free from water aft of the step. If we assume that aft of the step the free surface is not smooth, as in a train of waves, the question arises as to whether it becomes possible to predict at what speed this separation from the water will take place.

If this wave train surface is tangent to the bottom of the boat, the problem becomes that of finding under what conditions a wave train can be determined under a given direction and speed at a specified point of immersion.

We examine first the plane during the phase of normal boat movement, then the prevision of the moment of change from one phase to the other and finally, the phase of hydroplane movement.

In the resistance to movement of a seaplane during the first phase, there are a certain number of points in common with those met with in a study of the resistance of a boat of normal form.

It is known that a boat formed body in normal movement at the surface of the water gives rise to changes of level of diverse characters, forming systems or trains of waves. The conditions of propagation of such trains of waves depend on the force of gravity.

If we consider two ship-shaped forms, geometrically similar, and similarly placed with reference to the general water level, these various wave systems will not be similar nor similarly placed with regard to the ship nor with regard to the general water level, unless the Reech-Froude speed-length ratio, $\neg = V/\sqrt{g\Lambda}$, has the same value for both forms.

Theoretical studies of such trains of waves have been made for the case of movement in two dimensions, that is to say, for what may be called cylindrical propagation with right line crests at right angles to the direction of movement; and also for the case of movement in three dimensions; that is, for the general phenomena of changes of level in a free liquid surface with the forms of the crests arbitrarily specified.

For the case of two dimensions, the theory gives a satisfactory solution even when the changes of level are of finite amplitude.

For the case of three dimensions, the theory gives a solution valid only for very small changes in level. In very brief form, these two theories may be summarized as follows:

Two-dimensional Waves. The form of the free surface is a cylinder with trochoidal right section. Between the velocity and the length from crest to crest L is the relation,

$$\frac{V}{\sqrt{g L}} = \frac{1}{\sqrt{2\,\pi}} = 0.4$$

where V = velocity in meters per second,
L = length in meters,
g = 9.81 meters per second per second.

Such a system of waves cannot be used directly as representative of the transverse system formed by a ship shaped body, because they are periodic and continuous as a system from $-\infty$ to $+\infty$ while transverse waves as actually produced by a body of cylindrical form approach asymptotically the general level of the water at infinity.

The wave pattern and system formed by a ship is due to the progressive invasion of tranquil water by the system considered.

Such an invasion can be studied theoretically in the single case where the amplitude is small; but it may be assumed that the trochoidal system gives an approximate representation of the actual conditions about a body in movement on the water.

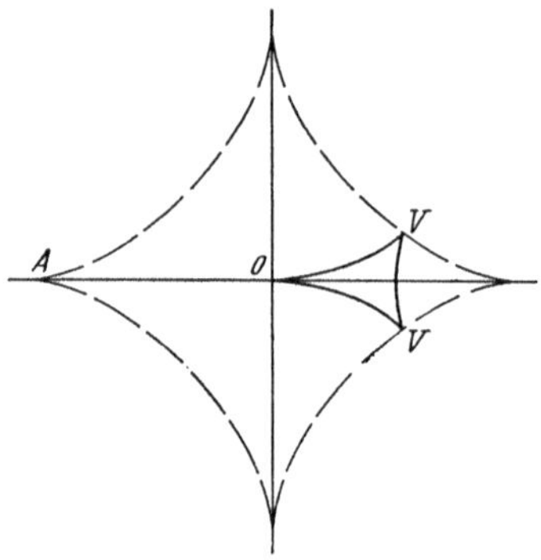

Fig. 13. Three-dimensional wave.

Three-dimensional System. Attention is directed to the form which the crests should take, considering a very small obstacle moving on the surface of the water and capable of creating all possible manner of wave systems, subject only to the condition that the system of crests shall be at rest relative to the obstacle; that is to say, that they accompany the obstacle in its movement.

If V is the speed of the movement, the speed of a train of waves moving in a direction inclined at an angle θ to the movement, will be $V \cos \theta$. To every value of θ will correspond a determinate wave length (assuming $L \sim V^2 \cos^2 \theta / K$). It is then readily shown that the equation to the line of the nth crest from the origin is

$$y \sin \theta - x \cos \theta = \frac{n V^2 \cos^2 \theta}{K^2}$$

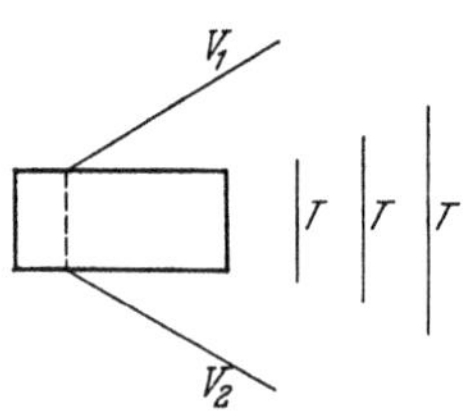

Fig. 14. Divergent and transverse wave system.

This is in the form

$$y \sin \theta - x \cos \theta = N \cos^2 \theta$$

It may be shown by well known methods that the envelop of these lines for N constant and θ variable is the arc of a curve which is, itself, the involute of a hypocycloid of four cusps.

Such a wave OVV, Fig. 13, created by the movement of a point O in the direction OA is readily observed with all sorts of water craft. It comprises a part VOV called the divergent system together with the part VV called the transverse system.

In the case of a seaplane in the gliding phase, these systems are reduced to their most simple form—two divergent waves V_1 V_2 and a transverse system TTT, see Fig. 14.

2. Trochoidal Waves. The equations (λ an auxiliary variable)

$$\left. \begin{aligned} x &= \lambda - A \sin \frac{2\pi}{L} \lambda \\ z &= \frac{\pi A^2}{L} + A \cos \frac{2\pi}{L} \lambda \end{aligned} \right\} \qquad (2.1)$$

represent the free surface of a trochoidal wave formed on a liquid originally at rest with its free surface in the plane $z = 0$. The vertical height is $2A$, the wave length L, the location of the center of the generating circle is at a height of $\pi A^2/L$ above the plane of still water, and the origin is at the point on $z = 0$ underneath the crest.

Let us now determine such a trochoid by the condition that at a given point at a depth h below the plane of repose $(z = -h)$, the slope of the surface shall be i. That is,

$$\frac{dz}{dx} = \tan i$$

To write this condition we take the derivatives $dx/d\lambda$ and $dz/d\lambda$ and then by division find the derivative dz/dx, thus:

$$-\frac{dz}{dx} = \frac{(2\pi A/L)\sin 2\pi\lambda/L}{1-(2\pi A/L)\cos 2\pi\lambda/L} = \frac{(2\pi/L)\sqrt{A^2-(z-\pi A^2/L)^2}}{1+(2\pi/L)(\pi A^2/L-2)} \tag{2.2}$$

The condition may then be written

$$-\tan i = \frac{2\pi\sqrt{A^2-(-h-\pi A^2/L)^2}}{L+2\pi(\pi A^2/L+h)} \tag{2.3}$$

If the velocity of the trochoidal wave is fixed, the length will be fixed likewise by the relation $L = (2\pi/g)\,V^2$ and in consequence (2.3) will determine A as a function of the given quantities i, h, and the velocity V.

No solution is acceptable unless the value found for A is less than $L/2\pi$, for otherwise the trochoid would have double points.

If i is very small, the equation for A gives

$$A = h + \frac{\pi A^2}{L}$$

The limiting case will be reached if this condition is satisfied for $A = L/2\pi$; that is, if $h = L/4\pi$. This is equivalent to the relation between h and the speed V, $V = \sqrt{2gh}$.

For speeds less than $\sqrt{2gh}$ there will be no trochoidal free surface tangent to the obstacle.

If i is not very small, the condition that $L/2\pi$ shall be a solution is found by making $A = L/2\pi$ in the equation

$$\tan^2 i\left[L+2\pi\left(\frac{\pi A^2}{L}+h\right)\right]^2 - 4\pi^2\left[A^2-\left(h+\frac{\pi A^2}{L}\right)^2\right] = 0$$

which gives

$$\frac{3}{4}L^2(1-3\tan^2 i) - 2\pi hL(1+3\tan^2 i) - 4\pi^2h^2(1+\tan^2 i) = 0$$

Let us apply this to an obstacle defined by

$$\tan i = 0.14$$

$$h = 3.45 \text{ cm.}$$

The equation in L becomes

$$L^2 - 32.6\,L - 680 = 0$$

An equation which has only a single positive root—a result holding for all cases of present interest for which $(1 - 3\,tan^2\,i)$ is positive. This root is $L = 47$ cm. The corresponding velocity

$$V = \sqrt{\frac{g\,L}{2\,\pi}} = 85.65 \text{ cm./sec}$$

and
$$A = \frac{L}{2\,\pi} = 7.48 \text{ cm.}$$

The tangent trochoid will then appear in the actual case when the speed of the boat reaches 86 cm./sec.

The horizontal distance from the first crest to the obstacle is found thus:

We first seek the distance $\alpha\alpha_1$, Fig. 15, between two points on the same level, assuming the curve to be practically trochoidal between these two points and hence symmetrical about a vertical through the point distant $L/2$ from the crest. The known value of z in (2.1) will give the angle $(2\,\pi\lambda/L)$. The length L being known, this will give λ and this will give x_1 for the point α_1. Then from the diagram it is clear that

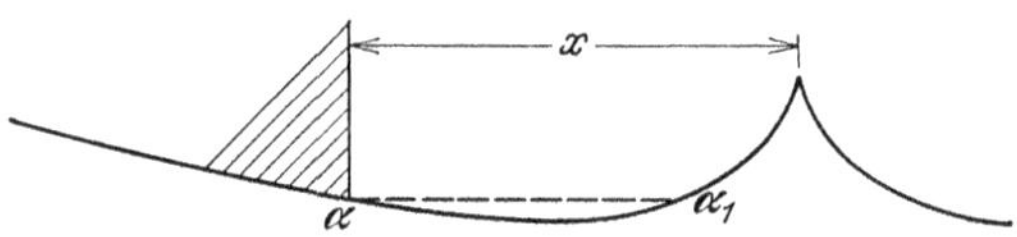

Fig. 15.

$$\frac{\alpha\,\alpha_1}{2} = \frac{L}{2} - x_1 \quad \text{or} \quad \alpha\,\alpha_1 = L - 2\,x_1$$

Thus for the preceding numerical case we find $\pi A^2/L = 3.74$ and $cos\ 2\pi\lambda/L = -.961$ or $2\pi\lambda/L = 164^0$. Then $\lambda = (164/360)\ L = 21.4$. Putting this in (1) we find $x_1 = 19.35$. Whence

$$\alpha\alpha_1 = 47 - 38.7 = 8.3 \text{ cm.}$$

The distance from the crest to the obstacle will then be

$$x_1 + \alpha\alpha_1 = 27.65 \text{ cm.}$$

The distance from the given immersion to the bottom of the wave hollow is $7.48 - (3.74 + 3.45) = .29$ cm.

3. Appearance of the Stern Wave. If we assume a plane with constant inclination to the horizontal and a depth of immersion relative to still water likewise constant, then over a considerable range of speeds, we should expect, *a priori*, to determine a point of brusque variation in the curve of resistance on speed, when the speed is reached where the wave at the stern begins to form clear of the boat. However, no such sudden change is found and it must be assumed that before the appearance of the stern wave, lining in with the obstacle, there is a brief period during which the wave exists though drowned, as it were, by a mass of entrained water.

4. Experiments on the Separation of the Stern Wave. These experiments, to the present time, are few in number, but sufficient to show that the order of magnitude of the speeds at which the wave separates at the stern is that given by the theory thus briefly summarized.

The question of the direction of movement of the water as it issues aft of the plane gliding surface is perhaps of no great importance for a seaplane full size. On the other hand it merits study for the case of a structure of model size. The separation aft of the step being of fundamental importance, it is necessary to make sure that the conditions of separation for the model are similar to those for the planes' actual size. This will give a lower limit to the length of the model.

In the case of models of water gliders with step, the experimental channels with travelling bridge usually employ models of small dimensions. The value of the speed-length ratio should in fact, be the same for the model and for the full-scale object. The maximum speed of the bridge is limited by the character of the installation, and in order to reach a value sufficiently high, it may become necessary to reduce the length. In such case, with very small models, it is found that the reduction of resistance secured for full scale boat by the use of the step, is not always found on the model. The explanation is that for the same value of the water separates from the hull of the full scale boat, but does not separate in the case of a very small model. It would appear that there is an effect of capillarity of which due note should be taken.

From the view point of agreement with theory, it is equally useful to know if, at the point of separation, the velocity is in the plane tangent to the surface of the body. Regarding this point, few observations seem to have been made. In a sketch found in a paper of Sottorf[1] a separation is shown aft of a thin gliding plane with a sharp break in direction. It seems rather in point of fact, that while the velocity is not in a direction tangent to the surface of the body it differs but little from it, the stern wave forming a hollow aft of the body, having an initial tangent a little less inclined to the horizon than the lower surface of the body itself.

5. Study of Take-Off From Diagrams. The detailed study of the three equations of equilibrium of the seaplane, as above, gives rise to certain difficulties, the character of which must be noted. Even with simple cases it is impossible to write down explicitly the various coefficients figuring in these equations, because they all depend upon the variables.

The thrust of the propeller, for example, depends at the same time on the speed, on its position with reference to the water and to the wings; co-efficients R_x, R_y, H_x, H_y depend on the same elements and likewise on the instantaneous orientation of the wings, and similarly for the others. It is possible, however, to determine certain things graphically regarding the conditions of equilbrium, and these diagrams have the

[1] Werft, Reederei, Hafen 1929.

advantage of showing the character and relation of the various difficulties.

Any position in general of the seaplane is here defined with reference to the general still water level by the aid of two parameters, of which one, θ, relates to its orientation; for example inclination of the wings to the horizontal, and the other to the immersion $(N - n_1)$, the number of vanes immersed in the first diagram, or x, length of plane immersed in the second diagram. Before treating the three equations of equilibrium it is necessary to know the values and positions of the resistances, both aerodynamic and hydrodynamic, as functions of the speed and of the two geometrical parameters $[\theta$ and $(N - n_1)$, or θ and $x]$. The first diagram is arranged in such manner that the equation of moments can be neglected. It is further assumed that the forces—thrust of the propeller, weight, aerodynamic and hydrodynamic resultants, all converge in the same point, the two latter having the same position. The problem is then one of constant incidence and the variable θ does not enter.

6. First Diagram. For a diagram thus constituted, the equations of equilibrium, neglecting the vertical forces of inertia, will be, see Fig. 16.

$$f(V) = \alpha\, n_1 V^2 + 800\, \alpha\, (N - n_1)\, V^2 + M \frac{dV}{dt} \tag{6.1}$$

$$M g = \beta\, n_1 V^2 + 800\, \beta\, (N - n_1)\, V^2 \tag{6.2}$$

The coefficient 800 recognizes the fact that the density of water is about 800 times that of air. The second equation shows that to each value of V there corresponds a determinate immersion characterized by the value of n_1. The immersion is independent of the function $f(V)$ which determines the law of thrust of the propeller. We then eliminate n_1 between the two equations, thus obtaining

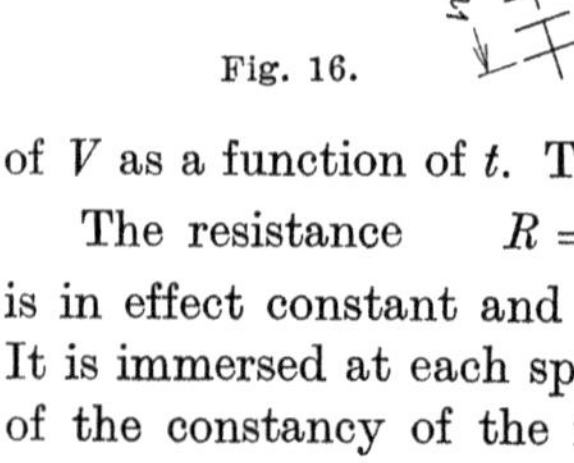

Fig. 16.

$$f(V) = \frac{\alpha}{\beta} M g + M \frac{dV}{dt}$$

An equation giving the law of variation of V as a function of t. This equation does not contain the factor 800.

The resistance $\quad R = \alpha n_1 V^2 + 800\, \alpha\, (N - n_1)\, V^2$

is in effect constant and equal to $(\alpha/\beta)\, M g$. The plane regulates itself. It is immersed at each speed in such manner as to satisfy this condition of the constancy of the resistance.

For each law determined for $f(V)$ there will be a law of variation of speed as a function of the time. If $f(V)$ is constant we have dV/dt constant and a uniform accelerated motion. If $f(V) = A - BV$ it would be necessary to integrate

$$A - BV = \frac{\alpha}{\beta} M g + M \frac{dV}{dt} \tag{6.3}$$

which gives

$$\left.\begin{aligned} V = \frac{1}{B}\left[A - \frac{\alpha}{\beta} M g - e^{-\frac{B}{M}t}\left(A - \frac{\alpha}{\beta} M g\right)\right] = \\ = \frac{1}{B}\left[A - \frac{\alpha}{\beta} M g\right]\left[1 - e^{-\frac{B}{M}t}\right] \end{aligned}\right\} \quad (6.4)$$

and supposing V equals 0 for t equals 0

$$t = -\frac{M}{B} \log \left[\frac{A - \frac{\alpha}{\beta} M g - B V}{A - \frac{\alpha}{\beta} M g}\right] = \frac{M}{B} \log \left[\frac{A - \frac{\alpha}{\beta} M g}{A - \frac{\alpha}{\beta} M g - B V}\right] \quad (6.5)$$

The speed of take-off will then be given by (6.2) when $n_1 = N$ or

$$V = \left(\frac{M g}{\beta N}\right)^{1/2}$$

and the time duration will be given by putting this value of V into (6.5). As will be noted, the factor 800 disappears from the final result. An analogous calculation can be carried out for other laws of variation of the thrust of the propeller. For example, assuming that this thrust varies as:

$$A \cos \frac{\pi}{2} \frac{V}{V_0}$$

where V_0 is the speed for which the thrust is 0, it will be necessary to consider the integral

$$\int \frac{d x}{\cos x - \alpha}$$

Here we may place x equals $2 \tan^{-1} u$ and reduce to a form directly integrable.

7. Second Diagram. Here the ensemble of the wings (wings and tail surfaces) is assimilated to a single plane of surface σ Fig. 17. The hull is assimilated to a sustaining plane of width b, of which the immersed length is x. The surface σ is inclined at an angle i relative to the surface S. The instantaneous position is defined by x and θ. The equations of vertical forces and moments are

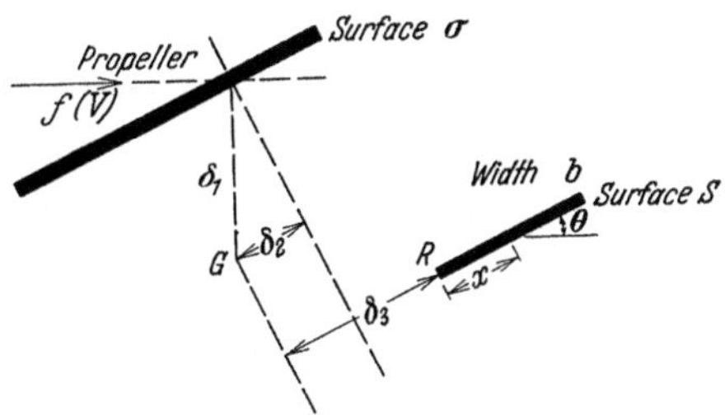

Fig. 17.

$$M g = \varrho \sigma V^2 (\theta + i) + 800\, \varrho b x V^2 \theta$$

$$f(V)\, \delta_1 = \varrho \sigma V^2 (\theta + i)\, \delta_2 + 800\, \varrho b x V^2 \theta\, (\delta_3 + 2\, x/3)$$

Eliminating θ we have the relation between x and V expressed in the form

$$u = \frac{M g - \varrho \sigma V^2 i}{f(V)\, \delta_1 - \varrho \sigma V^2 i\, \delta_2} = \frac{\sigma + 800\, b\, x}{\sigma\, \delta_2 + 800\, b\, x\, [\delta_3 + 2\, x/3]} \quad (7.1)$$

Is it then possible to have in the neighborhood of x equals 0 (take-off), V decreasing with increase of x? That is to say,

$$\frac{d V}{d x} < 0.$$

For x equals 0 we have

$$Mg - \varrho\sigma V^2 i = \varrho\sigma V^2\theta$$

$$f(V)\,\delta_1 - \varrho\sigma V^2 i\,\delta_2 = \varrho\sigma V^2\theta\,\delta_2$$

We then form the two derivatives du/dV and du/dx. The former will take its sign from

$$-2\varrho\sigma i\,[f(V)\delta_1 - \varrho\sigma V^2 i\delta_2]\,V - (Mg - \varrho\sigma V^2 i)\,[f'(V)\delta_1 - 2V\varrho\sigma i\delta_2]$$

Then for $x = 0$ with the above relations this reduces to $-f'(V)\delta_1$. Then since $f'(V)$ is always negative, the sign of δ_1 will evidently here control.

Similarly the derivative du/dx takes its sign from:

$$[\sigma\,\delta_2 + 800\,b\,x\,(2\,x/3 + \delta_3)]\,800\,b - (\sigma + 800\,b\,x)\,[(4/3)\,800\,b\,x + 800\,\delta_3\,b]$$

For $x = 0$ this reduces to $(\delta_2 - \delta_3)$. Hence the derivative dV/dx will have the sign of $(\delta_2 - \delta_3)/\delta_1$ and the condition sought will be satisfied if $\delta_3 > \delta_2$.

Numerical Example Given:

$Mg = 2400$ $\qquad b = 1.5$

$\sigma = 50$ $\qquad i = 0$

$\delta_1 = 1$ $\qquad f(V) = 2500 - 154\,V$ (propeller equation)

$\delta_2 = \delta_3 = 0.4$

The relation (7.1) reduces to

$$f(V) = 2400\,\frac{80\,x^2 + 48\,x + 2}{120\,x + 5}$$

For each value of x we have a value of $f(V)$ and consequently V. We may then calculate $\varrho\theta$ by the equation of vertical forces, giving

$$\varrho\theta = \frac{Mg}{(\sigma + 1200\,x)\,V^2}$$

Now taking the resistance to advance of the part in the air as proportional to $V^2\theta\sigma$, it is seen to follow from the equation for vertical forces (since ϱ and σ are both constants) that $V^2\theta\sigma$ is also proportional to $Mg/(\sigma + 1200\,x)$. Also we take the resistance of the hydroplane proportional to $V^2\theta x b \times 800$, that is, to $V^2\theta x \times 1200$. Since $\sigma = 50$, the ratio of these two resistance values is seen to be $24\,x$.

We may write, therefore,

$$\text{Resistance (air)} = k'\,\frac{Mg\cdot\sigma}{\sigma + 1200\,x} = k'\,\frac{50\,Mg}{\sigma + 1200\,x}$$

$$\text{Resistance (water)} = k'\,\frac{1200\,Mg\cdot x}{\sigma + 1200\,x}$$

Whence:

$$\text{Resistance (Total)} = k'\,Mg\,\frac{\sigma}{\sigma + 1200\,x} + \frac{1200\,x}{x + 1200\,x}$$

The results of the calculation are given in the following table, and are shown in Figs. 18a and 18b. The above analysis only applies during the period of hydroplaning. The parts of the curves in dotted line indicate the period for which the wing and float cannot yet be represented by the combination of two planes. It will be noted from the formulae and also

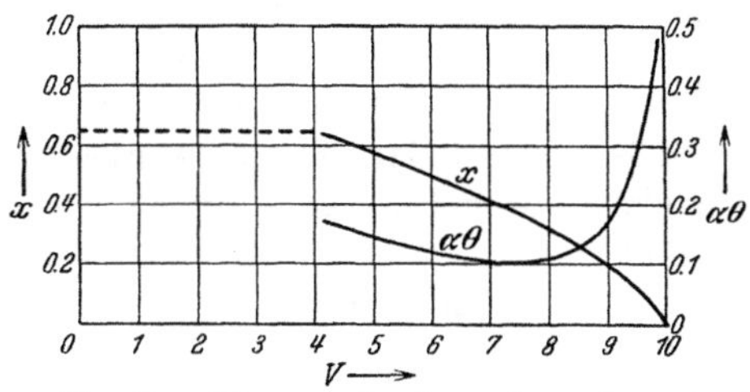

Fig. 18a. Diagram showing values of x and $\alpha\theta$ on speed.

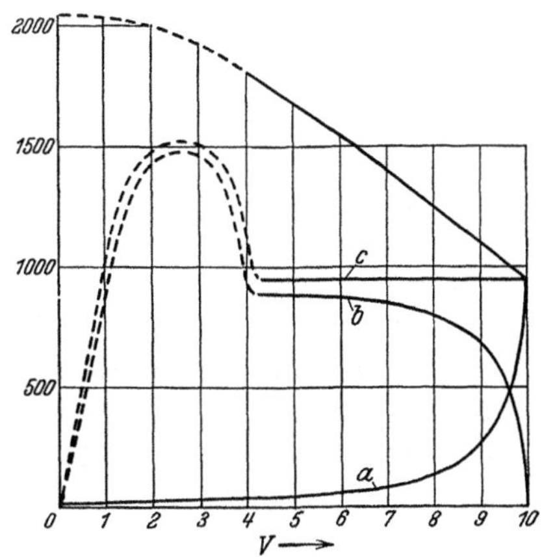

Fig. 18b. Diagram showing resistance on speed. a) Resistance of the air. b) Resistance of the water. c) Total resistance.

from the values in the table that, under the special assumptions made and for the period during which they are valid, the total resistance remains constant.

x [1]	$f(V)$ [2]	V [3]	$\sigma + 1200$ [4]	$\alpha\,\theta$ [5]	$\frac{47500}{\sigma + 1200\,x}$ [6]	Resistance (water) [7]	Resistance (total) [8]
0	960	10	50	0.480	950	0	950
0.1	1073	9.27	170	0.1645	279.4	670.5	950
0.2	1225	8.28	290	0.121	163.8	786.2	950
0.3	1381	7.26	410	0.111	115.9	834.1	950
0.4	1540	6.24	530	0.1166	89.65	860.35	950
0.5	1698	5.20	650	0.1366	73.1	876.91	950
0.6	1858	4.17	770	0.1794	61.7	888.3	950
0.7	2017	3.14	890	0.2741	53.5	896.6	950
0.8	2177	2.1	1010	0.539	47.1	902.9	950
0.9	2336	1.06	1130	1.876	42.1	907.9	950
1	2496	0.03	1250		38	912	950

[1] Values of the variable x.

[2] Values of $f(V) = 2400\,\frac{80\,x^2 + 48\,x + 2}{5 + 120\,x}$.

[3] Values of V deduced from $V = \frac{2500 - f(V)}{154}$.

[4] Values of $\sigma + 1200\,x = 50 + 1200\,x$.

[5] Values of $\alpha\,\theta = \frac{2400}{(\sigma + 1200\,x)\,V^2}$.

[6] Resistance of the part in the air $= \frac{47500}{\sigma + 1200\,x}$.

[7] Resistance of the part immersed $= (6) \times 24\,x$.

[8] Total resistance $= (6) + (7)$.

This second diagram is evidently applicable only for values of θ less than those for which the tail of the plane barely touches the water. The curve of θ as a function of V is, for the period of hydroplaning, not without some similarity with the curves obtained experimentally by the National Advisory Committee for Aeronautics[1].

The principal use of this diagram is to show the relation between the characteristics of the wings and of the propeller on the one hand with those of the hull on the other, and that it is impossible to treat separately a study of these different parts of the structure as a whole.

8. Third Diagram. Determination of Elements Relative to the Critical Speed. We assume for purposes of study a schematic plane formed by a prismatic float and a wing, Fig. 19. Let

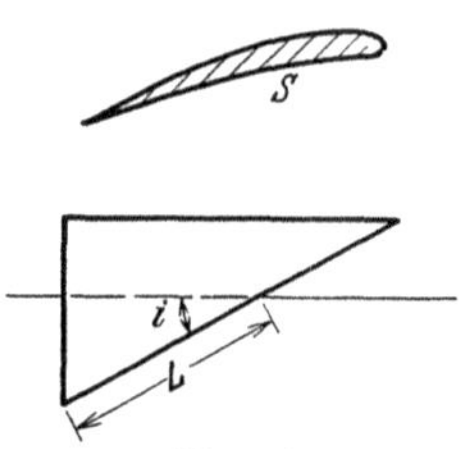

Fig. 19.

W = Weight of the plane in tons,
L = Length at rest of the immersed body along the slope in meters,
S = Surface of the wing in square meters,
V_0 = Minimum speed over the ground in kilometers per hour,
b = Width of the float in meters,
δ = Weight of unit volume of water.

When the quantities W, L, b, are given the slope i of the bottom follows from the condition of equilibrium at rest. We here assume i small and $\cos i$ = sensibly 1. We have then,

$$W = \frac{1}{2}\delta\, b\, L^2 \sin i \tag{8.1}$$

We shall not consider at present variations of the inclination during the take-off. While in movement the immersed length, l, will be variable between the values L at the start and 0 when the speed reaches V_0. The force of buoyancy will be a function of l alone and taken equal to

$$\frac{l^2}{L^2} W$$

The hydrodynamic sustentation will be taken proportional to the surface of support $b\,l$ to the square of the speed V^2 and to $\sin i$. We may write then, $\frac{1}{2}\delta\, K' b\, l\, V^2 \sin i = \frac{K' l\, W}{L^2} V^2$ [see (8.1)]

where K' is a coefficient of hydrodynamic sustentation.

The resistance to movement due to this sustentation will be sensibly proportional to the sustentation in the ratio $\sin i$, which, putting in $\sin i$ from (8.1), will give

$$K'' \frac{l}{b} \frac{W^2}{L^4} V^2 \quad \text{where} \quad K'' = 2\, K'/\delta$$

[1] Crowley, J. W., Jr., and Ronan, K. M., Characteristics of a Single Float Seaplane During Take-off, U.S. N.A.C.A. Technical Report No. 209, 1925. — Characteristics of a Boat Type Seaplane During Take-off, U.S. N.A.C.A. Technical Report No. 226, 1926.

At the speed V the aerodynamic sustentation will be $K_1 S V^2$, and the air resistance will be $K_2 S V^2$. The condition of equilibrium in flight at the moment of take off gives

$$W = K_1 S V_0^2$$

and consequently the air sustentation at the speed V due to the wings will be

$$W \frac{V^2}{V_0^2}$$

The equations of equilibrium will then be

$$W = W \frac{l^2}{L^2} + K' l \frac{W}{L^2} V^2 + W \frac{V^2}{V_0^2}$$

$$R = K'' \frac{l}{b} \frac{W^2}{L^4} V^2 + K_2 S V^2$$

This system of two equations determines the two unknowns l (position of the float with regard to the water) and R (resistance to movement) as functions of the speed. Eliminating l we shall have the law of resistance to movement as a function of speed. We have first for l

$$l = \frac{R - K_2 S V^2}{K'' W^2 V^2} b L^4$$

and the law of resistance to motion as a function of V is given by the relation between R and V.

$$1 = \frac{1}{L^2} \left[\frac{R - K_2 S V^2}{K'' W^2 V^2} b L^4\right]^2 + K' \frac{V^2}{L^2} \left[\frac{R - K_2 S V^2}{K'' W^2 V^2} b L^4\right] + \frac{V^2}{V_0^2}$$

The part of the resistance due to the hull alone in the series of conditions through which the plane passes is $R_1 = R - K_2 V^2 S$ and the law of variation of R_1 as a function of V is

$$1 = \frac{1}{L^2} \left[\frac{R_1 b L^4}{K'' W^2 V^2}\right]^2 + K' \frac{V^2}{L^2} \left[\frac{R_1 b L^4}{K'' W^2 V^2}\right] + \frac{V^2}{V_0^2} \qquad (8.2)$$

In this relation S, the area of wing surface, does not directly enter, but only indirectly, through the agency of V_0. We then place

$$x = \frac{V^2}{V_0^2} \quad (x \text{ lies between 0 and 1})$$

$$y = \frac{R_1 L^4}{W^2}$$

$$A = \frac{b}{L K'' V_0^2}$$

$$B = \frac{K'}{K''} \frac{b}{L^2}$$

The discussion of the relation between R_1 and V is then reduced to a discussion of the variations of y as a function of x when y and x are related by the equation

$$A^2 y^2 + B x^2 y + x^3 - x^2 = 0$$

This represents a cubic of the general form shown in Fig. 20. It will have a maximum value for y between $x = 0$ and 1. Let us seek the coordinates of this maximum which will give us the value of V^2/V_0^2 for which $R_1 L^4/W^2$ is maximum and in consequence the value of V for which R_1 will be a maximum. The cubic passes through the points $x = 0$, $y = 0$, and $x = 1$, $y = 0$. We desire then the coordinates of y maximum for x between 0 and 1. To this end we utilize a substitute conic having with the cubic a point in common at the origin, and a point in common at $x = 1$, $y = 0$. By putting $y = mx$ it is readily shown that the tangent of the slope at the origin is $y' = 1/A$.

We next form the two derivatives y' and y'' thus:

$$2A^2 y y' + 2Bxy + Bx^2 y' + 3x^2 - 2x = 0$$

$$2A^2 y y'' + 2A^2 y'^2 + 2Bxy' + 2By + 2Bxy' + Bx^2 y'' + 6x - 2 = 0$$

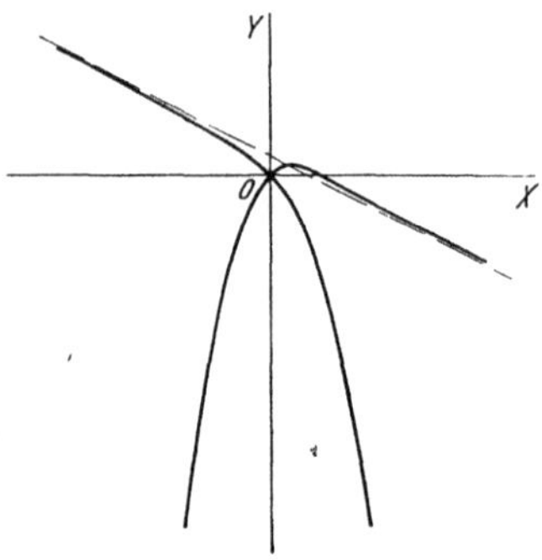

Fig. 20.

At the point $y = 0$, $x = 1$ these give

$$y' = -\frac{1}{B} \qquad y'' = -\frac{2A^2}{B^3}$$

We thus have five conditions—two points, two slopes and one second derivative. This will determine a general conic passing through the origin with five coefficients. It is known from analytical geometry that if $y = m_1 x + b_1$ and $y = m_2 x + b_2$ are any two lines, then, λ being a constant coefficient, the equation

$$\lambda y^2 + (y - m_1 x - b_1)(y - m_2 x - b_2) = 0$$

will give a conic crossing X at the same points as these lines and having at these points the same slope as the lines. Hence in the present case the equation to a conic passing through the origin and the point $y = 0$, $x = 1$ and having the same slope at these points as the cubic will be

$$\lambda y^2 + (Ay - x)(By + x - 1) = 0$$

We have now to apply the further condition for y''. To this end we form y' and y'' from this equation thus

$$2\lambda y y' + (Ay' - 1)[By + (x - 1)] + (Ay - x)[By' + 1] = 0$$

$$2\lambda y y'' + 2\lambda y'^2 + (Ay - x) By'' = 0$$

Whence putting in the values of y' and y'' for the point $y = 0$, $x = 1$ we find $\lambda = -A^2$. The conic sought is therefore

$$-A^2 y^2 + (Ay - x)[By + (x - 1)] = 0$$

We then seek the maximum using the equation giving y' and putting $y' = 0$. This gives

$$-[By + (x - 1)] + Ay - x = 0$$

Combining this equation with that of the conic we have

$$-A^2 y^2 + (Ay - x)^2 = 0$$

or

$$y = \frac{x}{2A}$$

Then replacing y by its value in the equation of condition $y' = 0$

$$-\frac{Bx}{2A} - x + 1 + \frac{x}{2} - x = 0$$

whence

$$x = \frac{2A}{3A + B}$$

$$y = \frac{1}{3A + B}$$

Returning now to the initial data and calling V_m the critical speed, and R_m the resistance of the hull for this speed we have

$$\frac{V_m^2}{V_0^2} = \frac{2}{3 + K' V_0^2/L}$$

$$R_m = \frac{W^2}{b L^3} \frac{K'' V_0^2}{3 + K' V_0^2/L}$$

If we consider the case in which we impose the weight W, the speed of take-off V_0 and length L of the part forward of the step, V_m will be independent of b, and R_m will decrease as b increases. This furthermore is readily seen from (8.2) which gives $R_1 b$ as a function of V.

These latter formulae may be written in such manner as to bring out the speed in relation to the square root of the dimensions. Thus, put

$$\frac{V_0}{\sqrt{L}} = \alpha \quad \text{and} \quad \frac{V_0}{\sqrt{b}} = \beta$$

Then

$$\frac{V_m^2}{V_0^2} = \frac{2}{3 + K' \alpha^2}$$

$$\frac{R_m}{W} = \frac{K'' W}{L^3} \cdot \frac{\beta^2}{3 + K' \alpha^2}$$

Under this form it is easy to determine the values of K' and K'' utilizing the results obtained by experiment either model or full scale.

$\frac{V_m}{V_0}$ is usually not far from 0.4

$\frac{W}{L^3}$ is in general not far from 0.050

Thus $W = 2.4\ T$ for $L = 3.6$ m.

α is usually not far from 53

Thus $V_0 = 100$ km./h. for $L = 3.60$ m.

β is usually not far from 83

Thus $V_0 = 100$ km./h. for $b = 1.45$ m.

This gives $3 + K'\alpha^2 = 12.5$ and K' for $\alpha = 52.7$ has the value 0.00343.

Again R_m/W is not far from 0.2 and hence

$$0.2 = K'' \cdot 0.0503 \cdot \frac{6960}{12.5}$$

whence $K'' = 0.00715$

It is interesting to trace the curve of total resistance as a function of the speed. To this end we examine again the cubic in x, y.

We have here

$$A = \frac{b}{L K'' V_0^2} = .00563 = 6 \cdot 10^{-3} \text{ app.}$$

$$B = \frac{K' b}{K'' L^2} = 0.054$$

The cubic is then

$$36 \cdot 10^{-6} y^2 + 0.054\, x^2 y + x^3 - x^2 = 0$$

This is readily traced by using polar coordinates after having put $y = 100\, y_1$. This gives

$$0.36\, y_1^2 + 5.4\, x^2 y_1 + x^3 - x^2 = 0$$

whence

$$r = \frac{\cos^2\theta - 0.36 \sin^2\theta}{5.4 \sin\theta \cos^2\theta + \cos^3\theta}$$

This curve having been traced, it is only necessary to change the distribution of the abscissae in order to have the curve of resistance as a function of the speed. Furthermore, the curve gives immediately the law of variation of l as a function of the speed, Fig. 21. In effect l is proportional to y/x.

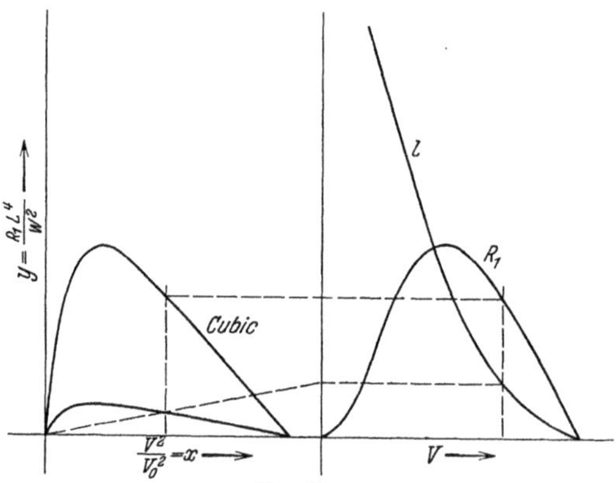

Fig. 21.

By the use of this method it appears that with the hyperbola which has been employed, the abscissae of the summit is less than that for the summit of the cubic, and that the summit of the hyperbola is higher than that of the cubic. If instead of using the hyperbola having three features in common with the cubic for $x = 1$, and two features for $x = 0$, use were made of a hyperbola having two points in common for $x = 0$, and three points for $x = 1$, the approximation would be somewhat better because the maximum is always nearer to $x = 0$ than to $x = 1$. The calculation in this case presents no difficulty but gives rise to formulae less simple than with the other substitution.

The formulae thus developed for the determination of the elements of the critical speed admit of diverse applications. In the first place they determine the length of immersed keel at the moment of maximum resistance, and thus furnish an indication regarding the minimum length to be given to the straight part of the keel before its upward curvature forward. Again they indicate the manner in which the critical speed varies with varying conditions. In a paper by Herrmann[1] there may be found for a single form of float the variations of elements V_m for varying speeds of take-off and for varying weight. It there appears that for the same weight and for varying speeds of take-off there will be found between R_m, V_m, and V_0 the following relations:

V_0 km./h.	V_m m./s.	R_m kgs.	$\sim V_m/V_0$	R_m/V_m^2
70	8.7	665	1.24	8.8
85	9.5	705	1.11	7.8
100	10.0	770	1.00	7.7

It thus appears, as indicated by the formula

$$\frac{V_m^2}{V_0^2} = \frac{2}{3 + K' V_0^2/L}$$

that as V_0 increases, V_m increases but with a decreasing ratio V_m/V_0.

The formula for R_m would indicate the ratio $R_m/V_m^2 = \text{const}$. This, however, is not exactly verified.

It is shown in the same paper that for different total weights the values of the speed of take-off, the maximum speed and the maximum resistance are as follows:

W kgs.	V_0 m./s.	V_m m./s.	R_m	$\sim \frac{R_m}{W^2 V_m^2}$
3000	22.0	8.8	887	127
2600	20.5	8.1	700	158
2200	18.5	7.9	512	170
1800	17.0	8.1	295	138

The formula would give for this case $V_m = \text{constant}$ and $R_m/W^2 V_m^2 = \text{constant}$. The agreement is not altogether satisfactory, but this result must be expected from a diagram as simple as the one which we have considered.

The practical application of the relation R_1 inversely proportional to b when the weight of the plane, the wing surface and the motor-propeller unit are fixed, is limited by the fact that an increase of b entails an increase in the weight of the hull and in its air resistance. The influence of the increase in air resistance can be neglected in most practical cases. It may be noted further that by the use of an under structure extending beyond the width of hull proper, the value of b may be increased without any corresponding departure from the general fineness of line. However, it is necessary to examine the effect of an increase in hull weight on the general economy of the structure as a whole.

In the comparison of different solutions involving the same general data, due account must be taken of the general schedule of design. The total weight W will represent weight of hull, weight of plane structure, weight of motor-propeller unit, weight of personnel and navigational equipment and useful load. Other elements remaining the same, any increase in hull weight will diminish in like amount the useful load and it might result in enlarging the value of b, starting from an initial design,

[1] Herrmann, H., Seaplane Floats and Hulls, Berichte und Abhandlungen der Wissenschaftlichen Gesellschaft für Luftfahrt, December 1926.

that a point would be reached where the increase in hull weight would equal the original useful load, thus resulting in a structure of no value as an agency of transport.

The problem might, however, present itself in an entirely different manner if, instead of an agency of transport, the design should be for racing purposes and without limiting conditions otherwise regarding useful load. We shall limit the present discussion to the case where the plane is to be of value as an agency of transport.

In order to permit of a simple calculation, let us assume as part of the initial data, a useful load eW and as width a value b. It may be assumed that the weight of hull structure as due to change in b alone will vary as b^2. As a matter of fact the evaluation of this variation should be based in part on an examination of distribution of weight in order to determine the part of the weight pertaining to the forward under structure, and in part on a consideration of the strength of the materials employed in order to determine the increase in local load at the instant of alighting on the water and during the hydroplaning period.

If, in the original design the weight of hull structure is βW, we may assume then, as a result of an increase in b, a weight of $\beta W\,(b_1^2/b^2)$. The combined weight, hull structure plus useful load, would in the original design for $(eW + \beta W)$ become, following the change, $e_1 W + \beta\,(b_1^2/b^2)\,W$ and the limit of b_1 will be reached when $e_1 = 0$. This gives

$$(e + \beta)\,W = \beta \frac{b_1^2}{b^2}\,W$$

or

$$(e + \beta) = \beta \frac{b_1^2}{b^2}$$

$$\frac{b_1}{b} = \sqrt{\frac{e + \beta}{\beta}}$$

If, for example, for the original design we have $\beta = 0.20$ and $e = 0.40$, the limit of b_1/b will be $\sqrt{0.60/0.20} = 1.732$.

Studies of this character make possible an estimate of the sacrifice to accept regarding useful load and the purpose of improving the conditions of take-off.

The problem may be approached from a different point of view by considering a reduction of resistance at the critical point, not for the purpose of improving the conditions of take-off, but in order to increase the load with which take-off can be realized. In this case the resistance at the critical point will be held as a constant, and we associate an increase in b with an increase in total weight as bearing upon the weight of hull structure and on the useful load.

The resistance R varying as W^2/b, it appears that, in this case, W^2/b should be held constant.

Let x denote the ratio b_1/b. Then the weight carried by the design as modified will be $W\sqrt{x}$. For the original design we have

$$W = C + U + D$$

where $C =$ weight of hull structure, U the useful load, and D the remainder of the total weight.

If the weight of hull structure varies with b according to an index m, we shall have, for the modified design

$$W\sqrt{x} = C\,x^m + U_1 + D$$

Comparing the two equations the gain in useful load is

$$G = U_1 - U = W\left(\sqrt{x} - 1\right) - C\left(x^m - 1\right)$$

From this we have

$$\frac{dG}{dx} = \frac{W}{2\sqrt{x}} - m\,C\,x^{m-1}$$

The limit beyond which increase of b would be useless is then given by $dG/dx = 0$.

If, in the original design there is no purpose in an increase of b, the above value of $dG/dx = 0$ with $x = 1$ will give

$$\frac{W}{2} = m\,C$$

This gives, for the ratio of weight of hull structure to total weight,

$$\beta = \frac{1}{2\,m}$$

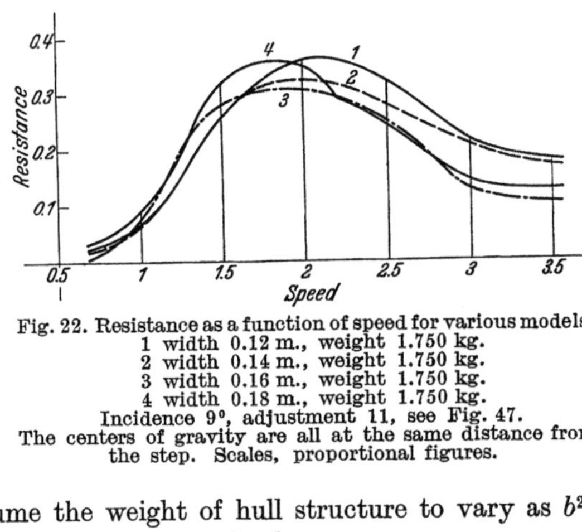

Fig. 22. Resistance as a function of speed for various models:
1 width 0.12 m., weight 1.750 kg.
2 width 0.14 m., weight 1.750 kg.
3 width 0.16 m., weight 1.750 kg.
4 width 0.18 m., weight 1.750 kg.
Incidence 9°, adjustment 11, see Fig. 47.
The centers of gravity are all at the same distance from the step. Scales, proportional figures.

If, as above, we assume the weight of hull structure to vary as b^2, it will then be advantageous to increase b whenever

$$\beta < 0.25$$

These considerations are limited in application by the fact that an increase in b carries with it a decrease in the ratio L/b, and during the period preceding that of hydroplaning, when the hull is moving through the water as an ordinary boat, its resistance may be disproportionately increased. In such case, forms may be found giving rise to a different maximum. An example of this is given, showing on curve 4, Fig. 22, corresponding to the largest model, a maximum different from that which we have considered and which will find its explanation in the study of the fourth curve sheet.

9. Fourth Diagram. Hydroplaning at a Constant Angle of Inclination. The theory of Wm. Froude regarding hydroplaning is briefly as follows.

Let A (Fig. 23) be the surface bearing upon the water (in square feet) and consequently A/b the length of water bearing, θ the angle of inclination to the horizontal, and V the speed in knots.

Froude assumes that the plane is subject to water forces as follows:

(1) A normal force:

$$F_1 = 3\,A\,V^2 \sin\theta \text{ (pounds)}$$

This gives a vertical component upward $3\,A\,V^2 \sin\theta \cos\theta$ and a horizontal resistance component $3\,A\,V^2 \sin\theta$.

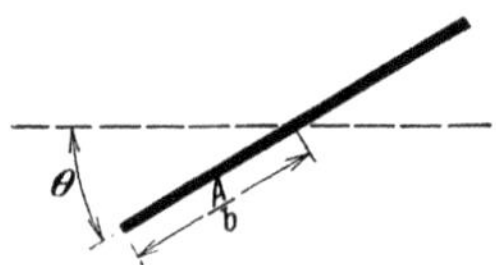

Fig. 23.

(2) A tangential force:

$$F_2 = 0.01\,A\,(V\cos\theta)^2 \text{ (pounds)}$$

This gives a vertical component downward $0.01\,A\,(V\cos\theta)^2 \sin\theta$ and a horizontal resistance component $0.01\,A(V\cos\theta)^2 \cos\theta$.

The sum of the vertical components must equal the weight W, thus

$$W = 3\,A\,V^2 \sin\theta\cos\theta - 0.01\,A\,V^2 \cos^2\theta\sin\theta \qquad (9.1)$$

The sum of the horizontal components, under steady conditions, must equal the resistance R. Thus

$$R = 3\,A\,V^2 \sin^2\theta + 0.01\,A\,V^2 \cos^3\theta \qquad (9.2)$$

In these values for W and R, the speed V enters only in the product $A\,V^2$ and hence the ratio R/W is a function solely of θ. Thus

$$\frac{R}{W} = \frac{3\sin^2\theta + 0.01\cos^3\theta}{3\sin\theta\cos\theta - 0.01\cos^2\theta\sin\theta}$$

For θ small we have

$$\frac{R}{W} = \frac{300\,\theta^2 + 1}{300\,\theta - \theta} = \frac{1}{299}\left(300\,\theta + \frac{1}{\theta}\right)$$

Finding the condition for a minimum of this expression we have

$$\theta = 0.0577 = 3^0.3$$

In (9.1) for a given value of θ it is clear that $A\,V^2$ must remain constant and hence the same in (9.2). It follows according to this theory that the resistance is independent of the speed. That is, A will decrease in the same ratio as V^2 increases and $A\,V^2$ remain constant. If, on the other hand we should introduce a supplementary term proportional to the square of the speed such as $B\,V^2$, recognizing for example the existence of parasitic resistance due to immersed parts, the equations will take the form

$$W = 3\,A\,V^2 \sin\theta\cos\theta - 0.01\,A\,V^2 \cos^2\theta\sin\theta \qquad (9.3)$$

$$R = 3\,A\,V^2 \sin^2\theta + 0.01\,A\,V^2 \cos^3\theta + B\,V^2 \qquad (9.4)$$

In this case the speed does not appear alone in the term $A\,V^2$. Assuming W known and constant, (9.3) and (9.4) give the values of the two unknowns A and R as functions of speed. In order to find the law of resistance (R as a function of V) A must be eliminated between the two equations.

Solving the first equation for A and putting this value in the second, we have

$$R = \frac{(3\,sin^2\,\theta + 0.01\,cos^3\,\theta)\,W}{3\,sin\,\theta\,cos\,\theta - 0.01\,cos^2\,\theta\,sin\,\theta} + B\,V^2$$

At constant speed, the value of θ for minimum R is the same as that found above. In both these cases the resistance is not zero at zero speed. The reason for this insufficiency in the theory arises from the assumption that the hydrodynamic components in V^2 alone equilibrate the weight of the plane, neglecting the part taken by the forces of buoyancy.

In the present state of hydrodynamic science it is not possible to give an exact evaluation of buoyant forces in the case of a body in motion at the surface of separation of two fluids. Even for the case of a body in motion wholly immersed in a fluid there exist solutions only in very special cases. All that can be said with assurance is that in the case where $V = 0$, the vertical buoyant force is equal to the weight of the volume of fluid displaced by the body, relative to the general fluid level,

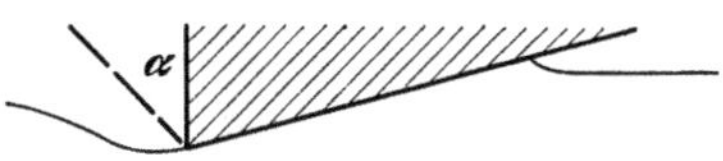

Fig. 24.

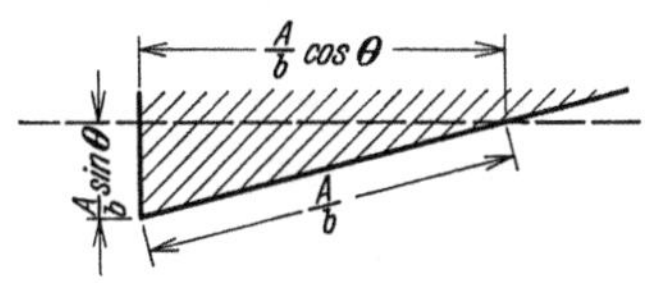

Fig. 25.

and if in movement the volume so displaced is zero, there exists a movement possible at which the buoyant force is zero. A value of the buoyant force satisfying these two limiting conditions will be given by assuming that during the motion the body is subject to two vertical forces, one as assumed by Froude and the other a force of buoyancy measured by the volume of the structure below the general water level plane.

This definition is evidently not correct for the intermediate phases of the movement, since the stern wave may separate from the hull and the addition of a hull volume such as α (see Fig. 24) does not really change the vertical forces in operation while it does change the value as calculated.

In the case of an ordinary ship, the introduction of this supplementary upward force will have no effect on the result. We may, however, admit it by assumption, as a working hypothesis for the present case of floats with step. Instead of representing the float in longitudinal section by a straight inclined line, let us assume it to be limited aft by a vertical line, see Fig. 25. The two equations then become:

$$W = 3\,A\,V^2\,sin\,\theta\,cos\,\theta - 0.01\,A\,V^2\,cos^2\,\theta\,sin\,\theta + 0.5\,\delta\,\frac{A^2}{b}\,sin\,\theta\,cos\,\theta$$

where δ is the weight in pounds of a cubic foot of water.

$$R = 3\,A\,V^2\,sin^2\,\theta + 0.01\,A\,V^2\,cos^3\,\theta$$

This system of two equations gives for each value of the speed V, the values of the two unknowns A and R. We shall have then R as a function of V by eliminating A between the two equations. This gives

$$W = \frac{[3 \sin\theta \cos\theta - 0.01 \cos^2\theta \sin\theta]\, R}{3 \sin^2\theta + 0.01 \cos^3\theta} + \frac{0.5\,(\delta/b) \sin\theta \cos\theta\, R^2}{V^4\,[3 \sin^2\theta + 0.01 \cos^3\theta]^2}$$

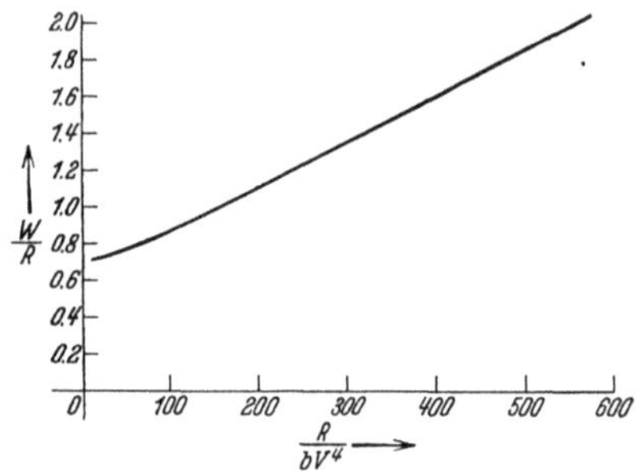

Fig. 26.

Such is the relation giving the resistance R as a function of the speed V. In this form, it is seen that for $V = 0$, $R = 0$, and further, that the relation corresponding to Froude's analysis is correct for $V = \infty$. The Froude theory alone then gives results correct only at the limit $V = \infty$.

In this equation the resistance R results as an implicit function of V. The law of resistance can, however, be represented by two simple methods which are directly indicated by the form of the relation between R and V.

If we plot the values of $R/b\,V^4$ as abscissae and those of W/R as ordinates, the points should fall on a straight line. If reference be made to the experiments of W. Froude as reported by Johns[1], it will be seen that points plotted and representing the general results of his experiments, do fall nearly on a straight line. See Table 1 and Fig. 26.

TABLE 1. *Froude's Experiments on the Ramus Step*[2].

Speed	V^4	Resistance	Proportional Values $\frac{1}{R}$	$\frac{R}{V^4}$
3	81	0.475	2.12	586
4	256	0.74	1.35	291
5	625	0.97	1.03	156
6	1300	1.17	0.856	91.0
7	2400	{1.24	0.805	52.0
		{1.28	0.780	53.1
8	4100	{1.28	0.780	31.6
		{1.35	0.740	33.2
9	6560	{1.30	0.770	19.9
		{1.40	0.713	21.3
10	10000	{1.30	0.770	13.0
		{1.45	0.690	14.8
11	14700	1.50	0.668	10.1
12	20700	1.60	0.627	7.7

Another form of graphical representation useful for this discussion is given by taking $x = V^4$ and $y = R$. Let us examine the relation between R and V, assuming θ small. We have

$$W = \frac{299\,R}{300\,\theta + \frac{1}{\theta}} + \frac{R^2}{V^4}\,\frac{\delta}{2\,b}\,\frac{\theta}{(3\,\theta^2 + 0.01)^2}$$

Replacing δ by its numerical value and putting $y = R$ and $x = V^4$ we have

$$\frac{W}{299} = \frac{y}{300\,\theta + \frac{1}{\theta}} + 1040\,\frac{\theta}{b}\,\frac{y^2}{x}\,\frac{1}{(300\,\theta^2 + 1)^2} \tag{9.5}$$

[1] Engineering, Vol. 110, London, September 24, 1920.

[2] The double values in this table refer to two conditions of the model, the larger when dragging along a body of dead water and the smaller when over-riding and free of such dead water.

This equation represents a hyperbola of which the horizontal asymptote is given by $y = \frac{W}{299}\left(300\,\theta + \frac{1}{\theta}\right)$ variable with θ and having a minimum value for $\theta = 0.0577$ (Fig. 27).

Let us examine the relation between R and V for a numerical case

$$W = 299$$
$$b = 30$$
$$\theta = 0.1$$
$$300\,\theta + 1/\theta = 40$$
$$1 = \frac{R}{40} + 3.47\frac{R^2}{V^4} \text{ [Eq. (9.5)]}$$

or

$$V^4 = \frac{138.8\,R^2}{40 - R}$$

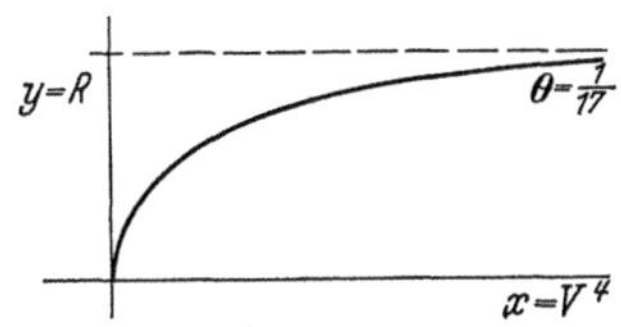

Fig. 27.

This gives the curve shown in Fig. 28. Actual experience gives curves of this same general character.

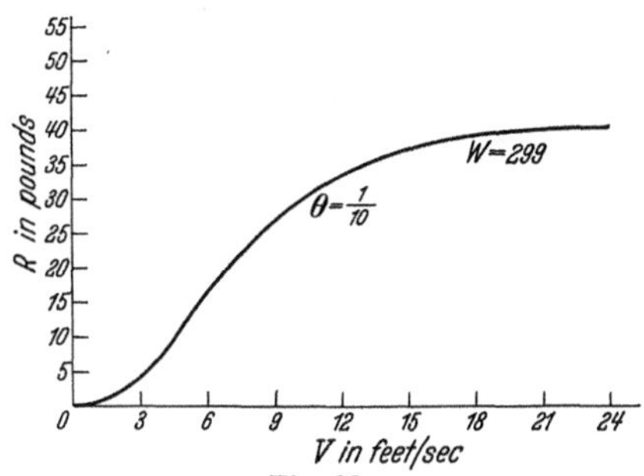

Fig. 28.

The case of a single inclined plane is indeed rather far removed from the form of an actual seaplane under-body. The forward part of the seaplane, in fact, is at a definitely steeper angle than the surface aft near the step and the analogy would be made more complete by the use of two planes in series with different inclinations.

In order to simplify the problem we assume a vertical plane B forming the stem and a plane moderately inclined forming the under surface, Fig. 29. Two cases will then develop, according as the vertical plane is or is not immersed. The second case (A variable with speed) is identical with the case examined above.

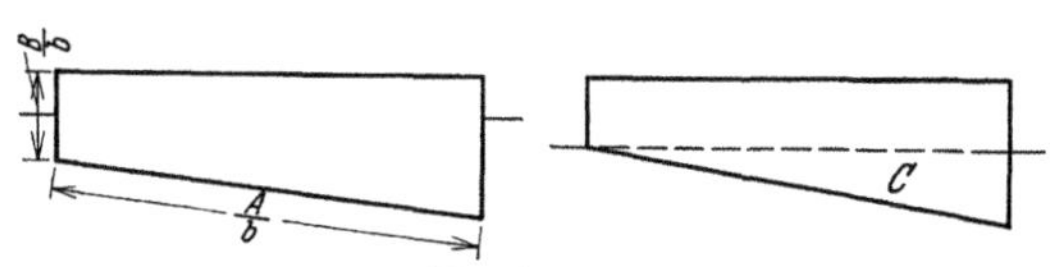

Fig. 29.

In the first case, A will be a constant and B will be the variable to eliminate. The angle of incidence for hydroplaning is here considered always the same.

A position of the float or boat with reference to the general water level is defined by the surface B. The resistance may then be taken in the form

$$R = \alpha A\,V^2 + \beta B\,V^2$$

while the equilibrium of the vertical forces will give

$$W = \delta\left[C + \frac{B}{b}\cdot\frac{A}{b}\cos\theta\right] + K' A\,V^2$$

where C is the volume of the immersed part when $B = 0$. B is essentially positive. If B were negative, the conditions would revert to those of the preceding case. For $B = 0$ the two cases coincide. In these two equations B is an unknown and A a constant. We may then find the law of R as a function of V by eliminating B. This will give

$$R = \left[\alpha A + \frac{\beta b^2}{A \cos \theta}\left(\frac{W}{\delta} - C - \frac{K' A V^2}{\delta}\right)\right] V^2$$

which is of the form $R = A_1 V^2 - A_2 V^4$

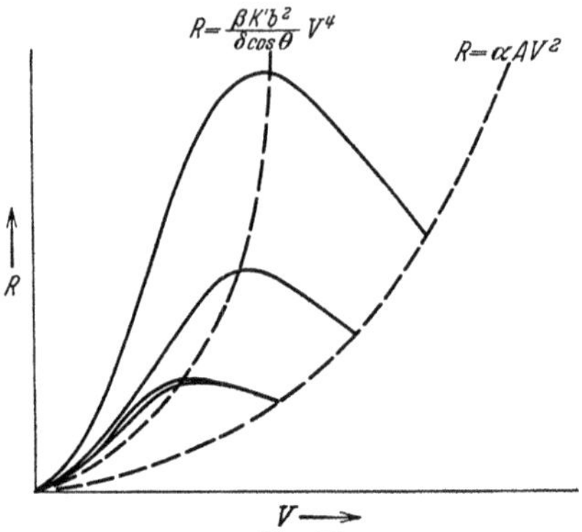

Fig. 30.

This formula is only applicable between $V = 0$ and the value for which $B = 0$, that is the value given by:

$$V^2 = \frac{W - \delta C}{K' A}$$

The value of R for this point will be

$$R = \alpha A V^2 = \alpha \left(\frac{W - \delta C}{K'}\right)$$

If we give to W all possible values, the ensemble of the end points will be on the curve given by eliminating W between these last two equations. This gives, of course, the same equation as above

$$R = \alpha A V^2$$

The end points are then on a parabola (see Fig. 30).

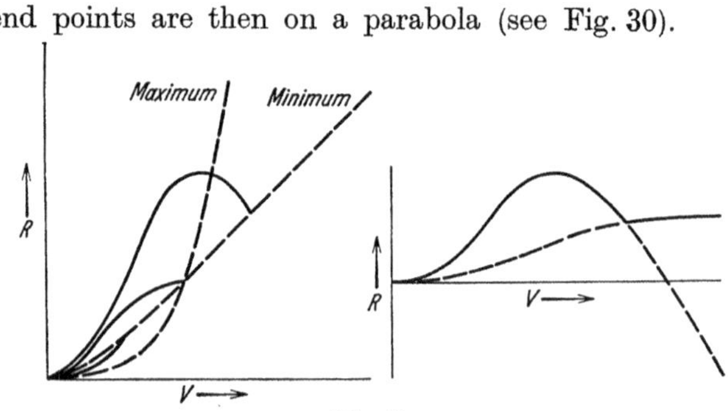

Fig. 31.

The maximum values of R will be given by the condition

$$\frac{dR}{dV} = 0 = 2 A_1 V - 4 A_2 V^3$$

whence $V^2 = A_1/2A_2$ and putting this in the value for R we find:

$$R = \frac{A_1^2}{4 A_2}$$

and using again the value of $V^2 = A_1/2A_2$, this becomes

$$R = A_2 V^4$$

If then we return to the expanded form of R for the value of A_2, we have

$$R = \frac{\beta K' b^2}{\delta \cos \theta} V^4$$

The results thus developed have been obtained by the aid of a doubtful working hypothesis. It is therefore necessary to check, by the aid of model experiments, in order to determine whether the results indicated by the formulae agree in satisfactory measure with those given by experiment. We give, therefore, as follows, certain results of model experiment.

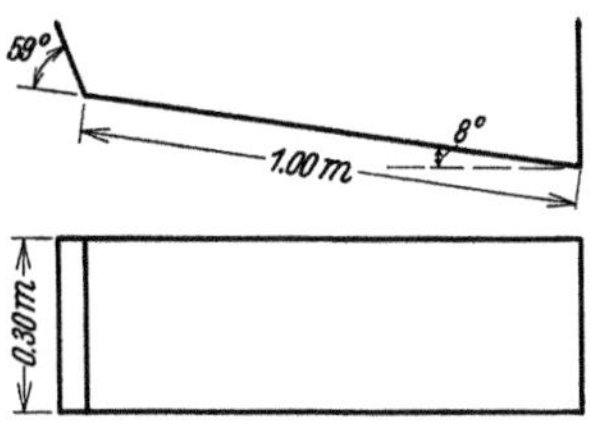

Fig. 32.

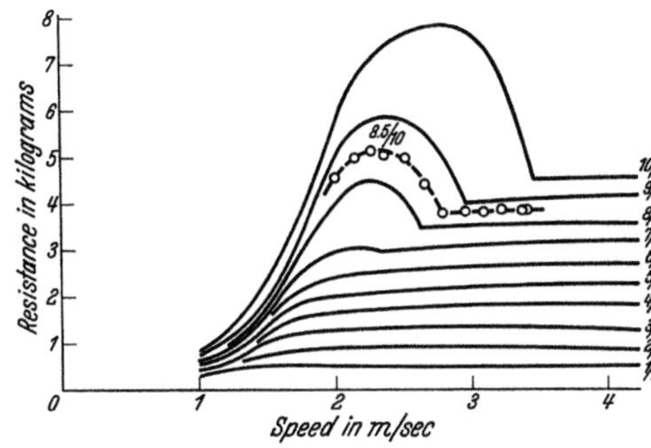

Fig. 33. Curves of resistance on speed.

10. Measures Taken on a Model Seaplane. Hydroplaning at Constant Angle of Incidence. The model Fig. 32 comprised a plane surface inclined 8° to the horizontal, together with a forward plane inclined 59° to the plane. The length of the first plane was one meter, the width 0.3 m. In these experiments while the model was compelled to maintain a constant attitude relative to the horizontal, it was free to move vertically. The model was run successively with a load of 24 kgs. (noted as 10/10) then with loads of 9/10, 8/10 and thus to the lightest loads. There was made also a similar series of runs with the intermediate series of loads, 9.5/10, 8.5/10 . . .

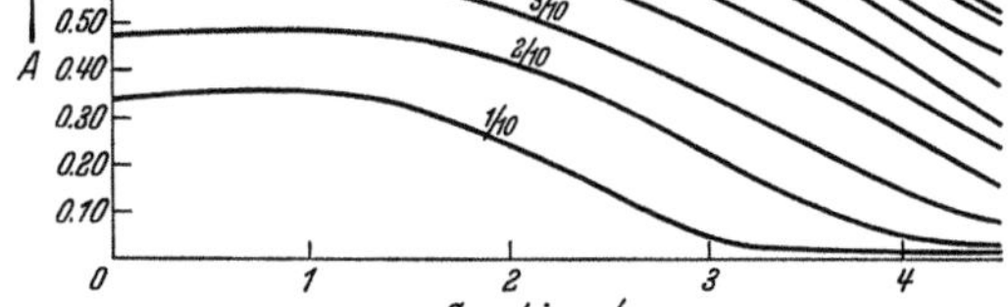

Fig. 34. Curves of location of model on speed.

Each run gave three curves. The first, Fig. 33, is the curve of resistance R as a function of the speed V. The two other curves, Fig. 34, relate to the position of the model relative to the undisturbed water level. One of these latter, B, gives the length of the forward plane below the undisturbed water level, and the second, A, the length of the lower plane below the same level. The curves A and B are given as functions of V.

In addition to these three sets of curves, three other curves are given, placed on the same sheet corresponding to the run with loading 9/10 (see Fig. 35). A comparison of this character brings to light a certain number of phenomena of a general nature.

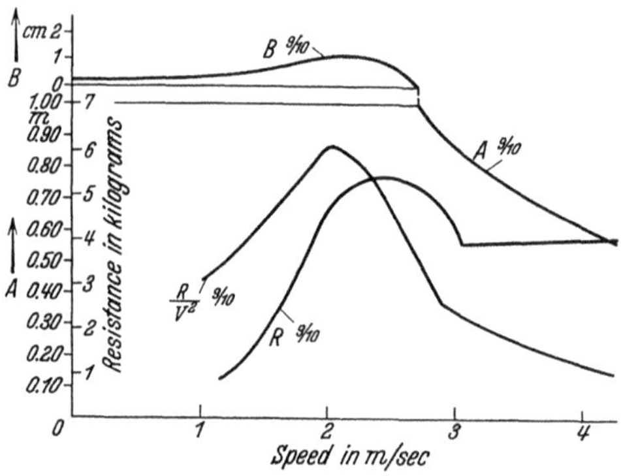

Fig. 35. Curves for loading 9/10.

(a) At low speeds for models starting with B not zero, the immersion increases; that is, B increases at the start.

(b) At low speeds for models starting with the edge between A and B out of water, A increases at the start.

(c) For models with B not zero, the maximum of resistance coincides with the maximum immersion.

(d) For the same models, the appearance of a resistance nearly constant corresponds to a progressive diminution of A.

(e) If the curve R/V^2 is traced, the maximum of this curve and its break, correspond with the maximum of B and with the disappearance of B.

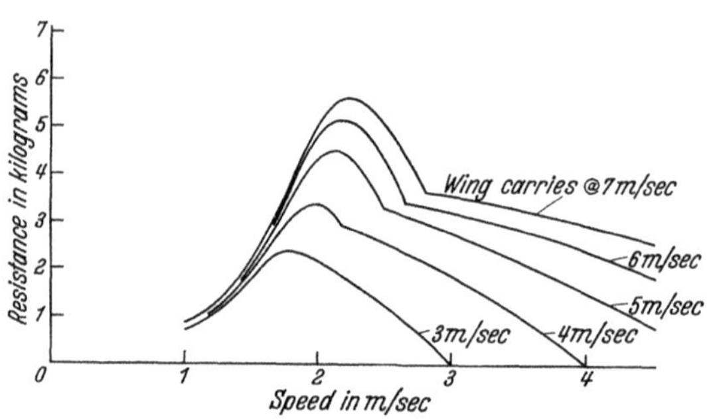

Fig. 36. Curves for resistance (water) assuming aerodynamic lift varying with the square of the speed.

(f) The general character of the curves of R is closely similar to that indicated by theory, but numerical verification is rendered difficult by the fact that, at the start, for example, B increases, a feature not included in the approximate theory above developed.

(g) By the aid of these curves, it is easy to obtain curves of variation of the resistance as a function of V when there exists an aerodynamic sustentation proportional to the square of the speed. If V_0 is the speed for which the aerodynamic sustentation is equal to the weight W of the plane, then for any speed x, the corresponding sustentation will be $W\,(x/V_0)^2$ and the weight carried by the water will be

$$W - W\left(\frac{x}{V_0}\right)^2 = W\left[1 - \left(\frac{x}{V_0}\right)^2\right]$$

The curve for $[1 - (x/V_0)^2]$ will then give the point on the abscissa x. The results of this method of representation are shown in Fig. 36.

Verduzio has given a simple method of showing in graphic form the values of the buoyant force required from the water wherein the incidence of the wings is variable. This graphic gives the values of the ratio, *sustentation from water* to the *square of the speed* wherein the incidence of the wings and the speed are known. It is assumed that the aerodynamic sustentation is of the form $L = \lambda i V^2$. If then the weight is W we have:

$$B = W - \lambda i V^2$$

where B = sustentation due to the water and L = that due to the air. This gives

$$\frac{B}{V^2} = \frac{W}{V^2} - \lambda i$$

These are straight lines for V = constant, each line for a particular value of V, Fig. 37. The ordinate at the origin is W/V^2. The point of intersection with the axis of abscissae gives the incidence corresponding to the speed of flight at altitude zero. In order to trace these lines, it is then sufficient to know, for a single speed, the value of i for flight at sea level.

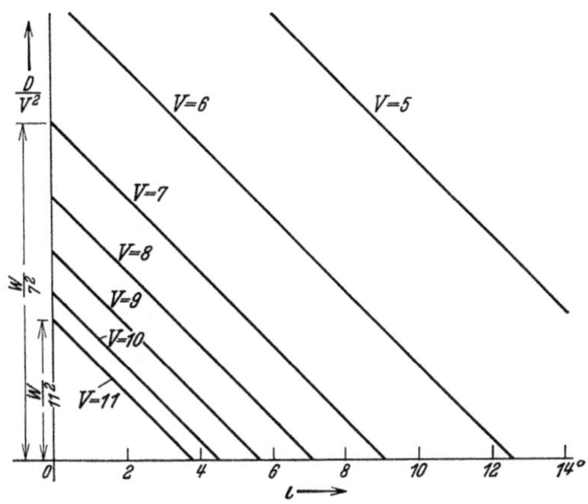

Fig. 37. Buoyant force required from water with varying incidence of the wings and varying speed.

11. Influence of the Angle of Incidence in Hydroplaning. If we consider singly the states of a seaplane at which the hydroplane regimen is established, it may be assumed that the resistance is given as a function of the weight carried p and of the angle of incidence, in the general form,

$$R = \alpha p \theta$$

Taking the angle of incidence of the wings, i, as variable, we may put this in the form,

$$R = \alpha p (e - i)$$

for the law of the resistance under all conditions of incidence and of weight carried. The wing data gives furthermore, the value of the aerodynamic sustentation as a function of speed and of the angle of incidence. This will have the form $\beta V^2 i$.

The weight carried by the reaction on the boat or float will then be

$$p = W - \beta V^2 i$$

and in consequence

$$R = \alpha (W - \beta V^2 i)(e - i)$$

This formula provides, for each value of the speed, for the study of the variation of R as a function of i. To represent these results graphically we employ the method given by Verduzio[1]. The abscissae represent the values of the wing incidence i and the ordinates y represent the values of $(R + R')/V^2$, R' being the resistance of the wings which may be taken in the form $K V^2 (i - i_1)^2$. We may then propose to trace the curves corresponding to a constant speed. We have thus

$$y = \frac{R + R'}{V^2} = \alpha \left(\frac{W}{V^2} - \beta i \right) (e - i) + K (i - i_1)^2$$

The curves corresponding to constant V will then be parabolas. The term in i^2 having a positive coefficient, the curves are concave upward. The value of the incidence i for which y is minimum is given by

$$i_m = \frac{\alpha W/V^2 + \alpha \beta e + 2 K i_1}{2 (\alpha \beta + K)}$$

The value of i_m will then decrease with increasing speed. The ordinate of the summits is found to be

$$y = - \frac{[\alpha W/V^2 + \alpha \beta e + 2 K i_1]^2}{4 (\alpha \beta + K)} + \frac{\alpha W}{V^2} e + K i_1^2$$

The locus of the summits is found by eliminating V between the values of y and of i. This gives likewise a parabola. These results are all in accord with those obtained graphically by Verduzio.

These results are only applicable to the period of hydroplaning. They show that, during this period, in order to reduce the power required to a minimum, the angle of incidence must slightly decrease as the speed increases.

During the critical period of mounting upon the step, the resistance of the air is small in comparison with that of the water and likewise the aerodynamic sustentation is weak. In consequence, during this period, there is no need of bringing in the incidence i of the wings.

It will be noted that thus far there has been no reference to changes of attitude due to the effect of the propeller. The influence due to the longitudinal stability has thus far been left entirely out of account.

This brief and approximate examination has provided means for a determination of the general character of the phenomena involved but can scarcely provide for the calculation of the more exact results required by the constructor. Independently of the fact that we have assumed the angle of incidence constant, a certain number of accessory phenomena have been completely neglected. These latter may be grouped under: phenomena entering in the two-dimensional problem, and, phenomena entering only in the three dimensional problem.

The first of these are: the immersion of the model at low speeds, and the formation of transverse waves (complicated by vortex phenomena).

[1] Hydrodynamic Congress, Innsbruck, 1922.

The second of these are: the divergence of fluid filaments on the sides of the body, the formation of divergent waves and friction of the lateral faces. Nevertheless, this preliminary examination has provided means for a determination of the general character of the conditions affecting the take-off of a seaplane.

12. Complete Study During the Hydroplane Period. The complete study of the plane during the hydroplane phase may be attacked by theory, by model experiment, and by full scale test. The method by theory with our present means hardly seems capable of giving useful results in cases other than where the flow is two-dimensional; that is to say, in the case where the width of the plane in the direction perpendicular to the line of movement is infinite. The method by model test has been employed by several experimenters. The most complete published results are those of Sottorf, which relate to the dynamometric examination of a model in movement under steady conditions.

Tests at full scale have thus far not given important results, principally because in the case of a full scale seaplane the phenomena due to the sustentation surface are confused with those due to parts of the hull aft of the step.

Before any analysis, observation on a model is necessary in order to permit the isolation of the diverse phenomena. We shall suppose that the situation is represented by a plane with small inclination to the horizontal and in uniform translation at high speed. Under these conditions there results a deformation of the free surface of the water similar to that caused by the movement of any body whatever. The free surface of the water takes a form which accompanies the body; that is to say, a form which observed from the body itself appears at rest. This may be expressed by saying that the speed of the system of waves formed by the body is equal to the speed of translation of the body itself.

Observed in longitudinal vertical section this deformation comprises a mass of water B elevated above the general level. This elevated mass forward seems to join with the horizon as an asymptote. At the stern there is formed a cavity, again joining ultimately with the general level of the water by the system of waves V (see Fig. 38). According to circumstances the separation of this system V may develop according to the form V_2, leaving the surface at the stern tangentially with concavity upward, according to form V_3 with concavity upward but not tangent to the under surface, or again according to form V_1 with concavity downward and not tangent to the under surface.

For an obstacle limited laterally, the observable portion of these wave forms is limited by the return of the water which the body has displaced laterally.

In plane form the deformation of the free surface is as follows. The water touches the plane following a surface nearly rectangular, limited forward by an arc pertaining to the elevated mass B. The material body having its lateral faces vertical, these faces are not in contact with the water (see Fig. 39). The free surface of the water is hollowed and

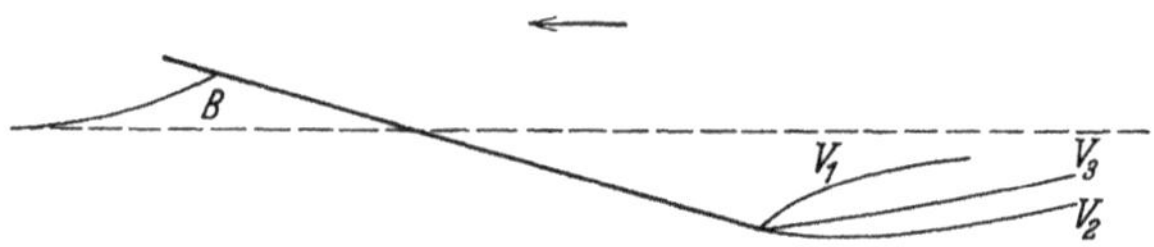

Fig. 38.

joins ultimately with the divergent waves W having their origin in B. The region at the stern is formed by a sort of basin, of which the bottom has the form nearly of a cylinder with horizontal generatrix, and the sides R are formed by the water returning along the longitudinal axis. At J the return of the water from both sides forms a sort of jet, Fig. 40.

Fig. 39.

The nature of the movement in contact with the body may be studied experimentally through the form of the curve B, through the form of the trajectories along the plane and through the value of the local pressure. The form of the curve B is readily observed through the use of a plane of glass. It is thus shown that this curve is convex forward. Its form, however, cannot be defined with any great precision because in this region steady conditions have not yet been fully established.

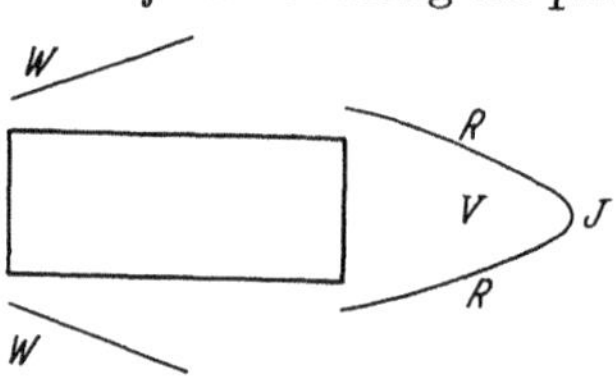

Fig. 40.

The form of the trajectories on the under surface of the body may be observed by the use of a plane of brass in which are fixed small steel rivets flush with the surface. Traces of rust may then be clearly observed

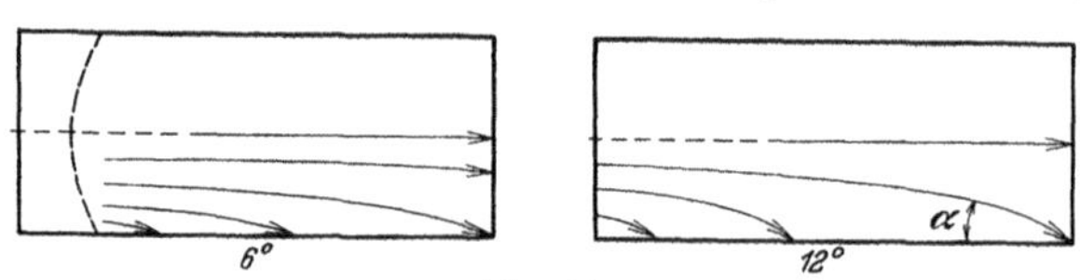

Fig. 41.

showing that a part of the water which has traversed the region B follows the body through to the after edge, while another portion escapes laterally, making an angle α with the direction of the lateral edges, Fig. 41. This angle α increases with the inclination of the plane to the horizon. For a plane inclined at an angle of 12^0 and a speed of two meters per second this angle reaches 60^0 (length of plane 58 cm. and width 14.4 cm.).

The values of the local pressure may be observed by the aid of manometers connecting with small holes pierced in the plane. Results of such measurements have been published by Sottorf[1].

These results relate to three planes having characteristics as shown in the following table.

Width cm.	Length below original level cm.	Wetted length cm.	Inclination	Speed m./sec.	Vertical component kgs.
30	17.5	24.5	8^0	6	18
30	37.0	45.0	6^0	6	18
30	80.0	87.0	4^0	6	18

The local pressure reaches its maximum value in region B and then decreases continuously to the after edge. At this edge Sottorf found for the first plane a positive pressure, and for the other two the pressure zero. It is a simple matter to pass from a knowledge of the pressure at a given point and of its immersion relative to the horizon, to a value of the local velocity. To this end it is only necessary to apply the Bernoulli principle. It thus appears that if at the after edge, which is below the general level, the pressure is equal to that of the atmosphere, the velocity at this point will be greater than the velocity of translation of the body. Applying the Bernoulli equation to the ensemble of the points for which measurements had been taken, Sottorf found for the three planes average velocities as follows:

5.57 m./sec., 5.81 m./sec., 5.92 m./sec.

A knowledge of these values is useful for the determination of that part of the resistance due to friction.

The form of the basin shaped cavity aft of the plane is important in regard to the practical determination of the form to be given to a seaplane aft of the step.

13. Zone of Water Contact Above General Level. Let us consider the part of the liquid forming the elevated mass forward and lying between the undisturbed level of the water and the forward portion of the plane. In consequence of the elevation of this mass above the general level a portion of the surface of the plane above this level is in contact with the water. It has the form indicated in Fig. 42. For a body with fixed attitude relative to the horizontal the dimension σ is a function of the velocity only. As the body is completely defined by b, λ and i, we should then have σ as a function of V, b, λ, i.

In each series of Sottorf's experiments V and b were constant. Hence σ is given as a function of λ, i.

[1] Werft, Reederei, Hafen 1929.

If we plot the values of σ as a function of λ for each series and place the curves according to the values of i, we shall obtain an ensemble of curves on which we may trace the lines for constant i.

It is also necessary to recognize the fact that the sweep of the air will modify the geometry of these phenomena. Experiments with the plane at rest are necessary in order to take account of the importance of this effect.

In a general way, however, it appears that this zone of contact depends principally on λ and i, but in a small degree upon the speed. In consequence, this zone will not be in relation with the system of longitudinal waves, but only with the flow of water laterally.

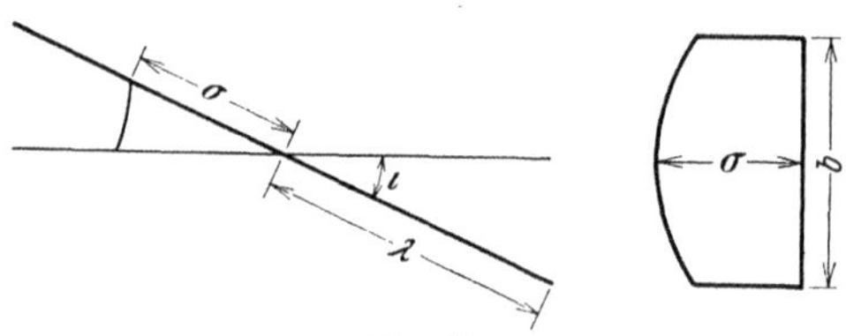

Fig. 42.

In order to definitely clear up this question it is necessary to run the model at all possible speeds, first without a forward air screen, and second with such screen. In such case, there develops an experimental difficulty which arises from the effect of the air on the region forward of the body. Let us consider the three following experimental arrangements relative to the same body (see Fig. 43).

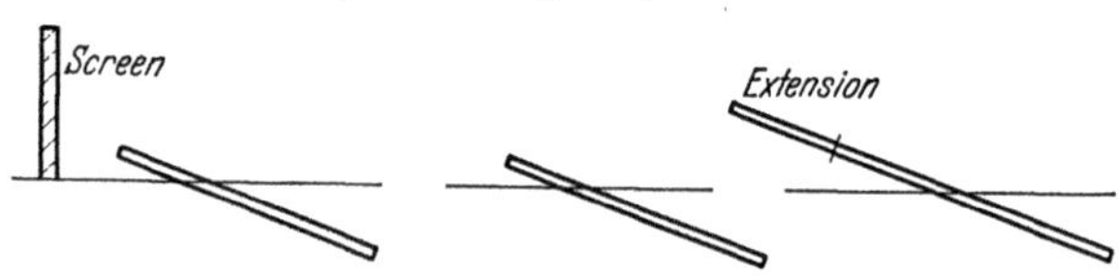

Fig. 43.

The results obtained in these three cases are different, and we must therefore conclude that the velocity of the body relative to the air has an influence on this zone of contact above the general level. This may have an effect of considerable importance on the position of the resultant. Sottorf found in certain cases the resultant passing through this zone.

If this zone were neglected the problem could then be treated by a method derived from that of Bobyleff by considering the body as the half of the Bobyleff body and imposing *a priori* the speed in the wake by the Bernoulli condition.

14. Study of Two Planes in Tandem with Constant Incidence. The curves marked x Fig. 44 give the resistance of a group of two bodies in tandem with an intervening space x. For all these bodies the slope of the under surface is 8^0 and the transverse width 30 cm. The load carried by the group is the same as in the preceding case, 9.6 kg.

The curve $x = \infty$ corresponds to the infinite separation of the two bodies. Its ordinates are therefore double those of the curve corresponding to a single body with load 4.8 kg.

On the curve $x = 80$ cm. it is seen that in comparison with the curve for $x = \infty$ the resistance of the group may be greater than the sum of the individual values for the two bodies. This arises from the fact that the wave produced by the forward body is projected with force on the after body, and meets it between its vertical walls. This situation develops in special degree for the case marked $V = 1.80$ m./sec. The association of the two bodies in this case is unfavorable, and the same condition holds through to $V = 2.4$ m./sec. where there is equivalence. For higher speeds the association becomes favorable and the resistance of the tandem combination passes through a minimum at about $V = 4.3$ m./sec.

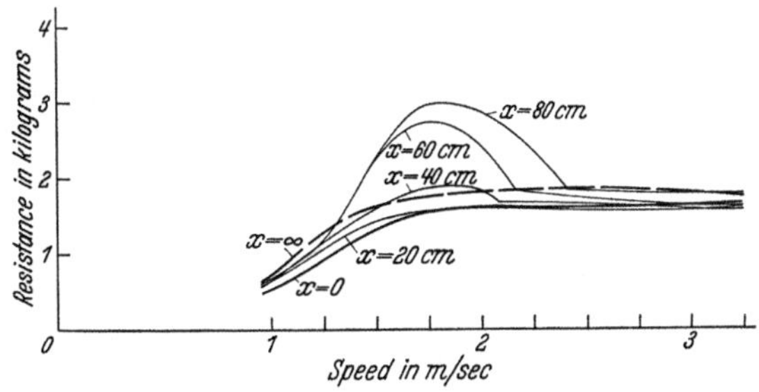

Fig. 44. Curves of resistance on speed for two planes in tandem, with varying distance between the planes.

For spacings larger than 30 cm. the same phenomena are found, with, however, maxima less and less definitely marked. The speeds of minimum resistance are related to the value x, as shown in the following table.

x	V m./sec.
80	4.30
70	4.18
60	3.87
50	3.75
40	3.60
30	3.25
20	2.70

The phenomena described here are susceptible of practical application for the prediction of the resistance of an ensemble of bodies. The description of these phenomena has been given solely for the purpose of showing the complexity of the actions intervening in the period during which the after part of a seaplane trails along in the wave formed by the portion forward of the step. The study of these phenomena through theory alone seems scarcely possible, especially because it would be necessary to take into account the closure of the water upon the after portion of the first step.

The experimental study of these phenomena presents as a principal difficulty the very great number of parameters. The present example is in effect relative to a single width and a single incidence. If such a study were to be undertaken systematically with a view to practical application, the number of parameters to be introduced would be six: the two values of the incidence, the two widths, a variable such as x giving the ratio between the longitudinal separation of the two bodies and the transverse width of one of them, and finally the weight of the ensemble. In such a study it would be further necessary to measure not alone the resistances to movement which have been mentioned here, but likewise the locations of their resultants. The extent of this program serves to explain why thus far such a study could be undertaken only in very special cases. In practice, such researches can only be carried on as variants of practical solutions, and the method by the use of models is the only one permitting in this way the improvement of a given initial design.

15. Comparison of Seaplane Under Water Forms with Variable Width. In these researches the following elements remain the same:

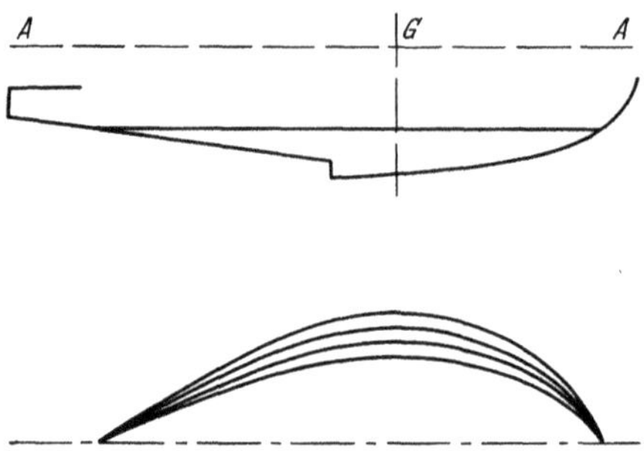

Fig. 45.

The wings and their characteristics, the weight of the model, the position of the wings with regard to the step, and the position of the center of gravity with relation to the step.

All the models have as geometrical form for the bottom, the same cylinder with horizontal generatrix following the form of the keel with step. The cylinder is limited by a vertical cylinder different for each model. All these vertical cylinders, however, are similar; that is to say, for the same transverse section the widths are in the same ratio as the values of b. As all of the models have the same weight, isolated they would not float at the same immersion, nor in the same attitude. The wider models have the reference line AA, Fig. 45, elevated toward the front. In effect the under water body being enlarged will take a lesser immersion for the same weight, and if the attitude of the model should remain unchanged the center of buoyancy would be displaced forward where the water lines are wider than aft.

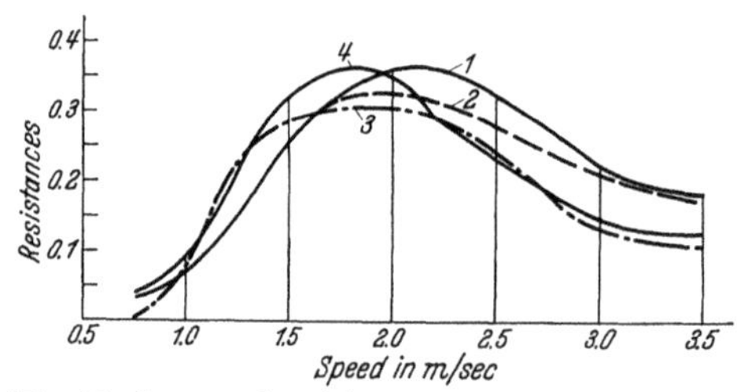

Fig. 46. Curves of resistance on speed for varying widths of model.

As the center of gravity is in the same longitudinal position for all cases, the model would tend to rise forward.

In consequence, at the start, the different models are more inclined upward at the bow as they are wider. In the case of the widest model the value of the inclination at the start is 9 mm. for a length of 900 mm. The value of the inclination at the start for the narrowest model is 0.

The values of b increase by successive equal values, the narrowest model having a width b, the widest model, No. 4, having a width 1.5 b. The constant weight was 1.75 kg. The resistance of these models plotted on speed is shown in Fig. 46.

16. Tests on Model with Varying Longitudinal Location of Wing. Figure 47 shows the resistance of a seaplane together with wing and suspension bars for a constant speed and a constant angle of incidence. There are altogether eleven tests differing with regard to the fixation of the model lengthwise with reference to the wing. The figure shows for each test the position of the model and its resistance. For Test No. 1, the wing being very far forward, the model remains supported on the water only by the part aft of the step, and the attitude is inclined upward in marked degree. For Tests 2, 3, 4, 5, the inclination decreases, but the general situation is the same. From 1 to 5 the resistance likewise decreases. For Test No. 6 the step begins to carry and the plane touches the water at two points. The resistance continues to diminish as well as the angle of inclination. For Tests 7, 8, 9, the step carries alone. The resistance still further diminishes, and for No. 9 the resistance is a minimum and the inclination negligible. For Test No. 10 the inclination begins again to increase as well as the resistance; the plane begins to lack longitudinal stability. For Test No. 11 the plane porpoises, the shift has gone too far. The minimum of resistance for these conditions of test was found for position No. 9. The distance between two consecutive positions of wing and model was 1/94 of the length of the model. The angle of incidence in all cases is 6^0 and the speed is 4.5 meters per second.

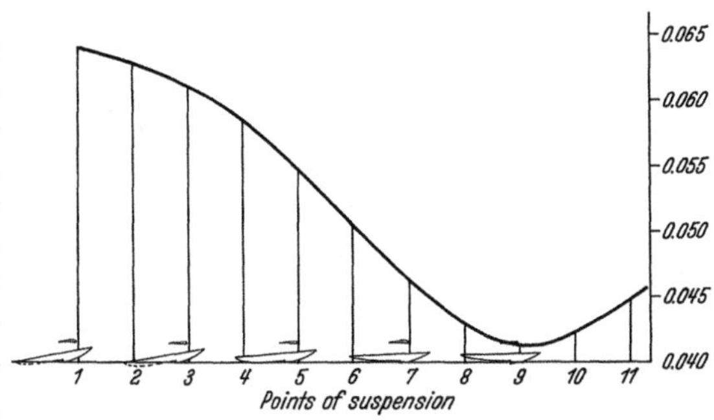

Fig. 47. Diagram showing effect of varying longitudinal location of wing.

CHAPTER IV

DIFFERENCES BETWEEN AIRPLANES AND SEAPLANES WITH REFERENCE TO THE AERIAL PORTIONS

In flight the general condition of equilibrium for any plane is the convergence of the three forces, weight, propeller thrust, aerodynamic reactions. The principal differences between airplanes and seaplanes is found in the distance between the general center of gravity and the propeller shaft. In the airplane this distance is small, and if the center of gravity is above the propeller axis the thrust of the propeller gives a couple tending to elevate the nose. In consequence, in case of a slowing of the motor the plane tends to go down by the nose, which permits it to pick up in speed.

With the seaplane, on the contrary, this distance is large, and the center of gravity is below the axis of the propeller, and in consequence the center of gravity is aft of the aerodynamic resultant R (wings and tail assembly). If, in flight, the plane should rise by the nose, due to any external cause, the resistance to movement increases, and in consequence the speed decreases. This results in an increase of thrust of the propeller, and in consequence an increase in the couple tending to turn the nose down, thus correcting the incipient perturbation. On the other hand, if in flight the motor weakens and stops, the plane rises forward since the opposing couple of the propeller decreases or becomes nothing. This involves a hazard of undue loss of speed. This defect can be corrected by an inclination of the axis of the propeller relative to the plane of the tail assembly in such fashion that in horizontal flight the propeller axis has an incidence positive, and the tail assembly an incidence negative.

The wash of the propeller gives then a downward reaction on the tail assembly and in consequence a moment tending to throw the nose of the plane upward. With a plane thus planned, the center of gravity should be carried forward in order to obtain proper centering in flight. If now the motor weakens, two effects are produced: one due to the decrease in propeller thrust—the effect tending to throw the nose upward; the other due to the decrease in the reaction of the tail assembly—the effect tending to throw the nose downward. In this manner compensation may be realized between these two effects. These consequences of a lack of equilibrium are especially important in small planes for which the distance between the center of gravity and the axis of the propeller is most important.

The study of the differences between airplane and seaplane should be completed by an examination of the stability of route in flight and of the transverse stability in flight and in turning.

(a) Regarding the stability of route, consideration must be given to the relative positions of the general center of gravity and the overall center of air drift. The latter should be located aft of the center of gravity.

The forward part of the boat or float of a seaplane being larger than the corresponding part of an airplane, the natural method of compensation would be found in a larger extent of the vertical drift-surface for the seaplane than for the airplane.

The vertical position of the center of gravity in relation to the center of drift is likewise different in the two forms. In the seaplane the center of drift may be still lower than the center of gravity of the structure as a whole, the result of which is a tendency to incline transversely in the direction of the drift.

(b) For stability among waves, consideration must be given on the longitudinal plan, to the position of the center of drift with reference to the axis of rolling and to the axis of yawing. In order that a seaplane may, at the same time, have stability of route and stability under the action of the waves, the center of air drift should be found in the angle hatched, see Fig. 48. This can only be realized by raising the general vertical drift-surface.

(c) In turning, the rudder carried by the tail of the plane must necessarily be above the water, and in consequence above the axis of rolling. The secondary effects due to pitching and rolling motions caused by turns, will then be more important in the seaplane than in the airplane.

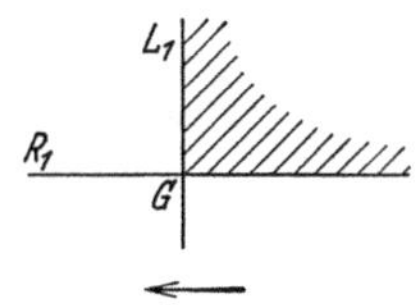

Fig. 48.

These differences between the seaplane and the airplane are not of basic importance with reference to the hydrodynamic study of the problem. There is need, however, to note briefly these relations in order to call to notice the sense in which advantage should be taken of such liberty of change as may develop from the hydrodynamic study, in order to improve the quality of the seaplane in flight.

CHAPTER V

DIFFERENCES AND ANALOGIES BETWEEN FORMS FOR HYDROPLANES AND FOR SEAPLANES

1. Introductory. The differences between seaplanes and ordinary boats have already been noted in the course of this study. They result chiefly from the difference in the values of the speed length ratio $\daleth = V/\sqrt{g\Lambda}$ in the two cases. Ships of normal form show values of this ratio less than 0.6. For example, even for a destroyer at 50 knots for a length of 100 m. at the water line, we should have $\daleth = 0.82$. On the contrary, for seaplanes, the speeds at the instant of maximum resistance are not far from 50 km./h. for lengths varying from 25 to 5 m. at the water line, giving values of the speed-length ratio between 0.89 and 2.0. Still higher values are to be met with for the period between the moment of maximum resistance and that of take off. The hull of a seaplane boat or float and that of an ordinary ship are then in conditions markedly different as regards the movements caused by their run over or at the free surface of a fluid. Nevertheless, there exist certain similarities between the conditions of operation of these two forms of hydro-glider. For very small boats, there have, indeed, been obtained values of the speed length ratio of 3.5 but these small boats are not suited for navigation in a troubled sea, and consequently this comparison is without significance. In order to realize more practical conditions, we may consider certain small boats described by Sir William

Thornycroft[1]. These boats have a length of about 17 m. for a speed of 40.5 knots, giving a value of the speed-length ratio of 1.6. These values give conditions similar to those for the hull of a seaplane, both as regards relative speed and the conditions for sustained movement in rough water. Table 2 gives a comparison between a boat of this character and a large modern seaplane.

TABLE 2.

	Boat	Seaplane
Ratio length to beam	5	5.5
Ratio $\frac{\text{distance of step to bow}}{\text{length}}$	0.5	0.67
Ratio $\frac{\text{distance of step to bow}}{\text{beam}}$	2.5	3.7
Ratio $\frac{\text{distance center of gravity from bow}}{\text{length}}$	0.54	0.57
Ratio $\frac{1000 \text{ volume}}{\text{length}^3}$	2.45	6.85
Angle of keel forward of the step with the water level	2^0	2^0
Break of the keel at the step	4^0	8^0

Length is here taken as length at the water level.

Although the two forms present the same general characters—a step, a sharp angle forward of the step, and similar forms of section for the forward body—the ratio length to beam is greater for the seaplane, which has, therefore, a somewhat finer form. For this there are several reasons.

In the first place, the boat in movement should be stable transversely due to its beam alone, while the seaplane is stabilized by the effect of its wings. Baker has indicated that if the length is less than 8.5 times the beam, the seaplane will be stable in movement[2].

Again, the boat is intended normally to move as a hydroplane, and for this case there is reason for an increase of beam, since the boat will have less to fear from jumps clear of the water due to an over development of hydroplane surface.

If instead of the ratio length to beam we consider the ratio 1000 volume $\div$ length^3, it is seen that from this point of view, the seaplane is much fuller in form than the boat. This difference results from the need of limiting the length of the hull of the seaplane in order to make possible the necessary changes in longitudinal attitude by a reasonable effort

[1] THORNYCROFT, SIR J. E., and LIEUT. BREMNER, Coastal Motorboats ("C.M.B."), Their Design and Service During the War, Transactions of the Institution of Naval Architects, Vol. 65, pp. 32—43, London, March 1923.

[2] BAKER, G. S., Flying Boats—the Form and Dimensions of their Hull, Engineering, Vol. 109, pp. 323—327, London, March 5, 1920.

on the controls. This condition is without force for the boat, but is imperative for the seaplane. A further reason is found in the need for an adequate stability of route.

In the seaplane, the length forward of the step should be the least compatible with the realization of a sufficient hydroplane area and of seaworthy qualities generally.

The longitudinal position of the step is very different in the two cases. On the seaplane the step is near the vertical containing the center of gravity. This is necessary in order to readily bring the plane upon the step. For the boat, on the contrary, the normal attitude is that of resting on two surfaces. The center of gravity is therefore, at nearly mid-distance between the center of pressure of the forward surface and that of the after surface, these surfaces being nearly of the same area and having each the optimum angle, determined in taking account of the speed of the water over the surface at each point. The advance of the center of gravity forward is here favorable to the speed but unfavorable from the point of view of seaworthy qualities.

The difference in the break of the keel arises from need, with the seaplane, to disengage completely the after part from the wave formed aft of the step—a condition which does not exist for the boat. The nearer we approach the stern the more the form of the hydro-glider differs from that of the seaplane.

Finally, if we examine, at the state of rest, the silhouette of the part above water, we shall find that the seaplane is much higher above the water and much more elongated aft. The first difference is due to the need, for the seaplane when alighting, of an approach to a rough water surface at a rather brusque angle, while the hydro-glider, always supported by the water, continuously adapts itself to the changes of the level on the free surface. The second difference is due to the conditions regarding the lever arm necessary for the movable air surfaces serving as controls for the seaplane.

If we should turn the comparison to the case of a racing hydroplane rather than a hydroplane boat, the differences would become more pronounced and would relate on the one hand, to the ratio between the resistance and the weight carried, and on the other, to the form of the transverse sections, which would no longer need to be limited at the bottom by a line rising from the mid-length toward each end fore and aft.

In the series of relative speeds where the extremes are occupied by the racing hydroplane and the ordinary boat, the seaplane occupies an intermediate position in which there should be combined a low value of the resistance and seaworthy qualities in good degree.

Historically the form of the seaplane has undoubtedly developed out of the hydro-glider. Practically it is a compromise among conflicting conditions as indeed are all projects in the art of naval architecture. In the case of the seaplane, the necessary nautical qualities must be paid for by a less favorable ratio between the resistance and the weight.

In order to properly understand the progressive evolution leading from the ship form to that of the seaplane hull, it is necessary to note the conditions in which the introduction of a step may be favorable or unfavorable. Dynamometric measures show within what limits of the speed-length ratio any given case may be found, but do not indicate the mechanism of the phenomenon.

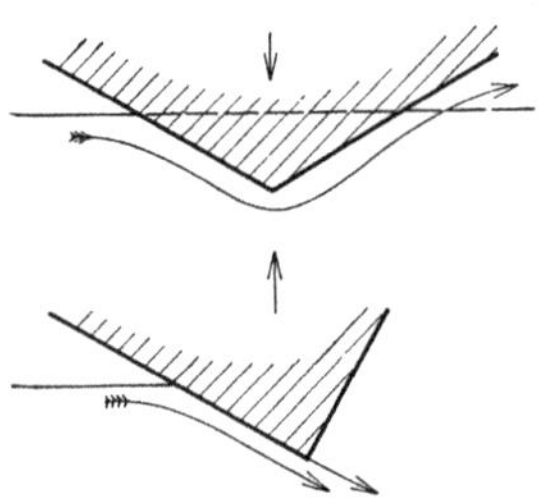
Fig. 49.

Theory can give to this subject some indication as follows:

Consider an obstacle, motionless in a current of water. The theorem of momentum indicates that if the water flows about the obstacle and leaves it with a speed and vertical component directed upward, the obstacle will be subject to a reaction from the water, directed downward; that is, it will tend toward a greater immersion. On the other hand, if the water leaves the obstacle with a velocity and a vertical component directed downward, the body will be subject to a reaction from the water, with vertical component directed upward; that is, to a force of sustentation Fig. 49. The theoretical problem would, then, be much advanced if we could determine the actual conditions under which separation of the plane from the water and take-off occurs.

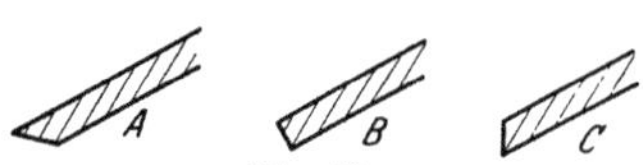

Fig. 50.

2. Experiments on Bevelled Planes. The importance of a brusque separation of the water at the after end of the body is readily shown by observations made in the experimental channel with planes having, either at the after end or on the side, a bevelled edge. For the same weight carried, the resistance is much greater with a following edge similar to A, Fig. 50 rather than to B or C. The same result is found, but in less marked degree, with regard to the lateral edges.

3. The Immersion of Ship-Formed Models and the Emergence of Seaplane Models. The consideration of an obstacle having a position fixed with reference to the level of undisturbed water is important from the point of view of a prevision of results by theory.

If observations are made on a model thus held in a fixed position in a current of water it will be found that for low values of the velocity, the model is subject, from the water to a force directed downward. The same effect is noted with models having one or two degrees of freedom as usually employed in naval experimental tanks. In such case, for low values of the speeds, the upward component is less than with the model at rest, and in consequence, the model settles lower; that is in movement, it takes a position lower than when at rest. If the speed increases, the phenomena for the free model become complicated in consequence of the change of attitude. However, in general the center of gravity of a ship under way is lower than when at rest.

These conditions do not hold for hydro-gliding planes, which beyond a certain speed, rise continuously with reference to the undisturbed water level.

If for a body of given form, fixed attitude and constant speed, the values of the horizontal and vertical components of the total water reaction for all immersions measured relative to the undisturbed level are plotted in curve form, each relative to the value of the immersion, we then have two curves H and V, Fig. 51. The curve H passes through the origin and the curve V below the origin. This is due to the fact that even for a weight zero, a body in horizontal motion will take a finite immersion as soon as it touches the liquid. The curve showing the ratio H/V will then have the form of a hyperbola. It is interesting to compare this curve V to that obtained from the same measurements made on a body at rest relative to the water. This new curve is well known to naval architects under the name of "displacement curve". It gives the total upward force at each immersion, in each case exactly equal to the weight producing that immersion. This curve naturally passes through the origin of coordinates.

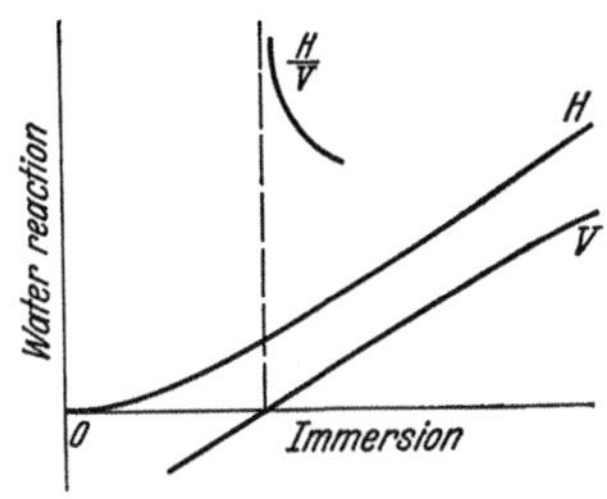

Fig. 51.

CHAPTER VI

CALCULATIONS OF DISPLACEMENT AND OF STABILITY

1. Displacement and Stability of the Seaplane at Rest. A floating body is in equilibrium at the surface of an undisturbed liquid when the hydrostatic pressures over the wetted surface have a single resultant, equal to the weight of the body, directed upward and through the center of gravity. The well known principles of hydrostatics show that this resultant is equal to the weight of the displaced liquid, or to the product of density of the liquid by the volume displaced by the body, and that it passes through the center of figure of this displaced volume.

For a seaplane, the problem for the conditions of equilibrium is usually presented in the following manner. Given the external form of the hull, the total weight of the plane and the position of its center of gravity, it is required to determine the immersion and the attitude of the plane at rest on the water.

The search for a position of equilibrium is, in consequence, a search for a water-plane section through the hull such that the volume cut off shall have a given value and that the perpendicular to this plane through the center of figure of this volume shall pass through a point fixed in advance (center of gravity of the structure). The solution of this problem depends on the manner in which the form of the hull is defined.

2. Determination of Form. As with ship lines, the form of the hull of a seaplane comprises a series of transverse sections, a series of horizontal or water-plane sections, and a series of longitudinal vertical sections (bow and buttock lines). A base line is taken passing through the lowest point of the step, AB on the various plans, and parallel to the water plane FL, Fig. 52. The lines of a seaplane hull differ from those for ships of normal form in two principal points. First the transverse sections are not equally spaced. This is due to the need for a specially accurate definition of the form in regions of rapid change and especially near the step or steps. If, for example, there are two steps, there will be a section at each step and there will be three distances, unequal in the general case, lying forward of, between and aft of these points. The second difference is due to the fact that the hull of a seaplane comprises longitudinal edges, and in consequence the lines will show a discontinuity whenever the plane cuts through such an edge.

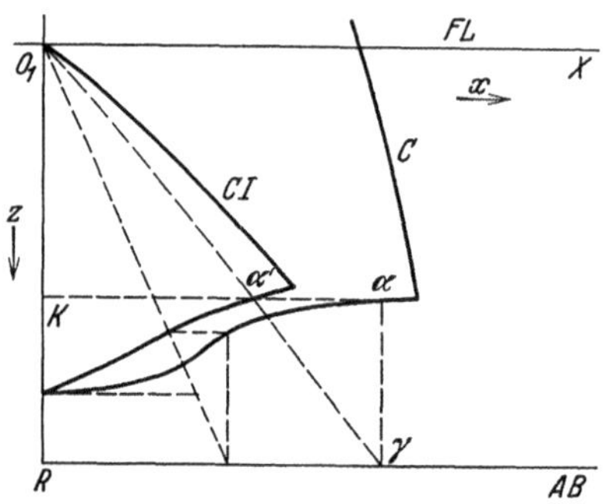

Fig. 52. Sections of form of seaplane body.

For these reasons, it is not possible to fully apply to the lines of a seaplane hull the methods commonly employed for normal ship forms. In particular the corrections at terminal points cannot be realized in the calculations for seaplanes. Constructors of seaplanes having, however, at the start followed the general practice in vogue for ship forms, the methods commonly employed are naturally based on the classic methods of naval architecture. On the contrary, the method here outlined is based on the special conditions of the actual problem. It is a planimetric method using simply a planimeter for areas rather than the more complex integrator giving areas, static moments and moments of inertia. The simple measurements of area may also be made without instrument and with almost equal rapidity by laying out the various diagrams and plans on cross-section paper.

The results required are first:

(a) The volume of hull below any given water-plane.

(b) The location of the center of figure of this volume, both longitudinally and vertically.

To obtain (a) and (b, vertical) there is traced at each section an auxiliary curve which may be called the section integral. The sections and section integrals are then measured by planimeter giving at each section the area and static moment about a reference axis. These areas and moments are then plotted on an axis of length and the curves drawn. These latter curves are then planimetered giving the volume and the moment of volume, and hence the center of volume relative to the reference water-plane. In order to find the longitudinal location of the center of volume the curve of longitudinal moments of section areas must be prepared and then planimetered in similar manner as for the sections.

We may now explain in the necessary detail the character of the various curves. The curve of section integral is found for any given section by the following construction, see Fig. 52. With each point α of the section, there is associated a point α' obtained by drawing $\alpha\alpha'$ and $\alpha\gamma$ respectively parallel and perpendicular to the water-plane and then drawing $O_1\gamma$. It is seen that, putting $\alpha K = x$, $\alpha' K = x'$ and $O_1 K = z$, we have from similar triangles,

$$x' = \frac{z\,x}{O_1 R}$$

The area of a curve with points found in this way is:

$$\int x'\,d\,z = \frac{1}{O_1 R}\int z\,x\,d\,z$$

which is the static moment of the section area about $O_1 X$, divided by $O_1 R$. This area multiplied by $O_1 R$ will then give the static moment of the section about $O_1 X$ (see Table 3).

In the same manner we may find the moment of inertia about $O_1 X$ by tracing a second integral curve. We should have thus

$$x'' = \frac{x'\,z}{O_1 R} = \frac{z^2\,x}{(O_1 R)^2}$$

whence
$$\int x''\,d\,z = \frac{1}{(O_1 R)^2}\int z^2\,x\,d\,z$$

The area of the curve of x'' multiplied by $(O_1 R)^2$ will therefore give the moment of inertia of the section about $O_1 R$.

The tracing of the section integral and the two measurements of area need occupy no more than five to ten minutes each, even when using section paper and without the use of a mechanical planimeter. With the latter the procedure is a little more rapid. In all, the number of sections being in the neighborhood of 20, this entire program can be carried through in about two hours.

TABLE 3. *Operations on the Curves of Section Areas.*

Sections	Water-line 1			Water-line 2		
	C_1	C_1I	M_1	C_2	C_2I	M_2
6	19.5	3.0	24.0	3.3	1.0	8.0
7	26.0	5.3	42.4	8.0	2.7	21.6
8	33.8	7.0	56.0	13.2	4.5	34.0
9	42.4	9.5	76.0	20.2	7.0	56.0
10	52.3	13.8	110.4	27.8	10.6	84.8
11	89.2	36.7	293.6	62.0	33.2	265.6
12	85.0	32.2	257.6	57.8	29.0	232.0
13	79.2	28.7	229.6	52.3	25.0	200.0
14	66.3	21.2	169.6	40.3	17.6	140.8
15	39.4	8.6	68.8	17.0	5.8	46.4
16	11.5	1.0	8.0	17.0	5.8	46.4

Depth $O_1R = 8$
Column headed C_1, areas of sections for water-plane 1
Column headed C_1I areas of section integrals as in Fig. 52
Column headed M, moments of section areas = col. $C_1I \times O_1R$
Similarly for water-plane 2

It remains to trace on length the curves of section areas and the curves of section moments vertically and longitudinally, and then to planimeter these two curves. This will occupy about one hour, after which we have the volume and the location of the center of volume.

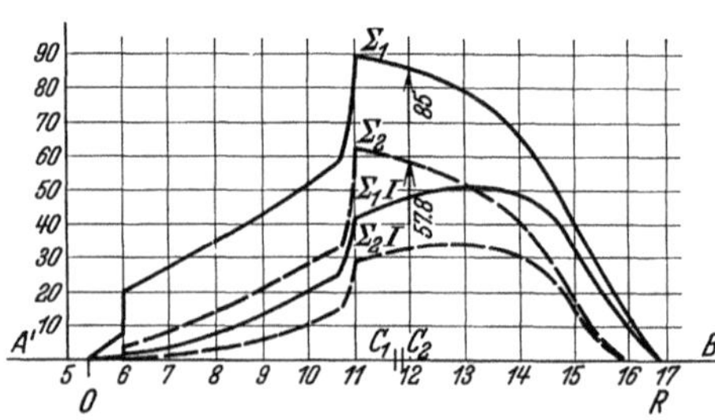

Fig. 53. Curves of section area and section moment on length.

Operating on two water-planes in succession, the total time required is about four hours and we have at our disposal two volumes and two centers of volume. It may be noted that the same section integrals will serve for the two water-plane locations, the area being properly limited in the second case.

The longest part of the work is that required for tracing the section integrals. This part of the program may be omitted when the forms employed are very similar to those already known, in which case the location of the center of volume may be fixed by a simple application of ratios.

It is evident that the curve of section areas will give the volume and the curve of longitudinal moments of these areas (see Fig. 53) will give the static moment longitudinally and thence by division the longitudinal location of the center of volume (see Table 4).

TABLE 4. Determination of the Center of Buoyancy, Longitudinally and Vertically for Water-Planes 1 and 2.

Curves relate to one-half body form.

Longitudinal (see Fig. 53).

Scales: Abscissae 1/6, Ordinates 1/10.

Area of curve Σ_1 = curve of section areas = 82.3 (cm.)^2
Area of curve $\Sigma_1 I$ = curve of section moments = 43.5 (cm.)^2
Area of curve Σ_2 = curve of section areas = 46.8 (cm.)^2
Area of curve $\Sigma_2 I$ = curve of section moments = 25.3 (cm.)^2
Distance $O_1 R$ = 17.65 cm.

$\text{Vol.}_1 = 82.3 \times 10 \times 6 \times 2 = 9870\ (\text{cm.})^3 = 9.87\ (\text{dm.})^3$
$\text{Vol.}_2 = 46.8 \times 10 \times 6 \times 2 = 5616\ (\text{cm.})^3 = 5.62\ (\text{dm.})^3$

$$\text{Abscissa of center of buoyancy } x_1 = \frac{43.5 \times 17.65 \times 6}{82.3} = 56 \text{ cm.}$$

$$\text{Abscissa of center of buoyancy } x_2 = \frac{25.3 \times 17.65 \times 6}{46.8} = 57.2 \text{ cm.}$$

Vertical (see Fig. 54).

Scales: Areas 1/10, Moments 1/80.

Area of curve Σ_1 as above = 82.3 (cm.)^2
Area of curve M_1 = curve of section integrals (Fig. 52) = 25.0 (cm.)^2
Area of curve Σ_2 as above = 46.8 (cm.)^2
Area of curve M_2 = curve of section integrals (Fig. 52) = 20.8 (cm.)^2

$$\text{Ordinate of center of buoyancy } y_1 = \frac{25 \times 8}{82.3} = 2.43 \text{ cm.}$$

$$\text{Ordinate of center of buoyancy } y_2 = \frac{20.8 \times 8}{46.8} = 3.55 \text{ cm.}$$

Similarly, the moments of the sections, as in Fig. 52, laid down as in Fig. 54 will give the moment of volume about the reference water-plane and hence the vertical location of the center of volume (see Table 4).

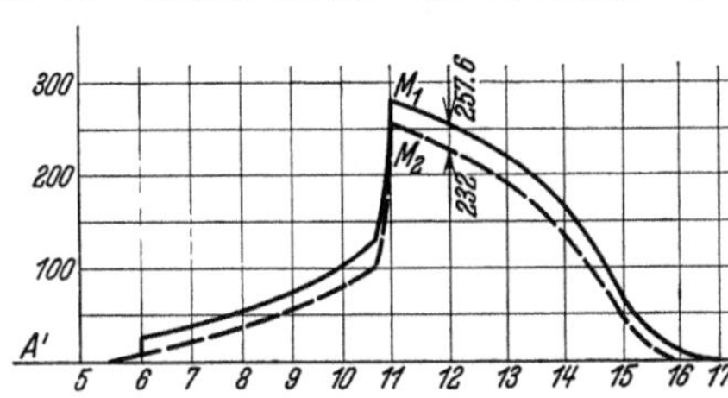

Fig. 54. Curves of section integral on length.

3. Interpolation of Volume. We have thus found, for two locations of the water-plane, the centers of volume both vertically and longitudinally. These are located on the longitudinal plan. One of the planes gives a volume greater than that specified, the other a volume smaller. By simple proportion, we may then locate, parallel to the base line, the water-plane and the volume specified, and also in like manner its center of figure.

This volume must, however, meet the further condition of equilibrium. To determine this condition we must find, for changing inclination with constant volume, the envelope of the water-planes and the locus of the resulting center of volume. To this end we must compute a radius of curvature $R = I/V$ and the derivative value $r = \triangle I / \triangle V$.

To this end we make use of the moments of inertia of the areas of the water-planes.

These moments must be taken about a gravity axis. Table 5 gives the form of calculation for the center of gravity of a water-plane area and for its moment of inertia about a transverse axis through this point. For the first, use is made of the curve FI and for the second, of the curve FII, Fig. 55. Since the moment of inertia directly resulting from this operation will not be about a gravity axis, use is made of the well known relation:

$$I_G = I - h^2 S$$

Where S is area and h is the distance between the first and the gravity axes. This correction, as shown in Table 5, thus gives the values needed, and thence R_1, R_2 and by interpolation R_v corresponding to C_v and r.

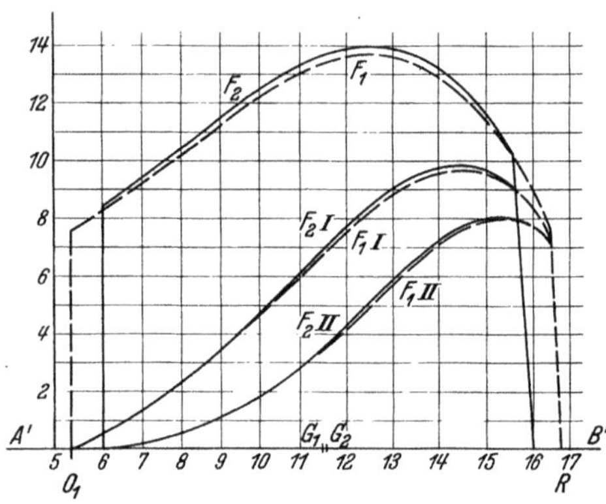

Fig. 55. Curves for two water-planes.

TABLE 5. *Calculation of the Center of Gravity of the Water-Plane Areas.*

Scales: Abscissae 1/6, Ordinates 1/4.

Planimetry $1/2\ WP_1 = 198.5$ sq. cm.

$$1/2\ WP_1 I = 100.4 \quad G_1 = d_1 = \frac{100.4 \times 17.65}{198.5} \times 6 = 53.58 \text{ cm.}$$

$$1/2\ WP_2 = 175$$

$$1/2\ WP_2 I = 89.5 \quad G_2 = d_2 = \frac{89.5 \times 17.65}{175} \times 6 = 54.18 \text{ cm.}$$

$$d_1^2 = 0.287 \text{ m.}^2 \qquad d_2^2 = 0.2935 \text{ m.}^2$$

Calculation of r and R.

Planimetry of $WP_1 II = 0.0063$ sq. m.

$WP_2 II = 0.00545$ sq. m.

$O_1 R = 0.1765$ m.

Inertia of WP_1 in relation to $O_1 = 0.0063 \times 17.65^2 \times 6^3 \times 2 = 0.0847$

Inertia of WP_2 in relation to $O_1 = 0.00545 \times 0.1765^2 \times 6^3 \times 2 = 0.0733$

$$I_G = I_{01} - d^2 S$$

$$S\ WP_1 = 0.1946 \times 2 \times 6 = 0.2335 \text{ sq. m.}$$

$$S\ WP_2 = 0.1773 \times 2 \times 6 = 0.213 \text{ sq. m.}$$

$$I_{1G} = 0.0847 - 0.287 \times 0.2335 = 0.0177$$

$$I_{2G} = 0.0740 - 0.2935 \times 0.213 = 0.0113$$

$$r = \frac{\triangle L}{\triangle V} = \frac{0.0064}{0.00425} = 1.505 \text{ m.}$$

$$V \text{ given} = 0.00774$$

$$I_V = 0.0145$$

$$R = \frac{I}{V} = \frac{0.0145}{0.00774} = 1.873$$

These calculations depend on the following geometrical properties which are demonstrated in any treatise of theoretical naval architecture.

(1) For small inclinations about a transverse axis, the envelope of the water-lines is a circle of radius $r = \triangle I/\triangle V$.

(2) Under these conditions the buoyant force passes through a point (sensibly stationary) called the metacenter. The distance from this point to the center of volume is given by I/V. The line joining the metacenter to the center of volume is $\perp$ to the water-line.

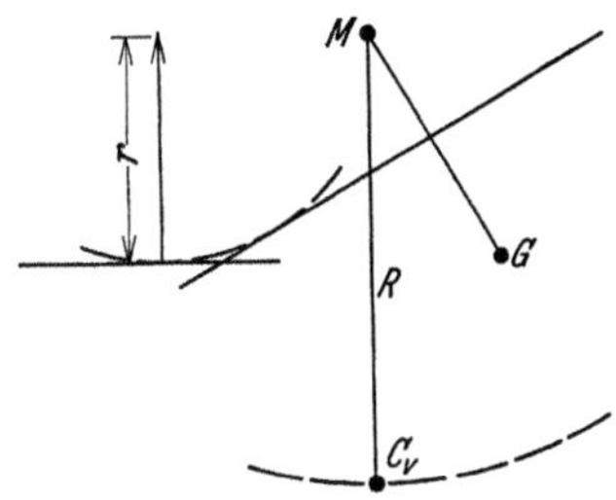

Fig. 56.

(3) The locus of the center of volume is the arc of a circle with the metacenter as center and I/V as radius.

For the actual problem we may then locate on the plan the center of volume, C_v, the center of gravity of the structure G and the metacenter M. The line GM will then be $\perp$ to the water-line. A line drawn $\perp$ to GM and tangent to the circle with $r = \triangle I/\triangle V$ as radius will then give the water-line sought, Fig. 56.

On the longitudinal section, Fig. 57, will be found in addition the centers of gravity of water-planes and the centers of volume for a series of other water-plane locations, with curves drawn through, showing the movement of these points with varying immersion. These curves show the character of the discontinuities which are to be anticipated in such cases.

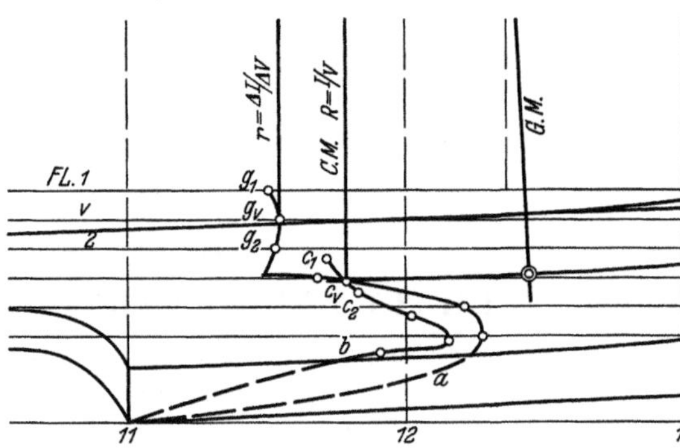

Fig. 57. Centers of figure: curve a of buoyancy, curve b of water-plane area.

4. Bonjean Scales. A procedure due to Bonjean is very useful for calculations of this character in the case where certain compartments of the hull may be invaded by water. In such case it is preferable to use methods of numerical integration.

For each section, the half widths are read for a series of equally spaced water-lines. We then write on each line (see Fig. 58) the half width (Column 1 of the table) and the sum to each line from the bottom upward, putting the results on the right (Column 2). These values, to suitable scale, are then laid off as abscissae giving a curve showing the section area from the base line up to any water-line at choice.

Each section is treated in this manner. If the water-line is drawn in, the corresponding abscissae at each water-line intersection will give the values for a curve of section areas on length. This form of calculation is useful for the determination of the heights to be given to water tight compartments.

In the application of this method to seaplane hulls, care must be taken with reference to the unequal spacing of the sections. In finding volume, for example, each section area must be used with the length which it represents, and then these parts summed.

Bonjean Scales.

Areas	Totals	Areas	Totals
19.0	374.6	9.0	79.7
19.0	355.6	9.0	70.7
19.1	336.6	9.0	61.7
19.1	317.5	9.0	52.7
19.1	298.4	9.0	43.7
19.2	279.3	9.0	34.7
19.3	260.1	6.6	25.7
19.4	240.8	5.0	19.1
19.5	221.4	4.0	14.1
19.5	201.9	3.1	10.1
19.6	182.4	2.5	7.0
19.7	162.8	1.9	4.5
19.8	143.1	1.3	2.6
19.9	123.3	0.9	1.3
19.9	103.4	0.4	0.4
20.0	83.5	0.0	0.0
15.3	63.5		
12.8	48.2		
11.0	35.4		
9.8	24.4		
8.2	14.6		
6.4	6.4		
0.0	0.0		

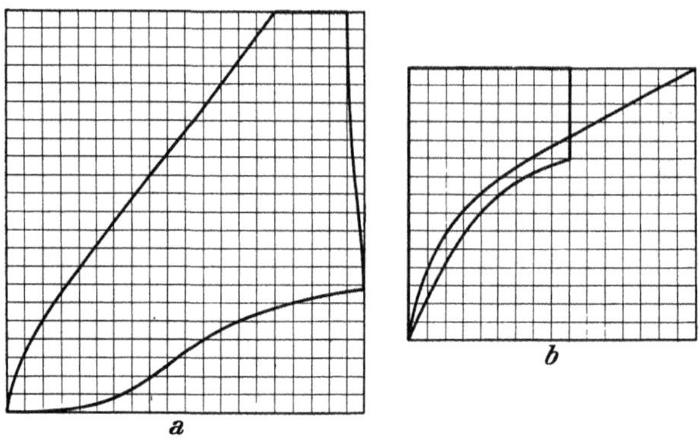

Fig. 58. Curves for Bonjean Scales.

CHAPTER VII

STABILITY UNDER VARYING CONDITIONS

1. Stability when Drifting with Motors Stopped. A seaplane assumed floating on the water with the motors stopped, will be subject to the action of the wind and waves. The effect of these actions will be different according to the orientation of the wind and that of the wave crests in relation to the longitudinal axis of the plane.

Let us first consider the effect of a head wind acting alone. The wind tends to push the plane sternward, which will, in consequence, pick up sternway. The water opposes this motion and develops thus a hydrodynamic resistance increasing with the speed until it equals the air force. At this moment there is equality between these two forces, but they will not be exactly opposed, and hence will form a couple. The nose of the plane rises, therefore, and a position is taken where the moment of longitudinal stability equals that due to the aero- and hydrodynamic forces.

To study this condition it is necessary to examine what happens when a small perturbation is introduced, and especially the effect produced by a change in the direction of the wind. Under these conditions, the combined air effects give a force sensibly horizontal and passing through the forward third of the part of the hull above water. This is the point $\triangle A$, the center of air drift, Fig. 59.

The plane receives, the same as before, a reaction from the water and this reaction passes sensibly through the after third of the immersed part of the hull. This results from the fact that the forces from the water are the same as though the water were in movement with reference to the plane, the current flowing from stern to bow. This is the center of water drift.

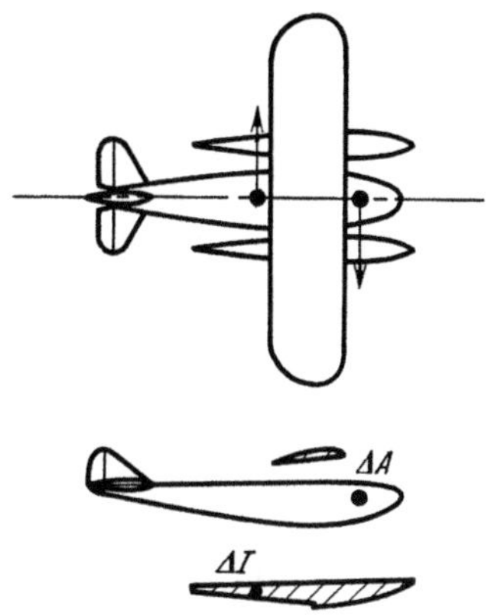

Fig. 59.

The situation of equilibrium with head to wind will not be stable unless the couple due to air and water forces tends to turn back the head into the wind, that is, if the center of air drift is aft of the center of water drift.

To realize this condition it is necessary to carry the center of air drift aft by the addition of a vertical surface toward the stern. With reference to the problem of stability under drift, the seaplane should be studied by model with reference to its parts above water (wings and non-immersed hull). These tests will serve to determine the vertical location of the air forces. The model should then be studied in a seaplane channel under horizontal traction by means of a cord attached at a suitable height at varying positions in length. In this way the location farthest from the stern is found for which the model will move in the direction of the thread. This gives the longitudinal position of the point $\triangle I$, Fig. 59.

When the point $\triangle I$ of the seaplane is aft of the point $\triangle A$ the plane tends of itself to turn head to wind. The effect of wind coming from the side and from aft is found in the same manner. For the case of a wind on the beam it will be necessary to carry out a calculation similar to that for the determination of the stability of a ship under sail—a procedure well known to the naval architect[1].

2. Stability of Route with Motors Running. This problem is treated in the same manner as for ships and as set forth in text books of naval architecture. In this case the resultant force on the plane is directed forward. The center of water drift is placed toward the forward third of the immersed body.

[1] Gagnotto, Luigi, Étude sur l'action du vent lateral sur les navires, *Bulletin de l'Association Technique Maritime et Aéronautique*, Vol. 33, pp. 53—74, Paris, 1929.

In the case of ships, the stability of route is estimated by comparing the positions of the general center of gravity and the center of water drift. The latter being always forward of the former, there is, properly speaking, no stability of route. There is nothing to oppose a continued turn initiated by any external cause. It is well known that the stability of route in such cases can be increased by the increase of immersed area toward the stern. This method may likewise be employed with seaplanes. It may be remarked finally that the increase of stability of route stands in opposition to the improvement of aptitude for evolutions. Too large a stability of route should therefore not be sought.

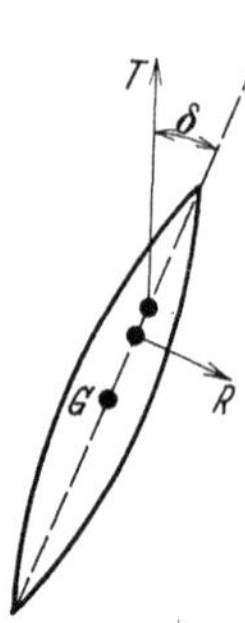

Fig. 60.

3. Stability under Tow. Differing from the conditions for the preceding cases, the problem of a seaplane under tow brings in an external force of which the direction is variable with reference to the plane. The length of the tow line is also a factor in this case. If the tow line is very short, it may be seen that the situation would develop as though the towing force passed through a fixed point of the plane. If this point of attachment is forward of the forward drift center there is stability.

In the case where the tow line is long, this condition is still necessary, but the plane no longer moves in a straight line. With a very long tow line we may assume that the direction of the line is constant. Under the influence of a small deviation δ of the bow, the water resistance will give a considerable transverse force R, which, relative to the center of gravity G, will have a moment greater than that of the traction T, Fig. 60. The angle δ will then increase, thus increasing the moment of T until an angle of yaw is reached where the two moments are equal. Due to the turning inertia, the bow will pass beyond this angle until brought back by the increasing value of the traction moment. The plane will thus have an oscillating movement about its mean path.

CHAPTER VIII

RULES OF EXTRAPOLATION

1. Introductory. The rules of extrapolation represent in a general way the collection of procedures making possible the determination of the principal characteristics of a design without passing through all the detailed calculations necessary for the design itself. Such rules are useful when there are a large number of variants. Rules of extrapolation should not be confused with rules drawn from the law of similitude. The latter relate to details while the former relate rather to the project as a whole. While such rules depend upon general considerations of mechanical similitude they include as well other diverse phenomena.

In order that rules of extrapolation may be established on a suitable foundation it is needful on the one hand that the number of previous cases be large and relative to varied characteristics; and on the other hand, that the project in hand be not too far removed in characteristics from those which have served as a foundation for the formulae employed. It will be understood, therefore, that in a branch of structural technique relatively young, as in the case of seaplane design, extrapolation is relatively difficult.

As an example, we may cite the proportions given by Rumpler. These propositions are based upon exact mechanical principles, but the formulae employed are based upon previous designs too far removed from the cases to which they are applied. Rumpler starts with the following considerations.

The principal difficulty in the increase of the dimensions of a design depends, for planes of present form, upon the fact that with such increase the useful load continually decreases. This results from an increase in the weight of construction due to the concentration of load in the central portion. By avoiding such concentration a percentage of useful weight will be maintained, the same for large planes as for small. This decentralization leads Rumpler to a design which is schematically the lateral juxtaposition of a series of small planes. This decomposition applies to the power plants as well as to the boats or floats. In order to justify such a type of construction Rumpler proposes to show that the limitation imposed by the weight of the wing structure of present designs is nearly realized. The foundation of these considerations is then a problem of extrapolation.

2. Rumpler's Method of Extrapolation. An initial design is assumed as follows:

Wing surface	S
Total Weight, W	2,000 kgs.
Weight of wing structure	300 kgs.
Residual Weight	1,000 kgs.
Total	1,300 kgs.
Useful Load	700 kgs.

It is then assumed that the design is increased in linear ratio n, holding constant the loading per unit area of wing surface. We have then

Wing surface $\Sigma = S\,n^2$

Total weight $Sn^2 \times W/S = 2000\,n^2$

Weight of wings $300\,(0.8\,n^3 + 0.2\,n^2) = 240\,n^3 + 60\,n^2$

Other weight (50 per cent of total) $= 1000\,n^2$

Remainder = useful load $= 2000\,n^2 - (240\,n^3 + 60\,n^2 + 1000\,n^2)$

$= 940\,n^2 - 240\,n^3$

The latter vanishes for $n = 3.92$

Under these conditions we should have

Wing surface $S \times (3.92)^2 = 15.37\,S$

Total weight $2000 \times (3.92)^2 = 30{,}740$ kgs.

For varying values of n the various elements considered by Rumpler may be computed, holding constant the total weight at 2000 kgs. For such a design the load per square meter would be about 53 kgs. The type design should then have

$$S = \frac{2000}{53} = 37.6 \text{ m.}^2$$

Consequently, for other designs:

$$\text{Wing surface} = 37.6 \times n^2$$

$$\text{Weight} = 2000 \times n^2$$

$$\frac{\text{Useful load}}{\text{Total Weight}} = \frac{940\,n^2 - 240\,n^3}{2000\,n^2} = 0.47 - 0.12\,n$$

The law of linear decrease indicated by Rumpler, and giving for the type design a ratio of 0.35 ($n = 1$) gives a ratio 0 for a limit design of 30,740 kgs.

Discussion. These results are based upon three hypotheses:

(1) The weight carried per square meter of wing area is constant (53 kgs.).

(2) Half of the weight carried is allotted to weights other than wing structure and useful load.

(3) The weight of wing construction per square meter of surface Σ is

$$\frac{240\,n^3 + 60\,n^2}{S\,n^2} = \frac{240\,n + 60}{37.6}$$

With $n = \sqrt{\Sigma/37.6}$, and an aspect ratio of 6, this weight of wing structure per unit area may be expressed as a function of the span by using the relation $\Sigma = 0.167\,b^2$.

The weight of wing per square meter as a function of span is then

$$\frac{240\,b\sqrt{0.167/37.6} + 60}{37.6} = 0.426\,b + 1.6$$

This weight of wing structure per unit area in the hypothesis is then a linear function of the span.

For $b = 10$ m. weight per unit area = 5.86 kg./m.^{2}

$b = 50$ m. weight per unit area = 22.90 kg./m.^{2}

We may now examine Rumpler's rules in the light of actual realization.

(1) Constant Load per Unit Area of Wing. This is inexact, see Fig. 61, giving the relation between R prediction and D realization.

(2) Half of the weight carried is allotted to weights other than wing and useful load. The Dornier designs show a distribution of weight carried as follows:

Total Weight. . . .	670	2860	6030	14100	51500
Weight of Wings . .	108	535	767	1899	7476
Useful load	80	962	2401	6411	25838
Sum.	188	1497	3168	8310	33314
Ratio to Total Weight	0.280	0.523	0.526	0.590	0.646

Instead of the constant ratio 0.5, it is seen that the useful load per unit of wing area continuously increases. Here again the relation between R and D as in Fig. 61 is shown in Figs. 62 and 63.

(3) The weight of the wing structure per unit area of wing $= 0.426b + 1.6$. This is likewise inexact; the relation is rather *Weight of wing structure per unit area of wing* $= 0.25b + 5$. See Fig. 64, Curve R' where again R and R' are according to these two formulae while D gives actual results.

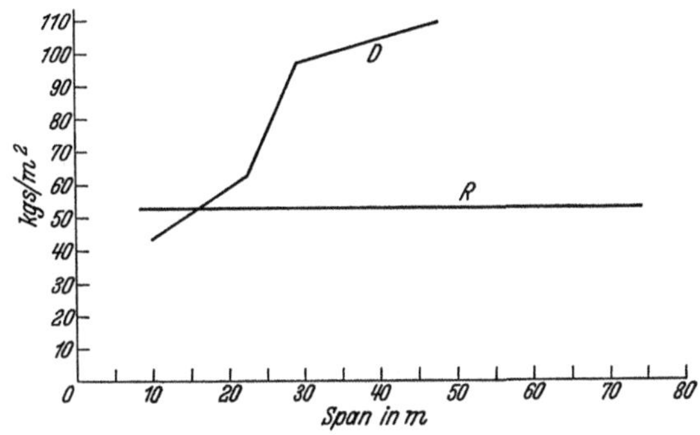

Fig. 61.

Figures 63 and 64 show the comparison between the prediction R and the realization D.

If we assume the following rules:

(1) $\frac{\text{Total Weight}}{\text{Wing Area}} = \text{constant} = 110$, for example.

(2) Useful load plus weight of wing structure equals one-half total weight.

(3) Weight carried per unit area of wing equals $5 + b/4$ we shall have

$$\text{Useful load} + S\left(5 + \frac{b}{4}\right) = 55\,S$$

The useful load will then become 0 for $5 + \frac{b}{4} = 55$ or $b = 200$.

By utilizing the most recent designs of Dornier, the extrapolation would lead to a construction X of 88 m. wing span with 36,000 hp., and of which the wing characteristics are given in Table 6 in comparison with the five Dornier designs indicated *Do.* and the two seaplanes $n = 1$ and $n = 3.92$ to which reference has been made above.

For the design X, the power would be 36,000 hp. and the useful load 110,000 kg., comprising 150 passengers and combustible for some ten hours flight.

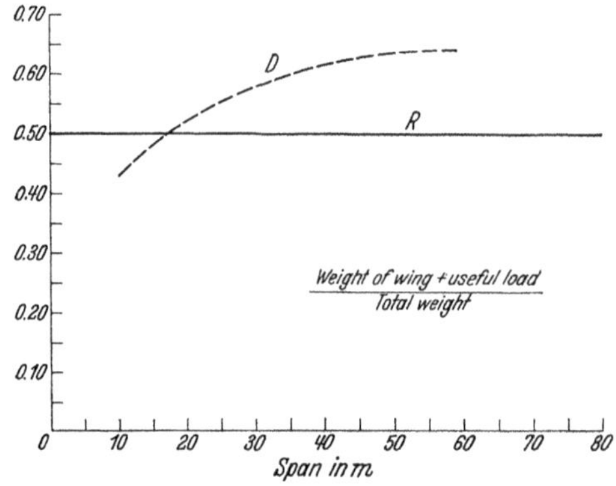

Fig. 62.

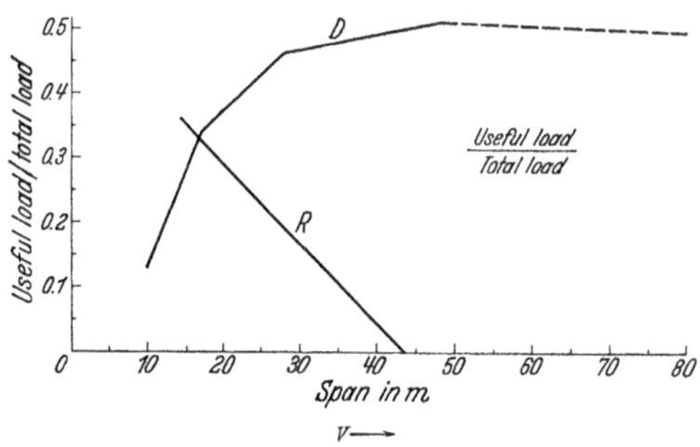

Fig. 63.

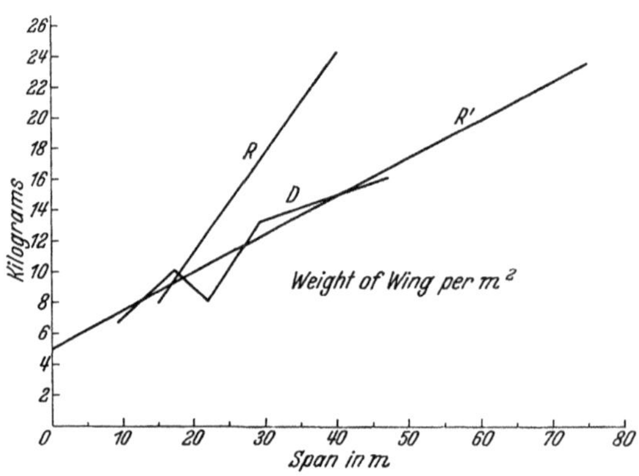

Fig. 64.

To sum up, there is certainly a difficulty in the fact that sustentation increases with the square of the span while the weight increases with the cube.

TABLE 6.

Type	Span b Meters	Wing Area Σ square meters	$10^3 \frac{\Sigma}{b^2}$	Total Load kg.	$\frac{\text{Total Load}}{\Sigma}$	Wt. Wings	$\frac{\text{Wt. Wings}}{\Sigma}$	Useful Load	$\frac{\text{Useful Load}}{\text{Total Load}}$	$\frac{\text{Useful Load}}{\text{Total Load—Wt. Wings}}$
Do.	9.80	15.51	162	670.	43.1	108	6.97	80	0.119	0.143
$n = 1.00$	15.00	37.60	167	2000.	53.0	300	8.00	700	0.350	
Do.	17.50	52.96	172	2860.	54.2	535	10.12	962	0.336	0.414
Do.	22.50	95.25	190	6030.	63.4	767	8.05	2401	0.400	0.456
Do.	28.60	143.86	178	14100.	98.0	1899	13.20	6411	0.457	0.526
$n = 3.92$	59.00	580.00	167	30740.	53.0	15318	26.50	0		
Do.	48.00	467.70	203	51500.	110	7476	15.97	25838	0.503	0.585
x	88.00	1781.00	230	240451.	135	44500	25.00	110451	0.460	0.564

However, the limit placed by Rumpler is too narrow. It may be remarked that there is a ratio nearly constant between the total weight and the weight empty. A simple rule such as

$$\frac{\text{Weight of wing structure}}{\text{Wing area}} = \frac{1}{6}\,\frac{\text{Total weight}}{\text{Wing area}}$$

combined with the assumption of weight varying as the cube of the span, and the supporting surface as the square would lead to a weight of wing construction per unit area proportioned to the span. However, the rule $a + cb$ is more in conformity with actual realization. Table 7 gives the numerical elements which have served as basis for the preceding discussion[1].

TABLE 7. *Weight in Kilograms.*

Span, meters	9.80	17.50	22.50	28.60	48.00	88.00
Wing	108.2	535.10	767.5	1898.9	7475.8	44500
Structure — (Wing and Hull)	67.8	203.0	497.7	878.6	2473.9	11500
Hull	106.7	416.0	1006.6	2023.3	7235.3	34000
Power Plant	194.1	491.1	995.5	2350.8	7143.5	34000
Motor and Control Accessories	33.6	93.0	121.9	217.1	613.7	3000
Crew	80.0	160.0	240.0	320.0	720.0	3000
Useful Load, Gasoline and Oil	80.0	962.0	2401.0	6411.0	25838.0	110451
Total Weight	670.4	2860.2	6030.2	14099.7	51500.2	240451
Horse Power	80	450	900	2000	6000	36000

CHAPTER IX

TESTS ON REDUCED SCALE MODELS

1. Introductory. The impossibility of deducing satisfactory predictions from theory alone, or by the extrapolation from results obtained on actual designs, has lead in the study of seaplane boats or floats, as in many other like problems, to the use of reduced scale models.

In the general technique of the use of models, it is not customary to represent all the minute peculiarities of form of the full scale object. The important question then arises regarding the characteristics which may be neglected, which are those which must be reproduced, and at what desirable scale. These same questions arise in connection with tests of models for ships, and it is known that for the study of resistance to uniform motion, the following conditions must be realized:

Geometrical similitude of form.

Weight of model proportional to the cube of linear dimensions.

[1] DORNIER, C., Notes on a Family of Similar Flying Boats, Journal of the Royal Aeronautical Society, Vol. 32, pp. 981—1020, London, December, 1928.

Fixed relative position of the center of gravity longitudinally.

Speed proportional to the square root of linear dimensions.

It is known also that corrections are necessary to allow for the influence of friction.

Likewise in tests of ordinary ship forms, whenever study is made of a non-permanent movement (tests of turning, oscillation among waves, etc.) it is necessary to realize in addition similitude of inertia. For model tests of seaplanes the question arises as to the need of realizing similitude of inertia as well as center of gravity. The question results from the fact that the models are usually small and not easily realized with suitable weight, and that in consequence it is not always easy to dispose of the weight needed to realize similitude of inertia.

The realization of the correct weight has lead in certain cases to ingenious measures; such, for example, as models of paper[1] but more frequently the excessive weight of the model is corrected by a counter weight acting over a pulley. This causes no trouble for steady movement, but may lead to discussion in the case of models for which it is desired to observe "porpoising". Efforts have therefore been made to realize inertia similitude about a transverse axis passing through the center of gravity.

In our present ignorance of hydrodynamic forces acting on a solid in irregular motion, it is not easy to give decisive arguments proving the necessity of employing models fulfilling the condition of inertia similitude.

In the larger number of the tests carried out thus far, the condition of inertia similitude not having been realized, it has seemed necessary to examine this point experimentally. To this end a model was made with rather heavy weight, a considerable proportion of which was adjustable relative to the center of gravity.

Within the limits between which the moment of inertia of an actual seaplane may be varied, the variations obtained on the model by changing the moment of inertia have been found small.

This question may therefore be considered at the present time as of secondary importance. It is, however, needful to note the point for it may later become of importance.

2. Models with Fixed Incidence. The study of a model with constant incidence is made in general by running the model with free angular movement at a series of speeds, and then over again the same series after a change in the distribution of the weight on the model, or in the forces due to the suspension. From the results of these various tests plotted in suitable curves, points corresponding to a given incidence can be collected.

---

[1] Denny, Sir Archibald, Transactions Institution Naval Architects, p. 194, 1915.

A more rapid method makes use of a system of articulated parallelograms, as in Fig. 65 which maintains the model with a fixed incidence. The various experimental results used in the present work have been found by means of a mounting of this character.

3. Tests with Unloading Proportional to the Square of the Velocity. Tests are often carried out in which the model is subject to a vertical force variable with the speed, and intended to represent the sustentation due to a wing. Various mechanical dispositions may be employed to this end. Thus on a balance a weight may be placed for each run corresponding to the speed for this run. Otherwise, by the use of a suitable device, the unloading may be readily effected by subjecting the model to the centrifugal force of a rotating counterweight properly connected to the driving mechanism of the carriage.

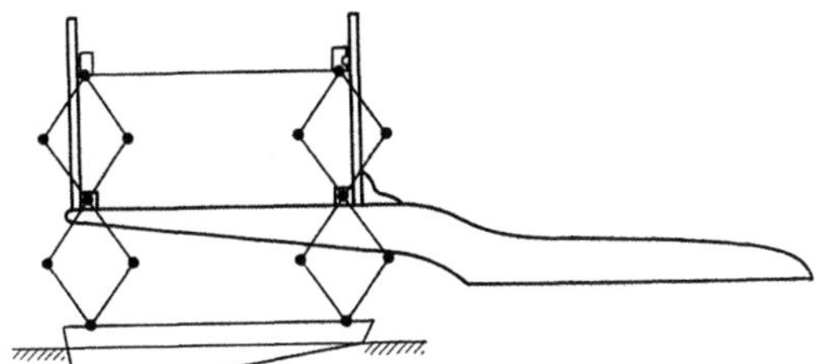

Fig. 65. Mounting for model tests with fixed incidence.

In the case where unloading should be proportional to the square of the speed the special regulation for each run may be avoided by utilizing the procedure suggested by Sottorf and employed in the Hamburg experimental canal[1]. This procedure is based on the use of a resisting body studied by Kempf, and giving at the same time good stability under tow, and a resistance proportional to the square of the speed. This body has schematically the form of a cone with circular base, a portion of the cone being removed as indicated in the diagram Fig. 66. The tow line is fixed on the generatrix opposite x. The displacement of the arc S with reference to the axis O permits a change of ratio between the resistance of the body under tow and the force of sustentation on the model. By the installation of an electric control for the displacement of S by the model itself the effect of variations of incidence on the sustentation due to the wings may be represented. An advantage of this arrangement is the possibility of locating the body under tow at a considerable distance from the model itself in order to avoid all effects of interference.

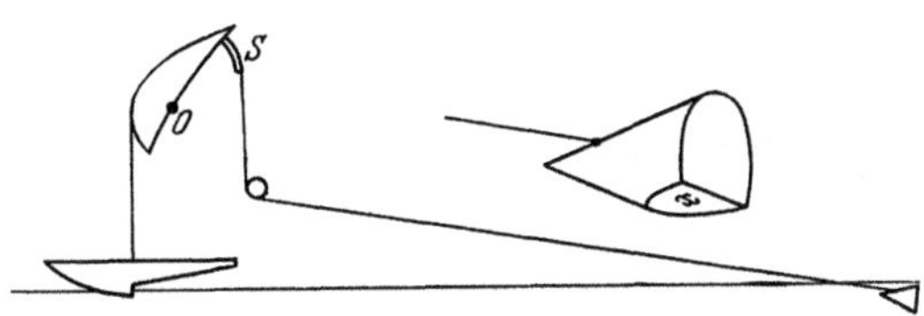

Fig. 66. Mounting for simulating hydrodynamic lift.

[1] Werft, Reederei, Hafen, 1929.

4. Tests with Free Incidence. The apparatus for the test of seaplanes with a representation of the effect due to a wing in the air by the use of an immersed wing[1] is composed of a double Roberval balance carried by the platform *P*, Fig. 67. The platform is movable vertically through the action of an electric motor driving four vertical screws *V*. The platform which carries the observer and the dynamometer carriage can thus be placed correctly relative to the water. And in this manner the vertical location of the model may be properly adjusted. The supporting arms of the double balance *A*, as well as their connection with the platform, are represented in the diagram. The forward balance *B* is connected by an articulation to the model *H*. This point of connection α may be varied from one test to another, but remains the same throughout a series of runs.

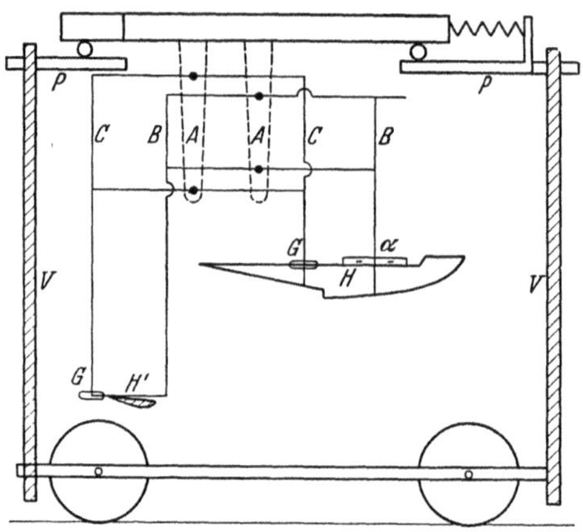

Fig. 67. Mounting for model tests with free incidence.

The forward balance is likewise connected to the forward part of the wing *H'*. The only difference for the after balance is that the connection to the model *H* or to the wing *H'* is made by means of a slide *G*. The model *H* is placed in its normal position. The model *H'* is, on the other hand, mounted reversed.

It is thus seen that when the platform is in motion relative to the water, the wing *H'* will be subject to a force directed downward, which, by the effect of the articulated parallelograms, will subject the model *H* to a force directed upward similar to the aerial sustentation due to the wing of an actual seaplane.

The wing *H'* naturally is not made to the same scale as the model *H*, since it is acted upon by water, while the model of the actual plane is in the air. We should therefore expect the wing *H'* to be made to a scale about 800 times smaller than the model *H*, and this indeed would be necessary if the wing *H'* were geometrically similar to the wing of the plane itself. Such a model wing would be practically very difficult to realize, and in consequence use is made of a wing very much reduced in span, and but slightly in chord. With the mounting, as indicated, the model *H* is free to incline, but at the same time it gives a like inclination to the wing *H'*.

[1] Commission Permanente d'Etudes aéronautiques, February 1927.

The apparatus and model being adjusted, a test consists in a series or runs carried out at uniformly spaced speeds and repeated three times under the following conditions.

In the first series everything is mounted as in the diagram. For each speed the resistance is measured and the position and attitude of the model are noted.

In the second series the model *H* is removed and instead of leaving the two balances free they are fixed to the supports *A* in the position taken for the preceding tests at each value of the speed. This second series gives, therefore, the resistance due to wing *H'* and supporting bars.

In the third series the wing is removed and runs are made giving the resistance due to suspension alone. These data will give then the resistance to which the model is subjected under the various conditions through which it passes between rest and the speed of take off.

Reference may again be made to Fig. 47 in which the results of tests with such a form of balance are given.

The runs differ only by the relative position of wing and hull indicated experimentally by the changing of the point α from one run to another. If the position of the wing is far forward, the model decreases its displacement, but the stern drags in the water. The wing being progressively moved aft, the stern drags less and less in the water, then the model rests on both the stern and the step, and then on the step only. Finally, if the wing is placed too far aft, the model begins to porpoise. If it is desired to study the same model with the same wing in the same longitudinal position, but with incidence variable relative to the hull, it is only necessary to change the length of the vertical connecting links. To this end the after links *C* are formed of two pieces permitting suitable adjustment in overall length.

5. Channel for Flowing Water. The apparatus used for measures in a current of water is formed of a balance with oscillating movement and an articulated parallelogram. The scale pan *I* is connected to the parallelogram. The scale pan *II* is connected to the oscillating bar *B*, which, with the suspension link and the model, is free to turn about the point Ω, Fig. 68.

With reference to Ω the model may assume four positions, 1, 2, 3, 4, designated respectively as left short, right short, left long, and right long. Four measures are made, one for each of these positions, and in each one the frame *C* supporting the apparatus is displaced in such manner that the model, for the four measures, shall be in the same position relative to the water.

For a given measurement a weight is placed in scale pan *I* equal to the vertical component *V*, which it is desired to obtain. Equilibrium is then reestablished by changing a part of the weight from pan I to pan II. The measure in each of the positions is the value of the weight contained in the pan II at the moment of equilibrium. Before each measurement the apparatus is brought into equilibrium out of the water. There is then no other compensating measure to be made. If w_1, w_2, w_3, w_4 are the measurements made, H and V the values of the horizontal and vertical reactions of the water, d the distance of the resultant R with regard to a reference point on the model, λ_1, λ_2 the two lengths of the vertical bars ($\lambda_2 > \lambda_1$), β the displacement of the vertical bar between the measurements right and left, π the length of the lever arm of the scale, it is seen that

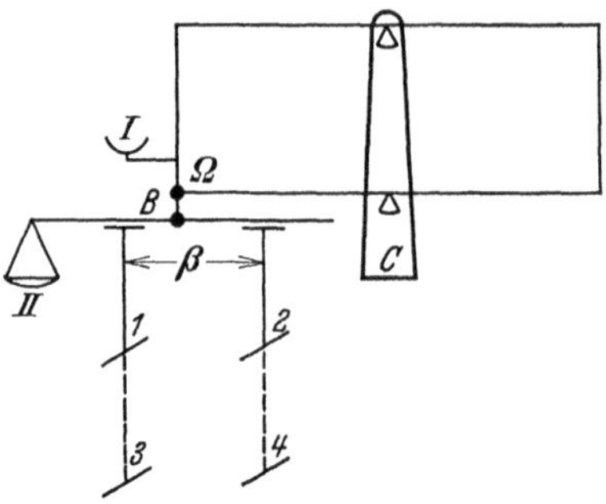

Fig. 68.

$$\pi (w_2 - w_1) = V\beta$$
$$\pi (w_4 - w_3) = V\beta$$
$$\pi (w_3 - w_1) = H (\lambda_2 - \lambda_1)$$
$$\pi (w_4 - w_2) = H (\lambda_2 - \lambda_1)$$

There are various verifications in these measures. Thus we should find $w_2 - w_1 = w_4 - w_3$ or otherwise, $w_3 - w_1 = w_4 - w_2$. Furthermore the two values found for V should be equal to the weight placed initially on the scale I.

The relations written above give immediately V and H and in consequence $R = \sqrt{V^2 + H^2}$ and the inclination $\alpha = tan^{-1} H/V$ of the resultant to the horizontal.

Since anyone of these measures will give the moment of this resultant, we have four determinations of the moment Rd and hence four determinations of d. The part of these measurements requiring the longest time being the balancing of the apparatus, it is preferable to make successively all the measurements No. 1 for a series of increasing values of V at the constant inclination of the model (or otherwise with V constant and i increasing), and then similarly all the measurements No. 2, No. 3, and No. 4. In this manner, with only four adjustments of the balance, the entire series of measurements are made permitting the complete tracing of curves of H, V, and moment for V variable and i constant (or otherwise i constant and V variable).

CHAPTER X

STRENGTH OF SEAPLANE HULL

For the same power and useful load the weight of a seaplane is greater than for a land machine. This increase is due chiefly to the hull structure which not only is much heavier than the normal landing gear, but in addition increases the resistance to movement and consequently the weight of power plant, all without any gain in sustentation.

This difference between the seaplane and the land machine was especially marked with the smaller sizes employed in the earlier years. For this there are several reasons: the dimensions of the human body remaining the same, the space and weight required for personnel is proportionally more important on a small machine. Furthermore the requirement of transverse stability for the seaplane afloat necessitated the addition of wing floats, thus accentuating still further the prejudicial effects due to the hull structure; and finally the necessity of a strength of construction suited to withstand the effects of the waves, leads, as in the case of ordinary boats, to a proportional weight of hull the greater as the structure is smaller.

In proportion as these constructions increase in overall dimension, these defects will become less and less important and would disappear if the dimensions became sufficient to permit of locating within the thickness of the wings all passenger accomodations. There would then remain to the disadvantage of the seaplane only the excess of weight of the hull in relation to the landing gear. It is furthermore by no means certain that for very large constructions, the need of landing on the sea will neccesitate weight greater than that of a corresponding landing gear for a land machine, and it is even possible that a large augmentation of useful load in relation to total load may be realized only by a seaplane design, for which the difficulties due to landing on the ground do not exist.

Under present conditions, it is, in any case, necessary to reduce to the possible minimum the weight of hull structure, always with due regard to the conditions which develop at take-off from and return to the surface of the water. Studies of the local surface loading under these conditions have been made both by model and full scale[1].

The conclusions drawn from these two studies are similar. The highest pressures develop in the vicinity of the step, immediately forward. These maximum pressures decrease rapidly with approach to the bow, but less rapidly along the line of the keel itself. The maximum pressure on the step is of the order of 400 g./cm.^{2}. Similar differences are found at the after step. Absolute values depend on the speed of contact with the water and on the fining out of the form.

These hydrodynamic pressures result in forces which must be balanced by the strength of the different parts of the structure. The role of the hull, considered as a structure calculated to resist these forces, is then

[1] BAKER, G. S., Flying Boats—the Form and Dimensions of their Hull, Engineering, Vol. 109, pp. 323—327, London, March 5, 1920.

THOMPSON, F. L., Water Pressure Distribution on a Seaplane Float, U.S. N.A.C.A. Technical Report No. 290, 1928.

to transmit and diffuse throughout the structure all local loads, operating thus by its mass and by the work required for its deformation.

The influence of an increase in size in a seaplane hull upon its weight, is difficult to determine at the present time, chiefly because the rules of design and factors of safety imposed on constructors are not uniform.

In some cases a load proportional to the weight of the plane is distributed over the bottom; in others, the assumption is made of a load constant per unit of area. By the first rule, an increase in the width will increase the distances over which these loads must be transported but will not augment the loads themselves. With the second rule, both the loads and the distances will be increased.

CHAPTER XI

GAPS BETWEEN THEORY AND PRACTICE

1. Introductory. It certainly would be very advantageous if we could have available a method permitting the use of the general equations of rational mechanics, to predict the results to be expected in service for any form of seaplane hull as defined by its plans. This objective has indeed attracted serious attention and it is useful to examine the degree of approach thus far realized and whether it may be possible to approach still more closely to this ideal.

Let us assume, for the moment, that we have a solution of the following problem—given a form of seaplane hull with a free water surface, the position of the hull being defined with reference to the plane of the undisturbed water level. To this body is given a motion of uniform rectilinear translation. It is required to determine the forces to which the surface of the hull is subject from the liquid.

The solution of this problem would be very useful for ship construction in normal forms, always assuming that the time required for the various calculations would not exceed a reasonable duration. It is certain that such a solution would permit a deep penetration into the mechanism of the production of fluid resistance and would furnish important indications regarding the search for forms of least resistance. But it is likewise certain that this problem has not as yet been solved for a ship form given *a priori*, and that the solutions obtained in special cases (and which have given occasion for remarkable experimental verification) involve numerical calculations of such a length that their practical application cannot be considered by the practitioner and, what is still more important, do not furnish, regarding the means for the improvement of a form, indications superior to those already furnished by experience direct.

Since, then, for ordinary ship forms such theoretical guides as we have at our disposal do not permit, for a ship in motion, a determination of the changes of position relative to the undisturbed water level, we should not expect, in the case of a seaplane hull for which the changes of incidence are of such great importance, to find in the present theories of rational mechanics, any adequate reply to the questions presented in practice. The difficulty is, in fact, still larger in the latter case and for a number of reasons. First, it is no longer a matter of continuous form, but rather of a form discontinuous longitudinally (step) and discontinuous transversely (abrupt angles). Furthermore, it is no longer a problem of a body subject to its weight and a single propulsive thrust and reacting fluid resistance, but of a body subject to external forces depending on the speed both in direction and magnitude and also on its attitude relative to the horizontal. Furthermore, it is not sure *a priori*, that a solution even if obtained, for uniform translation would be applicable to accelerated movement, either with or without pitching.

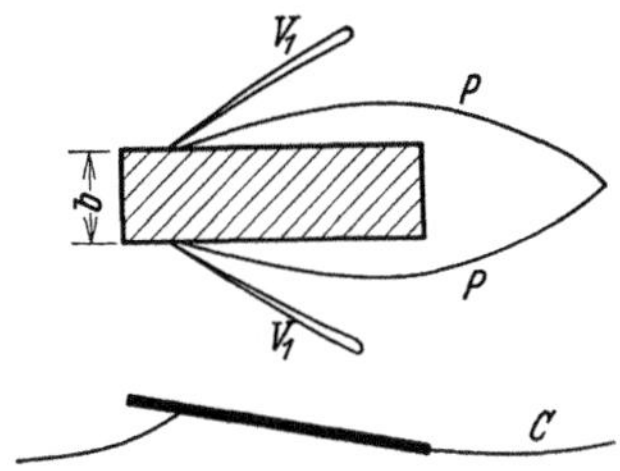

Fig. 69.

The hydrodynamic problem of the seaplane, in all its complexity, must then be considered as lying beyond our present means for theoretical treatment and it becomes then necessary to ask whether simplified solutions will be susceptible of useful applications.

The first simplification to be considered is that of considering uniform translation only. The most important phase in separation and take-off is that of maximum resistance, during which the horizontal acceleration is certainly very small. Such a simplification would, therefore, be acceptable.

A second simplification would consist in the substitution of a two-dimensional problem for one of three dimensions. This would be realized by assuming the hull defined by its vertical axial longitudinal section and generated by an indefinite right line perpendicular to this plane of section. In order to justify this simplification the following reasons might be adduced. The wave resistance is due to the formation of changes of level in the free liquid surface; we see first the diverging system V, Fig. 69, and then a hollow with walls P closing itself behind the body and from the bottom rising according to the form of a cylinder C with transverse horizontal generatrix. If then the principal part of the resistance is due to the formation of this wave form C, we shall have an approach possibly sufficient through a study of this formation as a problem in two dimensions.

Experiment furnishes an answer with reasonable assurance, that this is not the case, and that the part of the resistance due to the transverse wave form is not the principal term in the total value. This conclusion is based on the analysis made by Sottorf of his own experiments, permitting, by the aid of certain hypotheses, the separation of the total into three parts, termed by him, the part due to friction, the induced resistance, and the wave-making resistance. In this analysis, the induced resistance corresponds to the formation of the transverse wave and the wave-making resistance to what we have here called resistance due to the divergent wave system.

2. Sottorf's Analysis[1]. The three elements of the resistance are as noted above. The resistance due to friction having been calculated, subtraction from the total, as furnished by experiment, gives the sum of the induced and wave-making resistances. In order to separate these two, the former is calculated according to the following hypothesis.

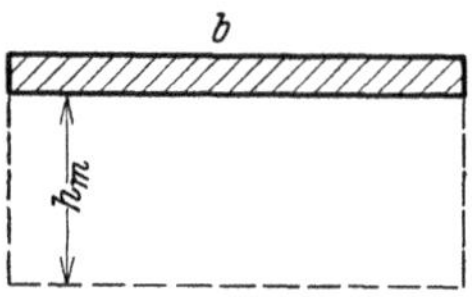

Fig. 70.

It is assumed that the mass of liquid acted upon is a prism $b \times h_m$ where b is the breadth of the body in contact with the water and h_m is unknown, Fig. 70. The mass acted on per second is then $\varrho b h_m V$, ϱ being the density and V the speed. The sustentation is then equal to the time rate of change of momentum in the vertical direction. This gives

$$L = \varrho\, b\, h_m V \cdot V_m \sin \alpha$$

$V_m \sin \alpha$ being the mean vertical component of the velocity of the water in contact with the body surface. This gives:

$$h_m = \frac{L}{\varrho\, b\, V \cdot V_m \sin \alpha} \tag{2.1}$$

Each experiment gives then a value for h_m. For example, with a value of $L = 18$ kgs. $= 18{,}000$ gms., $V = 6$ m./sec. $= 600$ cm./sec., $b = 30$ cm. and $\varrho = 1/981$ (whence $L/b\varrho V = 981$) the following results were found.

Test No.	Inclination of Under Water Surface	Horizontal Length cm.	V_m cm./s.	$V_m \sin \alpha$ cm./s.	h_m cm.
1	4^0	80.0	592	41.3	23.8
2	6^0	37.0	581	60.7	16.2
3	8^0	17.5	557	77.5	12.6

The resulting values of h_m are then as given in the last column of the table.

The kinetic energy expended per second is equal to the induced resistance R_i multiplied by the mean speed V_m. This energy again is equal to half the product of the mass acted on per second by the square of the speed $V_m \sin \alpha$.

Equating these two values we have

$$R_i V = \frac{1}{2}\, \varrho\, h_m\, b\, V\, (V_m \sin \alpha)^2$$

[1] Werft, Reederei, Hafen 1929.

In this equation if we put the value of h_m in (2.1) we have

$$R_i V = \frac{1}{2} L V_m \sin \alpha$$

Substituting the proper numerical values, values of R_i are found as in the following table.

Test No.	Resistance Due to Friction	Induced Resistance	Wave-Making Resistance	Total
1	1410	619	811	2840
2	708	912	1026	2646
3	352	1167	1374	2893

In a general way the sum of the induced and wave-making resistance equals $L \tan \alpha$. Furthermore, the induced resistance is nearly equal to $(L/2) \sin \alpha$; and V_m/V differs but little from unity. Hence the induced resistance will approximately be half the sum of induced resistance plus the wave-making. That is, the induced resistance would be approximately equal to the wave-making, whatever the ratio of width to length.

It should be noted that in this calculation no account is taken of the lateral deviation of the liquid filaments in contact with the body. This in effect, enters only indirectly in connection with the measurement of the mean velocity.

Comparing these conditions of resistance with those for airplane wings, Sottorf notes that for the conditions of minimum resistance, the resistance of a hydro-gliding plane is divided into three parts of the same order of magnitude and that the part due to friction is much greater than for the aerial wing. The question remains if a part of the induced resistance is not due to the formation of the divergent wave system. An opinion on this point must be based upon an examination of the system of waves formed by a gliding plane.

Although it is recognized that the two-dimensional problem will surely not serve by itself to give a satisfactory solution for the problem in three dimensions, it would be premature to conclude that the two-dimensional solution is without interest. It might be, for example, that the part due to the divergent wave system is constant, in which case some advantage could be drawn from the two-dimensional solution.

To examine this question, let us take three floats with widths 10 cm., 20 cm., 30 cm., with bottom inclined at an angle of 8^0 and moving at a speed of 5 m./sec. The floats are run at an incidence as chosen, and the weights are as 1, 2, 3. We shall then find resistances which are sensibly in arithmetical progression—that is, on a straight line when plotted as in Fig. 71. This line projected back to $b = 0$ gives an ordinate as shown which may be considered as the resistance due to the formation of the divergent wave system independent of the width b of the body. On this assumption, the remainder is due to friction and to the resistance due to the formation of the transverse wave system. The resistance due to friction being nearly proportional to b, the resistance due to the transverse wave system should also be proportional to b.

But it might otherwise be assumed that the resistance due to the divergent system comprises two terms, one a constant and one proportional to the width. To decide this question very exact examination would be required of the divergent system for a series of forms differing only in width and to the present time, no such examination has been made.

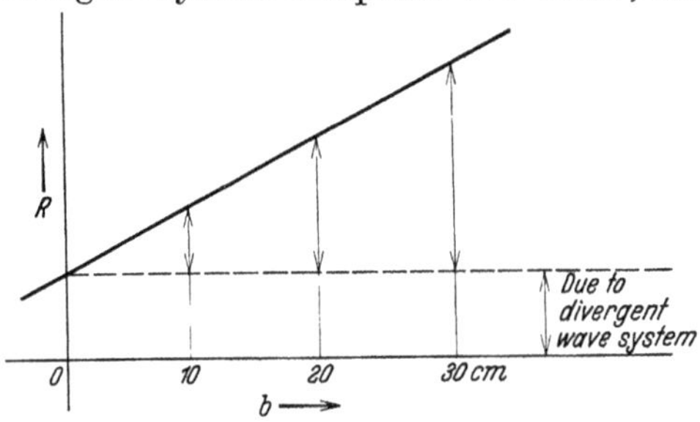

Fig. 71.

In the mean time, it may be agreed that a solution of the two-dimensional problem is very much to be desired, a solution which does not seem to lie beyond the possibilities of present mathematical analysis; and further to be desired likewise, is the solution for a case of three dimensions with a gliding plane.

With the solution of these two problems, there would be furnished a sound foundation for the discussion of experimental results. The solution of problems only slightly more complicated, as that of the comparison of different forms of transverse section, problems, the solution of which would be of immediate practical use, seem at present to be beyond the reach of theoretical treatment.

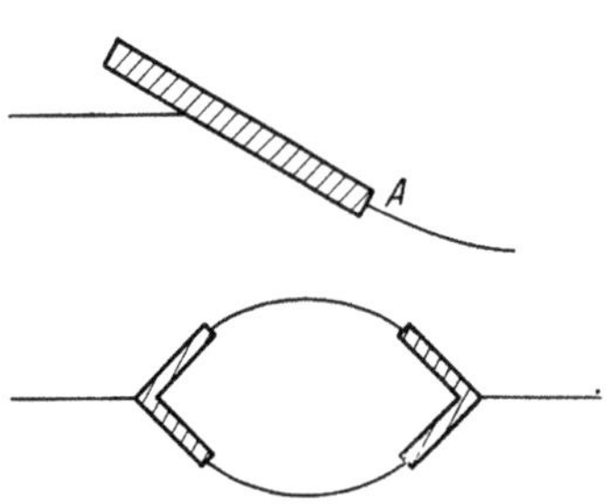

Fig. 72.

It is the same for the problems relating not alone to the forward gliding plane, but also to the part of the hull located aft of the step. Nevertheless, for these cases, the solution of the first problem noted, would permit, already, by a knowledge of the form of the free surface aft of the step, to determine *a priori*, the incidence to be given to the region aft which is to rest upon this free surface during the period preceding that of gliding on the step alone.

As regards the first of these fundamental problems, that of two dimensions, it may be approached in various ways—by seeking a solution in a similar manner to that employed by Lord Kelvin, especially in his paper "On the Front and Rear of a Free Procession"[1], or again by the method given by Lamb[2] for a circular obstacle deeply immersed.

This method of approach seems likely to give a useful result only in the case of a body of small dimensions.

Another method of approach to this problem would be analogous to that employed in the study of diversion dams, only introducing the gravity field for the purpose of determining, by Bernoulli's equation, the local velocity at the point of issue A, Fig. 72. The remainder of the problem would consist in finding a suitable hydrodynamic field, either by a suitable distribution of sources and sinks, or by assimilation to a problem with a surface of discontinuity. This general method of attack has the advantage of permitting the consideration of the case of an obstacle of finite dimensions, but would require, the same as for the case of the dam, an experimental verification in order to justify the hypotheses employed.

CONCLUSION

In the preceding pages, we have held especially in view the possibilities of approach between hydrodynamic theory and the phenomena observed with a seaplane during the period of contact with the water. In consequence, purely technical considerations have been entirely passed by and no note has been made of procedures permitting the avoidance of the maximum resistances due to the water, such as launching by catapult, independent towage, etc. All matters relating to purely structural problems, to economic considerations and to the future of the seaplane have also been entirely passed by.

It is nearly the same with regard to studies relative to the analysis of the results obtained from full scale experiments on actual seaplanes, from studies made on actual hull structures without the wing equipment, and from studies made on models. In this group of studies, theory alone is an insufficient guide and technical procedure takes therefrom only those general principles well known under the name of "rules of kinematic similitude".

At the present time, the principal difficulty in connecting theory with technical procedure is certainly the lack of results furnished by theory alone. The time when theory will be far enough advanced to permit the prediction of the forces to which a hull would be subjected in varying movement does not seem near. On the other hand, we may hope that theory will succeed in giving such results for the case of uniform horizontal translation.

[1] Kelvin, Lord (Sir William Thompson), "Mathematical and Physical Papers", Vol. IV, p. 307, Cambridge 1882—1911.

[2] Lamb, Horace, Ann. di Matematica, 1913.

Thus far, in order to deal with these forces and with the general behavoir of the plane, we are compelled to use diagrams and procedures in which hydrodynamic and hydrostatic forces are mingled and which are introduced by assimilation with the principle of Archimedes. It is part of the role of theory to show how, when the speed varies, the principle of Archimedes (strictly valid only for $V = 0$) may be brought, as a limiting case into the general equations for laws of fluid motion. At the present time, the vertical force on an immersed body in movement is given by theory only for the case where the immersion is very great[1].

Until theory has overcome this obstacle, we shall, in this domain as in that of ship design, be reduced to the necessity of direct observation, either full scale or on models, and to methods of analysis and interpretation of these measurements based upon subdivisions of the total resistance which find no justification in theory.

In procedures of this character, however, theory will have an important role, especially as a guide to experimental research. At the present time the best method of reducing the gap between theory and practice seems to lie in the direction of a study of models leading progressively from actual forms of seaplane hulls to thin plane surfaces inclined to the surface of the water. Thus, for example, a study of the hull itself, then the hull bounded by a horizontal cylindrical bottom and with vertical cylindrical sides, then the latter limited to the part forward of the step, and finally the thin plane alone. In the present work, several experimental results of this general character have been given. In this manner the most important steps in any practical problem are made evident; but general conclusions can only be reached after studies of this character have been applied to a great number of forms.

Questions of stability, so important for the seaplane, may be approached from two different points of view. We may attack the problem in its overall aspect, in which case theory should furnish a solution relative to non-permanent motion. In this direction, we have thus far no useful result. Or otherwise we may assume *a priori* that in irregular movement, the reactions from the water are the same as for permanent tangential movement; in other words, that a knowledge of the instantaneous movement of the solid will permit a determination of reactions from the water, or otherwise that the latter depend only on the instantaneous velocities.

In the case of a seaplane of which the longitudinal vertical plane remains fixed in space, each instantaneous state is defined by five parameters: two parameters of position (one giving the position of a point in the plane with reference to the undisturbed water level, and the other to fix the attitude) and then three parameters of instantaneous velocity (horizontal translation, vertical translation and rotation). By reasons

[1] Havelock, Proceedings Royal Society 1929.

of symmetry, the reactions from the water will, in this case, become reduced to a single force depending, then, on these five parameters.

In a similar manner as in the case of normal ship forms as dealt with in theoretical naval architecture, the consideration of states of equilibrium leads to a definition of metacenters. With the normal ship form and considering only longitudinal inclinations, there is a single metacenter, because the force of buoyancy is one of fixed magnitude and its location for varying inclination with constant displacement depends on one parameter only, the angle of inclination. Here, on the contrary, giving proper significance to the hydrodynamic forces in play, there will be five metacenters corresponding each to the change in a single one of these five parameters, the other four remaining constant. Furthermore, to each metacenter there will correspond two parameters of distribution—one for changes of intensity, the other for changes of direction. The conditions of a small change in state will therefore require for specification fifteen quantities, these being the partial derivatives of F_x, F_y and M with regard to these five parameter variables.

The facility of application of the metacentric method in the case of a floating body arises (other than for the reasons already noted) from the fact that the external force is always perpendicular to the water-plane and turns exactly with the inclination. There is, then, no need of considering a parameter for varying intensity, nor for varying inclination, and the knowledge of the single position of the metacenter defines the location of the known value of the buoyant force for angular positions near any given initial position of the body.

The application of the metacentric method to the hull of a seaplane, on the contrary, presents a very much more complicated problem. There are at present available only a few practical examples insufficient as a foundation for theory.

Happily, however, notable simplifications in practice are possible. Thus at the start the case may be restricted to a dependence of the force on three parameters—horizontal speed, vertical component force, and inclination. We then have three metacenters corresponding respectively to changes in speed, in load and in inclination. Experiment will then show that often certain metacenters are far separated or that the parameter of distribution of the intensities is of secondary importance. To forward these questions, the study of metacenters by model experiment is the only way which seems at present possible.

It is to be foreseen, therefore, that still for a long time, the study of the marine parts of a seaplane must be based principally on model tests the same as in the case of normal ship design. For the latter, in fact, notwithstanding a condition of theory relatively much farther advanced, it is still by way of experiment on reduced scale models, that projects are carried to realization.

APPENDIX

At the third International Congress of Applied Mechanics (Stockholm 1930) two papers were presented, bearing upon the subject of the Hydrodynamics of Boats and Floats and of which brief abstracts may be presented as follows:

The first paper, by Professor Herbert Wagner of Charlottenburg, relates to the theoretical computation of the loads supported by a V shaped bottom under shock contact with the water. In this paper the problem is considered under two dimensions in a vertical transverse plane, neglecting the effect of weight. It is then necessary to find a correct solution of nonpermanent flow.

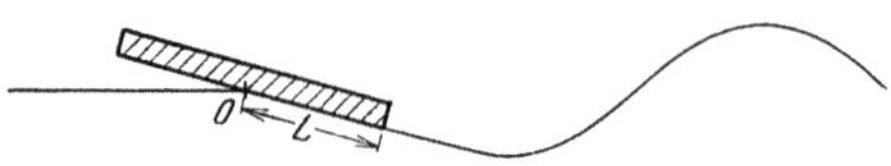

Fig. 73.

The author considers first an angle of the V very small, and computes the force and the distribution of pressures neglecting terms of the same order of magnitude as the angle. The energy expended in the shock is found in the energy of the water thrown laterally with high velocity. There results local pressures of high intensity on a restricted zone, and these pressures are not sensibly decreased by the elasticity of the structure of the hull.

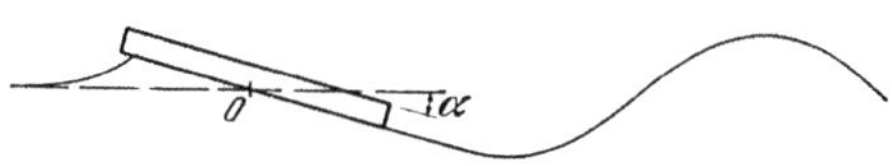

Fig. 74.

The author then considers a finite angle of the V shaped form and calculates for various cases the distribution of the pressure. Although these numerical computations are too long for practical application, the paper is to be especially noted not only by reason of the method employed, but also for the hope which it gives for the ultimate treatment of a problem which thus far has shown itself beyond the reach of theoretical methods.

The second paper, by Pavlenko of Leningrad deals with the theoretical computation of the loads supported by a plane surface in hydroplaning. The problem is considered again in two dimensions in the vertical longitudinal plane, and treated by the method proposed by Lord Kelvin and Lamb[1].

Let us place the origin of coordinates at the point O where the plane of quiet water at infinity meets the plane. The author then arrives, for the free surface, at a half plane passing through O forward of the obstacle, and a sinusoidal wave as it leaves the obstacle (see Fig. 73). He considers furthermore the development L of the plane in the direction of movement as small. It is very doubtful with these assumptions, whether the phenomena can be correctly represented, because no account is taken of the super-elevation of the water forward of the obstacle

[1] LAMB, H., "Hydrodynamics", 4th ed., p. 387, Cambridge, 1924.

(see Fig. 74). The conclusion of the paper gives an expression of the resistance for high velocities under the form

$$R = \frac{\varrho\, V^4\, tan^2\, \alpha}{4\, g}$$

It is readily seen that this expression might have been obtained by computing simply the height of the sinusoidal change of level satisfying the two conditions:

No. 1. At the altitude zero having the inclination α to the plane of the horizon.

No. 2. Having a speed equal to that of the obstacle.

The height of the change of level being known, the energy contained in the wave, or otherwise the energy expended in its formation, may be deduced immediately, and from this the resistance to the movement.

The problem of the theoretical determination of the form of free surface before and behind the obstacle still remains therefore to be solved.

Bibliography.

P. 5 Flying Boat No. 86, Impact Test—(Experiments with Full Size Machines. Third Series), Br. A.R.C. R. and M. 926, 1924.

Comparison of Atalanta and Model; Seaworthiness and Fore and Aft Angle; By the Staffs of the Marine Aircraft Experimental Establishment, Felixstowe, and the William Froude National Tank, N.P.L., Br. A.R.C. R. and M. 1076, 1926.

Rohrback, Adolf, Recent Experiments with Large Seaplanes, U.S. N.A.C.A. Technical Memorandum No. 353, 1926. From Berichte und Abhandlungen der Wissenschaftlichen Gesellschaft für Luftfahrt, July 1925.

Bénoit, E., Le Tracé des carènes d'hydravions, L'Aérotechnique, Vol. 4, No. 42; also L'Aéronautique, Vol. 8, No. 85, Paris, June 1926; also U.S. N.A.C.A. Technical Memorandum No. 376, 1926.

Crowley, J. W. Jr., and Ronan, K. M., Characteristics of a Boat Type Seaplane During Take-Off, U.S. N.A.C.A. Technical Report No. 226, 1926.

— Characteristics of a Twin-Float Seaplane During Take-Off, U.S. N.A.C.A. Technical Report No. 242, 1926.

Herrmann, H., Schwimmer und Flugbootskörper, Berichte und Abhandlungen der Wissenschaftlichen Gesellschaft für Luftfahrt E. V. (WGL), Vol. 14, pp. 126—154, München and Berlin, December 1926.

Magaldi, Giulio, Cenni sul problema dei grandi idrovolanti, L'Aerotecnica, Giornale ed Atti dell'Associazione Italiana di Aerotecnica, Numero straordinario, pp. 209—245, Pisa 1926.

Richardson, Holden Chester, Naval Development of Floats for Aircraft Engineering, Vol. 122, No. 3177, pp. 705—708, London, Dec. 3, 1926.

— The Trend of Flying Boat Development, Journal of the American Society of Naval Engineers, Vol. 38, No. 2, pp. 231—253, Washington, D.C., May 1926.

Eula, Antonio, Studio sistematica alla vasca di galleggianti a scafo per idrovolanti, Rendiconti Technici della Direzione General del Genio Aeronautico, Vol. 15, No. 3, pp. 33—52, Rome 1927.

Herrmann, H., Seaplane Floats and Hulls, U.S. N.A.C.A. Technical Memorandum No. 426, 1927 and 427, 1927.

Oswald, S. L., A Consideration of the Design of a Light Flying-Boat, Journal of the Royal Aeronautical Society, Vol. 31, No. 200, pp. 767—768, London, August 1927.

DORNIER, C., Notes on a Family of Similar Flying Boats, Journal of the Royal Aeronautical Society, Vol. 32, No. 216, pp. 981—1020, London, Dec. 1928.

GUIDONI, ALESSANDRO, Seaplanes. Fifteen Years of Naval Aviation, Journal of the Royal Aeronautical Society, Vol. 32, No. 205, pp. 25—64, London, January 1928.

GUSTOSA, CORRADO, Che cosa occorre per facilitare lo sviluppe degli idrovolanti plurimotori: 1º maggiore numero di idroscafi; 2º conche o bacini; 3º *redans* con assali e mozzi "Standard", IV Congresso Internazionale di Navigazione Aerea, Octo. 24—30, Rome, Vol. 3, pp. 268—279, Rome 1928.

RICHARDSON, HOLDEN CHESTER, Aircraft Float Design, The Ronald Press Company, New York 1928.

— The Reaction on a Float Bottom When Making Contact with Water at High Speeds, U.S. N.A.C.A. Technical Note. No. 288, 1928.

SALLE, M., Les essais de modèle en hydraviation, Bulletin de l'Association Technique Maritime et Aéronautique, No. 32, 1928.

SCHUBIGER, CLAUDE, L'aviation allemande: L'hydravion géant Dornier-Super-Wal, quadrimoteur, La Nature, No. 2782, pp. 312—313, Paris, April 1928.

THOMPSON, F. L., Water Pressure Distribution on a Seaplane Float, U.S. N.A.C.A. Technical Report No. 290, 1928.

COOMBES, L. P., and CROUCH, A. S., The Accelerations of a Fairey "Flycatcher" Seaplane During Acrobatic Manoeuvres, Br. A.R.C., R. and M. 1288, April 1929.

DORNIER, C., Mitteilungen über eine Familie ähnlicher Flugboote, Die Luftwacht, Vol. 4, pp. 169—180, Berlin, April 1929.

KÁRMÁN, TH. v., The Impact on Seaplane Floats During Landing, U.S. N.A.C.A. Technical Note No. 321, 1929.

LOWER, J. H., The Hydrodynamical Efficiency of Flying Boat Hulls and Seaplane Floats, Aircraft Engineer. suppl. Flight, Vol. 21, No. 26, pp. 522c—522f, London, June 27, 1929.

— The Static Longitudinal Stability of Seaplane Floats, Aircraft Engineer suppl. Flight, Vol. 21, No. 5, pp. 82a—82c, London, January 31, 1929.

P., C. W., Stability of Seaplanes Afloat, Aircraft Engineer suppl. Flight, Vol. 21, No. 33, pp. 446d—446e, London, May 30, 1929.

RICHARDSON, HOLDEN C., Design of a Large Flying Boat, Airway Age, Vol. 10, No. 1, pp. 28—33, New York, January 1929.

ROHRBACH, I. A. K., Flying Boat Design, Aero Digest, Vol. 15, No. 4, pp. 86—87, 274, 276, New York, October 1929.

THOMPSON, F. L., Water Pressure Distribution on a Twin-Float Seaplane, U.S. N.A.C.A. Technical Report No. 328, 1929.

CARTER, W. G., Flying Boat Development for Speed Research, The Aircraft Engineer, Flight Engineering Section, Suppl. to No. 1135, Vol. 22, No. 39, pp. (1068a—1068c) 65—67, London, Sept. 26, 1930.

COOMBES, L. P., The Testing of Seaplanes and Flying Boats, Journal of the Royal Aeronautical Society, Vol. 34, No. 230, pp. 190—209, London, February 1930.

COWLEY, W. L., and WARDEN, R., Tests of Models of High-Speed Seaplanes for the Schneider Trophy Contest of 1927. Section I., Br. A.R.C. R. and M. 1296 (Ae. 430), November 1927, London 1930.

— Tests on Quarter-Scale Models of High-Speed Seaplanes for the Schneider Trophy Contest of 1927. Section IV. Comparison with Full Scale and Conclusions, Br. A.R.C. R. and M. 1929 (Ae. 433), October 1929, London 1930.

GARNER, H. M., and COOMBES, L. P., The Determination of the Water Resistance of Seaplanes, Br. A.R.C. R. and M. 1289 (ae. 438), October 1929, London 1930.

NELSON, WILLIAM, Flying Boat Ranges as Influenced by Size, Airway Age, Vol. 11, No. 10, pp. 1302—1303, New York, October 1930.

PABST, WILHELM, Theorie des Landestoßes von Seeflugzeugen, Jahrbuch 1930 der Deutschen Versuchsanstalt für Luftfahrt, E. V., pp. 69—78, München and Berlin, 1930; also U.S. N.A.C.A. Technical Memorandum No. 580, 1930.

PARKINSON, H., Seaplane Take-Off Calculations, The Aircraft Engineer, Flight Engineering Section, Suppl. to No. 1148, Vol. 22, No. 52, pp. 1480—1481, London, December 26, 1930.

THOMPSON, F. L., Water Distribution on a Flying Boat Hull, U.S. N.A.C.A. Technical Report No. 333, 1930.

Manoeuvrability Tests of a Seaplane, Aeronautical Engineering, Suppl. to The Aeroplane, Vol. 39, No. 9, p. 500, London, August 27, 1930.

COLLINS, A. R., A Formula for the Buoyancy of the Wing Floats of Flying Boats and Single Float Seaplanes, Aircraft Engineer, Flight Engineering Section, Suppl. to No. 1178, Vol. 23, No. 30, London 1931.

DORNIER, C., Entwurf eines Dornier-Rennflugbootes, Luftschau, Vol. 4, No. 7, Berlin 1931.

GARNER, H. M., and COOMBES, L. P., The Determination of the Water Resistance of Seaplanes, Br. A.R.C. R. and M. 1289, 1931.

MITCHELL, R. J., Essais de maquettes d'hydravions au bassin des carènes, L'Aéronautique, Vol. 13, No. 142; L'Aérotechnique, Vol. 9, No. 99, 1931.

PABST, WILHELM, Über den Landestoß von Seeflugzeugen, Zeitschr. f. Flugt. u. Motorluftf. Vol. 22, No. 1, Berlin 1931, and Jahrbuch 1931 der Deutschen Versuchsanstalt für Luftfahrt, 1931; also U.S. N.A.C.A. Technical Memorandum No. 624, 1931.

RENNIE, J. D., The Development of the Long-Range Flying Boat, Flight, No. 1167, 1168, Vol. 23, Nos. 19, 20, London 1931.

SCHRÖDER, PAUL, Determination of Resistance and Trimming Moment of Planing Water Craft, U.S. N.A.C.A. Technical Memorandum No. 619, 1931.

— Ein Übergangsgesetz der Hydrodynamik und seine Anwendung bei der Untersuchung des Startes von Seeflugzeugen, Zeitschr. f. Flugtechnik u. Motorl. 22, 1, 1931; also U.S. N.A.C.A. Technical Memorandum No. 621, 1931.

SEEWALD, FRIEDRICH, Über Schwimmer und Schwimmerversuche, Zeitschr. f. Flugtechnik u. Motorl. 22, 9 1931.

TAUB, JOSEF, Beitrag zur Frage der Belastungsannahmen für den Landungsstoß von Seeflugzeugen, Zeitschr. f. Flugtechnik u. Motorl. 22, 14, 1931 and Jahrbuch 1931 der Deutschen Versuchsanstalt für Luftfahrt, 1931; also U.S. N.A.C.A. Technical Memorandum No. 643, 1931.

VERDUZIO, RODOLFO, Sollecitazioni alla partenza ed all'ammaramento negli idrovolanti, l'Aerotecnica, Vol. 11, No. 11, 1931; Abstract in English, pp. 1488—1489.

WAGNER, HERBERT, Über die Landung von Seeflugzeugen, Zeitschr. f. Flugtechnik u. Motorl. 22, 1, 1931; also U.S. N.A.C.A. Technical Memorandum No. 622, 1931.

WILLIAM, FROUDE NATIONAL TANK, Report on Experiments with Model of Seaplane Float Test with the S. 5 Duralumin Float. Model 807B. Br. A.R.C. R. and M. 1300, 1931.

COOMBES, L. P., and CUSHING, R. K., Full Scale Measurement of Lift and Drag of Large Seaplanes—Experiments on Blackburn "Iris", Br. A.R.C. R. and M. 1354, 1932.

GARNER, H. M., Porpoising Tests on a Model of a Flying Boat Hull, Br. A.R.C. R. and M. 1492, 1932.

KARTVELI, ALEXANDER, Stress Analysis of Flying Boat Hulls, Aviation Engineering, Vol. 6, No. 6, June 1932.

MAGALDI, GIULIO, Il raggio d'azione dell'idrovolante odierno, Revista Aeronautica, Vol. 8, No. 11, November 1932.

SCHRÖDER, P., Towing Tests of Models as an Aid in the Design of Seaplanes. U.S. N.A.C.A. Technical Memorandum No. 676, 1932.

SOTTORF, W., Über den Einfluß des Modellmaßstabes bei Untersuchung von Flugzeugschwimmern, Zeitschr. f. Flugtechnik u. Motorl. 23, 24, 1932.

VERDUZIO, RUDOLFO, Stresses Developed in Seaplanes While Taking Off and Landing, U.S. N.A.C.A. Technical Memorandum No. 677, 1932.

WIGLEY, W. G. S., Mathematical Investigation of Strength of Wooden Seaplane Hulls of the Linton-Hope Type of Construction, Br. A.R.C. R. and M. 1376, 1932.

BELL, JOE W., Tank Tests of Two Floats for High Speed Seaplanes, U.S. N.A.C.A. Technical Note No. 473, 1933.

BONIFACIO, FERDINANDO, L'evoluzione costruttiva dell'idrovolante, l'Aerotecnica, Vol. 13, No. 12, 1933.

COOMBES, L. P., and READ, R. H., The Effect of Various Types of Lateral Stabilizers on the Take-Off of a Flying Boat, Br. A.R.C. R. and M. 1411, 1933.

CROUCH, A. S., Full Scale Lift and Drag Curves of a Standard Seaplane, Br. A.R.C. R. and M. 1448, 1933.

DIEHL, WALTER S., The Estimation of Maximum Load Capacity of Seaplane and Flying Boats, U.S. N.A.C.A. Technical Report No. 453, 1933.

GARNER, H. M., Seaplane Research, Journal of the Royal Aeronautical Society, Vol. 37, No. 274, 1933.

GERARD, I. J., A Method of Testing the Strength of Aircraft Hulls, Br. A.R.C. R. and M. 1398, 1933.

HARDY, J. K., Experimental Comparison Between a Series of Turns of Different Diameter on a Gloster IV Seaplane, Br. A.R.C. R. and M. 1301, 1933.

KOHLER, M., Luftkräfte und Luftkraftmomente an einem Seeflugzeug auf dem Wasser, Zeitschr. f. Flugtechnik u. Motorl. 24, 16, 1933; also U.S. N.A.C.A. Technical Memorandum No. 728, 1933.

MEYER, M. L., Remarques sur les dimensions des flotteurs d'hydravions en catamaran, Bulletin de l'Association Technique Maritime et Aéronautique No. 37, 1933; also U.S. N.A.C.A. Technical Memorandum No. 719, 1933.

PERRING, W. G. A., and GLAUERT, H., Stability on the Water of a Seaplane in the Planing Condition, Br. A.R.C. R. and M. 1493, 1933.

RICHARDSON, HOLDEN C., Interpretation of Data of Model Basin Tests on Aircraft Floats, Aero Digest, Vol. 23, No. 2, New York 1933.

SHOEMAKER, JAMES M., A Complete Tank Test of a Model of a Flying Boat Hull—N.A.C.A. Model 16, U.S. N.A.C.A. Technical Note No. 471, 1933.

—, and PARKINSON, JOHN B., A Complete Tank Test of a Model of a Flying Boat Hull—N.A.C.A. Model 11, U.S. N.A.C.A. Technical Note No. 464, 1933.

SOTTORF, W., Scale Effect of Model in Seaplane-Float Investigations, U.S. N.A.C.A. Technical Memorandum No. 704, 1933.

TRUSCOTT, STARR, The N.A.C.A. Tank, a High-Speed Towing Basin for Testing Models of Seaplane Floats, U.S. N.A.C.A. Technical Report No. 470, 1933.

STRETTENSKY, On the Motion of a Glider on Deep Water, U.R.S.S. Academy 1933.

NELSON, WILLIAM, Seaplane Design, McGraw-Hill Book Co., New York 1934.

SHOEMAKER, JAMES M., Tank Tests of Flat and V-Bottom Planing Surfaces, U.S. N.A.C.A. Technical Note No. 509, 1934.

VON KÁRMÁN, Contributions to the Theory of Wave Resistances, Fourth International Congress for Applied Mechanics, Cambridge, 1934.

WAGNER, H.: Über das Gleiten von Körpern auf der Wasseroberfläche, Fourth International Congress for Applied Mechanics, Cambridge, 1934.

GRUSON, F., Sur la similitude dans les essais hydrodynamiques concernant l'hydroplanage, Journées de Mécanique, Lille, 1934.

BELL, JOE W., The Effect of Depth of Step on the Water Performance of a Flying-Boat Model, N.A.C.A. Model 11-C, U.S. N.A.C.A. Technical Note No. 535, 1935.

GOUGE, A., Flying-Boats and Their Possible Developments, Journal of the Royal Aeronautical Society, August 1935.

DIVISION T

AERODYNAMICS OF COOLING

By

H. L. Dryden,

Washington, D. C.

PREFACE

The purpose of this Division is to present a critical review of the fundamental aspects of the aerodynamics of cooling. Neither the theoretical nor the experimental aspects of this subject have been sufficiently developed to permit a logical presentation starting from a few basic assumptions and progressing to the mathematical solution of simple illustrative problems. Instead, it is found necessary to review the progress made along several independent lines of attack, to expose our ignorance of the essential characteristics of eddying flow, and to call attention to many implicit assumptions which are generally made to secure simple formulae. In addition there is a considerable body of theoretical knowledge not yet expressed in precise mathematical form, relating principally to the factors controlling the transition from laminar to eddying flow, which is of sufficient importance to record.

The point of view adopted is that of the physicist, rather than that of the engineer. The engineer will look in vain for design data and no attempt has been made to make exhaustive presentation of experimental results. The experimental data presented have been selected for their illustrative value in respect to basic phenomena.

The subject of heat transfer by forced convection has until recently been studied principally by investigators with training in heat flow, who have had little contact with recent developments in aerodynamics. This situation is rapidly being remedied. It is hoped that the present treatment, written from an aerodynamic background, may prove stimulating and helpful to those interested in the field of heat transfer.

CHAPTER I

FUNDAMENTAL IDEAS

1. Temperature, Heat Energy. Heat is a form of energy associated with irregular, random motions of molecules. Generally we are concerned only with changes in the heat energy of bodies or with the flow of heat from one body to another. The direction of the flow of heat energy between two bodies is determined by that physical property to which the name "temperature" has been given. Experiment has shown that the energy required to increase the temperature of a given body from a temperature t_1 to a higher temperature t_2 is independent of the source of the energy.

To give a number to temperature, some definite property of a definite body, which can be measured and which changes with the temperature, is selected, for example, the apparent volume of a quantity of mercury in a glass bulb with a fine stem. Then two standard thermal conditions are chosen, such as the freezing point and the boiling point of water at an atmospheric pressure of 76 cm. Hg. To these conditions are assigned arbitrarily selected values of the temperature, for example, 0 and 100 respectively on the Centigrade scale, 32 and 212 on the Fahrenheit scale. The selected physical property of the selected body is measured in the standard thermal conditions and in the condition for which the temperature is desired. The value of the temperature is determined by simple proportion. The relation between the Centigrade and Fahrenheit scales is obviously $t_F^0 = 32 + 1.8\, t_C^0$.

Heat energy can be measured in the ordinary mechanical units such as ergs or foot-pounds, but since heat effects are not as a rule produced by direct mechanical processes, it is not always convenient to do so. It is more common to choose as a unit the amount of energy required to raise the temperature of a unit mass of water by a known amount. Three such units in general use may be mentioned: the mean gram-calorie (cal.), one one-hundredth of the energy required to raise the temperature of one gram of water from 0^0 C to 100^0 C; the mean British Thermal Unit (B.T.U.), one one-hundred-and-eightieth of the energy required to raise the temperature of one pound of water from 32^0 F to 212^0 F; and the mean Centigrade Heat Unit (C.H.U.), one one-hundredth of the energy required to raise the temperature of one pound of water from 0^0 C to 100^0 C. The ratio of these units to the mechanical unit is known as the mechanical equivalent of heat. It has been determined by experiment with the results given in the following table of equivalents:

1 cal. $= 4.186 \times 10^7$ ergs $= 3.968 \times 10^{-3}$ B.T.U. $= 2.205 \times 10^{-3}$ C.H.U. $= 3.087$ foot-pounds.

1 B.T.U. $= 1054.8 \times 10^7$ ergs $= 252$ cal. $= 0.5556$ C.H.U. $= 778.7$ foot-pounds.

1 C.H.U. $= 1898.7 \times 10^7$ ergs $= 1.8$ B.T.U. $= 453.6$ cal. $= 1401.7$ foot-pounds.

The energy required to raise the temperature of a given mass of water one degree varies somewhat with the temperature, and the calorie has sometimes been defined for a particular 1^0 interval, say 15^0 C to 16^0 C. For engineering purposes this variation can be neglected. In view of the difficulty of accurately measuring the mechanical equivalent of heat in terms of any particular definition of the calorie, it has been suggested by some authorities that a value be arbitrarily selected for the mechanical equivalent and the calorie be defined as standing in this ratio to the mechanical unit. Thus the mean calorie would be defined as 4.186×10^7 ergs, the mean British Thermal Unit as 1054.8×10^7 ergs, etc.

The energy required to raise the temperature of a unit mass of any substance through 1^0 is called the specific heat. Obviously if either Calories, British Thermal Units, or Centigrade Heat Units are used in measuring the heat energy, the specific heat of water (average between freezing and boiling point) is unity; if mechanical units are used, the specific heat of water is equal numerically to the mechanical equivalent of heat. The specific heat is in general a function of the temperature, although for water and many other substances the variation is so small that it is usually neglected in engineering computations.

2. Transmission of Heat. Heat energy may be transferred from place to place by an actual motion of heated particles or portions of a fluid, a process termed convection. If the motion of the fluid is the result of differences in density resulting from temperature differences and the action of gravity, the process is known as natural convection. If the motion is produced by mechanical means, the process is that of forced convection, the only method of transmission to be discussed in any detail in the present Division.

Heat energy may also be transferred from one part of a body to another part of the same body, or from one body to another in physical contact with it, without appreciable motion of the particles of the body. This is the process of conduction. Even in natural and forced convection, the heat energy is transferred from the boundaries of solids to the fluid by this process. If we are dealing with the flow of heat energy in one direction, s, in a homogenous substance, the rate of flow of heat energy $dQ/d\theta$, θ being the time, is proportional to the temperature gradient dt/ds and to the area of cross section S at right angles to the direction of the heat flow.

Thus
$$\frac{dQ}{d\theta} = - k S \frac{dt}{ds} \tag{2.1}$$

the minus sign being introduced since the energy flows from a region of high to a region of low temperature. The coefficient k is known as the conductivity. It generally varies with the temperature, although in engineering computations it is usually regarded as a constant. For conduction in a steady state, $dQ/d\theta = Q/\theta =$ constant.

It is found that heat energy is also transferred from a hot body to a colder body even when there are no material bodies between. This process is called radiation and the energy is believed to be transferred by the same mechanism which conveys the energy of light or electromagnetic waves. The practical aspects of the transfer of heat by radiation are discussed in Chapter III of Heat Transmission by William H. McAdams (McGraw-Hill, 1933).

In nearly all actual cases of heat transmission, more than one process is involved. The separation of the total transfer into these component processes is often difficult, but only in this way can a logical basis of calculation be developed, since the laws governing the three processes differ considerably.

3. Laminar and Eddying Flow[1]. The fundamental aspects of the dynamics of fluids are treated at length in other volumes of this series. Nevertheless it seems advisable to review the fact that there exist in nature two radically different types of flow, commonly designated as laminar and turbulent. Whether one or the other type occurs depends among other things on the turbulence initially present in the on-coming fluid stream. It is necessary to distinguish between this "initial" turbulence and the "fully-developed" turbulent motion. For this reason the author prefers the use of a special term "eddying flow" as a synonym of fully-developed turbulent flow.

By laminar flow is meant a flow in layers, the transfer of momentum between layers being by molecular motions whose effect is integrated in the viscosity. The laminar flow satisfies the equations of Stokes for the flow of a viscous fluid. The laminar flow is not always a steady flow, fluctuations corresponding to a wavering of the laminae about a mean position often being observed. The components of the fluctuations of the speed at any point are not correlated and there is no resultant transfer of momentum by the fluctuations.

In eddying flow there is an additional transfer of momentum by the motion of small masses which is exhibited as rapid fluctuations of the velocity at any point. The components of the fluctuations show correlations. The mean motion does not satisfy the equation of Stokes.

For geometrically similar arrangements and a given initial turbulence, the transition from laminar to eddying flow is determined by the Reynolds number, which equals a reference speed times a reference length fixing the scale, divided by the kinematic viscosity. The Reynolds number at which transition begins is called the critical Reynolds number. Below the critical value, the flow is laminar. The critical Reynolds number is decreased if the initial turbulence is increased. There is in fact a functional relation between the critical Reynolds number and the initial turbulence.

The two radically different types of flow are found in practically all aerodynamic problems. In many cases, both types of flow are present in different parts of the field of flow, leading to complication in integrated

[1] See also Division G.

quantities such as the total friction or the total heat transfer. The theory of heat transfer will be treated separately for the laminar and the eddying flow.

4. Physical Properties of Air. For ready reference, there are listed below the specific heat, conductivity, viscosity, density, and kinematic viscosity of air at several temperatures. The values are taken directly or interpolated between those given in the International Critical Tables.

Specific Heat, c, of Air at Constant Pressure (1 atmosphere).

Temperature		ergs (gram)$^{-1}$ (^{0} C)$^{-1}$	cal. (gram)$^{-1}$ (^{0} C)$^{-1}$ or B.T.U. (lb)$^{-1}$ (^{0}F)$^{-1}$ or C.H.U. (lb)$^{-1}$ (^{0}C)$^{-1}$
^{0}C	^{0}F		
0	32	1.004×10^7	0.2398
100	212	1.006×10^7	.2403
200	392	1.010×10^7	.2413
400	752	1.017×10^7	.2430
600	1112	1.034×10^7	.2470

Conductivity of Air[1]*, k.*

Temperature		ergs (sec.)$^{-1}$ (cm.)$^{-2}$ [^{0} C (cm.)$^{-1}$]$^{-1}$	cal. (sec.)$^{-1}$ (cm.)$^{-2}$ [^{0} C (cm.)$^{-1}$]$^{-1}$	B.T.U. (sec.)$^{-1}$ (ft.)$^{-2}$ [^{0} F (ft.)$^{-1}$]$^{-1}$ or C.H.U. (sec.)$^{-1}$ (ft.)$^{-2}$ [^{0} C (ft.)$^{-1}$]$^{-1}$
^{0}C	^{0}F			
0	32	2230	5.33×10^{-5}	3.57×10^{-6}
50	122	2540	6.08×10^{-5}	4.07×10^{-6}
100	212	2850	6.82×10^{-5}	4.57×10^{-6}
150	302	3120	7.46×10^{-5}	5.00×10^{-6}
212.5	414.5	3460	8.26×10^{-5}	5.53×10^{-6}
430.6	807.1	5750	13.75×10^{-5}	9.21×10^{-6}
539	1002.2	8180	19.56×10^{-5}	13.09×10^{-6}

Viscosity of Air, μ.

Temperature		dyne (cm.)$^{-2}$ sec.	pound (ft.)$^{-2}$ sec.
^{0}C	^{0}F		
0	32	1.709×10^{-4}	3.569×10^{-7}
50	122	1.951×10^{-4}	4.075×10^{-7}
100	212	2.175×10^{-4}	4.543×10^{-7}
150	302	2.385×10^{-4}	4.981×10^{-7}
200	392	2.582×10^{-4}	5.393×10^{-7}
250	482	2.770×10^{-4}	5.785×10^{-7}
300	572	2.946×10^{-4}	6.153×10^{-7}
350	662	3.113×10^{-4}	6.502×10^{-7}
400	752	3.277×10^{-4}	6.844×10^{-7}
450	842	3.433×10^{-4}	7.170×10^{-7}
500	932	3.583×10^{-4}	7.483×10^{-7}

[1] According to I.C.T. the accuracy of the experimental values is probably not better than 7 per cent.

Density of Air, ϱ, at 760 mm. Hg. Pressure.

Temperature °C	Temperature °F	grams (cm.)$^{-3}$	pounds (sec.)^2 (ft.)$^{-4}$
0	32	1.293×10^{-3}	2.509×10^{-3}
50	122	1.093×10^{-3}	2.121×10^{-3}
100	212	$.946 \times 10^{-3}$	1.836×10^{-3}
150	302	$.834 \times 10^{-3}$	1.619×10^{-3}
200	392	$.746 \times 10^{-3}$	1.448×10^{-3}
250	482	$.675 \times 10^{-3}$	1.310×10^{-3}
300	572	$.616 \times 10^{-3}$	1.195×10^{-3}
350	662	$.567 \times 10^{-3}$	1.099×10^{-3}
400	752	$.525 \times 10^{-3}$	1.018×10^{-3}
450	842	$.488 \times 10^{-3}$	$.947 \times 10^{-3}$
500	932	$.457 \times 10^{-3}$	$.886 \times 10^{-3}$

Kinematic Viscosity, $\nu = \mu/\varrho$, of Air at 760 mm. Hg. Pressure.

Temperature °C	Temperature °F	cm.^{2}/sec.	ft.^{2}/sec.	Temperature °C	Temperature °F	cm.^{2}/sec.	ft.^{2}/sec.
0	32	0.1322	1.423×10^{-4}	300	572	.4784	5.149×10^{-4}
50	122	.1785	1.921×10^{-4}	350	662	.5496	5.916×10^{-4}
100	212	.2298	2.474×10^{-4}	400	752	.6246	6.723×10^{-4}
150	302	.2859	3.077×10^{-4}	450	842	.7034	7.571×10^{-4}
200	392	.3460	3.724×10^{-4}	500	932	.7846	8.446×10^{-4}
250	482	.4103	4.416×10^{-4}				

The Dimensionless Ratio $\sigma = c\mu/k$ at 760 mm. Hg. Pressure[1].

°C	°F	$\sigma = c\,\mu/k$	°C	°F	$\sigma = c\,\mu/k$
0	32	0.769	300	572	.719
100	212	.760	400	752	.621
200	392	.767	500	932	.521

CHAPTER II

THEORY OF HEAT TRANSFER IN LAMINAR FLOW

1. General Problem. Let us imagine a solid A (Fig. 1), with boundary surface B, which contains a steady source of heat and which is placed in an air stream of speed V. The speed V is assumed to be constant both in magnitude and direction at a great distance from the body. When equilibrium is reached, there will be a steady flow of heat energy by conduction through the solid to the boundary B, thence by conduction to the air close to the boundary, and finally by convection to the air at remote distances. In general the temperature and the heat flow will not be uniform along the boundary B. The final state of affairs

[1] The large decrease at 400 and 500° C is in all probability due to inaccurate values of the conductivity.

is fully determinate but depends on the thermal properties of the solid as well as on its external shape.

If now one attempts to divide the problem into two parts, the first dealing with the flow of heat within the solid A and the second with the flow of heat in the fluid, the result is complication. For the temperature gradients and the heat transfer within the solid depend on the rate of loss of heat at the boundary B and this is dependent on the processes occurring in the fluid. Likewise the flow of the fluid and the heat transfer within it depend on the boundary conditions, namely, the distribution of temperature over B which is not known until the first problem is solved.

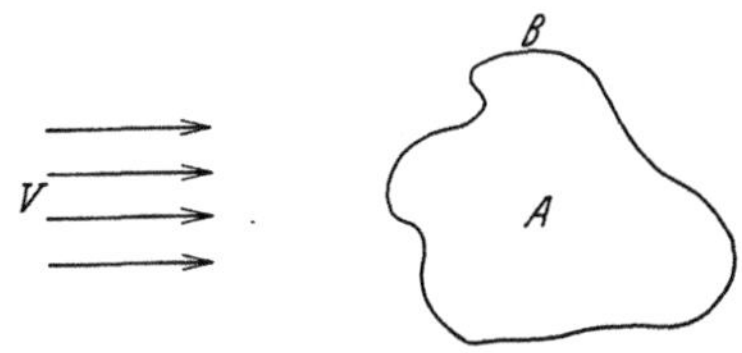

Fig. 1. General problem of forced convection.

The difficulty is analogous to that encountered in the simpler problem of determining the speed of rotation when a fan is placed on the shaft of a particular motor. Nature solves the problem of determining the speed for which the characteristics of motor and fan agree very simply and quickly. But to predict in advance of the experiment requires a knowledge of the speed of the motor with all possible torque loads and a knowledge of the torque of the fan at all possible speeds of rotation.

Likewise in the heat transmission problem, one would have to know the flow of heat within the solid for all possible rates of heat loss at the boundary and the flow of heat within the fluid for all possible rates of supply at the boundary. Then those solutions would be selected for which the boundary conditions agree, that is, for which the rate of loss from the solid equals rate of supply to the fluid at every point of B.

This fully general problem seems beyond our powers of analytical solution for some time to come. The usual procedure is either to assume that the conductivity of the material in A is so high that the temperature throughout is substantially uniform (for a given heat flow there must be a gradient inversely proportional to the conductivity, which is small if the conductivity is large); or to assume some simple distribution of temperature or heat flow based on observations in experiments on similar bodies. It must be borne in mind, however, when dealing either with experimental data or theoretical solutions that the application to any new technical problem must be studied in the light of the more general problem just outlined.

2. Equations of Motion of the Fluid. We restrict the general problem by assuming that the boundary conditions are known either in the form of the distribution of temperature or of the rate of heat flow over the boundaries. We shall likewise consider only problems of the steady state. With these simplifications we proceed to write down the equations of motion of the fluid and in the next section the equations for the flow of heat in the fluid.

By reference to other divisions of this work[1] it will be found that the equations for the steady motion of a viscous fluid are as follows:

$$u\frac{\partial u}{\partial x}+v\frac{\partial u}{\partial y}+w\frac{\partial u}{\partial z}=g_x-\frac{1}{\varrho}\left(\frac{\partial p_{xx}}{\partial x}+\frac{\partial p_{xy}}{\partial y}+\frac{\partial p_{xz}}{\partial z}\right) \tag{2.1}$$

$$u\frac{\partial v}{\partial x}+v\frac{\partial v}{\partial y}+w\frac{\partial v}{\partial z}=g_y-\frac{1}{\varrho}\left(\frac{\partial p_{xy}}{\partial x}+\frac{\partial p_{yy}}{\partial y}+\frac{\partial p_{yz}}{\partial z}\right) \tag{2.2}$$

$$u\frac{\partial w}{\partial x}+v\frac{\partial w}{\partial y}+w\frac{\partial w}{\partial z}=g_z-\frac{1}{\varrho}\left(\frac{\partial p_{xz}}{\partial x}+\frac{\partial p_{yz}}{\partial y}+\frac{\partial p_{zz}}{\partial z}\right) \tag{2.3}$$

where u, v, w are the components of the velocity at any point (x, y, z) along three Cartesian axes of reference, ϱ the density, g_x, g_y, and g_z the components of the acceleration of gravity along the three axes and p_{xx}, p_{yy}, p_{zz}, p_{xy}, p_{xz} and p_{yz} are the components of a stress tensor defined by the equations

$$p_{xx}=p+2\mu\frac{\partial u}{\partial x}-\frac{2}{3}\mu\left(\frac{\partial u}{\partial x}+\frac{\partial v}{\partial y}+\frac{\partial w}{\partial z}\right)$$

$$p_{yy}=p+2\mu\frac{\partial v}{\partial y}-\frac{2}{3}\mu\left(\frac{\partial u}{\partial x}+\frac{\partial v}{\partial y}+\frac{\partial w}{\partial z}\right)$$

$$p_{zz}=p+2\mu\frac{\partial w}{\partial z}-\frac{2}{3}\mu\left(\frac{\partial u}{\partial x}+\frac{\partial v}{\partial y}+\frac{\partial w}{\partial z}\right)$$

$$p_{xy}=\mu\left(\frac{\partial u}{\partial y}+\frac{\partial v}{\partial x}\right)$$

$$p_{xz}=\mu\left(\frac{\partial w}{\partial x}+\frac{\partial u}{\partial z}\right)$$

$$p_{yz}=\mu\left(\frac{\partial v}{\partial z}+\frac{\partial w}{\partial y}\right)$$

where p is the pressure and μ is the viscosity.

To these equations must be added the equation of continuity

$$\frac{\partial(\varrho u)}{\partial x}+\frac{\partial(\varrho v)}{\partial y}+\frac{\partial(\varrho w)}{\partial z}=0 \tag{2.4}$$

and the equation of state of the fluid

$$f_1(p, \varrho, t)=0 \tag{2.5}$$

where t is the temperature.

If we regard the fluid as incompressible so that ϱ may be considered independent of p and if the departures from Boyle's law may be neglected, equation (2.5) becomes

$$\varrho=\frac{A}{T} \tag{2.6}$$

where T is the absolute temperature and A is a constant.

Equations (2.1), (2.2) and (2.3) are essentially the equations of Stokes. The density ϱ and the viscosity μ are, however, not constant but vary from place to place as determined by the temperature. Consequently, the reduction to the form commonly given is not permissible.

[1] See Division G, also Bulletin U.S. National Research Council No. 84, Washington D.C. 1932.

3. Equation for the Flow of Heat. The equation for the flow of heat may be developed by considering the conservation of energy as applied to a small element of volume. Radiation of heat from the element and the generation of heat by friction within the element are assumed to be negligible.

Consider first the conduction of heat into the Cartesian volume element $dxdydz$ of Fig. 2. If t is the temperature at the center of the face $ABCD$, the gradient normal to the face is $\partial t/\partial x$ and the rate of loss of heat by conduction across the face is $k\,(\partial t/\partial x)\,dydz$. At the face $EFGH$ the rate of increase of heat by conduction will be

$$\left[k\frac{\partial t}{\partial x} + \frac{\partial}{\partial x}\left(k\frac{\partial t}{\partial x}\right)dx\right]dydz$$

Hence the net rate of increase of heat by conduction for these two faces is $\frac{\partial}{\partial x}\left(k\frac{\partial t}{\partial x}\right)dxdydz$. By similar considerations for the other four faces the total rate of increase by conduction is found to be

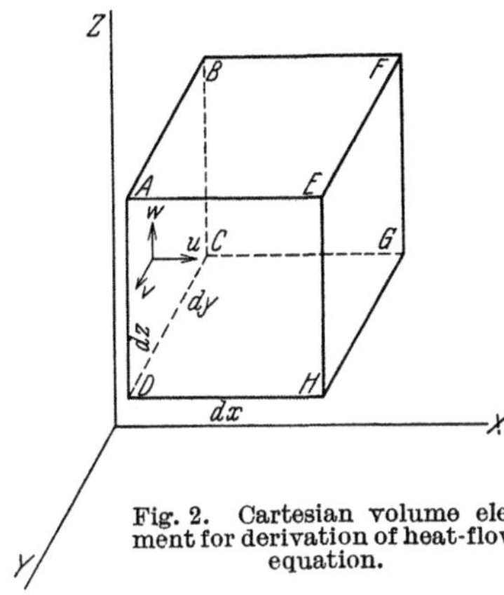

Fig. 2. Cartesian volume element for derivation of heat-flow equation.

$$\left[\frac{\partial}{\partial x}\left(k\frac{\partial t}{\partial x}\right) + \frac{\partial}{\partial y}\left(k\frac{\partial t}{\partial y}\right) + \frac{\partial}{\partial z}\left(k\frac{\partial t}{\partial z}\right)\right]dxdydz$$

Consider next the effect of convection. The fluid entering through the face $ABCD$ at temperature t leaves the face $EFGH$ with temperature $t + \frac{\partial t}{\partial x}dx$. The mass entering per unit time is $\varrho u dydz$. The increase in heat energy per unit time is $\varrho u dydz\, c\frac{\partial t}{\partial x}\,dx$ where c is the specific heat. By similar considerations for the other faces, the total rate of increase in heat energy within the volume element by convection is

$$\varrho c\left(u\frac{\partial t}{\partial x} + v\frac{\partial t}{\partial y} + w\frac{\partial t}{\partial z}\right)dxdydz$$

This rate of increase must equal that supplied by conduction to satisfy the principle of conservation of energy. Hence the temperature satisfies the equation

$$\varrho c\left(u\frac{\partial t}{\partial x} + v\frac{\partial t}{\partial y} + w\frac{\partial t}{\partial z}\right) = \frac{\partial}{\partial x}\left(k\frac{\partial t}{\partial x}\right) + \frac{\partial}{\partial y}\left(k\frac{\partial t}{\partial y}\right) + \frac{\partial}{\partial z}\left(k\frac{\partial t}{\partial z}\right) \qquad (3.1)$$

If the heating due to pressure changes is neglected, c is the specific heat at constant pressure.

To take account of the variation of k, c, and μ with temperature, we must add the equations expressing these relations

$$k = f_2(t) \qquad (3.2)$$

$$c = f_3(t) \qquad (3.3)$$

$$\mu = f_4(t) \qquad (3.4)$$

4. Common Assumptions to Facilitate Solution. The determination of the temperature distribution and flow of heat within the fluid requires the solution of the nine equations (2.1), (2.2), (2.3), (2.4), (2.6), (3.1), (3.2), (3.3), and (3.4) in the nine variables u, v, w, p, t, ϱ, k, c, and μ.

Little progress has been made in the solution of this formidable problem. The attention of many workers in the field of heat transmission has been centered on (3.1), usually with the simplification that ϱ, c and k are assumed to be independent of the temperature. The dynamic problem of the motion of the fluid is disposed of by simple assumptions, such as that the velocity or the mass flow is everywhere constant or that the mass flow varies linearly with distance from the boundary. These convenient assumptions seem so far from anything observed in the actual flow of fluids that the solutions can only be regarded as mathematical exercises. Those readers who are interested will find valuable summaries of these solutions in papers by Drew[1] and by Lévêque[2].

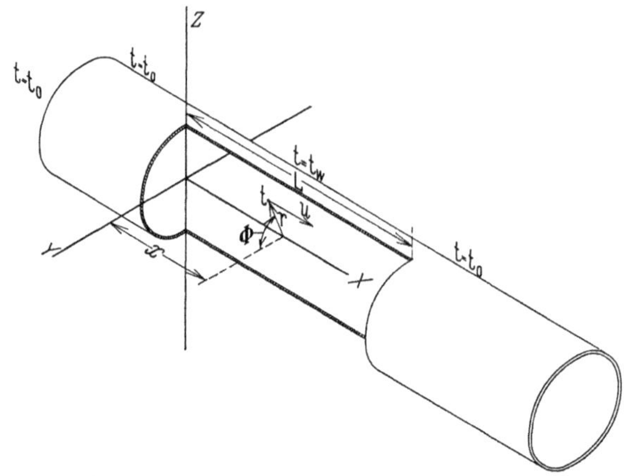

Fig. 3. Cylindrical coordinates for flow in a pipe.

A somewhat more advanced but bold assumption utilizes the results of computations of the isothermal flow of the fluid to obtain values of u, v, and w to be used in (3.1). It is perhaps worth while to illustrate this procedure by Graetz's solution for the flow of heat from the wall of a pipe to a fluid flowing through it in laminar flow, *i. e.* below the critical Reynolds number.

5. Laminar Flow in a Pipe. It is most convenient for this problem to use cylindrical coordinates x, r, and φ as shown in Fig. 3, the x axis coinciding with the axis of the pipe. By direct transformation of (3.1) we obtain

$$\left.\begin{aligned} \varrho c\left(u\frac{\partial t}{\partial x}+\dot r\frac{\partial t}{\partial r}+\dot\varphi\frac{\partial t}{\partial\varphi}\right)=\frac{\partial}{\partial x}\left(k\frac{\partial t}{\partial x}\right)+\frac{\partial}{\partial r}\left(k\frac{\partial t}{\partial r}\right)+\\ +\frac{1}{r^2}\frac{\partial}{\partial\varphi}\left(k\frac{\partial t}{\partial\varphi}\right)+\frac{k}{r}\frac{\partial t}{\partial r}\end{aligned}\right\}\quad(5.1)$$

where $\dot r$ and $r\dot\varphi$ are the radial and circumferential components of the velocity.

---

[1] Trans. Am. Inst. Chem. Engineers, Vol. 26, p. 26, 1931.
[2] Annales des Mines, Vol. 13, pp. 201, 305 and 381, 1928.

Assuming symmetry about the axis of the pipe, which means either that the pipe is vertical or that the convection currents are negligible, $\partial t/\partial \varphi$ is zero and (5.1) becomes

$$\varrho c\left(u\frac{\partial t}{\partial x}+\dot{r}\frac{\partial t}{\partial r}\right)=\frac{\partial}{\partial x}\left(k\frac{\partial t}{\partial x}\right)+\frac{\partial}{\partial r}\left(k\frac{\partial t}{\partial r}\right)+\frac{k}{r}\frac{\partial t}{\partial r} \tag{5.2}$$

The velocities u and $\dot{r}$ are assumed from the flow in the isothermal case, namely, $\dot{r}=0$, $\varrho u = 2G(1-r^2/R^2)$ where G is the average mass velocity over a cross section and R is the radius of the cylindrical pipe. Setting $r/R = y$, (5.2) reduces to

$$2cG(1-y^2)\frac{\partial t}{\partial x}=\frac{\partial}{\partial x}\left(k\frac{\partial t}{\partial x}\right)+\frac{1}{R^2}\frac{\partial}{\partial y}\left(k\frac{\partial t}{\partial y}\right)+\frac{k}{yR^2}\frac{\partial t}{\partial y} \tag{5.3}$$

Graetz made the additional assumptions that $\frac{\partial}{\partial x}\left(k\frac{\partial t}{\partial x}\right)$ could be neglected in comparison with $\frac{\partial}{\partial y}\left(k\frac{\partial t}{\partial y}\right)$ and that k and c are independent of the temperature. With these assumptions, (5.3) becomes

$$(1-y^2)\frac{\partial t}{\partial x}=A\left(\frac{\partial^2 t}{\partial y^2}+\frac{1}{y}\frac{\partial t}{\partial y}\right) \tag{5.4}$$

where $A = k/2cGR^2$, a constant.

As boundary conditions, the pipe wall was assumed to be heated to a constant temperature t_w from $x=0$ to $x=L$, and the initial temperature of the fluid was assumed constant and equal to t_0.

The type solution of equation (5.4) is of the form

$$\frac{t-t_w}{t_0-t_w}=\sum_m B_m\, e^{-a_m A x} f_m(y, a_m) \tag{5.5}$$

where B_m and a_m are constants to be determined and $f_m(y, a_m)$ satisfies the differential equation

$$\frac{d^2 f_m}{dy^2}+\frac{1}{y}\frac{d f_m}{dy}+a_m(1-y^2)f_m=0 \tag{5.6}$$

The constant a_m is a root of $f_m(1, a_m)=0$.

Equation (5.6) characterizes certain functions related to Bessel functions which must be evaluated by development in series. According to Drew's revision of Graetz's computation, the average rate of heat flow per unit surface area of the pipe, q_{av}, is given by the formula

$$q_{av}=\frac{cGR}{2L}(t_w-t_0)\left[1-8P_2(AL)\right] \tag{5.7}$$

where P_2 is the convergent infinite series

$$P_2 = 0.10238\, e^{-7.3136\, AL} + 0.01220\, e^{-44.61\, AL} + \text{etc.}$$

If we consider a very long heated section of pipe, it is obvious that the temperature of the fluid will approach the temperature of the wall of the pipe and the rate of transfer of heat will approach zero. Thus in (5.7), as L increases, the series P_2 approaches zero and because of the L in the denominator, q_{av} approaches zero.

The local rate of transfer of heat q_x at a distance x from the beginning of the heated section is given by the expression

$$q_x = \frac{2\,k\,(t_w - t_0)}{R}\,P_1\,(A\,x) \tag{5.8}$$

where P_1 is the convergent infinite series

$$P_1 = 0.74877\,e^{-7.3136\,A\,x} + 0.544\,e^{-44.61\,A\,x} + \text{etc.}$$

The results are often expressed in terms of a heat transfer number defined in terms of the average temperature of the fluid at the section x. Graetz and Lévêque used an average temperature t_{av} defined by the relation

$$t_{av} - t_0 = \frac{\int_0^1 c\,\varrho\,u\,(t - t_0)\,2\,\pi\,y\,d\,y}{\int_0^1 c\,\varrho\,u\,2\,\pi\,y\,d\,y} \tag{5.9}$$

It is found that $$t_w - t_{av} = 8\,(t_w - t_0)\,P_2\,(A\,x) \tag{5.10}$$

The local heat transfer number h_x, defined as $q_x/(t_w - t_{av})$ is given by the relation

$$h_x = \frac{k}{4\,R}\,\frac{P_1\,(A\,x)}{P_2\,(A\,x)} \tag{5.11}$$

Nusselt uses an average temperature t'_{av} defined by the relation

$$t'_{av} - t_0 = \frac{1}{\pi}\int_0^1 (t - t_0)\,2\,\pi\,y\,d\,y \tag{5.12}$$

and obtains a corresponding h'_x

$$h'_x = \frac{k}{2\,R}\,\frac{P_1\,(A\,x)}{N_1\,(A\,x)} \tag{5.13}$$

where $N_1 = 0.14525\,e^{-7.3136\,A\,x} + 0.0334\,e^{-44.61\,A\,x} + \text{etc.}$

The two formulations correspond to two different experimental procedures for determining the average temperature. Equation (5.9) gives the "mixing cup" temperature found by thoroughly mixing the fluid in a special chamber and measuring the temperature of the mixture. Equation (5.12) gives the average temperature found by making a traverse of the pipe with a thermocouple.

When $A\,x$ is greater than 0.2, h_x and h'_x are constant within 0.1 per cent and are given by the formulae

$$h_x = 1.83\,\frac{k}{R} \tag{5.14}$$

$$h'_x = 2.58\,\frac{k}{R} \tag{5.15}$$

Thus while q_x approaches zero as the length of the heated section is increased, the heat transfer number approaches a constant value.

This brief summary gives some idea of the mathematical complications even after many simplifying assumptions have been made.

It must be stated here that experimentally the heat flow is greater than that given by (5.7) by as much as 100 per cent for certain values of $A\,L$. This discrepancy will be discussed more fully in Chapter VII.

6. Laminar Flow in a Two-Dimensional Boundary Layer. When a fluid having a small coefficient of viscosity (for example, air) flows past a solid, the flow at a considerable distance approximates a potential flow and the effect of viscosity is confined to a thin boundary layer at the surface of the body. In a similar way, if the thermal conductivity of the fluid is small, the temperature gradient is large only in a thin layer at the surface of the body. Under these circumstances the general equations permit of certain simplifications which will be discussed for the case of two-dimensional flow. The effects of natural convection and the internal heating of the fluid by viscous friction will be neglected. Likewise the density, viscosity, thermal conductivity, and specific heat will be considered constant. In other words, the equations of motion for the isothermal case will be used to obtain values of the velocities for use in the equation for heat flow.

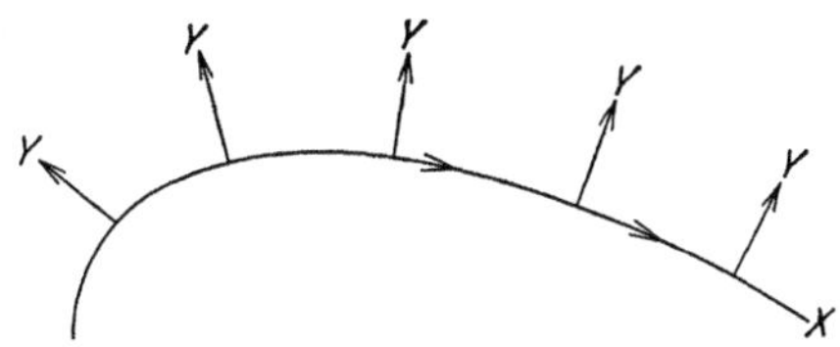

Fig. 4. Coordinates for flow in a boundary layer.

The assumption that a boundary layer exists means that we shall be interested in distances from the surface of the order of magnitude of some small distance ε which may be regarded as some conveniently defined "thickness" of the boundary layer. Actually there is no sharp boundary, but various exact definitions can be made. At present we will be concerned only with orders of magnitude. The X axis (Fig. 4) will be selected as the curvilinear boundary line of a section of the two-dimensional body, the radius of curvature of which is assumed to be sufficiently large in comparison with the thickness of the boundary layer that the effect of the curvature need not be considered. The Y axis will be taken as perpendicular to the X axis.

The simplifications which are made in the general equations for the case of a thin boundary layer are usually introduced by a consideration of orders of magnitude of terms in the general equations. The procedure cannot be considered in any sense as a mathematical proof. The justification for dropping certain terms must come ultimately from the experimental results. Nevertheless the discussion of order of magnitudes is instructive.

Two quantities are to be understood to be of the same order of magnitude when they do not differ by a factor of more than say ten, and of a different order when they differ by a factor of 100 or more. In dealing with quantities which vary throughout the layer, we are interested in maximum values since we desire to know what terms can be dropped. A difficulty arises when dealing with derivatives, since mathematically the value of a derivative is not related to the absolute values of the variables. Thus let us consider u to become of order of magnitude of 1 for x of order of 1 and y of the order of ε. The slope $\partial u/\partial y$ mathematically may lie anywhere between $-\infty$ and $+\infty$, but assuming no sharp discontinuities, we expect in physical problems a slope of the same order of magnitude (in the sense defined above) as the total change in u divided by the total change in y. That is we expect absolute values of $\partial u/\partial y$ at any point to lie between 0 and perhaps ten times the mean value of $\partial u/\partial y$.

Considering u to become of order of magnitude 1 for y of order of ε $\partial u/\partial y$ is of the order of $1/\varepsilon$ and $\partial^2 u/\partial y^2$ of the order of $1/\varepsilon^2$, while $\partial u/\partial x$ and $\partial^2 u/\partial x^2$ will remain of the order of 1. From the equation of continuity (2.4) which becomes, with the assumptions which have been made,

$$\frac{\partial u}{\partial x} + \frac{\partial v}{\partial y} = 0 \tag{6.1}$$

$\partial v/\partial y$ is of the order of 1 and hence v is of the order of ε and $\partial^2 v/\partial y^2$ of the order $1/\varepsilon$.

Equations (2.1), (2.2), (2.3) reduce for the present case to the following, where the orders of magnitude of the maxima of the several terms are indicated:

$$\left.\begin{array}{l} u\dfrac{\partial u}{\partial x} + v\dfrac{\partial u}{\partial y} = \dfrac{\mu}{\varrho}\left(\dfrac{\partial^2 u}{\partial x^2} + \dfrac{\partial^2 u}{\partial y^2}\right) - \dfrac{1}{\varrho}\dfrac{\partial p}{\partial x} \\ 1,\ 1 \qquad \varepsilon,\ \dfrac{1}{\varepsilon} \qquad\qquad 1, \qquad \dfrac{1}{\varepsilon^2} \qquad\quad 1 \end{array}\right\} \tag{6.2}$$

$$\left.\begin{array}{l} u\dfrac{\partial v}{\partial x} + v\dfrac{\partial v}{\partial y} = \dfrac{\mu}{\varrho}\left(\dfrac{\partial^2 v}{\partial x^2} + \dfrac{\partial^2 v}{\partial y^2}\right) - \dfrac{1}{\varrho}\dfrac{\partial p}{\partial y} \\ 1,\ \varepsilon \qquad \varepsilon,\ 1 \qquad\qquad \varepsilon, \qquad \dfrac{1}{\varepsilon} \end{array}\right\} \tag{6.3}$$

From (6.2) it follows that ε^2 must be of the order of μ/ϱ; for if it is greater, the terms involving viscosity drop out when ε is considered small with respect to 1, and if it is less, the second term on the right becomes infinite. Equation (6.2) becomes, when μ/ϱ is assumed of the order of ε^2 and terms of the order of ε are omitted,

$$u\frac{\partial u}{\partial x} + v\frac{\partial u}{\partial y} = \frac{\mu}{\varrho}\frac{\partial^2 u}{\partial y^2} - \frac{1}{\varrho}\frac{\partial p}{\partial x} \tag{6.4}$$

Equation (6.3) then states that $\partial p/\partial y$ is of the order of ε and hence that the pressure may be regarded as independent of y. It therefore equals the pressure in the potential flow outside the boundary layer or the pressure on the boundary.

Equation (3.1) becomes for the two-dimensional case

$$\varrho c\left(u\frac{\partial t}{\partial x} + v\frac{\partial t}{\partial y}\right) = \frac{\partial}{\partial x}\left(k\frac{\partial t}{\partial x}\right) + \frac{\partial}{\partial y}\left(k\frac{\partial t}{\partial y}\right) \tag{6.5}$$

or with constant k and c

$$\varrho c\left(u\frac{\partial t}{\partial x} + v\frac{\partial t}{\partial y}\right) = k\left(\frac{\partial^2 t}{\partial x^2} + \frac{\partial^2 t}{\partial y^2}\right) \tag{6.6}$$

We cannot necessarily conclude that the thermal boundary layer is of the same order of magnitude as ε and therefore we assume that the

y in (6.6) is of order ε_1. It follows that $\partial t/\partial y$ will be of order $1/\varepsilon_1$ and $\partial^2 t/\partial y^2$ of order $1/\varepsilon_1^2$ while $\partial t/\partial x$ and $\partial^2 t/\partial x^2$ are of order 1. It is readily seen that ε_1^2 must be of the order of $k/\varrho c$, in which case (6.6) reduces to

$$\varrho c \left(u \frac{\partial t}{\partial x} + v \frac{\partial t}{\partial y}\right) = k \frac{\partial^2 t}{\partial y^2} \tag{6.7}$$

If $k/\varrho c$ and μ/ϱ are of the same order of magnitude, ε and ε_1 are of the same order and equations (6.1), (6.4) and (6.7) constitute the equations for heat transfer in a laminar boundary layer. For air μ/ϱ is about 0.70 to 0.77 $k/\varrho c$ and hence the thermal and dynamic boundary layers are of the same order of magnitude.

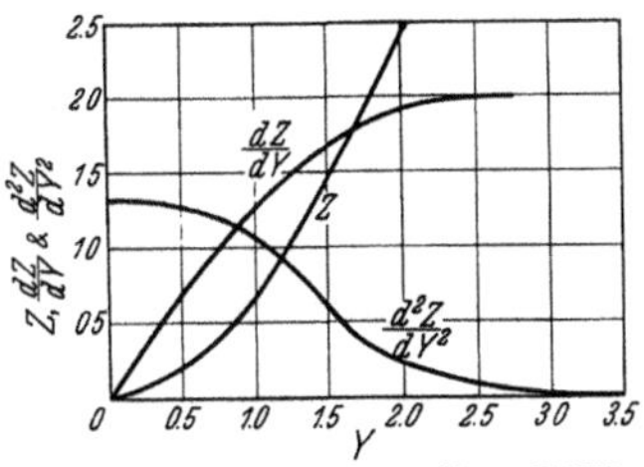

Fig. 5. Blasius' function $Z(Y)$ for flow near skin-friction plate.

7. Laminar Flow along a Thin Flat Plate. As an application of the boundary layer equations developed in the last section, we shall consider the heat transfer from a thin flat plate placed in an air stream of uniform velocity U_0, the plate being parallel to the flow and maintained at a constant temperature t_p. The temperature of the air stream at a distance from the plate is assumed constant and is designated t_0. Since the air stream is of uniform velocity, $\partial p/\partial x$ is zero. The three equations are therefore

$$\frac{\partial u}{\partial x} + \frac{\partial v}{\partial y} = 0 \tag{7.1}$$

$$u \frac{\partial u}{\partial x} + v \frac{\partial u}{\partial y} = \frac{\mu}{\varrho} \frac{\partial^2 u}{\partial y^2} \tag{7.2}$$

$$u \frac{\partial t}{\partial x} + v \frac{\partial t}{\partial y} = \frac{k}{\varrho c} \frac{\partial^2 t}{\partial y^2} \tag{7.3}$$

To satisfy (7.1), set

$$u = \frac{\partial \psi}{\partial y}; \quad v = -\frac{\partial \psi}{\partial x} \tag{7.4}$$

and introduce new variables

$$Y = \frac{1}{2}\sqrt{\frac{U_0 \varrho}{\mu}} \frac{y}{\sqrt{x}} \tag{7.5}$$

$$\psi = \sqrt{\frac{U_0 \mu x}{\varrho}}\, Z(Y) \tag{7.6}$$

$$t = t_p - (t_p - t_0)\,\theta(Y) \tag{7.7}$$

whence

$$u = \frac{1}{2} U_0 \frac{dZ}{dY}; \quad v = \frac{1}{2}\sqrt{\frac{U_0 \mu}{\varrho x}}\left(Y \frac{dZ}{dY} - Z\right)$$

The boundary conditions $t = t_p$, $u = v = 0$ at $y = 0$ and $t = t_0$, $u = U_0$ at $y = \infty$ become $Z = 0,\ Z' = 0,\ \theta = 0$ at $Y = 0$

and $Z' = 2,\ \theta = 1$ at $Y = \infty$

The partial differential equations reduce to total differential equations

$$\frac{d^3 Z}{dY^3} + Z \frac{d^2 Z}{dY^2} = 0 \tag{7.8}$$

and
$$\frac{d^2\theta}{dY^2} + \sigma Z \frac{d\theta}{dY} = 0 \tag{7.9}$$
where σ is the dimensionless ratio $\mu c/k$.

The function $Z(Y)$ was computed by Blasius by development in series. Figure 5 shows a graph of the function and its derivatives.

Assuming Z as known, the solution of (7.9) can be obtained by quadrature as
$$\theta = \alpha(\sigma) \int_0^Y e^{-\sigma \int_0^Y Z\,dY} dY \tag{7.10}$$
where
$$\alpha(\sigma) = \frac{1}{\int_0^\infty e^{-\sigma \int_0^Y Z\,dY} dY} \tag{7.11}$$

The temperature gradient
$$\frac{dt}{dy} = \frac{\partial t}{\partial \theta}\frac{\partial \theta}{\partial Y}\frac{\partial Y}{\partial y} = -\frac{1}{2}\sqrt{\frac{U_0 \varrho}{\mu x}}\frac{\partial \theta}{\partial Y}(t_p - t_0).$$

But $\partial\theta/\partial Y = \alpha(\sigma)\, e^{-\sigma \int_0^Y Z\,dy}$ and at the surface of the plate $(d\theta/\partial Y)_{Y=y=0} = \alpha(\sigma)$ and hence
$$\left(\frac{\partial t}{\partial y}\right)_{y=0} = -\frac{1}{2}\alpha(\sigma)\sqrt{\frac{U_0\varrho}{\mu x}}(t_p - t_0) \tag{7.12}$$
Hence the heat transfer per unit area, q_x, is given by
$$q_x = \frac{k}{2}\sqrt{\frac{U_0\varrho}{\mu x}}\,\alpha(\sigma)\,(t_p - t_0) \tag{7.13}$$
This is the local rate of flow per unit area and is a function of x. The average for a plate of length L is given by
$$q_{av} = \frac{1}{L}\int_0^L q_x\,dx = \frac{k}{L}\sqrt{\frac{U_0\varrho L}{\mu}}\,\alpha(\sigma)\,(t_p - t_0) \tag{7.14}$$
E. Pohlhausen found that α was approximately equal to $0.664\sqrt[3]{\sigma}$ for σ between 0.6 and 15.0. He also gives graphs of equation (7.10)[1] for various values of σ.

It may be noted that for $\sigma = 1$, the distribution of dZ/dY (hence u) and θ are identical. For air $\sigma = 0.70$ to 0.77, and the difference between the two distributions is not large, but for liquids the difference is very considerable, the thermal boundary layer being much thinner than the dynamic boundary layer.

CHAPTER III

THEORY OF HEAT TRANSFER IN EDDYING FLOW

1. Reynolds Theory of Eddying Flow. We shall now consider the theory of eddying flow, and we shall assume that the entire field of flow

[1] Pohlhausen, E., Zeitschrift für angewandte Mathematik und Mechanik, Vol. 1, p. 115, 1921.

is of this type. At any point the average speed and temperature are constant but the instantaneous values fluctuate with time. Reynolds regarded the flow as consisting of a mean flow and a superposed fluctuating motion. The possibility of clearly distinguishing the two rests upon the possibility of choosing time and space intervals, over which averages are to be taken to obtain the mean motion, which are large in comparison with the period and wave length of the fluctuations but which may be regarded as small in comparison to times and distances which are of significance in the mean motion.

We write $u = \bar{u} + u'$, $v = \bar{v} + v'$, $w = \bar{w} + w'$, $t = \bar{t} + t'$ where the bars relate to the mean motion and the primes to the fluctuations. The fluctuations of ϱ, μ, c and k will be neglected. The above values of u, v, w and t are substituted in the equations of laminar flow and mean values taken. The rules for taking these means were formulated by Reynolds. Thus the mean value of a quantity which has already been averaged is not changed by taking the new mean. The averaging obeys the distributive law. Symbolically, if a and b are two of the quantities for which mean values are to be taken and if the bars are used to denote mean values, $\bar{\bar{a}} = \bar{a}$, $\overline{\bar{a}\,b} = \bar{a}\,\bar{b}$, $\overline{a + b} = \bar{a} + \bar{b}$, $\overline{\partial a/\partial x} = (\partial/\partial x)\,\bar{a}$, etc.

By definition $\overline{u'} = 0$, $\overline{v'} = 0$, $\overline{w'} = 0$, $\overline{t'} = 0$ and hence $\overline{u\,u} = \bar{u}\,\bar{u} + \overline{u'u'}$, $\overline{u v} = \bar{u}\,\bar{v} + \overline{u'v'}$, $\overline{u t} = \bar{u}\,\bar{t} + \overline{u't'}$, etc.

When the above expressions for u, v, w and t are introduced in the general equations of motion (2.1), (2.2), (2.3), (2.4) and (3.1) of Chapter II with the terms occurring in non-steady flow added, mean values taken, and account taken of the fact that u, v, w and u', v', w' satisfy the equation of continuity, we obtain

$$\bar{u}\frac{\partial \bar{u}}{\partial x} + \bar{v}\frac{\partial \bar{u}}{\partial y} + \bar{w}\frac{\partial \bar{u}}{\partial z} = g_x - \frac{1}{\varrho}\left(\frac{\partial \bar{p}_{xx}}{\partial x} + \frac{\partial \bar{p}_{xy}}{\partial y} + \frac{\partial \bar{p}_{xz}}{\partial z}\right) \tag{1.1}$$

$$\bar{u}\frac{\partial \bar{v}}{\partial x} + \bar{v}\frac{\partial \bar{v}}{\partial y} + \bar{w}\frac{\partial \bar{v}}{\partial z} = g_y - \frac{1}{\varrho}\left(\frac{\partial \bar{p}_{xy}}{\partial x} + \frac{\partial \bar{p}_{yy}}{\partial y} + \frac{\partial \bar{p}_{yz}}{\partial z}\right) \tag{1.2}$$

$$\bar{u}\frac{\partial \bar{w}}{\partial x} + \bar{v}\frac{\partial \bar{w}}{\partial y} + \bar{w}\frac{\partial \bar{w}}{\partial z} = g_z - \frac{1}{\varrho}\left(\frac{\partial \bar{p}_{xz}}{\partial x} + \frac{\partial \bar{p}_{yz}}{\partial y} + \frac{\partial \bar{p}_{zz}}{\partial z}\right) \tag{1.3}$$

where
$$\bar{p}_{xx} = p + 2\mu\frac{\partial \bar{u}}{\partial x} - \frac{2}{3}\mu\left(\frac{\partial \bar{u}}{\partial x} + \frac{\partial \bar{v}}{\partial y} + \frac{\partial \bar{w}}{\partial z}\right) - \varrho\,\overline{u'u'}$$

$$\bar{p}_{yy} = p + 2\mu\frac{\partial \bar{v}}{\partial y} - \frac{2}{3}\mu\left(\frac{\partial \bar{u}}{\partial x} + \frac{\partial \bar{v}}{\partial y} + \frac{\partial \bar{w}}{\partial z}\right) - \varrho\,\overline{v'v'}$$

$$\bar{p}_{zz} = p + 2\mu\frac{\partial \bar{w}}{\partial z} - \frac{2}{3}\mu\left(\frac{\partial \bar{u}}{\partial x} + \frac{\partial \bar{v}}{\partial y} + \frac{\partial \bar{w}}{\partial z}\right) - \varrho\,\overline{w'w'}$$

$$\bar{p}_{xy} = \mu\left(\frac{\partial \bar{u}}{\partial y} + \frac{\partial \bar{v}}{\partial x}\right) - \varrho\,\overline{u'v'}$$

$$\bar{p}_{xz} = \mu\left(\frac{\partial \bar{w}}{\partial x} + \frac{\partial \bar{u}}{\partial z}\right) - \varrho\,\overline{u'w'}$$

$$\bar{p}_{yz} = \mu\left(\frac{\partial \bar{v}}{\partial z} + \frac{\partial \bar{w}}{\partial y}\right) - \varrho\,\overline{v'w'}$$

$$\frac{\partial(\varrho\bar{u})}{\partial x} + \frac{\partial(\varrho\bar{v})}{\partial y} + \frac{\partial(\varrho\bar{w})}{\partial z} = 0 \tag{1.4}$$

$$\left.\begin{aligned}\varrho\, c\left(\bar{u}\frac{\partial \bar{t}}{\partial x}+\bar{v}\frac{\partial \bar{t}}{\partial x}+\bar{w}\frac{\partial \bar{t}}{\partial z}\right)=\frac{\partial}{\partial x}\left(k\frac{\partial \bar{t}}{\partial x}\right)+\frac{\partial}{\partial y}\left(k\frac{\partial \bar{t}}{\partial y}\right)+\\+\frac{\partial}{\partial z}\left(k\frac{\partial \bar{t}}{\partial z}\right)-c\frac{\partial(\varrho\overline{u' t'})}{\partial x}-c\frac{\partial(\varrho\overline{v' t'})}{\partial y}-c\frac{\partial(\varrho\overline{w' t'})}{\partial z}\end{aligned}\right\}\qquad(1.5)$$

Equations (1.1), (1.2), (1.3) differ from the corresponding equations (2.1), (2.2) and (2.3) of Chapter II only in the presence of the derivatives of the six "eddy stresses"

$$-\varrho\overline{u' u'},\quad -\varrho\overline{u' v'},\quad -\varrho\overline{u' w'},\quad -\varrho\overline{v' v'},\quad -\varrho\overline{v' w'},\quad \text{and}\quad -\varrho\overline{w' w'}$$

Equation (1.4) is the same as (2.4) of Chapter II. Equation (1.5) differs from (3.1) of Chapter II in the presence of the derivatives of the "eddy heat transfer" terms which may be written

$$-c\varrho\overline{u' t'},\quad -c\varrho\overline{v' t'},\quad \text{and}\quad -c\varrho\overline{w' t'}$$

if c is regarded as independent of x, y, z.

If the fluctuations u', v', w', t' were perfectly random, $\overline{u' v'}$, $\overline{u' w'}$, $\overline{v' w'}$, $\overline{u' t'}$, $\overline{v' t'}$, $\overline{w' t'}$ would be zero, *i. e.* the eddy shearing stresses and the eddy heat transfer would be zero. There would remain only the eddy normal stresses $\varrho\overline{u' u'}$, $\varrho\overline{v' v'}$, $\varrho\overline{w' w'}$, which generally are quite insignificant in comparison with the pressure. The essential feature of eddying flow is then the correlation between the fluctuations of the several components of the velocity at any point.

The eddy shearing stresses show a certain parallelism with the viscous shearing stress. In laminar flow, the fluctuations are molecular fluctuations. The speeds u, v, w, are the mean speeds of many molecules. The effect of the molecular motions appears in the "smoothed" equations of motion in the form of the viscosity coefficient, and in the "smoothed" heat convection equation as the conductivity. These coefficients are functions of the temperature.

Reynolds general theory shows the influence of the fluctuations on the mean motion but does not give any information as to the fluctuations themselves. Little progress has been made on a fundamental theory of the fluctuations. However by more or less plausible theoretical considerations combined with a judicious use of empirical results obtained by experiment, notable advances have been made. The more important of these semi-empirical methods are (1) the use of the concept of eddy viscosity by Reynolds, Boussinesq, Stanton, Richardson, G. I. Taylor and many others; (2) the use of the concept of mixing length with momentum transfer by Prandtl; (3) the use of the concept of mixing length with vorticity transfer by G. I. Taylor; and (4) the use of the principle of similarity by von Kármán.

2. The Concept of Eddy Viscosity. Equations (1.1), (1.2) and (1.3) may be made to exhibit a formal similarity with II (2.1), (2.2) and (2.3) by defining an "apparent" or eddy viscosity which is assumed to satisfy simultaneously all of the following equations

$$-\varrho\,\overline{u'u'} = 2\,\varepsilon\frac{\partial \bar{u}}{\partial x} - \frac{2}{3}\,\varepsilon\left(\frac{\partial \bar{u}}{\partial x} + \frac{\partial \bar{v}}{\partial y} + \frac{\partial \bar{w}}{\partial z}\right) \tag{2.1}$$

$$-\varrho\,\overline{u'v'} = \varepsilon\left(\frac{\partial \bar{u}}{\partial y} + \frac{\partial \bar{v}}{\partial x}\right) \tag{2.2}$$

$$-\varrho\,\overline{u'w'} = \varepsilon\left(\frac{\partial \bar{v}}{\partial z} + \frac{\partial \bar{w}}{\partial y}\right) \tag{2.3}$$

$$-\varrho\,\overline{v'v'} = 2\,\varepsilon\frac{\partial \bar{v}}{\partial y} - \frac{2}{3}\,\varepsilon\left(\frac{\partial u}{\partial x} + \frac{\partial \bar{v}}{\partial y} + \frac{\partial \bar{w}}{\partial z}\right) \tag{2.4}$$

$$-\varrho\,\overline{v'w'} = \varepsilon\left(\frac{\partial \bar{w}}{\partial x} + \frac{\partial \bar{u}}{\partial z}\right) \tag{2.5}$$

$$-\varrho\,\overline{w'w'} = 2\,\varepsilon\frac{\partial \bar{w}}{\partial z} - \frac{2}{3}\,\varepsilon\left(\frac{\partial u}{\partial x} + \frac{\partial \bar{v}}{\partial y} + \frac{\partial \bar{w}}{\partial z}\right) \tag{2.6}$$

On introducing the above relations in (1.1), (1.2) and (1.3), equations identical with those for a viscous fluid are obtained except that the viscosity μ is replaced by $\mu + \varepsilon$. It must be noted however that ε varies from place to place and is a function of the mean motion (not merely of the temperature). Moreover, the use of the concept of eddy viscosity implies certain relations between u', v' and w' at a given point which are expressed in the equations (2.1) to (2.6) inclusive.

Attempts have been made to use a vector quantity ε_x, ε_y, ε_z by some investigators in place of the scalar ε, especially in meteorological problems.

A similar "eddy conductivity" β may be introduced in (1.5) by writing

$$-c\,\varrho\,\overline{u't'} = \beta\frac{\partial \bar{t}}{\partial x} \tag{2.7}$$

$$-c\,\varrho\,\overline{v't'} = \beta\frac{\partial \bar{t}}{\partial y} \tag{2.8}$$

$$-c\,\varrho\,\overline{w't'} = \beta\frac{\partial \bar{t}}{\partial z} \tag{2.9}$$

Here again (2.7), (2.8) and (2.9) involve further assumptions as to the relations between the fluctuations. The introduction of eddy viscosity and eddy conductivity is not a mere introduction of two new symbols, but the introduction of a physical hypothesis as to the relations between the fluctuations of the several components of the speed, the temperature at any point, the mean motion and the mean temperature gradient.

The eddy viscosity ε and the eddy conductivity β are usually large in comparison with μ and k, and hence μ and k are generally neglected in problems of eddying flow. The relation between β and ε cannot be determined without some more definite picture of the underlying mechanism. The rather obvious assumption that $\beta = k\varepsilon/\mu$ does not seem to have been often used. The more usual assumption is that $\beta = c\varepsilon$, based on considerations given in 4 of this Chapter. If $k/\mu c = 1$, which is nearly true for gases, the two assumptions are equivalent.

In spite of the formal simplification, we are still not in a position to solve either the dynamical or the thermal problem. The variation of ε and β from place to place must first be known. Some solutions of the thermal problem have been given by computing from experimental data for the isothermal flow, using one of the relations between β and ε given above. An example is the computation of heat transfer in eddying flow in a pipe by Latzko[1].

3. Eddying Flow in a Pipe. We shall consider a section of a long cylindrical pipe of radius R, diameter D, far from the entrance and suppose the speed to be sufficiently high that the flow is eddying throughout and the final speed distribution has been attained. The air will be assumed to enter the pipe at a uniform temperature t_0, and the pipe wall will be assumed to be heated to a uniform temperature t_w for a distance L. The system of cylindrical coordinates of Fig. 3 will be used as in II 5, symmetry will be assumed, $\frac{\partial}{\partial x}\left(k\frac{\partial t}{\partial x}\right)$ will be neglected in comparison with $\frac{\partial}{\partial r}\left(k\frac{\partial t}{\partial r}\right)$, and k will be replaced by β. The equation of heat flow then becomes

$$\varrho c u \frac{\partial t}{\partial x} = \frac{1}{r}\frac{\partial}{\partial r}\left(\beta r \frac{\partial t}{\partial r}\right) \tag{3.1}$$

This equation may be derived directly by considering the heat flow in a short cylinder of radius r concentric with the pipe, as shown by Latzko.

The computation of Latzko was made on the basis of the older formulation of the empirical laws of the isothermal eddying flow in which a power law was used. This representation is in good agreement with the experimental data for Reynolds numbers between 3000 and 100,000. Denoting the average velocity by $\bar{u}$, the empirical power law for the velocity distribution as used by Latzko is

$$u = \frac{8}{7}\bar{u}\left(1 - \frac{r^2}{R^2}\right)^{1/7} \tag{3.2}$$

The frictional shearing stress τ_0 at the wall is given by the formula of Blasius

$$\tau_0 = 0.03955\,\varrho\,\bar{u}^2\left(\frac{\bar{u}D\varrho}{\mu}\right)^{-1/4} \tag{3.3}$$

The shearing stress at any distance r is related to the velocity gradient $\partial u/\partial r$ by the formula developed by von Kármán

$$\tau = \frac{7}{8.82}\,\tau_0^{3/7}\,\varrho^{3/7}\,\mu^{1/7}\left(\frac{R^2 - r^2}{2R}\right)^{6/7}\frac{\partial u}{\partial r} \tag{3.4}$$

From these expressions the eddy viscosity ε is readily computed to be

$$\varepsilon = 0.199\,\bar{u}^{3/4}\,\varrho^{3/4}\,\mu^{1/4}\,D^{-3/28}\left(\frac{R^2 - r^2}{2R}\right)^{6/7} \tag{3.5}$$ [2]

---

[1] Zeitschrift für angewandte Mathematik und Mechanik, Vol. 1, p. 269, 1921.

[2] In (3.5) $\bar{u}^{3/4}$ signifies mean u with index 3/4 and similarly in (3.6) and (3.7).

Assuming that the eddy conductivity β equals $k\varepsilon/\mu$, β is given by the expression

$$\beta = 0.199\, k\, \bar{u}^{3/4}\, \varrho^{3/4}\, \mu^{-3/4}\, D^{-3/28} \left(\frac{R^2 - r^2}{2R}\right)^{6/7} \tag{3.6}$$

Introducing the value of u from (3.2) and β from (3.6) in (3.1) and setting

$$A = \frac{0.199\, k}{\frac{8}{7}\, \varrho^{1/4} c\, \bar{u}^{1/4}\, \mu^{3/4}\, D^{3/28}} \tag{3.7}$$

we find

$$r\left(1 - \frac{r^2}{R^2}\right)^{1/7} \frac{\partial t}{\partial x} = A \frac{\partial}{\partial r}\left[\left(\frac{R^2 - r^2}{2R}\right)^{6/7} r \frac{\partial t}{\partial r}\right] \tag{3.8}$$

As in II 5 the type solution is

$$\frac{t - t_w}{t_0 - t_w} = \sum_m B_m\, e^{-a_m A x} f_m(r, a_m) \tag{3.9}$$

where $f_m(r, a_m)$ satisfies the ordinary differential equation

$$\frac{d}{dr}\left[r\left(\frac{R^2 - r^2}{2R}\right)^{6/7} \frac{d f_m}{dr}\right] = -a_m f_m r \left(1 - \frac{r^2}{R^2}\right)^{1/7} \tag{3.10}$$

or placing $\left(1 - \frac{r^2}{R^2}\right)^{1/7} = y$ and $49\left(\frac{R}{2}\right)^{8/7} a_m = \omega$

$$\frac{d}{dy}\left[(1 - y^7)\frac{d f_m}{dy}\right] = -\omega\, y^7 f_m \tag{3.11}$$

This again is a differential equation which describes a class of functions which are not readily computed. Solutions satisfying the boundary conditions $f_m = 0$ at $y = 0$ (equivalent to $t = t_w$ at $r = R$) and $d f_m/dy$ finite at $y = 1$ (equivalent to $\partial t/\partial r = 0$ at $r = 0$) are obtained only for certain values of ω.

The approximate solution found by Latzko is as follows:

$$\left.\begin{aligned} \frac{t - t_w}{t_0 - t_w} &= 1.129\, e^{-a_1 (x/D)}\, (0.9544\, y - 0.0212\, y^3 + 0.0668\, y^5) \\ &\quad - 0.180\, e^{-a_2 (x/D)}\, (-0.7472\, y - 4.275\, y^3 + 6.022\, y^5) \\ &\quad + 0.048\, e^{-a_3 (x/D)}\, (20.34\, y - 54.80\, y^3 + 35.47\, y^5) \end{aligned}\right\} \tag{3.12}$$

where

$$a_1 = 0.151\left(\frac{\mu}{\bar{u} D \varrho}\right)^{1/4} \frac{k}{c\mu};\quad a_2 = 2.844\left(\frac{\mu}{\bar{u} D \varrho}\right)^{1/4} \frac{k}{c\mu};\quad a_3 = 29.42\left(\frac{\mu}{\bar{u} D \varrho}\right)^{1/4} \frac{k}{c\mu}$$

Latzko placed $k/c\mu = 1$, making the assumed relation $\beta = c\varepsilon$, rather than $\beta/k = \varepsilon/\mu$.

The rate of heat transfer q_x per unit area from the wall of the pipe can be computed from the relation

$$q_x = \left(\beta \frac{dt}{dr}\right)_{r=R} = -\left[\frac{2r}{7R^2}\left(1 - \frac{r^2}{R^2}\right)^{-6/7} \beta \frac{dt}{dy}\right]_{y=0}$$

There is obtained

$$\left.\begin{aligned} q_x = 0.0346\, \frac{k}{D}\left(\frac{\bar{u} D \varrho}{\mu}\right)^{3/4} (t_w - t_0)\, [1.078\, e^{-a_1 (x/D)} + \\ + 0.1345\, e^{-a_2 (x/D)} + 0.976\, e^{-a_3 (x/D)}] \end{aligned}\right\} \tag{3.13}$$

The average rate q_{av} for the heated section of length L is defined as

$$q_{av} = \frac{1}{L}\int_0^L q_x\,d\,x = 0.247\,c\,\mu\left(\frac{\bar{u}\,D\,\varrho}{\mu}\right)\frac{(t_w - t_0)}{L}\,[1.0113 - e^{-a_1(L/D)} - 0.0066\,e^{-a_2(L/D)} - 0.0047\,e^{-a_3(L/D)}] \tag{3.14}$$

The mean temperature difference $(t_w - t_m)$ between the pipe wall and the fluid may be computed from the relation

$$(t_w - t_m) = \frac{1}{\pi R^2}\int_0^R (t_w - t)\,2\pi\,r\,d\,r$$

whence

$$\frac{t_w - t_m}{t_w - t_0} = 0.970\,e^{-a_1(x/D)} + 0.024\,e^{-a_2(x/D)} + 0.006\,e^{-a_3(x/D)} \tag{3.15}$$

The heat transfer coefficient h_x is defined by Latzko as equal to $q_x/(t_w - t_m)$. We find from (3.13) and (3.15)

$$h_x = 0.0346\,\frac{k}{D}\left(\frac{\bar{u}\,D\,\varrho}{\mu}\right)^{3/4}\frac{1.078\,e^{-a_1(x/D)} + 0.1345\,e^{-a_2(x/D)} + 0.976\,e^{-a_3(x/D)}}{0.970\,e^{-a_1(x/D)} + 0.024\,e^{-a_2(x/D)} + 0.006\,e^{-a_3(x/D)}} \tag{3.16}$$

As x increases, h_x approaches a minimum value

$$h_{min} = 0.03846\,\frac{k}{D}\left(\frac{\bar{u}\,D\,\varrho}{\mu}\right)^{3/4} \tag{3.17}$$

Since Latzko assumed $k/c\mu = 1$, he obtained instead of (3.17)

$$h_{min} = 0.03846\,\bar{u}\,\varrho\,c\left(\frac{\mu}{\bar{u}\,D\,\varrho}\right)^{1/4} \tag{3.18}$$

The experimental results of Nusselt are in better agreement with (3.18) than (3.17) so that the assumption $\beta = c\varepsilon$ is apparently better than $\beta = k\varepsilon/\mu$.

4. The Concept of Mixing Length. The first progress toward a theory relating the eddy viscosity to the mean motion was made by Prandtl, who introduced the concept of mixing length. The mixing length plays the same part in the theory of the fluctuations as the mean free path plays in the kinetic theory of gases. It is the length of path described by a small mass of fluid before it loses its individuality by mixing with neighboring masses. The isolation of small masses and the mixing length itself can be conceived as existing only in a statistical sense.

Considering for simplicity the case of two-dimensional flow of an incompressible fluid in the direction of the X axis with a velocity gradient du/dy perpendicular to the direction of flow, we imagine a small mass of air having a lateral motion of its own of magnitude v' travelling through a lateral distance l before it mixes with its surroundings. If the x component of its speed is equal to the mean x component at the place from which it started, the difference between the speed of this mass and the mean speed $\bar{u}$ in its new position is to a first approximation $-l\,(d\bar{u}/dy)$. We can thus set the average fluctuation u' proportional to $-l\,(d\bar{u}/dy)$. The lateral motion can be imagined to arise as the result of collisions of fluid masses with different velocities u' and can therefore be placed proportional to u' and hence to $l\,(d\bar{u}/dy)$. The eddy shearing stress $\tau' = -\varrho\overline{u'v'}$ is accordingly proportional to $\varrho l^2 (d\bar{u}/dy)^2$. Since l is still unknown, the various constants of proportionality and the correlation factor involved in the product $-\overline{u'v'}$ may be included in l^2. Since τ' must change sign with $d\bar{u}/dy$, we write

$$\tau' = \varrho\, l^2 \left|\frac{d\bar{u}}{dy}\right| \frac{d\bar{u}}{dy} = \varepsilon\,\frac{d\bar{u}}{dy} \tag{4.1}$$

The eddy viscosity on this theory is equal to $\varrho l^2 |d\bar{u}/dy|$, where the mixing length l varies from point to point.

At first sight nothing has been gained, since an assumption as to the variation of ε from place to place has merely been replaced by an assumption as to the variation of l from place to place. However, experiment shows that at large Reynolds numbers the mixing length is practically independent of the magnitude of the velocity and simple assumptions as to spatial distribution give reasonably accurate results. Moreover, suitable assumptions can often be found by dimensional reasoning. For example, near a wall, it seems clear that the mixing length must be proportional to the distance from the wall.

In the Prandtl theory the initial velocity of the fluid mass remains constant during the "life" of the mass. In effect this assumption implies a neglect of the effect of pressure fluctuations. Any other property of the fluid such as color or temperature will also be exchanged between two fluid layers by the motion of the small fluid masses.

Thus if we consider the "eddy heat transfer" $-c\varrho\overline{v't'}$, we should have v' and t' as proportional to $l\,(d\bar{u}/dy)$ and $-l\,(d\bar{t}/dy)$. The "eddy heat transfer" may thus be assumed to be proportional to $c\varrho l^2 \dfrac{d\bar{u}}{dy}\dfrac{d\bar{t}}{dy}$. Here again the constants of proportionality and the correlation factor between v' and t' may be incorporated in the unknown l. This l is usually assumed to be the same as the mixing length involved in the transfer of shearing stress, in which case the eddy conductivity β is equal to $c\varrho l^2 \left|\dfrac{d\bar{u}}{dy}\right|$ which equals $c\varepsilon$.

The identity of the two mixing lengths is usually founded on the identity of the mechanism of transfer, a given mass being responsible for bringing about a certain u' and t' which are in phase and hence have the same correlation with v' and the same factor of proportionality with $d\bar{u}/dy$ and $d\bar{t}/dy$. It has been found experimentally that this identity does not hold in all cases, and the reason has been sought in the neglect of the pressure fluctuations or in the neglect of the rotational motion of the fluid masses.

The mixing length concept can be applied also to three-dimensional problems. The stress tensor is of the form given in (2.1) to (2.6) inclusive with

$$\left.\begin{aligned}\varepsilon^2 = \varrho^2\, l^4 \Big[\Big(\frac{d\bar{w}}{d\,y} + \frac{d\,\bar{v}}{d\,z}\Big)^2 + \Big(\frac{d\bar{u}}{d\,z} + \frac{d\,\bar{w}}{d\,x}\Big)^2 + \Big(\frac{d\,\bar{v}}{d\,x} + \frac{d\,\bar{u}}{d\,y}\Big)^2 \\ - 4\frac{d\,\bar{v}}{d\,y}\frac{d\bar{w}}{d\,z} - 4\frac{d\,\bar{u}}{d\,x}\frac{d\bar{w}}{d\,z} - 4\frac{d\,\bar{u}}{d\,x}\frac{d\,\bar{v}}{d\,y}\Big]\end{aligned}\right\} \tag{4.2}$$

The applications of the Prandtl theory have been largely to the dynamical problem of the isothermal eddying motion. Solutions of heat transfer problems have been of the type described in the preceding section, using the relation $\beta = c\,\varepsilon = c\,\varrho\, l^2 \left|\frac{d\,\bar{u}}{d\,y}\right|$, the values of l and of $d\bar{u}/d\,y$ being taken from the dynamical results for the isothermal case.

A similar theory of eddy stresses and eddy conductivity has been advanced by G. I. Taylor with the difference that the vorticity remains constant during the motion of a fluid mass, the momentum being modified in its journey by changes in pressure.

Consider again the two-dimensional flow of an incompressible fluid along the x axis, and neglect viscosity. The equation of motion is

$$\frac{d\,u}{d\,\theta} + u\,\frac{d\,u}{d\,x} + v\,\frac{d\,u}{d\,y} = -\frac{1}{\varrho}\frac{d\,p}{d\,x} \tag{4.3}$$

and writing $\eta = \frac{1}{2}\left(\frac{d\,u}{d\,y} - \frac{d\,v}{d\,x}\right)$ for the vorticity, (4.3) becomes

$$-\frac{d}{d\,x}\left(\frac{p}{\varrho} + \frac{1}{2}u^2 + \frac{1}{2}v^2\right) = \frac{d\,u}{d\,\theta} + 2\,v\,\eta \tag{4.4}$$

Taking mean values and supposing η is on the average independent of x, (4.4) becomes

$$-\frac{1}{\varrho}\frac{d\,\bar{p}}{d\,x} = 2\,\overline{v'\eta'} \tag{4.5}$$

If now the vorticity η is conveyed without change as is the momentum in Prandtl's theory, the correlation between v and η arises in the same manner as the correlation between v and u in Prandtl's theory. The vorticity of the mean motion is $1/2\ \frac{d\,\bar{u}}{d\,y}$ and hence the fluctuation η' is proportional to $-l\frac{d}{d\,y}\left(\frac{1}{2}\frac{d\,\bar{u}}{d\,y}\right)$ and the mean value

$$2\,\overline{v'\eta'} = -2\,l^2\frac{d\,\bar{u}}{d\,y}\frac{d}{d\,y}\left(\frac{1}{2}\frac{d\,\bar{u}}{d\,y}\right) = -\,l^2\frac{d^2\bar{u}}{d\,y^2}\frac{d\,\bar{u}}{d\,y}$$

Hence (4.5) becomes

$$\frac{1}{\varrho}\frac{d\,\bar{p}}{d\,x} = l^2\frac{d^2\bar{u}}{d\,y^2}\frac{d\,\bar{u}}{d\,y} \tag{4.6}$$

Considering the equilibrium of a small prism of the fluid, the term on the left may be recognized as $\frac{1}{\varrho}\frac{d\,\tau}{d\,y}$, where τ is the shearing stress.

Therefore

$$\tau = \tau_{y=0} + \int_0^y \varrho\, l^2 \frac{d^2\bar{u}}{d\,y^2}\frac{d\,\bar{u}}{d\,y}\,d\,y \tag{4.7}$$

The rate M at which momentum is communicated to a unit volume of the fluid is

$$\text{according to Prandtl} \quad M = \varrho \frac{d}{dy}\left[l^2 \left|\frac{d\bar{u}}{dy}\right| \frac{d\bar{u}}{dy}\right] \tag{4.8}$$

$$\text{according to Taylor} \quad M = \varrho\, l^2 \left|\frac{d\bar{u}}{dy}\right| \frac{d^2\bar{u}}{dy^2} \tag{4.9}$$

On both theories the rate of heat transfer Q per unit area perpendicular to y is

$$Q = c\,\varrho\, l^2 \left|\frac{d\bar{u}}{dy}\right| \frac{d\bar{t}}{dy} \tag{4.10}$$

For the particular case where l is independent of y, we find

$$\text{according to Prandtl} \quad M = 2\,\varrho\, l^2 \frac{d\bar{u}}{dy} \frac{d^2\bar{u}}{dy^2}$$

$$\text{according to Taylor} \quad M = \varrho\, l^2 \frac{d\bar{u}}{dy} \frac{d^2\bar{u}}{dy^2}$$

The generalization of Taylor's theory to three dimensions leads to quite complicated equations. Taylor proved however that Prandtl's momentum theory applies when u', the x component of the fluctuation, is zero.

It is clear from these considerations that the ratio between the eddy conductivity and the eddy viscosity is not constant but depends on the circumstances of the motion. Ruden has attempted to give the physical reasons for this fact[1]. He points out that a mass of fluid in a flow with a velocity gradient can maintain its individuality only if it rotates with an angular velocity $\frac{1}{2}\frac{d\bar{u}}{dy}$, and that without any translatory motion, there will be an exchange of properties of the fluid other than momentum, for example heat energy because of the rotational motion. The effect of the rotation will be negligible if the size of the fluid mass is small in comparison with the mixing length. In general, the effect may vary through wide limits.

The Taylor theory likewise does not make possible the direct computation of the mixing length. Solutions of heat transfer problems follow the same lines as in the preceding section, except for the different relations between β and ε.

5. Von Kármán's Principle of Similarity. A further advance in the theory of eddying flow was made by von Kármán by means of the principle of similarity. It was assumed that the fluctuations were similar to each other throughout the field of flow so that the conditions of flow in the neighborhood of two points differ only by a multiplicative factor of the magnitude of the fluctuations and by a length characteristic. The theory has been outlined in detail only for two-dimensional flow.

Let us suppose the average flow to be a parallel flow in the direction of the x axis and that the axis is chosen through the point to be

[1] Die Naturwissenschaften, Vol. 21, p. 375, 1933.

investigated. We then develop the average velocity near the point in powers of y

$$u = U_0 + U_0' y + U_0'' \frac{y^2}{2} + \ldots \tag{5.1}$$

If the stream function of the fluctuating motion is designated $\psi'(x, y)$ the stream function ψ of the instantaneous motion will be

$$\psi = U_0 y + U_0' \frac{y^2}{2} + U_0'' \frac{y^3}{6} \ldots + \psi' \tag{5.2}$$

We assume that the fluctuations have only a small extension in the y direction so that the development to the term in U_0'' is sufficiently accurate and we assume that the fluctuations may be regarded as a field of flow superposed on and carried along by the fundamental flow. The assumption of similarity means that if we set

$$\left.\begin{aligned} x &= l\,\xi \\ y &= l\,\eta \\ \psi' &= A\,f(\xi, \eta) \end{aligned}\right\} \tag{5.3}$$

only l and A are dependent on the particular point under investigation, that is on U_0' and U_0'', and $f(\xi, \eta)$ is independent of position in the field.

Th. von Kármán obtains the conditions for similarity by actually writing down the equation of motion, but by dimensional analysis it is immediately obvious that if l and A depend only on U_0' and U_0'', $U_0'' l / U_0'$ is constant, and $U_0''^2 A / U_0'^3$ is constant. Similarly, if the shearing stress τ depends only on ϱ, U_0' and U_0'', $\tau U_0''^2 / \varrho U_0'^4$ is constant. Likewise, if the rate of transfer of heat q per unit area depends only on the product ϱc [ϱ and c occurring only in this combination in the differential equation (1.5)], U_0', U_0'', and the temperature gradient t_0' in the mean flow, $q U_0''^2 / \varrho c t_0' U_0'^3$ are constant.

We may write, therefore,

$$l = \frac{a\,U_0'}{U_0''} \tag{5.4}$$

$$A = \frac{b\,U_0'^3}{U_0''^2} = b'\,l^2\,U_0' \tag{5.5}$$

$$\tau = \frac{d\,\varrho\,U_0'^4}{U_0''^2} = d'\,\varrho\,l^2\,U_0'^2 \tag{5.6}$$

$$q = \frac{e\,\varrho\,c\,t_0'\,U_0'^3}{U_0''^2} = e'\,\varrho\,c\,l^2\,U_0'\,t_0' \tag{5.7}$$

where a, b, b', d, d', e, e' are constants.

Since l, the length characteristic, has an undetermined factor, any one of the constants of proportionality may be chosen to be unity. von Kármán selected $d' = 1$ defining l such that

$$\tau = \varrho l^2 U_0'^2 \tag{5.8}$$

The result (5.8) is identical with Prandtl's result (4.1) and l is commonly spoken of as the mixing length in this theory also. The notable advance is the equation (5.4) which relates the mixing length to the mean flow.

The eddy viscosity ε is given by the same expression as before

$\varepsilon = \varrho l^2 U_0'$ or in the old notation,

$$\varepsilon = \varrho\, l^2 \frac{\partial \bar{u}}{\partial y}$$

The eddy conductivity β is not however $c\varepsilon$ but $e' c\varepsilon$ where e' is an undetermined constant. The principle of similarity therefore yields the result that the eddy conductivity is proportional to but not equal to $c\varepsilon$.

It is interesting to note that Taylor's theory of the transport of vorticity also leads to a ratio between β and ε which depends on the mean flow. By (4.7)

$$\varepsilon = \frac{\tau_{y=0} + \int\limits_0^y \varrho l^2 (d^2 u/d y^2)(d\bar{u}/d y)\, d y}{d\bar{u}/d y} \tag{5.9}$$

whereas $\beta = c\varrho l^2 |d\bar{u}/dy|$. Hence if $y = 0$ is chosen at $\tau = 0$,

$$\frac{\beta}{\varepsilon} = \frac{c\,\varrho\, l^2 \left|\frac{d\bar{u}}{dy}\right| \frac{d\bar{u}}{dy}}{\int\limits_0^y \varrho\, l^2 \frac{d^2\bar{u}}{d y^2} \frac{d\bar{u}}{d y}\, d y} \tag{5.10}$$

This ratio obviously depends on the variation of l, $d\bar{u}/dy$, and $d^2\bar{u}/dy^2$ with y. If l is independent of y, $\beta = 2c\varepsilon$. In general it appears from Taylor's formula that β/ε may be a function of y. The principle of similarity on the other hand leads to a ratio which is independent of y, β/ε being constant for a given flow.

The theory of von Kármán has not been extended to the case of three-dimensional flow. A rigorous computation was given only for the simple case of parallel flow, the results for the flow in pipes being obtained partly by a combination of plausible inferences based on the analogy between boundary layer thickness and radius of the pipe and partly by dimensional reasoning.

6. Present Status. The present position of the theory of heat transfer in eddying flow is very unsatisfactory. It appears quite certain that the exchange of quantities having scalar properties such as heat energy does not always obey the same law as the exchange of a quantity having vector properties such as momentum, the mixing lengths being different. The available experimental data indicate that near a wall the mixing lengths are the same for heat transfer and momentum transfer, whereas in the wake of a heated body the mixing length for the heat transfer is $\sqrt{2}$ times that for the momentum transfer (the eddy conductivity being twice the product of specific heat and eddy viscosity).

The laborious computations of heat transfer based on the formulae recently developed by von Kármán for the isothermal flow in pipes and near a wall have not yet been made. These formulae for the velocity distribution and the skin friction are in very good agreement with the experimental data.

CHAPTER IV

DIMENSIONAL ANALYSIS

1. Introductory. As in many other fields of technical physics, where the differential equations governing the phenomena are such that their solution is impractical if not altogether impossible, the method of dimensional analysis has proved a valuable tool in the treatment of heat-transfer problems. The method is particularly useful in the planning of experiments, and in the interpretation of experimental data.

The application of the theory of dimensions to problems in mechanics has been discussed at some length in Division A IV. Hence a brief review of the essentials and illustrations of the application to heat transfer problems will suffice for present purposes.

2. The Π Theorem. The methods of dimensional analysis depend on the fact that all of the terms of any correct and complete physical equation must have the same dimensions. Expressed in another way, the form of such an equation must be independent of the size of the units involved in the various terms of the equation, since the equation describes a relation which exists quite independently of the units selected to measure the several quantities.

By means of this principle it may be shown that any equation

$$F(Q_1, Q_2, \ldots . Q_n) = 0 \tag{2.1}$$

describing a relation among the n different kinds of quantity Q_1, Q_2, ... Q_n is always reducible to the form

$$f(\Pi_1, \Pi_2, \ldots . \Pi_{n-k}) = 0 \tag{2.2}$$

in which each of the variables Π represents a dimensionless product of the form

$$\Pi = Q_1^a, Q_2^b \ldots . Q_n^n \tag{2.3}$$

where k is the number of independent fundamental units required to specify the units of the n kinds of quantitiy and f is some unknown function to be found by experiment. If there are n separate kinds of quantities but more than one quantity of each kind, for example, a number of lengths all the quantities of any one kind may be represented by specifying a single one of that kind and the ratios r', r'' .... of the others to this one. Equation (2.2) then becomes

$$f(\Pi_1, \Pi_2, \ldots . \Pi_{n-k}, r', r'' \ldots .) = 0 \tag{2.4}$$

This is the basic theorem, usually called the Π theorem.

Equation (2.4) may be solved for one of the Π's, say Π_1 to give

$$\Pi_1 = \varphi(\Pi_2, \ldots. \Pi_{n-k}, r', r'' \ldots.) \tag{2.5}$$

The form of the function φ cannot be obtained except by experiment. In particular, the theory of dimensions does not yield the result sometimes attributed to it

$$\Pi_1 = \varphi_1(\Pi_2)\,\varphi_2(\Pi_3) \ldots. \varphi_{n-k-l}(\Pi_{n-k})\,\varphi'(r')\;\varphi''(r'') \ldots. \tag{2.6}$$

The validity of (2.6) in which the variables are separated can be established only as a result of experiment, not as a result of dimensional reasoning. Similarly, the approximation of the function φ by power laws can be established only on an experimental basis. The theory of dimensions alone does not permit us to go beyond (2.5).

The accuracy of the results of dimensional analysis is that of the original list of physical quantities which are involved in the problem. One cannot determine by the theory of dimensions what physical quantities are of importance in any given problem, but only that, *if* certain quantities are of importance, the relation must be of the form (2.5). The accuracy of the result must be tested in the last analysis by experiment.

3. Fundamental Units. In problems in mechanics three independent fundamental units suffice, the units of other physical quantities being expressed in terms of the three fundamental units. In heat-transfer problems it is necessary to add one additional fundamental unit, which may be conveniently the unit of temperature difference. Some writers add a unit of heat energy as a fundamental unit, which is permissible when there is no conversion of heat into mechanical energy and *vice versa,* or in any case if the mechanical equivalent of heat is added as one of the quantities on which the result depends. There is an essential difference between the unit of temperature difference and the unit of heat energy, for whereas a unit of heat energy can readily be derived from the mechanical units of mass, length, and time, the unit of temperature cannot be derived from mechanical units without a further act of arbitrary choice, the choice of the fixed points and of the numerical value to be assigned to the difference between the temperatures of the fixed points (see I **1**).

While any convenient number of fundamental units may be used as discussed by Buckingham[1] if proper account is taken of all the known relations between the units, a great deal of confusion is avoided by the use of the customary four fundamental units, mass, length, time, and temperature difference.

---

[1] Notes on the Method of Dimensions, Phil. Mag., Vol. 42, p. 696, 1921.

In terms of these fundamental units, we may write down the several quantities with which we are concerned in the problem of heat transfer by convection and their dimensions:

Quantity		Dimensions
Mass	Fundamental	M
Length	Fundamental	L
Time	Fundamental	θ
Temperature	Fundamental	d
Area		L^2
Density		$M L^{-3}$
Viscosity		$M L^{-1} \theta^{-1}$
Force		$M L \theta^{-2}$
Pressure		$M L^{-1} \theta^{-2}$
Energy, Heat or Mechanical		$M L^2 \theta^{-2}$
Power, or Rate of Flow of Heat Energy		$M L^2 \theta^{-3}$
Velocity		$L \theta^{-1}$
Acceleration		$L \theta^{-2}$
Specific Heat (per unit mass)		$L^2 \theta^{-2} d^{-1}$
Thermal Conductivity		$M L \theta^{-3} d^{-1}$
Coefficient of Thermal Expansion		d^{-1}
Coefficient of Heat Transfer, *i. e.* Rate of flow of heat energy per unit area divided by a temperature difference		$M \theta^{-3} d^{-1}$

4. Application to Convective Heat Transfer. Let us consider a simple idealized case of heat transfer in apparatus of one particular design so that the only quantity needed for specifying the apparatus is some linear dimension D. Let us suppose further that the density ϱ, viscosity μ, specific heat c, thermal conductivity k, and coefficient of thermal expansion β are constant, hence independent of the temperature. Let the temperature of the heated section be uniform and equal to t. Let the mean speed at a given section of the apparatus (or in some problems the speed at a great distance) be V and the mean temperature of the fluid (or in some problems the temperature at a great distance) be t_0. Suppose the temperature differences are sufficiently low that the heat transfer depends only on the difference $t - t_0$ and not on t and t_0 separately. For the present, we will not neglect the natural convection so that β, the coefficient of thermal expansion, and g, the acceleration of gravity, are considered of importance. The rate of flow q of heat energy per unit area is then determined by the quantities $D, V, \varrho, \mu, c, k, \beta, g, (t - t_0)$.

There is then a relation of the form

$$f\,[q, D, V, \varrho, \mu, c, k, \beta, g, (t - t_0)] = 0 \tag{4.1}$$

Since there are ten kinds of quantities and four fundamental units, there are six Π's. These may be built up in a great many different ways. By custom some of the non-dimensional quantities which can be formed bear special names as given in the following table:

Name	Symbol	Definition
Nusselt Number	N	$qD/k\,(t-t_0)$
Reynolds Number	R	$DV\varrho/\mu$
Prandtl Number[1]	σ	$c\mu/k$
Grashof Number	Gr	$gD^3\beta\varrho^2\,(t-t_0)/\mu^2$

There are two additional independent Π's which may be taken as $\beta\,(t-t_0)$ and $k\,(t-t_0)/\mu V^2$. Thus by the Π theorem (4.1), may be written

$$N = \varphi\left[R, \sigma, Gr, \frac{k\,(t-t_0)}{\mu V^2}, \beta\,(t-t_0)\right] \tag{4.2}$$

Consider first the case of natural convection, for which $V = 0$. The quantities R and $k\,(t-t_0)/\mu V^2$ are then absent and (4.2) becomes

$$N = \varphi'\,[\sigma, Gr, \beta\,(t-t_0)] \tag{4.3}$$

The quantity $\beta\,(t-t_0)$ is a measure of the volume changes in the fluid and may have considerable influence in special problems such as in the loss of heat from a hot wire of very small diameter. Generally, however, the expansion effects will be negligible except insofar as they alter the weight per unit volume and so set up gravity currents. In this case, β will occur only in combination with g in the Grashof number and $\beta\,(t-t_0)$ may be omitted.

Returning to forced convection, let us consider the case in which natural convection currents are negligible so that β and g have no influence on the heat transfer. The Grashof number and $\beta\,(t-t_0)$ are then absent and we have

$$N = \varphi''\left[R, \sigma, \frac{k\,(t-t_0)}{\mu V^2}\right] \tag{4.4}$$

This is as far as one can go by dimensional reasoning alone. There are two ways of proceeding to obtain the result commonly given. One is to introduce the assumption that there is no conversion of mechanical work into heat, *i. e.* that the heating due to viscous friction is negligible. Then the ratio $k\,(t-t_0)/\mu V^2$, which expresses the ratio of the flow of heat to the energy dissipated by viscous friction, can have no effect. An alternative procedure is to utilize the experimental fact that q is proportional to $t-t_0$ in which case $t-t_0$ cannot enter on the right hand side of (4.4). Either procedure leads to the result, $N = \varphi'''\,(R, \sigma)$ or

$$\frac{qD}{k\,(t-t_0)} = \varphi'''\left(\frac{DV\varrho}{\mu}, \frac{c\mu}{k}\right) \tag{4.5}$$

This equation could also be derived by using a unit of heat energy as an additional fundamental unit, but such a procedure involves the tacit assumption that there is no conversion of mechanical energy into heat, the same assumption as introduced above. It should be noted

[1] The reciprocal $k/c\mu$ is called the Stanton Number.

that the same assumption was introduced in Chapter II before writing down the differential equations.

5. Effect of Variation of Properties of the Fluid with Temperature. Equation (4.5) was derived for a fluid for which μ, ϱ, c, and k are independent of the temperature. For actual fluids, there is generally considerable variation of some or all of these quantities with the temperature so that the question immediately arises as to the values which should be used in (4.5). If the several temperature coefficients, for example β, the expansion coefficient, are included in the list of quantities governing the phenomena, one finds additional non-dimensional Π's, as in (4.2), of the general form $\beta\,(t - t_0)$. A general formula with such a large number of Π's is of little practical use.

This dilemma is usually met by the proposal to use "average" values, often without a clear definition of what is meant by average value. It is generally recognized that this average should be something more than the simple average throughout a volume defined by the integral $\iiint \varrho\, dx dy dz$ taken over the volume, divided by the volume. A more common proposal is to weight the local values according to the local temperature

$$\mu_{av} = \frac{\iiint \mu\,(t - t_0)\, d\,x\, d\,y\, d\,z}{\iiint (t - t_0)\, d\,x\, d\,y\, d\,z}$$

What is really desired are the values of ϱ, c, μ, and k for a fluid with properties independent of the temperature which would give the same heat transfer as the real fluid with properties which are dependent on the temperature. To define these "average" values with precision would require the solution of the complete problem.

If the range of variation is not large, the various types of averages will not differ greatly, and the temperature weighted average will be sufficiently accurate.

6. Modification by Introduction of Experimental Data. Equation (4.5) is the result of dimensional analysis plus two assumptions, (1) that the effects of natural convection are negligible and (2) that the heating due to viscous friction is negligible. By introducing other assumptions or information obtained from experiment, simpler formulae are obtained. Thus if the effect of viscosity is considered to be negligible (4.5) reduces to

$$\frac{q\,D}{k\,(t - t_0)} = \varphi'''' \left(\frac{D\,V\,c\,\varrho}{k}\right) \tag{6.1}$$

The combination $D\,V c \varrho / k$ is known as the Péclet number. This form of equation is that used by Boussinesq and deduced by Rayleigh from the principle of similitude.

The same equation may be obtained if the Prandtl number $c\mu/k$ is constant, as it approximately is for gases. For if $c\mu/k$ is constant it may be omitted and one may use either the Péclet number or the

Reynolds number as the argument of the function on the right hand side of the equation.

If now we find by experiment that q varies as V^n, it is obvious that

$$\varphi''''\left(\frac{D\,V\,c\,\varrho}{k}\right) = \left(\frac{D\,V\,c\,\varrho}{k}\right)^n \tag{6.2}$$

In a number of analyses of experimental data, power laws of this character have been used. The exponent n usually varies a little with the range of Péclet numbers under investigation. The function φ''' in (4.5) is often approximated by a product of powers of the Reynolds number and the Prandtl number. It should be remembered however that the power law and the separation of the function of two variables into the product of two functions each of a single variable do not result from dimensional theory alone. These specific results are approximations found experimentally to apply over certain ranges of the two variables.

CHAPTER V

THE ANALOGY BETWEEN HEAT TRANSFER AND SKIN FRICTION

1. The Reynolds Formulation. Another general method of attack on the heat transfer problem is by means of analogy between heat transfer and skin friction. The earliest formulation of the analogy was given by Reynolds for the flow in a pipe. He assumed that the mechanism for the transfer of heat was the same as the mechanism for the transfer of momentum, and concluded that the ratio of the momentum lost by skin friction between two cross sections distance $d\,x$ apart to the total momentum of the fluid is the same as the ratio of the heat actually supplied (by convection only) to the fluid between these sections and the heat which would have been supplied if the whole of the fluid flowing through the pipe had been carried up to the surface between the two sections.

Let us consider the case of fluid of specific heat c flowing with mean velocity V through a cylindrical pipe of radius R whose walls are maintained at constant temperature t_w. Let G be the mass of fluid passing through the pipe per second, t_{av} the average temperature between the two sections $d\,x$ apart, $d\,t/d\,x$ the temperature gradient, $d\,p/d\,x$ the pressure gradient. Then Reynolds' conclusion is that

$$\frac{-\pi\,R^2\,(d\,p/d\,x)}{G\,V} = \frac{G\,c\,(d\,t/d\,x)}{G\,c\,(t_w - t_{av})} \tag{1.1}$$

The rate of heat flow q_x per unit area at the surface of the pipe is given by the relation

$$2\,\pi\,R\,q_x = G\,c\,\frac{d\,t}{d\,x} \tag{1.2}$$

The heat transfer number h_x, which equals $q_x/(t_w - t_{av})$ is therefore

$$h_x = \frac{G\,c}{2\,\pi\,R\,(t_w - t_{av})}\,\frac{d\,t}{d\,x} \tag{1.3}$$

The skin friction F_x per unit area is given in terms of the pressure drop by the relation

$$2\pi R F_x = \pi R^2 \frac{dp}{dx} \tag{1.4}$$

The coefficient of skin friction, c_f is defined by the relation

$$F_x = c_f \frac{1}{2} \varrho V^2 \tag{1.5}$$

whence

$$c_f = \frac{R\,(dp/dx)}{\varrho V^2} \tag{1.6}$$

Substituting (1.6) and (1.3) in (1.1) we obtain

$$h_x = \frac{1}{2} \varrho c V c_f = \frac{1}{2} c G c_f \tag{1.7}$$

This is one form of the Reynolds equation. In terms of F and q_x

$$\frac{F_x}{V} = \frac{q_x}{c\,(t_w - t_{av})} \tag{1.8}$$

which is the form originally given by Reynolds.

Although derived for flow in a pipe, it is often assumed that the same relation applies to any flow for which the thermal and dynamic boundary conditions correspond.

Reynolds pointed out that ultimately it is by conductivity that the heat passes from the walls of the pipe to the fluid so that k should enter into the result.

In the derivation of the Reynolds equation, no hypothesis is made as to whether the flow is laminar or eddying. If the molecular motions in the laminar flow are similar to the molar motions in the eddying flow, the relation between heat transfer and momentum transfer will be the same in each case.

Let us examine the solution given in II 5 for laminar flow in a pipe. It is of course obvious that there is no relation between friction and heat transfer when the pipe is heated for only a short section. A relation can exist only at some distance from the beginning of the heated section, where the heat transfer number h_x is constant. For this case we found $h_x = 1.83\,(k/R)$. By simple transformation of the Poiseuille-Hagen formula,

$$c_f = \frac{8\mu}{\varrho V R} \tag{1.9}$$

By (1.7) we would expect from (1.9), that the heat transfer number would be equal to $4\mu c/R$ which is 2.185 $(\mu c/k)$ times the correct value.

Similarly for the laminar flow along a plate. From II (7.13), we have

$$h_x = \frac{k}{2} \alpha \left(\frac{c\mu}{k}\right) \sqrt{\frac{U_0 \varrho}{\mu x}}$$

By the Blasius formula for skin friction

$$c_f = 0.664 \sqrt{\frac{\mu}{U_0 \varrho x}} \tag{1.10}$$

By (1.7) we would expect from (1.10) that h_x would be equal to $0.332\, c\mu \sqrt{\frac{U_0 \varrho}{\mu x}}$ which is $\frac{0.664}{\alpha\,(c\mu/k)} \frac{c\mu}{k}$ times the correct value. Since $\alpha\,(c\mu/k)$

is approximately equal to 0.664 $(c\mu/k)^{1/3}$, the ratio of the predicted to the correct value is $(c\mu/k)^{2/3}$.

It is obvious that the formula (1.7) does not apply to laminar flow, that the ratio of the predicted to the correct value is a function of $c\mu/k$, and that this function is different for different geometrical arrangements.

If we examine the solution of III 3 for eddying flow in a pipe, we find that the Reynolds equation (1.8) is approximately though not exactly satisfied.

2. The Prandtl-Taylor Formulation. As experimental technique developed, it became clear that the flow through a pipe above the critical Reynolds number was not of the eddying type throughout the entire flow, but there was an approximation to laminar flow in a thin film adjacent to the walls of the pipe. Let us consider the hypothetical case of a thin layer of thickness ε adjacent to the wall of the pipe. Let the temperature at the outer boundary of the layer be t_1, the velocity V_1, and suppose that the layer is so thin that the distributions of velocity and temperature are linear. Then

$$F_x = \mu \frac{V_1}{\varepsilon} \tag{2.1}$$

$$q_x = \frac{k\,(t_w - t_1)}{\varepsilon} \tag{2.2}$$

For the turbulent core, the Reynolds equation (1.8) gives

$$\frac{F_x}{V - V_1} = \frac{q_x}{c\,(t_1 - t_{av})} \tag{2.3}$$

By eliminating t_1 and ε from (2.1), (2.2) and (2.3) we obtain

$$\frac{q_x}{F_x} = \frac{c\,(t_w - t_{av})}{V} \left(\frac{1}{1 + (V_1/V)\,(\mu\, c/k - 1)}\right) \tag{2.4}$$

This equation is usually known as the Prandtl-Taylor formulation of the heat transfer friction analogy. Before the equation can be used, V_1/V must be determined. Using the same power law formulation as was used in III 3, we may proceed as follows. From III (3.2)

$$\frac{\partial u}{\partial r} = -\frac{16}{49}\frac{\bar{u}\, r}{R^2}\left(1 - \frac{r^2}{R^2}\right)^{-6/7} = -\frac{16}{49}\frac{\bar{u}\, r}{R^2}\left(\frac{7}{8}\frac{u}{\bar{u}}\right)^{-6} \tag{2.5}$$

Corresponding to our previous assumption of a thin layer in which the flow is laminar, we assume that $-\mu\,(\partial u/\partial r)$ taken near the wall is equal to the shearing stress at the wall given by III (3.3). Hence setting

$$r = R \qquad 0.7276\,\frac{\bar{u}}{R}\,\mu\left(\frac{u}{\bar{u}}\right)^{-6} = 0.03955\,\varrho\,\bar{u}^2\left(\frac{\bar{u}\, D\, \varrho}{\mu}\right)^{-1/4}$$

$$\text{whence} \qquad \frac{u}{\bar{u}} = 1.82\left(\frac{\bar{u}\, D\, \varrho}{\mu}\right)^{-1/8} = 1.67\left(\frac{\bar{u}\, R\, \varrho}{\mu}\right)^{-1/8} \tag{2.6}$$

In the notation of the present section

$$\frac{V_1}{V} = 1.67\left(\frac{V R\, \varrho}{\mu}\right)^{-1/8} \tag{2.7}$$

Prandtl with a slightly different expression for the velocity distribution obtained the value 1.6 for the numerical coefficient, but remarks that its value is very uncertain, since actually there is no sharp transition between the laminar and eddying regions. He suggests a value 1.0 or 1.1 as best fitting the experimental data.

In this Division we are particularly interested in the flow of air, for which $c\mu/k$ is about 0.7. The difference between (2.4) and (1.8) is in this case not very large.

3. General Remarks. The analogy between skin friction and heat transfer has been discussed most extensively for the problem of flow in pipes. The results apply also to certain other cases, as for example the boundary layer flow near a skin friction plate, but the analogy is not universally applicable. To gain an insight into the conditions to which the analogy applies, let us consider the relations at some point very near to a heated surface (temperature t_w) past which a fluid stream flows. We suppose for simplicity that the flow is two-dimensional. If the speed and temperature at the point are u_1 and t_1 respectively and if the point is distant ε from the surface, we have the relations

$$F_x = \frac{\mu u_1}{\varepsilon} \quad \text{and} \quad q_x = \frac{k(t_w - t_1)}{\varepsilon}$$

whence eliminating ε,

$$\frac{F_x}{q_x} = \frac{\mu u_1}{k(t_w - t_1)} \tag{3.1}$$

Assuming that the effects of heating due to viscous friction and of natural convection are negligible, and that c, μ and k are independent of the temperature, we have by dimensional reasoning

$$\frac{u_1}{V} = f_1\left(\frac{c\mu}{k}, \frac{VD\varrho}{\mu}\right) \tag{3.2}$$

and

$$\frac{t_w - t_1}{t_w - t_0} = f_2\left(\frac{c\mu}{k}, \frac{VD\varrho}{\mu}\right) \tag{3.3}$$

where V, D and t_0 are reference speed, reference length, and reference temperature, respectively.

Substituting (3.2) and (3.3) in (3.1)

$$\frac{F_x}{q_x} = \frac{V}{c(t_w - t_0)} \frac{(c\mu/k) f_1}{f_2} \tag{3.4}$$

To obtain the Reynolds equation, we must have

$$\frac{c\mu}{k} f_1 = f_2$$

or returning to u_1 and t_1

$$\frac{u_1}{V} = \frac{k}{c\mu}\left(\frac{t_w - t_1}{t_w - t_0}\right)$$

The equations governing the distribution of u/V and $\frac{k}{c\mu}\frac{(t_w - t)}{(t_w - t_0)}$ must be identical *for points near the surface.* Since the flow is believed to be laminar very near the surface and the point may be selected as close as one pleases to the surface, we may use the equations for laminar flow in a boundary layer, namely

$$u\frac{\partial u}{\partial x} + v\frac{\partial u}{\partial y} = \frac{\mu}{\varrho}\frac{\partial^2 u}{\partial y^2} - \frac{1}{\varrho}\frac{\partial p}{\partial x} \tag{3.5}$$

$$\varrho c\left(u\frac{\partial t}{\partial x} + v\frac{\partial t}{\partial y}\right) = k\frac{\partial^2 t}{\partial y^2} \tag{3.6}$$

Placing $u/V = X$ and $(k/c\mu)(t_w - t)/(t_w - t_0) = Y$, (3.5) and (3.6) become

$$X\frac{\partial X}{\partial x} + \frac{v}{V}\frac{\partial X}{\partial y} = \frac{\mu}{\varrho V}\frac{\partial^2 X}{\partial y^2} - \frac{1}{\varrho V^2}\frac{\partial p}{\partial x} \tag{3.7}$$

$$X\frac{\partial Y}{\partial x} + \frac{v}{V}\frac{\partial Y}{\partial y} = \frac{k}{\varrho c V}\frac{\partial^2 Y}{\partial y^2} \tag{3.8}$$

The distributions of X and Y will be similar, provided $c\mu/k = 1$ and $dp/dx = 0$. Accordingly these are the conditions for which the Reynolds equation is valid.

Equation (3.6) and hence (3.8) were derived on the assumption that there was no source of heat within the fluid. If heat were generated at the rate S per unit volume, the term S would appear on the right hand side of (3.6) and the term $-\frac{k}{c\mu}\cdot\frac{S}{\varrho c V(t_w - t_0)}$ on the right hand side of (3.8). The distributions of X and Y would then be similar provided $k/c\mu = 1$ and $\frac{1}{V}\frac{\partial p}{\partial x} = \frac{S}{c(t_w - t_0)}$. If $t_w > t_0$, and $\partial p/\partial x$ is negative, S is negative and represents a sink. Suppose then that we consider a hypothetical case in which we have a source of heat at the wall of the pipe and a sink of strength S per unit volume extending throughout the pipe, where

$$S = \frac{c(t_w - t_0)}{V}\frac{\partial p}{\partial x} \tag{3.9}$$

Such a picture in which the heat generated equalled that absorbed by the sink would be a more or less reasonable approximation to the eddying flow in a pipe for which the temperature is approximately uniform over the central core. As suggested by Prandtl we can use (3.9) to compute the heat flow from the wall approximately. The rate of flow of heat q_x per unit area is related to S by the expression $2\pi R q_x = \pi R^2 S$ or $q_x = (R/2)S$. Since the skin friction $F_x = \frac{R}{2}\frac{\partial p}{\partial x}$, (3.9) becomes

$$q_x = \frac{c(t_w - t_0)}{V}F_x$$

From the nature of the approximate picture, t_0 would be selected as the average temperature and V as the average speed.

To conclude, the Reynolds analogy can be shown to be valid only for the case of flow near a skin friction plate of a fluid for which $c\mu/k$ is equal to unity. The application of the Reynolds equation or the Prandtl-Taylor equation to other cases is generally speaking, without other than empirical justification. Neither equation applies to flows in which separation occurs, as for example, the flow outside a cylindrical pipe normal to the wind.

CHAPTER VI

HEAT TRANSFER FROM A SKIN FRICTION PLATE

1. Introductory. In Chapters II and III we have reviewed that part of the theory of the aerodynamics of cooling which is capable of more or less orderly mathematical development. We have treated certain illustrative problems, usually with simplifying assumptions and always, with the assumption that the flow is either everywhere laminar or everywhere eddying. In practice, both types exist in different parts of the field of flow, and the conditions which determine the transition between the two types of flow have been the subject of much investigation, both theoretical and experimental. Some knowledge has been accumulated, but in a form not susceptible as yet of exact mathemetical treatment. This information is of the greatest importance in its bearing on many practical problems of heat transfer.

The distinction between laminar and eddying flow is usually approached on the basis of Reynolds' original experiments on the flow in pipes, corresponding to the historical development of the subject. In some respects, however, the two-dimensional flow of a fluid near a thin flat plate with sharp symmetrical leading edge, the plate being set parallel to the flow, illustrates the basic phenomena somewhat better, and furnishes a more logical beginning. Such a plate is usually known as a "skin-friction" plate.

The heat transfer from a skin friction plate is of considerable interest in aeronautics since the ratio of heat transfer to drag for such a surface represents an ideal maximum which cannot be exceeded. The performance of the plate sets the standard of excellence.

We shall discuss at some length both the dynamical and thermal problem. Experimental data on heat transfer from skin friction plates are not very plentiful, but because of the possibility of applying the Reynolds formula [see V (1.8)], the data on skin friction may be used to compute the heat transfer when dealing with the flow of air. The chief interest, however, is not the data on the total heat transfer or total friction, but the development of certain basic phenomena illustrated in the distribution of speed and temperature near a heated plate. It will be necessary to review certain information which is presented in greater detail in other divisions of this work.

2. Distribution of Speed in Isothermal Flow. The pioneer measurements of the distribution of speed in isothermal flow near a skin friction plate were made by J. M. Burgers and B. C. van der Hegge Zijnen in

1924 with the aid of a hot-wire anemometer. The results are presented in the dissertation of van der Hegge Zijnen in the form of numerous tables and curves giving the observed speeds at several hundred points, whose x and y coordinates with respect to the leading edge of the plate are tabulated, for five speeds of the approaching air stream. Fortunately, it is possible to utilize the methods of dimensional analysis to devise a method of plotting which enables one to obtain a general view of the hundreds of measurements. The speed V at any point (x, y) at distance x from the leading edge and at distance y

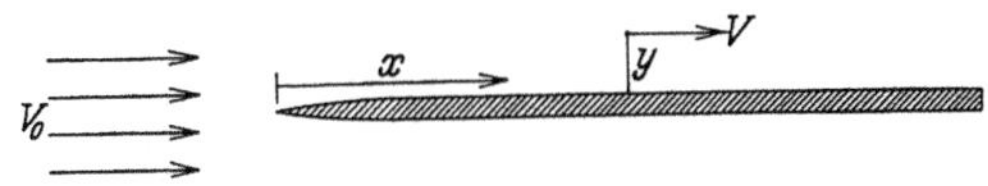

Fig. 6. Coordinates for skin-friction plate.

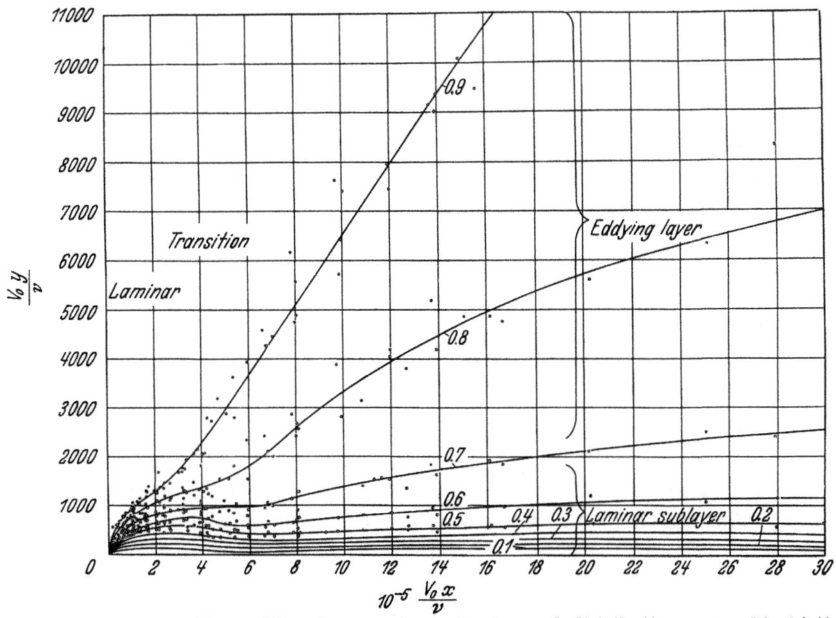

Fig. 7. Van der Hegge Zijnen's measurements of speed distribution near skin-friction plate. See Fig. 6 for x and y. V_0 is speed of free stream. ν is kinematic viscosity of the fluid. Numbers on contours are values of $\frac{V}{V_0}$, where V is the local speed.

from the plate (Fig. 6) is a function of the speed V_0 of the approaching air stream, of the density ϱ and the viscosity μ of the fluid, and of x and y. By the methods of Chapter IV we find at once that

$$\frac{V}{V_0} = \left(\frac{V_0\, x\, \varrho}{\mu}, \frac{V_0\, y\, \varrho}{\mu}\right)$$

The results can therefore be represented by a three-dimensional model, or more conveniently by a contour diagram of the three-dimensional model.

A contour diagram of this type for the measurements of van der Hegge Zijnen is given in Fig. 7. The contours are for values of V/V_0

in steps of 0.1, the corresponding x and y being found by interpolation in the original tables, from which $V_0 x/\nu$ ($\nu = \mu/\varrho$) and $V_0 y/\nu$ were computed. For convenience the scale of $V_0 y/\nu$ has been magnified 200 times. If one wishes to think in terms of x and y, we may imagine a flow of air at a speed of 200 ft./sec., in which case the numbers along the abscissae represent distances in inches and each square along the ordinates, a distance of one thousandth of an inch. Or at a speed of 20 ft./sec., the numbers along the abscissae are in tens of inches and each square along the ordinates a distance of one hundredth of an inch.

The diagram contains data for five speeds and in general the results are very consistent. The deviations correspond to about 0.02 in V/V_0 or 0.005 inches on the average in y. When examined on a large scale there are certain systematic differences between the results at different speeds, which are to be ascribed to the influence of a slight pressure gradient in the air stream in which the measurements were made.

Near the leading edge the contour lines are approximately parabolic in shape and correspond approximately to the theoretical result of Blasius for laminar flow (see II 7). For this reason, the flow in this part of the field is labeled "laminar".

At a $V_0 x/\nu$ of about 300,000, the contours for small values of V/V_0 approach the axis of abscissae, indicating an increasing speed along the plate while the contours for large values bend away from the axis, indicating a rapid thickening of the layer. The process continues over the range from 300,000 to about 500,000, a region usually designated as the "transition" region.

There follows a different type of speed distribution which resembles very closely that found in eddying flow in pipes. In the part marked "eddying" layer, there is at any x a logarithmic relation between V and y. The relations are different near the wall, a region commonly termed the "laminar sub-layer" because the distribution resembles that in the laminar layer. It should be noted that the laminar sub-layer accounts for only a small part of the thickness of the layer but for two-thirds of the fall in speed.

The contour for $V/V_0 = 1$ is not shown, for the reason that V approaches V_0 asymptotically. Various unambiguous procedures can be used to define the "thickness" of the layer of fluid affected by the presence of the plate. We may perhaps think of the distribution as approximated by some specific mathematical expression, and the thickness δ as the value of y which, substituted in that expression, gives $V = V_0$. Or we may make use of the "reduced thickness" δ^* defined

by the relation
$$\delta^* = \frac{1}{V_0} \int_0^\infty (V_0 - V)\, d\, y$$

3. The Concept of Initial Turbulence. In air streams, especially those produced by artificial means as in wind tunnels, the motion is never absolutely steady, and there are always present small ripples or fluctuations which do not usually exceed a few per cent of the average speed. It is difficult to believe that the presence of these fluctuations, usually of frequencies of the order of 20 to perhaps 1000 per second, could play any part in determining the nature of the flow around an object placed in the stream. Yet it has been found experimentally that these fluctuations exert a comparatively large influence in many cases. The basic effect in all these cases is believed to be the effect on the transition from laminar to eddying flow in the boundary layer, which is best illustrated by experiments on flow near a skin friction plate.

Before proceeding with these experiments, it is desirable to review briefly the methods by which this property of air streams, the "initial turbulence", commonly designated simply the turbulence, is evaluated. It is possible to measure directly the mean fluctuation of the speed at any point with time by means of a special form of hot-wire anemometer, with a wire of small diameter, an amplifier, an electrical network to compensate for lag of the wire, and an alternating-current milliammeter. The speed fluctuation is converted into an alternating electric current whose intensity is measured. The turbulence may then be defined as the ratio of the average fluctuation to the mean speed and is usually expressed as a percentage.

Another method of comparing the turbulence of different air streams is by observation of the resistance of spheres or by measurements of the pressure differences between holes at the front and rear of a sphere.

These methods have been compared and both have been found to give a suitable basis for the correlation of data on the effect of turbulence on aerodynamic properties as measured in wind tunnels. A summary of the status of the knowledge of the effects of turbulence as of December 1934 is given in a paper by the author[1]. Experiments completed since the publication of that summary show that some length characteristic as well as an intensity characteristic is required to characterize the effect of the fluctuations on the air resistance of spheres. The method by which the turbulence should be numerically evaluated is therefore still in process of development. This difficulty of measurement and the absence of a satisfactory theory does not alter the fact of the existence of this property of an airstream which has a profound effect on the transition from laminar to eddying flow.

4. The Effect of Turbulence on the Transition from Laminar to Eddying Flow. The qualitative nature of the effect of turbulence was demonstrated by van der Hegge Zijnen by placing a wire screen ahead of his plate to

[1] Journal of the Aeronautical Sciences, p. 67, April 1934.

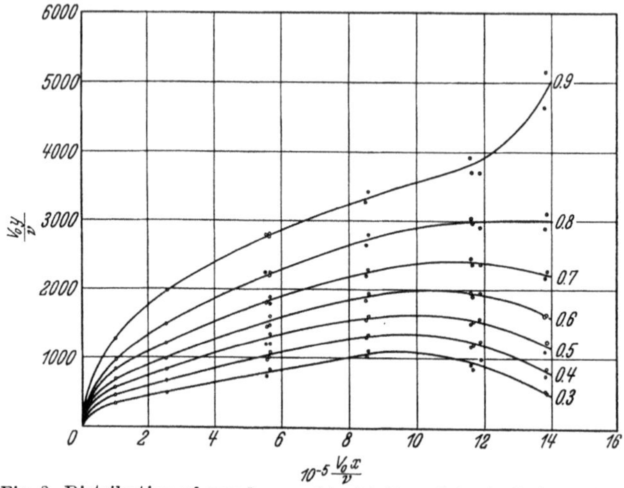

Fig. 8. Distribution of speed near skin-friction plate, turbulence 0.5 per cent Measurements at National Bureau of Standards.

produce greater turbulence. The transition then occurred for $V_0 x/\nu$ of about 100,000. To give some idea of the possible magnitude of the effect, the results of some experiments at the National Bureau of Standards are shown in Figs. 8 and 9. These are contour diagrams similar to Fig. 7. The results of Fig. 8 were obtained in a wind tunnel having a turbulence of 0.5 per cent as measured by the hot-wire method. The transition occurs at a value of $V_0 x/\nu$ of about 1,100,000. For a given speed, the transition begins at a distance from the leading edge nearly four times that found by van der Hegge Zijnen or by Hansen at Aachen.

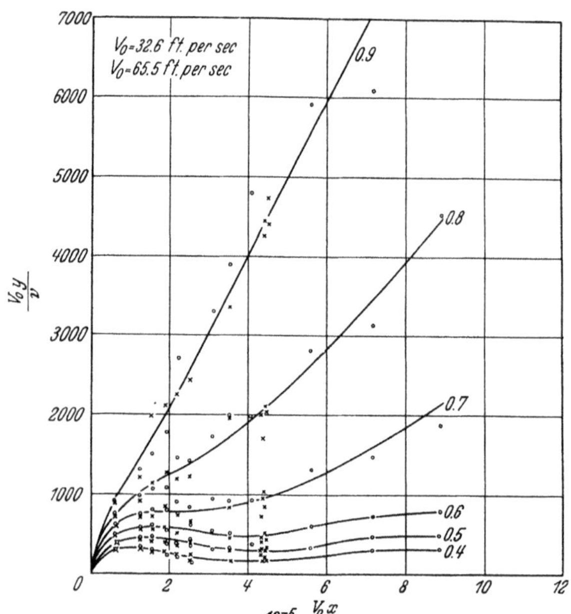

Fig. 9. Distribution of speed near skin-friction plate, turbulence 3.0 per cent.

The results of Fig. 9 were obtained by introducing a screen in the wind tunnel ahead of the plate which increased the turbulence as measured by the hot-wire method to 3.0 per cent. The transition occurs very much earlier, namely at $V_0 x/\nu$ of about 100,000.

These experiments may be summarized as follows. So long as the turbulence of the air stream is unchanged, the transition occurs at a fixed value of the Reynolds number formed from the distance to the leading edge and the speed in the free air stream. If, however, the turbulence is increased, the Reynolds number for transition is decreased and *vice versa*. There is thus a functional relation between the Reynolds number for transition and the turbulence of the air stream, which has not yet been experimentally determined.

5. Distinction Between Laminar and Eddying Flow. The distinction between laminar and eddying flow, made on the basis of the type of distribution of mean speed, is often extended to suggest that the laminar flow is a steady flow (since that is assumed in the theory of laminar flow) and that the eddying flow is not steady, except on the average. The same hot-wire equipment used to measure the turbulence of the air stream can be used to measure the fluctuations near the plate. Contour diagrams of the fluctuations corresponding to Figs. 8 and 9 are given in Figs. 10 and 11. It is seen that the fluctuations in the laminar part of the layer are three times as large as those in the free stream. Large fluctuations develop in the transition region, but these do not persist in the eddying flow.

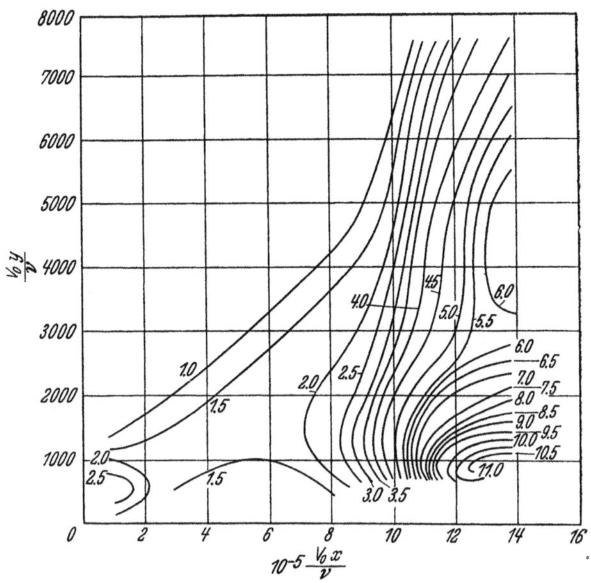

Fig. 10. Distribution of fluctuations of speed corresponding to Fig. 8. Numbers on contours are values of $\frac{100\,\Delta V}{V_0}$ where ΔV is the root-mean-square fluctuation of the local speed. In the free stream $\frac{100\,\Delta V}{V_0} = 0.5$.

The distribution of mean speed indicates a gradual transition. Actually, however, the transition is sudden as shown by the oscillographic records of the fluctuations in Fig. 12 (Plate II). It will be noted that the fluctuations are much more rapid when the layer is eddying than when it is laminar. Likewise, the records in the transition region show a sudden and intermittent alteration of the flow from laminar to eddying, occurring at infrequent intervals near the beginning of the transition region and at more and more frequent intervals as the end of the transition region is approached. From another point of view, the point of transition wanders back and forth erratically and this wandering is responsible for the very large fluctuations in the transition region. These observations are entirely analogous to Reynolds observations of the behaviour of bands of color in the flow of water through a pipe.

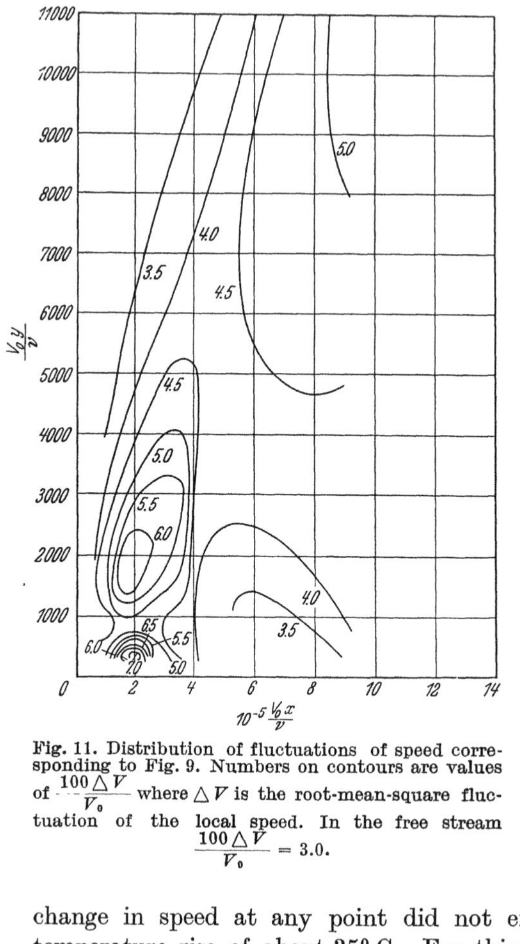

Fig. 11. Distribution of fluctuations of speed corresponding to Fig. 9. Numbers on contours are values of $\frac{100 \triangle V}{V_0}$ where $\triangle V$ is the root-mean-square fluctuation of the local speed. In the free stream $\frac{100 \triangle V}{V_0} = 3.0$.

As emphasized in I 3, the true distinction between laminar and eddying flow must rest not on the presence or absence of fluctuations with time, but on the presence or absence of correlation between the three components of the fluctuations of the speed at a point. The values in Figs. 10 and 11 represent only one component of the fluctuations, approximately that parallel to the plate.

6. Effect of Heat Transfer on the Velocity Field. Comparatively little information is available as to the effect of the heating of the plate on the velocity field. Éliás[1] found that the change in speed at any point did not exceed 2 or 3 per cent for a temperature rise of about 35° C. For this temperature range the kinematic viscosity changes by about 20 per cent. Éliás apparently uses the value corresponding to the arithmetic mean of the plate and air temperatures.

When the results of Éliás are plotted in the manner described in 2, there is evidence of a slightly earlier transition for the heated plate. The effect is not very large, however, and may be neglected.

[1] Zeitschrift für angewandte Mathematik und Mechanik, Vol. 9, p. 434, 1929.

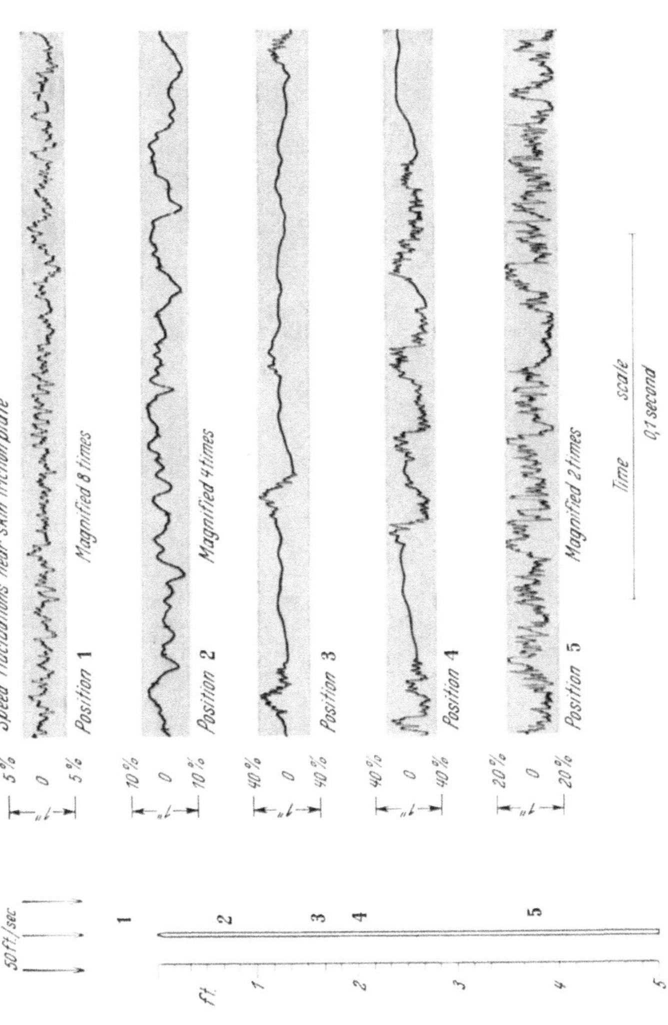

Fig. 12. Oscillographic records of fluctuations of speed at several points near skin-friction plate.

7. Distribution of Temperature. An extensive series of measurements on the temperature distribution near a heated plate in an air stream was made by Éliás. These results may be plotted in the same manner as the measurements of speed, using x and y Reynolds numbers as abscissae and ordinates, and contours of equal temperatures instead of

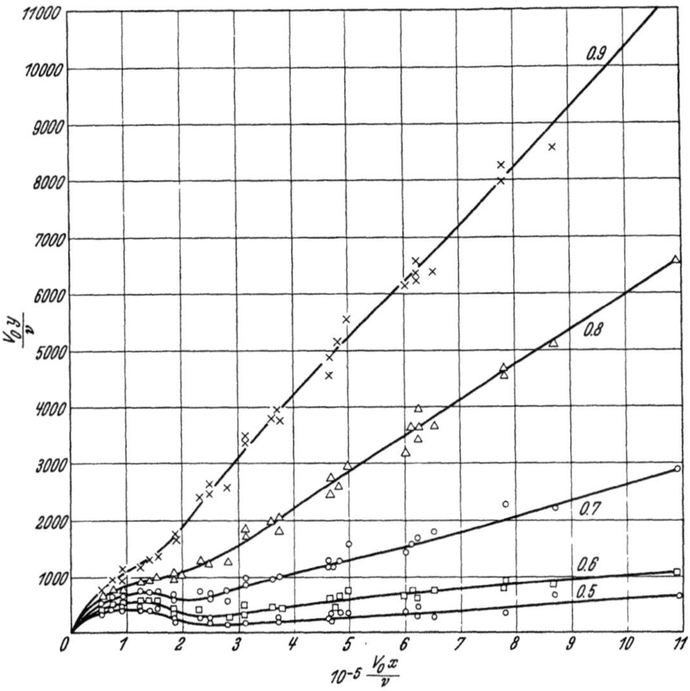

Fig. 13. Distribution of temperature near heated skin-friction plate according to Éliás' measurements. Numbers on contours are $\frac{t_w - t}{t_w - t_0}$, where t_0 is the temperature of the air stream at a distance, t_w is the temperature of the plate, and t is the temperature at points along the contour.

contours of equal speeds. To secure a closer correspondence, the quantity $\frac{t_w - t}{t_w - t_0}$ is used, where t_w is the temperature of the plate, t_0 the temperature of the air stream at a considerable distance from the plate, and t the temperature at points along the contour.

It was not practicable in the experiments of Éliás to heat the plate to a constant temperature, beginning at the sharp leading edge. As a consequence, the thermal boundary layer began at a point 10 cm. behind the leading edge of the plate, at which the dynamic boundary layer begins. The boundary conditions for the speed and temperature were accordingly not quite the same. At distances of 30 cm. and more from the leading edge, the effect of this difference in the boundary conditions appears to have disappeared. We shall therefore confine our attention to the measurements at distances of 30 cm. or more from the leading edge of the plate.

Figures 13 and 14 show the distribution of temperature and the corresponding distribution of speed which was also measured by Éliás. The general similarity of the two diagrams is remarkable. In the laminar region the deviation is in the direction required by the theory of II 7, for $c\mu/k < 1$, *i. e.* the thermal boundary layer is thicker than the dynamic

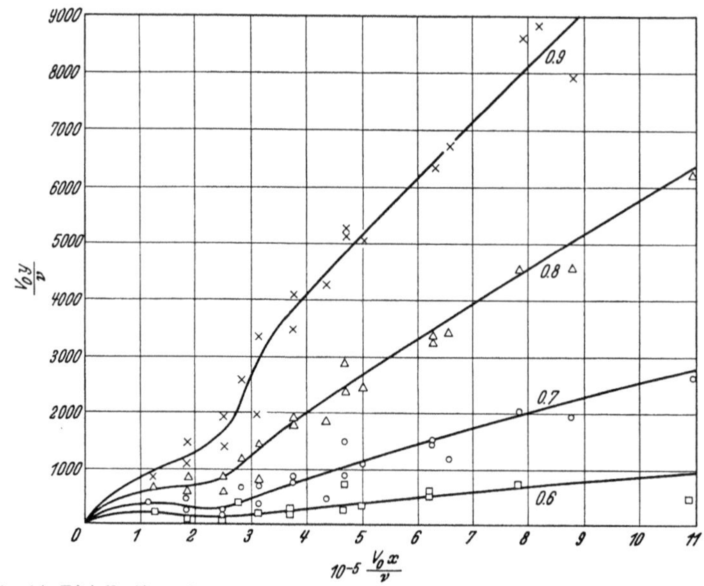

Fig. 14. Distribution of speed near heated skin-friction plate according to Éliás' measurements. See Fig. 6 for x and y. V_0 is speed of free stream. ν is kinematic viscosity of the fluid. Numbers on contours are values of $\frac{V}{V_0}$, where V is the local speed.

boundary layer. In the eddying region, the distributions are substantially identical.

The results of Éliás are applicable to flows for which the turbulence is the same as that present in the wind tunnel at Aachen. It is to be inferred that, corresponding to the speed distributions of Figs. 8 and 9, there are temperature distributions which resemble the speed distributions as Fig. 13 resembles Fig. 14. In general terms, all of the discussion of 2—5 may be applied to the heat transfer problem with temperature substituted for speed. To anticipate a generalization, measurements of heat transfer in air streams, for which no measurement of turbulence has been made, cannot be applied to predict the heat transfer in other air streams, without risk of serious error.

8. Heat Transfer from Skin Friction Plate. It should be clear from the previous discussion that the relations between total skin friction or total heat transfer, the properties of the fluid, and the speed, are likely to be very complex, and that the turbulence must be considered as a variable of primary importance. The skin friction problem is treated in another division of this work. Because of the small practical interest, the heat transfer from plates has not often been experimentally studied.

Éliás' results for the laminar region are in fair agreement with the Pohlhausen formula [II (7.13)], and that formula may be safely used to compute the local heat transfer coefficients in the laminar region. The loss by natural convection and radiation should be investigated, if the speed of the forced air stream is low.

For the eddying region, Latzko by the methods of III 3 obtained the following result for q_x, the local rate of transfer of heat per unit area

$$q_x = 0.0285\,(t_w - t_0)\,V_0\,\varrho\,c\left(\frac{\mu}{V_0\,\varrho\,x}\right)^{1/5} \qquad (8.1)$$

which Éliás finds in fair agreement with his results if x is measured from the beginning of the heated section.

The power law formulation of the skin friction theory has been superceded by the logarithmic law

$$\frac{1}{c_f^{1/2}} = 1.7 + 4.26\,log_{10}\,(R_x\,c_f) \qquad (8.2)$$

where c_f is the local skin friction coefficient and $R_x = V_0\,\varrho\,x/\mu$. For air flow, the local rate of transfer of heat may be estimated from V (1.7)

$$q_x = h_x\,(t_w - t_0) = \frac{1}{2}\,\varrho\,c\,V_0\,c_f\,(t_w - t_0) \qquad (8.3)$$

A comparison of (8.1) and (8.3) is given in the following table:

The power law formulation is a good approximation for Reynolds numbers not exceeding 5,000,000. The experimental results of Éliás extend only to 1,000,000. Formula (8.3) is the more reliable for extrapolation.

R_x	c_f	$\frac{q_x}{\varrho\,c\,V_0\,(t_w - t_0)}$ by (8.1)	$\frac{q_x}{\varrho\,c\,V_0\,(t_w - t_0)}$ by (8.3)
200000	0.00480	0.00240	0.00248
500000	.00402	.00201	.00207
1000000	.00357	.00179	.00180
2000000	.00317	.00159	.00157
5000000	.00274	.00137	.00130
10000000	.00245	.00123	.00113
20000000	.00222	.00111	.00099
50000000	.00196	.00098	.00082
100000000	.00179	.00090	.00072
200000000	.00163	.00082	.00062
500000000	.00145	.00073	.00052
1000000000	.00135	.00068	.00045

To compute the heat transfer from a plate, the Reynolds number of transition and the approximate limits of the transition region must first be estimated. The local rates of transfer in the laminar and eddying regions are then to be computed, and a reasonable transition curve assumed in the transition region. By graphical integration of these local rates of transfer, the total heat transfer may be obtained.

9. Effect of Pressure Gradients. The theoretical treatment of the flow near a skin friction plate is based on the assumption that the pressure gradient along the plate is zero. Both experiment and theory show that for laminar flow the effect of even a very small pressure gradient is quite large, changing the shape of the speed distribution curve, the thickness of the boundary layer, and the local skin friction coefficient. This factor must be considered in the heat flow problem as well.

The existence of a small pressure gradient greatly modifies the Reynolds number of transition for a given turbulence. There is some evidence that a transition Reynolds number defined in terms of the "reduced" thickness (see 2) is less affected by pressure gradients.

There is also evidence that the distribution of speed in an eddying boundary layer depends to some extent on the pressure gradient.

These effects arise naturally in the solution of the boundary layer equations. They are mentioned here only because of the sensitiveness of the boundary layer flow to small pressure gradients.

CHAPTER VII

HEAT TRANSFER FROM A PIPE TO A FLUID STREAM WITHIN THE PIPE

1. Flow near the Entrance. The flow near the entrance of a pipe is analogous to the flow near the front of a skin friction plate. In fact, the plate may be considered as a limiting case of a pipe of infinite radius. Just as the front edge of the plate must be sharp and symmetrical to avoid disturbances which would be similar in effect to increased initial turbulence, the entrance of the tube must be rounded to avoid entrance eddies and the formation of a vena contracta. The fluid enters the pipe with approximately uniform velocity and a boundary layer develops along the wall of the pipe. When however the thickness of the boundary layer equals the radius of the pipe, the process cannot continue further. If the Reynolds number formed from the radius of the pipe and the speed at the center of the pipe does not exceed a critical value determined by the turbulence of the incoming fluid, the flow will remain laminar. If, however, the critical Reynolds number is exceeded, a transition to eddying flow will occur.

The pipe flow differs from the flow about a plate in that the average speed over a cross section must remain constant, the same quantity of fluid passing each cross section, and a pressure gradient is necessarily present. Nevertheless, the phenomena are very similar, and the critical Reynolds numbers are of the same order of magnitude, if that for the plate is expressed in terms of the thickness of the boundary layer rather than in terms of distance from the leading edge. It is possible to have in the entrance region of the pipe laminar flow near the mouth, transition further downstream, and eddying flow as the final type.

Studies of the entrance conditions for a flow which remains laminar throughout have been made by Nikuradse, and approximate theories have been given by Boussinesq[1] and by Schiller[2]. Neither theory is in exact accord with the results of experiments. The final distribution, $u = 2\,\bar{u}\,(1 - y^2/R^2)$, where $\bar{u}$ is the average speed, y the radial distance at which the speed is u and R the radius, is approached asymptotically. The value of x_1, the distance from the entrance for which the actual distribution differs from the parabolic distribution by one per cent, is given by the relation

$$\frac{x_1}{R} = 0.26\,\frac{\bar{u}\,R}{\nu} \tag{1.1}$$

If we consider a tube 1/4 inch in diameter through which air flows at an average speed of 20 feet per second, the entrance length is about 40 inches. The tubes of most aircraft radiators, for example, are not sufficiently long to get out of the entrance length flow.

The entrance length for eddying flow is materially shorter than that for laminar flow and is less dependent on Reynolds number. The estimates range from 20 to 200 radii.

In the entrance length, it is necessary to have a larger pressure gradient than that which is found some distance from the entrance, in order to provide for the increased kinetic energy of the final distribution as compared with that of the initial uniform distribution of speed. In addition, it is probable that the friction is greater. Observations of this increased pressure gradient in short tubes, for example in radiator tubes, have occasionally been interpreted as indicating the presence of eddying flow. Such an interpretation is not always correct.

2. Heat Transfer in the Entrance Length. No accurate measurements of heat transfer in the entrance length of a pipe are known. It is possible to make approximate computations following the methods used by Schiller and Latzko, but the results have not as yet been tested by experiment. The general nature of the results is obvious without computation. At the entrance, both the velocity and temperature gradients are very large, infinite in fact if the velocity and temperature at the entrance are uniform across the section. The gradients decrease rapidly. Hence the local rate of heat transfer per unit area is very high at the entrance but decreases rapidly. If a transition to eddying flow occurs in the entrance section, the gradients of speed and temperature at the wall increase, and hence the local rate of heat transfer increases. Beyond the transition region, the local rate again falls. When the boundary layers coalesce at the center of the pipe at the end of the entrance length,

[1] Comptes rendus, Vol. 113, pp. 9 and 49, 1891.

[2] Zeitschrift für angewandte Mathematik und Mechanik, Vol. 2, p. 96, 1922.

the temperature at the center increases and the average temperature of the fluid increases at much faster rate than in the entrance section. The temperature difference between wall and fluid decreases and hence the local rate of heat transfer decreases. The rate per unit temperature difference, however, approaches a constant value asymptotically. Unfortunately we cannot as yet make accurate computations for this most important practical case of flow in the entrance length, illustrated by oil coolers and water radiators.

3. Heat Transfer for a Short Section of an Infinitely Long Pipe. A problem which has received much attention is that of the transfer of heat from a short section of an infinitely long pipe which is maintained at a uniform temperature. The flow may be either of the laminar or eddying type. The transition type of flow cannot usually be maintained in a very long pipe, although there may be an intermittent alternation between the two types of flow.

The theoretical formulae have already been given in II 5 for laminar flow and in III 3 for eddying flow. In each case, the local rate of transfer is infinite at the beginning of the heated section and decreases to a limiting value as the distance from the beginning of the heated section increases. The formulae are as follows:

$$\text{Laminar case:}\quad h'_x \frac{2R}{k} = f_1\left(\frac{k}{c\mu}\,\frac{\mu}{2R\varrho\bar{u}}\,\frac{x}{2R}\right) \tag{3.1}$$

$$\text{Eddying case:}\quad h'_x \frac{2R}{k}\left(\frac{2R\varrho\bar{u}}{\mu}\right)^{-3/4} = f_2\left[\frac{k}{c\mu}\left(\frac{\mu}{2R\varrho\bar{u}}\right)^{1/4}\frac{x}{2R}\right] \tag{3.2}$$

where R is the radius of the pipe, x is the distance along the pipe from the beginning of the heated section, k, μ and ϱ are the thermal conductivity, viscosity, and density of the fluid, $\bar{u}$ is the average speed defined by $\frac{2}{R^2}\int_0^R u r\,dr$ and h'_x is the local heat transfer number based on the difference between the temperature t_w of the wall and the average temperature t_{av} of the fluid, defined by

$$t_w - t_{av} = \frac{2}{R^2}\int_0^R (t_w - t)\,r\,dr$$

The local rate of heat transfer, q_x, per unit area is given by the relation $q_x = h'_x\,(t_w - t_{av})$.

For large values of $x/2R$ the formulae approach the following values:

$$\text{Laminar case:}\quad \frac{h'_x\,2R}{k} = 5.16 \tag{3.3}$$

$$\text{Eddying case:}\quad h'_x \frac{2R}{k}\left(\frac{2R\varrho\bar{u}}{u}\right)^{-3/4} = 0.03846 \tag{3.4}$$

It is not practicable to measure these local coefficients experimentally. The experimental data usually consist of data on the average rate of transfer of heat per unit area for the whole heated section. It is better then to return to II (5.7) and III (3.14), which may be written as follows:

$$\text{Laminar case:} \quad \frac{2\, q_{av}\, x}{c\, \varrho\, \bar{u}\, R\, (t_w - t_0)} = \psi\left(\frac{k}{c\,\mu}\, \frac{\mu}{2\, R\, \varrho\, \bar{u}}\, \frac{x}{2\, R}\right) \tag{3.5}$$

$$\text{Eddying case:} \quad \frac{2\, q_{av}\, x}{c\, \varrho\, \bar{u}\, R\, (t_w - t_0)} = \varphi\left[\frac{k}{c\,\mu}\left(\frac{\mu}{2\, R\, \varrho\, \bar{u}}\right)^{1/4} \frac{x}{2\, R}\right] \tag{3.6}$$

The left hand members of these equations have a simple physical interpretation. For, if we call t_2 the average temperature with which

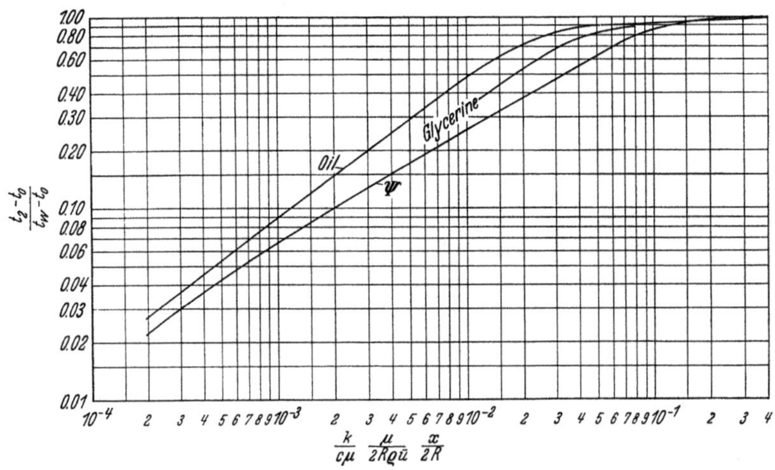

Fig. 15. Heat transfer from a short heated length of an infinitely long pipe to a fluid in laminar flow. t_2 is the average "mixing-cup" temperature of the fluid as it leaves the heated section, the wall of which is at temperature t_w. t_0 is the initial temperature of the fluid. k, c, μ, and ϱ are the thermal conductivity, specific heat, viscosity, and density of the fluid. R is the radius of the pipe, x the length of the heated section, $\bar{u}$ the average speed of the fluid. Curve ψ is the theoretical curve, natural convection negligible. Curves marked oil and and glycerine are experimental curves for those fluids.

the fluid leaves the heated section as determined by the mixing cup method; *i. e.*

$$t_2 - t_0 = \frac{2}{c\, \varrho\, \bar{u}\, R^2} \int_0^R c\, \varrho\, u\, (t - t_0)\, r\, dr$$

the total heat energy transferred to the fluid is

$$c\, \varrho\, \bar{u}\, (t_2 - t_0)\, \pi\, R^2$$

and hence

$$q_{av} = \frac{c\, \varrho\, \bar{u}\, (t_2 - t_0)\, \pi\, R^2}{2\, \pi\, R\, x} = \frac{c\, \varrho\, \bar{u}\, (t_2 - t_0)\, R}{2\, x} \tag{3.7}$$

Hence the left hand members of (3.5) and (3.6) reduce to the simple temperature ratio $\frac{t_2 - t_0}{t_w - t_0}$, which represents the actual temperature rise divided by the maximum possible rise. The values of the functions ψ and φ are shown in Figs. 15 and 16.

The theoretical equations were derived for a constant temperature along the heated section, such as is approximated when the pipe wall is steam heated. There are no reliable data on the transfer of heat to air in laminar flow in a pipe. There is plotted in Fig. 15 the empirical curve for transfer of heat to oil flowing in a heated pipe given by Drew, Hogan, and Mc.Adams[1]. It is seen that the actual transfer is greater than that given by (3.5). The nature of the departures suggests that the discrepancy may be due to the failure of the theory to consider the effects of convection currents in promoting a more thorough mixing

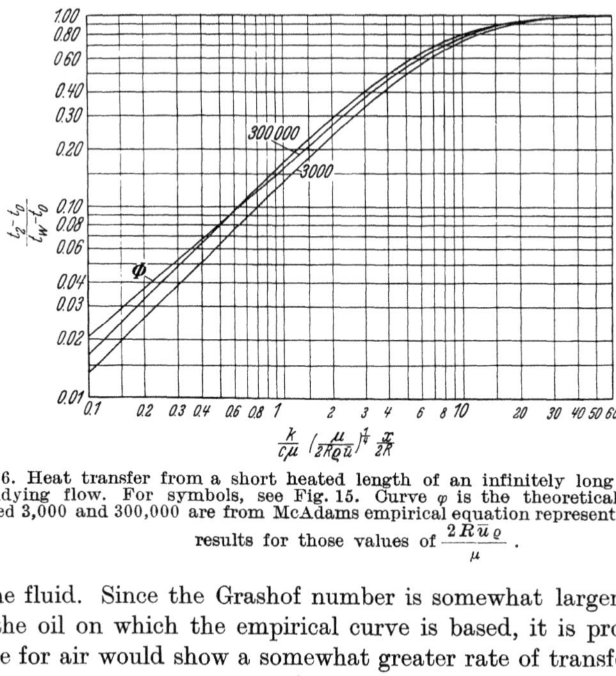

Fig. 16. Heat transfer from a short heated length of an infinitely long pipe to a fluid in eddying flow. For symbols, see Fig. 15. Curve φ is the theoretical curve. Curves marked 3,000 and 300,000 are from McAdams empirical equation representing experimental results for those values of $\frac{2 R \bar{u} \varrho}{\mu}$.

of the fluid. Since the Grashof number is somewhat larger for air than for the oil on which the empirical curve is based, it is probable that a curve for air would show a somewhat greater rate of transfer. An upper limit is set by the fact that $\frac{t_2 - t_0}{t_w - t_0}$ cannot exceed unity.

Some additional evidence that the discrepancy between theory and experiment is due to the effects of natural convection is provided by the experiments of Drew on glycerine[2] in which the Grashof number was about one one-hundredth that in the experiments on oil shown in Fig. 15. The results for glycerine lie closer to the theoretical curve.

For eddying flow McAdams concludes that the best empirical equation for representing the experimental results is as follows:

$$h \frac{2R}{k} = 0.0225 \left(\frac{2 R \varrho u}{\mu}\right)^{0.8} \left(\frac{c \mu}{k}\right)^{0.4} \tag{3.8}$$

[1] Trans. Am. Inst. Chem. Eng., Vol. 26, p. 81, 1931.

[2] Trans. Am. Inst. Chem. Eng., Vol. 27, p. 171, 1931.

The heat transfer number h in this equation is based on an average temperature difference $\triangle t_m$ which is the logarithmic mean of the "mixing cup" temperature of the incoming and outgoing fluid streams. By simple transformations, (3.8) may be written in a form parallel to (3.6) as

$$\frac{2\,q_{av}\,x}{c\,\varrho\,\bar{u}\,R\,(t_w - t_0)} = \frac{t_2 - t_0}{t_w - t_0} = 0.09\left(\frac{2\,R\,\varrho\,\bar{u}}{\mu}\right)^{-0.2}\left(\frac{c\,\mu}{k}\right)^{-0.6}\frac{x}{2\,R}\,\frac{\triangle t_m}{t_w - t_0} \qquad (3.9)$$

If we write for $\triangle t_m$ its equivalent value

$$\frac{(t_2 - t_0)}{\log_e \dfrac{t_w - t_0}{t_w - t_2}}$$

and set $0.09\left(\frac{2\,R\,\varrho\,\bar{u}}{\mu}\right)^{-0.2}\left(\frac{c\,\mu}{k}\right)^{-0.6}\frac{x}{2\,R} = A$, it is easily seen that

$$\frac{t_2 - t_0}{t_w - t_0} = \frac{e^A - 1}{e^A} \qquad (3.10)$$

Since the various quantities do not enter with the same exponents as in (3.6), a general comparison is difficult. If we assume $c\mu/k$ equal to 1, we can plot in Fig. 16 the values from (3.10) for the extreme Reynolds numbers covered by the experiments (3,000 and 300,000). This has been done, and we see that (3.6) and (3.10) agree within about 15 per cent,

when
$$\frac{k}{c\,\mu}\left(\frac{\mu}{2\,R\,\varrho\,\bar{u}}\right)^{1/4}\frac{x}{2\,R}$$
is greater than 1.

4. Remarks on Comparisons with Experimental Data. In the general case we may have to consider the heat transfer in a pipe in which the beginning of the heated section falls at some point within the entrance length. From general considerations we may infer that the local coefficients will be intermediate between those appropriate to the case where the heated section begins at the entrance of the pipe and those appropriate to a short section of an infinitely long pipe. These local coefficients cannot, however, be measured with precision and only overall average values are determined. These average coefficients are functions of the turbulence of the incoming fluid, of the ratio of the length of the heated section to the diameter of the tube, and of the ratio of the length of the unheated entrance to the diameter as well as of the Reynolds number and Prandtl number.

The experimental data available are of a most unsystematic nature as regards these first three important variables and most of the published summaries ignore them. Sufficient data of a systematic nature are not available to determine the magnitude of their influence.

The expression of the data in terms of a heat-transfer number should not mislead those unfamiliar with the subject to infer that the heat-transfer for cases in which the wall temperature varies can be computed from experimental data or theoretical formulae applicable to a constant wall temperature. The local heat-transfer number depends not only on the local wall temperature but also on the conditions elsewhere, especially upstream.

It should be noted that even in the infinitely long pipe, the local rate of transfer of heat varies considerably from upstream to downstream end of the heated length. When the upstream end of the heated section is not sharply defined, disturbing end effects are to be expected. It is not legitimate to determine the average rate of transfer from a limited portion of the heated length.

One frequently finds in the literature the logarithmic mean temperature difference. This type of mean value is correct only if the local rate of transfer is constant. It therefore has no legitimate claim to exactness in the case of heat transfer in a pipe, since the local rate is not constant but varies considerably.

These remarks give some idea of the clear distinctions that must be made and the care that must be taken in the comparison of experimental data from various sources with each other or with theoretical formulae. There is need of a carefully planned experimental investigation in which the influence of the many variables is recognized, and their effects are experimentally determined.

CHAPTER VIII

HEAT TRANSFER FROM CYLINDERS IMMERSED IN A FLUID STREAM

1. Dynamic Boundary Layer. In order to complete the picture of that body of theoretical knowledge not susceptible of mathematical treatment, but which is of importance in heat transfer problems, the final chapter of this Division will be devoted to the heat transfer from circular cylinders immersed in a fluid stream with axes at right angles to the flow. The purpose is primarily to illustrate the influence of pressure gradients and the phenomenon of separation rather than to cover exhaustively the experimental data on cylinders.

The nature of the flow about a cylinder depends on the value of the Reynolds number $V_0 D/\nu$, V_0 being the speed of the fluid stream at a distance from the cylinder, D the diameter and ν the kinematic viscosity. When the Reynolds number is very small, say of the order of 1, the fluid closes in completely behind the cylinder and the flow is everywhere of a laminar character. At a Reynolds number of about 3, a stationary eddy pair develops behind the cylinder. As the Reynolds number is increased, the eddies move away from the cylinder and become unstable. At a Reynolds number of 100, eddies form periodically in the wake, arranging themselves in the well known Kármán vortex street. Over the forward part of the cylinder the flow is still of a laminar character. The fluid stream however does not close in behind the cylinder but separates from the surface. There is evidence that the flow remains of a laminar character for some distance beyond the separation point before eddying motion develops. When the Reynolds number is as large as 30,000, the flow remains laminar up to the point of separation but becomes eddying almost immediately afterward. When the Reynolds number reaches values of the order of 200,000 or more, the flow becomes eddying before separation and the process of separation is delayed, the drag coefficient falling rapidly. The exact values at which the transition to eddying flow occurs in the wake after separation or in the flow ahead of the point of separation depend on the turbulence of the approaching stream.

We shall for the present confine our attention to the flow at values of the Reynolds number between 30,000 and 200,000, for which the drag coefficient is constant. There is a well defined thin boundary layer in which the flow is laminar up to the point of separation, and the flow in the separated boundary layer becomes eddying immediately after separation. The flow outside the boundary layer but not in the wake may be regarded as without viscous friction. The speed just outside the boundary layer increases from zero at the upstream stagnation point to a maximum of about 1.55 times the speed of the approaching stream. The boundary layer is accordingly subjected to a pressure gradient in the direction of flow, associated with the increasing speed. The presence of this pressure gradient has a profound influence on the development of the boundary layer, which is illustrated in the following table. This table gives certain data on the boundary layer flow about a cylinder (taken from Technical Report No. 497 of the National Advisory Committee for Aeronautics) and the thickness which would have been found if there were no pressure gradient, *i. e.* if the speed remained equal to the speed of the approaching stream. R denotes the Reynolds number, D the diameter, x the distance along the surface from the upstream stagnation point, V the local speed in the potential flow just outside the boundary layer, δ the "thickness" as defined by a particular power-series approximation to the speed-distribution curve within the boundary layer, and θ the azimuthal angle.

θ	$\frac{x}{D}$	$\frac{V}{V_0}$	$R\frac{\delta^2}{D^2}$ (cylinder)	$R\frac{\delta^2}{D^2}$ pressure drop absent, $V/V_0 = 1$
0	0	0	1.125	0
10^0	0.0872	0.307	1.140	2.745
20^0	.1744	.633	1.233	5.490
30^0	.2616	.928	1.377	8.236
40^0	.3488	1.175	1.500	10.981
50^0	.4360	1.384	1.731	13.726
60^0	.5232	1.522	2.340	16.471
65^0	.5668	1.549	2.880	17.844
69^0	.6015	1.554	3.600	18.942

It is obvious that the pressure drop reduces the thickness of the boundary layer very materially, (except near the stagnation point) and hence as shown in many experiments[1], increases the skin friction by a factor of two or more as compared with that on a thin flat plate set parallel to the flow.

2. Thermal Boundary Layer. From the equations for the flow in a laminar boundary layer [see II (6.1), (6.4) and (6.7)], it is seen that there is no term in the equation for the flow of heat corresponding to the pressure drop term in the dynamic equations. Thus the pressure drop affects the growth of the thermal boundary layer only indirectly through changes of u and v. While the mathematical solution of this problem has not been worked out in detail, it is perhaps obvious that the thermal boundary layer will develop substantially as it would along a plane surface, *i. e.*, thickening at a fairly rapid rate with consequent decrease in temperature gradient at the surface and in the local rate of transfer of heat, with increasing distance from the upstream stagnation point. By contrast, the dynamic layer remains of nearly constant thickness for some distance and because of the increasing value of V, the skin friction increases with increasing distance from the nose.

This general picture applies to a cylinder whose external surface is maintained at a constant temperature and not to the case where a small part of the cylindrical wall is heated. Many experimental studies have been made on the loss of heat from a heated strip on a cylinder as a function of the azimuthal angle. While these studies are of interest in exploring the nature of the air flow and because of their bearing on certain theoretical investigations, they give no information on the local rate of transfer of heat from a cylinder whose entire outer surface is maintained at a constant temperature.

3. The Phenomenon of Separation. A well-defined boundary layer is observed adjacent to the surface of the cylinder only over the forward part of the cylinder to an angular distance of about 70^0 from the upstream stagnation point. In this vicinity the flow separates from the surface and immediately behind the point of separation it may be demonstrated by smoke that the air near the surface is moving in a direction opposite to that of the main stream. Separation occurs only when the pressure increases in the downstream direction. The process is usually described in the following way. The particles near the wall are dragged along by the friction of the neighboring faster moving particles but are retarded by the pressure. As the layer thickens, the retarding effect predominates and this finally causes a reversal of the flow near the surface. The reversal of flow, on account of the

[1] For example Br. A.R.C. R. and M. 1369.

consequent accumulation of fluid, separates the flow from the surface[1].

When the Reynolds number of the cylinder is increased above about 200,000, a transition from laminar to eddying flow occurs within the boundary layer before separation and the point of separation moves to a larger azimuthal angle. In the eddying flow there is a more thorough mixing of the air particles, and the driving action of the outer layers on the fluid near the surface is greater. The fluid near the surface can accordingly proceed farther against the adverse pressure gradient.

An exact description of the flow near the surface in the separated region is not available. It is known that the average pressure is substantially uniform and that the velocity fluctuates both in magnitude and direction within comparatively wide limits. The average speed is rather low.

4. Local Rate of Heat Transfer as a Function of Azimuthal Angle. Drew and Ryan[2] describe some measurements made under their direction by Paltz and Starr on the average rates of heat transfer from longitudinal strips of a cylinder, the wall of which was maintained at a uniform temperature. The Reynolds number was 39,600, *i. e.*, within the range in which the boundary layer flow is laminar up to the point of separation and becomes eddying immediately after separation. The cylinder was 11.3 diameters long but spanned the air stream so that the flow was approximately two-dimensional. Each strip was 20° wide. The approximate ratios of the local rates to the average rate are given in the table below:

Azimuthal angle at center of strip	$\frac{\text{Local rate per unit area}}{\text{Average rate per unit area}}$
10°	1.39
30°	1.29
50°	1.12
70°	.83
90°	.54
110°	.64
130°	.88
150°	1.02
170°	1.29

By contrast with the behavior of the local skin friction, the local rate of transfer of heat decreases up to and beyond the separation point. In the region of separated flow the local rate of transfer of heat increases and is practically as great at the rear as at the front of the cylinder. The significance of this increase is not definitely known.

These results have been indirectly checked by Lohrisch by diffusion experiments, utilizing the similarity of the differential equations governing

[1] See also Division G **15, 16.**
[2] Trans. Am. Inst. Chem. Eng., Vol. 26, p. 118, 1931.

diffusion and heat transfer. It is obvious that there is no immediate correlation between heat transfer and skin friction because of the effect of the pressure gradient in the dynamic problem. This effect is not present in the diffusion or heat transfer problem.

5. Average Rate of Heat Transfer. Many summaries of the experimental data on the average rate of heat transfer from cylinders in approximately two-dimensional flow (long cylinders extending entirely across the air stream) have been published. Empirical equations, usually power law approximations, are found to be useful representations of the data over limited ranges of Reynolds numbers. McAdams, for example, suggests

$$\frac{q_{av} D}{k (t_w - t_0)} = 0.45 + 0.33 \left(\frac{V_0 D}{\nu}\right)^{0.56} \tag{5.1}$$

This equation is satisfactory for engineering computations when the Reynolds number is between 1 and say 100,000. Extrapolation is probably safe to the point where the boundary layer becomes eddying before separation.

It should be noted that results on cylinders in two-dimensional flow may not be applicable to short cylinders, where the fluid flows around the ends. Experimental data are not available to indicate the magnitude of the end effects.

6. Concluding Remarks. It is perhaps clear that while it is not yet possible to compute either the air flow or the heat transfer for a heated body such as a cylinder, we can recognize certain basic phenomena which are encountered in the flow about bodies of any shape for certain values of the Reynolds number. Thus the establishment of thin boundary layers may be expected over the upstream portions. The flow below a certain critical Reynolds number is laminar. A transition to eddying flow occurs when the Reynolds number reaches a critical value which is dependent on the initial turbulence. It is generally believed but not completely proved that if the thickness of the boundary layer is used as the characteristic length in the Reynolds number, the critical Reynolds number will depend only on the initial turbulence, not on the pressure gradient, shape of the body, etc. With either laminar or eddying flow, the boundary layer will under certain circumstances separate from the surface. It has not yet been possible to devise a universally applicable criterion to indicate when separation will occur.

The use of the concepts of boundary layer, laminar and eddying flow, transition, initial turbulence, separation, has been illustrated by the experimental data for skin friction plates, flow in pipes, and flow about cylinders. The same concepts may be applied to the description of the air flow and the study of the heat transfer in other problems.

Bibliography.

The following references supplement those given in the text with respect to the theoretical aspects of the heat-transfer problem. Other divisions of the work should be consulted for references to papers on fluid mechanics.

Books.

REIHER, H., Wärmeübergang durch Mitführung (Handbuch der Experimentalphysik, herausgeg. von W. Wien u. F. Harms, Vol. 9, part 1, Wärmeübertragung, Chapter 2, p. 268), Akad. Verlagsgesellschaft, Leipzig, 1929.

MCADAMS, W. H., Heat Transmission, McGraw-Hill, New York, 1933.

JAKOB, M., Wärmeleitung (Handbuch der Physik, herausgeg. von H. Geiger u. K. Scheel, Vol. XI, Anwendung der Thermodynamik, Chapter 2, Section VII, p. 132ff.), Julius Springer, Berlin, 1926.

TEN BOSCH, M., Die Wärmeübertragung, Julius Springer, Berlin, 1927.

GROBER, H., Die Grundgesetze der Wärmeleitung und des Wärmeüberganges Julius Springer, Berlin, 1921.

The references to papers are listed under the chapters to which the subject matter is most closely related.

Chapter II.

OBERBECK, A., Über die Wärmeleitung der Flüssigkeiten bei Berücksichtigung der Strömungen infolge von Temperaturdifferenzen, Ann. der Phys., Vol. 243 (3 series, Vol. 7), p. 271, 1879.

GRAETZ, L., Über die Wärmeleitungsfähigkeit von Flüssigkeiten, Ann. der Phys., Vol. 254 (3 series, Vol. 18), p. 79, 1883 and Vol. 261 (3 series, Vol. 25), p. 337, 1885.

BOUSSINESQ, J., Sur le pouvoir refroidissant d'un courant liquide ou gazeux, Comptes Rendus, Vol. 133, p. 257, 1901.

NUSSELT, W., Das Grundgesetz des Wärmeüberganges, Gesundheitsingenieur, Vol. 38, p. 477, 1915.

Chapter IV.

DAVIS, A. H., Convection of Heat and Similitude, Phil. Mag. (6), Vol. 40, p. 692, 1920; Cooling Power of a Stream of Viscous Fluid, Vol. 44, p. 940, 1922.

Chapter V.

REYNOLDS, O., On the Extent and Action of the Heating Surface of Steam Boilers, Proc. Manchester Lit. and Phil. Soc., Vol. 14, p. 9, 1874; also Sci. Papers, Vol. II, p. 81.

STANTON, T. E., On the Passage of Heat between Metal Surfaces and Liquids in Contact with them, Phil. Trans. Roy. Soc., Vol. 190, p. 67, 1897.

PRANDTL, L., Eine Beziehung zwischen Wärmeaustausch und Strömungswiderstand der Flüssigkeit, Phys. Zeitschrift, Vol. 11, p. 1072, 1910.

LANCHESTER, F. W., Surface Cooling and Skin Friction, Br. Aer. Research Com., Repts. and Memo. No. 94, 1913 with appendix by Stanton, T. E., Note on the Relation between Skin Friction and Surface Cooling.

TAYLOR, G. I., Conditions at the Surface of a Hot Body Exposed to the Wind, Br. Aer. Research. Com., Repts. and Memo. No. 272, 1916.

LORD RAYLEIGH (with appendix by T. E. Stanton), On the Suggested Analogy between the Conduction of Heat and Momentum During the Turbulent Motion of a Fluid, Br. Aer. Research Com., Repts. and Memo. No. 497, 1917.

DAVIS, A. H., The Heat Loss by Convection from Wires in a Stream of Air, and its Relation to the Mechanical Resistance, Phil. Mag. (6), Vol. 41, p. 899, 1921.

TEN BOSCH, M., Die Wärmeübergänge in tropfbaren Flüssigkeiten, Zeitschrift des Vereins deutscher Ingenieure, Vol. 70, p. 911, 1926; also Vol. 75, p. 40 and 468, 1931.

LOHRISCH, W., Bestimmung von Wärmeübergangszahlen durch Diffusionsversuche, Forschungsarbeiten a. d. Geb. d. Ingenieurwesens, No. 322, p. 46, 1929.

NUSSELT, W., Wärmeübergang, Diffusion und Verdunstung, Zeitschrift für angew. Math. u. Mech., Vol. 10, p. 105, 1930.

COLBURN, A. P., A Method of Correlating Forced Convection Heat Transfer Data and a Comparison with Fluid Friction, Trans. Am. Inst. Chem. Eng., Vol. 29, p. 174, 1933.

Chapter VI.

NUSSELT, W., and JÜRGES, W., Die Kühlung einer ebenen Wand durch einen Luftstrom. Gesundheitsingenieur, Vol. 45, p. 97 and 641, 1922.

Chapter VII.

NUSSELT, W., Der Wärmeübergang in Rohrleitungen, Forschungsarbeiten a. d. Geb. d. Ingenieurwesens, No. 89, p. 1, 1910.

Chapter VIII.

REIHER, H., Wärmeübergang von strömender Luft an Röhren und Röhrenbündel in Kreuzstrom, Forschungsarbeiten a. d. Geb. d. Ingenieurwesens, No. 269, p. 1, 1925 or Zeitschrift des Vereins deutscher Ingenieure, Vol. 70, p. 47, 1926.

INDEX

Printed by Universitätsdruckerei H. Stürtz A.G., Würzburg.

GPSR Compliance

The European Union's (EU) General Product Safety Regulation (GPSR) is a set of rules that requires consumer products to be safe and our obligations to ensure this.

If you have any concerns about our products, you can contact us on ProductSafety@springernature.com

In case Publisher is established outside the EU, the EU authorized representative is:

Springer Nature Customer Service Center GmbH
Europaplatz 3
69115 Heidelberg, Germany

Batch number: 10682323

Printed by Printforce, the Netherlands

Droplet Wetting and Evaporation

Droplet Wetting and Evaporation

From Pure to Complex Fluids

Edited by

David Brutin

Aix-Marseille University
Marseille
France

ELSEVIER

AMSTERDAM • BOSTON • HEIDELBERG • LONDON
NEW YORK • OXFORD • PARIS • SAN DIEGO
SAN FRANCISCO • SINGAPORE • SYDNEY • TOKYO
Academic Press is an imprint of Elsevier

Academic Press is an imprint of Elsevier
32 Jamestown Road, London NW1 7BY, UK
525 B Street, Suite 1800, San Diego, CA 92101-4495, USA
225 Wyman Street, Waltham, MA 02451, USA
The Boulevard, Langford Lane, Kidlington, Oxford OX5 1GB, UK

ISBN: 978-0-12-800722-8

British Library Cataloguing-in-Publication Data
A catalogue record for this book is available from the British Library

Library of Congress Cataloging-in-Publication Data
A catalog record for this book is available from the Library of Congress

For information on all Academic Press publications
visit our website at http://store.elsevier.com/

Typeset by MPS Limited, Chennai, India
www.adi-mps.com

Publisher: Joe Hayton
Acquisition Editor: Brian Guerin
Editorial Project Manager: Natasha Welford
Production Project Manager: Lisa Jones
Designer: Maria Ines Cruz

Contents

Section II
Heat and Mass Transfer

Section IV
Complex Fluids Drying

List of Contributors

Daniel Bonn Institute of Physics, University of Amsterdam, Amsterdam, The Netherlands

Edward Bormashenko Department of Physics, Ariel University, Ariel, Israel

Wassim Bou Zeid Mechanical Engineering Department, Aix-Marseille University, Marseille, France

David Brutin Mechanical Engineering Department, Aix-Marseille University, Marseille, France

Florian Carle Mechanical Engineering Department, Aix-Marseille University, Marseille, France

Odile Carrier Institute of Physics, University of Amsterdam, Amsterdam, The Netherlands

Andrea Cavalli Physics of Complex Fluids, Department of Science and Technology, University of Twente, Enschede, The Netherlands

Pierre Colinet Université Libre de Bruxelles—TIPs Fluid Physics, Brussels, Belgium

Jean Colombani Institute of Light and Matter, Université de Lyon, Université de Lyon 1 and CNRS, Villeurbanne, France

Paolo Di Marco DESTEC, Engineering Thermodynamics and Heat Transfer, University of Pisa, Pisa, Italy

Stéphane Dorbolo Université de Liège—GRASP Laboratory, Liège, Belgium

Fei Duan School of Mechanical & Aerospace Engineering, Nanyang Technological University, Singapore

David Fairhurst Department of Physics and Mathematics, Nottingham Trent University, Nottingham, UK

Tatiana Gambaryan-Roisman Institute for Technical Thermodynamics and Center of Smart Interfaces, Technische Universität Darmstadt, Darmstadt, Germany

Didier Laux France Institut of electronics and Systems Department, University Montpellier II, Montpellier, France

Marcus Camarinha Lopes Mechanical Engineering Department Technical University Darmstadt, Darmstadt, Germany

Alexandra Mailleur Institute of Light and Matter, Université de Lyon, Université de Lyon 1 and CNRS, Villeurbanne, France

Laurent Maquet Université de Liège—GRASP Laboratory, Liège, Belgium

Marc Medale Mechanical Engineering Department, Aix-Marseille University, Marseille, France

Florian Moreau PPrime ENSMA, Poitiers, France

Frieder Mugele Physics of Complex Fluids, Department of Science and Technology, University of Twente, Enschede, The Netherlands

Ludovic Pauchard Université Paris-Sud, CNRS. Laboratoire FAST, Bat 502, Campus Universitaire, Orsay, France

Aaron H. Persad Thermodynamics and Kinetics Lab, Department of Mechanical and Industrial Engineering, University of Toronto, Toronto, ON, Canada

Christophe Pirat Institute of Light and Matter, Université de Lyon, Université de Lyon 1 and CNRS, Villeurbanne, France

Alexey Rednikov Université Libre de Bruxelles—TIPs Fluid Physics, Brussels, Belgium

Fabrice Rigollet Mechanical Engineering Department, Aix-Marseille University, Marseille, France

Ramon G. Rubio Complutense University of Madrid, Department of Physical Chemistry I, Madrid, Spain

Sergey Semenov Mechanical Engineering Department, Aix-Marseille University, Marseille, France

Benjamin Sobac Mechanical Engineering Department, Aix-Marseille University, Marseille, France; Université Libre de Bruxelles, TIPs—Fluid Physics, Bruxelles, Belgium

Victor Starov Loughborough University, Department of Chemical Engineering, Loughborough, UK

Valérie Vancauwenberghe Mechanical Engineering Department, Aix-Marseille University, Marseille, France

Joachim Venzmer Evonik Industries AG, Essen, Germany

Charles A. Ward Thermodynamics and Kinetics Lab, Department of Mechanical and Industrial Engineering, University of Toronto, Toronto, ON, Canada

Tao Wei School of Mechanical & Aerospace Engineering, Nanyang Technological University, Singapore

Guiping Zhu School of Mechanical & Aerospace Engineering, Nanyang Technological University, Singapore

Biographies

BRUTIN David, Contributor & Editor
Aix-Marseille University, France, david.brutin@univ-amu.fr

David Brutin has been an associate professor at Aix-Marseille University since 2005. He obtained his PhD in the field of mechanical engineering in 2003 at the University of Provence. He got his master degree in mechanical engineering in 2000. David is now working on fluid mechanics, soft matter, energetic, and phase change heat transfer with pure and complex fluids (refrigerants, blood, nanofluids...). His applications include the fields of space, aeronautics, medical diagnosis, forensic science, and the printing industry... David has published over 40 research articles in recognized journals. In 2013, he co-organized the first International Workshop on Wetting and Evaporation: Droplets of Pure and Complex Fluids.

This work has been carried out in the framework of the Labex MEC (ANR-10-LABX-0092) and of the A*MIDEX project (ANR-11-IDEX-0001-02), funded by the *Investissements d'Avenir* French Government program managed by the French National Research Agency (ANR). The work were also funded by the French Space Agency (CNES), the European Space Agency (ESA) and the Carnot Star Institute.

BONN Daniel, Contributor
Institute of Physics, University of Amsterdam, Amsterdam, The Netherlands, d.bonn@uva.nl

Daniel Bonn is director of the Institute of Physics of the University of Amsterdam, where over 200 researchers work. He is also group leader of the "Soft Matter Group", which totals about 30 people, including 4 permanent scientific staff. He is currently supervisor of 15 doctoral students and in his career accompanied to date 26 successful PhD students. Before coming to Amsterdam recently, he was a CNRS research director at the Laboratoire de Physique Statistique of the Ecole Normale Supérieure in Paris, where he led the "complex fluids" group. He published more than 200 papers on wetting phenomena, complex fluids, hydrodynamics, mechanics and the statistical mechanics of glasses of which several in Nature, Science and Physical Review Letters; he was invited more than 40 times as a speaker at international conferences In the past five years.

BORMASHENKO Edward, Contributor
Ariel University, Israel, edward@ariel.ac.il
Edward Bormashenko is a professor of physics and materials science and is the head of the Laboratory of Polymers in Ariel University in Israel. His research interests include wetting phenomena, surface science, superhydrophobicity, wetting transitions, processes of self-assembly, polymer science, soft matter physics, and interaction of plasma with organic materials. He is the author of more than 140 publications in these fields, including the monograph *Wetting of Real Surfaces*. In the last decade, he has devoted his research to interfaces with prescribed wettability, including nonwetted surfaces and nonstick droplets.

CAVALLI Andrea & and MUGELE Frieder, Contributors
University of Twente, f.mugele@utwente.nl
Frieder Mugele is a full professor of physics of complex fluids at the University of Twente. He is currently heading a group of approximately 25 researchers who are studying the fundamental and applied aspects of fluids at interfaces, on scales ranging from molecular dimensions to millimeter. Mugele's research interests include electrowetting, wetting of complex surfaces, dynamic wetting, drop dynamics, two-phase flow microfluidics, AFM spectroscopy in fluids, confined liquids, nanofluidics, optofluidics, and colloids, (micro)rheology. His application areas include oil recovery, inkjet printing, immersion lithography, and lab-on-a-chip systems. Andrea Cavalli has been a postdoctorate in PCF since 2013. His core interests are wetting on textured surfaces, electrowetting, and numerical simulations of two-phase flows, with applications to microfluidics and display technology.

CARRIER Odile, Contributor
Institute of Physics, University of Amsterdam, Amsterdam, The Netherlands, O.M.H.Carrier@uva.nl
With a degree in Chemical Engineering and a PhD in Process and Product Engineering from Université de Lorraine (France), Odile Carrier has been working on emulsification in microfluidic systems and stabilisation of interfaces implying complex fluids, surfactants and crystallization processes. Further developing her interest for interfacial sciences and physical-chemistry, she is currently working on wetting phenomena at the Institute of Physics in the University of Amsterdam (The Netherlands). She took part in several international and/or applied projects for cosmetics, petroleum industry and forensic science.

COLINET Pierre, Contributor
Université Libre de Bruxelles, Belgium, pcolinet@ulb.ac.be

Pierre Colinet is a Senior Research Associate of the Fonds de la Recherche Scientifique (FNRS) and a Professor at the Université Libre de Bruxelles, where he heads the Fluid Physics Research Unit in the TIPs (Transfers, Interfaces, and Processes) laboratory. He currently teaches physical chemistry and modeling of multiphase systems, and his main research interests are in capillarity, wetting, surface-tension-driven flows, phase change (e.g., evaporation and boiling), heat/mass transfer, thin films and droplets, pattern formation, nonlinear dynamics, and micro- and nanofluidics. The authors gratefully acknowledge the financial support of ESA & BELSPO (PRODEX projects), Fonds de la Recherche Scientifique—FNRS (of which PC is a Senior Research Associate) and BELSPO (IAP 7/38μ-MAST project). This research was also carried out under COST Action MP1106's umbrella.

DORBOLO Stéphane, Contributor
University of Liège, Belgium, s.dorbolo@ulg.ac.be

Stéphane Dorbolo obtained his PhD in 2000 at the University of Liège under the supervision of Dr. M. Ausloos in the field of thermal properties of superconductors. He joined N. Vandewalle's group (GRASP) at the University of Liège in 2003. He oriented his research in the field of granular material and particularly the Branly's effect. He sojourned in several places, including the University of Cambridge, Clark University, Collège de France, ENS-Lyon, and Paris 7 for projects in the fields of granular materials, antibubbles, and droplets. Dorbolo became an FNRS research associate in 2006.

DUAN Fei, Contributor
Nanyang Technological University, Singapore, feiduan@ntu.edu.sg

Fei Duan is an associate professor in Nanyang Technological University, Singapore. His research areas include droplet evaporation, thermofluids, and thermal management. Currently, Guiping Zhu is a research fellow, and Tao Wei is a PhD student in Duan's group at Nanyang Technological University.

FAIRHURST David, Contributor
Nottingham Trent University, UK, david.fairhurst@ntu.ac.uk

David Fairhurst studied physics at Imperial College, London, including a year at Ecole Supériere de Physiques et Chimie Industrielles in Paris, where he studied motion in asymmetric profiles with Professor Armand Ajdari. He acquired his

PhD in Edinburgh, studying phase behavior of colloidal systems and the importance of variations in particle size with Professors Wilson Poon and Peter Pusey. After working for Crocodile Clips, an educational software company, he spent 3 months in Professor Paul Callaghan's NMR lab in the University of Victoria, Wellington, New Zealand. Fairhurst returned to the University of Edinburgh for a postdoctorate in nonequilibrium soft matter before taking up his current position at Nottingham Trent University. His research interests include simple and complex fluids, including polymers; surfactants and colloids; the dynamic behavior of droplets, including evaporation; and the stability and dynamics of bubbles and foams.

GAMBARYAN-ROISMAN Tatiana, Contributor
Technische Universität Darmstadt, Germany, gtatiana@ttd.tu-darmstadt.de
Tatiana Gambaryan-Roisman is a professor (apl. Prof.) and a research group leader at the Institute for Technical Thermodynamics, Technische Universität Darmstadt, in Germany. In 1998, she earned a DSc degree in mechanical engineering from Technion, Haifa, Israel. In 2003, Gambaryan-Roisman founded an Emmy Noether-Junior Research Group, "Evaporation of Thin Films on Structured Surfaces," at TU Darmstadt. In 2007, she founded with colleagues a new interdisciplinary research center, the Center of Smart Interfaces. Since January 2014, she has been coordinating an EU Marie Curie Initial Training Network, "Complex Wetting Phenomena" (CoWet). Her research interests include heat and mass transfer, interfacial phenomena in liquids and solids, complex wetting phenomena, multiphase flows, phase change, micro- and nanoscale transport phenomena, and transport phenomena in porous media.

LAUX Didier, Contributor
University Montpellier II, France, didier.laux@univ-montp2.fr
Didier Laux obtained his PhD in 2002 for mechanical properties of high burn-up irradiated nuclear fuel evaluation with ultrasonic methods. He worked 1 year in CEA in Cadarache, France, on homogenization and relocalization approaches in the thermomechanics of irradiated advanced nuclear fuels. Since 2004, he has been an associate professor at the University of Montpellier in a CNRS laboratory (IES—UMR CNRS 5214). He has conducted research on ultrasound and complex media, from the measurement of the Young modulus of nuclear ceramics to the evaluation of the viscosity of biological fluids. Currently, he teaches theoretical and applied physics at the University of Montpellier. He is an expert in high-frequency ultrasonic waves propagation in complex and heterogeneous media. He has published more than 26 papers in international journals and three book chapters.

LOPES Marcus, Contributor
Technische Universität Darmstadt, Germany, marcuscamarinha@gmail.com

Marcus Lopes completed his PhD and undergraduate studies in mechanical engineering at the Technical University Darmstadt in Germany. During his PhD, he conducted research on drop evaporation of liquid mixtures, solutions, emulsions, solvents, and dispersions on partially soluble and deformable surfaces. He has worked in cooperation with the Max-Planck-Institute for Polymer Research and was a visiting researcher at the University of Genoa and Carnegie Mellon University. In 2014, he joined the R&D team of DuPont de Nemours in Luxemburg.

MEDALE Marc, Contributor
Aix-Marseille University, France, marc.medale@univ-amu.fr

Marc Medale is a professor of mechanical engineering at Aix-Marseille University in France. His research is mainly focused on computational fluid dynamics, especially incompressible or thermal flow instability problems, for which he has obtained some reference results and benchmarks. He has authored or coauthored more than 40 journal papers and one book chapter in this field.

PAUCHARD Ludovic, Contributor
University Paris-Sud, France, pauchard@fast.u-psud.fr

Ludovic Pauchard is a CNRS director of research in Fluides, Automatique et Systèmes Thermique laboratory—University Paris-Sud. He completed his PhD in Laboratoire de Physique Statistique (Ecole Normale Supérieure) in 1998. His scientific interests are at the frontier between soft matter and mechanics, including drops, morphogenesis, and mechanical instabilities in out-of-equilibrium systems. He is an expert in studying crack morphologies in paintings to deduce the physical properties of pictorial matter (including the *Mona Lisa*) and judging the authenticity of paintings.

PIRAT Christophe, COLOMBANI Jean, MAILLEUR Alexandra, Contributors
University of Lyon, France, christophe.pirat@univ-lyon1.fr

Christophe Pirat is assistant professor of physics. His interests include multiscale interactions at liquid—solid interfaces in diverse forms, especially the dynamics of wetting and evaporation of drops, impact dynamics with vanishing interfacial friction, and confined flows past superhydrophobic surfaces. Jean Colombani is a professor of physics. His research is devoted to the physics of liquids at mineral interfaces, including porous media, suspensions, granular matter, mineral materials, and particularly the coupling of dissolution with mechanical stress, hydrodynamic flow, and evaporation. Alexandra Mailleur is currently working on her PhD under the supervision of Pirat and Colombani.

SEMENOV Sergey, Contributor
Aix-Marseille University, France, sergey.semenov@univ-amu.fr
Sergey Semenov graduated from Tyumen State University in Russia in 2006. He worked there as a junior researcher in the Laboratory of Liquid Microgravity Technologies for 3 years. In 2009, he began his PhD studies at Loughborough University in the United Kingdom. At that time, he worked as a research associate as part of the Marie Curie Initial Training Network "Multiscale Complex Fluid Flows and Interfacial Phenomena (MULTIFLOW)." Currently, he holds a CNES postdoctorate position at Aix-Marseille University, IUSTI Laboratory. His interests include numerical modeling and computer simulations of evaporating liquid layers and sessile droplets.

SOBAC Benjamin Contributor
Université Libre de Bruxelles, Belgium, bsobac@ulb.ac.be
Benjamin Sobac studied mechanical engineering at the Polytechnic School of Aix-Marseille University in France, where in 2009 he received a masters of engineering and a masters of science in the physics of fluids. He earned his PhD in 2012 from Aix-Marseille University, where he was awarded a prize for the best PhD thesis in the field of engineering sciences. Since 2012, he has been doing postdoctorate studies at the Fluid Physics group of the Transfers, Interfaces, and Processes Lab at the Université Libre de Bruxelles in Belgium. His research concentrates mainly on hydrodynamics at interfaces, wetting, phase change, and heat and mass transfer with pure and complex liquids.

VENZMER Joachim, Contributor
Evonik Industries AG, Germany, joachim.venzmer@evonik.com
Joachim Venzmer studied chemistry at the Johannes Gutenberg University Mainz in Germany and received his education in polymer science at the University of Massachusetts in Amherst, Massachusetts, in the United States. After 6 months as a visiting researcher at the Xerox Research Center of Canada in Mississauga/Ontario, Canada, and obtaining his PhD at the University of Mainz in 1991, he joined Central Research of Th. Goldschmidt AG, which became part of Degussa and then Evonik Industries. Currently, he is the head of Research Interfacial Technology of the Nutrition & Care Segment of Evonik. His main interest is the behavior of amphiphiles at interfaces.

WARD Charles, Contributor
University of Toronto, Canada, charles.ward@utoronto.ca

Charles Ward received his PhD from the University of Toronto in 1967. In 2009, he was awarded the Jules Stachiewicz Medal of the Canadian Society of Mechanical Engineering in recognition of his contributions to the study of heat transfer. He has published over 150 research papers and supervised over 25 doctoral students. His research applies quantum mechanics to the study of energy transport during phase-change processes. His research interests include thermodynamics and kinetics.

Preface

It is really a nice and fascinating book that is published here by David Brutin and some of his colleagues and friends, on "droplets considered as a small lab for wetting and evaporation." As David says, droplets spreading or evaporating on a substrate have of course motivated a huge number of studies, especially in the wetting field (de Gennes, 1985), where it is one of the cardinal case that one can imagine and realize, and also in diverse fields of coating (Kistler and Schweizer, 1997) and more generally engineering. However, it is only recently that the qualitative idea progressively emerges that a so small and fascinating object can become a very simple and nice lab inside which various secret laws of nature could be investigated simply and deeply under very well controled conditions.

The perhaps first sign of this evolution was, to my opinion, a series of remarkable papers from the Chicago group (Deegan, Nagel, Witten, and others) on the so-called "coffee stain" problem (Deegan et al., 1997). These authors realized that very surprising things were occuring in the vicinity of the contact line of a drying drop. Unsual conditions induced by the combined effect of evaporation and hydrodynamics led to an unexpected organization of the solute, and surprising self-organization patterns. This had a great impact and gave rise to a number of other ones on different aspects of this problem and of its generalizations. I was very lucky to be involved in this motion, during a rather short time, by co-organizing a small meeting in Leiden in 2010, with Tom Witen and Vincenzo Vittelli, in which I was surprised by the potentialities of this still exapanding subject, where three singularities were interacting at the same place (evaporation, hydrodynamics, and concentration). The complete understanding of this requires to master hydrodynamics, transferts, chemistry, soft matter, and evaporation, in a very interdisciplinary manner. This first branch of droplet studies was also progressively interacting with another emerging one: a growing number of works also adressed drop dynamics in condition of non-wetting, which led to surprising behaviors (superhydrophobic substrates, Leidenfrost levitation, drop impact and rolling, etc.). The accumulation of all these studies on droplets leads to another sign of this evolution, which was the multiplication of sessions "droplets" in several congresses (APS-DFD, Euromech, etc.), and that were federating an increasing number of research teams.

In some sense DROPLET 2013 was perhaps one of the first attempts to federate all these ones in a single motion, and arrived very timely. Very well organized by David and his collaborators, this meeting was really an oustanding success, in front of "Vieux Port de Marseille," which gives the occasion to many teams to exchange, discuss, federate themselves; and this book is a result among valuable ones of this great event. David has been able to identify and enroll famous collaborators with whom he had well coordinated, which gives this nice book. The great merit of the team is to have avoided to do some "Encyclopedia of Droplets" that would have been a never ending work... Instead, after giving to the reader some general principles on hydrodynamics, evaporation, and contact lines, the book focusses on several and fascinating hot topics, such as convection patterns in drops, self organization of drying induced fractures, drops on soft matter compounds, new repulsive substrates, investigations on biological compounds by drop drying, where, by the way, David realized remarkable and difficult things applicable to criminology (Brutin et al., 2011), and so on. It is rather a walk around the Droplet County that is proposed to the reader, and I wish to everyone to have the same pleasure as I had when discovering these new horizons.

I have no doubt that this book will be a very helpful tool for everyone wanting to learn in this new "Droplet field," and that it will catalyse new progresses in this expanding field. A great and nice book that I recommend to everyone!

Laurent Limat
Paris, April 12, 2015.

References

de Gennes, P.G., 1985. Wetting: statics and dynamics. Rev. Mod. Phys. 57, 827.

Kistler, S.F., Schweizer, P.M., 1997. Liquid Film Coating. Chapman Hall.

Deegan, R.D., Bakajin, O., Dupont, T.F., Huber, G., Nagel, S.R., Witten, T.A., 1997. Capillary flow as the cause of ring stains from dried liquid drops. Nature 389 (6653), 827–829.

Brutin, D., Sobac, B., Loquet, B., Sampol, J., 2011. Pattern formation in drying drops of blood. J. Fluid Mech. 667, 85–95.

Introduction

In the early 1800s, Thomas Young (1805) and Pierre-Simon Laplace (1806) conducted the first investigations on the wetting of droplets. They examined the wetting issues, the role of the contact angle, and the liquid/solid coupling nature driving the droplet problems. While a sessile droplet is a simple geometry, it is also a complex system to solve for real-life situations (metallic inks for inkjet printing, spreading of pesticides on leaves, drops of whole blood or blood serum spreading and drying for medical applications). By taking into account its wetting and its evaporation, this simple case becomes a very complex problem that has been studied by several teams worldwide. The complexity is mainly due to the physics involved, the full coupling with the substrate on which the drop sits (the Latin root of the word *sessile* means "on which one can sit"), the atmosphere, and the fluid nature (pure fluid, bi- or multiphases, or even containing colloids).

The idea for a book addressing the issues of sessile droplet wetting and evaporation came after the first workshop DROPLETS, organized in June 2013, at Aix-Marseille University, Marseille, France. Before the DROPLETS events, each team's work was presented in general conferences: the American Physical Society, Division of Fluid Dynamics (APS, DFD), or the European Colloids and Interface Society (ECIS). Before then, researchers working on droplets in the fields of chemical engineering, mechanical engineering, bioengineering, and surface science had no place to assemble. The workshop has successfully gathered more than 160 scientists, all of whom work in the field of sessile droplets and study the processes of wetting and evaporation. After the event, it was clear that a book on this topic was sorely needed for new scientists, masters and PhD students, and even established researchers who wanted to learn more. Thus, I asked some workshop participants to help me write this book, which covers most of the aspects of droplet wetting and evaporation.

Today, the topic of "droplet" and "evaporation" is the subject of more than 6700 papers published in the past 50 years (1964–2014) in referenced journals archived in the Scopus database. Because the total number of scientific publications also sharply increased in the same period for all areas of research in the life, health and physical sciences fields, the results were normalized and clearly show a linear increase in the percentage of works written by the scientific community. The number of publications dealing with "droplets" and "evaporation" has increased 10-fold in the last 50 years (Figure I.1).

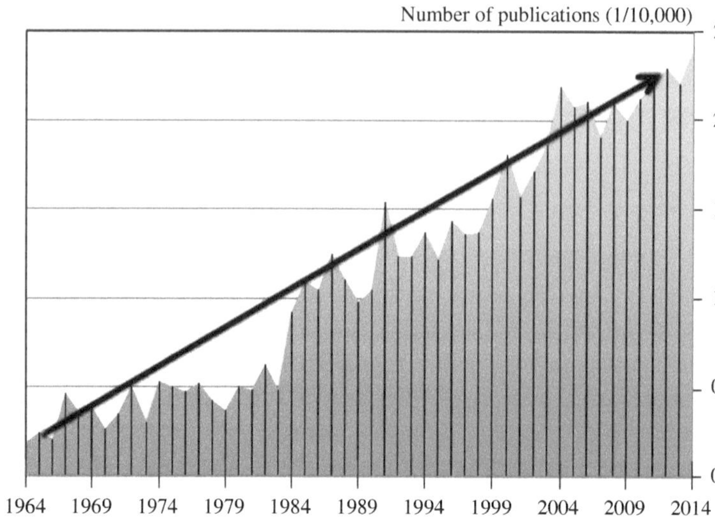

FIGURE I.1 Number of articles published between 1964 and 2014 having "droplet" and "evaporation" in their abstract, key words list, or title that are also referenced in the Scopus® database. The results are normalized by the total number of articles in the categories of "Life + Health + Physical" Sciences. *Scopus®*.

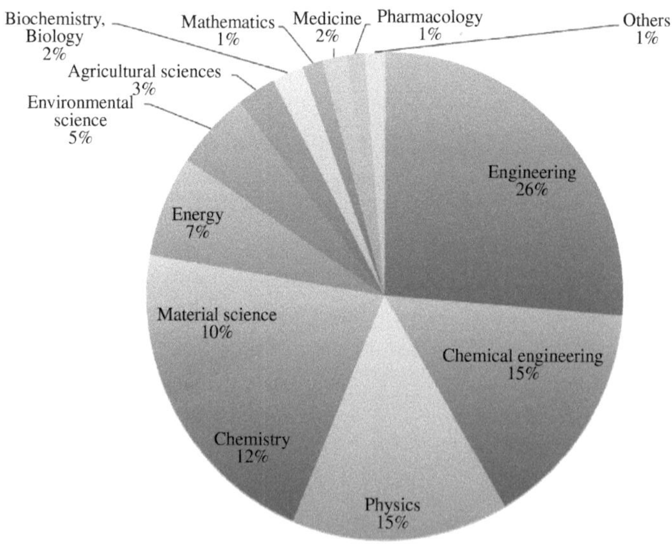

FIGURE I.2 Subject areas regarding the topics of "droplet" and "evaporation." *Scopus®*.

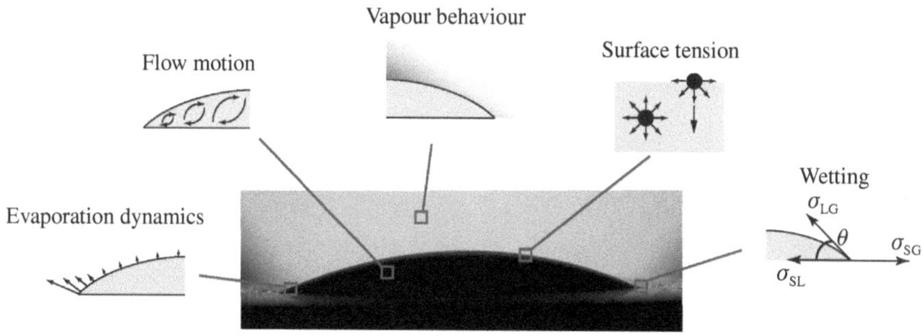

FIGURE I.3 Ways to address the topic of "Droplet Wetting and Evaporation."

Several domains are affected by this increase in publications. The distribution in 2014 shows a strong interest in engineering of more than 40%, fundamental physics and chemistry of 27%, material science of 10%, and energy of 7%. Energy is frequently discussed in published papers dealing with "droplet" and "evaporation," especially because of the applications (printing of complex fluids, spray cooling, DNA analysis) (Figure I.2).

This book contains five chapters that cover wetting and evaporation issues applied to sessile droplet configurations working with pure and complex fluids:

- Chapter 1 addresses wetting and spreading issues and provides the theoretical concepts.
- Chapter 2 addresses heat and mass transfer and presents the driving parameters.
- Chapter 3 addresses flow instabilities and details them either inside the droplet or in the vapor phase.
- Chapter 4 addresses drying complex fluids and soft matter applications and issues.
- Chapter 5 addresses the influence of external forces and presents unusual parameters.

Because the domain of sessile droplets is quite vast, it was not easy to structure the book, present the theoretical concepts, and analyze the work performed by the scientific community. A droplet can be viewed with the eyes of a chemist looking for wetting issues, colloid motions, and their influence on spreading and evaporation. The problem can also be addressed from a mechanical engineering point of view by looking at the internal flow motion or the vapor flow behavior around the evaporating droplet. There are several ways to address this topic, and they are summarized in Figure I.3. Almost all of them are covered in this book.

The mind map of the book is made of the book key words which can be found in the index. Each branch corresponds to a book chapter, but cross-links exist because the physics involved are strongly coupled (Figure I.4).

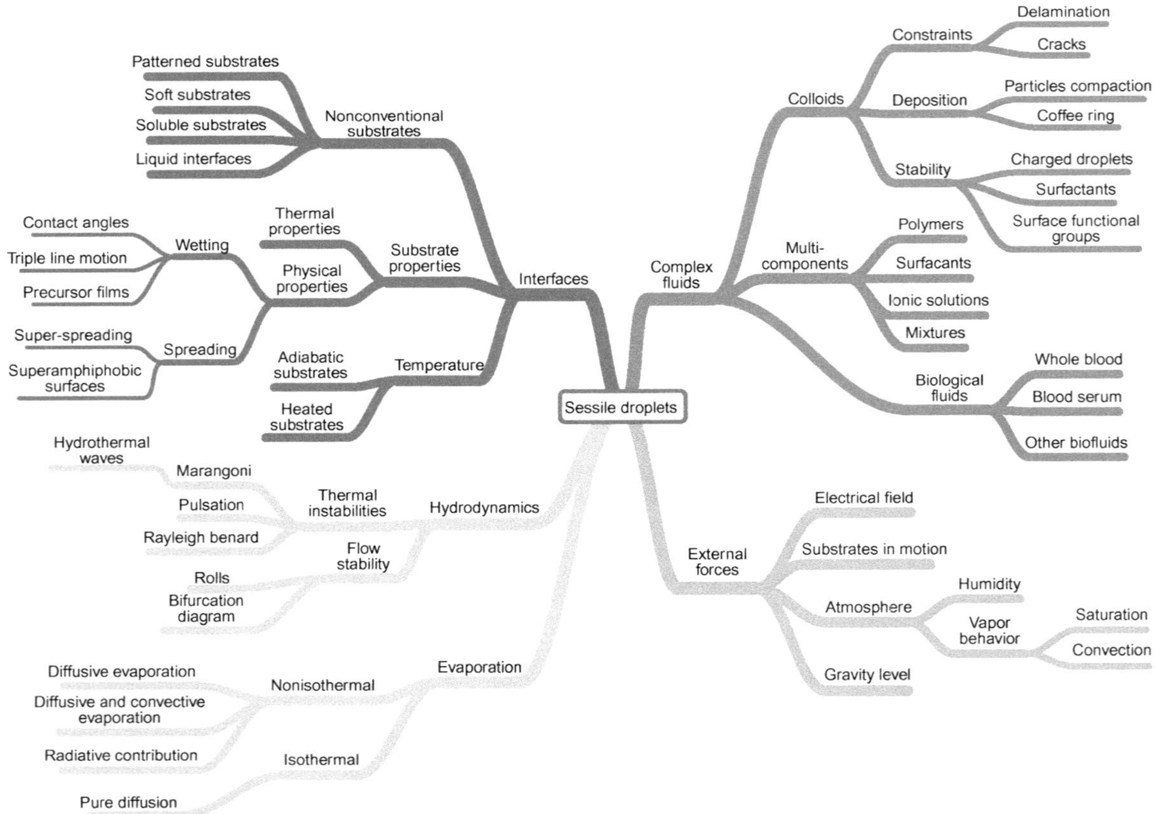

FIGURE I.4 Mind map and key words in the book.

The book is the result of several years of working on droplets as part of a stimulating community that is growing faster and faster all the time. It is the result of collaborative work. I owe much to my alumni who worked on studying droplets: Benjamin Sobac, Florian Carle and Wassim Bou-Zeid. I am also grateful to all other alumni who brought their bricks to this work. I extend my sincere thanks to Natasha Welford and Lisa M. Jones from the Elsevier team for their help and fruitful advices. I also thank all the members of the production team for their work. Finally, I owe much to my wife. Nathalie, you were a great support during this past year while I spent time on this book instead of you.

David Brutin

Aix-Marseille University, Marseille, France

References

Thomas, Y., 1805. An essay on the cohesion of fluids. Phil. Trans. R. Soc. Lond. 95, 65–87.

Pierre-Simon, L., 1806. De l'adhésion des corps à la surface des fluides. J. Phys. LXIII.

Section I

WETTING AND SPREADING

Chapter 1

LIQUID SPREADING

Odile Carrier and Daniel Bonn
Institute of Physics, University of Amsterdam, Amsterdam, The Netherlands

This section addresses the physics of liquid spreading from an energetics point of view. The different modes of spreading are presented and several fluid natures (pure or complex) are investigated.

1.1 Theoretical concepts: equilibrium surface thermodynamics

Capillary phenomena have been studied for centuries, and the key quantity to consider in equilibrium situations is the excess free energy per unit area associated with the formation of an interface between two phases. In order to define the excess free energy, or surface tension, it is common to work in the grand canonical ensemble, fixing the chemical potential rather than the total number of molecules present in the system. This is done because the interface can be viewed as an "open" system; molecules can move between the bulk and interfacial regions, so the number of particles in the interfacial region is not fixed. The particles in the interfacial region must, in addition, be in equilibrium—at a fixed chemical potential—with the particles of the same species in either of the bulk phases (Figure 1.1).

Droplet Wetting and Evaporation. DOI: http://dx.doi.org/10.1016/B978-0-12-800722-8.00001-1

FIGURE 1.1 Droplets on plant leaves. Because the plant leaves clearly repel water, it is difficult to get the droplets to stick to the leaves so, for example, pesticides can be applied.

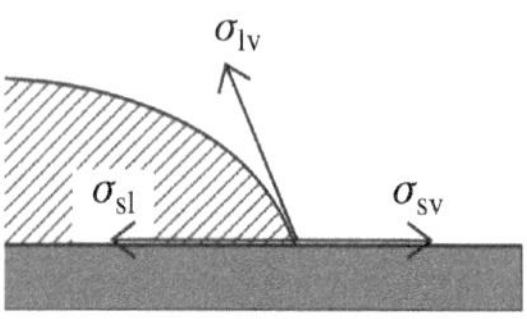

FIGURE 1.2 Force balance at the three-phase contact line. The mechanical equilibrium for this system is equivalent to the Young equation.

We thus employ the grand potential Ω; for a one-component system, $\Omega \equiv F - \mu N$, with F the Helmholtz free energy, μ the chemical potential, and N the number of molecules. The surface tension σ follows from the differential:

$$d\Omega = \sigma \, dA - S \, dT - p \, dV - n \, d\mu \tag{1.1}$$

with A the surface area, S the entropy, T the temperature, p the pressure, and V the volume. The surface tension is consequently defined as the excess free energy per unit area: $\sigma = \left(\frac{\partial \Omega}{\partial A}\right)_{T,V,\mu}$.

Consider, now, a liquid drop on a solid substrate, as depicted in Figure 1.2. Because three different phases are present, three surface tensions must be considered: solid–liquid σ_{sl}, liquid–vapor σ_{lv}, and solid–vapor σ_{sv}. The relation between the cosine of the so-called "contact angle" θ and the three surface tensions is given by the Young equation (1805):

$$\sigma_{sv} = \sigma_{sl} + \sigma_{lv} \cos \theta \tag{1.2}$$

As shown in Figure 1.2, this equation can be interpreted as a force balance for the line of three-phase contact. It simply expresses the mechanical equilibrium of the contact line. The unit of the surface tension ($\mathrm{J\,m^{-2}}$) is an energy per unit area, but it can also be interpreted as a force per unit length ($\mathrm{N\,m^{-1}}$) that acts on the three-phase contact line.

Wetting can be viewed as a direct consequence of the Young equation. The situation where a droplet is present on the surface is commonly called the *partial wetting* state, because the surface is only partially covered with the liquid, as depicted in Figure 1.3A. If we now change some thermodynamic variable—say, the temperature—the three surface tensions will change, and, consequently, the contact angle will also change. At a certain point, it may be that the sum of the solid–liquid and the liquid–vapor surface tensions becomes equal to the solid–vapor interfacial tension. The contact angle will then be zero, and the system will be in equilibrium when a uniform film covers the whole solid surface. In the usual terminology, we say that the liquid *completely wets* the surface. This is depicted in Figure 1.3B, where the *complete wetting* state is shown.

Analogously, it may be that, by changing the temperature, at a certain point the contact angle reaches 180°—that is, the sum of the solid–vapor and the liquid–vapor equals the solid–liquid interfacial tension. Then, the liquid does not wet the solid and, in equilibrium, remains separated from it by a film of vapor. This situation is depicted in Figure 1.3C and is, quite logically, called the

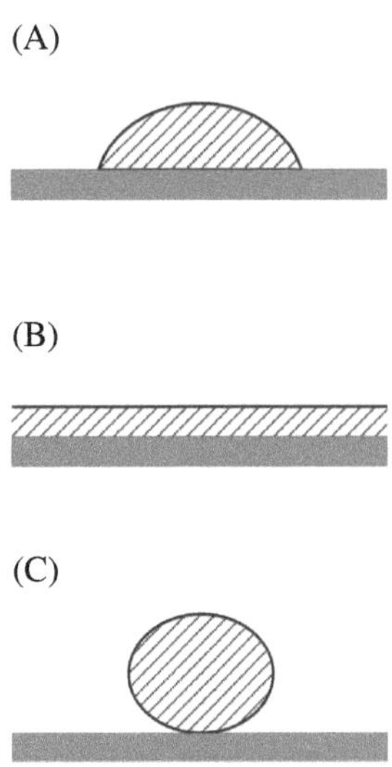

FIGURE 1.3 Wetting of a liquid drop on a solid substrate. (A) Partial wetting state. (B) Complete wetting state. (C) Complete drying state.

complete drying state. Because complete drying is in fact completely analogous to *complete wetting*—but with the roles of the liquid and vapor reversed—in the following we will just consider wetting.

The *wetting transition* is a transition between a partial state and a complete wetting state. If the system is studied as a function of temperature, the transition temperature is called the wetting temperature T_w. The order of the wetting transition is determined—in the same manner as for a bulk phase transition—by the discontinuities in derivatives of the *surface* free energy. If a discontinuity occurs in the first derivative of the free energy, the transition is said to be of first order and will take place in a discontinuous way. On the other hand, if the first derivative of the free energy is continuous through a phase transition point, then that indicates that it is a higher-order or critical phase transition.

The distinction between the different wetting states is usually made by considering the *spreading coefficient* S, which represents the surface free energy difference between partial and complete wetting: $S = \sigma_{sv} - (\sigma_{sl} + \sigma_{lv})$. The spreading coefficient is thus negative for the partial wetting state. Because the sum of the solid–liquid and the liquid–vapor tensions is larger than the solid–vapor surface tension, it is unfavorable to replace the solid–vapor interface with the two others.

An important remark is that *for complete wetting, the equilibrium spreading coefficient is zero*. This is because, at complete wetting, the solid–vapor interface, at equilibrium, is covered with a macroscopically thick wetting layer, so its tension is equal to the sum of the solid–liquid and liquid–vapor surface tensions. Simply said, the solid–vapor interface ceases to exist, because it is no longer stable thermodynamically. Consequently, positive spreading coefficients, sometimes reported in the literature, by definition consider nonequilibrium situations. A much-studied example of this is the spreading of a droplet (see, for instance, the review of de Gennes, 1985, and references therein). Although the *initial spreading coefficient* may be positive in this situation, because it is the driving force of the spreading, once the droplet has spread to cover the whole surface, the equilibrium spreading coefficient will be zero.

It is important to note here that the partial wetting state does not correspond to the complete absence of molecules in the fluid that does the wetting besides the droplet. This is, of course, because of entropy: the presence of a few molecules of the wetting fluid at the substrate—for instance, as a two-dimensional gas—lowers the free energy by $k_B T$ per molecule. Even if the interactions do not favor the presence of many of these molecules at the surface, as is the case for partial wetting, a few molecules will always be present (although it should be kept in mind that it is possible to have a depletion from the surface and thus a negative surface excess density); therefore, we speak of the presence of a microscopic film.

1.2 Spreading dynamics: pure fluids

If a drop is placed on a solid surface, it will in general be far from its equilibrium state—that is, in terms of the extent to which the free energy in the initial spreading coefficient is different from the equilibrium one. In this case, a hydrodynamic flow occurs until the equilibrium contact angle is reached, provided the drop is not trapped in some metastable state. The hydrodynamics of this problem has been studied extensively. The simplest case is that of complete wetting ($S_{eq} = 0$); here, the drop will spread almost forever, presumably until it reaches a thickness set by its own volume and possibly also the van der Waals forces. If the equilibrium contact angle is finite but smaller than the one just after deposition of the drop, the initial stage of the dynamics is similar: the drop will spread out but will eventually go to its equilibrium state given by the Young equation with a finite contact angle.

The dynamics of droplets have attracted a great deal of attention over the past few years and have been studied by many different groups. The spreading of drops is a prototypical singularity problem: a hydrodynamic singularity is at the moving contact line, where the viscous stresses diverge and need to be regularized in the lubrication equations in order to be able to predict the spreading rate. The problem is complicated by the fact that the form of the droplet during the spreading is a priori unknown. For a perfectly wetting liquid on a perfectly flat surface, the effects of the anchoring of the contact line are small. When you deposit a drop, because the liquid is perfectly wetting, the drop starts to spread on the surface and its radius increases in time. Figure 1.4 shows a few typical curves of the radius as a function of time for different alkanes spreading. These experiments were done by gently depositing drops on a glass surface and photographing the time evolution of the drop's radius with Kruss Easydrop, which performs automated image recognition and fits the drop image to a spherical cap. The volume of the droplets was about 1 μL. The largest radius these droplets attained was about 2 to 3 mm, somewhat larger than the capillary length (1 mm).

The key observation from this (and a lot of earlier) experiment is that the increase in drop radius is very small and is strongly sublinear in time. One might have (naively) anticipated a linear dependence of the radius in time. A simple dimensional analysis would (wrongly) assume that the spreading coefficient is the driving force, which is countered by viscous forces. This would lead to a linear increase of the drop radius in time, with a slope on the order of the capillary speed, which is on the order of $\sim 10\ \mathrm{m\ s^{-1}}$ for low-viscosity liquids such as the alkanes discussed above.

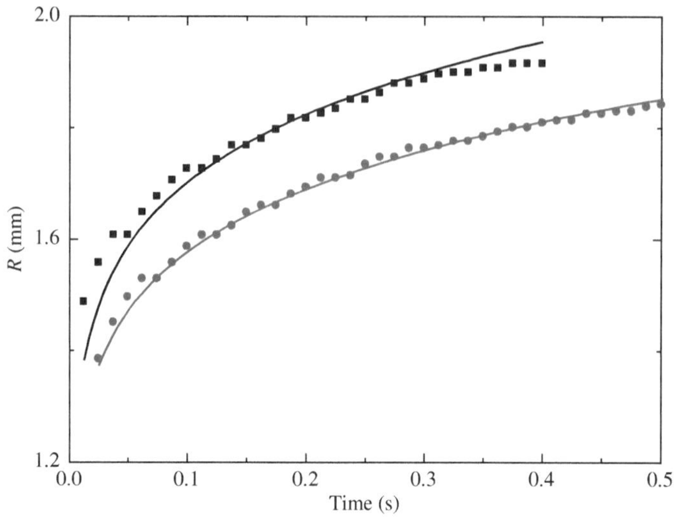

FIGURE 1.4 Spreading of 1 μL nonane (black squares) and dodecane (red dots) droplets on a glass microscope slide. The lines are fits to a power-law following $t^{0.1}$. (For interpretation of the references to color in this figure legend, the reader is referred to the online version of this book.)

Clearly, this is not the case, and the spreading is much, much slower. This happens for two reasons. First, the driving force is not the spreading coefficient (or, more precisely, the difference between the initial and equilibrium one), because this surface energy is lost in the development of the microscopic film discussed in Section 1.1. The existence of this film, called precursor film for the case of spreading, changes notably the solid–vapor surface energy. It does so in such a way that if, for instance, the equilibrium state is complete wetting (zero contact angle), there is no driving force for the spreading from the free energy difference between the partial and complete wetting states. The driving force for the spreading is then only the (much weaker) capillary pressure within the drop, given by the surface tension of the liquid–vapor interface multiplied by the curvature of that same interface.

The second reason for the slow spreading is the hydrodynamic singularity at the contact line. For a detailed review on this issue, we refer to the review of Bonn et al. (2009). For our purposes here, it is sufficient to note that the hydrodynamics of spreading droplets are commonly modeled using the lubrication approximation. For a droplet of radius R and height h:

$$\eta\frac{\partial h}{\partial t} - \gamma\nabla^2\left(h^3\frac{1}{r}\partial_r(r\partial_r h)\right) = 0 \tag{1.3}$$

where η is the viscosity, and γ is the surface tension of the liquid–vapor interface. A simple dimensional analysis, assuming $\nabla \approx \frac{1}{R}$, $\frac{\partial}{\partial r} \approx \frac{1}{R}$, $r \approx R$, and $\frac{\partial h}{\partial t} \approx \frac{h}{t}$, immediately gives:

$$\eta \frac{h}{t} \approx \gamma \frac{h^4}{R^4} \tag{1.4}$$

In addition, from mass conservation we have $h \sim 1/R^2$, so that from Eq. (1.4), it follows directly that $R \sim t^{1/10}$, which is commonly known as Tanner's law. The latter indeed follows from a complete and more rigorous resolution of what leads to the complete Tanner's law: $R = \text{constant}\ V^{3/10} \left(\frac{\gamma t}{\eta}\right)^{1/10}$, where V is the drop volume, γ the surface tension of the liquid–vapor interface, and η the viscosity of the spreading fluid. This agrees with the data shown above and in general relatively well with the available data in the literature. However, it should be noted that the increase in drop radius in experiments is usually only a factor of 2 or 3, and it is therefore difficult to unambiguously show that the very low 1/10th power of Tanner's law is really observed experimentally. Most experiments to date appear to agree with Tanner's law.

1.3 Spreading with complex fluids

Efficient spreading is critical to most painting, coating, and solution delivery applications. However, little work has been devoted to understanding the influence of complex fluids on the dynamics of spreading. The most obvious way to improve spreading for aqueous solutions is to add surfactants to the droplets to decrease the liquid–vapor interfacial tension and thus increase the initial spreading coefficient. It is, however, not clear how surfactants influence the two other surface tensions that intervene in the Young equation, so even though surfactants are widely used, the detailed mechanisms of surfactant action remain at least in part to be explored. Even less explored is the use of complex fluids with a complex rheology in order to influence spreading behavior. Here, we discuss some of the experiments that have been done on spreading with surfactants and non-Newtonian liquids and answer some questions.

For the important case of aqueous solutions on hydrophobic surfaces, one effect of surfactants is to make the spreading coefficient positive; for this, the surfactant concentration must typically be above the critical micellar concentration (cmc). For surfactants, the pertinent question appears to be whether an aqueous surfactant solution spreads faster or slower than a simple liquid with the same surface tension and viscosity. The starting point of the experiments of Rafaï et al. (2002) is therefore a reference system without surfactants on the same substrate.

In order to have similar liquid–vapor interfacial tensions as the surfactant solutions, they used silicon oils with a low liquid–vapor surface tension. Different oils allow varying the viscosity over several orders of magnitude, and the results were well described by the classical Tanner's law presented in Section 1.2:

$$R(t) \propto V^{3/10}\left(\frac{\gamma t}{\eta}\right)^{1/10}.$$

The spreading of aqueous solutions of the surfactant dioctyl sodium sulfosuccinate (AOT) solutions turns out to be very similar to that of the oils, at least for high surfactant concentration where the surface tension is similar to that of the oils (Figure 1.5). In this case, $S > 0$, and the evolution of the radius can be fitted to Tanner's law, but the prefactor is typically 25% smaller than a Newtonian reference system with the same viscosity and equilibrium tension. This implies that *the surfactant slows down the spreading* in comparison with the reference system. Viscosity measurements show that increase due to the addition of surfactant is negligible. Also, because all of the measurements are above the cmc, where the equilibrium tension depends only very weakly on the bulk surfactant concentration, the observed variations cannot be attributed to a variation in the equilibrium surface tension. One possible explanation for the reduction in the rate of movement is that when new interface is created during the spreading, surfactant needs to be transported to the interface in order to lower the tension. Because the creation of new area is faster than the arrival of surfactant molecules to the interface (Roché et al., 2009), this can decrease the dynamic spreading coefficient and slow down the droplet spreading dynamics. This behavior has since been verified for a number of different surfactants, and it appears to be generic (Aytouna, 2012, 2013).

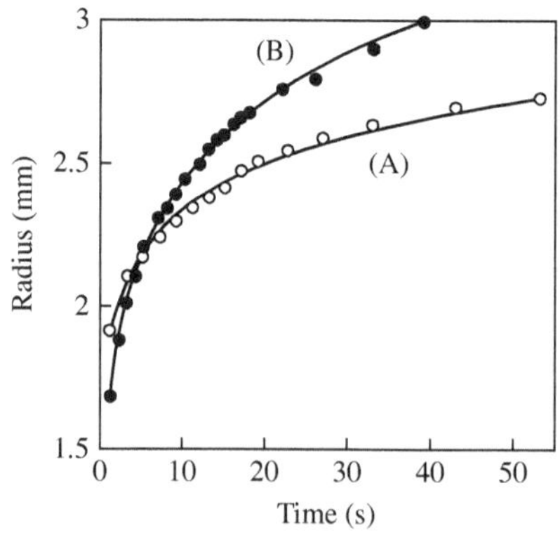

FIGURE 1.5 (A) Dynamics of an aqueous AOT drop spreading over a hydrophobic surface. The line is a fit to a power-law evolution $R = C\,t^{1/10}$, yielding the prefactor C. (B) Dynamics of aqueous trisiloxane drops. The line is fit to a power-law evolution $R = C\,t^{0.16}$ (Rafaï et al., 2002).

There is, however, one exception: trisiloxane surfactants are completely different. For all concentrations investigated (both above and slightly below the cmc), the spreading is faster than that of the corresponding simple fluid with the same equilibrium surface tension and viscosity. The spreading becomes much faster than $R \propto t^{0.1}$, with the power of t obtained from fitting a power-law behavior to $R(t)$ increasing proportionately to increasing surfactant concentration. For the highest bulk concentrations, a linear relation $R \propto t$ is obtained, a power one order of magnitude larger than that predicted by Tanner's law (Figure 1.6). Therefore, whereas the AOT appeared to slow down the spreading, *the trisiloxane system accelerates the spreading relative to the reference system.* Again, viscosity measurements show that the viscosity change is at most a few percent. This so-called superspreading suggests that the motor driving the spreading is different. Explanation of this effect poses a challenge because both dynamic surface tension effects and inertial forces can only slow down the dynamics. Although Marangoni effects are generally believed to be the cause for superspreading, the mechanism remains elusive and merits further theoretical and experimental attention.

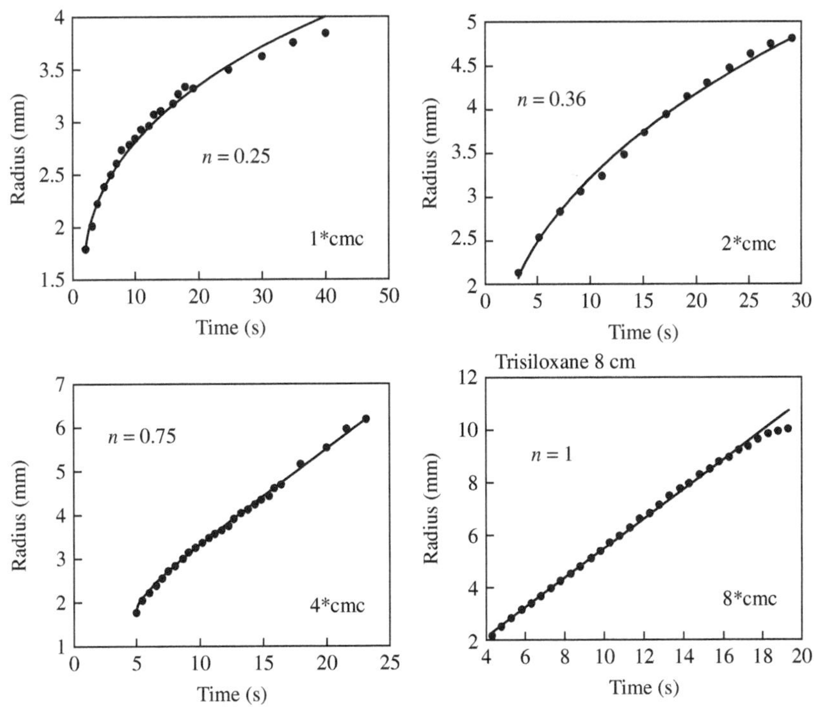

FIGURE 1.6 Spreading dynamics of trisiloxane surfactant solutions with increasing surfactant concentration (Rafaï et al., 2002).

A less obvious way to intervene in the spreading dynamics is to change the bulk rheology of the fluid. This is mostly done by using polymeric additives. The dynamics of such non-Newtonian liquids is far less understood than their Newtonian counterpart; the main challenge is that the constitutive equation is in general nonlinear. For a non-Newtonian bulk rheology, most theoretical studies have used simple constitutive equations, focusing on shear-thinning behavior where the viscosity decreases with increasing velocity gradient in the fluid. It was shown that within the lubrication approximation, the contact line singularity disappears due to the shear-dependent viscosity. One would—again perhaps naively—expect this to speed up the spreading.

However, this is again not what is observed in the experiment. Experimentally, the spreading of a shear-thinning fluid is found to be slightly slower than that of a Newtonian fluid: the power in the spreading law is found to be slightly smaller than 0.1, and decreases with increasing shear thinning—that is, increasing non-Newtonian effect. The fact that the singularity does not disappear is in fact expected. The calculations that show this are done for theoretical power-law models in which the fluid viscosity can become vanishingly small in the limit where the shear rate becomes very large. In reality, the viscosity of a polymer solution can never be smaller than that of the solvent, and thus the singularity persists. The reason why the spreading is slower is more subtle and follows from a lubrication calculation in which the dissipation is calculated near the moving contact line and matched to other regions of the flow (Figure 1.7).

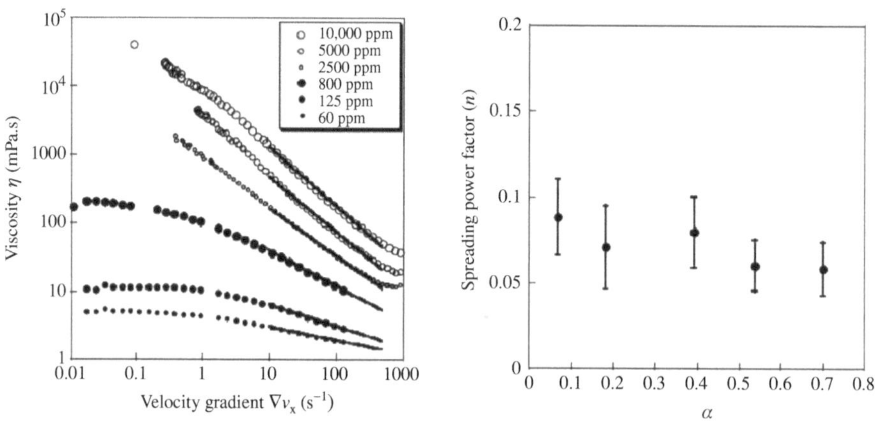

FIGURE 1.7 Rheology of xanthane polymer solutions at different concentrations (indicated in weight ppm in the legend) as a function of the shear rate (=velocity gradient in a simple shear flow). The high-shear behavior is fitted to a power-law behavior $\eta \sim (\text{grad } v)^{\alpha}$. The right figure shows the spreading exponent from the droplet experiment as a function of the shear-thinning exponent α (Rafaï et al., 2004, 2005).

If normal stresses are present, again only small deviations from Tanner's law are found in the experiment. This can be understood by comparing viscous and normal stress contributions to the spreading. It turns out that only logarithmic corrections to Tanner's law survive, which are nonetheless visible in the experiment.

The main conclusion of different experiments using different types of complex fluids (mainly surfactant and polymer solutions) is therefore the perhaps surprising robustness of Tanner's law; most deviations from it are very small, the only exception being "superspreading" surfactants. For the spreading of non-Newtonian liquids, the two most common non-Newtonian flow properties—a shear-rate dependence of the viscosity and the existence of normal stresses—have been considered both experimentally and theoretically. Again, the spreading behavior is found not to deviate strongly from Tanner's law.

References

Aytouna, M., 2012. Droplet Dynamics (Ph.D. thesis). University of Amsterdam.

Aytouna, M., Paredes, J., Shahidzadeh-Bonn, N., Moulinet, S., Wagner, C., Amarouchene, Y., et al., 2013. Drop formation in non-Newtonian fluids. Phys. Rev. Lett. 110, 034501.

Bonn, D., Eggers, J., Indekeu, J., Meunier, J., Rolley, E., 2009. Wetting and spreading. Rev. Mod. Phys. 81 (2), 739.

De Gennes, P.G., 1985. Wetting: statics and dynamics. Rev. Mod. Phys. 57 (3), 827.

Rafaï, S., Bonn, D., 2005. Spreading of non-Newtonian fluids and surfactant solutions on solid surfaces. Physica A 358, 58.

Rafaï, S., Bonn, D., Boudaoud, A., 2004. Spreading of non-Newtonian fluids on hydrophilic surfaces. J. Fluid Mech. 513, 77.

Rafaï, S., Sarker, D., Bergeron, V., Meunier, J., Bonn, D., 2002. Superspreading: aqueous surfactant drops spreading on hydrophobic surfaces. Langmuir 18, 10486.

Roché, M., Aytouna, M., Bonn, D., Kellay, H., 2009. Effect of surface tension variations on the pinch-off behavior of small fluid drops in the presence of surfactants. Phys. Rev. Lett. 103, 264501.

Young, T., 1805. An essay on the cohesion of fluids. Philos. Trans. R. Soc. Lond. 95, 65.

Chapter 2

CONTACT ANGLES AND THE SURFACE FREE ENERGY OF SOLIDS

Odile Carrier and Daniel Bonn
Institute of Physics, University of Amsterdam, Amsterdam, The Netherlands

This section presents the contact angle, its determination, hysteresis, and the different approaches to correct it, depending on the substrate roughness.

2.1 Equilibrium contact angle: Zisman plots

As presented in Chapter 1, the cosine of the equilibrium contact angle θ is linked to the three surface tensions according to the Young equation (1805) for pure liquids on an ideal solid substrate (smooth and homogeneous):

$$\sigma_{sv} = \sigma_{sl} + \sigma_{lv} \cos\theta \tag{2.1}$$

The contact angle is then used to characterize the wetting properties of solids: a droplet is deposited on the substrate, and the equilibrium contact angle is measured. To do it directly from only the contact angle and surface tension measurements, Fox and Zisman (1950) introduced a characteristic parameter of the solid: the critical surface tension. They wrote an empirical relation between $\cos\theta$ and σ_{lv} for homologous pure liquids. The authors assumed that the difference $\Delta\sigma = \sigma_{sv} - \sigma_{sl}$ is a property of the solid surface only; that is, it does not depend

Droplet Wetting and Evaporation. DOI: http://dx.doi.org/10.1016/B978-0-12-800722-8.00002-3

on the liquid used. A "Zisman plot" represents the cosine of the contact angle as a function of the liquid–vapor surface tension, which is changed by using different liquids. A linear extrapolation of the curve to $\cos\theta = 1$ then gives the critical surface tension σ_{cr} of the substrate. For example, classical values obtained for polyethylene are around 30 mN m^{-1} and for PTFE 20 mN m^{-1}. This critical surface tension characterizes the wetting properties of the substrate: all fluids whose interfacial tension is below σ_{cr} should completely wet the substrate. Above σ_{cr}, liquids do not spread, and there is only partial wetting or nonwetting.

σ_{cr} is only a characteristic of the solid and is independent of the pure fluids. However, this does not stay valid for mixtures and complex fluids such as alcohol solutions, surfactant solutions, or polymer solutions. Even if the linear relation can stay valid for some aqueous solutions, they can also show different behaviors leading to different values for the critical surface tension. Figure 2.1A shows a typical representation of a Zisman plot for a surfactant solution. There is no linear relationship between $\cos\theta$ and σ_{lv}. There is, however, a linear relationship between $\cos\theta$ and $1/\sigma_{lv}$ as suggested by Eq. (2.1); this is shown in Figure 2.1B.

A difference of critical surface tension for different liquids on one substrate means a difference of adsorption at the interfaces. Zisman plots were drawn for many different substrates and surfactants, such as cationic and anionic surfactants, polymeric substrates, and so on. Due to their different relative adsorbing abilities at interfaces, critical surface tension determinations based on solutions should then be considered carefully.

This characterization of substrates supposes that there is just one contact angle at equilibrium, as presented by the Young equation. However, there is not always only one single contact angle measured for a given fluid/substrate pair; a hysteresis of the contact angle exists.

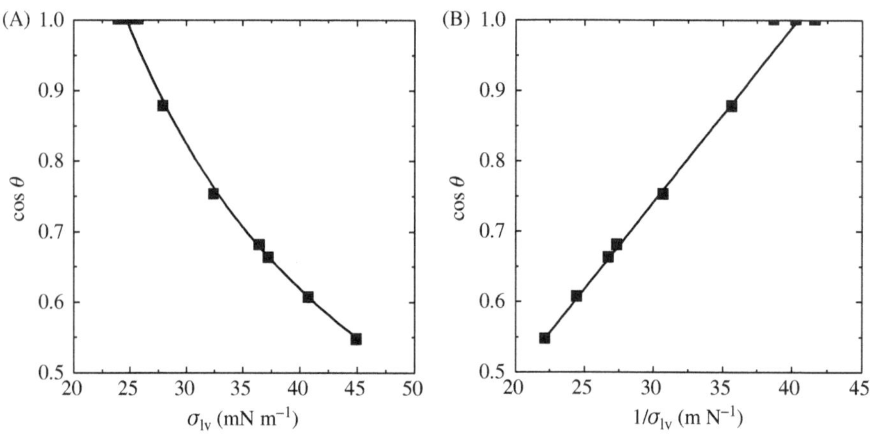

FIGURE 2.1 Zisman plots of cetyltrimethylammonium bromide solutions on a terphene substrate. (A) Classical representation. The line corresponds to a −1 power-law fit. (B) The line corresponds to a linear fit.

2.2 Contact angle hysteresis

2.2.1 Advancing and Receding Contact Angles

The Young–Laplace equation gives only one equilibrium contact angle for a homogeneous pure liquid on a perfectly flat, rigid, and smooth substrate without any impurity or heterogeneity. This is, however, hardly ever the case for real systems. Even on thoroughly cleaned and smooth surfaces, several contact angles can indeed be measured. Their values are between two extremes: the advancing contact angle θ_a and the receding contact angle θ_r. The triple-line is often pinned on the substrate and immobile, resulting in the multiple values of the contact angle. The contact angle hysteresis can be defined as the difference between these two angles (Good, 1979):

$$\Delta\theta = \theta_a - \theta_r \tag{2.2}$$

The advancing contact angle is determined for an increase in the contact area between the liquid and the solid, while the receding contact angle corresponds to a decreasing contact area. The wetting conditions for both configurations can then be seen as different: the advancing contact angle corresponds to a dynamic on a completely dry substrate—if we ignore the precursor film—while the receding contact angle corresponds to a retraction on a wet substrate.

In practice, the value of $\Delta\theta$ is of the order of several tens of degrees and can even be as high as 50° (Johnson and Dettre, 1993). It is largely above the error measurement of contact angles and cannot be considered negligible.

Different techniques can be used to measure the receding and advancing contact angles. The three most common techniques are presented in Figure 2.2 and find their origin in techniques to measure the equilibrium contact angle.

The first one consists in observing a droplet deposited on an inclined plan. It is probably the most direct technique because both angles can be seen on one droplet at the same time. Up to a certain inclination the drop can stay at equilibrium on the substrate without sliding. When the droplet starts to slide on the plate, the advancing contact angle can be measured at the front of the droplet, and the receding contact angle can be measured at its tail, as shown in Figure 2.2A. The balance between gravity and interfacial forces gives the hysteresis force per unit length (Furmidge, 1962):

$$H = \frac{mg \sin\alpha}{w} = \gamma(\cos\theta_r - \cos\theta_a) \tag{2.3}$$

with m the mass of the drop, w the width of the drop, α the minimum tilting angle to make the drop slide, and γ the liquid/air surface tension. It represents the required force to anchor the droplet on the substrate. If the surface was very smooth, perfectly flat, and completely homogeneous, without any hysteresis, the droplet would slide as soon as the plate was tilted. The hysteresis keeps the drop on the plate. The drop slides only when the gravity force is equal to *H*.

The dynamic sessile drop method can also be used to measure advancing and receding contact angles. However, while for the equilibrium contact angle, the drop is just deposited on the substrate, in this case, liquid is pumped in and out of the drop at a given constant low speed. It is done either with a syringe coming from above the droplet (see Figure 2.2B) or with a feeding line coming from under the droplet through the substrate. The contact area between the liquid and the solid is either increased or decreased, giving access, respectively, to the advancing and receding contact angles. The advancing contact angle corresponds to the measured contact angle just before the contact line begins to move—that is, θ_a while the contact line speed is 0—as can be seen in Figure 2.3.

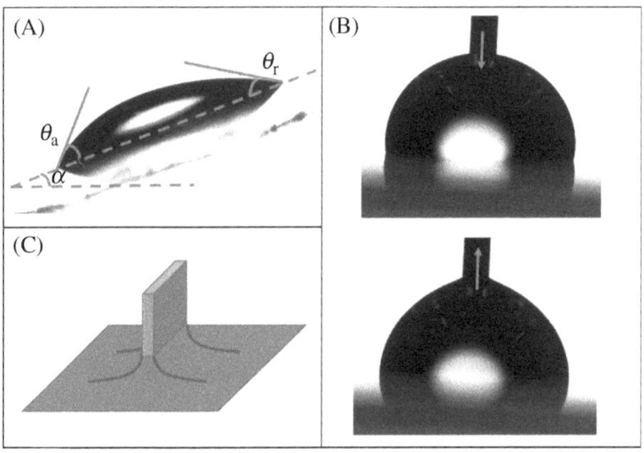

FIGURE 2.2 (A) Water drop on a tilted substrate ($\theta_a = 47°$ and $\theta_r = 30°$). (B) Sessile drop with pumped-in fluid (top, $\theta_a = 100°$) and pumped-out fluid (bottom, $\theta_r = 87°$). (C) Wilhelmy plate.

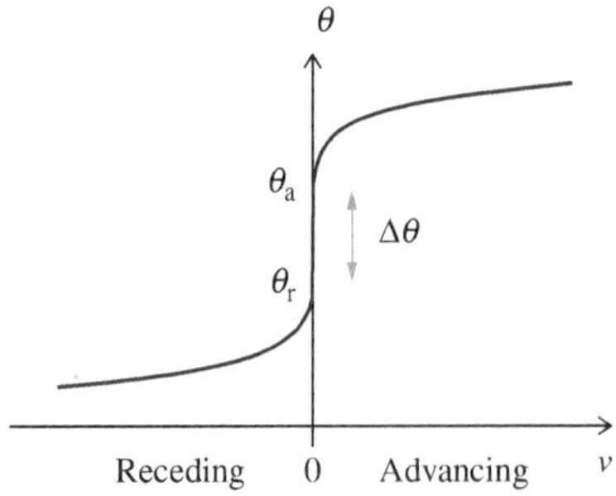

FIGURE 2.3 Dynamic behavior of the contact angles.

The third method is the Wilhelmy plate method (Figure 2.2C). In this method, the drop is not deposited on the solid, but a solid plate is immersed in the liquid of known surface tension. The force F required to pull or push the plate in the liquid gives the contact angle according to Wilhelmy (1863):

$$\cos\theta = \frac{F}{P\gamma} \tag{2.4}$$

with P the wetted perimeter of the solid. Unlike the first two methods, which directly provide the values of the contact angles, with this method, it is necessary to know the liquid/air surface tension.

2.2.2 Causes of Hysteresis

Many causes of hysteresis can be found. They mainly come from the nonrespect of hypothesis advanced by the Young–Laplace equation—that is, a flat, smooth, rigid, and homogeneous substrate in particular. A result of this deviation is that it brings the contact line to some metastable states (locally stable) but not globally stable states, and these distortions result in contact angle hysteresis. Many models were developed to describe the contact angle hysteresis based on the distributions of defects on the substrate (Decker and Garoff, 1997). We will not provide extensive descriptions of all these models, but we will discuss some key points and the assumptions made concerning the origin of the contact angle hysteresis.

The main causes often invoked for the contact angle hysteresis are the surface roughness (Shuttleworth and Bailey, 1948) and heterogeneities. Without any hysteresis considerations, their influence on the value of contact angles was modeled by Wenzel (1936) and Cassie and Baxter (1944). It increases the hydrophobic or hydrophilic characteristic of a substrate: if rough, a hydrophobic substrate will be even more hydrophobic than the same smooth substrate, and reciprocally, a hydrophilic substrate will be even more hydrophilic.

The contact angle of a liquid on a rough surface differs from that on a smooth surface. Wenzel (1936) was the first one to explain the deviation from the Young–Laplace contact angle with the roughness r of the solid, defined as the ratio of the real surface area by the geometric (apparent) surface area. As the liquid spreads and wets all the asperities of the surface (see Figure 2.4A), the apparent contact angle θ_{app} is

$$\cos\theta_{app} = r\cos\theta_{eq} \tag{2.5}$$

However, the possible wetting transition implied by this model for $r = 1/\cos\theta_{eq}$, θ_{eq} is the equilibrium contact angle, was never observed experimentally, so it should be taken as a way to describe what happens qualitatively, not quantitatively.

The model presented by Cassie and Baxter (1944) considers the trapping of a gas pocket in the asperities of a substrate, leading to a heterogeneous surface with which the liquid is in contact (see Figure 2.4B). Advancing and receding contact angles are then defined according to

$$\cos \theta_{\mathrm{app}} = -\frac{E_{\mathrm{D}}}{\gamma_{\mathrm{lv}}} = f \cos \theta_{\mathrm{eq}} - (1 - f) \tag{2.6}$$

with E_{D} the energy expanded in the forming area of the interface, f the area of the solid–liquid interface, and $1 - f$ the area of the gas–liquid interface induced by the gas pockets stuck in the roughness asperities.

However, both models do not take clearly into account and explain the hysteresis of the contact angle, only a difference from the Laplace–Young equilibrium contact angle. Depending on the state of the drop—Wenzel or Cassie—the hysteresis can be either very high or very low, respectively, for the same equilibrium contact angle.

Figure 2.5 presents a simplified way to see the effect of roughness on a contact angle and the deriving hysteresis. If we consider a rough surface without any other heterogeneity involved, the contact angle determined by the Laplace–Young equation is the microscopic contact angle θ_{eq}. From a macroscopic point of view, the measured contact angle θ can be very different from that value. For the extreme case presented in Figure 2.5, it can be seen as hydrophilic or hydrophobic. The slope of the asperities is an important factor influencing the advancing and receding contact angles (Eick et al., 1975; Figure 2.5). Depending on how the drop is deposited on the

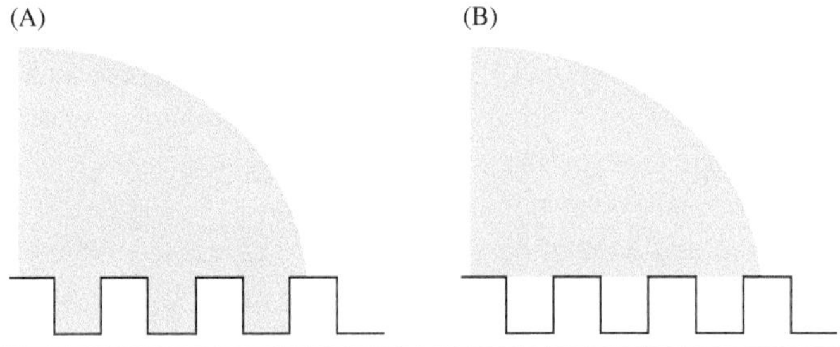

FIGURE 2.4 (A) Wenzel configuration. (B) Cassie configuration.

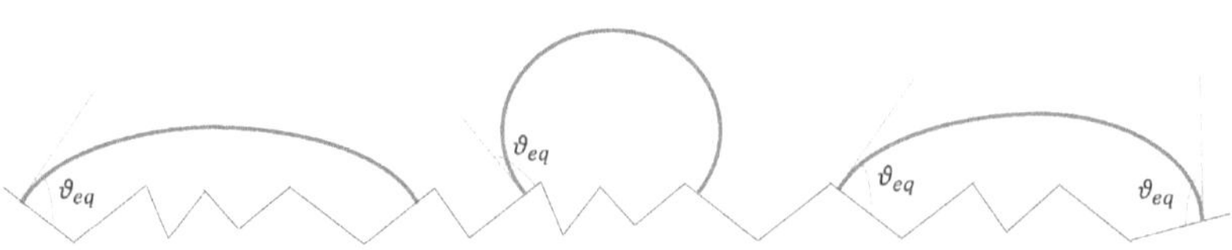

FIGURE 2.5 Simplified representation of the possible roughness effect on the contact angle.

substrate, non circular drops can also be observed and different contact angles can be mesured on one drop (third case on Figure 2.5).

Either heterogeneities or contaminations (defects on the surface) can also lead to a pinning of the contact line on the substrate and then to the hysteresis. These defects can present very different sizes and strengths, and as a result entail different effects on contact angles. For micrometer-sized defects, pinning and wiggling of the contact line can be directly observed, while for nanometric defects, the effect manifests itself as a hysteresis. Theoretical models are available for describing the response of the contact angle to such defects (Joanny and de Gennes, 1984).

The link between heterogeneities and contact angle hysteresis was studied from a thermodynamic point of view on heterogeneous regular and patterned surfaces (strips, concentrical rings of different wettability response; Johnson and Dettre, 1964; Neumann and Good, 1972). The advancing contact angle can be linked to the wettability on the low-energy area of the surface, while the smaller receding contact angle can be linked to the high-energy area. The different metastable states of drop configurations separated by energy barriers prevent the system from reaching the minimum free energy—that is, a unique equilibrium contact angle.

Considering nonperiodic defects covering the substrate, the dynamics of a contact line moving is similar to those of systems where an elastic boundary moves in a random medium. Thus, models have also been used to study the contact line as an elastic boundary: close to depinning, a transition occurs that can be characterized by a few critical exponents.

Other sources of hysteresis come from the chemical interaction between the liquid and the solid. If the liquid contains other species (surfactants, etc.), there can be a layer that deposits on the substrate surface. This adsorption can be a source of heterogeneities, leading to a possible modification of the solid surface tension locally.

Penetration of liquids in the microporosity of a condensed monolayer can also lead to hysteresis of the contact angle. It then depends on the molecular size of the liquid and can be eliminated for molecules larger than the pores (Timmons and Zisman, 1966).

Finally, the molecular interactions between the fluid and the substrate can induce hysteresis, especially at the contact line. Extrand (2004) thus observed contact angle hysteresis on a smooth substrate due to hydrogen bonding of water and polyamide substrate.

2.2.3 Hysteresis Consequences

As a result of all these possible explanations, almost all real systems involve that hysteresis. It is very hard to get a perfectly smooth and clean surface with no contamination of the surface or the liquid. The measure of a unique contact angle can then be theleast important concern: is it really the equilibrium contact angle? As a result of defects in the surface and pinning the contact line on this point, drops can also be asymmetric: the measured contact angle will not be the same in every point of the drop. Measures of contact angles should then be carried out very carefully. Moreover, measuring the receding contact angle can be difficult to reproduce: adsorption of the fluid on the substrate can lead to modification of the substrate properties. Furthermore, when attempting to predict wettability behavior, you must choose the correct model, depending on what you are trying to find out: is the hysteresis caused by chemical interactions, roughness, and/or defects?

The hysteresis of the contact angle can intervene in every case that involves wetting, and there are numerous processes where it has to be taken into account, such as coating processes or flows in porous media. As we said before, roughness tends to strengthen the hydrophilic or hydrophobic characteristic of a substrate. Moreover, the drops that present a low contact angle hysteresis can easily slide on a surface—that is, with only a slight tilt of the substrate. As a result, a drop with a high contact angle (on a hydrophobic substrate) and a low hysteresis contact angle will hardly wet a substrate and can be easily removed. An increase in the highly controlled roughness of a substrate can then be used as a superhydrophobic substrate. Hysteresis of the contact angle can then be used to design surfaces with specific wetting properties. Water-repellent surfaces in particular should have low contact angle hysteresis for water.

It is then necessary to better understand the origin of hysteresis. However, determining all the different possible causes for hysteresis can be tricky. It would be helpful to have theoretical explanations and experimental studies of these different phenomena on model substrates (Bonn et al., 2009).

References

Bonn, D., Eggers, J., Indekeu, J., Meunier, J., Rolley, E., 2009. Wetting and spreading. Rev. Mod. Phys. 81, 739–805.

Cassie, A.B.D., Baxter, S., 1944. Wettability of porous surfaces. Trans. Faraday Soc. 40, 546–551.

Decker, E.L., Garoff, S., 1997. Contact angle hysteresis: the need for new theoretical and experimental models. J. Adhes. 63, 159–185.

Eick, J.D., Good, R.J., Neumann, A.W., 1975. Thermodynamics of contact angles II. Rough solid surfaces. J. Colloid Interface Sci. 53, 235–248.

Extrand, C.W., 2004. Contact angles and their hysteresis as a measure of liquid-solid adhesion. Langmuir 20, 4017–4021.

Fox, H.W., Zisman, W.A., 1950. The spreading of liquids on low energy surfaces. I. Polytetrafluoroethylene. J. Colloid Interface Sci. 5, 514–531.

Furmidge, C.G.L., 1962. Studies at phase interfaces I. The sliding of liquid drops on solid surfaces and a theory for spray retention. J. Colloid Sci. 17, 309–324.

Good, R.J., 1979. In: Good, R.J., Stromberg, R.R. (Eds.), Contact angles and the surface free energy of solids, Surface and Colloid Science, vol. 11. Plenum Press, New York, NY.

Joanny, J.F., de Gennes, P.G., 1984. A model for contact angle hysteresis. J. Chem. Phys. 81, 552–562.

Johnson, R.E., Dettre, R.H., 1964. Contact angle hysteresis III. Study of an idealized heterogeneous surface. J. Phys. Chem. 68, 1744–1750.

Johnson, R.E., Dettre, R.H., 1993. Wetting of low energy surfaces. In: Berg, J.C. (Ed.), Wettability. Dekker, New York, NY, pp. 1–74.

Neumann, A.W., Good, R.J., 1972. Thermodynamics of contact angles. I. Heterogeneous solid surfaces. J. Colloid Interface Sci. 38 (2), 341–358.

Shuttleworth, R., Bailey, G.L.J., 1948. The spreading of a liquid over a rough solid. Discuss. Faraday Soc. 3, 16–22.

Timmons, C.O., Zisman, W.A., 1966. The effect of liquid structure on contact angle hysteresis. J. Colloid Interface Sci. 22, 165–171.

Wenzel, R.N., 1936. Resistance of solid surfaces to wetting by water. Ind. Eng. Chem. 28, 988–994.

Wilhelmy, J., 1863. Uber die Abhängigkeit der Capillaritäts-Constanten des Alkols von Substanz und Gestalt des benetzten festen Korpers. Ann. Phys. 119, 177–217.

Chapter 3

TRIPLE LINE MOTION AND EVAPORATION

David Brutin[1] and Benjamin Sobac[1,2]
[1]*Mechanical Engineering Department, Aix-Marseille University, Marseille, France,*
[2]*Université Libre de Bruxelles—TIPs Fluid Physics, Brussels, Belgium*

3.1 Triple line dynamics

Sessile droplet evaporation for contact angles $\theta \leq 90°$ have been widely studied in the past (Picknett and Bexon, 1977; Birdi et al., 1989; Shanahan and Bourgès, 1994; Bourgès-Monnier and Shanahan, 1995; Rowan et al., 1995; McHale et al., 1998; Deegan et al., 2000; Hu and Larson, 2002). However, it is only recently, with the emergence of structured surfaces, the study was extended to hydrophobic and super-hydrophobic situations, i.e when $\theta > 90°$ (McHale et al., 2005; Reyssat et al., 2008; Kulinich and Farzaneh, 2009; Gelderblom et al., 2011). These studies showed that different modes of evaporation are possible, depending on the triple line (TL) dynamics (Picknett and Bexon, 1977; Bourgès-Monnier and Shanahan, 1995; Shanahan, 2001a,b) and thus depending on the hysteresis of the contact angle (Kulinich and Farzaneh, 2009). A drop can evaporate with a constant contact area overtime, as illustrated in Figure 3.1. Then the triple line is pinned and the evaporation is characterized by a decrease of the contact angle during the evaporation. A drop can also evaporate with a constant contact angle overtime. Then the evaporation occurs with a recession of the TL. More complex evaporation modes also exit, when the two previous modes are either alternated or combined.

Droplet Wetting and Evaporation. DOI: http://dx.doi.org/10.1016/B978-0-12-800722-8.00003-5

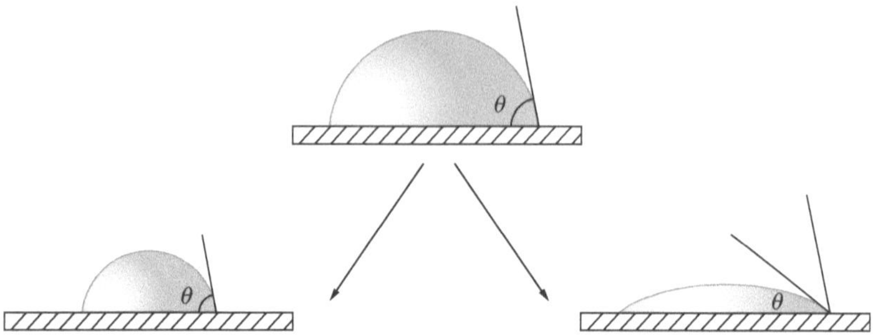

FIGURE 3.1 Two contact line evaporation modes: (left side) constant contact angle mode and (right side) constant contact area mode (Sobac, 2012).

In general, the studies that have investigated sessile drop evaporation with a pinned TL evidence a global evaporation rate proportional to the wetting radius (Birdi et al., 1989; Deegan et al., 2000; Grandas et al., 2005). Evaporation rate dynamics are strongly related to the contact angle. Birdi and Vu (1993) and Hu and Larson (2002) reported constant evaporation rates for very low initial contact angles, while Bourgès-Monnier and Shanahan (1995) and Gelderblom et al. (2011) found a decrease in the evaporation rate dynamics for larger values of the contact angle. The TL length directly affects the evaporation rate, because for the low contact angle, the evaporation rate is linked to the droplet wetting radius. The decrease in the radius is related to the square root of time, as showed by Erbil et al. (2002) and Shahidzadeh-Bonn et al. (2006), and a nonlinear evolution of the evaporation rate (Birdi and Vu, 1993). It is clear that droplet evaporation is directly linked to TL dynamics and wettability. These two mechanics are related to the substrate material properties (roughness) and physico-chemical properties (surface energy).

3.2 Pinned TLs

Only a few studies have investigated the influence of the substrate's physico-chemical properties on the evaporation process and its associated transport mechanisms. Grandas et al. (2005) and Shin et al. (2009) studied the evaporation of a water drop on different surfaces: Polytétrafluoroéthylène (PTFE) (hydrophobic) and aluminum (hydrophilic) in the Grandas study, and glass (hydrophilic), Octadecyltrichlorosilane (OTS) (hydrophobic), and alkylketene dimmer (AKD) (superhydrophobic) in the Shin study. The dynamics of the TL and the evaporation time change with the substrate type. However, in these two studies, the entire substrate was changed, and thus the physico-chemical properties (surface energy), thermo-physical properties (diffusivity and effusivity), and surface state (roughness) of the evaporation system were modified at once.

Sobac and Brutin (2011) described the influence of the substrate surface properties on sessile drop evaporation in the configuration of pinned TL. They covered substrates with various nano- and microcoatings: PFC, PTFE, SiOC, and SiOx. This method allows modifying the surface properties of the substrates (such as the roughness and surface energy) while maintaining identical thermal properties. Experiments were conducted under atmospheric conditions using five fluids: methanol, ethanol, propanol, toluene, and water. The various combinations of these fluids and coatings allow for a wide range of drop evaporation properties to be studied: the dynamics of the TL and a large range of wettabilities (from 17° to 135°). The influence of these parameters on the kinetics and dynamics of evaporation is studied in order to highlight the distinct and combined roles of the TL and the wettability of the drops on the sessile drop evaporation process. When the contact angle is less than 40°, the influence of the contact angle on the dynamics of evaporation is negligible. Thus, the experiments show that the evolution of the evaporative power is linearly related to the evolution of the wetting radius, regardless of the behavior of the TL. When the contact angle is greater than 40°, the influence of the contact angle is no longer negligible. All experimental data are in good quantitative agreement with the models of Hu and Larson (2002) (valid for $0° < \theta < 90°$) and Popov (2005) (valid for $0° < \theta < 180°$), showing that the evaporation process is quasi-steady and driven by diffusion. A universal relationship for the evolution of the drop mass, independent of the drop size and initial contact angle, is obtained, which confirms the recent results of Gelderblom et al. (2011). Ultimately, the experimental results reveal an increase of 50% of the evaporation time between drop evaporations occurring with a pinned and a receding TL, as well as an increase of 58% when the contact angle increases from 63° to 135°. These data clearly show the importance and effects of surface properties on sessile drop evaporation.

3.3 TL depinning during evaporation

During evaporation, the free energy balance at the contact is modified, which can induce a depinning of the TL. In 2009, Shanahan and Sefiane provided an explanation for this, which is presented below. Assuming that gravitational flattening is negligible, the drop is represented by a spherical cap of contact angle θ and contact radius R (Figure 3.2).

Drop volume, V, and liquid–vapor surface area, A, are given by

$$V = \frac{\pi R^3}{3} \frac{(1-\cos\theta)^2(2+\cos\theta)}{\sin^3\theta} \tag{3.1}$$

$$A = \frac{2\pi R^2}{(1+\cos\theta)} \tag{3.2}$$

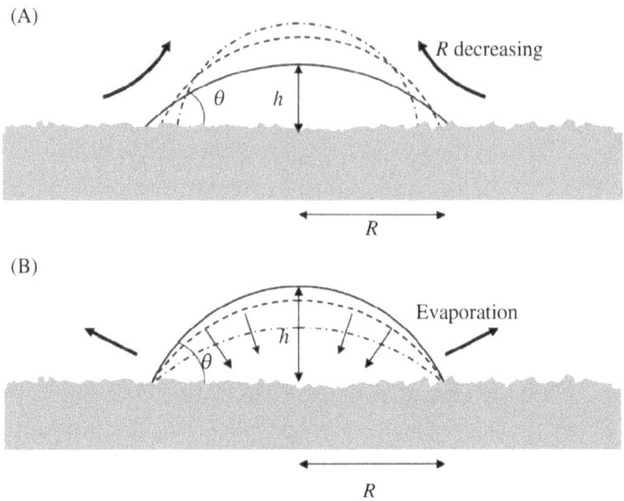

FIGURE 3.2 Sketch representing evaporation leading to stick–slip motion of triple line (TL) of a sessile drop. (A) Initial evaporation leads to decreased height, *h*, and contact angle, *θ*, at constant contact radius, *R*, due to TL pinning. (B) When the TL depins, it recedes, leading to increased *h* and *θ*, the latter tending toward its equilibrium value. The process then repeats itself with further evaporation, giving successive periods of "stick" and "slip" (Shanahan and Sefiane, 2009).

The Gibbs free energy associated, G, is given to within an additive constant by

$$\begin{aligned} G &= \gamma A + \pi R^2(\gamma_{SL} - \gamma_{SV}) \\ &= \gamma \pi R^2 \left[\frac{2}{(1+\cos\theta)} - \cos\theta_0 \right] \end{aligned} \tag{3.3}$$

where the Young equation has been used to replace $(\gamma_{SL} - \gamma_{SV})$ by $-\gamma\cos\theta_0$, θ_0, representing the equilibrium contact angle, γ, γ_{SL}, and γ_{SV} being liquid surface free energy, solid–liquid free energy, and solid–vapor interfacial free energy, respectively. For a given volume, V, we have an equilibrium contact radius, R_0, corresponding to contact angle θ_0, but if the drop is out of equilibrium with, say, contact radius $R = R_0 + \delta R$ (and therefore contact angle $\theta_0 - \delta\theta$), its excess free energy, δG, is given by

$$\begin{aligned} \delta G &= G(R) - G(R_0) \\ &= \delta R \left[\frac{dG}{dR} \right]_{R=R_0} + \frac{(\delta R)^2}{2} \left[\frac{d^2 G}{dR^2} \right]_{R=R_0} + O\left[(\delta R)^3\right] \end{aligned} \tag{3.4}$$

Because $[dG/dR]_{R=R0} = 0$—as R_0 is the equilibrium value—we may evaluate the excess free energy per unit length of TL, $\delta\tilde{G} = \frac{\delta G}{2\pi R}$ to $O[(\delta R)^2]$ from Eqs. (3.1)–(3.4):

$$\delta\tilde{G} = \frac{\gamma \sin^2 \theta_0 (2 + \cos \theta_0)(\delta R)^2}{2R} \tag{3.5}$$

Alternatively, we may use a Taylor expansion in powers of $\delta\theta$, rather than δR, as in Eq. (3.4), to obtain

$$\delta\tilde{G} = \frac{\gamma R (\delta\theta)^2}{2(2 + \cos \theta_0)} \tag{3.6}$$

During evaporation, G increases for a pinned TL (this is clearer from Eq. (3.6) than from Eq. (3.5), because $\delta\theta$ physically decreases during pinned evaporation), until it just exceeds the potential energy barrier, U, physical or chemical, for depinning. At this moment, the TL may depin and recede, leading to a smaller value of R, which is assumed to be a new equilibrium value of R_0 corresponding to the recently diminished drop volume. The new R_0 will be smaller than the previous one, but because the height of the drop, h, will increase during the (assumed rapid) TL motion, the original θ_0 may be regained, at least in this simplified model. G should again become zero. As discussed below, whether the original value of θ_0 should be expected to be attained after the initial depinning is doubtful. It should be noted that a further simplification assumed here is that TL motion occurs over a sufficiently short period for concomitant mass loss by evaporation to be considered negligible. Also, explicit time dependence is omitted in this preliminary model.

Glossary

PTFE polytétrafluoroéthylène
AKD alkylketene dimmer
OTS octadecyltrichlorosilane
PDMS polydiméthylsiloxane
SiOC PDMS highly carbonated
SiOx PDMS lowly carbonated
PFC Perfluorocarbone

References

Birdi, K.S., Vu, D.T., 1993. Wettability and the evaporation rates of fluids from solid surfaces. J. Adhes. Sci. Technol. 7 (6), 485–493.

Birdi, K.S., Vu, D.T., Winter, A., 1989. A study of the evaporation rates of small water drops placed on a solid surface. J. Phys. Chem. 93 (9), 3702–3703.

Bourgès-Monnier, C., Shanahan, M.E.R., 1995. Influence of evaporation on contact angle. Langmuir 11 (7), 2820–2829.

Deegan, R.D., Bakajin, O., Dupont, T.F., Huber, G., Nagel, S.R., Witten, T.A., 2000. Contact line deposits in an evaporating drop. Phys. Rev. E 62, 756–765.

Erbil, H.Y., McHale, G., Newton, M.I., 2002. Drop evaporation on solid surfaces: Constant contact angle mode. Langmuir 18 (7), 2636–2641.

Gelderblom, H., Marin, A.G., Nair, H., Van Houselt, A., Lefferts, L., Snoeijer, J.H., et al., 2011. How water droplets evaporate on a superhydrophobic substrate. Phys. Rev. E 83, 026306.

Grandas, L., Reynard, C., Santini, R., Tadrist, L., 2005. Experimental study of the evaporation of a sessile drop on a heat wall. Int. J. Therm. Sci. 44, 137.

Hu, H., Larson, R.G., 2002. Evaporation of a sessile droplet on a substrate. J. Phys. Chem. B 6 (106), 1334–1344.

Kulinich, S.A., Farzaneh, M., 2009. Effect of contact angle hysteresis on waterdroplet evaporation from super-hydrophobic surfaces. Appl. Surf. Sci. 255, 4056–4060.

McHale, G., Rowan, S.M., Newton, M.I., Banerjee, M.K., 1998. Evaporation and the wetting of a low-energy solid surface. J. Phys. Chem. B 102 (11), 1964–1967.

McHale, G., Aquil, S., Shirtcliffe, N.J., Newton, M.I., Erbil, H.Y., 2005. Analysis of droplet evaporation on a superhydrophobic surface. Langmuir 21 (24), 11053–11060.

Picknett, R.G., Bexon, R., 1977. The evaporation of sessile or pendant drops in still air. J. Colloid Interface Sci. 61 (2), 336–350.

Popov, Y.O., 2005. Evaporative deposition patterns: Spatial dimensions of the deposit. Phys. Rev. E 71, 036313.

Reyssat, M., Yeomans, J.M., Qu'er'e, D., 2008. Impalement of fakir drops. Europhys. Lett. 81, 26006.

Rowan, S.M., Newton, M.I., McHale, G., 1995. Evaporation of microdroplets and the wetting of solid surfaces. J. Phys. Chem. 99 (35), 13268–13271.

Shahidzadeh-Bonn, N., Rafai, S., Azouni, A., Bonn, D., 2006. Evaporating droplets. J. Fluid Mech. 549, 307–313.

Shanahan, M.E.R., 2001a. Condensation transport in dynamic wetting. Langmuir 17, 3997–4002.

Shanahan, M.E.R., 2001b. Spreading of water: condensation effects. Langmuir 17, 8229–8235.

Shanahan, M.E.R., Bourgès, C., 1994. Effects of evaporation on contact angles on polymer surfaces. Int. J. Adhes. Adhes. 14, 201–205.

Shanahan, M.E.R., Sefiane, K., 2009. Kinetics of triple line motion during evaporation in contact angle. In: Mittal, K.L. (Ed.), Wettability and Adhesion, vol. 6. CRC Press, Koninklijke Brill NV, Leiden.

Shin, D.H., Lee, S.H., Jung, J., Yoo, J.Y., 2009. Evaporating characteristics of sessile droplet on hydrophobic and hydrophilic surfaces. Microelectron. Eng. 86, 1350–1353.

Sobac, B., Brutin, D., 2011. Triple-line behavior and wettability controlled by nanocoated substrates: influence on sessile drop evaporation. Langmuir 27 (24), 14999–15007.

Sobac, B., 2012. Evaporation de gouttes sessiles: des fluides purs aux fluides complexes (Ph.D. thesis). Aix-Marseille Université, Marseille, France.

Chapter 4

PRECURSOR FILMS AND CONTACT LINE MICROSTRUCTURES

Pierre Colinet and Alexey Rednikov
Université Libre de Bruxelles—TIPs Fluid Physics, Brussels, Belgium

This chapter focuses on the small-scale structure of a contact line formed by a pure liquid, a gas (in particular, the vapor of the liquid considered), and a smooth and chemically homogeneous solid substrate. On the basis of the lubrication approximation (small slopes of the liquid/gas interface), various situations are considered, including volatile or nonvolatile liquid, partial or complete wetting, and advancing or receding contact lines. Two types of microfilms ahead of the contact line are considered: extended (i.e., covering the substrate everywhere) and truncated (i.e., formally ending at some "dry" spot). On the macroscopic side, results useful for the study of many configurations, such as droplet evaporation and boiling, are provided in terms of the Cox Voinov relationship, relating the scale-dependent apparent contact angle to the velocity of the contact line. Parameters entering this relationship—namely, the apparent contact angle of the steady microstructure and the microlength below which singularities of the contact line are regularized—are calculated in various cases. Particular attention is devoted to discussing the pertinence of disjoining pressure and the Kelvin effect (curvature dependence of saturation conditions) at small scales in relation with the resolution of contact line singularities. The selection between microstructures with extended and truncated films is also discussed.

Droplet Wetting and Evaporation. DOI: http://dx.doi.org/10.1016/B978-0-12-800722-8.00004-7

4.1 Introduction

This chapter covers the mathematical description of the small-scale structure of a contact line, particularly the possible existence of liquid microfilms covering the apparently dry portions of the substrate not covered by macroscopic liquid objects such as droplets, rivulets, etc. Such microfilms, often called precursor films since the works of de Gennes and colleagues (de Gennes, 1985), are now commonly invoked in so-called precursor film models as a way to avoid the viscous singularity associated with contact line motion, occurring when the film thickness truly vanishes at some point (Huh and Scriven, 1971). Even though they are certainly useful in this respect (e.g., for numerical simulations), the point of view followed here is that the existence of such *extended* microfilms (and the related conceptual question of whether or not a dry point exists) cannot be postulated but should depend on various effects such as the actual disjoining pressure isotherm of the considered fluid (i.e., thin-film forces), evaporation/condensation, the Kelvin effect (here generically understood as the influence of excess pressure in the liquid on the chemical equilibrium conditions), contact line motion, and possibly other factors. Hence, the question of the existence of precursor films is rather intricate, and this chapter does not claim to be exhaustive but rather discusses only a few cases that illustrate this complexity. It is worth emphasizing in this respect that there is still a lot of debate nowadays on which effects are pertinent in the microscale vicinity of a contact line, moving relatively to the substrate or not (see the recent review of Bonn et al., 2009). At the smallest (molecular) scales, the discrete nature of matter certainly plays a role, and this is typically studied using molecular dynamics (see, e.g., De Coninck and Blake, 2008) and molecular kinetic theory (Blake and Haynes, 1969). Coarse-grained approaches such as density functional theory (see, e.g., Nold et al., 2014) or phase-field methods (see, e.g., Anderson et al., 1998; Pismen and Pomeau, 2000; Sibley et al., 2013) also enable studying microscale properties of the contact line and avoiding their singularities, due to a non-vanishing thickness of the interfaces between contacting fluids. Yet, in this chapter we focus on macroscopic descriptions where interfaces have zero thickness, for on the one hand they are more "classical", and on the other hand because they have also proven to be adequate and successful in avoiding singularities provided that relevant microscale (yet continuum) effects are properly accounted for. Another advantage is that these "thermo-hydrodynamic" approaches make it easier to bridge the gap between the smallest scales, where microfilms possibly exist, and larger scales, where apparent contact angles are determined, as also seen in this chapter. Moreover, we also ignore molecular slip, which is often expected to be very small except for super-hydrophobic situations not dealt with here (Huang et al., 2008). We rather consider complete or partial wetting situations where the typical slope the liquid/gas interface makes with the substrate is not too large, such that the lubrication assumption may be used. In practice, this is valid in an appreciable range of slopes (sometimes up to 20–30°), while keeping the analysis rather easy compared to the original set of thermo-hydrodynamic equations and boundary conditions. We also ignore interface relaxation effects (Shikhmurzaev, 1993; Sibley et al., 2012), focusing on effects pertinent at larger length (and time) scales, such as evaporation (condensation) of the considered pure fluid into (from) its pure-vapor atmosphere (hence, the nonvolatile case here appears as a particular one within the general formulation).

In Section 4.2, we first establish a general thin-film evolution equation incorporating all the effects considered here (namely, capillarity, disjoining pressure, nonisothermal phase change, and the Kelvin effect) and applying to both cases of complete and partial wetting for advancing or receding contact lines. In Section 4.3, we recall basic results on the balance between viscous stresses and capillary pressure, leading to the Cox–Voinov asymptotics, which is valid at intermediate scales between the macroscopic meniscus and the microstructure of the contact line (whatever the microscale effects considered). In Section 4.4, some equilibrium flat microfilms and their stability are considered briefly. In Section 4.5, we consider the case of nonvolatile liquids at scales where the disjoining pressure acts and show how the precursor film preceding a moving contact line depends on its velocity (using the approach proposed by Hervet and de Gennes, 1984). Section 4.6 covers the effect of evaporation/condensation, first using a minimalist model without disjoining pressure but with the Kelvin effect (Section 4.6.1), and then in combination with disjoining pressure within the complete model, allowing us to study the transition between extended and truncated microfilm profiles (Section 4.6.2). In Section 4.7, we draw some conclusions.

Note that throughout the chapter, when appropriate, any dimensional quantity f^* (with an asterisk) will be written as $f^* = [f]f$—that is, the product of the chosen scale $[f]$ and of the corresponding dimensionless quantity f.

4.2 The lubrication approximation

Considering a sufficiently close vicinity of the contact line (the microregion), the geometry may be assumed two-dimensional (coordinates x^* along the film and z^* orthogonal to it), and under the lubrication approximation (small ratio $\boldsymbol{\varepsilon} = [h]/[x]$ of the typical film thickness scale $[h] = [z]$ and of the spatial scale $[x]$), the general evolution equation describing the evolution of the film thickness $h^* = h^*(x^*, t^*)$ reads

$$\frac{\partial h^*}{\partial t^*} + \frac{\partial q^*}{\partial x^*} + \frac{j^*}{\rho_1^*} = 0 \tag{4.1}$$

where q^* is the horizontal volumetric (or, in 2D, rather areal) flux

$$q^* = -\frac{h^{*3}}{3\eta_l^*}\frac{\partial \tilde{p}_l^*}{\partial x^*} \tag{4.2}$$

in which η_l^* is the liquid dynamic viscosity and $\tilde{p}_l^*$ is its pressure excess over the pressure in the gas (the latter assumed constant)—that is,

$$\tilde{p}_l^* = -\left(\gamma^* \frac{\partial^2 h^*}{\partial x^{*2}} + \Pi^*(h^*)\right) \tag{4.3}$$

where γ^* is the surface tension and $\Pi^*(h^*)$ is the disjoining (Derjaguin) pressure, due to long-range van der Waals forces acting between liquid and solid molecules, in addition with other components in general (Starov et al., 2007).

In the last term of Eq. (4.1), ρ_l^* is the liquid density, while j^* is its local evaporation flux density (in kg m^{-2}s^{-1}). The latter quantity should generally be expressed as a functional of h^*, which depends on the evaporation regime considered. In case evaporation is limited by the diffusion of vapor through an inert gas such as air, one generally has $j^* \sim -D_v^* \partial\rho_v^*/\partial z^*$, where D_v^* is the vapor/air diffusion coefficient, and ρ_v^* is the vapor density field typically obtained by solving the Laplace equation $\nabla^2\rho_v^* = 0$ with appropriate boundary conditions far away and on the film surface (e.g., phase equilibrium). This diffusion-limited regime will not be considered here, however, because a comprehensive theory of contact line microstructures is still lacking in this case (particular cases are dealt with in, e.g., Berteloot et al., 2008; Pham et al., 2010; Colinet and Rednikov, 2011). Rather, we focus here on the more extensively studied case of evaporation into a pure-vapor atmosphere maintained at constant pressure p_v^*, for which j^* is obtained by solving a much simpler thermal problem in the liquid only (one-sided model). Namely, denoting vapor and liquid chemical potentials by μ_v^* and μ_l^*, the interface temperature by T_Σ^*, and using a kinetic (linear irreversible thermodynamics) expression of the type $j^* = L_{ww}^*[\mu_l^*(p_v^* + \tilde{p}_l^*, T_\Sigma^*) - \mu_v^*(p_v^*, T_\Sigma^*)]/T_\Sigma^*$ (assuming local temperature continuity at the interface and defining a "conductance" L_{ww}^* as in, e.g., Kjelstrup and Bedeaux, 2008), it can be shown by linearizing this expression around the macroscopic equilibrium condition $\mu_l^*(p_v^*, T_0^*) = \mu_v^*(p_v^*, T_0^*)$ (which defines the boiling temperature T_0^* as a function of the vapor pressure p_v^*), and assuming the liquid to be incompressible, that

$$j^* = \frac{L_{ww}^*\mathcal{L}^*}{T_0^{*2}}\left[T_\Sigma^* - T_0^*\left(1 - \frac{\tilde{p}_l^*}{\rho_l^*\mathcal{L}^*}\right)\right] \tag{4.4}$$

where $\mathcal{L}^*$ is the latent heat of vaporization, and the other symbols have been defined before. Note the factor correcting the boiling temperature T_0^* in Eq. (4.4), actually accounting for the Kelvin effect—that is, the shift of the saturation conditions due to the excess of pressure in the liquid (here due to capillary and disjoining pressures; see Eq. (4.3)). Now, in the lubrication approximation, the liquid temperature profile $T_l^*(z^*)$ is simply linear, varying from an imposed temperature T_w^* (supposed uniform) at the substrate $z^* = 0$ to the interface temperature T_Σ^* at $z^* = h^*$. If heat conduction from the vapor is omitted (small thermal conductivity), combining the interfacial energy balance $-\lambda_l^*[\partial T_l^*/\partial z^*]_{z^*=h^*} = j^*\mathcal{L}^*$ (where λ_l^* is the liquid thermal conductivity) with Eq. (4.4) yields T_Σ^* (not given here because it is not needed further), as well as the final expression of the evaporation flux density

$$j^* = \mathcal{L}^{*-1} \frac{\Delta T^* + \dfrac{T_0^*}{\rho_l^* \mathcal{L}^*} \tilde{p}_l^*}{\dfrac{h^*}{\lambda_l^*} + \dfrac{T_0^{*2}}{L_{ww}^* \mathcal{L}^{*2}}} \tag{4.5}$$

the numerator of which is proportional to the driving force for evaporation (the superheat $\Delta T^* = T_w^* - T_0^*$ modulated by the Kelvin effect), while the two terms in the denominator can be seen as resistances to evaporation. The first one indeed represents the thermal resistance of the liquid film, while the second amounts to a kinetic resistance (usually associated with the finiteness of the number of molecules hitting the interface per unit time). Note finally that j^* can a priori be positive (evaporation) or negative (condensation).

Equation (4.1), together with Eqs (4.2), (4.3) and (4.5) is thus a fourth-order PDE in space, requiring a certain number of boundary conditions. Typically, as far as the microregion is concerned, its solutions will be of two types: (i) extended microfilms—that is, no dry points anywhere, a microfilm extending to, say, $x^* \to -\infty$ and the macroscopic zone being reached formally at $x^* \to +\infty$; and (ii) truncated microfilms—that is, the film thickness h^* formally vanishing at some "dry" point $x^* = x_0^*$ and the macroscopic region still corresponding to $x^* \to +\infty$. While the boundary conditions at infinity will be specified later on the basis of the asymptotics of the solution in these limits, we will here give some details about an important boundary condition applying at any (possibly moving) dry point $x^* = x_0^*(t^*)$. Namely, we require the film there to be in a state of local equilibrium, hence assuming instantaneous relaxation of the microcontact zone to a minimum of the relevant free energy. The resulting boundary condition, derived by Rednikov and Colinet (2011) in the framework of an approach similar to that of Yeh et al. (1999), reads

$$\frac{\gamma^*}{2}\left(\frac{\partial h^*}{\partial x^*}\right)^2 - P^*(h^*) \xrightarrow{x^* \to x_0^*} -S^* \tag{4.6}$$

where $S^* = \gamma^* + \gamma_{sl}^* - \gamma_{sg}^*$ is the spreading parameter (where γ_{sl}^* and γ_{sg}^* are interfacial energies of solid–liquid and solid–vapor surfaces, respectively), and

$P^*(h^*)$ is the excess film energy due to thin-film forces, related to the disjoining pressure by

$$P^*(h^*) = \int_{h^*}^{+\infty} \Pi^*(h^*)\mathrm{d}h^* \tag{4.7}$$

Two important remarks are at hand here. First, the condition (4.6) is written in the form of an asymptotic condition for x^* tending to x_0^*, in order to allow dealing with the usual Hamaker-like form $\Pi^*(h^*) = A^*/h^{*3}$ of the disjoining pressure for nonpolar liquids down to $h^* \to 0$, as seen later (see also Colinet and Rednikov (2011) for a general discussion of such an approach). Yet, it also makes it possible to handle nonsingular forms of $\Pi^*(h^*)$, for which $P^*(0)$ is finite (the microcontact angle at x_0^* is then $\theta_0 = [2(P^*(0) - S^*)/\gamma^*]^{1/2}$, according to Eq. (4.6)—that is, a Young's angle modified by disjoining pressure (see Yeh et al. (1999)). Second, it is often argued (in order to justify the existence of precursor films) that adsorption of vapor molecules on the substrate is unavoidable, so that the latter is never truly dry (see e.g. Starov and Velarde, 2009). The approach followed hereafter is that such (sub-)monomolecular films, escaping the hydrodynamic formalism adopted here, should be handled by using (in S^*) the appropriate value of the surface energy γ_{sg}^*, obtained from Gibbs adsorption theory (see, e.g., Defay and Prigogine, 1966).

To conclude this section, it should also be emphasized that the above theoretical framework is strictly valid under the lubrication approximation, relying on the condition $\varepsilon = [h]/[x] \ll 1$, where the length scales $[h]$ and $[x]$ still need to be selected, depending on the particular case considered. Importantly, the smallness parameter ε thus represents the typical order of magnitude of the film slopes—that is, in particular the apparent contact angles in which we are interested.

4.3 Cox–Voinov solution and viscous bending

Consider a contact line moving with velocity $c^* > 0$ to the left (i.e., advancing, as the macroscopic region is here always assumed to be on the right), and assume a steady film profile in the co-moving frame—that is, $h^* = h^*(x^* + c^*t^*)$. The first term of Eq. (4.1) may then be replaced by $c^*\partial h^*/\partial x^*$. At large scales, a balance between this term and the fourth-order capillary term can occur, while the effects of disjoining pressure and evaporation (including the Kelvin effect) are negligible. Integrating Eq. (4.1) once and assuming that there is no flux in the moving frame (i.e., $c^*h^* + q^* = 0$), the film thickness equation can then be cast in the form

$$3Ca + h^{*2}\frac{\partial^3 h^*}{\partial x^{*3}} = 0 \tag{4.8}$$

where the capillary number $Ca = c^* \eta_1^* / \gamma^* > 0$ has been introduced. Note the self-similarity of Eq. (4.8)—that is, its invariance with respect to a change of scales identical for h^* and x^*. On the contrary, the capillary number could be scaled out by using different scales $[h]$ and $[x]$ and defining the "slope parameter" by $\varepsilon = [h]/[x] = (3Ca)^{1/3}$—that is, the typical scale of purely motion-induced angles (viscous bending). Note that the capillary number should clearly be much smaller than unity for the lubrication approximation to hold. At large enough positive x^*, looking for solutions to Eq. (4.8) in the form $h^* \sim x^{*m} \log^n(x^*/\ell^*)$, and requiring that the curvature asymptotically vanishes (i.e., the macroscopic curvature is much smaller than at the scales considered in the present section), one finds the Cox–Voinov asymptotics (Voinov, 1976; Cox, 1986)

$$h^* = x^* \left[9Ca \log\left(\frac{x^*}{e\ell^*} \right) \right]^{\frac{1}{3}} = x^* \left[\theta_{\text{mic}}^3 + 9Ca \log\left(\frac{x^*}{e\ell_{\text{mic}}^*} \right) \right]^{\frac{1}{3}} \quad \text{for } \log\left(\frac{x^*}{\ell^*} \right) \gg 1 \quad (4.9a)$$

where given that the length ℓ^* in general depends on the capillary number Ca, the second expression of h^* will sometimes be preferred in what follows, in which θ_{mic} is taken to correspond to the apparent contact angle in the zero-velocity case $Ca = 0$ (note that the dependency of ℓ_{mic}^* with Ca should therefore be weaker than exponential, contrary to that of $\ell^* = \ell_{\text{mic}}^* \exp[-\theta_{\text{mic}}^3/9Ca]$, which is exponentially smaller than ℓ_{mic}^* for $Ca \to 0$). Note also the factor e within the logarithmic term, which is introduced in order to recover the most common form of the expression of the cube of the slope (see, e.g., Bonn et al., 2009)—that is,

$$\left(\frac{dh^*}{dx^*} \right)^3 = 9Ca \log\left(\frac{x^*}{\ell^*} \right) = \theta_{\text{mic}}^3 + 9Ca \log\left(\frac{x^*}{\ell_{\text{mic}}^*} \right) \quad \text{for } \log\left(\frac{x^*}{\ell^*} \right) \gg 1 \qquad (4.9b)$$

On the other hand, it is clear that Eqs (4.9a) and (4.9b) cannot be valid down to the microscale, where they cannot satisfy equilibrium conditions such as Eq. (4.6) at points where $h^* = 0$, even without disjoining pressure (in which case, Eq. (4.6) reduces to the small angle version of Young's law). Moreover, Eq. (4.9a) yields a capillary pressure increasing as x^{*-1} when x^* decreases—that is, nonintegrably. This can also be shown to be the case for the stress exerted on the substrate, thus indicating that the smallest scales provide a nonphysically large contribution to the total force exerted on the substrate (and to the corresponding viscous dissipation). As seen hereafter, a small-scale cutoff for these divergences will actually be obtained by reincorporating some of the missing effects (e.g., evaporation, disjoining pressure, or the Kelvin effect). This will also make it possible to determine the two as yet unknown quantities in Eqs (4.9a) and (4.9b)—namely, θ_{mic} and ℓ_{mic}^* (or ℓ^*)—which will turn out to be "inherited" from this so-called microstructure of the contact line (hence the subscript "mic").

4.4 Extended flat microfilms

As part of the analysis of such microstructures of the contact line, it is worth briefly examining the existence and stability of uniform thin-film solutions of Eq. (4.1) in order to determine under which conditions extended "precursor" films could possibly exist ahead of the contact line instead of a dry substrate. We will not consider here the limit of very large kinetic resistance—that is, L^*_{ww} very small in Eq. (4.5)—leading to vanishing evaporation flux j^*, for which Eq. (4.1) in principle admits flat film solutions of arbitrary thickness. Instead, requiring the numerator of Eq. (4.5) to vanish (chemical equilibrium between the film and the vapor), and with Eq. (4.3), we see that the steady flat film solutions have a thickness h^*_{f} such that

$$\Pi^*(h^*_{\mathrm{f}}) = \frac{\rho^*_{\mathrm{l}}\mathcal{L}^*\Delta T^*}{T^*_0} \tag{4.10}$$

In other words, excess forces due to molecular interactions with the substrate must compensate the driving force for phase change. This can a priori hold both for evaporation ($\Delta T^* > 0$) and condensation ($\Delta T^* < 0$)—depending, of course, on the disjoining pressure curve $\Pi^*(h^*)$ characterizing the liquid/solid pair. Note, however, that a linear stability analysis of such flat film solutions in the framework of Eq. (4.1) shows that the latter are stable to infinitesimal fluctuations provided $\Pi^{*'}(h^*_{\mathrm{f}}) < 0$. Otherwise, fluctuations grow, and the film either breaks or evolves to another type of solution.

In the following, we will restrict ourselves to the most classical form $\Pi^*(h^*) = A/h^{*3}$ of the disjoining pressure, with A^* positive. In that case, if $\Delta T^* > 0$, a stable flat film is possible, with thickness

$$h^*_{\mathrm{f}} = \left(\frac{A^*T^*_0}{\rho^*_{\mathrm{l}}\mathcal{L}^*\Delta T^*}\right)^{\frac{1}{3}} \tag{4.11}$$

as found by Potash and Wayner (1972), and often supposed to extend all over the parts of the substrate not covered by macroscopic liquid portions (see, e.g., Moosman and Homsy, 1980; Stephan and Busse, 1992; Morris, 2001; Ajaev, 2005). However, it has been more recently demonstrated by Rednikov and Colinet (2011) that the absolute stability of such film is not guaranteed, because this should also depend on the short-scale tendency to wetting, characterized by the spreading parameter S^*. To understand this, consider a system at constant external pressure p^*_{v} (but variable total volume V^*) and temperature $T^*_{\mathrm{w}} = T^*_0 + \Delta T^*$, for which the free energy $Z^* = F^* + p^*_{\mathrm{v}}V^*$ must be minimal at equilibrium (F^* is the Helmholtz free energy of the system). We can then compare the free energy Z^*_{f} of the extended flat film state of thickness given by

Eq. (4.11) with the corresponding free energy Z_b^* of the "bare" substrate. Rednikov and Colinet (2011) have demonstrated, in a rather general setting accounting for possible Gibbs adsorption on the substrate, that the difference between both free energies (per unit surface) is given by

$$\frac{Z_f^* - Z_b^*}{\text{Surf}} = P^*(h_f^*) + h_f^*\Pi^*(h_f^*) - S^* = \frac{3A^*}{2h_f^{*2}} - S^* \tag{4.12}$$

Hence, the extended liquid film is absolutely stable if $Z_f^* < Z_b^*$—that is, only if the spreading parameter exceeds the critical positive value

$$S_{cr}^* = \frac{3A^*}{2h_f^{*2}} > 0 \tag{4.13}$$

Conversely, for a given positive value of S^*, the extended flat film becomes metastable ($Z_f^* > Z_b^*$) when its thickness h_f^* decreases below the critical value

$$h_{cr}^* = h_p^* = \left(\frac{3A^*}{2S^*}\right)^{\frac{1}{2}} \tag{4.14}$$

which is just the thickness of the "pancake" solution (see, e.g., de Gennes et al., 2004). From Eq. (4.11), it turns out that this occurs when the superheat ΔT^* exceeds some critical value, beyond which truncated films are preferred (in a thermodynamic sense) over extended ones (see also Section 4.6.2).

4.5 Microstructure of the moving contact line for a nonvolatile liquid

We now consider a first possible small-scale resolution of the Cox–Voinov asymptotics described in Section 4.3, following Hervet and de Gennes (1984), whose analysis is here recovered by omitting any effects of phase change, including the Kelvin effect (i.e., $j^* = 0$ in Eq. (4.1)). In the absence of any slip on the substrate, only disjoining pressure can then possibly regularize the viscous singularity (we here consider the apolar form $\Pi^*(h^*) = A^*/h^{*3}$ with A^* positive), and it is natural to select the length scales $[x]$ along the film and $[h]$ across the film to have all terms (moving-frame term, capillarity, and disjoining pressure) of the same order in the thin-film Eq. (4.1). Integrating it once, again with zero flux in the moving frame and for advancing contact lines (Ca > 0), the dimensionless generalization of Eq. (4.8) then reads

$$1 + h^2 h''' - h^{-2} h' = 0 \tag{4.15}$$

and the scales are given by

$$[h] = \frac{3^{1/6} a^*}{Ca^{1/3}}, \quad [x] = \frac{a^*}{3^{1/6} Ca^{2/3}} \tag{4.16}$$

where numerical factors are used for convenience, and we have introduced the molecular length scale (extensively used by de Gennes and collaborators)

$$a^* = \left(\frac{A^*}{\gamma^*}\right)^{1/2} \tag{4.17}$$

and we see that the "slope parameter" is indeed $\varepsilon = [h]/[x] = (3Ca)^{1/3}$, as noted in Section 4.3 (motion-induced angle). We will be looking for solutions starting from $h = 0$ at $x = 0$ (the microedge), where the dimensionless form of the boundary condition (4.6) applies—that is,

$$h \to 0, \; h'^2 - \frac{1}{3h^2} \to -\Sigma \quad \text{for } x \to 0 \tag{4.18}$$

in which the dimensionless spreading coefficient Σ has been defined by

$$\Sigma = \frac{2}{(3Ca)^{2/3}} \frac{S^*}{\gamma^*} \tag{4.19}$$

On the macroscopic side $x \to +\infty$, we must match with the Cox–Voinov film profile (Eq. (4.9a)), whose dimensionless form here reads

$$h = x\left[3 \log\left(\frac{x}{e\ell}\right)\right]^{\frac{1}{3}} = x\left[\vartheta_{\text{mic}}^3 + 3 \log\left(\frac{x}{e\ell_{\text{mic}}}\right)\right]^{\frac{1}{3}} \quad \text{for } x \to +\infty \tag{4.20}$$

where $\ell = \ell^*/[x]$, $\ell_{\text{mic}} = \ell^*_{\text{mic}}/[x]$, and $\vartheta_{\text{mic}} = \theta_{\text{mic}}/(3Ca)^{1/3}$, in which we must remember that θ_{mic} is taken as the static ($Ca = 0$) contact angle that here should be equal to zero for $S^* > 0$ (perfect wetting) and to $(-2S^*/\gamma^*)^{1/2}$ for $S^* < 0$ (partial wetting, small angle approximation). The corresponding *equilibrium* film shapes can easily be found by expressing that the excess liquid pressure (Eq. (4.3)) is a constant. This yields the dimensionless equation $3h'' + h^{-3} = -\Omega$, to be solved with the boundary condition (4.18), and leading to the pancake solution for $\Sigma > 0$ (perfect wetting), and to "Young's wedge" for $\Sigma < 0$ (partial wetting); see, for example, Colinet and Rednikov (2011). Hence, the rescaled angle appearing in Eq. (4.20) is given by $\vartheta_{\text{mic}} = 0$ for $\Sigma > 0$ and by $\vartheta_{\text{mic}} = (-\Sigma)^{1/2}$ for $\Sigma < 0$. It is now necessary to calculate the dimensionless microlength ℓ_{mic} as a function of Σ by solving numerically Eq. (4.15) with Eq. (4.18) and looking for solutions matching Eq. (4.20) to the right. This has been done by Colinet and Rednikov (2011), but only $\ell(\Sigma)$ was given (and seen to quickly vanish for

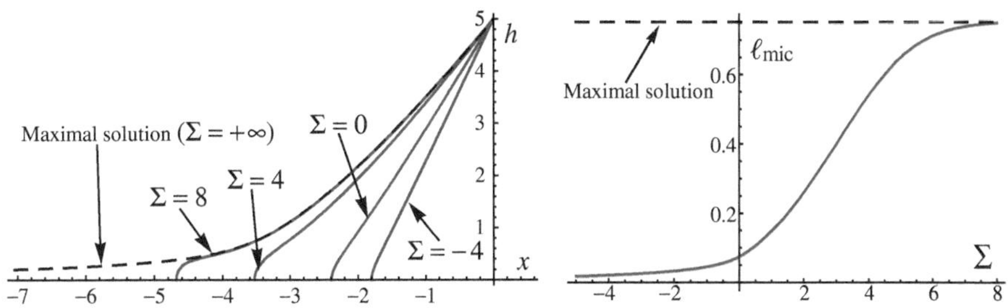

FIGURE 4.1 (Left): Some film thickness profiles in the microregion for several dimensionless spreading coefficients Σ. Note that each profile is horizontally shifted so it has thickness $h = 5$ at the same abscissa. (Right): The dimensionless microlength ℓ_{mic} appearing in the Cox–Voinov asymptotics as a function of Σ.

$\Sigma \to -\infty$), so we here provide the curve $\ell_{\text{mic}}(\Sigma) = O(1)$ instead (Figure 4.1, together with some typical microstructure profiles). Its dimensional counterpart $\ell^*_{\text{mic}} = a^* \ell_{\text{mic}}(\Sigma)/3^{1/6} Ca^{2/3}$ thus represents the sought cutoff length for the viscous singularity, which can clearly exceed the molecular scale a^* by orders of magnitude at sufficiently small velocities.

Note that we have here assumed $Ca > 0$—that is, an advancing contact line. However, the result (4.20), when rewritten in the dimensional form (Eqs (4.9a) and (4.9b)), is actually applicable to receding contact lines ($Ca < 0$), as well, as long as the logarithmic term is somewhat smaller than the static contribution θ^3_{mic}—that is, for partial wetting ($\Sigma < 0$) for a velocity that is small enough, and for sufficiently but not too large x^* (i.e., in an intermediate asymptotic sense). When this is not the case, we expect transitions to other regimes (e.g., Landau–Levich films, not described here; see, e.g., Delon et al., 2008; Bonn et al., 2009).

From Figure 4.1 (left) and Eq. (4.19), it is seen that the dimensionless length of the foot of the microstructure (hereafter classically called precursor film) elongates when the spreading coefficient S^* is positive and the velocity decreases (i.e., Σ increases). For $\Sigma \to +\infty$ and $x \to -\infty$, Eq. (4.15) is satisfied by the asymptotic shape $h_{\max} = 1/|x|$, the so-called "maximal solution" (as named by Hervet and de Gennes, 1984) represented as a dashed line in Figure 4.1 (left). For large but finite Σ, the film profile is actually truncated at some value of $|x|$, above which the film gets thinner than the pancake film thickness (Eq. (4.14)). In the scales (Eq. (4.16)) used in this section, the dimensionless pancake thickness is simply $h_{\text{p}} = \Sigma^{-1/2}$, such that requiring $h_{\max} \sim h_{\text{p}}$ yields the dimensionless precursor film length $|x_{\text{pf}}| \sim \Sigma^{1/2}$—that is, a dimensional length $|x^*_{\text{pf}}| \sim a^*(S^*/\gamma^*)^{1/2} Ca^{-1}$. Note that this scaling law has been validated by careful microscopic interferometry experiments on silicon oils of various viscosities by Kavehpour et al. (2003). Finally, remark that, as mentioned at the beginning of this section, disjoining pressure is indeed sufficient to regularize the viscous stress divergence associated with the motion of the contact line. This was demonstrated by Hervet and de Gennes (1984) and in a more general formulation considering both partial and complete wetting by Colinet and Rednikov (2011). It is shown there by considering the asymptotics near the microedge $x = 0$ that both the viscous stress and the dissipation now diverge more slowly as $\sim x^{-1/2}$,—that is, *integrably*—while the pressure diverges only logarithmically. It was also argued in Colinet and Rednikov (2011) that, although still problematic in some respects (multivaluedness of the velocity field and breakdown of the lubrication approximation in a molecular vicinity of the microedge), there is no real reason in our opinion to think this should invalidate experimentally verifiable predictions made on the basis of this (de Gennes) "paradigm."

4.6 Liquids with evaporation/condensation into/from their pure vapor

We now incorporate possible evaporation/condensation effects—that is, the last term in Eq. (4.1). As a preamble, it is instructive to estimate whether this phase change term is indeed negligible at the scales (Eq. (4.16)) used in the previous section. First, the effect of superheating (i.e., the first term at the numerator of Eq. (4.5)) can be omitted against the highest-order capillary flow term in Eq. (4.1), provided

$$\frac{[h]^5}{[x]^4} \gg \frac{\eta_l^* \lambda_l^* \Delta T^*}{\rho_l^* \mathcal{L}^* \gamma^*} \to Ca^{1/3} \longrightarrow \left(\frac{\eta_l^* \lambda_l^* \Delta T^*}{\rho_l^* \mathcal{L}^* \gamma^* a^*}\right)^{1/3} = \theta_{\text{evap}}^D \tag{4.21}$$

where the scales (Eq. (4.16)) have been used for the second inequality, and the notation θ_{evap}^D has been introduced to denote a typical scale of evaporation-induced contact angles (here with an essential influence of disjoining pressure, hence the superscript "D"). Thus, this condition can in principle be satisfied for sufficiently small values of ΔT^* and not too small Ca, when the motion-induced contribution to the contact angle—that is, $Ca^{1/3}$ (see Eqs (4.9a) and (4.9b))—dominates over the evaporation-induced contribution θ_{evap}^D.

Next, in order to estimate the influence of the Kelvin effect (i.e., the second term at the numerator of Eq. (4.5)), it is useful to introduce the new length scale

$$\delta^* = \frac{(\eta_l^* \lambda_l^* T_0^*)^{1/2}}{\rho_l^* \mathcal{L}^*} \tag{4.22}$$

which, for many liquids, incidentally turns out to be of the same order as the molecular length scale a^* defined by Eq. (4.17)—typically a few Ångströms.

Then, it is readily found that in order to omit the Kelvin effect against capillarity (and disjoining pressure) in Eq. (4.1), we should have

$$\frac{[h]^2}{[x]} \longrightarrow \delta^* \to a^* \longrightarrow \delta^* \qquad (4.23)$$

which, importantly, does not depend on any control parameter such as ΔT^* or Ca, and is generally not satisfied because, as we just said, we have $a^* \sim \delta^*$ instead. As a consequence, for liquids in contact with their own vapor, it is likely that the Kelvin effect might have a noticeable influence on de Gennes' picture of moving contact lines presented in the last section (unless the kinetic resistance, i.e. the second term at the denominator in Eq. (4.5), is particularly large—in other words, $[h]$ is much smaller than the "kinetic length" $\ell^*_{\mathrm{kin}} = \lambda_1^* T_0^{*2} / L^*_{\mathrm{ww}} \mathcal{L}^{*2}$).

In Section 4.6.1, we highlight the role of the Kelvin effect by omitting the disjoining pressure contribution altogether—that is, considering a hypothetical case where $\delta^* \gg a^*$. Then, in Section 4.6.2, we reincorporate disjoining pressure and focus on the effect of evaporation near the contact line on the selection between truncated and extended microfilm profiles, as already touched upon at the end of Section 4.4.

4.6.1 A Minimalist Model Without Disjoining Pressure but Accounting for the Kelvin Effect

Following Rednikov and Colinet (2012, 2013), we now describe how a minimalist approach based on the Kelvin effect alone is sufficient to fully resolve all singularities of moving contact lines, including evaporation-related ones (the latter due to the fact that the expression (4.5) of the evaporation rate a priori diverges as $h^* \to 0$ when the kinetic resistance term is omitted). We consider first the case where phase change is not externally imposed (vanishing superheat—i.e., saturated vapor atmosphere), and then turn to the effect of forced evaporation and kinetic resistance for slowly moving contact lines.

4.6.1.1 Moving contact lines in a saturated vapor atmosphere

Instead of balancing the capillary and disjoining pressure terms in Eq. (4.1), as was done in Section 4.5, we omit the latter (which is a priori valid when $\delta^* \gg a^*$) and select scales so both the moving-frame term and the Kelvin effect have the same order as the capillary flow term in Eq. (4.1). Specifically, we choose the scales

$$[h] = \frac{\delta^*}{(3Ca)^{1/3}}, \quad [x] = \frac{\delta^*}{(3Ca)^{2/3}} \qquad (4.24)$$

that is, again with $\varepsilon = [h]/[x] = (3Ca)^{1/3}$ but now with δ^*, defined by Eq. (4.22), instead of $\sqrt{3}a^*$ at the numerator. This scaling yields the dimensionless equation

$$h' + (h^3 h''')' - 3h^{-1}h'' = 0 \tag{4.25}$$

where the kinetic resistance term has been omitted—that is, assuming that $\delta^*/Ca^{1/3} \gg \ell^*_{\mathrm{kin}}$—together with the effect of superheat (i.e., the bulk liquid is in equilibrium with the vapor, $\Delta T^* = 0$). These two effects will be considered in Section 4.6.1.2. Because we here ignore any effect of the disjoining pressure, the dimensionless boundary condition (4.6) simply reads

$$h' = \vartheta_{\mathrm{Y}} = \frac{\theta_{\mathrm{Y}}}{(3Ca)^{1/3}} \quad \text{at } x = 0 \tag{4.26}$$

where θ_{Y} is the Young's angle—that is, 0 for $S^* > 0$ and $(-2S^*/\gamma^*)^{1/2}$ for $S^* < 0$. Note that contrary to the singular form $h \sim \sqrt{x}$ of the film profile encountered in Section 4.5 (devoted to "de Gennes' paradigm"), the microedge asymptotics here starts with a finite slope ϑ_{Y}, according to Eq. (4.26). More precisely, from Eq. (4.25) it can be shown (Rednikov and Colinet, 2012, 2013) that, up to some exponentially small eigenfunction, this behavior in fact corresponds to the leading-order term of a Taylor expansion $h = \vartheta_{\mathrm{Y}} x + \vartheta_{\mathrm{Y}}^2 x^3/18 + O(x^5)$ for $\vartheta_{\mathrm{Y}} > 0$ (partial wetting). For $\vartheta_{\mathrm{Y}} = 0$ (complete wetting), interestingly, such a "truncated microfilm" solution starting at $x = 0$ cannot be found, and the film instead extends to $x \to -\infty$ with the asymptotics $h = -6x^{-1} + O(x^{-7})$, hence another type of "extended microfilm" or "maximal solution" (similar to that encountered in Section 4.5 but with quite a different physics). Importantly, whatever the case considered, is it easy to check that neither the pressure $\tilde{p}_1 \sim h''$ nor the stress exerted on the substrate $\tau \sim h\tilde{p}_1'$ nor the viscous dissipation density $\dot{\varepsilon} \sim h^3(\tilde{p}_1')^2$ diverges; in other words, the viscous stress singularity is fully resolved by the sole consideration of the Kelvin effect (more details can be found in the original papers of Rednikov and Colinet). It also appears that the leading order balance at the microedge in Eq. (4.25) is between the first (moving-frame) and the third (Kelvin effect) terms, while the second one (capillarity) turns out to be of higher order, again for both partial and complete wetting. As demonstrated in detail in the original papers, this exact compensation (actually valid for receding contact lines as well, for $\vartheta_{\mathrm{Y}} > 0$) shows that even though motion of the fluid generally exists at some distance of the moving contact line, the velocity locally vanishes at the microedge, where the contact line moves only due to phase change (Figure 4.2).

On the other hand, far from the microedge, a Cox–Voinov behavior of the type (Eq. (4.20)) is again expected, where ϑ_{mic} is here also taken as the steady microstructure contact angle—that is, ϑ_{Y} (indeed, without the moving-frame term, the trivial solution of Eqs (4.25) and (4.26) is the straight wedge $h = \vartheta_{\mathrm{Y}} x$).

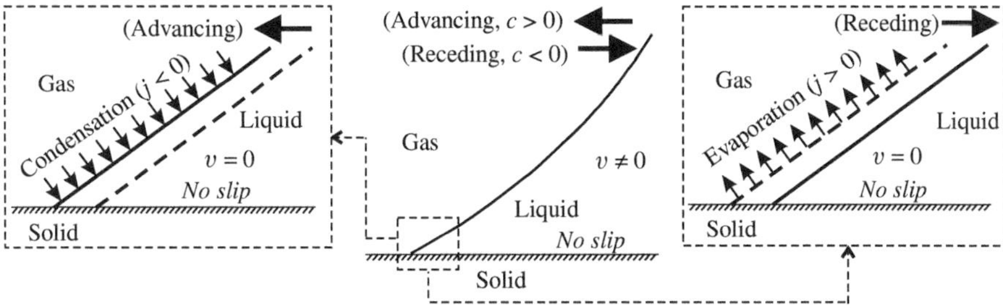

FIGURE 4.2 The physical mechanism behind the resolution of the moving contact line singularities by the Kelvin effect: the curvature-induced change of local saturation conditions near the microedge allows the contact line to recede by evaporation and to advance by condensation without any liquid velocity relative to the substrate. *From Rednikov and Colinet (2012, 2013).*

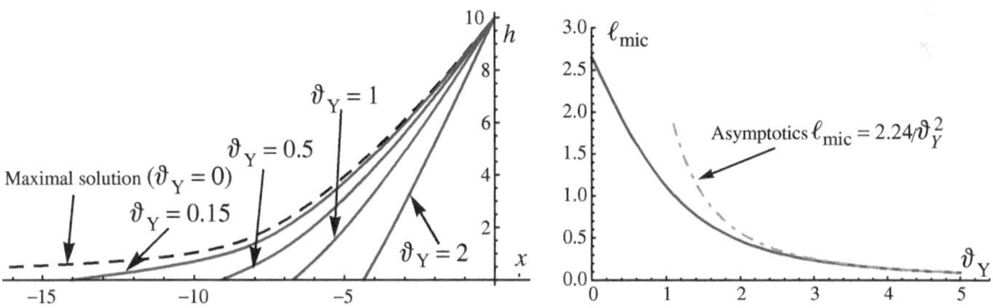

FIGURE 4.3 Microstructure of a contact line advancing into its pure vapor, with the Kelvin effect and no disjoining pressure. (Left): Some film thickness profiles for several rescaled Young's angles ϑ_Y. Note that each profile is horizontally shifted such as to have thickness $h = 10$ at the same abscissa. (Right): The dimensionless microlength ℓ_{mic} of the Cox–Voinov asymptotics as a function of ϑ_Y.

As for the microlength $\ell_{mic}(\vartheta_Y)$, it needs to be obtained by (numerically) solving Eq. (4.25) starting from Eq. (4.26). The result is represented in Figure 4.3, together with some typical microstructure profiles. The actual cutoff length to be used in the Cox–Voinov law (Eqs (4.9a) and (4.9b)), with $\theta_{mic} = \theta_Y$, then reads $\ell^*_{mic} = \delta^* \ell_{mic}(\vartheta_Y)/(3Ca)^{2/3}$. In particular, one has $\ell^*_{mic} \approx 1.27\, \delta^*/Ca^{2/3}$ when $\vartheta_Y \to 0$ (in view of the definition (4.26) of ϑ_Y, this corresponds to the case of a small Young's angle θ_Y compared to the motion-induced angles $Ca^{1/3}$, including complete wetting). At the other extreme, $\vartheta_Y \to \infty$ (the Young's angle is much larger than motion-induced ones), the semi-analytical result represented as a dot-dashed curve in Figure 4.3 (right) instead yields $\ell^*_{mic} \approx 2.24\,\delta^*/\theta_Y^2$.

To conclude this section, we emphasize that although the idea that phase change can regularize moving contact line singularities is not new in itself (Wayner, 1993; Pomeau, 2000, 2011; Shanahan, 2001), to our knowledge its first rigorous demonstration within a classical thermo-hydrodynamic formalism highlighting the central role of the Kelvin effect is that proposed by Rednikov and Colinet (2012, 2013) and summarized above. Somewhat later, other authors have arrived at similar conclusions, but they limited their analysis of the contact line microstructure to partial wetting and a motionless contact line (Janeček and Nikolayev, 2012) and subsequently extended it to complete wetting (or large superheat), and treating contact line motion as a small perturbation (Janeček et al., 2012, 2013), in contrast with the case $Ca^{1/3} \sim \theta_Y$ considered in this section.

4.6.1.2 Slowly moving contact lines with nonvanishing superheat

We now consider the role of an externally imposed evaporation—that is, a positive superheat ΔT^* in the expression (4.5) of the evaporation flux—while regarding the effect of contact line velocity as secondary. A different scaling is thus at hand, and we will here choose to balance the capillary flow term in Eq. (4.1) with both superheat and Kelvin effects. Doing so, we arrive at the scales

$$[h] = \frac{\delta^*}{\varepsilon}, [x] = \frac{\delta^*}{\varepsilon^2} \quad \text{with } \varepsilon = \left(\frac{3\eta_l^{*1/2}\lambda_l^{*1/2}\Delta T^*}{\gamma^* T_0^{*1/2}}\right)^{1/3} \sim \theta_{\text{evap}}^K \tag{4.27}$$

where, as in Eq. (4.21), we introduced the notation θ_{evap}^K for the scale of evaporation-induced angles (the present form being appropriate when the Kelvin effect dominates over the disjoining pressure). With this scaling, the dimensionless Eq. (4.1) then reads

$$ch' + (h^3 h''')' + \frac{1 - 3h''}{h + K} = 0 \tag{4.28}$$

where c is the dimensionless velocity of the contact line

$$c = \frac{3Ca}{\varepsilon^3} = \frac{c^* \eta_l^{*1/2} T_0^{*1/2}}{\lambda_l^{*1/2} \Delta T^*} \tag{4.29}$$

in which c^* is its actual dimensional velocity, and a dimensionless kinetic resistance number has been defined by

$$K = \frac{\ell_{\text{kin}}^*}{[h]} = \varepsilon \frac{\ell_{\text{kin}}^*}{\delta^*} \tag{4.30}$$

in which we have used the earlier introduced (typically nanometric) "kinetic length" $\ell_{\text{kin}}^* = \lambda_l^* T_0^{*2} / L_{\text{ww}}^* \mathcal{L}^{*2}$. As for the boundary conditions, we again consider a given (Young's) angle at the microedge—that is,

$$h' = \vartheta_Y = \frac{\theta_Y}{\varepsilon} \text{ and } h = 0 \quad \text{at } x = 0 \tag{4.31}$$

and at infinity we still look for solutions matching the Cox–Voinov behavior (Eq. (4.9a)), which in the new scaling adopted here reads

$$h = x\left[\vartheta_{\text{mic}}^3 + 3c\log\left(\frac{x}{e\ell_{\text{mic}}}\right)\right]^{\frac{1}{3}} \quad \text{for } x \to +\infty \tag{4.32}$$

with values of ϑ_{mic} and ℓ_{mic} again inherited from the microstructure region, and thus $\vartheta_{\text{mic}} = \vartheta_{\text{mic}}(\vartheta_Y, K)$ and $\ell_{\text{mic}} = \ell_{\text{mic}}(c, \vartheta_Y, K)$ in general. A parametric analysis of this problem is presented here for two important limiting cases: (i) $K = 0$ (local interfacial equilibrium) and ϑ_Y varied with $c \ll 1$ (i.e., velocity will be considered as a perturbation); and (ii) $\vartheta_Y = 0$ (complete wetting) and K varied, while $c \ll 1$ as well. In the latter case, note that we will not consider the particular case $K \to \infty$ (kinetically-limited evaporation regime), implied, for example, by Todorova et al. (2012).

Because $c \ll 1$ in both cases, we look for solutions of the form

$$h = h_0 + ch_1 + O(c^2) \tag{4.33}$$

where the leading-order term h_0 is the solution of $(h_0^3 h_0''')' + (h_0 + K)^{-1}(1 - 3h_0'') = 0$, with boundary conditions $h_0 = 0$ and $h_0' = \vartheta_Y$ at $x = 0$. It can be shown—see, e.g., Rednikov et al. (2009) for a more general case—that the solutions of this problem with vanishing curvature at infinity have the asymptotics

$$h_0 \to \vartheta_{\text{mic}} x - \frac{\log^2[x/k_0]}{4\vartheta_{\text{mic}}^4} + O(1) \quad \text{for } x \to +\infty \tag{4.34}$$

where the microstructure contact angle ϑ_{mic} and the length scale k_0 can be calculated numerically and here depend on ϑ_Y and K (in particular, ϑ_{mic} is represented in Figure 4.4 for case (i) and in Figure 4.5 for case (ii)). Remember that $\vartheta_{\text{mic}} = O(1)$ is the rescaled contact angle of the steady microstructure, the actual one being given by $\theta_{\text{mic}} = \vartheta_{\text{mic}}\varepsilon \ll 1$, with ϵ given by Eq. (4.27). In view of the definition of $\vartheta_Y = \theta_Y/\varepsilon$ and of Figure 4.4 (left), it can be seen that $\theta_{\text{mic}} > \theta_Y$—that is, evaporation indeed always induces an increase in the apparent contact angle compared to the equilibrium (Young's) value. Clearly, this is also true for case (ii), where $\theta_Y = 0$, and the apparent angle is entirely evaporation-driven, as seen in Figure 4.5. Film thickness profiles for case (i) are represented in Figure 4.4 as well for some values of ϑ_Y. They actually look similar to those presented in Figure 4.3 for the case of an advancing contact line, apart from the important difference that there is no maximal (or extended) solution here; in other words, the film is always truncated, even for perfect wetting (due to evaporation).

Now, for both cases and at the first order in c, Eq. (4.28) yields the equation (not reproduced here for conciseness) to be satisfied by the correction h_1 in

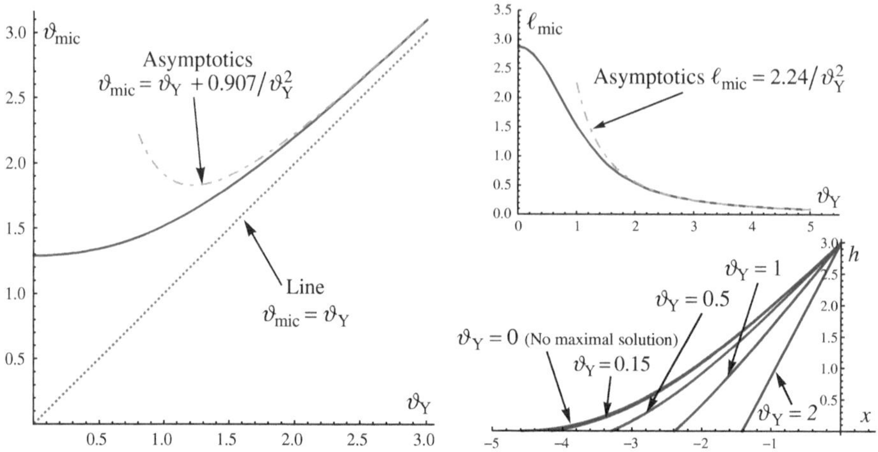

FIGURE 4.4 Parameters ϑ_{mic} (rescaled microstructure contact angle, on the left) and ℓ_{mic} (dimensionless microlength, on the upper right) of the Cox–Voinov asymptotics as a function of the rescaled Young's angle ϑ_{Y}, for a slowly moving contact line evaporating into its own vapor with negligible kinetic resistance and no disjoining pressure. Film profiles are shown on the lower right for some values of ϑ_{Y}.

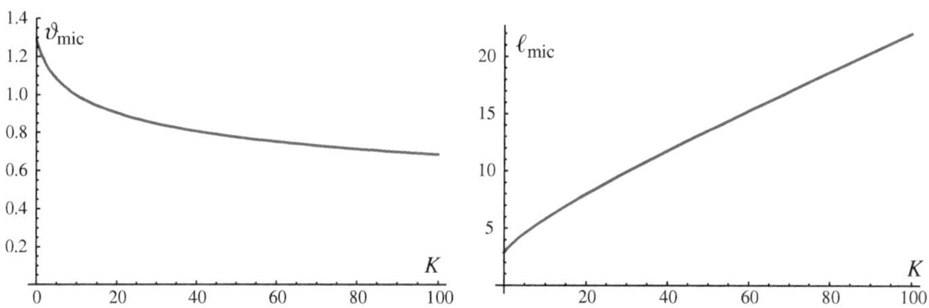

FIGURE 4.5 Influence of the kinetic resistance K on microstructure characteristics for a slowly moving contact line of a perfectly wetting liquid evaporating into its pure vapor, with Kelvin effect and no disjoining pressure. (Left): The rescaled contact angle ϑ_{mic} of the microstructure as a function of K. (Right): The dimensionless microlength ℓ_{mic} in the Cox–Voinov asymptotics, as a function of K.

Eq. (4.33). Examining its vanishing-curvature asymptotics for $x \to +\infty$ quite generally yields the behavior

$$h_1 \to \frac{x \log[x/k_1]}{\vartheta_{\text{mic}}^2} + O(\log^3[x/k_1]) \quad \text{for } x \to +\infty \tag{4.35}$$

where k_1 is another length scale to be calculated numerically and thus again dependent on ϑ_{Y} and K in general. Expanding Eq. (4.32) for small values of

$c \log(x/\ell_{mic})$, it is easy to check that the result asymptotically matches with Eqs (4.33)–(4.35), provided $\ell_{mic} = k_1(\vartheta_Y, K)/e$. This function is also represented in Figure 4.4 for case (i) and in Figure 4.5 for case (ii), and upon multiplication by the scale $[x]$, it yields the dimensional length scale $\ell^*_{mic} = \delta^* \ell_{mic}(\vartheta_Y, K)/\varepsilon^2$ to be used in combination with $\theta_{mic} = \vartheta_{mic}(\vartheta_Y, K)\varepsilon$ in the Cox–Voinov asymptotics (Eqs (4.9a) and (4.9b)). In particular, from Figure 4.4 (i.e., for $K = 0$), one obtains $\theta_{mic} \approx 1.29\varepsilon$ and $\ell^*_{mic} \approx 2.88\, \delta^*/\varepsilon^2$ for $\vartheta_Y \to 0$ (small Young's angle θ_Y compared to evaporation-induced ones $\theta^K_{evap} \sim \varepsilon$). In the opposite limit of a small evaporation-induced contribution compared to the Young's angle $\vartheta_Y \to \infty$, the asymptotic results represented in Figure 4.4, together with Eq. (4.27), again yield $\ell^*_{mic} \approx 2.24\, \delta^*/\theta_Y^2$ (as in the previous section), while

$$\theta_{mic} \approx \theta_Y + 2.72 \frac{\eta_1^{*1/2} \lambda_1^{*1/2} \Delta T^*}{\gamma^* T_0^{*1/2} \theta_Y^2}$$

in which the second term indeed represents a small correction to the Young's angle (see also Janeček and Nikolayev, 2012), and which is actually valid both for evaporation ($\Delta T^* > 0$) and condensation ($\Delta T^* < 0$).

To conclude this section, we point out from Figure 4.5 (left) that, as expected, an increase in the kinetic resistance leads to a decrease in the apparent contact angle (especially in the situation of complete wetting considered in Figure 4.5, where the contact angle is entirely due to evaporation). Moreover, it is worth emphasizing that whatever the importance of these kinetic effects, evaporation apparently always prevents the existence of extended microfilms (as for $K = 0$)—that is, the film always ends at some "dry" spot. There (at $x = 0$), it can be shown that the solution of Eq. (4.28) satisfying boundary conditions (4.31) has the regular asymptotics

$$h = \vartheta_Y x + \frac{1}{6} x^2 (1 + c\, K \vartheta_Y) + \frac{c}{54} x^3 (K + c\, K^2 \vartheta_Y + 3\vartheta_Y^2) + O(x^4) \quad \text{for } x \to 0$$

which not only shows that truncated film solutions indeed always exist (even for perfect wetting, in contrast with the case of vanishing superheat dealt with in Section 4.6.1.1), but also that no singularities occur, still thanks to the Kelvin effect. Indeed, with this asymptotics, it is straightforward to check that for $x \to 0$, the pressure $\tilde{p}_1 \sim h'' = O(1)$, the stress exerted on the substrate $\tau \sim h\, \tilde{p}_1' = O(x)$, and the viscous dissipation density $\dot{e} \sim h^3 (\tilde{p}_1')^2 = O(x^3)$, hence a fully analytical (and integrable) behavior. The evaporation flux $j = (1 - 3h'')/(h + K)$ itself tends to the constant value $-c\, \vartheta_Y$ independently of K—that is, it is never singular (even for the "extreme case" $K = 0$), and in fact it exactly balances the moving-frame term ch' in Eq. (4.28) at the leading order. This actually points to the general validity of the compensation mechanism sketched in Figure 4.2 (microedge motion entirely due to evaporation/condensation), even for $c = O(1)$—that is, when the motion-induced angle is comparable to the evaporation-induced (and Young's) angles, see Eqs (4.27), (4.29), and (4.31). When $c = O(1)$, however, note that ℓ_{mic} (but not ϑ_{mic}) should also depend on c (whereas values given in Figures 4.4 and 4.5 were obtained for $c \ll 1$).

4.6.2 Full Model: Extended Versus Truncated Microfilms

In this section, we reincorporate the disjoining pressure contribution (in the usual form $\Pi^*(h^*) = A^*/h^{*3}$ with $A^* > 0$), in view of the fact that small length scales a^* and δ^*, defined by Eqs (4.17) and (4.22), respectively, are generally comparable. In addition, in case the microstructure of the contact line ends at some "dry" point (i.e., when the film is truncated), we will consider the boundary condition (4.6), thereby incorporating the spreading coefficient S^* into the analysis ("de Gennes' paradigm"; see Section 4.5). The scaling used here is that adopted by Rednikov et al. (2009), Rednikov and Colinet (2011), and Colinet and Rednikov (2011)—that is, reflecting a balance not only between capillary and disjoining pressures but also between the driving force for evaporation and the Kelvin effect in Eq. (4.5). Specifically, we select

$$[h] = 3^{1/2}\frac{a^*}{\varepsilon} = h_f^*, [x] = 3^{1/2}\frac{a^*}{\varepsilon^2} = \frac{h_f^*}{\varepsilon} \quad \text{where } \varepsilon = 3^{1/2}\left(\frac{a^* \rho_l^* \mathcal{L}^* \Delta T^*}{\gamma^* T_0^*}\right)^{1/3} \sim \theta_{\text{evap}} \tag{4.36}$$

where the equilibrium film thickness h_f^* has already been defined by Eq. (4.11) and indeed turns out to be the scale used here for h^*. Then, the complete thin-film equation in the moving reference frame reads

$$ch' + (h^3 h''' - h^{-1}h')' + E\frac{1 - 3h'' - h^{-3}}{h + K} = 0 \tag{4.37}$$

where the dimensionless velocity, evaporation number, and kinetic resistance are defined by

$$c = \frac{3Ca}{\varepsilon^3}, \quad E = \frac{\eta_l^* \lambda_l^* T_0^*}{3(a^* \rho_l^* \mathcal{L}^*)^2} = \frac{\delta^{*2}}{3a^{*2}}, \quad K = \frac{\ell_{\text{kin}}^*}{[h]} = \frac{\ell_{\text{kin}}^*}{h_f^*} \tag{4.38}$$

Interestingly, the evaporation number E is of the order of the squared ratio of small scales δ^* and a^* ; that is, it will be considered $O(1)$ hereafter, contrary to the case $E \gg 1$ (i.e., $\delta^* \gg a^*$) assumed in Section 4.6.1 and to the case $E \ll 1$ (i.e., $\delta^* \ll a^*$) not considered here (see Colinet and Rednikov, 2011). Note that for $E = O(1)$, the scale of evaporation-induced angles θ_{evap} introduced in Eq. (4.36) is of the same order as θ_{evap}^D (defined by Eq. (4.21)) and θ_{evap}^K (defined by Eq. (4.27) and to be used when $E \gg 1$).

That said, we will here consider steady microstructures only—that is, $c = 0$, keeping in mind that Cox–Voinov asymptotics (Eqs (4.9a) and (4.9b)) are expected to be applicable (here with $\theta_{\text{mic}} \sim \varepsilon$ and $\ell^*_{\text{mic}} \sim \varepsilon^{-2} a^* \sim \varepsilon^{-1} h^*_{\text{f}}$) if the contact line slowly advances or recedes. Outlining the detailed study of Rednikov and Colinet (2011), the main goal of this section will be to examine the competition between extended and truncated (precursor) film solutions and to present some typical values for the rescaled contact angle $\vartheta_{\text{mic}} = \theta_{\text{mic}}/\varepsilon$, as a function of the governing dimensionless parameters. Note that for truncated microfilm solutions, the microedge equilibrium condition (4.6) reads, in dimensionless form,

$$h \to 0,\ h'^2 - \frac{1}{3h^2} \to -\Sigma \quad \text{for } x \to 0 \tag{4.39}$$

in which the dimensionless spreading coefficient Σ is here defined by

$$\Sigma = \frac{2}{\varepsilon^2}\frac{S^*}{\gamma^*} = \frac{2}{3}\frac{S^*}{A^*} h^{*2}_{\text{f}} = \frac{h^{*2}_{\text{f}}}{h^{*2}_{\text{p}}} \tag{4.40}$$

where h^*_{p} is the pancake thickness defined by Eq. (4.14), in the case $S^* > 0$.

In Section 4.4, we shortly discussed the competition between two possible equilibrium states—namely, the extended flat microfilm state with thickness $h^* = h^*_{\text{f}}$ (i.e., $h = 1$ in the present scaling) and the bare substrate state. The latter was mentioned to be the preferred state (in a thermodynamic sense) when $h^*_{\text{f}} < h^*_{\text{p}}$, which, in view of Eq. (4.40), corresponds to $\Sigma < 1$. On the contrary, the flat microfilm solution $h = 1$ has the lowest free energy when $\Sigma > 1$. To confirm this thermodynamic argument, Rednikov and Colinet (2011) have actually considered the dynamics of a kink-like front connecting these two states on the basis of Eq. (4.37). The results, reproduced in Figure 4.6, show that the front advances (i.e., the flat film "grows" onto the bare substrate) when $\Sigma > 1$, while it recedes (i.e., the precursor film "shortens") when $\Sigma < 1$.

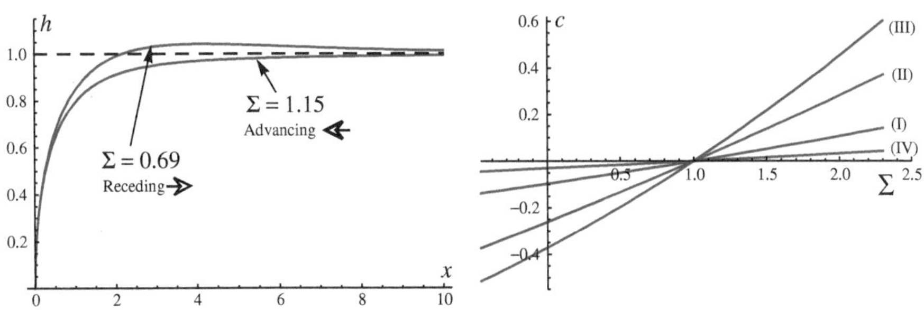

FIGURE 4.6 Dynamics of a front between the bare substrate ($h = 0$, $x < 0$) and the flat film ($h = 1$, $x \to +\infty$). (Left): The flat film advances (i.e., $c > 0$) for $\Sigma > 1$ and recedes (i.e., $c < 0$) for $\Sigma < 1$. (Right): The dimensionless front velocity c versus the dimensionless spreading coefficient Σ for four representative cases: (I) $E = 0.124$, $K = 5.74$; (II) $E = 7$, $K = 50$; (III) $E = 0.124$, $K = 0.01$; (IV) $E = 0.01$, $K = 5.74$. Left plots are obtained for case I.

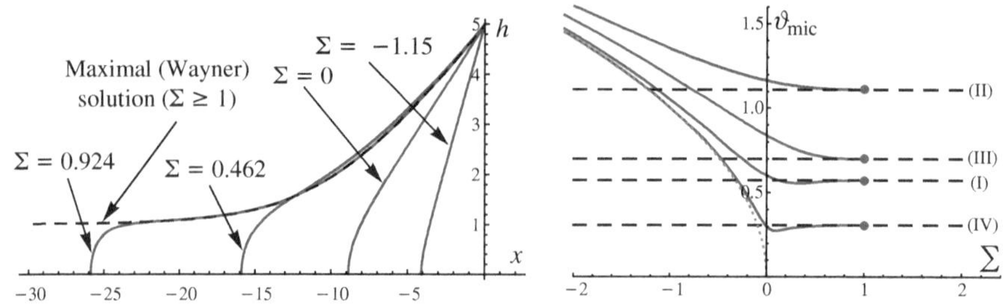

FIGURE 4.7 Truncated microfilm solutions for contact lines with evaporation into pure vapor (full model). (Left): Some typical microstructures for various values of the dimensionless spreading coefficient Σ, and $E=0.124$, $K=5.74$. (Right): Rescaled apparent contact angle ϑ_{mic} as a function of Σ. Case (I): $E=0.124$, $K=5.74$; case (II): $E=7$, $K=50$; case (III): $E=0.124$, $K=0.1$; case (IV): $E=0.01$, $K=5.74$. Plain curves correspond to microstructures with a truncated microfilm, and dashed lines to microstructures with an extended microfilm (for which the result does not depend on Σ). The dotted parabola corresponds to (rescaled) Young's law $\vartheta_{\text{mic}}=(-\Sigma)^{1/2}$.

While this indeed confirms the thermodynamic stability argument presented in Section 4.4, we still have to analyze corresponding contact line microstructures. Actually, it is well known already that contact lines with extended microfilms can be found as solutions of Eq. (4.37) with $c=0$ and boundary conditions

$$h \to 1 \quad \text{for } x \to -\infty, \quad h \to x\,\vartheta_{\text{mic}}(E,K) \quad \text{for } x \to +\infty \tag{4.41}$$

These extended microfilm solutions, which do not depend on Σ because there is obviously no dry point anywhere, have been studied since the 1970s by various authors (see, e.g., Potash and Wayner, 1972; Stephan and Busse, 1992; Morris, 2001; Ajaev, 2005). A rather extensive parametric study as a function of E and K has been published more recently by Rednikov et al. (2009).

The second type of solution, with a truncated microfilm, and thus an additional dependence on the spreading parameter, was analyzed even more recently by Rednikov and Colinet (2011) by solving Eq. (4.37) together with boundary condition (4.39) and $h \to x\,\vartheta_{\text{mic}}(E,K,\Sigma)$ for $x \to +\infty$. Typical truncated microfilm profiles are represented in Figure 4.7, along with values of $\vartheta_{\text{mic}}(E,K,\Sigma)$ for a few representative cases and the corresponding values (independent on Σ) for extended microfilm ("Wayner") solutions. Most importantly, it is seen that while the latter exist for any Σ, this is not the case for truncated microfilm solutions, which do not exist for $\Sigma>1$—that is, when the bare substrate state is thermodynamically metastable. Below this positive critical value $\Sigma_{\text{cr}}=1$—that is, according to Eq. (4.40) when the extended microfilm is thinner than the pancake—truncated solutions do exist and are expected to be preferred over extended solutions. In particular, when $\Sigma<0$, these truncated solutions look like Young's wedges modified by evaporation (although not significantly when Σ is strongly negative). More details about the competition between extended and truncated microfilms may be found in the original paper of Rednikov and Colinet (2011).

4.7 Summary and concluding remarks

In this chapter, we attempted to illustrate the complexity inherent in the question of the existence of precursor films ahead of the contact line formed by a pure liquid, a gas (in particular, the vapor of the liquid considered), and a smooth and chemically homogeneous solid substrate. Using the lubrication assumption throughout (small slopes of the liquid/gas interface), a variety of cases have been treated (volatile or nonvolatile liquids, partial or complete wetting, advancing or receding contact lines, etc.). The focus has been on the microstructure of the contact line (i.e., the liquid thickness profile) at scales that are large compared to molecular dimensions yet much smaller than the size of the macroscopic setup considered (be it a sessile droplet, a growing vapor bubble, a meniscus in a capillary, or a dewetting film). Results deemed useful for the study of such macroscopic configurations have been provided in terms of the Cox–Voinov relationship, which is generically valid at sufficiently large scales (yet smaller than macroscopic ones), and relating the scale-dependent apparent contact angle to the velocity of the contact line. Two important parameters entering this "effective boundary condition" (often used as a subgrid model in macroscopic numerical codes; see, e.g., Sui et al., 2014) have been calculated for several cases: the apparent contact angle of the steady microstructure and the microlength below which singularities of the contact line are regularized. Even though it enters a logarithmic term in the Cox–Voinov relationship, we have shown that the latter microlength can in principle vary in a wide range of scales (hence in some cases significantly affecting the value of this logarithmic term), depending on the physics at play at the smallest scales (e.g., disjoining pressure, Kelvin effect, evaporation kinetics, etc.). In particular, when the Kelvin effect dominates all other small-scale effects, a minimalist model has shown that this microlength is fully determined as a function of well-known macroscopic quantities, such as viscosity, latent heat, thermal conductivity, etc. Yet, beyond the conceptual question of resolving the singularities, other effects such as the disjoining pressure (or kinetic effects of molecular origin, slip, etc., depending on the fluid/solid pair considered) might need to be included in order to achieve a fully quantitative comparison with experiments, free of any fitting parameters (which is key to future progress in the field).

As for the question of whether precursor films exist, generally speaking, it was shown that the microfilms formed ahead of the contact line may be of two types: extended (i.e., covering the smooth substrate everywhere) or truncated (i.e., formally ending at some "dry" spot, within the continuum hydrodynamic description adopted here). The selection between these two types of microstructure depends on various effects, among which the intermolecular forces between solid and liquid (here quantified by the disjoining pressure), the tendency to wetting (measured by the spreading coefficient or by Young's contact angle, depending on the model considered), the driving force for phase change (i.e., the superheat or, correspondingly, the undersaturation of the vapor phase), the Kelvin effect, the velocity of the contact line, and possibly other effects not included in our analysis. Hence, we will not attempt to draw any specific conclusion about the preference for such extended films (or maximal films) here, apart from mentioning that wetting obviously favors them, while evaporation instead appears to favor truncated film microstructures.

Glossary

Disjoining pressure pressure which arises in thin films, due to long-range forces between solid and liquid molecules

Lubrication theory approximation based on the fact that length scales along a film are typically much larger than its thickness

Kelvin effect modification of the thermodynamic equilibrium (saturation) conditions by the curvature of the phase boundary

Precursor film microscopic thin film ahead of a contact line and covering the seemingly "dry" portions of the substrate

References

Ajaev, V.S., 2005. Spreading of thin volatile liquid droplets on uniformly heated surfaces. J. Fluid Mech. 528, 279–296.

Anderson, D.M., McFadden, G.B., Wheeler, A.A., 1998. Diffuse-interface methods in fluid mechanics. Ann. Rev. Fluid Mech. 30, 139–165.

Berteloot, G., Pham, C.-T., Daerr, A., Lequeux, F., Limat, L., 2008. Evaporation-induced flow near a contact line: consequences on coating and contact angle. Europhys. Lett. 83, 14003.

Blake, T.D., Haynes, J.M., 1969. Kinetics of liquid/liquid displacement. J. Colloid Interface Sci. 30, 421.

Bonn, D., Eggers, J., Indekeu, J., Meunier, J., Rolley, E., 2009. Wetting and spreading. Rev. Mod. Phys. 81, 739–805.

Colinet, P., Rednikov, A., 2011. On integrable singularities and apparent contact angles within a classical paradigm. Partial and complete wetting regimes with or without phase change. Eur. Phys. J. Spec. Top. 197, 89–113.

Cox, R.G., 1986. The dynamics of the spreading of liquids on a solid surface. Part 1. Viscous flow. J. Fluid Mech. 168, 169.

De Coninck, J., Blake, T., 2008. Wetting and molecular dynamics simulations of simple liquids. Annu. Rev. Mater. Res. 38, 1–22.

Defay, R., Prigogine, I., 1966. Surface Tension and Adsorption, English ed. John Wiley & Sons, New York, NY.

de Gennes, P.-G., 1985. Wetting: statics and dynamics. Rev. Mod. Phys. 57, 827–863.

de Gennes, P.-G., Brochart-Wyart, F., Quéré, D., 2004. Capillarity and Wetting Phenomena: Drops, Bubbles, Pearls, Waves, English ed. Springer, New York, NY.

Delon, G., Fermigier, M., Snoeijer, J.H., Andreotti, B., 2008. Relaxation of a dewetting contact line. Part 2: experiments. J. Fluid Mech. 604, 55.

Hervet, H., de Gennes, P.-G., 1984. The dynamics of wetting: precursor films in the wetting of "dry" solids. C. R. Acad. Sci. Paris 299, 499.

Huang, D.M., Sendner, C., Horinek, D., Netz, R.R., Bocquet, L., 2008. Water slippage versus contact angle: a quasiuniversal relationship. Phys. Rev. Lett. 101, 226101.

Huh, C., Scriven, L.E., 1971. Hydrodynamic model of steady movement of a solid/liquid/fluid contact line. J. Colloid Interface Sci. 35, 85–101.

Janeček, V., Nikolayev, V., 2012. Contact line singularity at partial wetting during evaporation driven by substrate heating. Europhys. Lett. 100, 14003.

Janeček, V., Andreotti, B., Prazak, D., Barta, T., Nikolayev, V.S., 2012. Moving contact line of a volatile fluid. arXiv.org: <http://arxiv.org/abs/1212.3717>.

Janeček, V., Andreotti, B., Prazak, D., Barta, T., Nikolayev, V.S., 2013. Moving contact line of a volatile fluid. Phys. Rev. E 88, 060404.

Kavehpour, H.P., Ovryn, B., McKinley, G.H., 2003. Microscopic and macroscopic structure of the precursor layer in spreading viscous drops. Phys. Rev. Lett. 91, 196104.

Kjelstrup, S., Bedeaux, D., 2008. Non-equilibrium Thermodynamics of Heterogeneous Systems. World Scientific, Singapore.

Moosman, S., Homsy, G.M., 1980. Evaporating menisci of wetting fluids. J. Colloid Interface Sci. 73, 212–223.

Morris, S.J.S., 2001. Contact angles for evaporating liquids predicted and compared with existing experiments. J. Fluid Mech. 432, 1–30.

Nold, A., Sibley, D.N., Goddard, B.D., Kalliadasis, S., 2014. Fluid structure in the immediate vicinity of an equilibrium three-phase contact line and assessment of disjoining pressure models using density functional theory. Phys. Fluids 26, 072001.

Pham, C.-T., Berteloot, G., Lequeux, F., Limat, L., 2010. Dynamics of complete wetting liquid under evaporation. Europhys. Lett. 92, 54005.

Pismen, L.M., Pomeau, Y., 2000. Disjoining potential and spreading of thin liquid layers in the diffuse-interface model coupled to hydrodynamics. Phys. Rev. E 62, 2480–2492.

Pomeau, Y., 2000. Représentation de la ligne de contact mobile dans les équations de la mécanique des fluides. C. R. Acad. Sci. Ser. IIB 328, 411.

Pomeau, Y., 2011. Contact line moving on a solid. Eur. Phys. J. Spec. Top. 197, 15–31.

Potash, M., Wayner, P.C., 1972. Evaporation from a two dimensional extended meniscus. Int. J. Heat Mass Transf. 15, 1851–1863.

Rednikov, A.Y., Colinet, P., 2011. Truncated versus extended microfilms at a vapor-liquid contact line on a heated substrate. Langmuir 27, 1758–1769.

Rednikov, A.Y., Colinet, P., 2012. Moving contact lines in a pure-vapor atmosphere: a singularity-free description in the sole framework of classical physics. arXiv.org: <http://arxiv.org/abs/1202.0273>.

Rednikov, A.Y., Colinet, P., 2013. Singularity-free description of moving contact lines for volatile liquids. Phys. Rev. E 87, 010401.

Rednikov, A.Y., Rossomme, S., Colinet, P., 2009. Steady microstructure of a contact line for a liquid on a heated surface overlaid with its pure vapor: parametric study for a classical model. Multiphase Sci. Technol. 21, 213–248.

Shanahan, M.E.R., 2001. Condensation transport in dynamic wetting. Langmuir 17, 3997–4002.

Shikhmurzaev, Y.D., 1993. The moving contact line on a smooth solid surface. Int. J. Multiphase Flow 19, 589–610.

Sibley, D.N., Savva, N., Kalliadasis, S., 2012. Slip or not slip? A methodical examination of the interface formation model using two-dimensional droplet spreading on a horizontal planar substrate as a prototype system. Phys. Fluids 24, 082105.

Sibley, D., Nold, A., Savva, N., Kalliadasis, S., 2013. On the moving contact line singularity: asymptotics of a diffuse-interface model. Eur. Phys. J. E 36, 1–7.

Starov, V.M., Velarde, M.G., Radke, C.J., 2007. Wetting and spreading dynamics. CRC Press, Boca Raton, FL.

Starov, V.M., Velarde, M.G., 2009. Surface forces and wetting phenomena. J. Phys.: Condens. Matter 21, 464121.

Stephan, P.C., Busse, C.A., 1992. Analysis of the heat transfer coefficient of grooved heat pipe evaporator walls. Int. J. Heat Mass Transf. 35, 383–391.

Sui, Y., Ding, H., Spelt, P.D.M, 2014. Numerical simulations of flows with moving contact lines. Ann. Rev. Fluid Mech. 46, 97–119.

Todorova, D., Thiele, U., Pismen, L.M., 2012. The relation of steady evaporating drops fed by an influx and freely evaporating drops. J. Eng. Math. 73, 17–30.

Voinov, O.V., 1976. Hydrodynamics of wetting. Fluid Dyn. 11, 714.

Wayner, P.C., 1993. Spreading of a liquid film with a finite contact angle by the evaporation/condensation process. Langmuir 9, 294–299.

Yeh, E.K., Newman, J., Radke, C., 1999. Equilibrium configurations of liquid droplets on solid surfaces under the influence of thin-film forces. Part I. Thermodynamics. J. Colloids Surf. A 156, 137–144.

Chapter 5

SUPERAMPHIPHOBIC SURFACES

Andrea Cavalli and Frieder Mugele
Physics of Complex Fluids, Department of Science and Technology, University of Twente, Enschede, The Netherlands

This chapter discusses recent trends in the development, fabrication, and characterization of superamphiphobic surfaces. An amphiphobic surface repels both polar liquids, such as water, and nonpolar (oily) liquids, and therefore has useful technological application in microfluidics devices, protective coatings, and the medical and food industries. Superamphiphobic surfaces are defined in analogy with superhydrophobic ones: on a superamphiphobic surface, both polar and nonpolar drops exhibit large (>150°) static contact angles and small contact angle hysteresis (limited to a few degrees). Different approaches can be used to achieve these properties. First, we describe the use of overhanging microstructures. These patterns support a heterogeneous Cassie–Baxter wetting state, even if the intrinsic contact angle of the liquid with the substrate is smaller than 90°, which is the case for many common oils. We then discuss liquid-infused porous films, a new concept for nonadhesive, liquid-repellent surfaces that became popular under the acronym SLIPS, which stands for Slippery Liquid Infused Porous Surfaces. They are characterized by a microscopic, rough texture imbued with a lubricant film, which significantly reduces the adhesion of drops.

Droplet Wetting and Evaporation. DOI: http://dx.doi.org/10.1016/B978-0-12-800722-8.00005-9

5.1 Introduction

This chapter covers surfaces that have remarkable wetting properties—namely, the ability to repel both polar (water-like) and nonpolar (oil-like) fluids. Aside from the scientific interest in such materials, these substrates could have immediate technological applications in many engineering fields, ranging from food processing and protective coatings to microfluidics and biomedical and electronic devices. The exciting properties of some prototype surfaces are immediately apparent even to readers with no previous experience in the field, and they have attracted significant media coverage (Varanasi, 2012).

Different authors have used different names to describe these substrates; the most common are *oleophobic* (Tuteja et al., 2007), *omniphobic* (Tuteja et al., 2008), and *hygrophobic* (Marmur, 2008) surfaces. This variety probably reflects the relative novelty of this research field. Because all these conventions are fairly common in the research literature, the reader should be aware of them. In this Chapter, we use the term *amphiphobic* surface, which loosely translates from the Greek word for "repelling both."

The "super-" prefix is added in the spirit of the better-known "superhydrophobic" surfaces—that is, substrates that exhibit extreme water repellency, with an apparent contact angle substantially higher than the intrinsic value for a flat surface of the same material. A more quantitative definition of superamphiphobic surfaces could then be "a substrate on which both polar and nonpolar liquids exhibit a large ($>150°$) apparent contact angle (De Gennes et al., 2004) and very low ($<5°$) contact angle hysteresis" (Lafuma and Quéré, 2003). This definition is not completely unambiguous (it depends on a choice of contact angle thresholds), but it gives a phenomenological description of the behavior of droplets on these substrates, similar to water on leaves from the Lotus plant (Neinhuis and Barthlott, 1997). We will show later how the condition of a large apparent contact angle can actually be relaxed—that is, there are surfaces on which water and other fluids exhibit very little adhesion and hysteresis, consistent with the intuitive notion of "liquid repellence," although their equilibrium contact angle is modest.

We will consider two families of superamphiphobic surfaces. First, we describe patterned substrates with overhanging texture and how they enable a suspended wetting state even for low surface tension liquids, which results in the required low hysteresis and large apparent contact angle. We then discuss Slippery Liquid Infused Porous Surfaces (SLIPS), which consist of an oil-impregnated rough texture that dramatically reduces adhesion and contact angle hysteresis of liquids and other contaminants. Our focus will be on the relevant physical parameters needed to assess the efficiency and robustness of the surface and a description of experimental designs and some naturally occurring examples of these antiwetting surfaces.

5.2 Overhanging textures

The design goal for "conventional" superhydrophobic surfaces is to reduce the contact between a droplet and the solid substrate. On a textured surface (an array of micron-sized posts is a good example), this can happen when the drop sits on top of the posts, while an air film is maintained between the posts below the drop. This configuration is known as "fakir" or Cassie–Baxter state. The reduced contact between the liquid and solid phase results in low adhesion of the liquid phase to the substrate. Consequently, the droplet shows a large contact angle and reduced hysteresis (De Gennes et al., 2004) However, if it is energetically favorable for the liquid phase to wet the substrate, it will be thermodynamically more convenient for the liquid to penetrate the texture, leading to an enhanced adhesion of the liquid to the substrate and a loss of the super-repellent properties. To characterize the texture, we define the parameters roughness r and solid fraction ϕ_s as the ratios of the real surface area and the area of the tops of posts over the apparent surface area of the material (for a flat surface, $r=1$ and $\phi_s=1$. In general, $r\geq 1$ and $\phi_s\leq 1$). The suspended state is then energetically favorable for the drop only if (De Gennes et al., 2004)

$$\cos\theta_Y < \frac{\phi_s - 1}{r - \phi_s} \tag{5.1}$$

where $\cos\theta_Y$ is the intrinsic (or Young) contact angle of the liquid on the substrate—that is, the equilibrium contact angle shown by a drop on a smooth surface of the same material. Because $\phi_s - 1 \leq 0$ by definition, this condition is satisfied only if $\theta_Y \geq \frac{\pi}{2}$. This is not the case for many oils on common substrates, making the Cassie state energetically more expensive. However, by carefully designing the shape of the posts, it is still possible to make the Cassie state a robust *local* energy minimum, thus extending the superphobic behavior to low surface tension fluid and achieving superamphiphobicity. We will now discuss how overhanging structures are critical to this purpose.

We introduce a simplified, two-dimensional picture that explains the main properties of overhanging superamphiphobic substrates. Let us consider a series of cylindrical protrusions of radius R, with spacing L between them, supported by a stem of height H, as shown in Figure 5.1. The top of these posts is a circular segment of central angle $2\varphi_{max}$. We consider equilibrium configurations for a drop sitting on such posts. In most cases, the characteristic size of the drop (in the

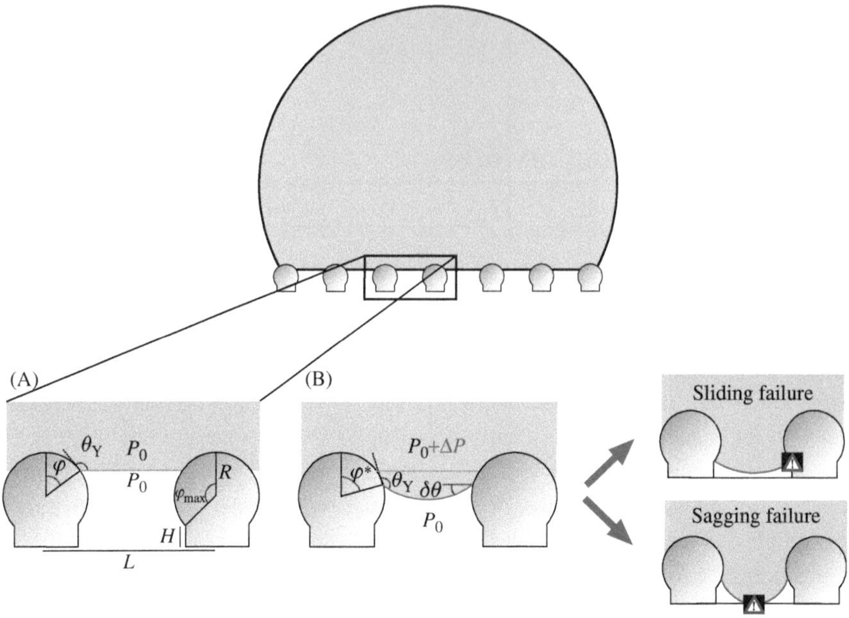

FIGURE 5.1 Sketch of an idealized amphiphobic texture that relies on overhanging protrusions to achieve a stable Cassie wetting configuration. (A) For small or zero pressure difference, the two dimensional interface is approximately flat, and sits at the height prescribed by the contact angle θ_Y. (B) An applied pressure difference across the interface results in the displacement of the contact line and bulging of the interface. Depending on the geometry and wettability of the system, either of these factors can trigger a transition to a fully wetted state.

range of millimeters) is much larger than the texture characteristic length. We can therefore, in first approximation, focus on a unit cell in our pattern and treat the macroscopic droplet as an "infinite" reservoir. We will also consider droplets small enough to neglect the effect of gravity. If the pressure difference across the interface is small, we can assume that the interface is flat (Figure 5.1A). The equilibrium position for the liquid–air interface can then be obtained by considering an infinitesimal angular displacement $\mathrm{d}\varphi$ of the contact line. The correspondent change in free energy $\mathrm{d}E$ for the system will be

$$\mathrm{d}E = (\gamma_{sl} - \gamma_{sg})R\,\mathrm{d}\varphi - \gamma_{lg}R(\sin(\varphi + \mathrm{d}\varphi) - \sin\varphi) \approx (\gamma_{sl} - \gamma_{sg})R\,\mathrm{d}\varphi - \gamma_{lg}R\cos\varphi\,\mathrm{d}\varphi, \tag{5.2}$$

where γ_{sl}, γ_{sg}, and γ_{lg} are the solid–liquid, solid–gas, and liquid–gas surface tensions, respectively. By imposing the equilibrium condition $\frac{\mathrm{d}E}{\mathrm{d}\varphi} = 0$, we then obtain

$$-\frac{(\gamma_{sg} - \gamma_{sl})}{\gamma_{lg}} = \cos\varphi \rightarrow -\cos\theta_Y = \cos\varphi \tag{5.3}$$

From which we conclude $\varphi = \pi - \theta_Y$. This means that, for any contact angle between π and $\pi - \phi_{max}$, there is an equilibrium height at which the liquid–air interface will sit, realizing a suspended, Cassie-like wetting state. The lower the contact angle, the further the interface will slide down. It is important to notice that without overhanging, there would be no equilibrium position for $\theta_Y < \frac{\pi}{2}$— that is, fluids with a low intrinsic contact angle would spread through the texture. Because many oils fall in this category, we therefore see the need of employing a overhanging geometry to hold them in the suspended Cassie state.

We can further check the stability of this equilibrium position with respect to small displacements by observing that $\frac{d^2E}{d\varphi^2} = \gamma_{lg} R \sin \varphi \geq 0$ for all $\varphi \in [0, \pi]$. The obtained equilibrium state is therefore stable with respect to small angular displacements. It can be shown (Nosonovsky, 2007; Marmur, 2008) that the stability property holds, more generally, for any convex protrusion. Conversely, concave profiles (cylindrical grooves, for example) would result in unstable equilibria unless some other stabilizing factor occurs.

In many circumstances, there can be a significant pressure difference across the liquid–air interface. An evaporating droplet in the suspended state, for example, develops an increasingly high Young–Laplace pressure, resulting in a significant curvature of the interface (Kusumaatmaja et al., 2008). A significant pressure at the interface will also appear when a drop impacts on a patterned surface (Bartolo et al., 2006), which could result in the impalement of the drop on the texture. A deformation of the interface can also be induced by electric fields (Manukyan et al. 2011). In two dimensions, the interface then assumes the shape of a circular arc of central angle $2\delta\theta$ (Figure 5.1B). Because the Young condition on the contact angle still holds, the interface will slide further to an angular position $\varphi^* = \pi - \theta_Y + \delta\theta$ to accommodate the contact angle. At equilibrium, the vertical component of the surface tension acting at the contact line is $F_{int} = \gamma_{lg} \sin \delta\theta$. This contribution is balanced by the pressure acting across the interface, $F_{\Delta P} = \Delta P\left(\frac{L}{2} - R \sin(\theta_Y - \delta\theta)\right)$.

We therefore get the following relation between the interface deflection $\delta\theta$ and the pressure difference ΔP:

$$\Delta P = \frac{\gamma_{lg} \sin \delta\theta}{\frac{L}{2} - R \sin(\theta_Y - \delta\theta)} \tag{5.4}$$

Once the interface gets deflected, two different failure modes can occur, resulting in a transition to a wetted state (Reyssat et al., 2008; Tuteja et al., 2008; Rathgen and Mugele, 2010; Butt et al., 2013). In the sagging mode, the interface touches the substrate between pillars, triggering a transition to a wetted state. In our simplified geometry, this happens at a critical deflection angle $\delta\theta_{sagging}$, for which the bulging of the interface $h_1 = \frac{\gamma_{lg}}{\Delta P}(1 - \cos \delta\theta)$ is equal to the height of the contact line over the substrate, $h_2 = H - R(\cos(\theta_Y - \delta\theta) + \cos \varphi_{max})$. In the sliding mode, the interface reaches the lowest point along the overhanging profile, usually corresponding to the stem of the post. The equilibrium contact angle is therefore not achievable anymore, and the interface continues to slide until the texture is fully impregnated. For the considered setup, this corresponds to a critical deflection angle $\delta\theta_{sliding} = \varphi_{max} + \theta_Y - \pi$. The pressure difference for which either mode triggers therefore depends on the geometry and wetting properties of the substrate, as Eq. (5.4) shows. It is worth stressing that the same analysis applies to posts with sharp edges (see, e.g., the "nanonails" in Figure 5.2A). As a matter of fact, sharp corners can be seen as round edges with small radii of curvature, for which our analysis still holds. Conversely, rounded features can be seen as "smeared-out" corners, which exhibit sliding instead of pinning of the contact line.

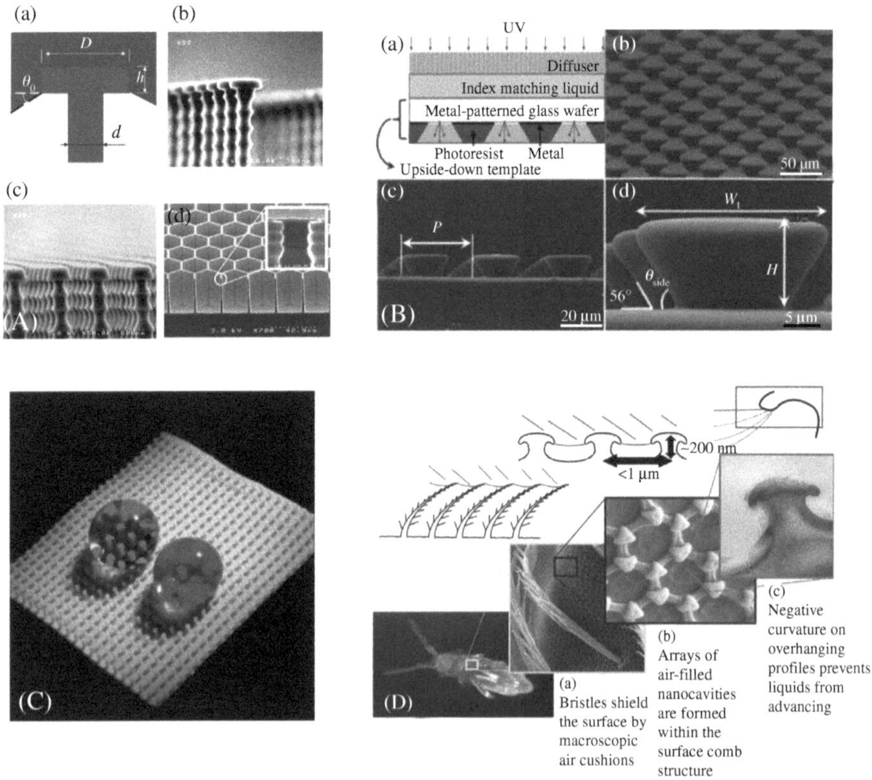

FIGURE 5.2 Different examples of overhanging textures with amphiphobic properties. (A) Nanonails by Ahuja et al. (2008). (B) Inverse trapezoids by Im et al. (2010). (C) Drops of different oils achieve a suspended wetting state on micro-hoodos, as described in Tuteja et al. (2008). (D) Microtextured skin of a springtail at different resolutions, from Helbig et al. (2011).

Some observations follow naturally from the previous analysis. For example, on any overhanging texture, the larger the intrinsic contact angle, the higher the pressure required to trigger the sliding mode. Moreover, the larger φ_{max}, the more robust the interface will be with respect to the sliding failure. Finally, supporting the overhanging features with higher stems makes the sagging transition increasingly harder.

Supported by similar arguments, several research groups have been developing overhanging amphiphobic surfaces. Tuteja et al. introduced their microhoodos surfaces (Figure 5.2C) in 2008 and provided a clear analysis of the two failure modes associated to this geometry. Chhatre et al. (2009) extended their analysis to include amphiphobic meshes—woven cylindrical fibers with a characteristic size in the order of tens to hundreds of μm. Their work shows how many kinds of textiles can achieve amphiphobic properties, provided the geometrical and chemical properties are adequate. Ahuja et al. (2008) describe substrates patterned with nanonails (Figure 5.2A). These structures are constituted by a silicon stem with an oxide nailhead. This design makes the substrate suitable for electrowetting: if voltage is applied between the silicon substrate and a drop in the suspended state, the contact angle is effectively reduced, and a transition to a fully wetted state can be triggered. Because the wetted configuration is thermodynamically more stable, however, the process is not reversible. Other interesting designs are the inverse trapezoids by Im et al. (2010) for the unusual choice of materials. Their structures are realized in PDMS through a molding process, thus providing extensive, flexible, and transparent amphiphobic coatings (Figure 5.2B).

Quite interestingly, the inspiration for overhanging superamphiphobic surfaces followed a different path compared to superhydrophobic ones. In the case of extreme water repellency, the inspiration stemmed from the investigation of several organisms that exhibit superhydrophobic properties. In particular, Neinhuis and Barlott (1997) presented an extensive survey of the microtexture on several water-repellent leaves that induces a suspended wetting state. In the animal realm, a well-known example of superhydrophobic surfaces are the legs of the water strider. As reported by Gao and Jiang (2004), these insects' legs have an intrinsic hydrophobic character and are covered in thin hairs with nanogrooves that trap air inside them. Thus far, however, no animal or plant that exhibits amphiphobic properties had been observed.

A naturally occurring example of overhanging texture with amphiphobic properties has only recently been observed in springtails (Helbig et al., 2011). These arthropods live in soil and decaying organic matter and have therefore developed a complex skin pattern that employs "all the tricks in the book" (Figure 5.2D). Micron-sized bristles pin the interface of high surface tension liquids and capture and retain a macroscopic air cushion. If low surface tension fluids overcome this barrier, they meet a second level of texture: nanocavities with overhanging walls arranged in a honeycomb lattice. The optimal placement, trapped air, and large overhanging make this surface extremely resistant to wetting. Indeed, Helbig and colleagues (2011) observed how a plastron (air cushion) appears around the springtails upon immersion in many polar and nonpolar liquids, even for applied pressure up to 3.5 atm.

Before we conclude this section, it is worth mentioning a well-known drawback of overhanging superamphiphobic surfaces. First, as even our two-dimensional stability analysis suggests, maintaining a stable air layer between the posts is challenging. Any kind of inhomogeneity in the chemical or topographical properties of the substrate could act as a trigger for a transition to a wetted state or as a pinning site for a drop sliding on the substrate, thus hindering the amphiphobic properties. Another drawback comes from the complex geometry required to achieve amphiphobicity. Overhanging designs are typically harder to fabricate than conventional micro-posts and thus more expensive. An interesting alternative to such complex fabrication techniques was recently demonstrated by Deng et al. (2012), who used candle soot as a template for growing three-dimensional silica networks with characteristic dimensions of the order of 100 nm. The same team also demonstrated excellent amphiphobic surfaces based on "raspberry" colloidal particles with multiscale roughness (Deng et al., 2012). These approaches demonstrate that simple and robust fabrication processes for large-scale coatings are increasingly becoming available. As a final remark, the robustness of these surfaces to mechanical wear is a major concern. Yet, also here, recent progress with hierarchically structured polymer surfaces provides hope for future developments (Groten and Rühe, 2013). In the following section, we will see how SLIPS are alternative candidates in overcoming these drawbacks and achieving robust and durable amphiphobicity.

5.3 Slippery liquid-infused porous surfaces (SLIPS)

As we saw in the previous section, overhanging superamphiphobic surfaces were originally developed as an extension of artificial superhydrophobic surfaces. The case is different for liquid-infused porous surfaces. Wong et al. (2011), in one of the seminal papers on this topic, specifically refer to the *Nepenthes* pitcher plants (Bohn and Federle, 2004) as the inspiration for their surfaces. The pitcher rim (peristome) of these plants (Figure 5.3A) presents a characteristic micro-roughness that traps natural humidity and a nectar secreted by the plant itself. As long as the texture is impregnated, most insects are not able to adhere to it and slide inside the pitcher. Wong pictured this system as a

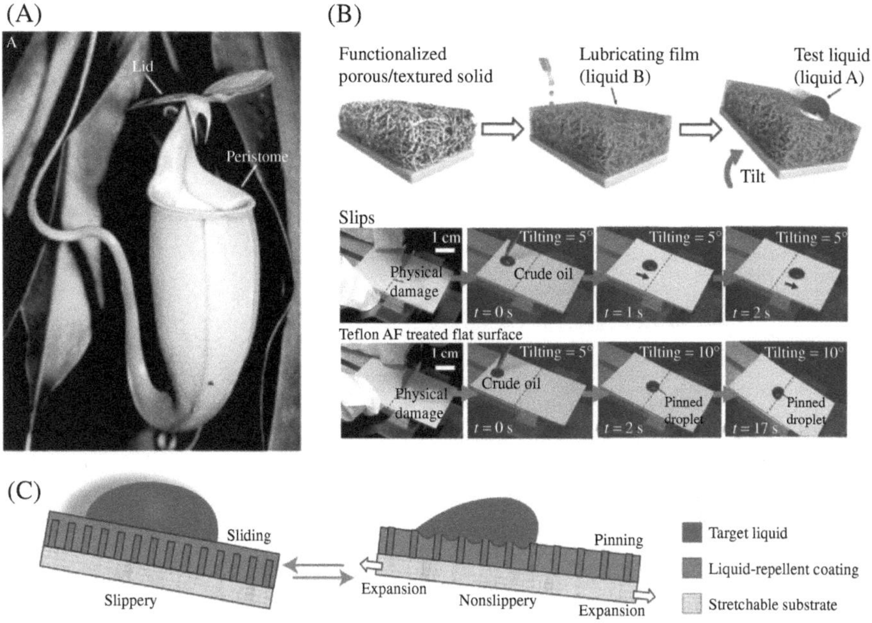

FIGURE 5.3 (A) A *Nepenthes* pitcher and peristome morphology, from Bohn and Federle (2004). (B) Sketch of the working principles of SLIPS, from Wong et al. (2011). In the bottom layer, a comparison between SLIPS and regular superhydrophobic surfaces is shown, highlighting the self-healing properties of the former substrates. (C) Sketch describing the working principle of elastic SLIPS, which show reversible wetting transitions, from Yao et al. (2013).

network of Teflon™ nanofibers imbued with a low surface tension perfluorinated liquid (Figure 5.2B), observing the same extreme slippery properties. In an almost contemporary paper,[1] Lafuma and Quéré (2011) describe the same behavior on a regular microtexture imbued with different kinds of alkanes and silicon oils.

The working principle is the same in both cases: a liquid drop deposited on such a substrate effectively sits on a mixed solid–liquid substrate, which can result in a significant reduction in friction. The lubricating oil is retained by the texture through capillary forces, and it is therefore much harder to displace in comparison to the air cushion on regular superhydrophobic/superamphiphobic substrates.

1. While Wong focuses more on the properties and applications of SLIPS, Lafuma and Quéré focus on the conditions for achieving such substrates.

Interface	Configuration	Total interface energy per unit area	Equivalent criteria		
Oil–solid–air 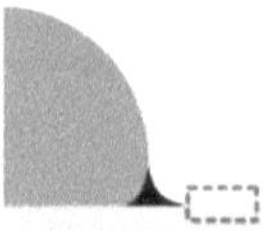	A1 Dry 	$E_{A1} = r\gamma_{sa}$	$E_{A1} < E_{A2}, E_{A3}$	$S_{os(a)} < -\gamma_{oa}\frac{r-1}{r-\phi_s}$	$\theta_{os(a)} > \theta_c$
	A2 Impregnated, emerged 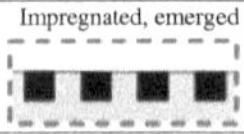	$E_{A2} = (r-\phi_s)\gamma_{os} + \phi_s\gamma_{sa} + (1-\phi_s)\gamma_{oa}$	$E_{A2} < E_{A1}, E_{A3}$	$-\gamma_{oa}\frac{r-1}{r-\phi_s} < S_{os(a)} < 0$	$0 < \theta_{os(a)} < \theta_c$
	A3 Encapsulated 	$E_{A3} = \gamma_{oa} + r\gamma_{os}$	$E_{A3} < E_{A2}, E_{A1}$	$S_{os(a)} \geq 0$	$\theta_{os(a)} = 0$
Oil–solid–water 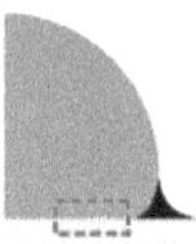	W1 Impaled 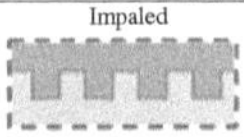	$E_{W1} = r\gamma_{sw}$	$E_{W1} < E_{W2}, E_{W3}$	$S_{os(w)} < -\gamma_{ow}\frac{r-1}{r-\phi_s}$	$\theta_{os(w)} > \theta_c$
	W2 Impregnated, emerged 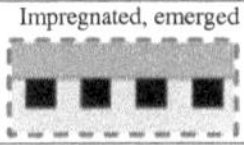	$E_{W2} = (r-\phi_s)\gamma_{os} + \phi_s\gamma_{sw} + (1-\phi_s)\gamma_{ow}$	$E_{W2} < E_{W1}, E_{W3}$	$-\gamma_{ow}\frac{r-1}{r-\phi_s} < S_{os(w)} < 0$	$0 < \theta_{os(w)} < \theta_c$
	W3 Encapsulated 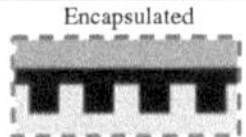	$E_{W3} = \gamma_{ow} + r\gamma_{os}$	$E_{W3} < E_{W1}, E_{W2}$	$S_{os(w)} \geq 0$	$\theta_{os(w)} = 0$

FIGURE 5.4 Possible wetting configurations for SLIPS. *From Smith et al. (2013).*

Once a drop is deposited on SLIPS, several thermodynamic equilibria can be established, depending on the chemical and physical properties of the four different phases in the system: the ambient gas, the substrate, the imbuing liquid, and the drop itself.

Smith et al. (2013) provide a thorough description of all accessible wetting states on SLIPS, which are shown in Figure 5.4 and are briefly summarized here. For simplicity, we consider the case of a water droplet on a lubricating oil film: this analysis can, however, be applied to any choice of liquids, provided that the relevant surface tensions are employed. In the following, we use the labels "o," "w," "s," and "a" to refer to the oil, water, solid, and air phases, respectively (e.g., γ_{ow} is the surface tension between the oil and water phases).

The exact geometry of the system is typically not relevant, because the wetting properties can be captured by the geometric parameters roughness r and solid fraction ϕ_s previously introduced. The first relevant condition is whether the imbuing oil will actually impregnate the surface. The spreading condition for a two-dimensional porous material is intermediate between the one for porous, three-dimensional media ($\gamma_{so} < \gamma_{sa}$, or $\theta_Y < 90°$) and the one for smooth solid surfaces ($\theta_Y < 0°$). To see this, let us compare the energy per unit area of a dry substrate and an impregnated one (configurations A1 and A2 in Figure 5.4). The liquid film will then form if

$$dE_{A2A1} = (\gamma_{os} - \gamma_{sa})(r - \phi_s) + \gamma_{lv} + (1 - \phi_s) < 0 \rightarrow \cos\theta_{os(a)} > \frac{(1-\phi_s)}{(r-\phi_s)} \quad (5.5)$$

where $\cos\theta_{os(a)} = \frac{\gamma_{sa} - \gamma_{os}}{\gamma_{oa}}$. Thus, the lubricant will spread if $0 < \theta_{os(a)} < \theta_c$. It can be seen that the film will fully cover the texture (configuration A3) only if $\theta_{os(a)} = 0$. While this scenario would avoid a direct contact between the droplet and the solid substrate, restricting the choice of lubricant to fully wetting ones would be a severe constraint on the design of SLIPS. An alternative strategy is tuning the solid fraction ϕ_s in such a way that the contact between the drop and the solid substrate is minute, while the lubricant is fully entrained in the texture.

Another relevant consideration is whether or not the water drop will be able to displace the lubricant film. Once again, we can compare the energy of the "emerged" and "impaled" scenarios (configurations W1 and W2 in Figure 5.4) so that the drop will be supported if

$$dE_{W2W1} = (\gamma_{os} - \gamma_{sw})(r - \phi_s) + \gamma_{ow}(1 - \phi_s) < 0 \rightarrow \cos\theta_{os(w)} > \frac{(1 - \phi_s)}{(r - \phi_s)} \tag{5.6}$$

where $\cos\theta_{os(w)} = \frac{\gamma_{sw} - \gamma_{os}}{\gamma_{ow}}$. A transition to the impaled state W1 would, of course, result in a strong adhesion of the drop to the substrate and a loss of the nonwetting behavior. In analogy to the A3 scenario, the posts below the drop will be encapsulated by oil only if $\theta_{os(w)} = 0$. An ideal SLIP will therefore exhibit a combination of the A2/A3–W2/W3" configuration, according to the labels in Figure 5.4, with the further caveat to keep the solid fraction ϕ_s as small as possible if the posts are surfacing. A final remark concerns situations in which the spreading coefficient of the lubricating oil on water, $S_{ow(a)} = \gamma_{wa} - \gamma_{wo} - \gamma_{oa}$, is positive. In this case, the water droplet will be cloaked in a lubricant oil film. This scenario is generally unfavorable for technical application, because shedded drops will deplete the lubricant oil and be contaminated by it.

From a dynamic point of view, it is not surprising to see that low-viscosity lubricants result in a faster sliding velocity for droplets. Smith and colleagues also suggest that the dissipation is mainly localized at the rim of the drop, where the lubricant film is locally displaced, similar to what is observed on soft substrates (Smith et al., 2013).

Even accounting for the constraints coming from the thermodynamic analysis, the design space for SLIPS is still large enough to consider other technologically relevant parameters in the development of these substrates. An example suggested by Lafuma and Quéré (2011) is choosing a lubricant oil that is also a solvent for contaminants in situations in which the surface is expected to be in contact with aerosols (such as kerosene in airports). Other relevant parameters are the optical properties of the lubricating oil, which can be chosen to make the substrate transparent or opaque. An interesting concept recently introduced is the use of elastic substrates for dynamic control of the surface properties, as shown by Yao et al. (2013). By depositing a Teflon™ nanofibers network over an elastic PDMS array, they developed a deformable SLIPS. If the elastic substrate is stretched

(Figure 5.3C), the lubricating oil level drops, leading to a change in the optical properties and an enhanced trapping of drops as the solid features emerge from the liquid bath. By releasing the strain, however, the surface recovers its enhanced slippery properties.

A final remark concerns the durability of these substrates. The capillary forces that keep the imbuing liquid in place give SLIPS a degree of self-healing property (Figure 5.3B): occasional defects in the solid texture will be covered by the lubricating film, thus keeping the substrate smooth. This is a major advantage over overhanging superamphiphobic surfaces for which mechanical wear is a serious issue, because the removal of one or more microstructures results in a strong pinning point for drops and a trigger for a wetting transition to a Wenzel/fully wetted state.

Glossary

Biomimetic A human device or technology whose function is based on or inspired by how other organisms solve a similar problem.

Superhydrophobic A substrate on which water exhibits apparent contact angles greater than 150° and small (few degrees) contact angle hysteresis.

Oleophobic An oil-repellent substrate on which nonpolar liquids exhibit a contact angle larger than 90°.

Superamphiphobic A substrate on which both polar and nonpolar liquids exhibit apparent contact angles greater than 150° and small (few degrees) contact angle hysteresis.

Nomenclature

θ_Y Young contact angle

r Surface roughness

ϕ_s Substrate solid fraction

φ_{max} Maximum overhang angle of micropatterned amphiphobic substrates

γ_{ab} Surface tension between phases a and b

References

Ahuja, A., Taylor, J., Lifton, V., Sidorenko, A., Salamon, T., Lobaton, E., et al., 2008. Nanonails: a simple geometrical approach to electrically tunable superlyophobic surfaces. Langmuir 24 (1), 9–14.

Bartolo, D., Bouamrirene, F., Verneuil, É., Buguin, A., Silberzan, P., Moulinet, S., 2006. Bouncing or sticky droplets: impalement transitions on superhydrophobic micropatterned surfaces. Europhys. Lett. 74 (2), 299.

Bohn, H.F., Federle, W., 2004. Insect aquaplaning: Nepenthes pitcher plants capture prey with the peristome, a fully wettable water-lubricated anisotropic surface. Proc. Natl. Acad. Sci. U.S.A. 101 (39), 14138–14143.

Butt, H.-J., Semprebon, C., Papadopoulos, P., Vollmer, D., Brinkmann, M., Ciccotti, M., 2013. Design principles for superamphiphobic surfaces. Soft Matter 9 (2), 418–428.

Chhatre, S.S., Choi, W., Tuteja, A., Park, K.-C., Mabry, J.M., McKinley, G.H., et al., 2009. Scale dependence of omniphobic mesh surfaces. Langmuir 26 (6), 4027–4035.

De Gennes, P.-G., Brochard-Wyart, F., Quéré, D., 2004. Capillarity and Wetting Phenomena: Drops, Bubbles, Pearls, Waves. Springer, Berlin.

Deng, X., Mammen, L., Butt, H.-J., Vollmer, D., 2012. Candle soot as a template for a transparent robust superamphiphobic coating. Science 335 (6064), 67–70.

Gao, X., Jiang, L., 2004. Biophysics: water-repellent legs of water striders. Nature 432 (7013), 36.

Groten, J., Rühe, J., 2013. Surfaces with combined microscale and nanoscale structures: a route to mechanically stable superhydrophobic surfaces? Langmuir 29 (11), 3765–3772.

Helbig, R., Nickerl, J., Neinhuis, C., Werner, C., 2011. Smart skin patterns protect springtails. PloS one 6 (9), e25105.

Im, M., Im, H., Lee, J.-H., Yoon, J.-B., Choi, Y.-K., 2010. A robust superhydrophobic and superoleophobic surface with inverse-trapezoidal microstructures on a large transparent flexible substrate. Soft Matter 6 (7), 1401–1404.

Kusumaatmaja, H., Blow, M., Dupuis, A., Yeomans, J., 2008. The collapse transition on superhydrophobic surfaces. Europhys. Lett. 81 (3), 36003.

Lafuma, A., Quéré, D., 2003. Superhydrophobic states. Nat. Mater. 2 (7), 457–460.

Lafuma, A., Quéré, D., 2011. Slippery pre-suffused surfaces. Europhys Lett. 96 (5), 56001.

Manukyan, G., Oh, J.M., van den Ende, D., Lammertink, R.G.H., Mugele, F., 2011. Electrical switching of wetting states on superhydrophobic surfaces: a route towards reversible Cassie-to-Wenzel transitions. Phys. Rev. Lett. 106 (014501), 1–4.

Marmur, A., 2008. From hygrophilic to superhygrophobic: theoretical conditions for making high-contact-angle surfaces from low-contact-angle materials. Langmuir 24 (14), 7573–7579.

Neinhuis, C., Barthlott, W., 1997. Characterization and distribution of water-repellent, self-cleaning plant surfaces. Ann. Bot. 79 (6), 667–677.

Nosonovsky, M., 2007. Multiscale roughness and stability of superhydrophobic biomimetic interfaces. Langmuir 23 (6), 3157–3161.

Rathgen, H., Mugele, F., 2010. Microscopic shape and contact angle measurement at a superhydrophobic surface. Faraday Discuss. 146, 49–56.

Reyssat, M., Yeomans, J., Quéré, D., 2008. Impalement of fakir drops. Europhys. Lett. 81 (2), 26006.

Smith, J.D., Dhiman, R., Anand, S., Reza-Garduno, E., Cohen, R.E., McKinley, G.H., et al., 2013. Droplet mobility on lubricant-impregnated surfaces. Soft Matter 9 (6), 1772–1780.

Tuteja, A., Choi, W., Ma, M., Mabry, J.M., Mazzella, S.A., Rutledge, G.C., et al., 2007. Designing superoleophobic surfaces. Science 318 (5856), 1618–1622.

Tuteja, A., Choi, W., Mabry, J.M., McKinley, G.H., Cohen, R.E., 2008. Robust omniphobic surfaces. Proc. Natl. Acad. Sci. U.S.A. 105 (47), 18200–18205.

Varanasi, G., 2012. LiquiGlide Ketchup Bottle. Available at the following link: https://www.youtube.com/watch?v=djwahGRi5iE

Wong, T.-S., Kang, S.H., Tang, S.K., Smythe, E.J., Hatton, B.D., Grinthal, A., et al., 2011. Bioinspired self-repairing slippery surfaces with pressure-stable omniphobicity. Nature 477 (7365), 443–447.

Yao, X., Hu, Y., Grinthal, A., Wong, T.-S., Mahadevan, L., Aizenberg, J., 2013. Adaptive fluid-infused porous films with tunable transparency and wettability. Nat. Mater. 12 (6), 529–534.

Chapter 6

SUPERSPREADING

Joachim Venzmer
Evonik Industries AG, Essen, Germany

Superspreading is a fascinating phenomenon that was first observed more than 20 years ago with dilute solutions of trisiloxane surfactants on hydrophobic substrates. It has attracted much attention mainly for two reasons: (i) the practical benefits of the effect in agrochemical applications and (ii) the scientific challenge to explain both the physics behind the phenomenon and the mode of action of the surfactants. Despite the work of many groups all over the world that has contributed significantly to the understanding of this phenomenon, the reasons why only some trisiloxane surfactants promote superspreading, whereas others of similar chemical structure behave more like ordinary surfactants, is still somewhat of a mystery. Therefore, it will continue to be an active field of research in interface science to elucidate the exact role of the surfactants in the superspreading process. This chapter not only covers the different approaches to explain superspreading, but also discusses the general, more philosophical aspects of these attempts; basically, it is all about how to discover the truth about an—on a molecular scale—unobservable phenomenon.

Droplet Wetting and Evaporation. DOI: http://dx.doi.org/10.1016/B978-0-12-800722-8.00006-0

6.1 Definition of superspreading—trisiloxane surfactants

The term *superspreading* as it is used in this chapter was defined over 20 years ago (e.g., Zhu et al., 1994). It refers to the quite surprising action of a droplet of a diluted solution of some special—so-called superspreading or superwetting—trisiloxane surfactants on hydrophobic substrates. Typically, a 50 μL droplet of a 0.1 wt% surfactant solution placed on a hydrophobic substrate yields a wetted area of about 70 to 80 mm in diameter within 1 min, as shown in Figure 6.1C. Compared to the behavior of a droplet of pure water (Figure 6.1A), this means an increase of the wetted area by a so-called spreading factor of about 100; this is about 25 times more effective than a nonsuperspreading surfactant of similar chemical structure (Figure 6.1B). However, the real fascination of the phenomenon can best be shared by watching the process in real time (see the video in the supplementary material of Venzmer, 2011).

The superspreading phenomenon was first observed using trisiloxane surfactants. These special silicone surfactants are commonly denoted as M(D′R)M, where M represents the trimethylsiloxy group $(CH_3)_3SiO_{1/2}-$, D′ stands for $-O_{1/2}Si(CH_3)(R)O_{1/2}-$, and R is a polyether group attached to the silicon. R can consist of either poly(ethylene oxide) ($R = -(CH_2)_3-O-(CH_2-CH_2-O-)_nR'$, or E_n for short), or it can be obtained from a mixture of ethylene oxide and propylene oxide

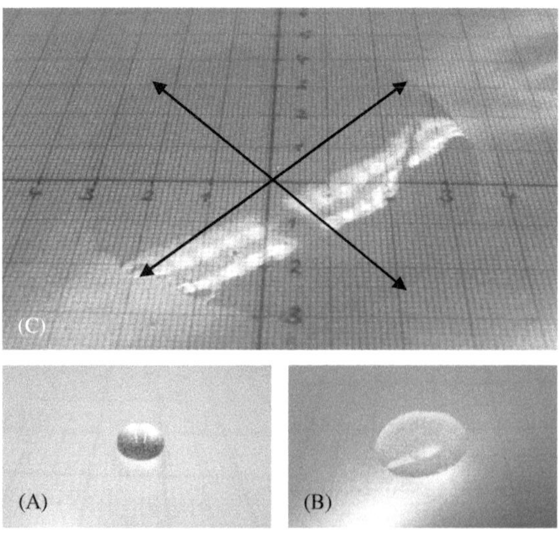

FIGURE 6.1 Photos taken 1 minute after placing a 50 μL droplet onto polypropylene film (with a cm scale underneath). (A) Water. (B) Nonsuperspreading trisiloxane surfactant ($M(D'E_{10}P_2OH)M$): 15 mm diameter. (C) Superspreading trisiloxane surfactant ($M(D'E_6P_3OH)M$): 70 mm diameter. *Figure from Venzmer (2011).*

($R = -(CH_2)_3-O-(CH_2-CH_2-O-)_n-(CH_2-CH(CH_3)-O-)_m R'$, or E_nP_m for short). R' stands for the end-capping group; mostly it is either $-H$ or $-CH_3$ (Me). For simplicity, the trisiloxane surfactants used in Figure 6.1 are denoted as $M(D'E_nP_mOH)M$. In most academic research, the commercial superspreader SILWET® L-77 (originally Union Carbide, now Momentive) has been used, which is a trisiloxane surfactant carrying a methyl-endcapped polyether of on average 7.5 EO ($M(D'E_{7.5}OMe)M$).

When you do a literature search on "superspreading" today, it can get confusing because other phenomena are now also termed *superspreading*—and only some of them are related to the behavior of droplets on solid substrates (e.g., electrowetting; see McHale et al., 2013). Most superspreaders found in the literature today are in the context of epidemiology: a super(-)spreader refers to an individual who comes into contact with a lot of people and thus contributes strongly to the spreading of infectious diseases. I am tempted to use this definition of the term in relation to some recent papers and review articles that are used to spread hypotheses, but this will be the topic of the following sections.

6.2 Papers and reviews on superspreading—the good, the bad, or the ugly?

In the past 25 years or so, a large number of papers have been published on the physics behind the phenomenon that try to elucidate the mode of action of silicone surfactants. The early works on superspreading have been reviewed twice by Hill (1998, 2002). More recently, on the occasion of the twentieth anniversary of the first physicochemical paper on superspreading (Ananthapadmanabhan et al., 1990), I attempted to compile all the information published as of June 2010 (Venzmer, 2011). Since then, three more reviews on superspreading have been published by Nikolov and Wasan (2011, 2013, 2014). So what is the point of adding yet another review on the very same topic? As you will soon see, this chapter is not just another summary of what has already been published on superspreading. It also covers some more general—maybe even philosophical—aspects of how to explain a phenomenon that cannot be observed at a molecular scale. The main challenge is that superspreading is a macroscopic, phenomenological manifestation of what a special kind of trisiloxane surfactant is doing on a molecular scale. Up to now, this "molecular action" could not be observed directly by using any analytical tools. This is not too surprising, considering that during the superspreading process, the three-phase contact line moves with a velocity of about 500 $\mu m\ s^{-1}$ for about 1 minute. This unusual movement of the three-phase contact line just indicates that the behavior of the superspreading surfactants must lead to an extraordinary force driving the contact line, which is absent in the case of non-superspreading trisiloxane surfactants.

In the last few years, for some reason, there were almost more reviews than "good" original publications with significant new results on superspreading. There are several reasons why one should have serious doubts about the relevance of some recent papers when it comes to providing new insights into the superspreading phenomenon. Experimentally, superspreading is much harder to study than a droplet having a relatively large finite contact angle, especially when utilizing a side-view setup typically used to measure contact angles. This seems to be one reason why some researchers have chosen experimental conditions at which no superspreading happens. This prevention of superspreading can be achieved by using surfactant concentrations that are too low (i.e., below the so-called cwc, or critical wetting concentration; Chengara et al., 2007) and/or a substrate that is too hydrophobic (e.g., perfluoroalkane; Ivanova et al., 2009). Also, it does not help to call a PTFE substrate with a water contact angle of 120° "moderately hydrophobic" (Radulovic et al., 2010, 2011). It cannot be ruled out that such studies could provide valuable information about the interfacial behavior of superspreaders; however, the usefulness of the findings to solve the mystery of superspreading seems to be rather limited. Another reason for the questionable value (in terms of being useful in explaining the superspreading phenomenon!) of some recent studies is a suboptimal choice of the surfactants used: either using superspreading surfactants only—that is, without a comparative example (e.g., Chengara et al., 2007; Ping et al., 2009; Ivanova et al., 2010; Radulovic et al., 2011)—or using comparative surfactants of very different chemical natures and thus much-reduced surface activity (e.g., Kumar et al., 2006; Ivanova et al., 2009). In case of a surfactant with a much lower surface activity, the surface tension σ_{lg} of the surfactant solution could well be higher than the surface tension σ_{sg} of the solid. As a consequence, the spreading coefficient ($S = \sigma_{sg} - \sigma_{sl} - \sigma_{lg}$) will always be negative, even if the interfacial tension σ_{sl} between the solid substrate and the liquid is zero. Therefore, if no spreading occurs under such conditions, what does that tell us about the mode of action of superspreading? This brings us to the chemical structures of the surfactants typically used and their correlation with spreading performance, which will be discussed in the next section.

6.3 Chemical structures of trisiloxane surfactants—correlation with spreading performance?

More interesting than classical surfactants such as sodium dodecyl sulfate (SDS) or fatty alcohol ethoxylates as a comparison would be systems that—from the standpoint of thermodynamics—have a chance to (super)spread because their surface tension is sufficiently low but—for some reason—do not like to do this. In other words, the real challenge for any hypothesis, model, or theory is explaining the different spreading behaviors between these two chemically pretty similar classes of trisiloxane surfactants: superspreading trisiloxanes like M(D′$E_{7.5}$OMe)M or M(D′E_6P_3OH)M, which form bilayer aggregates (L_α, L_3, vesicles) in aqueous solution, and more hydrophilic, micelle-forming, nonsuperspreading trisiloxanes that carry a little longer polyether (e.g., M(D′$E_{10}P_2$OH)M, M(D′E_{12}OH)M). Whether or not this correlation between phase behavior and spreading performance is just a coincidence is discussed in Section 6.11.

Some researchers did use a series of trisiloxane surfactants that carried polyether chains of different lengths, like M(D′E_nOH)M with $n = 4$, 6, or 8 (Ivanova et al., 2009), or $n = 4$, 5, 6, 7, 8, or 9 (Ritacco et al., 2010). Unfortunately, none of these polyether chains is sufficiently long to render the corresponding trisiloxane surfactant sufficiently hydrophilic to be micelle-forming and nonsuperspreading. In other words, despite all the efforts, a comparison with a trisiloxane surfactant that is not superspreading but behaving just as a "normal" surfactant (Figure 6.1B) is still not possible.

6.4 Data, hypotheses, models, and theories—how to prove the unprovable?

As mentioned already, up to now, it has not been possible to observe the superspreading process with molecular resolution—and it could well be that this will never be achievable. Therefore, one can ask the more or less general question of how we can explain such a macroscopic phenomenon on the basis of molecular processes at the relevant interfaces—that is, especially at the three-phase contact line. The situation at a moving contact line is already quite challenging even for pure liquids and solutions of "normal" surfactants, with a number of conflicting theories and models, giving rise to much debate. In an editorial in a recent "Discussion & Debate" issue of the *European Physics Journal* on "Wetting and Spreading Phenomena," Manuel Velarde stated that despite the numerous publications dealing with this topic, only a few have actually contributed to a consensus on what facts experiments and theories provide (Velarde, 2011). In a discussion note in the same issue, Terry Blake encouraged the protagonists to identify areas of potential agreement rather than just their differences (Blake, 2011). Superspreading is a brilliant example of how the situation can be even worse in terms of conflicting hypotheses and some pitfalls in science.

Let's go back to the question of how to explain such an unobservable phenomenon like superspreading. In *Star Trek VI: The Undiscovered Country*, Spock states that if you have eliminated the impossible, whatever remains—however improbable—must be the truth. Actually, this is a quote from Sherlock Holmes (Doyle, 1890), who also said, "It is a capital mistake to theorize before one has data. Insensibly one begins to twist facts to suit theories, instead of theories to suit facts" (Doyle, 1891). However, it is only human—but not necessarily advantageous for the advancement of science—for every researcher to consider certain results as being the most relevant: the results that are in line with their model. Conflicting experiments are considered irrelevant or are even ignored, even though they are the crucial ones. According to Karl Popper, scientific testing cannot prove theories but can only disprove them (Popper, 1959). However, often experimental data are used to "prove" that a hypothesis is correct; one should rather say that if the results are in line with a theory, this is an indication that the theory is still a valid option. Therefore, to advance the state of knowledge, both unbiased observation and the creativity to come up with unusual ideas, as well as the willingness to accept falsification, are needed—not only in the field of superspreading.

6.5 Mode of action of superspreading surfactants—too many possibilities?

The following sections briefly examine the different approaches to elucidating the mode of action of superspreading found in the literature. For further details concerning individual attempts and an in-depth discussion, I refer the reader to Venzmer (2011) and Nikolov and Wasan (2011).

From the standpoint of thermodynamics, the requirement for any spreading process is a positive spreading coefficient ($S = \sigma_{sg} - \sigma_{sl} - \sigma_{lg}$). Therefore, for a continuing process like superspreading, it should be crucial to maintain a positive spreading coefficient—that is, an efficient transport of the surfactant molecules to the freshly created interfaces is needed. Comparative spreading experiments on liquid oil as a hydrophobic substrate have shown that the surfactant transport to the air/water interface does not seem to be the limitation (Stoebe et al., 1997b). Hence, the limitation rather seems to be the supply of surfactant to the substrate/water interface, which could be facilitated by aggregate disintegration.

Thermodynamics, however, might not be the most relevant factor to explain the highly kinetic phenomenon of superspreading. As an alternative driving force, a Marangoni flow caused by a surface tension gradient between the apex of the drop and the expanding contact area has been proposed (Nikolov et al., 2002). A preadsorbed water layer should be detrimental to maintaining a positive spreading coefficient, but it could be crucial to maintaining a surface tension gradient. Also, such a preadsorbed water layer could explain why superspreading does not happen on substrates that are too hydrophobic.

To sum it all up, Figure 6.2 shows a collection of concepts that have been proposed to explain the mode of action of superspreading surfactants. Because they all involve the surfactants present as bilayer aggregates, they even suggest why there should be a difference in performances between the bilayer-forming superspreaders and the micelle-forming nonsuperspreaders (see also Section 6.11).

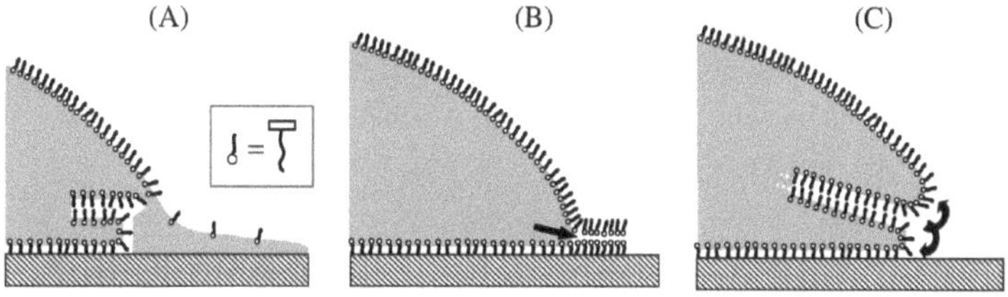

FIGURE 6.2 Various options to depict the mode of action of superspreading surfactants. (A) Disintegrating bilayer aggregates in combination with preexisting water layer (according to Stoebe et al., 1997a). (B) Bilayer spreading (according to Ruckenstein, 2012). (C) Unzippering of bilayer aggregates (Venzmer, 2011; see also Maldarelli, 2011). (Note: The surfactants are drawn in the typical hydrophilic head/hydrophobic chain fashion and not as hammer-like trisiloxanes, because the molecular shape does not seem to be crucial.)

6.6 Different stages of the spreading process—when does the spreading start to become "super"?

One simple question arises when comparing the spreading behavior of the classes of surfactants discussed here: when does the differentiation between superspreading and nonsuperspreading happen, or, in other words, when does the spreading start to become super? Recently, high-speed video microscopy with a time resolution of 0.019 ms (54,000 fps) has been used to observe what happens when a droplet of a surfactant solution is placed on a polypropylene film (Wang et al., 2013). During the first 12 ms, wetting was mainly controlled by inertia. All of the drop rims advanced with more or less the same power law (exponent between 0.3 and 0.45); neither the type of surfactant nor its concentration played a role. For the solutions of the conventional ionic surfactants CTAB and SDS, only this first wetting stage was observed before the drops reached a finite stable contact angle. Both trisiloxane surfactants used showed a second—viscosity-dominated—slower wetting stage that lasted from about 12 ms to about 1 s; the exponents of the power law describing the wetting were pretty similar: between 0.1 (for $M(D'E_{10}P_2OH)M$) and 0.15 (for $M(D'E_6P_3OH)M$). At the end of this stage, the nonsuperspreading $M(D'E_{10}P_2OH)M$ approached equilibrium with a well-defined contact angle. Only the superspreading $M(D'E_6P_3OH)M$ showed a third, and again faster, stage. The power law exponents derived from fits to the experimental data in this late superspreading stage (from 1 to 10 s) varied with concentration and ranged from 0.25 to 0.55. Therefore, it seems like the additional driving force for superspreading takes effect only after a characteristic time of over 1 s; before, superspreading and nonsuperspreading trisiloxane behaved similarly. Further insights into what initiates this superspreading stage are currently being gained by using the same high-speed video microscopy setup, except adding a droplet of concentrated trisiloxane surfactant solutions on the apex of a sessile water drop, sitting on a hydrophobic substrate. It seems like sessile drop spreading is kicked off after 1.6 ms, approximately when the surfactant first reaches the rim of the drop (Wang et al., 2015).

6.7 An equation containing substrate wettability and surface tension gradient

The conclusion of a very recent series of review papers by Nikolov and Wasan is an equation expressing the initial spreading velocity with substrate wettability (contact angle of water) and surface tension gradient (Nikolov and Wasan, 2013, 2014). Is this the ultimate solution to the superspreading phenomenon? Can such an equation elucidate the mode of action of the surfactants or predict the behaviors of different surfactants? Unfortunately, there is no chemistry-related parameter in these considerations that could explain the lack of spreading of the more hydrophilic, nonsuperspreading surfactants—which have somewhat similar surface tension dynamics. Therefore, it is still unclear what the superspreading molecules are actually doing to promote superspreading.

6.8 Surfactants jumping ahead of the contact line?

Another remarkable phenomenon is the spreading of the concept of "autophilicity": surfactant molecules—including trisiloxane superspreaders—are assumed to be able to transfer onto the bare hydrophobic substrate in front of the contact line in a tail-down/head-up fashion, hydrophilizing the substrate to render it wettable by the advancing water drop (Starov et al., 2010). The problem is that it is just taken as a given that surfactants will escape from the liquid phase to increase σ_{sg}, the substrate/air surface tension ahead of the contact line. Hence, all data in agreement are used to "prove" the existence of autophilicity, even if they can be explained by other effects, and all of the other results that could cast some doubt are not considered to be relevant. A detailed discussion on this can be found in Milne and Amirfazli (2010). Moreover, if the spreading kinetics do not fit a power law without assuming the existence of the autophilic effect, is this "proof" of the autophilic effect or an indication that the applicability of the power law is questionable? Be that as it may, this autophilic effect cannot play a role here because the superspreading process is way too fast for surfactants to "jump" ahead of the contact line.

6.9 Unusual concentration dependence—a lot does not always help a lot!

The relative humidity has a drastic effect on the concentration dependence of the spreading performance of superspreading trisiloxane surfactants such as M(D′$E_{7.5}$OMe)M (Figure 6.3). Because this can potentially provide some insights into what happens at the three-phase contact line, it is discussed here in a little more detail. At 100% relative humidity, the initial spreading velocity is more or less identical for concentrations above 0.1 wt%, and the maximum spreading area is proportional to the amount of surfactant molecules. This is to be expected, because the final arrangement is a "pancake" of water with one monolayer of surfactant each at the air/water and the substrate/water interfaces (Zhu, 1992). The situation becomes much more interesting when the spreading experiment is performed at 50% relative humidity. Although the spreading performance is not influenced by humidity at a surfactant concentration of 0.1 wt%, a drastic effect occurs at higher concentrations: the spreading process—although it has started with the same velocity—stops after some time (e.g., at 10 s for 0.25 wt%). The higher the surfactant concentration, the sooner the spreading process stops, and the lower the achievable spreading area. This maximum in spreading velocity at an intermediate concentration has often been interpreted as an indication that Marangoni flow is a driving force for superspreading (e.g., Nikolov et al., 2002). The driving force, however, does not seem to be the problem, because the

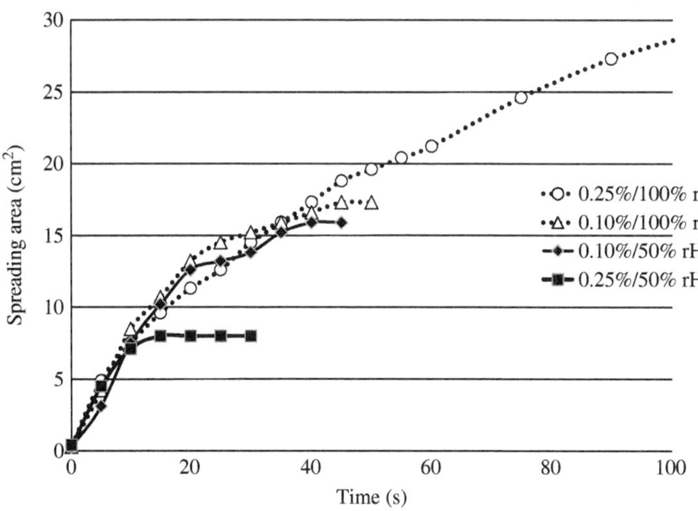

FIGURE 6.3 Spreading kinetics of M(D′$E_{7.5}$OMe)M (10 μL on polypropylene) as a function of concentration (0.1 and 0.25 wt%) and humidity (50% and 100% relative humidity). *Data replotted from Venzmer and Wilkowski (1998).*

spreading takes off without any hindrance, even at higher concentrations. When we performed these spreading experiments on a polypropylene film placed on an analytical balance, we found that by the time the spreading process stops, 10 wt% of the water has already been lost due to evaporation. Considering that evaporation preferentially takes place at the three-phase contact line (Starov and Sefiane, 2009), this could lead to a local increase in concentration at the moving contact line and thus the formation of a—highly viscous—lamellar phase stopping the spreading process. This is an alternative explanation for the peculiar concentration dependence—that is, maximum spread at an intermediate concentration—so this effect can hardly be used as proof of Marangoni flow as the main driving force.

6.10 Kinetic models to describe the moving contact line

Several models quantitatively describe the physics of a moving contact line, as discussed by Blake (2006). Attempts to fit experimental spreading data of $M(D'E_6P_3OH)M$ and $M(D'E_{10}P_2OH)M$ using the Molecular Kinetic Theory (MKT) have confirmed that there must be differences in surfactant transfer at the three-phase contact line between these two trisiloxane surfactants. Whereas the spreading kinetics of the "normal," nonsuperspreading $M(D'E_{10}P_2OH)M$ can be described reasonably well using MKT, this model is inappropriate to describe the wetting of solutions of the superspreading $M(D'E_6P_3OH)M$ (Wang et al., 2013). Again, this indicates that during the superspreading process, something "special" must be going on at the three-phase contact line, whatever it may be.

6.11 Specific phase behavior—coincidence or crucial?

This "something" that causes unusual behavior of the surfactants at the three-phase contact line could well be related to the phase behavior of the surfactants. As already mentioned, the superspreading trisiloxane surfactants form bilayer aggregates (L_α, L_3, vesicles) in aqueous solution, whereas the nonsuperspreading trisiloxane surfactants form spherical micelles—just like "normal" surfactants (e.g., He et al., 1993). Some authors believe that this correlation between phase behavior and spreading performance is just coincidence (e.g., Nikolov et al., 2002). However, there are indications that the phase behavior could be rather crucial for the spreading performance. First, the individual components of the typical homologue mixtures of commercial superspreading surfactants—that is, $M(D'E_nOMe)M$ with $n = 3, 4, 5, 6, 7, 8,$ or 9—exhibit an L_3-phase in a very narrow temperature range, which pretty much coincides with the temperature at which the highest initial spreading rates was observed (Wagner et al., 1999). The second strong indication that a specific phase behavior is crucial for superspreading comes from studying standard (i.e., nonsilicone), micelle-forming surfactants (Kennedy et al., 1998). At certain surfactant/cosurfactant ratios, mixtures of such surfactants with middle-chain alcohols as cosurfactants exhibit a similar phase behavior (L_α, L_3) in dilute solution as the superspreading trisiloxane surfactants. Interestingly, such mixtures exhibit the very same superspreading characteristics as those shown in Figure 6.1C. Therefore, somehow the physics seems to be the same, independently of the chemical structure of the surfactant or even the presence of a trisiloxane group, as long as the required phase behavior is there.

6.12 New approaches—experimental techniques and simulation

Studies addressing the comparison between superspreading and nonsuperspreading surfactants, especially if used at relevant concentrations, could provide valuable information to explain the superspreading phenomenon, even if the experiments are not related to spreading on hydrophobic substrates. One example of this is the gravitational drainage experiments of films supported by a wire frame reported recently (Sett et al., 2014). This method allows the determination of the disjoining pressure, which means looking into the liquid/air interface way beyond the typical (dynamic) surface tension measurements. The results revealed both systematic and dramatic differences in the behavior between superspreading and nonsuperspreading trisiloxane surfactants. The films of the two superspreading surfactants showed significant disjoining pressures leading to a stabilization of the films at a thickness of 35 nm, which has been attributed to the presence of "fluffy surfaces"—that is, long-hanging bilayers attached to the air/water interface. In contrast, the gravitational drainage of two more hydrophilic, nonsuperspreading surfactants proceeded uninhibited, just like in the case of "normal" surfactants. Interestingly, the fluffy bilayer protrusions of the superspreaders are assumed to be in the same size range as the "objects" that Brewster Angle Microscopy discovered in the air/water interface of superspreaders (Ritacco et al., 2010). Unfortunately, the latter authors did not use a more hydrophilic nonsuperspreader for comparison to check whether the aggregates observed in the air/water interface are somehow a special feature of the superspreading surfactants.

It is a real challenge to capture superspreading by computer simulation or molecular modeling, because this phenomenon is a macroscopic effect that happens over the course of dozens of seconds. Also, the systems have to be sufficiently large to describe the phase/aggregation behavior of the surfactants reasonably well; otherwise, it is difficult to draw conclusions concerning structure/performance relationships. Such large systems have to be coarse-grained—with the risk of being too coarse and therefore only leading to more or less trivial conclusions. Therefore, it will be interesting to see the future results of the multiscale approach currently undertaken in the group of Omar Matar at the Imperial College London (EPSRC Grant EP/J010502/1).

6.13 Conclusion

So where are we today? After all these years of research and debate, it is still not clear what is really driving the superspreading process and why other trisiloxane surfactants that have similar behaviors at the air/water interface do not show this exceptional spreading performance. For the time being, we have a choice between (i) quantitative models that are in agreement with the kinetics of superspreading, but not involving or even ignoring the chemical structures of the surfactants, or (ii) conceptual sketches that explain the phenomena encountered in the application of trisiloxane surfactants, including the known structure/property relationships, and are even useful in predicting the application performance of new surfactants or surfactant formulations. It is more or less a matter of personal preference which one of those alternatives is considered relevant. It will be interesting to see to which extent future research will be able to provide answers to the questions that have gone unanswered for decades.

Glossary

Disjoining pressure Surface force per unit area of a plane-parallel film
Lamellar phase Surfactant bilayers dispersed in water
L_3-phase Sponge phase, surfactant bilayer between two continuous water phases
Micelle Aggregate of surfactant molecules, typically spherical
Phase behavior Miscibility and formation of aggregates—for example, by surfactants in water
Superspreading Unusually fast and complete spreading by solutions of some trisiloxane surfactants
Superwetting See superspreading
Surface activity Consequence of the amphiphilic nature of surfactants; ability to accumulate at the air/water interface to reduce surface tension
Surfactant Surface active agent, a molecule with both hydrophilic and hydrophobic parts, therefore also called amphiphile
Vesicle Spherically closed surfactant bilayer

Nomenclature

CTAB Cetyl trimethyl ammonium bromide
fps Frames per second
$\mathbf{L}_{\alpha}$ Lamellar phase
$\mathbf{L}_{3}$ Sponge phase
rH Relative humidity
S Spreading coefficient
σ_{sg} Interfacial tension solid/gas; surface tension of a solid
σ_{sl} Interfacial tension solid/liquid
σ_{lg} Interfacial tension liquid/gas; surface tension of a liquid
SDS Sodium dodecyl sulfate

References

Ananthapadmanabhan, K.P., Goddard, E.D., Chandar, P., 1990. A study of the solution, interfacial and wetting properties of silicone surfactants. Colloids Surf. 44, 282–297.

Blake, T.D., 2006. The physics of moving wetting lines. J. Colloid Interface Sci. 299, 1–13.

Blake, T.D., 2011. Discussion notes: a more collaborative approach to the moving contact-line problem? Eur. Phys. J. Spec. Top. 197, 343–345.

Chengara, A., Nikolov, A.D., Wasan, D.T., 2007. Spreading of a water drop triggered by the surface tension gradient created by the localized addition of a surfactant. Ind. Eng. Chem. Res. 46, 2987–2995.

Doyle, A.C., 1890. The Sign of the Four. <www.gutenberg.org/ebooks/2097>.

Doyle, A.C., 1891. Adventures of Sherlock Holmes I: A scandal in Bohemia, in The Strand Magazine, London, vol. II, pp. 61–75.

He, M., Hill, R.M., Lin, Z., Scriven, L.E., Davis, H.T., 1993. Phase behaviour and microstructure of polyoxyethylene trisiloxane surfactants in aqueous solution. J. Phys. Chem. 97, 8820–8834.

Hill, R.M., 1998. Superspreading. Curr. Opin. Colloid Interface Sci. 3, 247–254.

Hill, R.M., 2002. Silicone surfactants—new developments. Curr. Opin. Colloid Interface Sci. 7, 255–261.

Ivanova, N., Starov, V., Johnson, D., Hilal, N., Rubio, R., 2009. Spreading of aqueous solutions of trisiloxanes and conventional surfactants over PTFE AF coated silicone wafers. Langmuir 25, 3564–3570.

Ivanova, N., Starov, V., Rubio, R., Ritacco, H., Hilal, N., Johnson, D., 2010. Critical wetting concentrations of trisiloxane surfactants. Colloids Surf. A: Physicochem. Eng. Asp. 354, 143–148.

Kennedy, M.W., Venzmer, J., Wilkowski, S.P., 1998. Organic superspreading surfactant blends vs. trisiloxane surfactants—spreading, wetting and active ingredient uptake. In: McMullan, P.M. (Ed.), Proceedings of Fifth International Symposium on Adjuvants for Agrochemicals, Memphis, TN, pp. 113–118.

Kumar, N., Maldarelli, C., Couzis, A., 2006. An infrared spectroscopy study of the hydrogen bonding and water restructuring as a trisiloxane superspreading surfactant adsorbs onto an aqueous-hydrophobic surface. Colloids Surf. A: Physicochem. Eng. Asp. 277, 98–106.

Maldarelli, C., 2011. On the microhydrodynamics of superspreading. J. Fluid Mech. 670, 1–4.

McHale, G., Brown, C.V., Sampara, N., 2013. Voltage-induced spreading and superspreading of liquids. Nat. Commun. 4, 1605.

Milne, A.J.B., Amirfazli, A., 2010. Autophilic effect: wetting of hydrophobic surfaces by surfactant solutions. Langmuir 26, 4668–4674.

Nikolov, A., Wasan, D., 2011. Superspreading mechanisms: an overview. Eur. Phys. J. Spec. Top. 197, 325–341.
Nikolov, A., Wasan, D., 2013. Superspreading: the role of the substrate surface energy. In: Rubio, R. G., et al., (Eds.), Without Bounds: A Scientific Canvas of Nonlinearity and Complex Dynamics. Springer-Verlag, Berlin, Heidelberg, pp. 301–314.
Nikolov, A., Wasan, D., 2014. Current opinion in superspreading mechanisms. Adv. Colloid Interface Sci. Available from: http://dx.doi.org/10.1016/j.cis.2014.03.006.
Nikolov, A.D., Wasan, D.T., Chengara, A., Koczo, K., Policello, G.A., Kolossvary, I., 2002. Superspreading driven by Marangoni flow. Adv. Colloid Interface Sci. 96, 325–338.
Ping, C.C.W., Ivanova, N.A., Starov, V.M., Hilal, N., Johnson, D., 2009. Spreading behaviour of aqueous trisiloxane solutions over hydrophobic polymer substrates. Colloid J. 71, 391–396.
Popper, K.R., 1959. The Logic of Scientific Discovery. Routledge, London.
Radulovic, J., Sefiane, K., Shanahan, M.E.R., 2010. Dynamics of trisiloxane wetting: effects of diffusion and surface hydrophobicity. J. Phys. Chem. C 114, 13620–13629.
Radulovic, J., Sefiane, K., Shanahan, M.E.R., 2011. The role of diffusion in superspreading. J. Adhes. Sci. Technol. 25, 1361–1370.
Ritacco, H.A., Ortega, F., Rubio, R.G., Ivanova, N., Starov, V.M., 2010. Equilibrium and dynamic surface properties of trisiloxane aqueous solutions. Part 1. Experimental results. Colloids Surf. A Physicochem. Eng. Asp. 365, 199–203.
Ruckenstein, E., 2012. Superspreading: a possible mechanism. Colloids Surf. A Physicochem. Eng. Asp. 412, 36–37.
Sett, S., Sahu, R.P., Sinha-Ray, S., Yarin, A.L., 2014. Superspreaders versus "cousin" non-superspreaders: disjoining pressure in gravitational film drainage. Langmuir 30, 2619–2631.
Starov, V., Sefiane, K., 2009. On evaporation rate and interfacial temperature of volatile sessile drops. Colloids Surf. A: Physicochem. Eng. Asp. 333, 170–174.
Starov, V., Ivanova, N., Rubio, R.G., 2010. Why do aqueous surfactant solutions spread over hydrophobic substrates? Adv. Colloid Interface Sci. 161, 153–162.
Stoebe, T., Lin, Z., Hill, R.M., Ward, M.D., Davis, H.T., 1997a. Enhanced spreading of aqueous films containing ethoxylated alcohol surfactants on solid substrates. Langmuir 13, 7270–7275.
Stoebe, T., Lin, Z., Hill, R.M., Ward, M.D., Davis, H.T., 1997b. Superspreading of aqueous films containing trisiloxane surfactant on mineral oil. Langmuir 13, 7282–7286.
Velarde, M.G., 2011. Editorial in spreading and wetting phenomena. Eur. Phys. J. Spec.Top. 197, 1–2.
Venzmer, J., 2011. Superspreading—20 years of physicochemical research. Curr. Opin. Colloid Interface Sci. 16, 335–343.
Venzmer, J., Wilkowski, S.P., 1998. Trisiloxane surfactants—mechanism of spreading and wetting. In: Nalewaja, J.D., Goss, G.R., Tann, R.S. (Eds.), Pesticide Formulations and Application Systems: Eighteenth Volume, American Society for Testing and Materials ASTM STP 1347, 140–151.
Wagner, R., Wu, Y., Czichocki, G., Berlepsch, H.v., Rexin, F., Perepelittchenko, L., 1999. Silicon-modified surfactants and wetting: II. Temperature-dependent spreading behaviour of oligoethylene glycol derivatives of heptamethyltrisiloxane. Appl. Organomet. Chem. 13, 201–208.
Wang, X., Chen, L., Bonaccurso, E., Venzmer, J., 2013. Dynamic wetting of hydrophobic polymers by aqueous surfactant and superspreader solutions. Langmuir 29, 14855–14864.
Wang, X., Bonaccurso, E., Venzmer, J., 2015. manuscript in preparation.
Zhu, X., 1992. Surfactant Fluid Microstructure and Surfactant Aided Spreading (Dissertation). University of Minnesota.
Zhu, S., Miller, W.G., Scriven, L.E., Davis, H.T., 1994. Superspreading of water-silicone surfactant on hydrophobic surfaces. Colloids Surf. A: Physicochem. Eng. Asp. 90, 63–78.

Chapter 7

LEIDENFROST DROPS

Benjamin Sobac[1,3], Alexey Rednikov[1], Stéphane Dorbolo[2] and Pierre Colinet[1]
[1]Université Libre de Bruxelles—TIPs Fluid Physics, Brussels, Belgium, [2]Université de Liège—GRASP Laboratory, Liège, Belgium, [3]Mechanical Engineering Department, Aix-Marseille University, Marseille, France

This chapter is concerned with drop levitation on a vapor layer when a volatile liquid is brought into contact with a very hot solid. This phenomenon is known as the Leidenfrost effect. A simple theory of a Leidenfrost drop covering the full range of stable shapes—that is, from small quasi-spherical droplets to larger puddles floating on a pocket-like vapor film—is presented. This modeling permits an accurate description of the geometry of the drop and of the underlying vapor film. It also provides deeper insights into heat transfer, evaporation, and flow that develop inside the vapor cushion. Finally, the model presented here can serve as a basis for studying other aspects or recent discoveries related to Leidenfrost drops. For instance, using a complement to the basic model, we show how to predict the lifetime of Leidenfrost drops.

7.1 Introduction

In 1756, Johann Gottlob Leidenfrost, a German physician, reported in his *Treatise on the Properties of Common Water* (Leidenfrost, 1756), "At the instant when the drop touches the glowing iron, it is spherical. It does not adhere to the spoon, as water is accustomed to do, which touches colder iron." He further added that when he placed

Droplet Wetting and Evaporation. DOI: http://dx.doi.org/10.1016/B978-0-12-800722-8.00007-2

a candle behind the drop, he could see light passing between the drop and the hot solid, revealing the existence of a vapor layer under the drop. Leidenfrost was probably not the first observer of this phenomenon, but he was the first to document it in detail. Therefore, the phenomenon bears his name. Figure 7.1A illustrates Leidenfrost's observation. A millimetric water drop "levitates" over a metallic substrate at 300°C. A small gap is indeed visible between the drop and the solid surface.

Indeed, a liquid that is placed on a solid surface at a temperature sufficiently above the saturation temperature does not come to a boil but rather levitates over a thin layer of its own vapor. The pressure exerted by the vapor emanating from the drop is able to sustain the weight of the drop. This levitation has mainly two consequences. First, the vapor layer preventing contact between the liquid and the heated surface ensures a higher degree of thermal insulation, conferring large lifetime to these objects. Second, this nearly frictionless state provides an extreme mobility to the drop, making it easy to manipulate but difficult to immobilize.

The Leidenfrost effect has been intensively investigated in the past (Bernardin and Mudawar, 1999; Biance et al., 2003; Quéré, 2013). However, many questions still remain open or under debate, in particular as far as the modeling of the phenomenon is concerned. A theoretical prediction of the Leidenfrost temperature is still unknown. This transition point depends on the nature of the liquid, as well as on the substrate's properties, such as surface roughness, thermal properties, and so on (Bernardin and Mudawar, 1999). Most experimental studies have concluded that the Leidenfrost temperature increases with the degree of roughness, the latter being detrimental to the stability of the vapor film. Accordingly, the generally accepted idea is that the higher the thermal diffusivity of the substrate and the smoother its surface (Bernardin and Mudawar, 1999), the closer the Leidenfrost point is to the saturation (boiling) temperature. However, a recent study also highlighted that a micropatterned surface can significantly reduce the Leidenfrost point (del Cerro et al., 2012). No present-day theory is able to explain these effects to the best of our knowledge.

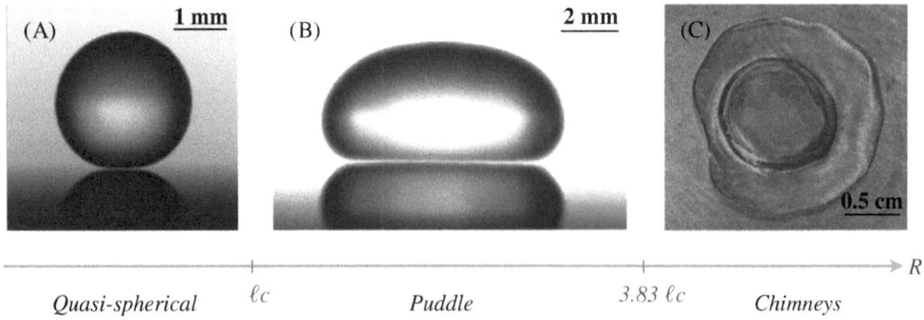

FIGURE 7.1 Typical shapes of Leidenfrost water drops of various sizes. A flat metallic substrate is maintained at 300°C. (A) and (B) Side view, courtesy of R. Thévenin and D. Soto. (C) Top view, taken from Biance et al. (2003), courtesy of A.-L. Biance.

The understanding and an accurate modeling of the Leidenfrost phenomenon are therefore crucial. Such a "film boiling" regime plays an important role in cooling devices. Indeed, if a vapor layer is present between the liquid and the device to be cooled, the energy transfer is drastically reduced, and the cooling becomes inefficient. This effect is, for example, undesirable from the point of view of rapid cooling of overheated components in high-power-density systems, such as nuclear reactors during possible accidents. On the other hand, the extreme mobility of the Leidenfrost drops and their nonwetting-like behavior make it possible to manipulate liquid mixtures without contact with any solid. This is a major advantage compared with regular micro- or millifluidic systems, where contact with walls is a potential source of pollution.

Recently, the Leidenfrost effect has been subject to a renewed interest following the discovery by Linke et al. (2006). Placed on an asymmetrically structured surface (called a ratchet), Leidenfrost drops can self-propel in a preferential direction with velocities of the order of $10\ \mathrm{cm\ s^{-1}}$. These findings opened new doors to the control and manipulation of drops. It can indeed be particularly useful in systems where drops need to be transported rapidly and accurately. Playing on the geometry of the structure, it is then possible to control their velocities and trajectories and even to immobilize them (Ok et al., 2011; Cousins et al., 2012). Taking advantage of this new control, such drops could, for example, be used to transport a component. Leidenfrost drops in mixtures is addressed in Chapter 24, but here, we discuss pure liquids.

This chapter is not intended to be a review of the subject. For a wider overview, the reader is referred to the recent review by Quéré (2013). Here, we concentrate on a recent model of a Leidenfrost drop. We use a simple theory that permits an accurate description of the geometry of the drop and the underlying vapor layer (Sobac et al., 2014). Because the evaporation of Leidenfrost drops is typically long compared to thermal and viscous relaxation times, only (quasi-) steady shapes are considered. This modeling appears to nicely match experimental data and in particular provides a good picture of the influence of the superheat and drop size on the Leidenfrost drop geometry. Moreover, it gives us deeper insights into the heat transfer, evaporation, and flow developing inside the vapor cushion. Finally, we show that the modeling can be extended to predict the lifetime of Leidenfrost drops and is readily amenable to further generalization.

7.2 Shape of Leidenfrost drops

Because there is no contact between the Leidenfrost drop and the surface, the apparent contact angle is 180°. A Leidenfrost drop therefore resembles a perfect superhydrophobic drop. Figure 7.1 shows the various shapes that a Leidenfrost drop can adopt, depending on its size. The smaller drops are quasi-spherical (Figure 7.1A), whereas the larger ones are flattened by gravity (Figure 7.1B). This transition between the spheres and puddles occurs when the gravity forces overcome the capillary ones. Typically, the effect of gravity on the drop shape can be disregarded when the drop radius R (as seen from the top of the drop) is smaller than the capillary length $\ell_c = (\gamma/\rho_\ell g)^{1/2}$, where g, γ, and ρ_ℓ are the gravity acceleration, surface tension, and liquid density, respectively. In a Leidenfrost situation, the temperature inside the drop is at the saturation (boiling) temperature (Biance et al., 2003). For water, this implies a density $\rho_\ell = 960$ kg m^{-3} and a surface tension $\gamma = 59$ mN m^{-1}, and thus a capillary length $\ell_c = 2.5$ mm. When the size of the drop is too large, the shape of the drop becomes unstable. The underlying vapor layer destabilizes, and vapor chimneys form, rising and eventually bursting through the center of the puddle (Figure 7.1C). This phenomenon has been interpreted as a Rayleigh–Taylor instability that develops as soon as $R > 4\ell_c$ (Biance et al., 2003; Snoeijer et al., 2009). Under certain conditions, drops spontaneously start to oscillate and develop "star shapes" or "faceted shapes" (Brunet and Snoeijer, 2011). The model presented here is limited to axisymmetric stable shapes of Leidenfrost drops—that is, to $R < 4\ell_c$.

7.2.1 Equilibrium Drop Shape (Outer Drop Surface)

The upper part of the drop, above the vapor cushion, is assumed to be of equilibrium shape. At equilibrium, the Laplace pressure $\gamma\kappa$ (with κ the curvature of the drop surface) balances (up to a constant) the hydrostatic pressure $-\rho_\ell g z$, where z is the vertical coordinate directed upward. Scaling all lengths by ℓ_c and counting z from the top of the drop, the dimensionless equation for the equilibrium shape then simply reads

$$\kappa + z = \kappa_{top} \tag{7.1}$$

where κ_{top} is the curvature at the top of the drop. Numerically integrating this differential equation for a given value of κ_{top} yields equilibrium shapes and in particular the value of the radius $R = R(\kappa_{top})$. The integration can be continued up to a horizontal local slope, beyond the envisaged applicability of an equilibrium shape, yielding the associated superhydrophobic drop (whose upper part coincides with that of our Leidenfrost shape); see Figure 7.2. The height and the contact radius of this superhydrophobic drop are $H = H(\kappa_{top})$ and $R_c = R_c(\kappa_{top})$, respectively. The height varies from $H \approx 2R$ for small drops (which are quasi-spherical, with the curvature $\kappa_{top} \approx 2/R$) to $H \approx 2.1$ for large drops (the top of the puddle being almost flat, the curvature $\kappa_{top} \approx 0$). Note also that in the limit $R \ll 1$, the contact radius is given by $R_c = \sqrt{2/3}\,R^2$ (Mahadevan and Pomeau, 1999; Burton et al., 2012), whereas in the limit of $R \gg 1$, R_c scales as R.

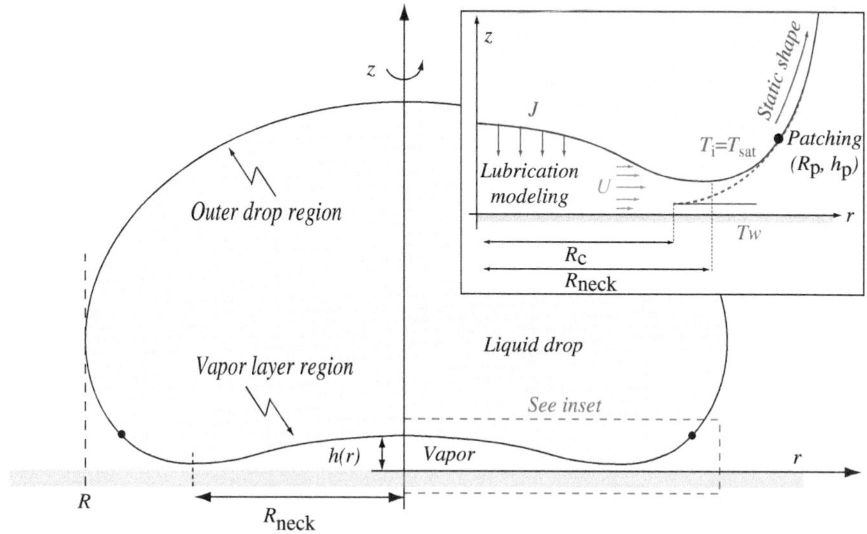

FIGURE 7.2 Schematic illustration of a Leidenfrost drop. The drop levitates on a vapor cushion of nonuniform thickness profile $h(r)$, composed by a pocket surrounded by an annular neck. Due to evaporation, an outward radial flow develops in the vapor layer. R is the drop radius as seen from above, while R_{neck} is the radius of the neck (at its thinnest section). The outer drop region (assumed to be an equilibrium shape) and the inner vapor layer region (described using lubrication theory, including evaporation) are solved separately and numerically matched at the "patching" point with coordinates (R_p, h_p). R_c is defined from the outer drop shape as the "contact" radius of the associated superhydrophobic drop at equilibrium.

In the case of a Leidenfrost state, however, the drop does not contact the substrate but rather levitates on a vapor cushion. It is unlikely that the bottom remains flat, especially for $R>1$ when gravity becomes stronger than capillarity. Indeed, we then rather expect the formation of a vapor pocket below the drop similarly to what happens in film boiling regimes. In the framework of our model of the Leidenfrost drop, the equilibrium shape is applied only until a point located at $r=R_p$, between R_c and R (Figure 7.2), where nonequilibrium effects of evaporation and viscous pressure losses in the underlying vapor layer come into play. Let us now discuss the shape of the liquid/vapor interface under the drop.

7.2.2 Vapor Film Geometry

In the vapor layer region ($0<r<R_p$), the gas is assumed to be pure vapor and incompressible, and its properties are taken as constant. Assuming that heat is only transferred by conduction across the film, the local evaporation flux at the

interface is expressed as $J = L^{-1}\lambda_v(T_w - T_i)/h$, where λ_v is the vapor thermal conductivity, T_i is the liquid–vapor interface temperature equal to the saturation temperature T_{sat}, T_w is the substrate temperature, and L is the latent heat of vaporization. The drop squeezing the vapor exerts an excess of pressure P_v, forcing the vapor to escape laterally. Ignoring possible motion inside the drop as it is typically the case in the literature (Snoeijer et al., 2009), the balance of normal forces at the drop interface gives $P_v = -(\rho_\ell gh + \gamma\kappa)$ up to a constant. The thickness of the vapor film and its slope can be considered small enough so as to use the lubrication approximation. Then, this pressure drives a Poiseuille flow with a volumetric flux $\vec{q} = -\frac{h^3}{12\mu_v}\vec{\nabla}P_v$, where μ_v is the vapor dynamic viscosity. Note the coefficient 1/12 in the mobility factor, typical of no-slip conditions imposed both at the drop surface and at the substrate. Finally, the classical expression of vapor mass conservation under the lubrication hypothesis (at steady state) reads $\vec{\nabla}\cdot\vec{q} - J/\rho_v = 0$, where ρ_v is the vapor density. Combining these equations, scaling all lengths by ℓ_c, and assuming the axial symmetry yields the following equation for the film thickness:

$$\frac{1}{12}\frac{1}{r}\frac{\partial}{\partial r}\left(rh^3\frac{\partial}{\partial r}(h+\kappa)\right) - \frac{E}{h} = 0 \tag{7.2}$$

with an evaporation number E defined by

$$E = \frac{\lambda_v\mu_v\Delta T}{\gamma\rho_v\ell_c L} \tag{7.3}$$

where $\Delta T = T_w - T_{sat}$ is the superheat. As for the curvature κ, it is here taken in a general form

$$\kappa = \frac{\frac{\partial^2 h}{\partial r^2} + \frac{1}{r}\left[1 + \left(\frac{\partial h}{\partial r}\right)^2\right]\frac{\partial h}{\partial r}}{\left[1 + \left(\frac{\partial h}{\partial r}\right)^2\right]^{3/2}} \tag{7.4}$$

It is interesting to note that E is hereby the only parameter remaining in the problem, apart from the size R of the drop, and depends on the fluid properties and the superheat ΔT. While the liquid properties are taken at T_{sat}, the vapor properties are here evaluated at the mean temperature $(T_w + T_{sat})/2$ of the vapor film and thus they vary with the substrate temperature. Four boundary conditions are needed to solve Eqs (7.2) and (7.4): symmetry conditions at $r = 0$—that is, $h'(0) = 0$ and $\kappa'(0) = 0$—while at $r = R_p$ the solution must match with the outer equilibrium shape of the drop—that is, the continuities of $h'(r)$ and $\kappa(r)$ are required. The continuity of $h(r)$ itself here merely amounts to a vertical shift of the outer shape. This problem is numerically solved using a standard second-order finite-difference method, and it is checked a posteriori that the choice of the "patching" point R_p has an insignificant influence on the results (up to a few percent).

Figure 7.3 shows typical shapes of Leidenfrost drops computed with the present model for various drop sizes. It is seen that the vapor layer forms a concave depression in the drop interface that becomes increasingly marked with the drop size. The vapor layer is composed of a vapor pocket at its center, surrounded by an annular neck. The film thickness at the neck location appears to increase with the drop size at a given E (inset of Figure 7.3). Figures 7.4A–C describe the Leidenfrost drop shape by means of the values of the neck thickness h_{neck}, the depth of the vapor pocket $\Delta h = h_{\mathrm{center}} - h_{\mathrm{neck}}$, and the neck radius R_{neck} as a function of the drop size. It can be seen that the vapor cushion thickness increases with the drop size. In the situation where $E = 7.16 \times 10^{-7}$ (i.e., a water drop on a substrate at $T_{\mathrm{w}} = 300°\mathrm{C}$), the neck thickness can reach 60 μm for puddles. The drop bottom is almost flat $\Delta h \approx 0$ for small drops, but the depth of the vapor pocket strongly increases with the drop size ($\Delta h \sim H$ when $R \approx 3.9$). Finally, the neck radius R_{neck} increases linearly with R when $R > 1$. Recently, Burton et al. (2012) experimentally studied the bottom of Leidenfrost drops by observing water drops on a hot transparent surface from below. Then, using interferometry, they characterized the shape of the vapor layer for various drop sizes and substrate temperatures. A comparison between these experimental data and the present model showed a good agreement (Sobac et al., 2014). It appears, among other things, that the position of the neck practically does not depend on the superheat and that the geometry of the vapor layer is much more affected by the drop size than by the applied superheat (Figure 7.4). The influence of the evaporation number on the vapor cushion characteristics is reported in Figures 7.4D and E. While the neck thickness turns out to scale as $h_{\mathrm{neck}} \sim E^{1/3}$ for all the drop sizes considered in the figure, the pocket thickness grows more slowly as $h_{\mathrm{center}} \sim E^{1/6}$, except for sufficiently small droplets, where the pocket levels with the neck and the exponent gets larger (closer to 1/3). The superheat value thus appears to have a more appreciable effect on the film in the neck region. These scalings can be directly obtained from Eq. (7.2) by an asymptotic

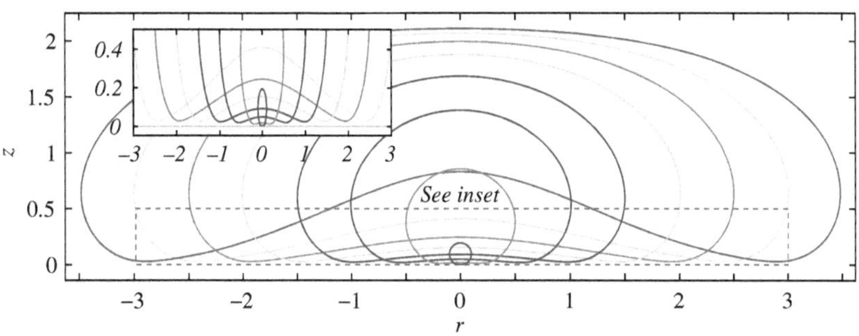

FIGURE 7.3 Numerically determined shapes of Leidenfrost drops for an evaporation number $E = 1.21 \times 10^{-6}$ ($T_{\mathrm{w}} = 370°\mathrm{C}$ for water). The lengths are scaled by the capillary length (i.e., 2.5 mm for water).

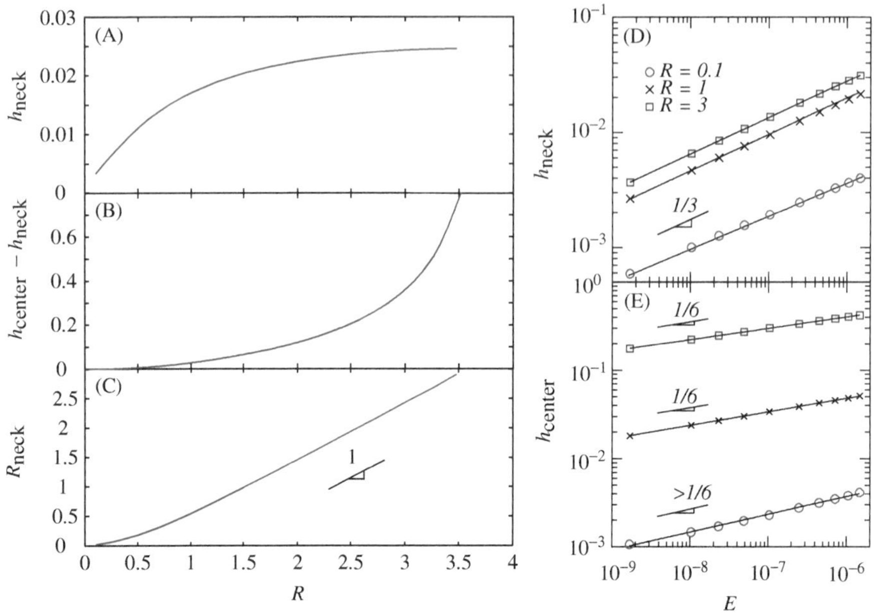

FIGURE 7.4 Geometrical parameters of the vapor layer under a Leidenfrost water drop over a substrate at $T_w = 300°C$ (i.e., $E = 7.16 \times 10^{-7}$). Neck thickness (A), depth of the vapor pocket (B), and neck radius (C) are plotted as a function of the drop size. All lengths are scaled by the capillary length $\ell_c = 2.5$ mm. The neck thickness (D) and the vapor-film thickness in the center (E) are also plotted versus the evaporation number for some drop sizes.

analysis on account of E being small (see Sobac et al., 2014). A slightly different scaling exponent for h_{center} was proposed by Pomeau et al. (2012) (although these authors also find $h_{neck} \sim \Delta T^{1/3}$)—namely, 1/5 was obtained instead of 1/6, against which we argued elsewhere (Sobac et al., 2014). We will see in the following section that this result is related to a rather important conceptual difference concerning the physics of the vapor layer. Understanding the shape and physical mechanisms at play in the film is crucial to predicting the heat transfer between the solid and the liquid, a practically important quantity, as well as the global evaporation rate of the drop or the drop lifetime (see Section 7.4).

7.3 Physics of the vapor film

In this section we look more closely at the physical mechanisms at play inside the vapor layer. It has already been mentioned that radial Poiseuille flow is driven in the vapor film underneath the drop. Given the no-slip conditions both at the plate and the interface, the maximum vapor velocity at a given radial location is thus located in the middle of the vapor layer, at $z = h/2$. Figure 7.5A shows the radial profile of this velocity U_{max} for various sizes of the drop. It is worth mentioning that U_{max} can attain values of the order of several meters per second in the vicinity of the neck (where U_{max} reaches its maximum, denoted as U_{neck}, even though it may not be exactly at the neck itself). The velocity difference between the pocket and the neck regions is more pronounced for puddles. Moreover, the magnitude of the velocity decreases with the increase in the drop size, the effect being apparently determined by the associated increase in the layer thickness. The local Reynolds number in the neck region, defined as $Re = \rho_v U_{neck} h_{neck}^2 / \mu_v \ell_{neck}$ and comparing the inertial and viscous effects therein, is then found to be of the order of 0.1 (inset of Figure 7.5A). This justifies a

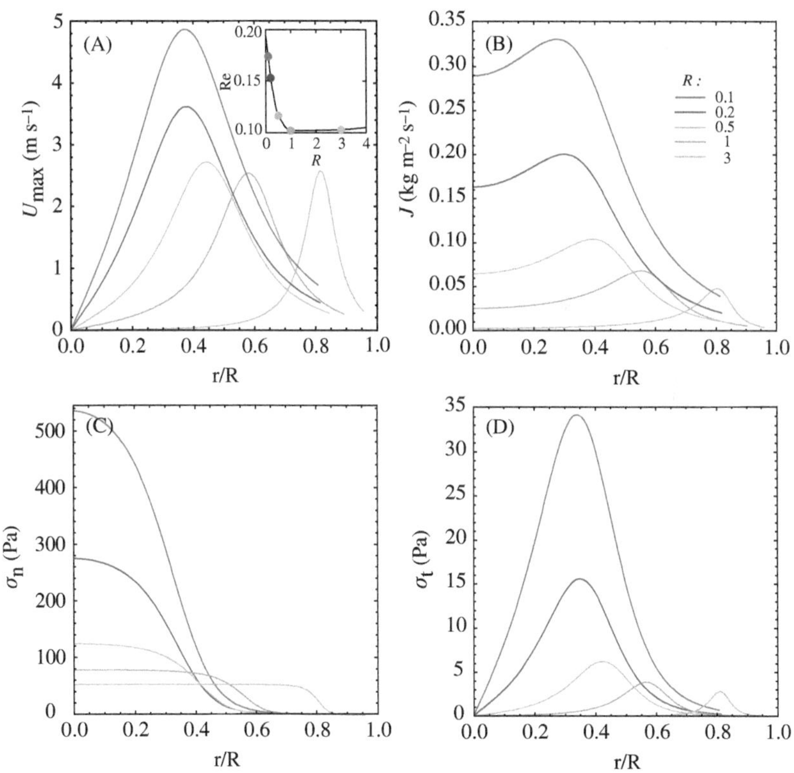

FIGURE 7.5 Some relevant flow quantities inside the vapor layer underneath a Leidenfrost water drop, the substrate being at $T_w = 300°C$ (i.e., $E = 7.16 \times 10^{-7}$). Radial profile of the maximum local velocity (A), the local evaporation flux (B), and the (excess) normal (C) and tangential (D) stresses acting on the bottom surface of the drop for various drop sizes. The profiles end at the patching points.

posteriori the dominant role of viscous forces in the vapor film assumed in the present analysis. Here, ℓ_{neck} is the length of the neck region as it appears at the height $2h_{neck}$ of the film profile.

Similar observations can be made for the profile of the local evaporation flux J (Figure 7.5B). Considering conduction in the vapor as the main mechanism of heat exchange between the substrate and the drop, the profile of J scales as $1/h$. Evaporation is thus more intense in the neck region, and an appreciable effect of evaporation on the film geometry in this place is not surprising. On the contrary, it turns out that the shape of the central pocket is almost unaffected by evaporation (it is an equilibrium shape governed by capillarity and gravity, just like the top of the drop (Sobac et al., 2014)). This contrasts with the analysis of Pomeau et al. (2012), who (unrightfully in our opinion) assumed that evaporation affects the shape of the vapor pocket already to the same order of magnitude as gravity and capillarity (see Sobac et al., 2014; it is also the basis of the earlier mentioned different scaling for h_{center} obtained in Pomeau et al., 2012). The Péclet number Pe = Pr Re (where $\mathrm{Pr} = \nu_v/\alpha_v \sim 0.7$ is the Prandtl number, and α_v is the thermal diffusivity of vapor) comparing heat transfer by convection and conduction is estimated to be of the order of 0.1, as well, supporting a posteriori the hypothesis of a mostly conductive heat transfer in the vapor film. A scaling argument reveals that the dimensionless rate of vapor production over the pocket is of the order of $(E/h_{center})\pi R_{neck}^2 \sim E^{5/6}$ where $\ell_{neck} \sim \mathrm{E}^{1/6}$ and $\mathrm{R} \sim 1$, which is comparable to the global evaporation rate in the neck region $(E/h_{neck})2\pi\ell_{neck}R_{neck} \sim E^{5/6}$ (Sobac et al., 2014). Thus, even though the local rate of evaporation is much lower in the pocket region than in the neck, it occurs on a larger surface, so that both regions contribute by the same order of magnitude to the global evaporation rate. Note that the conclusions of Pomeau et al. (2012) also differ from ours in that respect; that is, unlike the present study, they found that the global evaporation rate is mainly determined by the pocket region.

Finally, Figures 7.5C and D present the tangential and (excess) normal stresses acting on the drop. The normal stresses sustain the levitation of the drop and are (in the lubrication approximation) equal to the vapor pressure, for which we here plot the excess value (relative to the ambient one). These stresses are maximum in the vapor pocket and quickly vanish through the neck region. Furthermore, they are quite homogeneous in the pocket underneath large puddles. The integral of the excess normal stresses over the bottom surface of the drop must be equal to the weight of the drop, which has been verified numerically. Note that these excess normal stresses in the vapor film underneath the drops can be as large as 5×10^{-3} atm. The values of the tangential (viscous) stresses $\sigma_t = \mu_v \partial u/\partial z|_{z=h}$ are about an order of magnitude smaller. These tangential stresses are obviously most important in the neck region. Needless to say, the integral of these tangential viscous stresses over the bottom surface of the drop is equal to zero by symmetry, which is compatible with the drop remaining static.

7.4 Global evaporation rate and lifetime

An integration of the local evaporation flux J all over the drop surface allows us to evaluate the quantity of vapor produced per unit time ($-\dot{M}$, the global evaporation rate). The model developed thus far yields us right away the global evaporation rate in the underlying film (from $r=0$ to $r=R_p$): $(-\dot{M}_{\text{film}}) = \frac{2\pi\lambda_v \Delta T}{L}\int_0^{R_p}\frac{r\,dr}{h}$. This evaporation rate is typically of the order of 1 mg s^{-1}, and it depends on the size of the drop. We can now numerically confirm the scaling we obtained before that shows that the effects of the vapor pocket and neck regions on the overall evaporation rate are of the same order of magnitude. Even though evaporation also occurs on the rest of the drop surface, $(-\dot{M}_{\text{film}})$ should be the main contribution into $(-\dot{M})$ in view of the thinness of the film. Nonetheless, we shall here be interested in accounting for evaporation over the entire surface of the drop. To this purpose, the following simplified approach is used. Having already obtained the overall Leidenfrost drop geometry in the framework of our model, we can now simply solve the stationary heat transfer problem for this fixed geometry. In principle, if the convective (maybe including buoyancy) contribution is important, it should also include a resolution of the Navier–Stokes equations in conjunction with the energy equation. In this chapter, however, we limit ourselves to the case of pure conduction when the gas temperature field T_g satisfies just the Laplace equation $\nabla^2 T_g = 0$. The drop is still assumed to be at the boiling temperature—that is, $T_g = T_{sat}$ at the drop surface. At the substrate, we still have $T_g = T_w$. As for the ambient temperature, we assume that the vicinity of the drop is heated up to the temperature of the substrate, hence $T_\infty = T_w$ for the temperature that is formally imposed for T_g at infinity. The formulated problem for T_g (which is 2D axisymmetric) is here solved using the COMSOL Multiphysics™ software. The local evaporation rate at the drop surface is then calculated as $J = \frac{\lambda_v}{L}\frac{\partial T_g}{\partial n}$, whereas the global one is $(-\dot{M}) = \frac{\lambda_v}{L}\int \frac{\partial T_g}{\partial n}\,dS$, by integrating over the entire surface area of the drop, where n is the external normal. As before, λ_v is here evaluated at the "mean" vapor temperature—that is, $(T_w + T_{sat})/2$. The results for the gas temperature field and the local and global evaporation rates are shown in Figure 7.6. As previously observed in Figure 7.5B, the local evaporation rate increases from the center of the vapor pocket to reach a maximum at the neck. Then, the local evaporation rate decreases down to a minimum at the top of the drop. Evaporation is thus more intense in the neck region, as it has been previously pointed out. Evaporation at the top of the drop appears to be relatively weak compared to the contribution of the vapor layer. However, considering the global evaporation rate (Figure 7.6D), the contribution of the upper part (i.e. above the patching point) to the evaporation rate does not appear to be negligible, at least not for the small drops. Indeed, the inset in Figure 7.6D shows the quantity $((-\dot{M}) - (-\dot{M}_{\text{film}}))/(-\dot{M}) \equiv \varepsilon_{\dot{M}}$, which represents the relative contribution of the upper part of the drop into the global evaporation rate.

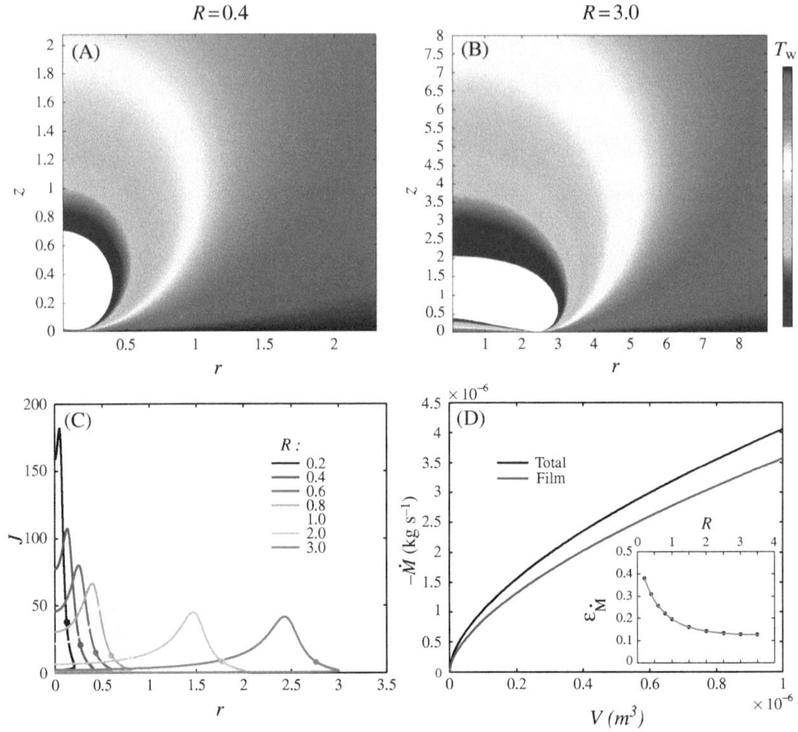

FIGURE 7.6 Evaporation of a Leidenfrost water drop over a surface at 300°C. (A and B) Temperature fields around drops of two different sizes ($R = 0.4$ and $R = 3.0$). (C) Dimensionless profile of the local evaporation flux at the drop surface for drops of various sizes (the open circles indicate the location of the patching points). (D) Global evaporation rate as a function of the drop volume, both the one obtained all over the drop surface and the one in the vapor film only. Inset: The relative difference between the two.

It reveals that for puddles ($R \gg 1$), the drop evaporates mainly via the vapor film; the upper part contributes only about 10% to drop evaporation. But this is not the case for smaller drops ($R \leq 1$), where evaporation takes place over the entire surface, rather than mostly in the film. The upper surface then contributes about 40% to drop evaporation with a dimensionless radius $R = 0.2$. These conclusions are found to be in qualitative agreement with the results of Biance et al. (2003). On the other hand, as it can be appreciated from Figure 7.6C, a big part of the upper-surface contribution comes in fact from its lower part, between the point of patching and the turning point $r = R$. Note finally that the global evaporation rate $\dot{M}$ turns out to scale approximately as $V^{0.6}$ (Figure 7.7A).

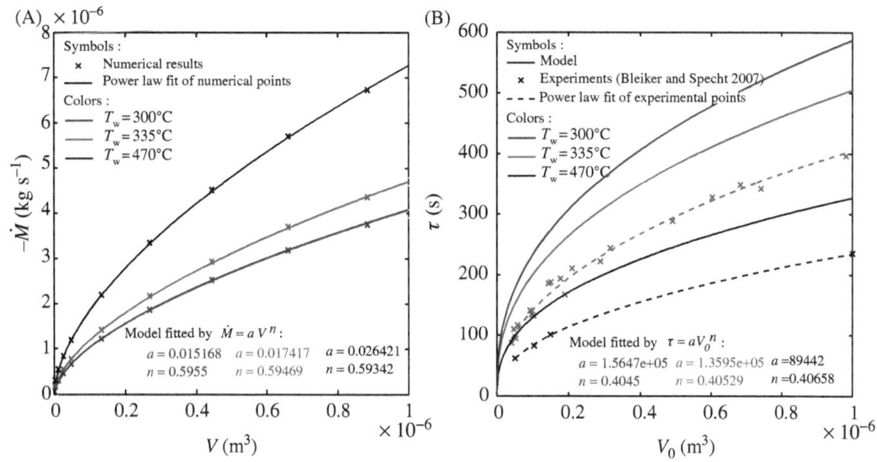

FIGURE 7.7 Global evaporation rate (A) and lifetime (B) of a water drop as a function of its size for several substrate temperatures. The predicted lifetime are compared with experimental data extracted from Bleiker and Specht (2007).

We can now estimate the lifetime of an evaporating Leidenfrost drop $\tau = \rho_\ell \int_0^{V_0} \mathrm{d}V/{-\dot{M}}$, where V_0 is the initial drop volume. A water Leidenfrost puddle of a few millimeters ($V_0 = 0.1$ mL) evaporates over a substrate at 300°C in 230 s according to our prediction, including evaporation all around the drop. Assuming that the drop only evaporates through the film, we instead get a lifetime estimate of 295 s. This simplification clearly leads to an overestimation of the lifetime by about 30%. The smaller the drop, the higher the overestimation due to this assumption. Therefore, evaporation all around the drop appears crucial for a correct lifetime estimate, even for puddles. Figure 7.7 shows the influence of the substrate temperature on the evaporation rate and on the lifetime of an evaporating Leidenfrost drop. The predicted evaporation times are compared to experimental data extracted from Bleiker and Specht (2007). The model prediction appears to overestimate the experimental lifetime of the drop. The overestimation is of the order of 25% for puddles over a substrate at 335°C and 38% over a substrate at 470°C. Part of this discrepancy is deemed to be due to convective effects, neglected in the present study, which would tend to increase the evaporation rate over the upper part of the droplet. On the other hand, the gas was here considered as pure vapor even all around the upper part of the drop which must not be exactly the case in experiments.

7.5 Conclusions and perspectives

The theory presented here provides an accurate description of the overall geometry of the Leidenfrost drop and of the underlying vapor layer, which proves to be composed of a vapor pocket at its center, surrounded by an annular neck. This enables us to examine the roles of heat transfer, evaporation, and flow developing inside the vapor layer. The model reveals in particular that the vapor flow developing in the film is dominated by viscous forces ($Re \ll 1$) and that the heat is mostly transferred by conduction ($Pe \ll 1$). Moreover, it turns out that evaporation has a more appreciable effect on the film geometry in the neck region and that the contributions of the pocket and neck regions to the global evaporation rate are comparable. This simple theory of the Leidenfrost effect can serve as a basis for further studies. For instance, it could be generalized to account for substrate nonflatness (roughness, geometrical structures), internal flow in the drop, and so on. As for this chapter, our basic model has in particular been supplemented with a consideration of the evaporation flux over the upper part of the drop, yielding the overall evaporation flux and permitting us to estimate the lifetime of the drop. The relative contribution of the upper part of the drop to the global evaporation rate reveals that for puddles ($R \gg \ell_c$), the drop evaporates mainly via the vapor film (pocket and neck), whereas for smaller drops ($R \leq \ell_c$), evaporation takes place over the entire surface, rather than mostly in the film. However, examining the evaporation that occurs all around the drop is crucial for estimating the lifetime accurately, even for puddles, because this regime contributes more to the total evaporation time than the spherical drop regime.

References

Bernardin, J.D., Mudawar, I., 1999. The Leidenfrost point: experimental study and assessment of existing models. J. Heat Transf. 121, 894–903.

Biance, A.-L., Clanet, C., Quéré, D., 2003. Leidenfrost drops. Phys. Fluids 15 (6), 1632–1637.

Bleiker, G., Specht, E., 2007. Film evaporation of drops of different shape above a horizontal plate. Int. J. Therm. Sci. 46, 835–841.

Brunet, P., Snoeijer, J., 2011. Star-drops formed by periodic excitation and on an air cushion—a short review. Eur. Phys. J. Special Topics 192, 207–226.

Burton, J.C., Sharpe, A.L., van der Veen, R.C.A., Franco, A., Nagel, S.R., 2012. Geometry of the vapor layer under a Leidenfrost drop. Phys. Rev. Lett. 109, 074301.

Cousins, T.R., Goldstein, R.E., Jaworski, J.W., Pesci, I.A., 2012. A ratchet trap for Leidenfrost drops. J. Fluid Mech. 696, 215–227.

del Cerro, D.A., Marín, A.G., Römer, G.R.B.E., Pathiraj, B., Lohse, D., Huis in't Veld, A.J., 2012. Leidenfrost point reduction on micropatterned metallic surfaces. Langmuir 28, 15106–15110.

Leidenfrost, J.G., 1756. On the fixation of water in diverse fire (Trans. C. Wares, 1966). Int. J. Heat Mass Transf. 9, 1153–1166.

Linke, H., Alemàn, B., Melling, L., Taormina, M., Francis, M., Dow-Hygelund, C., Narayanan, V., Taylor, R.P., Stout, A., 2006. Self-propelled Leidenfrost droplets. Phys. Rev. Lett. 96 (15), 154502.

Mahadevan, L., Pomeau, Y., 1999. Rolling droplets. Phys. Fluids 11, 2449–2453.

Ok, J., Lopez-Ona, E., Nikitopoulos, D., Wong, H., Park, S., 2011. Propulsion of droplets on micro- and sub-micron ratchet surfaces in the Leidenfrost temperature regime. Microfluid. Nanofluid. 10, 1045–1054.

Pomeau, Y., Le Berre, M., Celestini, F., Frisch, T., 2012. The Leidenfrost effect: from quasi-spherical droplets to puddles. C. R. Mecanique 340, 867–881.

Quéré, D., 2013. Leidenfrost dynamics. Ann. Rev. Fluid Mech. 45, 197–215.

Snoeijer, J., Brunet, P., Eggers, J., 2009. Maximum size of drops levitated by an air cushion. Phys. Rev. E 79, 036307.

Sobac, B., Rednikov, A., Dorbolo, S., Colinet, P., 2014. Leidenfrost effect: accurate drop shape modeling and refined scaling laws. Phys. Rev. E 90, 053011.

Section II

HEAT AND MASS TRANSFER

Chapter 8

PURE DIFFUSION

Benjamin Sobac[1,2] and David Brutin[1]
[1]*Mechanical Engineering Department, Aix-Marseille University, Marseille, France,*
[2]*Université Libre de Bruxelles, TIPs—Fluid Physics, Bruxelles, Belgium*

The evaporation phenomenon is a phase change transition from a liquid to its own vapor. The molecular scale description is a transfer of molecules from the liquid to the surrounding gas, while the macroscopic scale description is a mass flux crossing the liquid/vapor interface. Evaporation occurs as soon as the vapor partial pressure in the gas is lower than the saturation pressure. This basic mechanism is called diffusive evaporation, and it always exists as long as the saturated conditions are not reached. Later, other supplementary heat and mass transfer mechanisms can appear, such as convections (thermo-gravitational, thermo-capillary, and thermo-solutal) or radiation. These additional mechanisms are discussed in Section 2.2. This section examines the theory of purely diffusive evaporation from analytical solution to approximated models. Indeed, droplet evaporation is usually considered a process that is controlled by the vapor diffusion into the environment. This is considered the basic modeling of droplet evaporation. This description appears to be very satisfactory for water evaporating into air at room temperatures (Gelderblom et al., 2011, Sobac and Brutin, 2011).

Consider small drops with a contact radius R of less than the capillary length $L_c = (\sigma/\rho_l g)^{1/2}$, where σ is the surface tension, ρ_l the liquid density, and g the gravitational constant. Then, the fluid adopts a spherical cap shape because gravitational forces are negligible compared to the capillary ones. The droplet volume V can then

Droplet Wetting and Evaporation. DOI: http://dx.doi.org/10.1016/B978-0-12-800722-8.00008-4

be expressed using only two of the following geometrical parameters: the contact line radius R, spherical radius R_s, contact angle θ, and drop height h. As a function of the contact line radius and the contact angle, the volume can be expressed as follows:

$$V(R,\theta) = \frac{\pi R^3}{3} \frac{(1-\cos\theta)^2(2+\cos\theta)}{\sin^3\theta} \tag{8.1}$$

8.1 Theory

The evaporation phenomenon occurs at the liquid/vapor interface as long as the saturated conditions are not reached. At the molecular scale, one observes the exchange of molecules going from the liquid to the gas phase and vice versa (Figure 8.1). If the gas phase is saturated, the same number of molecules are both entering and leaving the liquid phase. It means that the equilibrium condition is reached. If the same number of molecules are not entering and leaving the liquid phase, then the equilibrium is not satisfied. For example, more molecules reach the vapor phase than the liquid phase in the case of evaporation. The macroscopic scale description is thus a mass flux crossing the interface. Therefore, evaporation exists as soon as the vapor partial pressure in the gas is lower than the saturation pressure. In these conditions a droplet of volatile liquid will evaporate into air. The transition for a molecule from the liquid to the gas state is related to energy consumption, which can be thermodynamically quantified by the phase change enthalpy of vaporization L_V. The evaporation is thus an endothermic process at the opposite of condensation that is exothermic.

The natural evaporation of a droplet is described as a process controlled by the vapor diffusion into the surrounding gas phase. When a droplet evaporates, the vapor concentration around the droplet is not homogeneous. While the

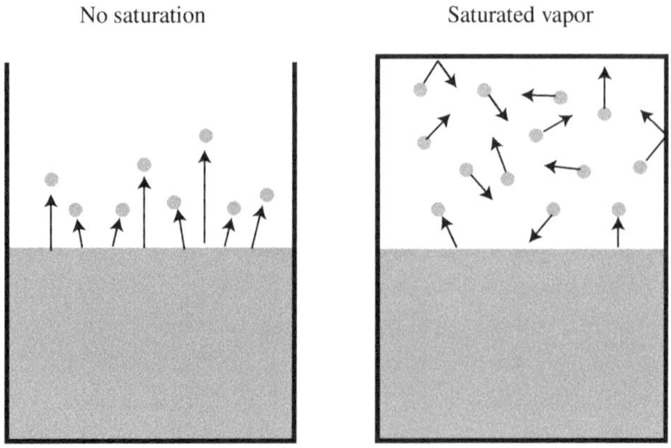

FIGURE 8.1 Schematic illustration of the diffusive evaporation of a liquid into a gas phase with or without saturated conditions.

evaporation mechanism ensures saturated conditions at the interface—that is, the vapor concentration at the interface c_0 is thus equal to the saturated concentration c_S—the vapor concentration far from the droplet reaches the value $c_\infty \approx Hc_S$, where H is the gas phase humidity. The vapor concentration difference in the surrounding air $\Delta c = c_S - c_\infty$ is then the leading phenomenon driving the evaporation. The field of concentration is thus driven by the mass diffusion balance (Eq. (8.2)):

$$\frac{\partial c}{\partial t} = D\Delta c \tag{8.2}$$

where D is the diffusion coefficient of the vapor in the gas phase. By definition, droplet evaporation is a nonsaturation process. Droplets evaporation is classically considered as a quasi-steady-state process. This condition is satisfied when the diffusive time of the vapor into the gas phase t_D is small compared to the evaporation time t_F (Eq. (8.3)):

$$\frac{t_D}{t_F} = \frac{R^2}{D\, t_F} \approx \frac{c_S - c_\infty}{\rho_l} \approx \frac{c_S(1-H)}{\rho_l} \ll 1 \tag{8.3}$$

In this case, the vapor concentration field quickly adjusts compared to the time needed to evaporate. This ratio of times can be simplified as a ratio of densities where the vapor concentration at the drop interface c_0 is much smaller than the liquid density ρ_l. As an example, for a microliter water droplet evaporating into air at room temperature, the diffusion time is of the order of magnitude of 10^{-2} s, while the evaporation time is much more important, of the order of magnitude of 10^2 s. As presented in Table 8.1, the ratio of characteristic times for water is between 10^{-6} and 10^{-9} s, depending on the surrounding air humidity. Table 8.1 informs that even for very volatile liquids like HFEs, the quasi-steady assumption remains valid at room temperature.

TABLE 8.1 Quasi-Steady Assumption for Several Binaries under 25°C and 1 atm

Liquid	Surrounding Gas	P_{sat} (hPa)	c_S (kg m^{-3})	H	ρ_l (kg m^{-3})	t_D/t_F
Water	Humid air (95%)	2.3	1.71E − 03	0.95	1000	8.5E − 08
Water	Humid air (50%)	2.3	1.71E − 03	0.50	1000	8.5E − 07
Water	Dry air	2.3	1.71E − 03	0	1000	1.7E − 06
Ethanol	Air + ethanol vapor (50%)	5.8	1.10E − 02	0.50	789	7.0E − 06
Ethanol	Air	5.8	1.10E − 02	0	789	1.4E − 05
Methanol	Air	13	1.72E − 02	0	791	2.2E − 05
HFE-7100	Air	267	2.76	0	1482	1.9E − 03
HFE-7000	Air	642	5.30	0	1401	3.8E − 03

Under these conditions, the vapor concentration field is the solution of the Laplace equation:

$$\Delta c = 0 \tag{8.4}$$

The evaporation mass flux at the droplet interface can be modeled by the diffusion of molecules at the liquid/gas interface to the surrounding air. The vapor transport is thus only a mass diffusion. Fick's law describes the outward mass flux called j in any point of the interface:

$$j = -D[\vec{\nabla}\rho]_i \cdot \vec{n} \tag{8.5}$$

where $\vec{n}$ is the normal vector to the interface and $[\vec{\nabla}\rho]_i$ is the vapor concentration gradient normal to the interface. The subscript i points out that the gradient is defined from the interface. Because the droplet mass variation is the consequence of the mass flux evaporated, the evaporation rate $\mathrm{d}m/\mathrm{d}t$ is the spatial integral of the local mass flux j over a surface element $\mathrm{d}s$ (Eq. (8.6)):

$$\frac{\mathrm{d}m}{\mathrm{d}t} = \iint j \, \mathrm{d}s \tag{8.6}$$

This description considers droplet evaporation as quasi-steady and controlled by the vapor diffusion in the gas phase. It also implies the isothermia of the system. This modeling was first proposed by Maxwell, in 1877, when he investigated aerosol droplets. His study is the basis of the next modeling used by other authors for a sessile droplet configuration (Deegan et al., 2000; Hu and Larson, 2002; Popov, 2005). Assuming a spherical droplet of radius R_s and evaporating into inert gas, Maxwell in 1877 and Langmuir in 1918 solved the problem with different approaches. Maxwell used an analogy with electrostatics, while Langmuir used an analogy with heat conduction. In both case, the evaporation mass can be expressed as follows:

$$-\frac{\mathrm{d}m}{\mathrm{d}t} = 4\pi R_s D(c_S - c_\infty) \tag{8.7}$$

The local evaporation mass flux is uniform at the droplet interface. It is worth noting that the global evaporation evaporation mass flux is proportional to the droplet radius R_s and not to the droplet surface area, as one could have first intuited. Solving Eq. (8.7), assuming all parameters are constant, leads to the drop radius evolution that is found to be proportional to the root square of time $R_s = \sqrt{t_F - t}$. This result is well-known as the R^2 law. One can deduce then that the droplet surface evolves linearly with time. We can deduce then that the droplet surface evolves linearly with time.

When a droplet evaporates when it comes into contact with a substrate, the solution seems more complicated because of the loss of symmetry and the existence of a triple line. Consider small drops that satisfy the spherical cap assumption (i.e., $R < L_c$). In that situation, most of the authors used the electrostatic analogy, as Maxwell did in 1877, to solve the problem (Picknett and Bexon, 1977; Deegan et al., 2000; Popov, 2005). Indeed, when evaporation is assumed to be stationary, the diffusion equation is similar to Poisson's equation (Eq. 8.4). The evaporation mass flux j and the vapor concentration c are, respectively, the electrical field E and the electrical potential V (Figure 8.2).

The local evaporation mass flux is then expressed as integrals of the Legendre function in toroidal coordinates (Lebedev, 1965). The general relation for the evaporation mass flux is complex, but an approximation near the triple line exists (Jackson, 1962):

$$j(r) \approx j_0 (R - r)^{-\lambda}$$
$$\lambda = \frac{\pi - 2\theta}{2\pi - 2\theta} \tag{8.8}$$

Although for a spherical droplet, the local evaporation mass flux profile at the interface was perfectly uniform, the presence of a substrate changes it considerably. Equation (8.8) shows that, the local evaporation mass flux changes along the interface (r being the distance from the drop center) as a function of the contact angle θ. This is illustrated in Figure 8.3 where the normalized local evaporation mass flux is plotted as a function of the normalized radius for different contact angle values from 0° to 157°.

The evaporation mass flux is normalized by the value of the evaporation mass flux at the apex $j(r = 0) = j_0$, while the radius is normalized by the contact line radius R. It appears that the local evaporation mass flux is mostly nonuniform, except when the contact angle is 90°. For that specific contact angle value, the

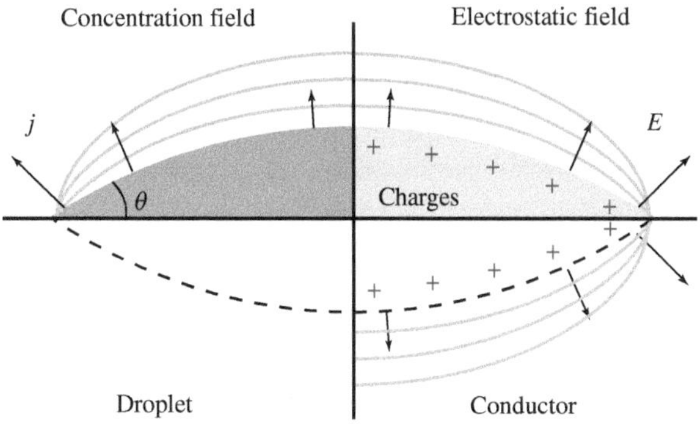

FIGURE 8.2 Analogy of a sessile droplet with an evaporation mass flux j and a conductor with the same shape that has an electric field E and the symmetry with the substrate used to solve the divergence at the triple line (Carle, 2014).

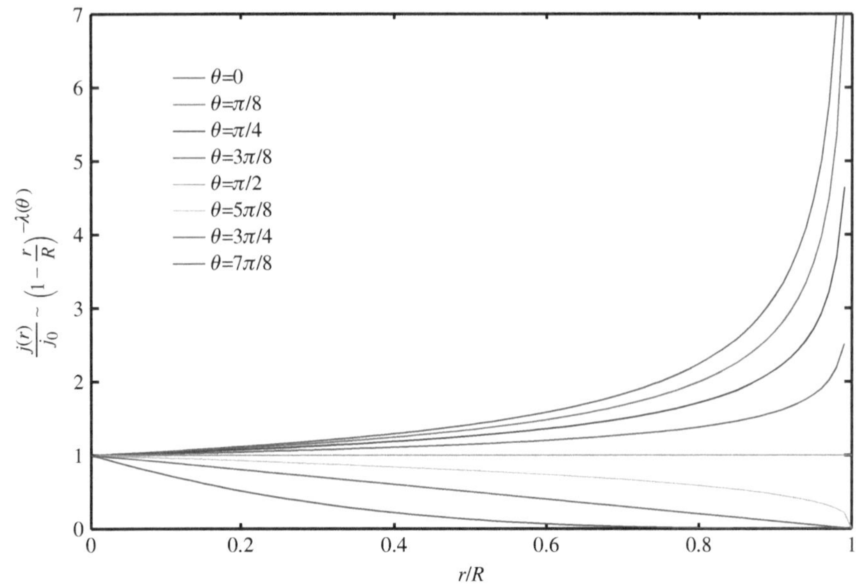

FIGURE 8.3 Dimensionless profile of the local diffusive evaporation mass flux as a function of the contact angle.

evaporative mass flux is perfectly uniform around the droplet. This is in accordance with the profile obtained for a spherical droplet. Three situations for the diffusive evaporation mass flux profile can be extracted from Figure 8.2:

- $0° < \theta < 90°$: The evaporation mass flux is maximum at the triple line and minimum at the droplet apex. Indeed, the evaporative mass flux increases near the triple line to diverge when $r = R$, as observed in Eq. (8.8). This mass flux divergence can be interpreted as a "pointe" effect.
- $\theta < 90°$: the evaporation mass flux is perfectly uniform along the drop interface.
- $90° < \theta < 180°$: the evaporation mass flux is maximum at the apex and minimum at the triple line. This can be understood as a "corner" effect. When θ tends to 180°, the droplet is almost spherical and sits on the substrate. The molecules cannot easily leave the triple line due to the very tiny gap between the interface and the surface.

The integration of the local evaporation mass flux at the spherical cap interface allows us to obtain the evaporation rate as follows:

$$-\frac{\mathrm{d}m}{\mathrm{d}t} = \int_0^R j(R)\sqrt{1 + (\partial_r h)^2}\, 2\pi r \mathrm{d}r \tag{8.9}$$
$$= \pi R D(c_S - c_\infty) f(\theta)$$

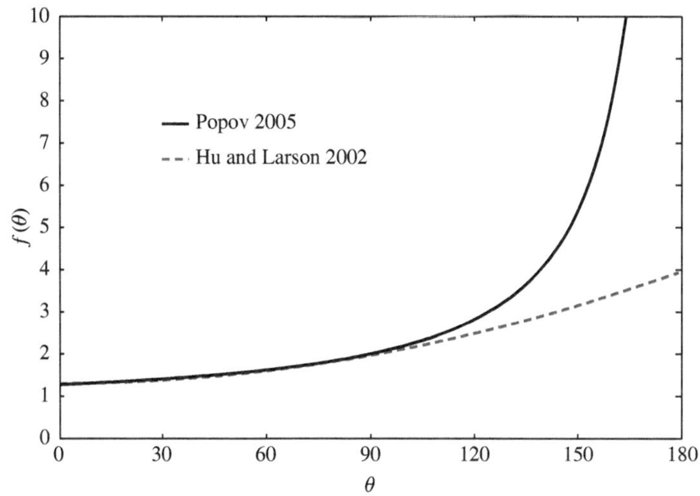

FIGURE 8.4 Comparison of two expressions of the function of the contact angle $f(\theta)$.

where $h(r, t)$ is the droplet profile, t is the time, r is the radial coordinate, and $f(\theta)$ is a function of the contact angle:

$$f(\theta) = \frac{\sin\theta}{1+\cos\theta} + 4\int_0^\infty \frac{1+\cosh 2\theta\tau}{\sinh 2\pi\tau}\tanh((\pi-\theta)\tau)\mathrm{d}\tau \tag{8.10}$$

The analytical expression of $f(\theta)$ is valid whatever the value of the contact angle between 0° to 180° (Popov, 2005). In 2002, Hu and Larson, numerically obtained an approximated polynomial expression of this function for contact angles between 0° and 90° as follows:

$$f_{\mathrm{HL}}(\theta) = 1.3 + 0.27\theta^2 \tag{8.11}$$

Both expressions of $f(\theta)$ are plotted in Figure 8.4 and show the quantitative agreement below 90°. Above 90°, the Hu and Larson polynomial expression is no longer valid and underestimates $f(\theta)$.

In the situation of a droplet evaporating with a constant wetting radius (i.e., in pinning mode), it is possible to combine Eqs (8.1) and (8.9) to obtain the contact angle evolution as follows:

$$\frac{\mathrm{d}\theta}{\mathrm{d}t} = -\frac{D(c_{\mathrm{S}} - c_\infty)}{\rho R^2}(1+\cos\theta)^2 f(\theta) \tag{8.12}$$

By integrating this equation, the purely diffusive evaporation time is obtained where θ_0 is defined as the initial contact angle (Eq. (8.13)):

$$t_{\mathrm{F}} = \frac{\rho R^2}{D\Delta c}\int_0^{\theta_0} \frac{\mathrm{d}\theta}{(1+\cos\theta)^2 f(\theta)} \tag{8.13}$$

In the situation where the contact angle is small enough—that is, $\theta \ll 1$ rad ($\theta \ll 57°$)—several previous equations can be simplified using Taylor expansion. Then, the contact angle appears to be linearly related to time (Eq. (8.14)) such as the droplet volume (Eq. (8.15)), where t_F is the total time of evaporation (Eq. (8.16)):

$$\theta = \theta_0 \left(1 - \frac{t}{t_F}\right) \tag{8.14}$$

$$V = \frac{\pi R^3 \theta_0}{4} \left(1 - \frac{t}{t_F}\right) \tag{8.15}$$

$$t_F = \frac{\pi \rho R^2 \theta_0}{16 D \Delta c} \tag{8.16}$$

Based on Eq. (8.15), we can see that in the case of a wetting droplet with $\theta < 57°$ and with a pinned triple line (constant radius), the evaporation rate remains constant during the entire evaporation because the volume decreases linearly.

8.2 Application

Let's consider experimentally now the evaporation of a bidistilled water droplet into air at room humidity, pressure and temperature (H = 47.5 ± 1%, P_a = 1015 HPa, T_a will be mentioned later). The droplet sits on an aluminum substrate that can be covered by two different nanocoatings in order to exhibit two different wetting conditions. With a SiOx coating, the initial contact angle is 67 ± 1° (hydrophilic condition), while with a PFC coating, the contact angle is 133 ± 1° (hydrophobic condition). The surface roughness in both cases is identical and enables a pinning of the triple line during almost all of the evaporation. The droplet has an initial volume of 3.64 μL ± 2% and evaporates with a constant radius of 1.44 mm ± 2%. The substrate is equipped with a heating system associated with a thermal regulation in order to be able to impose a temperature.

In the experiments presented here, we investigate the evaporation of a drop on a substrate at different temperatures, from the room temperature (without heating) to about 45°C. The experimental data are compared with the model of drop evaporation presented in Section 8.1, assuming that the evaporation is a quasi-steady state process controlled by the diffusion of vapor into air and drop isothermia. The assumption of drop isothermia consists in the modeling to take the vapor concentration at the interface equals to the saturated concentration at the substrate temperature. The comparison between the experiments and the model permits to investigate the validity of the model at room temperature and with a slight increased of the substrate temperature.

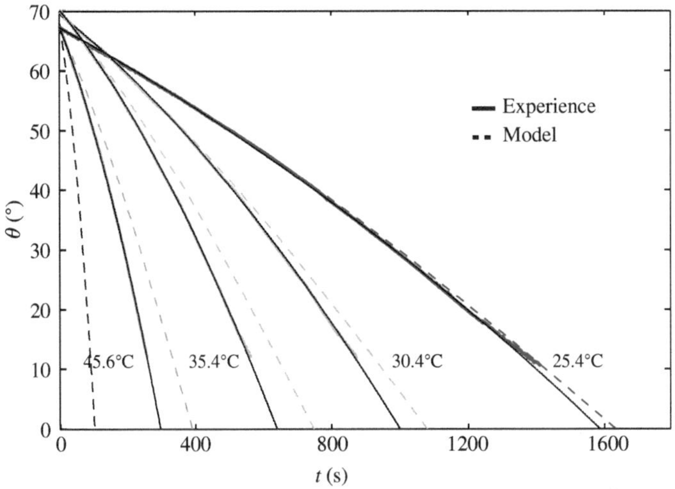

FIGURE 8.5 Hydrophilic case—evaporation of a pinned water droplet on an aluminum substrate coated with a nanolayer of SiOx as a function of the substrate temperature: contact angle variation versus time.

The contact angle evolution for the hydrophilic case is plotted in Figure 8.5. The surrounding air temperature is 25.4°C. Because the initial contact angle is above 57°, the evolution of the contact angle is not linear. Based on Eq. (8.12), the theoretical evolution based on a purely diffusive model is plotted. As evidenced for a substrate that is not heated, the theoretical model perfectly fits the experimental evolution. As soon as the substrate temperature is increased, the experimental evolutions start to shift from the model. With a substrate at room temperature, the gap between the experimental total time of evaporation and the theoretical one predicted by Eq. (8.13) is only 3.5%. For substrates heated at 30.4°C, 35.4°C, and 45.6°C, the gaps are 9%, 17%, and 32%, respectively. It is worth noting that the purely diffusive model still succeeds to quantitatively describe the experiments at higher substrate temperatures. One observes that the purely diffusive model underestimates the experimental drop evaporation rate. Moreover, the difference between the model and the experiments increases with the substrate temperature. Indeed, some assumptions are relaxed when increasing the substrate temperature. The buoyancy convection in the gas phase is becoming more and more important as the substrate temperature increases (Saada, 2010, Sobac, 2012). The consideration of a purely diffusive model is no longer valid at high substrate temperatures. The incorporation of the effect of buoyancy convection in the vapor phase is necessary if a more accurate modeling is needed. This additional mechanisms will be discussed in Chapter 9. However, the assumption of drop

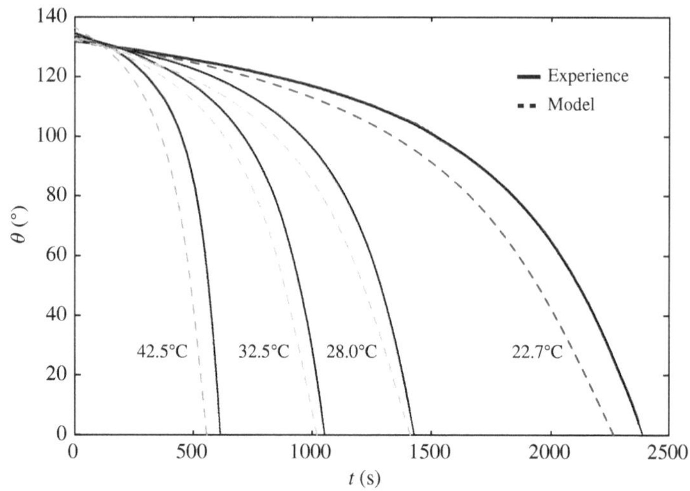

FIGURE 8.6 Hydrophobic case—evaporation of a pinned water droplet on an aluminum substrate coated with a nanolayer of PFC as a function of the substrate temperature: contact angle variation versus time.

isothermia still remains valid in the hydrophilic situation as numerically showed by Girard et al. (2006).

The contact angle evolution for the hydrophobic case is plotted in Figure 8.6. The surrounding air temperature is 22.7°C. For this second case as well, based on Eq. (8.12), the theoretical evolution based on a purely diffusive model can be plotted. As for the hydrophilic case, at room temperature without heating the substrate, agreement is very good, with a 4% difference in the total time of evaporation. Then, heating the substrate at 28.0°C, 32.5°C, and 42.5°C, the differences between the experimental and theoretical times are 2%, 9%, and 16%, respectively. One can note that the purely diffusive model still quantitatively succeeds to describe the drop evaporation at higher substrate temperatures in the hydrophobic situations. However, even if the deviation between experiences and theory increases with the substrate temperature, the deviation is less important compared to the hydrophilic situation. Moreover, one observes in this case, that the theory overestimates the evaporation rate.

Previously, to accurately describe the evaporation process, the hydrophilic situation showed that is was necessary to account for the buoyant convection of transport in the vapor phase because the diffusion-driven evaporation model globally underestimates the global evaporation rate. In the hydrophobic situation, the convective transport is still missing in the model, and yet the model overestimates the global evaporation rate. Hence, the hydrophobic situation highlights another mechanism acting during droplet evaporation that is not included in the model.

Indeed, the assumption of droplet isothermia is no longer justifiable in this situation as it was in the hydrophilic one. The thickness of the drop increases, and the temperature gradient at the interface is no longer negligible. Considering the local evaporative flux at the liquid–air interface, maximum evaporative flux occurs along the interface where the temperature is below the substrate temperature. Hence, the prediction, calculated for an interfacial temperature equal to the substrate temperature, overestimates the real contribution of diffusion transport during evaporation. The deviation due to this assumption appears to be even more important than the contribution of convective transport in the vapor phase. It clearly reveals the importance of the evaporative cooling effect on the evaporation rate even in natural evaporation (without heating the substrate) compared with forced evaporation (when the substrate is heated).

Glossary

Fick's law Law that relates the diffusive flux to the concentration under the assumption of steady state. It postulates that the mass flux goes from regions of high concentration to regions of low concentration with a magnitude proportional to the concentration gradient.

Partial pressure In a mixture of ideal gases, each gas has a partial pressure, which is the pressure the gas would have if it completely occupied the volume. The total pressure of a gas mixture is the sum of the partial pressures of each individual gas in the mixture.

Saturation A gas can be saturated in its own vapor; a liquid is saturated when a compound cannot be dissolved anymore.

Sessile Droplets resting on a support.

Spherical cap Portion of a sphere cut off by a plane.

References

Carle F., 2014. Flow Motion in Sessile Droplets: Evaporation and Nanoparticles Assembly (Ph.D. thesis). Aix-Marseille University.

Deegan, R.D., Bakajin, O., Dupont, T.F., Huber, G., Nagel, S.R., Witten, T.A., 2000. Contact line deposits in an evaporating drop. Phys. Rev. E 62 (1), 756.

Gelderblom, H., Marín, Á.G., Nair, H., van Houselt, A., Lefferts, L., Snoeijer, J.H., et al., 2011. How water droplets evaporate on a superhydrophobic substrate. Phys. Rev. E 83, 026306.

Girard, F., Antoni, M., Faure, S., Steinchen, A., 2006. Evaporation and marangoni driven convection in small heated water droplets. Langmuir 22 (26), 11085–11091.

Hu, H., Larson, R.G., 2002. Evaporation of a sessile droplet on a substrate. J. Phy. Chem. B 6 (106), 1334.

Jackson, J.D., 1962. Classical Electrodynamics, first ed. Wiley, New York, NY.

Langmuir, I., 1918. The evaporation of small sphere. Phys. Rev. 12, 368.

Lebedev, N.N., 1965. Special Functions and Their Applications. Prentice-Hall, Englewood Cliffs, NJ.

Maxwell, J.C., 1877, Diffusion Collected Scientific Papers, Encyclopaedia Brittanica. Cambridge, UK.

Picknett, R.G., Bexon, R., 1977. The evaporation of sessile or pendant drops in still air. J. Colloid Interface Sci. 61 (2), 336.

Popov, Y.O., 2005. Evaporative deposition patterns: spatial dimensions of the deposit. Phys. Rev. E 71, 036313.

Saada, M.A., Chick, S., Tadrist, L., 2010. Numerical investigation of heat and mass transfer of an evaporating sessile drop on a horizontal surface. Phys. Fluid. 22, 112–115. Available from: http://dx.doi.org/10.1063/1.3488676.

Sobac, B., Brutin, D., 2011. Triple-line behavior and wettability controlled by nanocoated substrates: influence on sessile drop evaporation. Langmuir 27, 14999–15007.

Sobac, B., Brutin, D., 2012. Thermal effects of the substrate on water droplet evaporation. Phys. Rev. E 86, 021602.

Chapter 9

CONVECTION

Florian Carle and David Brutin
Mechanical Engineering Department, Aix-Marseille University, Marseille, France

While in chapter 8, only diffusion is making the droplet evaporate, in this chapter, convection exists. Its origin can be thermal or capillary. The substrate can or cannot be heated, and gravity can be suppressed by performing experiments in microgravity for the sake of the explanation. Thus, in addition of diffusion, the flow in the surrounding gas is considered (Stefan flow), such as the flow inside the droplet (Marangoni flow). They can originate thermo-gravitational convection, thermo-capillary convection, and/or thermo-solutal convection. The complexity of the situation is that all these contributions can add themselves. In this chapter, each contribution is explained separately.

9.1 Basis

An interface between two phases (here a liquid and its vapor) is often thought of as a sharp infinitesimal boundary in the direction of its normal state. However, at the molecular scale, an interface is a transition region for physical values—that is, density—whose wideness depends on the temperature. At low temperature, the interface thickness is on the order of nanometers, but it can reach micrometer values near the fluid critical point.

Droplet Wetting and Evaporation. DOI: http://dx.doi.org/10.1016/B978-0-12-800722-8.00009-6

In the liquid phase, adjacent molecules attract one another with different interactions, such as van der Waals forces or hydrogen bonds. These interactions characterize a condensed phase. However, near the interface, fewer molecules attract a molecule, and the interaction is not isotropic. This leads to a tension in the direction of the interface: the surface tension. In a sense, surface tension can be seen as the force needed to bring a molecule from the inside of the liquid to the interface.

The surface tension depends on the nature of the liquid as well as on the temperature and pressure conditions. For a pure compound, the surface tension σ almost always decreases with an increase in temperature T, and it becomes zero at the critical point:

$$\sigma = \sigma_0 - \gamma(T - T_0) \tag{9.1}$$

$$\gamma = -\frac{\partial\sigma}{\partial T} \tag{9.2}$$

Marangoni effects are a consequence of surface tension and occur at the interface, inducing a flow. The most famous Marangoni effects can usually be observed in a glass of wine as tears along the side of the glass. Capillary forces make the wine rise on the side of the glass. During the rise, evaporation occurs, and the wine, which is composed of water and alcohol, evaporates. Due to vapor pressure differences, the alcohol (which has a lower surface tension than water) evaporates faster than water and locally changes the surface tension. The local increase in surface tension draws more fluid to the side of the glass to form droplets that will fall once gravity overcomes the capillary forces.

Flow induced by Marangoni effects can been induced by local changes in concentration as described above, but it can also be caused by temperature gradients, as is the case for droplets. Surface tension is strongly influenced by temperature, as seen in Eq. (9.1). Therefore, a temperature gradient along the droplet interface brings up tangential stress for the latter. The surface tension forces are not in equilibrium and will make the fluid flow from the hot area, which has low surface tension, to the cold area, which has high surface tension, until equilibrium is reached.

The tangential stress at the interface of a fluid in contact with a gas phase is zero. Therefore, the stress due to surface tension at the droplet interface is compensated by viscous friction at the interface (i.e., the surface flow will diffuse inside the droplet). Overall, in order to conserve mass through a vertical plane from the interface to the substrate, lower layers of fluids will have speeds that are the opposites of those in the upper layer (right side of Figure 9.1).

The whole phenomenon creates a complex flow inside the droplet. To this is added another flow pattern, such as thermo-gravitational flow (left side of Figure 9.1), which occurs due to the buoyancy forces and evaporation-induced outward flow. The flow motion that is observed in a droplet results from

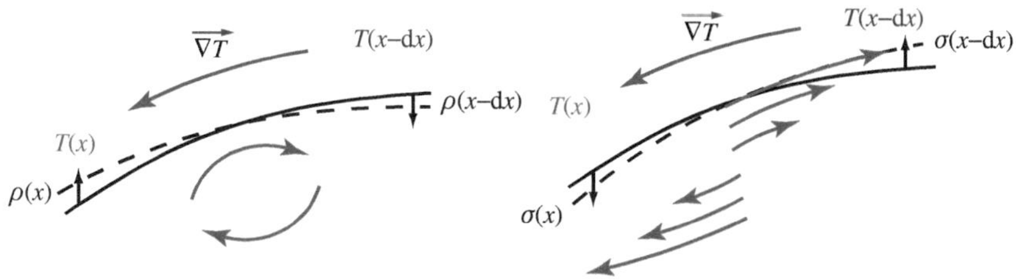

FIGURE 9.1 Schematic representation of two types of flows induced by temperature gradient in a droplet with different origins: gravitational flow (left) due to density variation and capillary flow (right) due to surface tension variation.

competition among all of the phenomena just mentioned. Thermo-gravitational flow tends to raise the fluid in the warm area, whereas thermo-capillary flow pushes the fluid in the cold area. The capillary length (Eq. (9.3)) can be used as the critical size; if the droplet size is more than twice this value, thermo-gravitational flows predominate:

$$L_C = \sqrt{\frac{\sigma_{LV}}{g(\rho_L - \rho_V)}} \tag{9.3}$$

For lower values, thermo-capillary flow occurs (Garnier, 2000). This rule only specifies the type of instability that can occur. Moreover, the dynamic Bond number Bd compares the Marangoni effects (described by the Marangoni number Ma) and natural convection (described by the Rayleigh number):

$$\mathrm{Ra} = \frac{\beta g \Delta T L^3}{\nu \alpha} \tag{9.4}$$

$$\mathrm{Ma} = \frac{-\frac{d\sigma}{dT}\Delta T L}{\mu \alpha} \tag{9.5}$$

$$\mathrm{Bd} = \frac{\mathrm{Ra}}{\mathrm{Ma}} = \frac{\beta g \rho}{-\frac{d\sigma}{dT}} \tag{9.6}$$

where β is the thermal expansion coefficient, L is the characteristic length, α is the thermal diffusivity, ν is the kinematic viscosity, and μ is the dynamic viscosity.

9.2 Spalding model

The previous models only take into account the concentration gradient as the cause of vapor diffusion away from the droplet; they ignore the vapor/air convection (see sub-chapter 2.1) and Stefan flow. Because the vapor density is lower than the liquid density, the volume of the liquid droplet is much less than the volume of the same quantity of vapor. Therefore, in addition to diffusion, the vapor that is created will push the saturated vapor further away to continue the evaporation. This phenomenon will create a flow outward from the droplet that is not taken into account by the purely diffusive models. This flow can be expressed by a variant of Maxwell's equation:

$$\begin{aligned}\frac{\mathrm{d}m}{\mathrm{d}t} &= 4\pi R^2\left(D_\mathrm{V}\frac{\mathrm{d}\rho_\mathrm{V}}{\mathrm{d}R} - \rho_\mathrm{V}U\right)\\ &= 4\pi R^2\left(D_\mathrm{V}\frac{\mathrm{d}\rho_\mathrm{V}}{\mathrm{d}R} - D_\mathrm{G}\frac{\rho_\mathrm{V}}{\rho_\mathrm{G}}\frac{\mathrm{d}\rho_\mathrm{G}}{\mathrm{d}R}\right)\end{aligned} \tag{9.7}$$

The first term inside the brackets is the expression of the diffusion (similar to the purely diffusive model), and the second term takes into account the vapor flow, where U is the flow velocity. This equation can be simplified (Sazhin, 2005) by assuming that the vapor and the gas diffusive coefficients, D_V and D_G, are equal:

$$\frac{\mathrm{d}m}{\mathrm{d}t} = -4\pi RD(\rho_\mathrm{V} + \rho_\mathrm{G})\ln(1 + B_\mathrm{M}) \tag{9.8}$$

$$B_\mathrm{M} = \frac{\rho_\mathrm{VS} - \rho_{\mathrm{V}\infty}}{\rho_\mathrm{GS}} = \frac{Y_\mathrm{VS} - Y_{\mathrm{V}\infty}}{1 - Y_\mathrm{VS}} \tag{9.9}$$

$$Y_\mathrm{VS} = \left(1 + \left(\frac{p}{p_\mathrm{sat}} - 1\right)\frac{M_\mathrm{G}}{M_\mathrm{V}}\right)^{-1} \tag{9.10}$$

This model, known as the Spalding model, is based on the Spalding mass number B_M, which compares the mass fraction of vapor/gas at the droplet interface $Y_\mathrm{V/G}$. $M_\mathrm{G/V}$ is the molar mass of air and vapor, and p/p_sat is the ambient pressure and the pressure of saturated fuel vapour.

Estimating the evaporation flux rate can lead to the time of evaporation and the quantity of vapor produced by a droplet. This knowledge is highly valuable for combustion applications in which fuel is sprayed into a combustion chamber in spherical droplets, so this model is commonly used in the combustion community. The model is based on similar assumptions to the purely diffusive models: isothermal spherical droplets vapor/liquid equilibrium at the droplet surface with a gas boundary layer quasi-steady and pure diffusion following Fick's law. The droplet temperature can also been obtained by using numerical simulation to solve the mass and heat transfer equations for each phase.

9.3 Stefan flow

Stefan flow occurs in the surrounding gas, assuming a concentration difference. It's driven by the solutal convection. In order to distinguish the contribution of the Stefan flow from the purely diffusive one on the evaporation mass flux, we have to be able to remove all convection induced by gravity. In order to achieve this, experiments were performed in weightlessness.

Pure ethanol droplets are allowed to evaporate onto heated substrates from ambient temperature to +35°C with similar experimental setups. The two sets of experiments were conducted under terrestrial ($1g$) and reduced gravity (μg) conditions. The results of these experiments are plotted in Figure 9.2, which shows the global evaporation flux rate plotted as a function of the difference in temperatures of the substrate and the ambient air for both gravity levels. Each data point is the evaporation rate given by one single droplet.

These experimental data are compared to the quasi-steady, diffusion-controlled evaporation model (Hu and Larson, 2002; Popov, 2005) already presented in subchapter 2.1 by Eq. (9.9) and are plotted in red in Figure 9.2, defined as follows:

$$-\frac{dm}{dt} = \pi R D (c_S - c_\infty) f(\theta) \tag{9.11}$$

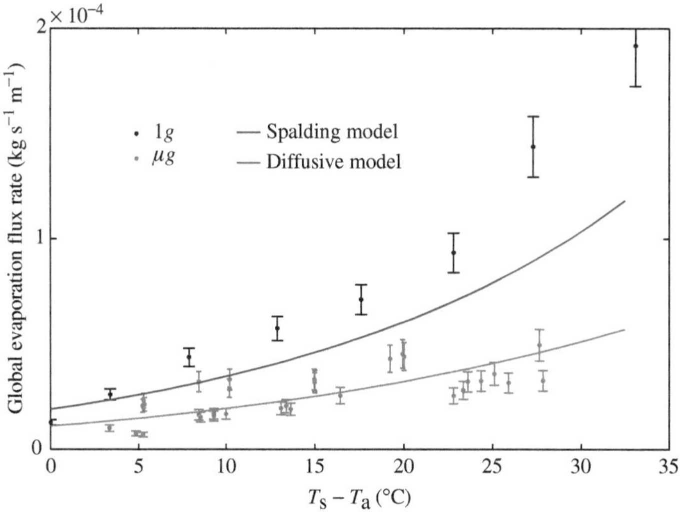

FIGURE 9.2 Global evaporation flux rate of ethanol droplets under normal and reduced gravity conditions as a function of the temperature difference between substrate and air at 20°C. Colored dots are experimental data with their error bars. The red line is the purely diffusive model, and the blue line is the Spalding model. (For interpretation of the references to color in this figure legend, the reader is referred to the online version of this book.)

The data are also compared with the Spalding model (Spalding, 1953; Bertoli and Migliaccio, 1999; Sazhin, 2006), which takes into account diffusion and Stefan flow and is shown in blue:

$$S = \pi\left(\frac{R^2}{4} + h^2\right) \tag{9.12}$$

$$\frac{\mathrm{d}m}{\mathrm{d}t} = SD(\rho_V + \rho_G)\ln(1 + B_M) \tag{9.13}$$

The first obvious observation is that the temperature strongly influences the evaporation flux rate. The flux rate increases with substrate temperature following a power law trend $\frac{1}{R}\frac{dm}{dt} = a(Ts - Ta)^b$. Under terrestrial gravity conditions (black dots) and at ambient temperature, the experimental value is nearly equal to the diffusive model value, consistent with other published results (Deegan, 2000; Gelderblom et al., 2011; Sobac and Brutin, 2011). However, when the substrate temperature differs from the ambient temperature, the evaporation rate becomes greater than the diffusion-controlled rate, and the difference between the two values increases with the substrate temperature. The error between the experimental evaporation rate and the diffusive model rapidly diverges according to a power law trend, and the deviation reaches up to 70% at $\Delta T = 35°\mathrm{C}$. Thus, this model globally underpredicts the experimental evaporation rate under normal gravity conditions.

The Spalding model (blue line) gives estimates of the evaporation flux rate closer to the experimental data compared to the purely diffusive model, but it still underpredicts the observed values. Stefan flow could explain the higher value of the evaporation flux rate compared to the theoretical value of purely diffusive evaporation. If, around the ambient temperature, a correct estimation of the value is obtained, then the deviation between the model and the experimental data reaches 38% at $\Delta T = 35°\mathrm{C}$, and Stefan flow is not the only component of evaporation.

Under reduced gravity conditions (red dots), Figure 9.2 shows that the diffusion-controlled evaporation model correctly predicts the experimental data, despite the significant dispersion of the data due to perturbations caused by the aircraft flight. This agreement highlights that, for the first time to our knowledge, this model has been shown to be valid and can account for the variations in the substrate temperature. In the absence of gravity effects, the evaporation can then be correctly assumed to be a quasi-steady, diffusion-controlled process, regardless of the substrate temperature and the isothermal temperature of the droplet. Indeed, the thermal gradient that develops inside an evaporating droplet driven by the latent heat of vaporization appears to be negligible when the droplet is sufficiently thin and when it is evaporated on a highly thermally conductive substrate (Girard et al., 2008; Sefiane and Bennacer, 2011; Sobac and Brutin, 2011). In this case, the model does not need to be corrected to include Stefan flow as the evaporation rate estimation is correct. However, Stefan flow will be taken into account in the numerical simulations to have a good representation of the vapour behaviour in the cell.

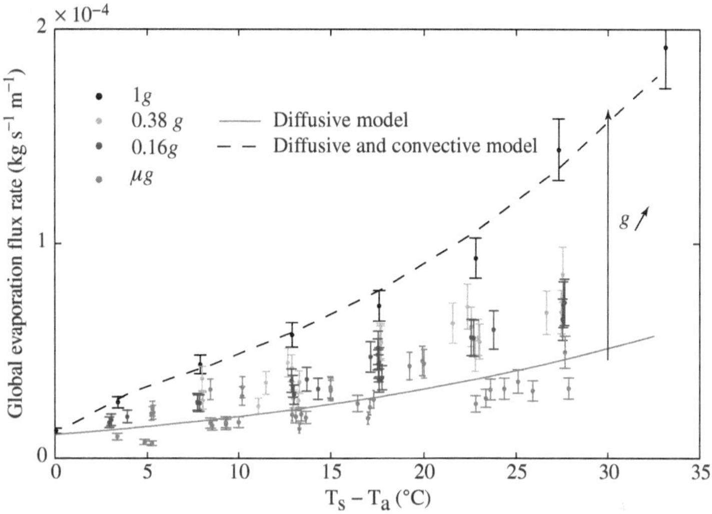

FIGURE 9.3 Global evaporation flux rate of ethanol droplets under various gravity levels—Earth (1*g*), Mars (0.38*g*), the moon (0.16*g*), and microgravity ($10^{-2}g$)—as a function of the temperature difference between substrate and air. Colored dots are experimental data with their error bars, and lines are the diffusive model (Eq. 9.14) and diffusive and convective models (Eq. 9.19). (For interpretation of the references to color in this figure legend, the reader is referred to the online version of this book.)

Consequently, the deviation noticed under the 1*g* conditions can only be due to the development of another mechanism accompanying the evaporation under the influence of gravity; this additional mechanism that contributes to the diffusion increases the evaporation rate. The contribution of this mechanism is greater when the substrate temperature is higher. These observations indicate that the mechanism is the natural convection in the vapor phase that develops due to buoyant forces. Experimental evidence can be found in Figure 9.3, which shows the global evaporation flux rate of an ethanol droplet as a function of the temperature difference for several gravity levels: Earth (1*g*), Mars (0.38*g*), the Moon (0.16*g*), and microgravity ($10^{-2}g$). Despite the data dispersion, a clear increase in the global evaporation flux rate can be seen as the gravity level increases, indicating an increase in natural convection contribution (Figure 9.3).

9.4 Influence of thermo-gravitational convection

In order to visualize the effects of convection, the profile in the vapor/air mixture surrounding the droplet has to be numerically investigated for a given configuration with identical boundary conditions.

The software COMSOL Multiphysics has been used with the Finite Element Method. All of the equations used in this code are expressed in their weak form, and the boundary conditions are fulfilled using constraints with Lagrange multipliers. The center of the computational domain is the droplet heart, and it is assumed that the problem is axis-symmetrical.

Figure 9.4 shows the diffusive evaporation of a droplet with Stefan flow in its vicinity and Marangoni flow on the inside, and the right side of the figure shows the results for evaporation plus natural convection. The colored contours represent the concentration in $mol \cdot m^{-3}$, and the streamlines show the velocity field. With pure diffusion only, the air velocity around the droplet is uniform and directed outward, except near the triple line, and the Stefan flow is strong due to the singularity of the evaporation profile. The vapor concentration is homogeneous and spherical around the droplet until it reaches the cell sides. When natural convection takes place in the test cell, the air motion around the droplet is far more complicated than in the diffusive case. Natural convection increases the mass transfer in the vapor phase, modifying the vapor field around the droplet by creating a flow motion and renewing the gas around the droplet. This phenomenon is clearly visible in the bottom picture, where a convective cell drags the vapor away from the interface. This naturally leads to an evaporation flux rate increase, as the evaporation is controlled by the difference in the vapor concentration at the interface and at infinity. Because the purely diffusive-controlled evaporation model does not take into account the natural convection, it naturally underestimates the experimental evaporation rate. However, it correctly describes the evaporation flux rate under microgravity, so the diffusive model can be used as a basis for the development of an empirical model that takes convection into account (Figures 9.4 and 9.5).

9.5 Empirical model

We demonstrated that at ambient temperature, the diffusive model is in good agreement with the experimental data, but at higher temperatures, the model underpredicts the evaporation flux rate with important deviations. To correctly describe the evaporation process is questionable when thermal effects related to evaporation are no longer negligible—that is, when the substrate plays the role of an insulator (David et al., 2007; Sefiane and Bennacer, 2011; Sobac and Brutin, 2011) or when it is heated. In such cases, the deviation has been shown to result from atmospheric convection around the thermal scene (Carle et al., 2013). Air, heated by the dry side of the substrate, rises due to buoyancy forces and modifies the concentration gradient that drives the evaporation.

Both diffusion and convection are significant here. Because these phenomena are linear, the evaporation flux rate E can be expressed as the sum of

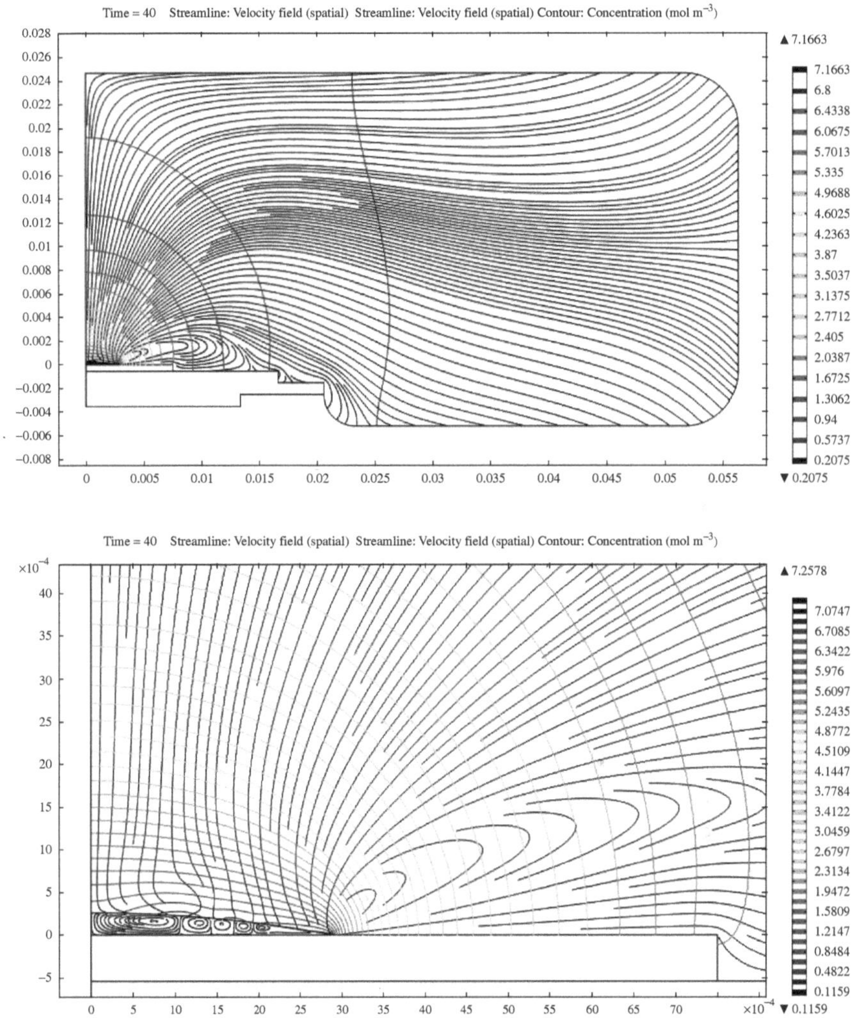

FIGURE 9.4 Numerical simulations performed on COMSOL Multiphysics of the vapor/air mixture around an evaporating droplet under reduced ($10^{-2}g$—diffusion, Stefan flow, Marangoni flow). The streamlines show the velocity field and contours concentrations (Semenov et al., 2010).

diffusive evaporation E_D, described by Eq. (9.14), and convective evaporation, E_C (Kelly-zion et al., 2011). Therefore, the dimensionless evaporation flux rate can be defined as follows:

$$E_D = -\frac{1}{R}\frac{dm}{dt} = \pi D \Delta c f(\theta) \approx 4D\Delta c \tag{9.14}$$

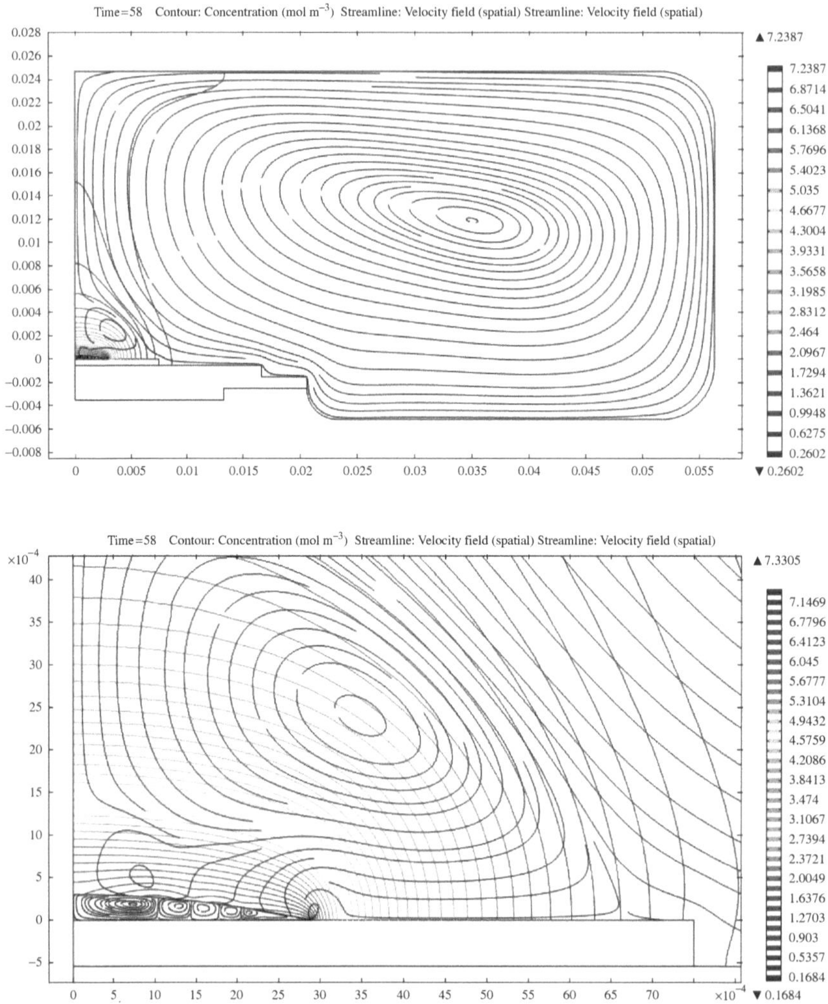

FIGURE 9.5 Numerical simulations performed on COMSOL Multiphysics of the vapor/air mixture around an evaporating droplet under normal gravity conditions (1-*g*-diffusion, Stefan flow, Marangoni flow, natural convection). The streamlines show the velocity field and contours concentrations (Semenov et al., 2010).

$$E = E_D + E_C \tag{9.15}$$

$$E^* = \frac{E}{E_D} = 1 + E_C^* \tag{9.16}$$

where E^*_C is the dimensionless evaporation rate due to convection. For this last part, the Rayleigh number, which compares thermal diffusivity over convection, is used to describe the convective contribution to evaporation:

$$\mathrm{Ra} = \frac{g\beta_T \Delta T R^3}{\nu_V \alpha} \tag{9.17}$$

$$\beta_T = \frac{1}{\rho_V}\frac{\partial \rho}{\partial T} \tag{9.18}$$

Rayleigh calculations are based on the vapor properties of the fluid and used to take into account convection into the vapor phase, diffusion being the limiting phenomenon of evaporation.

The Rayleigh number decreases as the dimensionless evaporation term E^*_C increases. To integrate the convective part into the model, a power law evolution for E_C as a function of the Rayleigh number is proposed based on experimental results provided in the literature (Carle, 2014). The coefficients C_1 and C_2 depend on the fluid evaporating. By integrating these coefficients inside the expression of the dimensionless global evaporation and by using the ideal gas law, the equation of the model can be expressed as:

$$\begin{aligned} E &= 4Dc_0(1 + C_1 \mathrm{Ra}^{C_2}) \\ &= \frac{4RDM_V P_V}{\hat{R}T_S}\left(1 + C_1\left[\frac{g\frac{\partial\rho}{\partial T}\Delta T R^3}{\rho_V \nu_V \alpha}\right]^{C_2}\right) \end{aligned} \tag{9.19}$$

where M is the molar mass, P is the pressure (considered at saturation as the substrate temperature for the vapor), the subscript V specifies the consideration of the vapor, the subscript S specifies the substrate, and $\hat{R}$ is the ideal gas constant.

Note that the developed model is similar to the one proposed in 2011 by Kelly-Zion et al. at ambient temperature only and tested on droplets of various sizes and various liquid volatilities (3MP, hexane, cyclohexane, and heptane). The authors based their model on the Grashof number, which compares the buoyant forces with the viscous forces. In their study, the authors mentioned the necessity to refine their model for small droplets at ambient temperature.

9.6 Community presents concerns

The vapor distribution around a sessile droplet has been very recently studied by experimental (Kelly-Zion et al., 2013; Dehaeck et al., 2014) and numerical

(Ait Saada et al., 2013; Pan et al., 2013) approaches in the case of purely diffusive evaporation. In 2013, Kelly-Zion et al. showed that vapor from sessile drops behaved differently than the commonly accepted diffusion-limited model. It turns out that the isotropic representation of the vapor concentration gradient around the droplet is not accurate, and models based only on diffusion are limited to describing the behavior under certain experimental conditions.

Models taking into account convection are developed in the combustion community to predict the evaporation time in order to obtain the best reagent concentration for combustion (Bertoli and Migliaccio, 1999; Sazhin, 2006). The Spalding evaporation model is based on diffusion, but it also takes into account Stefan flow. New models that can correctly predict diffusive and convective evaporation are emerging (Kelly-Zion et al., 2011; Carle et al., 2013), but their use is limited to particular fluids.

9.7 Applications

In sub-chapter 2.1, an application case with water droplets evaporation was presented. A few substrate temperatures were investigated in order to evidence the agreement between the experimental results and the theoretical modeling at room temperature but also the gap increasing as soon as the substrate is heated. In this chapter, we see the convection contribution to the evaporation rate that is also applied following for water droplet evaporation in the same conditions as those in sub-chapter 2.1.

For the sake of the application presented here with water droplets, the substrate temperature is investigated from room temperature up to 75°C. Figure 9.6 illustrates the normalized evolutions of the contact angle as a function of the temperature of the substrate for the two situations previously presented in sub-chapter 2.1: the hydrophilic case (A) and the hydrophobic case (B). The experimental data are compared with the dimensionless evolution of the contact angle predicted by the quasi-steady, diffusion-driven evaporation model (black dashed line) previously presented. The results show good agreement between the model and the experiments at ambient temperature.

However, one observes deviation of the curves with an increase in the temperature of the substrate: the higher the temperature of the substrate, the more deviate the curves. An increase in the deviated is more visible in the hydrophobic case compared with the hydrophilic case. A deviation of about 10% is reached between the contact angle at ambient temperature and at 75°C, $t/t_F = 0.8$ in the hydrophobic situation. Hence, the results reveal the influence of the thermal-linked mechanisms to the contact angle evolution. This influence is not accounted for in the model and could be due to the temperature gradient at the free surface

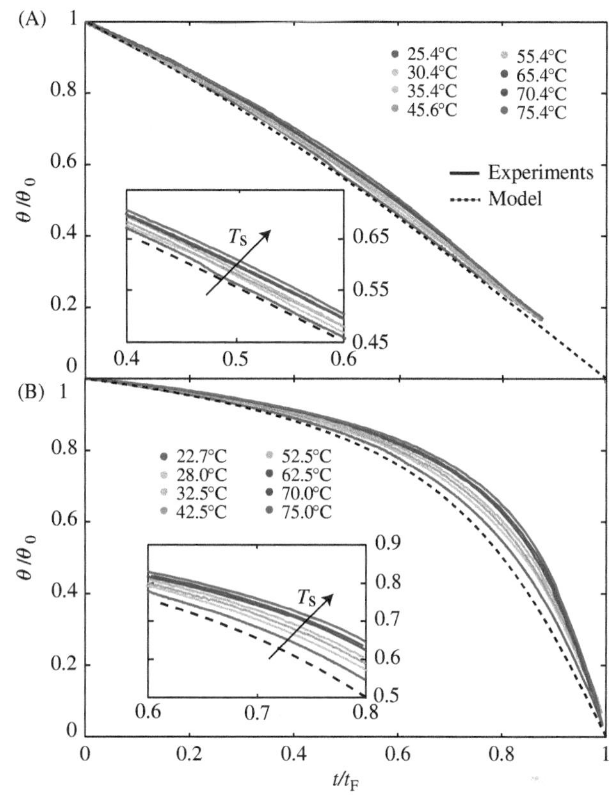

FIGURE 9.6 Influence of convective transport on the contact angle evolution—normalized contact angle θ/θ_0 versus the normalized time t/t_F as a function of the temperature of the substrate for the hydrophilic case (A) and the hydrophobic case (B).

induced by evaporative cooling, which leads to a variation of the saturated vapor concentration along the interface: the temperature gradient is a function of the wettability. This result affects both the dynamic contact angle and the atmospheric convection.

Glossary

Marangoni convection Mass transfer along an interface between two fluids due to surface tension gradient.

Stefan flow Transport phenomenon concerning the movement of chemical species by a flowing fluid that is induced to flow by the production or removal of the species at an interface. Stefan flow is distinct from Fick's law, but diffusion almost always also occurs in multispecies systems that are experiencing Stefan flow.

Buoyancy Upward force exerted by a fluid that opposes the weight of an immersed object.

References

Ait Saada, M., Chikh, S., Tadrist, L., 2013. Evaporation of a sessile drop with pinned or receding contact line on a substrate with different thermophysical properties. Int. J. Heat Mass Transf. 58, 197–208.

Bertoli, C., Migliaccio, M., 1999. A finite conductivity model for diesel spray evaporation computations. Int. J. Heat Fluid Flow 20 (5), 552–561.

Carle, F., 2014. Flow Motion in Sessile Droplets: Evaporation and Nanoparticles Assembly (Ph.D. thesis). Aix-Marseille University.

Carle, F., Sobac, B., Brutin, D., 2013. Experimental evidence of the atmospheric convective transport contribution to sessile droplet evaporation. Appl. Phys. Lett. 102 (6), 061603–061603–4.

David, S., Sefiane, K., Tadrist, L., 2007. Experimental investigation of the effect of thermal properties of the substrate in the wetting and evaporation of sessile drops. Colloids Surf. A Physicochem. Eng. Asp. 298 (1–2), 108–114.

Deegan, R.D., 2000. Pattern formation in drying drops. Phys. Rev. E 61 (1), 475.

Dehaeck, S., Rednikov, A., Colinet, P., 2014. Vapor-based interferometric measurement of local evaporation rate and interfacial temperature of evaporating droplets. Langmuir 30 (8), 2002–2008.

Garnier, N., 2000. Ondes non-linéaires à une et deux dimensions dans une mince couche de fluide (Ph.D. thesis). Université Paris 7 Diderot.

Gelderblom, H., Marín, Á.G., Nair, H., van Houselt, A., Lefferts, L., Snoeijer, J.H., et al., 2011. How water droplets evaporate on a superhydrophobic substrate. Phys. Rev. E 83 (2), 026306.

Girard, F., Antoni, M., Faure, S., Steinchen, A., 2008. Influence of heating temperature and relative humidity in the evaporation of pinned droplets. Colloids Surf. A Physicochem. Eng. Asp. 323 (1–3), 36–49.

Hu, H., Larson, R.G., 2002. Evaporation of a sessile droplet on a substrate. J. Phys. Chem. B 106 (6), 1334–1344.

Kelly-Zion, P.L., Pursell, C.J., Vaidya, S., Batra, J., 2011. Evaporation of sessile drops under combined diffusion and natural convection. Colloids Surf. A Physicochem. Eng. Asp. 381 (1–3), 31–36.

Kelly-Zion, P.L., Pursell, C.J., Hasbamrer, N., Cardozo, B., Gaughan, K., Nickels, K., 2013. Vapor distribution above an evaporating sessile drop. Int. J. Heat Mass Transf. 65, 165–172.

Pan, Z., Dash, S., Weibel, J.A., Garimella, S.V., 2013. Assessment of water droplet evaporation mechanisms on hydrophobic and superhydrophobic substrates. Langmuir, 15831–15841.

Popov, O., 2005. Evaporative deposition patterns: spatial dimensions of the deposit. Phys. Rev. E 71 (3), 036313.

Sazhin, S., 2005. Modelling of heating, evaporation and ignition of fuel droplets: combined analytical, asymptotic and numerical analysis. J. Phys. Conf. Ser. 22, 174.

Sazhin, S., 2006. Advanced models of fuel droplet heating and evaporation. Prog. Energy Combust. Sci. 32 (2), 162–214.

Sefiane, K., Bennacer, R., 2011. An expression for droplet evaporation incorporating thermal effects. J. Fluid Mech. 667, 260–271.

Semenov, S., Starov, V.M., Rubio, R.G., Velarde, M.G., 2010. Instantaneous distribution of fluxes in the course of evaporation of sessile liquid droplets: computer simulations. Colloids Surf. A Physicochem. Eng. Asp. 372 (1–3), 127–134.

Sobac, B., Brutin, D., 2011. Triple-line behavior and wettability controlled by nanocoated substrates: influence on sessile drop evaporation. Langmuir 27 (24), 14999–15007.

Spalding, D.B., 1953. The combustion of liquid fuel, 4th Symposium (Int.) on Combustion, 847–864, The Combustion Institute, Pittsburgh.

Chapter 10

LIQUID AND VAPOR PROPERTIES

Florian Carle and David Brutin
Mechanical Engineering Department, Aix-Marseille University, Marseille, France

The influence of the liquid and vapor properties is evidenced in two types of fluids: linear carbon chains, which were selected to be evaporated on heated substrates—alcohols from CH_3OH (methanol) to $C_7H_{15}OH$ (heptanol)—and alkanes from C_5H_{12} (pentane) to $C_{10}H_{22}$ (decane). The fluids, obtained from Sigma Aldrich, are anhydrous with a purity of 99% or higher and contain only one compound, except for hexane, that is a mixture of isomers. The fluids have been used without any treatments prior to experiments. All fluid properties can be found in Table 10.1. Properties are given at medium temperature ($T_m = 40°C$) unless otherwise indicated. The temperature dependencies of properties have been taken into account in all calculations. Moreover, throughout this chapter, properties either have been measured in our laboratory or are accompanied by references.

The droplet must have the same volume in order to compare only the fluid properties' influence of the evaporation dynamics. The droplets are deposited onto a substrate using a motorized dispenser that is controlled by a computer to create perfect spherical cap droplets with similar volume and similar initial geometric conditions (Kruss DSA30). Alcohols and alkanes are prone to moderate wetting due to their low surface

Droplet Wetting and Evaporation. DOI: http://dx.doi.org/10.1016/B978-0-12-800722-8.00010-2

TABLE 10.1 Fluids Properties at $T = 40°C$ and $P = 1013.25$ hPa, Except for P_{SAT} and $T = 20°C$

Fluid Properties		M_W	ρ_V	ν_V	L_V	T_{SAT}	γ	L_C	θ
Name	Formula	g mol^{-1}	kg m^{-3}	μm^2 s^{-1}	kJ kg^{-1}	K	mN m^{-1}	mm	(°)
Methanol	CH_4O	32	0.486	21.7	1111	333	22.5	1.70	21
Ethanol	C_2H_6O	46	0.322	28.5	857	347	22.3	1.70	20
Propanol	C_3H_8O	60	0.159	51.8	664	370	23.7	1.74	25
Butanol	$C_4H_{10}O$	74	0.082	81.3	423	390	24.6	1.78	28
Pentanol	$C_5H_{12}O$	88	0.059	119	505	411	25.6	1.81	29
Hexanol	$C_6H_{14}O$	102	0.238	27.3	486	430	24.5	1.73	36
Heptanol	$C_7H_{16}O$	116	0.010	611	440	488	26.2	1.82	37
Pentane	C_5H_{12}	72	3.37	2.17	357	309	16	1.65	18
Hexane	C_6H_{14}	86	1.27	5.47	335	342	18.4	1.72	20
Heptane	C_7H_{16}	100	0.48	13	317	371	19.3	1.72	21
Octane	C_8H_{18}	114	0.18	33	303	399	21.7	1.80	23
Nonane	C_9H_{20}	128	0.07	81.7	295	424	22.9	1.82	26
Decane	$C_{10}H_{22}$	142	0.03	208	263	447	23.9	1.85	28

The molar mass M_w, vapour density ρ_v, vapour viscosity η_v, latent heat L_v saturated temperature T_{sat} are obtained from (Sefiane et al., 2010), while surface tension gamma is from (NIST, 2013) exceptfor heptane and decane (CAMEO, 2013).

tension, and they therefore induce small contact angles. The lowest contact angle was $\theta_{pentane} = 18°$, and the highest contact angle was $\theta_{heptanol} = 37°$. Alcohol droplets are allowed to evaporate onto PFC substrates, whereas alkane droplets are deposited onto PTFE substrates (Table 10.1). Substrates are chosen so that their surface energies induce good wetting. They had low initial contact angles, and there was pinning of the triple line during a large part of the evaporation process. However, the dissolution of the PFC substrates by alkanes forces us to use PTFE substrates instead for this type of liquid. Due to the small thickness of the substrates, the substrates thermal conductivity difference does not influence the temperature gradient imposed by the heater and the ambient air (negligible 0.6% temperature decrease at the interface heater/air for a set point at 30°C for the PTFE substrate compared to the PFC substrate) and therefore the evaporation rate.

Once all of the fluids were evaporated at temperatures from ambient to +35°C above ambient, the evaporation flux rate was measured for each case and divided by the initial radius R_0 to compare all experimental data. This quantity is hereafter referred to as the global evaporation flux rate. Figure 10.1 shows the global evaporation flux rate as a function of the number of carbon atoms in each molecular chain for various temperatures. The evaporation flux rate decreases as the length of the molecular chain increases. This observation is valid for alcohols and alkanes and, to some extent, for all temperatures. It is well known that most of the evaporation occurs at the triple line for sessile droplets with low contact angles (Hu and Larson, 2002). If hemispheric droplets ($\theta = 90°$) have a uniform evaporation flux profile along the interface,

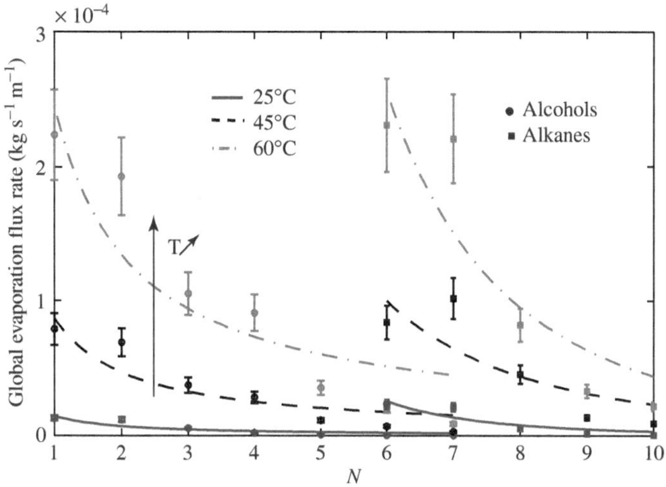

FIGURE 10.1 Chain length dependence of the evaporation flux rate for alcohols (circles) and alkanes (squares) at 25°C, 45°C, and 60°C in an air at 25°C, where N is the number of carbons in the chain. The lines fitting the data follow the power law ax^b.

droplets with small contact angles ($\theta < 40°$) present a singularity in the evaporation flux profile near the contact line. In this case, the evaporation rate is proportional to the droplet radius, and the evaporation flux rates of droplets with various radii can be compared by dividing these values by the initial radius (Eq. (10.1)):

$$\begin{aligned} E_C &= -\frac{1}{R}\frac{\mathrm{d}m}{\mathrm{d}t} \\ &= \pi D \Delta c f(\theta) \\ &\approx 4 D \Delta c \end{aligned} \tag{10.1}$$

$$f(\theta) = 1.3 + 0.27\theta^2 \tag{10.2}$$

The sessile droplet evaporation flux singularity can be explained by the increased probability of a molecule escaping the liquid phase if it is located near the contact line when driven only by a random walk (Deegan, 2000). Therefore, a substantial part of the liquid is evaporating at this location, and in order to conserve mass, a liquid flow is induced inside the droplet from the apex to the triple line. If the size of the molecule increases, the random walk will be influenced by the reduced freedom of the molecule (Huang et al., 2005), leading to a lower probability of escaping the liquid phase.

This effect is also enhanced by the fact that longer carbon chains have less freedom and tend to entangle (Fox and Loshaek, 1955) more easily, which increases the viscosity (Table 10.1). Indeed, the viscosity will decrease the liquid flow inside the droplet, thereby reducing the evaporation. The empirical model needs to take into account these changes in fluid properties.

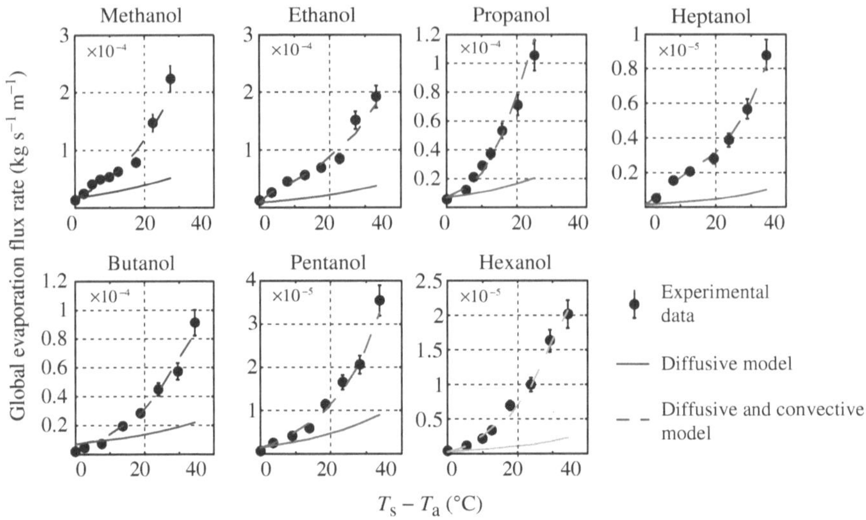

FIGURE 10.2 Global evaporation flux rates of alcohols droplets as a function of temperature difference between ambient air at 25°C and substrate.

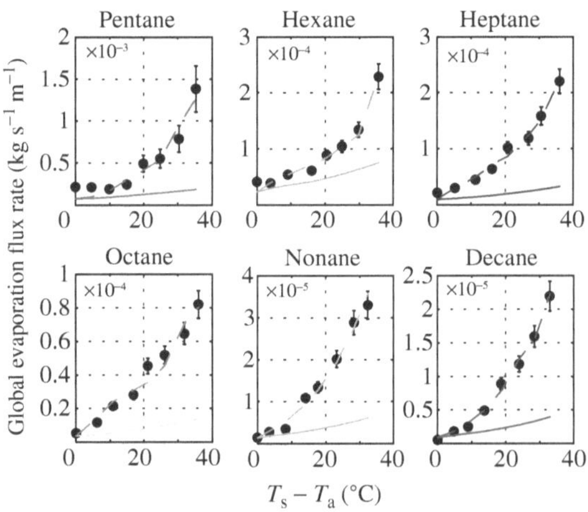

FIGURE 10.3 Global evaporation flux rates of alkanes droplets as a function of temperature difference between ambient air at 25°C and substrate. The legend is the same as Figure 10.2.

Figures 10.2 and 10.3 provide a closer look at each individual fluid. The global evaporation flux rate is plotted as a function of the temperature between the substrate and the ambient air. The error bars on the experimental data have been calculated as the quadratic sum of all of the measurement errors. All experiments can actually be considered quasi-steady because the calculated diffusion time is smaller than the measured evaporation time, $t_D/t_E = L^2/\alpha t_E \approx 10^{-3}$. Plain color lines plotted in the figure represent estimates of the global evaporation flux rate by the quasi-steady diffusion-controlled evaporation model. In the case of sessile droplets with contact angles below 40°, the evaporation flux rate can be expressed as in Eq. (10.1).

D is the coefficient of diffusion of molecules inside air. For the most volatile compound $D_{\text{pentane}} = 8.7 \times 10^{-6}\ \text{m}^2\ \text{s}^{-1}$ in an air at 20°C, and for the less volatile compound $D_{\text{heptanol}} = 6.7 \times 10^{-6}\ \text{m}^2\ \text{s}^{-1}$. D is the coefficient of diffusion of molecules inside air. This coefficient increases linearly with temperature, with an average increase of 17% at 60°C compared to ambient temperature. $\Delta c = c_0 - c_\infty$ is the vapor concentration difference between the interface, which is assumed to be saturated at the substrate temperature $[c_0 = c_V(T_S)]$ and a point far from the droplet. The vapor concentration far from the droplet is considered to be zero due to the large characteristic length of the test cell. The last term, $f(\theta)$, is a function of the contact angle that can be approximated as 1.3 in the limit of small contact angles (Hu and Larson, 2005). This diffusion-driven evaporation model implemented with the temperature assumes that the droplets are isothermal at the substrate temperature. This assumption is proved to be correct because the temperature gradient that develops inside the droplet, which is driven by the latent heat of vaporization, can be ignored if the droplet is thin and evaporates on a highly thermally convective substrate, such as a thick aluminum heater with thin substrates (Girard et al., 2008).

At ambient temperature, the diffusive model is in good agreement with the experimental data (black dots), but at higher temperatures, the model underpredicts the evaporation flux rate with deviations of up to 100% at +35°C. The model's ability to correctly describe the evaporation process is questionable when thermal effects related to evaporation are no longer negligible—that is, when the substrate plays the role of an insulator (David et al., 2007; Sefiane and Bennacer, 2011; Sobac and Brutin, 2011) or when it is heated. In such cases, the deviation has been shown to result from atmospheric convection around the thermal scene (Carle et al., 2013). Air, heated by the dry side of the substrate, rises due to buoyancy forces and modifies the concentration gradient that drives the evaporation.

Both diffusion and convection are significant in our problem. Because these phenomena are linear, the evaporation flux rate E can be expressed as the sum of diffusive evaporation E_D, described by Eq. (10.1), and convective evaporation, E_C (Kelly-Zion et al., 2011). Therefore, the dimensionless evaporation flux rate can be defined as follows:

$$E = E_D + E_C \tag{10.3}$$

$$E^* = E/E_D = 1 + E_C^* \tag{10.4}$$

where E_C^* is the dimensionless evaporation rate due to convection. For this last part, the Rayleigh number, which compares thermal diffusivity to convection, is used to describe the convective contribution to evaporation:

$$\mathrm{Ra} = \frac{g\beta_T \Delta T R^3}{\nu_V \alpha} \tag{10.5}$$

$$\beta_T = \frac{1}{\rho_V}\frac{\partial \rho}{\partial T} \tag{10.6}$$

Rayleigh calculations are based on the vapor properties of the fluid and used to take into account convection into the vapor phase, diffusion being the limiting phenomenon of evaporation.

Figure 10.4 shows the dimensionless convective evaporation flux rate E_C^* as a function of the Rayleigh number for the alcohol and the alkanes. The Rayleigh number ranges from less than 1 for heptanol to 23,000 for pentane. Figure 10.4 shows that the convective regime of each compound vapor is laminar. The critical Rayleigh number for turbulent flow induced by a horizontal heated plate is $\mathrm{Ra_C} = 2 \times 10^7$ (Incropera et al., 2011). In the case of short molecules, the motion of the vapor around the droplet is far from being purely diffusive ($\mathrm{Ra} > 200$ for alcohols and $\mathrm{Ra} > 1500$ for alkanes), and convection induces a strong concentration gradient between the droplet interface and the surrounding air. For the longest molecular chains, the Rayleigh numbers are low (less than 1), but there is still a significant deviation between the purely diffusive model and the global evaporation flux rate (Figure 10.2).

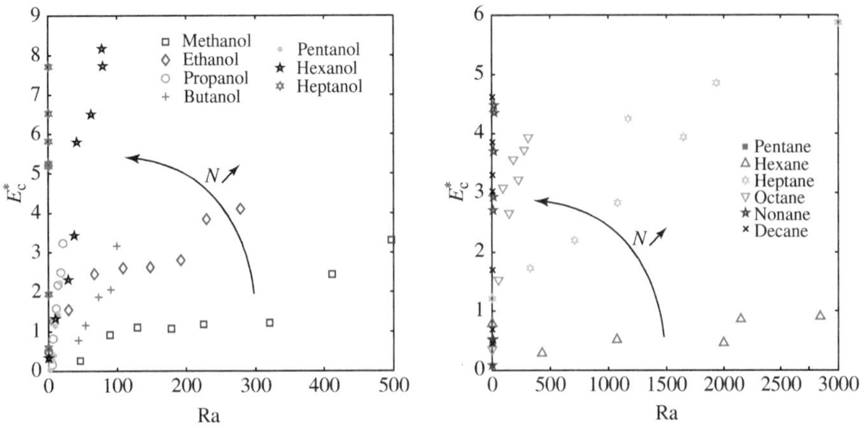

FIGURE 10.4 Variation of the dimensionless evaporation term E_c^* as a function of the Rayleigh number for alcohols (left) and alkanes (right).

In a general way, the Rayleigh number decreases with the number of carbons, N, in the chain as the dimensionless evaporation term E_{C}^{*} increases. This is true for alkanes and most alcohols (except pentanol and hexanol). To integrate the convective part into the model, each set of data was fitted with a power law ($E_{\mathrm{C}} = C_1\ \mathrm{Ra}^{C2}$), giving a pair of coefficients C_1 and C_2 per fluid that can be related to the molar mass.

The first coefficient, C_1, strongly increases with the number of carbon atoms and can be fitted by an exponential law ($C_1 = \varphi M^{\psi}$). In contrast, the coefficient C_2 undergoes a lesser variation and can be assumed to be constant close to 1. These coefficients depend only on the physical properties because no other experimental parameters have been changed. This gives us the coefficients for C_1 for alcohols and alkanes are presented in Table 10.2.

By integrating these coefficients inside the expression of the dimensionless global evaporation and by using the ideal gas law, the equation of the model can be expressed as

$$E = 4Dc_0(1 + C_1\mathrm{Ra}) = \frac{4RDM_{\mathrm{V}}P_{\mathrm{V}}}{\hat{R}T_{\mathrm{S}}}\left(1 + \frac{\varphi M^{\psi} g\frac{\partial \rho}{\partial T}\Delta T R^3}{\rho_{\mathrm{V}}\nu_{\mathrm{V}}\alpha}\right) \tag{10.7}$$

where M is the molar mass, P is the pressure (considered at saturation as the substrate temperature for the vapor), the subscript V specifies the consideration of the vapor, the subscript S specifies the substrate, and $\hat{R}$ is the ideal gas constant. φ and ψ are two coefficients dependent on the fluid.

A good agreement between the experimental data and the diffusive and convective model is observed. At ambient temperature, the model only takes into account the convective part because the temperature gradient is null. However, as soon as the substrate temperature increases, the convective part is added to the diffusive part, with the Rayleigh number determined by the empirical coefficients. Overall, the good agreement between the model and the experimental data can be estimated using Figure 10.5, where the percentage of error (calculated between the experimental data and the use of the fit) is plotted as a function of the temperature difference between the substrate and the ambient air. For most of the fluids, the error is less than 10% (dashed black lines). However, larger errors can be seen at ambient and low

TABLE 10.2 Coefficients for Alcohols and Alkanes Assuming a Power.

Coefficients for C_1	Alcohols	Alkanes
φ	1.72×10^{-9}	2.50×10^{-32}
ψ	4.13	1.48

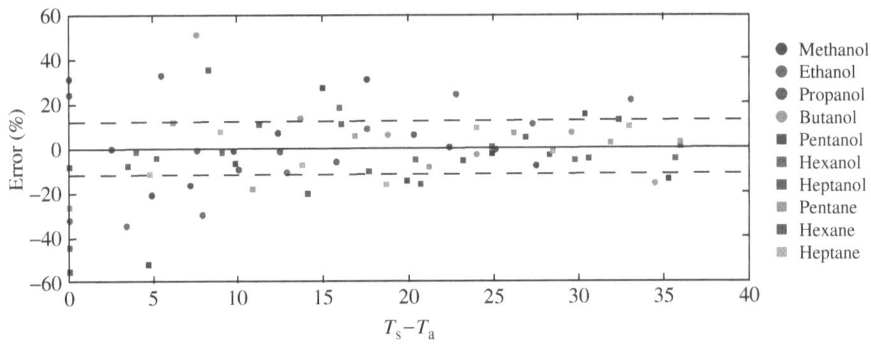

FIGURE 10.5 Percentage of error in the model as a function of the temperature difference between the substrate and the ambient air for alcohols and alkanes.

temperatures. At these temperatures, the droplets are in a transitory phase between diffusive and convective evaporation, and the model is not accurate enough to provide a good estimation. Moreover, the diffusive model does not take into account the convection induced either by the Marangoni forces in the vicinity of the droplet or by the vapor diffusion. However, when the temperature increases, the average error decreases, and the model can be accurately used.

Note that the developed model is similar to the one proposed in 2011 by Kelly-Zion et al. at ambient temperature only and tested on droplets of various sizes and various liquid volatilities (3MP, hexane, cyclohexane, and heptane). The authors based their model on the Grashof number, which compares the buoyant forces with the viscous forces. In their study, the authors also mentioned the necessity to refine their model for small droplets at ambient temperature.

The empirical model developed in this chapter takes into account only diffusion and convection, while ignoring all the flow dynamics that can occur inside the droplet and that can influence the evaporation dynamics. Work is still ongoing on this macroscopic approach to address this issue.

References

Carle, F., Sobac, B., Brutin, D., 2013. Experimental evidence of the atmospheric convective transport contribution to sessile droplet evaporation. Appl. Phys. Lett. 102 (6), 061603–061603–4.

CAMEO 2013. Database chemicals v. 2.4.1. <http://cameochemicals.noaa.gov/>.

David, S., Sefiane, K., Tadrist, L., 2007. Experimental investigation of the effect of thermal properties of the substrate in the wetting and evaporation of sessile drops. Colloids Surf. A Physicochem. Eng. Asp. 298 (1–2), 108–114.

Deegan, R.D., 2000. Pattern formation in drying drops. Phys. Rev. E 61 (1), 475.

Fox, T.G., Loshaek, S., 1955. Isothermal viscosity–molecular weight dependence for long polymer chains. J. Appl. Phys. 26 (9), 1080–1082.

Girard, F., Antoni, M., Faure, S., Steinchen, A., 2008. Influence of heating temperature and relative humidity in the evaporation of pinned droplets. Colloids Surf. A Physicochem. Eng. Asp. 323 (1–3), 36–49.

Hu, H., Larson, R.G., 2002. Evaporation of a sessile droplet on a substrate. J. Phys. Chem. B 106 (6), 1334–1344.

Hu, H., Larson, R.G., 2005. Analysis of the microfluid flow in an evaporating sessile droplet. Langmuir 21, 3963–3971.

Huang, D., Simon, S.L., McKenna, G.B., 2005. Chain length dependence of the thermodynamic properties of linear and cyclic alkanes and polymers. J. Chem. Phys. 122 (8), 084907.

Incropera, F.P., Lavine, A.S., DeWitt, D.P., 2011. Fundamentals of Heat and Mass Transfer. John Wiley & Sons, Hoboken, NJ.

Kelly-Zion, P.L., Pursell, C.J., Vaidya, S., Batra, J., 2011. Evaporation of sessile drops under combined diffusion and natural convection. Colloids Surf. A Physicochem. Eng. Asp. 381 (1–3), 31–36.

NIST 2013. Database number 69. <http://webbook.nist.gov/chemistry/>.

Sefiane, K., Steinchen, A., Moffat, J.R., 2010. On hydrothermal waves observed during evaporation of sessile droplets. Colloids Surf. A 365 (1-3), 95–108.

Sefiane, K., Bennacer, R., 2011. An expression for droplet evaporation incorporating thermal effects. J. Fluid Mech. 667, 260–271.

Sobac, B., Brutin, D., 2011. Triple-line behavior and wettability controlled by nano-coated substrates: influence on sessile drop evaporation. Langmuir 27 (24), 14999–15007.

Chapter 11

SOLID SUBSTRATE PROPERTIES

Tatiana Gambaryan-Roisman
Institute for Technical Thermodynamics and Center of Smart Interfaces, Technische Universität Darmstadt, Darmstadt, Germany

Evaporation of sessile droplets is accompanied by cooling of the liquid–gas interface. The evaporation-induced heat transport is governed by the droplet geometry and by the thermal properties of the liquid, ambient gas, and substrate. As a result, thermal properties of the substrate influence the temperature distribution in the droplet and at the liquid–gas interface, Marangoni-induced flow patterns in the droplets, and, under certain conditions, the evaporation rate. The influence of evaporative cooling on the overall evaporation rate is strong for poorly conductive substrates and quickly evaporating drops. The surface substrate properties influence the heat transfer through the contact angle and its hysteresis. Modification of the substrate surface properties through texturing and coating may significantly affect the behavior of evaporating droplet and influence the heat and mass transfer.

11.1 Thermal properties

Two basic phenomena are associated with the influence of temperature at the liquid–gas interface on the droplet behavior. The first one is the decrease in the

Droplet Wetting and Evaporation. DOI: http://dx.doi.org/10.1016/B978-0-12-800722-8.00011-4

evaporation rate caused by a decrease of the interface temperature. The second phenomenon is the nonuniformity of the interface temperature, leading to the development of Marangoni convection. These phenomena can be affected by the thermal properties of the substrate.

In the early studies of drop evaporation, the effect of the substrate thermal properties was not considered to be an important issue, although the thermal properties of the substrates used in droplet evaporation experiments varied widely. The substrate materials included glass covered by a thin polytetrafluoroethylene (PTFE) layer in the work of Picknett and Bexon (1977), poly(methyl methacrylate) (PMMA) with typical thermal conductivity of around 0.2 $W(m\ K)^{-1}$ in the work of Rowan et al. (1995), and silicon with thermal conductivity of around 140 $W(m\ K)^{-1}$ in the work of Poulard et al. (2005). Beginning in the middle of the last decade, the effect of the substrate thermal properties on the evaporation rate became a focus of many experimental and theoretical/numerical investigations. The dedicated experiments have been performed both for spontaneous evaporation of droplets (David et al., 2007; Dunn et al., 2009; Lopes et al., 2013) and for evaporation of droplets on heated substrates (Sobac and Brutin, 2012). Table 11.1 summarizes the substrates and liquids used in experiments, the experimental conditions, and the main findings of these works. The thermal properties of the substrates tested in these works are summarized in Table 11.2. The listed thermal properties include the thermal conductivity, k_s, which governs the temperature field in the substrate in steady state; the thermal diffusivity, $a_s = k_s/(\rho_s c_s)$, which determines the characteristic time of heat conduction; and the thermal effusivity, $e_s = k_s/\sqrt{a_s}$, which governs the transient evolution of the temperature field.

In these works, measures have been taken to keep the influencing parameters, except for the substrate thermal properties, constant. In particular, the surface properties of the substrates have been kept constant using an appropriate surface treatment. David et al. (2007) and Dunn et al. (2009) coated different substrates with an aluminum layer, providing an initial contact angle of 60° for water droplets. Sobac and Brutin (2012) used two types of nanocoating: the SiO_x coating that provided an initial contact angle of $67 \pm 5°$ (hydrophilic situation) and the perfluorocarbon coating that provided an initial contact angle of $133 \pm 2°$ (hydrophobic situation) for water droplets. Still et al. (2014) sputtered a chromium layer of a submicron thickness over the substrates, which resulted in an initial contact angle of 73° for water on a SrF_2 substrate and 76° for water on a sapphire substrate immediately after deposition. In the work of Lopes et al. (2013), the substrate's wettability was controlled with a grafted layer of polystyrene. In this case, the contact angle for water was $82 \pm 3°$, and the constant contact angle (CCA) regime dominated during the most of the droplet evaporation time.

The experiments of different groups listed in Table 11.1 were performed at different thermal boundary conditions. In the works of Lopes et al. (2013) and Still

TABLE 11.1 Experimental Investigations of Influence of Substrate Thermal Properties on Droplet Evaporation Dynamics

Reference	Substrate Materials	Liquids	Regime	Geometry, Conditions	Findings
David et al. (2007)	Aluminum Titanium Macor PTFE	Water Methanol Acetone	CCR	$h_s = 1$ mm $p = 10^4 – 10^5$ Pa (for water)	Evaporation rate is decreased for substrates with low k_s (PTFE), especially at low p. Drop is strongly cooled on substrates with low k_s.
Dunn et al. (2009)	Aluminum Titanium Macor PTFE	Water Methanol Acetone	CCR	$h_s = 10^{-3}$ m	Evaporation rate for a PTFE substrate is lower than that for aluminum substrate (e.g., 18% difference for methanol).
Sefiane et al. (2009)	Aluminum Titanium Macor PTFE	Water	CRR before unpinning	$h_s = 10^{-3}$ m $p = 4 \cdot 10^3 – 10^5$ Pa Gas: He, N_2, CO_2	Up to threefold factor difference in evaporation rate between the Al and PTFE substrates at low p.
Sobac and Brutin (2012)	Aluminum Copper Brass Bronze POM	Water	CCR	$h_s = 10^{-2}$ m Heated substrate, $25 \leq T_b \leq 75$°C	Evaporation rate is higher for metallic substrates. A difference of 14% in global evaporation rate at ambient temperature; 53.9% difference at 75°C between copper and POM substrates.
Lopes et al. (2013)	Glass Silicon	Water	Initially CCR followed by CCA	$h_s = 5 \times 10^{-4}$ m back side of substrate exposed to air	Evaporation rate is higher for metallic substrates. The difference in evaporation time exceeds 10%.
Still et al. (2014)	Sapphire SrF_2 (infrared-transparent substrates)	Water	CCR	$h_s = 10^{-3}$ m back side of substrate exposed to air	Evaporation rate is higher for sapphire substrate. The difference in evaporation time is around 25%.

et al. (2014), the droplet evaporated spontaneously, and the back side of the substrate was exposed to the quiescent gas. In Sobac and Brutin's (2012) paper, the substrate was heated from below to a prescribed temperature. The substrate thickness was 10 mm, which significantly exceeded the droplet contact radius (1.44 mm). In the other works, the substrate thickness was comparable to the drop contact radius, which is typically on the order of 1 mm. In spite of the differences,

TABLE 11.2 Thermal Properties of Substrate Materials

Reference	Material	k_s, W(m K)$^{-1}$	a_s, m^2 s^{-1}	e_s, W s$^{1/2}$(m^{2}K)$^{-1}$
Dunn et al. (2009)	Aluminum	237	9.58×10^{-5}	2.42×10^4
	Titanium	21.9	9.22×10^{-6}	7.21×10^3
	Macor	1.46	7.33×10^{-7}	1.70×10^3
	PTFE	0.25	1.08×10^{-7}	7.60×10^2
Sobac and Brutin (2012)	Aluminum	237	9.75×10^{-5}	2.40×10^4
	Copper	401	1.18×10^{-4}	3.68×10^4
	Brass	117	3.68×10^{-5}	1.99×10^4
	Bronze	50	1.51×10^{-5}	1.29×10^4
	POM	0.31	1.5×10^{-7}	8.0×10^2
Lopes et al. (2013)	Glass	1.05	5.8×10^{-7}	1.38×10^3
	Silicon	139.4	8.0×10^{-5}	1.56×10^4
Still et al. (2014)	Sapphire	40.81	1.33×10^{-5}	1.12×10^4
	SrF_2	1.42	5.40×10^{-7}	1.93×10^3

all of the studies found the same qualitative result: the droplet evaporation rate was higher and the evaporation time was shorter on substrates with higher thermal conductivity. The differences in droplet evaporation rates among different metallic substrates with thermal conductivity of between 20 and 400 W(m K)$^{-1}$ was small. A substantial reduction in evaporation rate was observed on the substrates with thermal conductivity on the order of 1 W(m K)$^{-1}$ and below. Sobac and Brutin (2012) hypothesized that the droplet evaporation on thick heated substrates, for which the characteristic heat conduction time was comparable to the drop evaporation time, was governed by transient heat transport processes. Therefore, the evaporation kinetics was determined by the thermal effusivity of the substrate material and liquid rather than by thermal conductivity. It was also observed that the influence of thermal properties was especially significant at high evaporation rates, caused by reducing the ambient pressure, accompanied by increasing the vapor diffusion coefficient (David et al., 2007; Sefiane et al., 2009), or by heating the substrate to high temperatures (Sobac and Brutin, 2012).

In order to explain the influence of thermal properties on droplet evaporation rate, we consider the evaporation of droplets in a still ambient gas. This gas is a mixture of a noncondensable gas such as air and the vapor of evaporating fluid. The schematic of the droplet evaporating from the substrate, including relevant dimensions, is shown in Figure 11.1. It is assumed that the droplet is perfectly axisymmetric. In the most important applications, the droplet size is below the capillary length so that the sessile

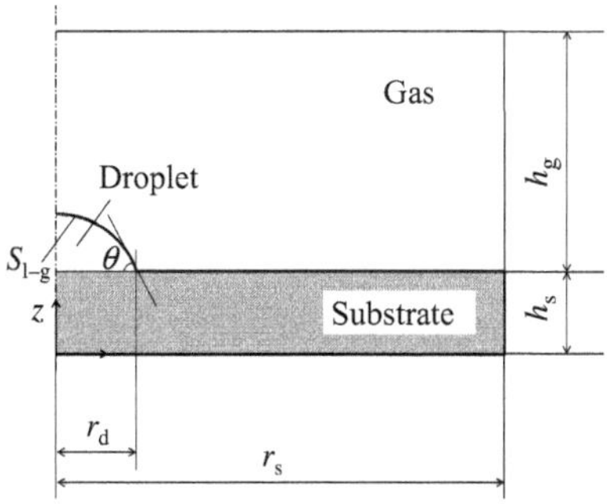

FIGURE 11.1 Schematic representation of an evaporating sessile droplet, along with the coordinate system.

droplet has the shape of a spherical cap. The evaporation rate of the droplet is governed by the rate of diffusive transport of vapor in the gas. The stationary diffusion in the gas phase is described by the Laplace equation for the concentration of vapor, C. Using the stationary diffusion equation is justified by the fact that the characteristic diffusion time has an order of r_d^2/D, where D is the vapor diffusivity in the gas. For the diffusion of water vapor in air at room temperature, $D = 2.1 \times 10^{-5} \mathrm{m}^2/\mathrm{s}$ (VDI-GVC, 2010). For a droplet radius of 1 mm, the diffusion time has the order of 0.05 s. This is much lower than the droplet evaporation time, which is typically several hundreds of seconds for droplets of different fluids, including water (David et al., 2007; Girard et al., 2008; Sobac and Brutin, 2012; Lopes et al., 2013), octane (Cazabat and Guéna, 2010), and blood (Brutin et al., 2012). If the density of the vapor significantly differs from that of the noncondensable gas, free convection may contribute to the vapor transport (Kelly-Zion et al., 2011; Dehaeck et al., 2014).

It is assumed that a local equilibrium is held at the liquid–gas interface so that the partial vapor pressure at the liquid–gas interface is equal to the saturation pressure at the interface. The discussion of the cases where the departure from equilibrium may play a significant role is given by Cazabat and Guéna (2010), Murisic and Kondic (2011), and Semenov et al. (2012). The evaporation rate, which is equal to the rate of droplet mass reduction, is given by

$$\frac{\mathrm{d}m}{\mathrm{d}t} = \rho_l \frac{\mathrm{d}V}{\mathrm{d}t} = -J_{\mathrm{ev}} = 2\pi D \int_{S_{l-g}} \nabla C \cdot \mathbf{n} r \mathrm{d}S_{l-g}. \tag{11.1}$$

If the liquid–gas interface is at a uniform temperature T_i, the evaporation rate can be computed from the equation

$$\frac{\mathrm{d}m}{\mathrm{d}t} = -J_{\mathrm{ev}} = -2\pi r_d D[C_{\mathrm{sat}}(T_i) - C_\infty]\frac{f(\theta)}{\sin\theta}, \tag{11.2}$$

where $C_{sat}(T_i) = p_{sat}(T_i)M/(RT_i)$, C_∞ is the vapor concentration far away from the droplet, and T_∞ is the temperature of the ambient gas. In the case of evaporation of water droplets, $C_\infty = p_{sat}(T_\infty)HM/(RT_\infty)$, where H is the relative humidity far from the droplet. The function $f(\theta)$ originally introduced by Picknett and Bexon (1977) as a result of the solution to the Laplace equation for the distribution of vapor concentration at different values of the contact angle θ can be found in the literature, including Picknett and Bexon, 1977; Hu and Larson, 2002; and Sefiane and Bennacer, 2011. According to simulations reported by Semenov et al. (2010), in the cases where the temperature distribution along the liquid–gas interface is nonuniform, the evaporation rate still can be computed with a good accuracy using Eq. (11.2) with T_i replaced by the average interface temperature $\overline{T}_i$. The expression for evaporation rate can be rewritten in the form

$$\frac{dm}{dt} = -J_{ev} = -\rho_l r_d \beta \frac{f(\theta)}{\sin\theta}, \quad \beta = 2\pi \frac{D[C_{sat}(\overline{T}_i) - C_\infty]}{\rho_l} \tag{11.3}$$

The parameter β characterizes the volatility of the fluid at the given thermodynamic conditions and is independent of the system geometry (Semenov et al., 2011).

The evaporation rate depends on the interface temperature through the term $C_{sat}\ \overline{T}_i$ in Eq. (11.3). Generally, this term decreases with time due to evaporative cooling. We estimate the temperature difference that would lead to a significant decrease of the evaporation rate. The evaporation rate decreases by 10% if the following inequality holds:

$$\frac{\Delta J_{ev}}{J_{ev}} = \frac{\Delta[C_{sat} - C_\infty]}{C_{sat} - C_\infty} = \frac{dC_{sat}}{dT}\frac{\Delta \overline{T}_i}{C_{sat} - C_\infty}$$
$$= \frac{p_{sat}(\overline{T}_i)M}{R\overline{T}_i^2}\left(\frac{h_{lv}M}{R\overline{T}_i} - 1\right)\frac{\Delta \overline{T}_i}{C_{sat}(\overline{T}_i) - C_\infty} > 0.1,$$

or

$$\Delta \overline{T}_i > 0.1 \frac{[C_{sat}(\overline{T}_i) - C_\infty]R\overline{T}_i^2}{p_{sat}(\overline{T}_i)M}\left(\frac{h_{lv}M}{R\overline{T}_i} - 1\right)^{-1}.$$

If $C_\infty = 0$, then the preceding inequality for the temperature difference takes the following form:

$$\Delta \overline{T}_i > 0.1\overline{T}_i\left(\frac{h_{lv}M}{R\overline{T}_i} - 1\right)^{-1}. \tag{11.4}$$

In the preceding expressions, h_{lv} is the latent heat of evaporation, M is the molar mass of the liquid, and R is a universal gas constant. The expression for dC_{sat}/dT has been derived using the Clausius–Clapeyron relation (Clapeyron, 1834;

Clausius, 1850) under an assumption that the specific volume of the vapor is much higher than the specific volume of the liquid and that the vapor is an ideal gas. For water at atmospheric pressure and room temperature, the minimum interface temperature reduction that affects the evaporation rate is, according to Eq. (11.4), around 1.7 K. For acetone and methanol these values are 2.2 K and 2.0 K, respectively (the fluid properties for these estimations have been taken from the work of Dunn et al. (2009)).

Let us estimate the effect of substrate properties on interface temperature for the case of an unheated substrate of infinite extent. The assumption of an infinitely large substrate is justified for the case if the droplet is much smaller than the substrate thickness and much smaller than any substrate lateral dimension. We evaluate the difference between the temperature at the substrate-droplet interface and the temperature of the substrate far away from the evaporating droplet. The substrate-droplet interface is a disk of radius r_{d}, at which a cooling rate of $\dot{Q}_{\mathrm{ev}} = h_{\mathrm{lv}} J_{\mathrm{ev}}$ is applied. After a transient period with a duration of order $r_{\mathrm{d}}^2/a_{\mathrm{s}}$, a stationary temperature distribution in the substrate is established, with a characteristic difference $\Delta T_{\mathrm{sd}} = \dot{Q}_{\mathrm{ev}}/(4 r_{\mathrm{d}} k_{\mathrm{s}})$ between the temperature far from the droplet base and the temperature at the solid–liquid interface (VDI-GVC, 2010). Evidently, this temperature difference increases as the thermal conductivity of the substrate decreases. Moreover, because $\dot{Q}_{\mathrm{ev}} = h_{\mathrm{lv}} J_{\mathrm{ev}} \sim r_{\mathrm{d}}$, this temperature difference is independent of the drop size. For a water droplet evaporating at room temperature and atmospheric pressure on an aluminum substrate, this temperature difference is equal to $\Delta T_{\mathrm{sd}} = 4.0 \times 10^{-3}$ K for a titanium substrate $\Delta T_{\mathrm{sd}} = 4.3 \times 10^{-2}$ K, whereas for a PTFE substrate, $\Delta T_{\mathrm{sd}} = 3.8$ K. In these cases a contact angle of 25° has been assumed. This simple estimation shows that the metallic supports stay nearly isothermal during the water droplet evaporation. This explains why the droplet evaporation rates on different metallic substrates are similar.

If the substrate thickness is comparable to the droplet contact radius, the temperature distributions in the substrate and in the droplet depend not only on the thermal properties of the phases but also on the substrate size and on the boundary conditions at the back side of the substrate. In this context the concept of the substrate thermal resistance has been used in the literature (Dunn et al., 2009; Sefiane and Bennacer, 2011). Due to the system complexity, a simple analytical estimation of the temperature variation in this case is impossible.

Several numerical works have been devoted to the modeling of heat and mass transfer during the droplet evaporation on a finite-size substrate with the aim of predicting the droplet evaporation rate as a function of substrate properties (Dunn et al., 2008, 2009; Sefiane et al., 2009; Semenov et al., 2010, 2011; Sefiane and Bennacer, 2011; Lopes et al., 2013; Saada et al., 2013). In all of these works, the droplet has been assumed to have a spherical shape. In addition, a certain evaporation regime (constant contact radius (CCR) or CCA) was prescribed. The evolution of the droplet volume is predicted on the basis of the rate of volume change computed from Eq. (11.1). Most of the models (except for the model developed in the work of Lopes et al. (2013)) are based on the assumption of quasi-steady heat and mass transport in the substrate, in the droplet, and in the gas. This assumption is justified for relatively thin substrates if the characteristic time of the transient heat transport processes in the substrate is given by h_s^2/a_s. For a substrate thickness of $h_s = 1$ mm and a thermal diffusivity of $a_s = 10^{-7}$ m^2 s^{-1}, this characteristic time is equal to 10 s, which is much shorter than the total evaporation time. The preceding estimation of the characteristic time of a transient process is valid if the constant temperature boundary condition can be posed at the back side of the substrate. The constant temperature at the back side of the substrate ($z = 0$) is a frequent assumption in the drop evaporation models.

Dunn et al. (2008, 2009) developed a numerical model describing a steady heat conduction in a substrate and in a droplet as well as diffusion of vapor in the gas phase. The effect of heat transfer on the evaporation rate has been taken into account through the dependence $C_{sat}(T_i)$, which is linearized around $C_{sat}(T_\infty)$. In the earlier work (Dunn et al., 2008), one-dimensional heat conduction through a thin substrate and a thin droplet were modeled. In the later work (Dunn et al., 2009), the Laplace equation for the temperature field in the substrate and in the gas was solved using the Finite Element Method (FEM). In the same work, the authors estimated the influence of vapor buoyancy on the droplet evaporation rate. They also tested the influence of boundary conditions at the solid–gas interface on the transport processes and the evaporation rate. Dunn et al. (2009) substituted the boundary condition of an isothermal solid–gas interface at an ambient temperature T_∞ with a more realistic boundary condition of the third kind. However, the assumed heat transfer coefficients in the range 100–10^4 W(m^2 K)$^{-1}$ were significantly higher than the expected heat transfer coefficient for free convection, which is below 10 W(m^2 K)$^{-1}$ (the estimation has been performed based on correlations and air properties from VDI-GVC (2010)). The model has further improved by Sefiane et al. (2009) by using a quartic approximation of the function $C_{sat}(T_i)$. The computational results showed a significant reduction in evaporation rate on substrates with low thermal conductivity, which is in agreement with the experimental trend.

Semenov et al. (2010) developed a FEM model of droplet evaporation that described the steady vapor diffusion in the gas phase, heat conduction in the solid and the gas phases, and Marangoni-driven flow and heat convection in the droplet. The computational results confirmed that the evaporation rate decreases as the substrate thermal conductivity decreases. It has been shown that Marangoni convection leads to an increase in the evaporation rate. This result differs from the findings of Girard et al. (2008), who claimed that Marangoni convection did not affect the evaporation rate, at least in the case of a heated substrate. A further important result reported by Semenov et al. (2010) is the already mentioned dependence of the total evaporation rate on the average interface temperature rather than on the temperature distribution over the interface. In their later work, Semenov et al. (2011) developed a universal model describing the drop evaporation dynamics during two stages of the evaporation process: the CCR stage, in which the contact angle changes from the advancing contact angle to the receding contact angle, and the CCA stage, during which the contact angle stays constant and the contact radius decreases. The universal model, which has been validated for a wide range of substrate properties, is based on a time scale $r_{\mathrm{b\ init}}^2/\beta$, where $r_{\mathrm{b\ init}}$ is the initial contact radius of the droplet.

Sefiane and Bennacer (2011) modeled the ratio between the actual drop evaporation rate and the maximal drop evaporation rate possible for the same droplet geometry, same ambient temperature and pressure, and same ambient gas composition. For $C_\infty = 0$, this ratio is equal to $C_{\mathrm{sat}}(\overline{T}_{\mathrm{i}})/C_{\mathrm{sat}}(T_\infty)$. This ratio has been represented as a function of a single dimensionless parameter, depending on the characteristic temperature difference in the droplet, the ratio of thermal resistances of the liquid layer and the solid layer, and the expression $\frac{1}{C_{\mathrm{sat}}}\frac{\mathrm{d}C_{\mathrm{sat}}}{\mathrm{d}T}$. A wide range of experimental data that have been collected for different liquids, substrate materials, and gas pressures could be indicated in a universal curve in this representation.

In the work of Lopes et al. (2013), the influence of transient heat conduction on the droplet evaporation dynamics has been studied using FEM for a silicon substrate and a glass substrate. The authors have solved the heat conduction problem with the condition of thermal insulation at the solid–gas interface exposed to air. Preliminary computations have shown that incorporation of the boundary condition of the third kind with a heat transfer coefficient of 10 $\mathrm{W(m^2\ K)^{-1}}$ or below does not influence the evaporation rate during the whole evaporation time. The problem is intrinsically transient, and the temperature field is mainly governed by the thermal effusivity of the materials. A good agreement between the experiment and the numerical predictions has been shown. The evaporation rate is higher in a silicon substrate that has a higher thermal effusivity.

Thermal properties of the substrate determine not only the droplet evaporation rate but also the direction of the temperature gradient at the liquid–gas interface and, therefore, the direction of Marangoni convection. In the absence of Marangoni convection, the high evaporation rate in the vicinity of the triple contact line leads to the emergence of flow in the direction of the contact line (Deegan et al., 2000) and to the deposition of particles from the droplet in the vicinity of the triple contact line ("coffee stain effect"). Marangoni convection in evaporating droplets manifests itself in a flow from hotter to colder areas at the liquid–gas interface and affects the particles' deposition patterns on the substrate.

Most experimental and numerical works report that, despite the fact that the local evaporation flux is higher in the vicinity of the triple contact line than at the droplet apex, the apex is the coldest point of the liquid–gas interface, whereas the triple contact line is the warmest location (David et al., 2007; Dunn et al., 2009; Semenov et al., 2010; Lopes et al., 2013; Saada et al., 2013; Still et al., 2014). The reason is the higher local thickness of the liquid layer. However, the flow direction changes for certain combinations of the thermal properties of the substrate and of the liquid and the contact angle.

Hu and Larson (2005) modeled the steady heat conduction in both the evaporating water droplet and glass substrate, the vapor diffusion in the gas, and Marangoni- and evaporation-induced creeping flow in the droplet, using FEM. The glass substrate thickness was 0.15 mm, and the droplet base radius was 1.3 mm. The authors applied the boundary condition of constant temperature at the back side of the substrate. The simulations have been performed for the contact angles from 10° to 40°, with a step of 5°. For the contact angles from 15° to 40°, the droplet apex was the coldest point at the liquid–gas interface, and Marangoni flow was directed from the periphery of the droplet to the apex. However, for the contact angle of 10°, the apex was the warmest location of the interface, and Marangoni flow was directed from the apex to the triple contact line.

Ristenpart et al. (2007) developed a criterion for the flow reversal, depending on the contact angle and the thermal conductivities of the liquid and the substrate. They considered the steady heat transfer in the immediate vicinity of the three-phase contact line. The evaporation flux was approximated as $j_{ev} = j_{ev0}(2r \cos \theta)^{-1/2+\theta/\pi}$, where r denoted the distance from the contact line. The direction of the temperature gradient in the vicinity of the three-phase contact line was determined from the solution of the heat conduction problem. The analysis brought the following results: if $k_s/k_l > 2$, the drop is warmest at the triple contact line; if $1.45 < k_s/k_l < 2$, then the triple contact line is the warmest point for $k_s/k_l > \tan \theta \cot(\theta/2 + \theta^2/\pi)$ and the coldest point if the ratio k_s/k_l is below this value; if $k_s/k_l < 1.45$, then the three-phase contact line is the coldest point. These theoretical criteria are supported by experimental data.

Xu et al. (2010) developed a model for determining the direction of the temperature gradient, depending on the thermal properties of the liquid and the substrate, the contact angle, and the substrate thickness. They considered the steady heat transfer in the vicinity of the droplet axis. The back side of the substrate was assumed to be isothermal and kept at the ambient temperature. Xu et al. (2010) derived the following criterion for determination of the temperature gradient direction: if $\frac{h_s k_l}{r_b k_s} < \frac{\sin 2\theta - 4(1/2 - \theta/\pi)\tan(\theta/2)}{4(1/2 - \theta/\pi) + 2\sin^2\theta}$, then the droplet apex is locally the coldest point. If the opposite is true, then the droplet apex is locally the warmest point. The criterion does not reduce to the criterion of Ristenpart et al. (2007) in the case of an infinitely thick substrate.

At present, no general criterion exists that can predict the direction of Marangoni convection on the scale of the entire drop.

11.2 Surface properties

It is evident from Eq. (11.2) that the evaporation rate and the associated heat and mass transfer depend on the substrate surface properties through the contact angle, θ. The function $J_{ev}/J_{ev}|_{\theta=\pi/2}$ relating the droplet evaporation rate at a contact angle θ to the evaporation rate at the contact angle $\pi/2$ for the same thermodynamic conditions, the same interface temperature, and the same contact radius is represented graphically in Figure 11.2. At a fixed contact radius the evaporation rate increases as the contact angle increases, whereas for $\theta < 40°$, the dependence of J_{ev} on θ is rather weak. Moreover, the substrate surface properties govern the regime of evaporation: CCR, CCA, or stick–slip behavior, which determine the evolution of the droplet volume and, therefore, the droplet evaporation time and average heat and mass transfer rate associated with the drop evaporation. In the CCA regime and under the condition that the expression $C_{sat}(\overline{T}_i) - C_\infty$ depends only weakly on time, Eq. (11.3) leads to the conclusion that r_d^2 or, equivalently, $V^{2/3}$ decreases linearly with time (Cazabat and Guéna, 2010; Semenov et al., 2011). In the CCR regime the evolution of the contact angle is given by the expression $t = \frac{r_d^2\pi}{\beta}\int_\theta^{\theta_{init}} \frac{\sin\tilde{\theta}\, d\tilde{\theta}}{f(\tilde{\theta})(1+\cos\tilde{\theta})^2}$, where θ_{init} is the initial contact angle (Schönenfeld et al., 2008; Gelderblom et al., 2011). For small values of the contact angle, the value of the expression $f(\theta)/\sin\theta$ is close to 0.64. Therefore, the evaporation rate is nearly constant, and the droplet volume decreases linearly with time (Schönenfeld et al., 2008; Sefiane and Bennacer, 2011). It has been shown that the values of the advancing/receding contact angles, as well as the degree of contact angle hysteresis, depend on both the substrate surface energy and the substrate roughness and heterogeneity (de Gennes et al., 2003; Bormashenko, 2013).

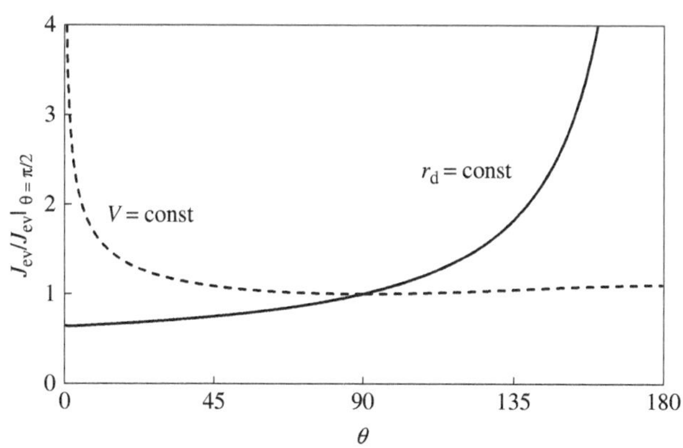

FIGURE 11.2 Scaled droplet evaporation rate at constant contact radius and constant drop volume.

The effect of substrate surface properties on the droplet evaporation rate, as well as on heat and mass transfer induced by the droplet evaporation, have been systematically studied by Sobac and Brutin (2011) and Brutin et al. (2012).

Sobac and Brutin (2011) studied spontaneous evaporation of five different liquids from a massive aluminum cylinder equipped with a heat flux meter and coated with four different materials characterized by different surface energies and roughnesses. In this way the influence of substrate surface properties on evaporation rate could be determined independently of the substrate thermal properties. Evidently, the liquid–solid interface temperature was very close to the ambient temperature. For evaporation of ethanol droplets with a constant initial volume from substrates characterized by different initial contact angles below 40°, the initial evaporation rate increased as the contact angle decreased. This result agrees with the curve $V = \text{const}$ in Figure 11.2. Moreover, the evaporation rate stayed constant as long as the three-phase contact line stayed pinned ($r_d = \text{const}$), which is also in agreement with Figure 11.2. Water droplets formed initial contact angles from 63° to 135° with different substrates. The water droplet evaporation rate in the pinned mode decreased as the contact angle decreased, which is also in agreement with the theory illustrated in Figure 11.2: curve $r_d = \text{const}$. It is also seen that the evaporation time significantly increases in an unpinned mode compared with the pinned mode. In addition, the droplet evaporation time significantly increases as the initial contact angle in a pinned mode increases (a 58% increase in water droplet evaporation time, with an increase in the initial contact angle from 63° to 135°).

Brutin et al. (2012) studied transport processes during evaporation of blood droplets. Three substrates were tested: an aluminum substrate with a thickness of 0.5 mm and initial contact angle of 95.7°, a glass substrate with a thickness of 1 mm and initial contact angle of 20.5°, and a gold-coated glass substrate with a thickness of 1 mm and initial contact angle of 91.9°. The substrates were mounted on a heat flux meter fixed on an aluminum block 8 mm in length. The resulting average evaporation rate was nearly identical for the gold-plated glass substrate and the aluminum substrate that had close initial contact angles with the droplet and very different thermal properties. The droplet average evaporation rate on a glass substrate with a lower initial contact angle was more than twice as much higher. The thermal properties of the substrate did not have any influence on evaporation behavior. This can be explained by the fact that for two substrates forming a contact angle of around 90°, the thermal resistance of the substrates was significantly lower than that of the droplet, and the temperature at the liquid–gas interface was practically independent of the temperature drop within the substrate.

Peculiar phenomena of transport associated with droplet evaporation on superhydrophobic surfaces are addressed by Gelderblom et al. (2011), Dash and Garimella (2013), and Pan et al. (2013). Gelderblom et al. (2011) conducted experiments on spontaneous evaporation of water droplets from substrates covered with carbon nanofibers possessing a chaotic three-dimensional microstructure. The substrates showed a superhydrophobic behavior with an initial contact angle of 150° after droplet deposition. The droplets stayed pinned during evaporation, except for the final moments of their lifetime. It has been found that the evolution of the contact angle during the droplet evaporation followed the analytical solution of Popov (2005), which is equivalent to the claim that the evaporation rate is described by Eq. (11.3), with β not changing with time. The scaled curve $\theta(\hat{t})$, where $\hat{t} = \frac{t(C_{sat} - C_\infty)D}{\rho_l r_d^2} = \frac{t\beta}{2\pi r_d^2}$, has shown a universal behavior for all initial drop volumes, which agrees with the conclusions of Semenov et al. (2011). The authors conclude that the thermal effects do not play a role in droplet evaporation kinetics. Unfortunately, it was impossible to estimate the thermal contact between the liquid and the substrate beneath the carbon nanofiber coating.

Dash and Garimella (2013) fabricated double-roughness structures made of 23 μm-high silicon pillars with a width of 45 μm, a pitch of 48 μm, and a roughness of 2.93 μm playing a role of the second-level texture. The structured surfaces have spin-coated with a Teflon-AF layer. The resulting substrates showed a contact angle of 160°, with hysteresis of 1°. The spontaneous drop evaporation took place in the constant contact angle regime. Evaporation times computed on the basis of the diffusion model (Eq. (11.3) with $\beta =$ const) significantly underpredicted the experimental data. The evaporation rate is overestimated by 20%. This discrepancy has been attributed to thermal effects—that is, to the significant reduction in the droplet interface temperature due to evaporation. In a following numerical work (Pan et al., 2013), the authors performed a numerical study in which stationary convective heat transfer in the droplet, heat conduction in the substrate, and convective heat transfer and convective-diffusive vapor transport in the gas phase were taken into account. The authors assumed that the Cassie–Baxter wetting state prevailed during evaporation and modeled the effective thermal conductivity of the layer of pillars based on an assumption that the gap between the pillars was filled with air. The improved model predictions agreed well with experimental data. Due to a very high contact angle with negligible hysteresis and due to a Cassie–Baxter wetting state, the thermal contact area between the liquid and the solid was very small. As a result, around 53% of the thermal energy was supplied to the droplet from the gas phase.

In this chapter, we have shown that the substrate surface properties affect heat and mass transport during droplet evaporation and the drop evaporation rate through the contact angle, contact angle hysteresis, and liquid–solid contact area. In reality, the substrate texture may induce more complex liquid–solid interactions governing heat and mass transport. For example, Bormashenko et al. (2007) imply that the complex shape of the liquid–gas interface under an evaporating droplet on textured surfaces induces Marangoni convection, which affects the mass transport and formation of deposits.

The substrates with textured surfaces may exert a capillary action on a liquid. This action is especially pronounced for substrates that have porous coatings.

FIGURE 11.3 Water droplet on a heated steel foil covered with a PAN nanofiber mat. *From Weickgenannt et al. (2010).*

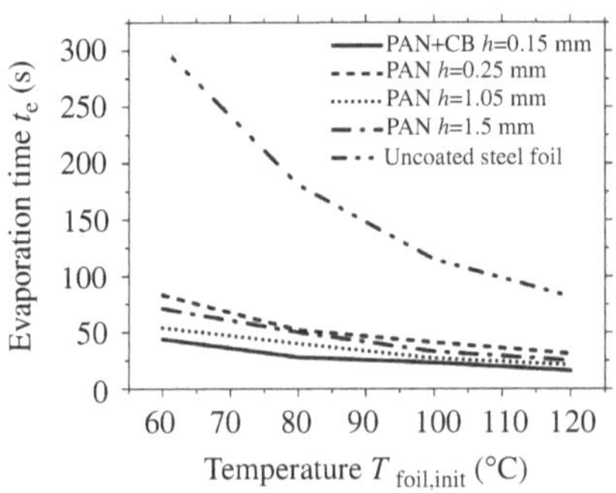

FIGURE 11.4 Cooling efficiency of nanofiber mats compared with the uncoated steel foil: the evaporation time versus the initial foil temperature. CB, carbon black. *From Weickgenannt et al. (2010).*

Examples of such coatings are nanofiber mats, which can be produced by electrospinning. The spontaneous evaporation of droplets or the evaporation of droplets facilitated by heating such substrates is accompanied by imbibition of liquid into the porous layer (Lembach et al., 2009; Weickgenannt et al., 2010). Figure 11.3 shows a water droplet evaporating from steel foil coated with PAN [poly(acrylonitrile)] nanofiber mats with a thickness of 1.05 mm. The steel foil is heated to a temperature of 60°C. The dark area around the droplet corresponds to a region of porous coating partially filled with water. In this case evaporation of liquid takes place not only from the droplet surface but also from the regions of coating around the droplet. This phenomenon significantly affects heat and mass transport and strongly reduces the lifetime of the droplets (Figure 11.4).

It can be concluded that the influence of the contact angle, as well as the contact angle hysteresis, on heat and mass transfer during droplet evaporation is well understood. On the other hand, if the textured substrate or a substrate with a porous coating exerts an influence on the microscale convection in the droplet or induces the imbibition of the liquid into the capillary-porous surface layer, the heat and mass transfer processes accompanying the droplet evaporation are not sufficiently understood. Due to the significance of these phenomena, it is expected that the efforts of the scientific community in the next several years will be focused on the development of numerical tools capable of dealing simultaneously with transport processes on the scale of a whole droplet and with the transport on the scale of elements of surface texture or unit cells of porous coatings.

Nomenclature

a	thermal diffusivity (m^2 s^{-1})
C	vapor concentration in gas (kg m^{-3})
D	diffusivity of vapor in gas (m^2 s^{-1})
e	thermal effusivity (W $s^{1/2}$ m^{-2} K^{-1})
H	ambient relative humidity
h	thickness (m)
h_{lv}	latent heat of evaporation (J kg^{-1})
J_{ev}	evaporation rate (kg s^{-1})
j_{ev}	evaporation flux (kg s^{-1} m^{-2})
k	thermal conductivity (W m^{-1} K^{-1})
M	molar mass of the fluid (kg mol^{-1})
m	mass of the droplet (kg)
$\mathbf{n}$	unit normal vector
p	pressure (Pa)
$\dot{Q}_{ev}$	cooling rate due to evaporation (W)
R	universal gas constant (J mol^{-1} K^{-1})
r	radial coordinate or radius (m)
r_d	drop contact radius (m)
S_{l-g}	liquid–gas interface
T	temperature (K)
t	time (s)
V	volume of the droplet (m^3)
z	axial coordinate (m)

Greek Symbols

θ	contact angle
ρ	density (kg m^{-3})

Subscripts

g gas
l liquid
sat saturation
s substrate

Abbreviations

CCA constant contact angle
CCR constant contact radius

References

Bormashenko, E., Bormashenko, Y., Pogreb, R., Stanevsky, O., Whyman, G., 2007. Droplet behaviour on flat and textured surfaces: co-occurance of Deegan outward flow with Marangoni solute instability. J. Colloid Interface Sci. 306, 128–132.

Bormashenko, E.Y., 2013. Wetting of Real Surfaces. Walter de Gruyter GmbH, Berlin/Boston (De Gruyter Studies in Mathematical Physics 19).

Brutin, D., Sobac, B., Nicloux, C., 2012. Influence of substrate nature on the evaporation of a sessile drop of blood. ASME J. Heat Transf. 134, 061101.

Cazabat, A.-M., Guéna, G., 2010. Evaporation of macroscopic sessile droplets. Soft Matter 6, 2591–2612.

Clapeyron, E., 1834. Mémoire sur la puissance motrice de la chaleur. J. École. Polytech. 14, 153–190.

Clausius, R., 1850. Über die bewegende Kraft der Wärme und die Gesetze, die sich daraus für die Wärmelehre selbst ableiten lassen. Annal. Phys. 155, 500–524.

Dash, S., Garimella, S.V., 2013. Droplet evaporation dynamics on a superhydrophobic surface with negligible hysteresis. Langmuir 29, 10785–10795.

David, S., Sefiane, K., Tadrist, L., 2007. Experimental investigation of the effect of thermal properties of the substrate in the wetting and evaporation of sessile drops. Colloids Surf. A Physicochem. Eng. Asp. 298, 108–114.

De Gennes, P.G., Brochard-Wyart, F., Quéré, D., 2003. Capillarity and Wetting Phenomena. Springer, Berlin, Germany.

Deegan, R.D., Bakajin, O., Dupont, T.F., Huber, G., Nagel, S.R., Witten, T.A., 2000. Contact line deposits in an evaporating drop. Phys. Rev. E 62, 756–765.

Dehaeck, S., Rednikov, A., Colinet, P., 2014. Vapor-based interferometric measurement of local evaporation rate and interfacial temperature of evaporating droplets. Langmuir 30, 2002–2008.

Dunn, G.J., Wilson, S.K., Duffy, B.R., David, S., Sefiane, K., 2008. A mathematical model for the evaporation of a thin sessile liquid droplet: comparison between experiment and theory. Colloids Surf. A Physicochem. Eng. Asp. 323, 50–55.

Dunn, G.J., Wilson, S.K., Duffy, B.R., David, S., Sefiane, K., 2009. The strong influene of substrate conductivity on droplet evaporation. J. Fluid Mech. 623, 329–351.

Gelderblom, H., Marín, Á.G., Nair, H., van Houselt, A., Lefferts, L., Snoeijer, J.H., et al., 2011. How water droplets evaporate on a superhydrophobic substrate. Phys. Rev. E 83, 026306.

Girard, F., Antoni, M., Sefiane, K., 2008. On the effect of Marangoni flow on evaporation rates of heated water drops. Langmuir 24, 9207–9210.

Hu, H., Larson, R.G., 2002. Evaporation of a sessile droplet on a substrate. J. Phys. Chem. B 106, 1334–1344.

Hu, H., Larson, R.G., 2005. Analysis of the effects of Marangoni stresses on the microflow in an evaporating sessile droplet. Langmuir 21, 3972–3980.

Kelly-Zion, P.L., Pursell, C.J., Vaidya, S., Batra, J., 2011. Evaporation of sessile drops under combined diffusion and natural convection. Colloids Surf. A Physicochem. Eng. Asp. 381, 31–36.

Lembach, A.N., Tan, H.-B., Roisman, I.V., Gambaryan-Roisman, T., Zhang, Y., Tropea, C., et al., 2009. Drop impact, spreading, splashing and penetration into electrospun nanofiber mats. Langmuir 26, 9516–9523.

Lopes, M.C., Bonaccurso, E., Gambaryan-Roisman, T., Stephan, P., 2013. Influence of the substrate thermal properties on sessile droplet evaporation: effect of transient heat transfer. Colloids Surf. A Physicochem. Eng. Asp. 432, 64–70.

Murisic, N., Kondic, L., 2011. On evaporation of sessile drops with moving contact line. J. Fluid Mech. 679, 219–246.

Pan, Z., Dash, S., Weibel, J.A., Garimella, S.V., 2013. Assessment of water droplet evaporation mechanisms on hydrophobic and superhydrophobic substrates. Langmuir 29, 15831–15841.

Picknett, R.G., Bexon, R.J., 1977. Evaporation of sessile or pendant drops in still air. J. Colloid Interface Sci. 61, 336–350.

Popov, Y., 2005. Evaporative deposition patterns: spatial dimensions of the deposit. Phys. Rev. E 71, 036313.

Poulard, C., Guéna, G., Cazabat, A.M., 2005. Difusion-driven evaporation of sessile drops. J. Phys. Condens. Matter 17, S4213–S4227.

Ristenpart, W.D., Kim, P.G., Domingues, C., Wan, J., Stone, H.A., 2007. Influence of substrate thermal conductivity on circulation reversal in evaporating drops. Phys. Rev. Lett. 99, 234502.

Rowan, S.M., Newton, M.I., McHale, G., 1995. Evaporation of microdroplets and the wetting of solid surfaces. J. Phys. Chem. 99, 13268–13271.

Saada, M.A., Chikh, S., Tadrist, L., 2013. Evaporation of a sessile drop with pinned or receding contact line on a substrate with different thermophysical properties. Int. J. Heat Mass Transf. 58, 197–208.

Schönenfeld, F., Graf, K.-H., Hardt, S., Butt, H.-J., 2008. Evaporation dynamics of sessile liquid drops in still air with constant contact radius. Int. J. Heat Mass Transf. 51, 3696–3699.

Sefiane, K., Bennacer, R., 2011. An expression for droplet evaporation incorporating thermal effect. J. Fluid Mech. 667, 260–271.

Sefiane, K., Wilson, S.K., David, S., Dunn, G.J., Duffy, B.R., 2009. On the effect of the atmosphere on the evaporation of sessile droplets of water. Phys. Fluids 21, 062101.

Semenov, S., Starov, V.M., Rubio, R.G., Velarde, M.G., 2010. Instantaneous distribution of fluxes in the course of evaporation of sessile liquid droplets: computer simulations. Colloids Surf. A Physicochem. Eng. Asp. 372, 127–134.

Semenov, S., Starov, V.M., Rubio, R.G., Agogo, H., Velarde, M.G., 2011. Evaporation of sessile water droplet: universal behavior in presence of contact angle hysteresis. Colloids Surf. A Physicochem. Eng. Asp. 391, 135–144.

Semenov, S., Starov, V.M., Rubio, R.G., Agogo, H., Velarde, M.G., 2012. Computer simulations of evaporation of pinned sessile droplets: influence of kinetic effects. Langmuir 28, 15203–15211.

Sobac, B., Brutin, D., 2011. Triple-line behavior and wettability controlled by nanocoated substrates: influence on sessile drop evaporation. Langmuir 27, 14999–15007.

Sobac, B., Brutin, D., 2012. Thermal effects of the substrate on water droplet evaporation. Phys. Rev. E 86, 021602.

Still, M., Gambaryan-Roisman, T., Stephan, P., 2014. Experimental investigation of interfacial temperature evolution during evaporation of sessile droplet. In: Proceedings of the 15th International Heat Transfer Conference, IHTC15-9151, Kyoto, Japan.

VDI-GVC, 2010. VDI Heat Atlas. Springer, Berlin, Germany.

Weickgenannt, C.M., Zhang, Y., Lembach, A.N., Roisman, I.V., Gambaryan-Roisman, T., Yarin, A.L., et al., 2010. Non-isothermal drop impact and evaporation on polymer nanofiber mats. Phys. Rev. E 3, 036305.

Xu, X., Luo, J., Guo, D., 2010. Criterion for reversal of thermal Marangoni flow in drying drops. Langmuir 26 (3), 1918–1922.

Chapter 12

SOFT SUBSTRATES

Marcus Camarinha Lopes
Mechanical Engineering Department Technical University Darmstadt, Darmstadt, Germany

Performed experimental investigations show that soft substrates (Young's modulus below 10 MPa) are able to directly influence the dynamics of the triple-phase-contact-line (TPCL) of a sessile drop due to deformation caused by interfacial forces. By fine-tuning the Young's modulus of the substrate underneath an evaporating droplet, it is possible to control the evaporation mode, and consequently, evaporation time of the droplet. Further investigations on particle deposition during the evaporation of water-silica suspension drops showed that the TPCL velocity increases with decreasing Young's modulus of the substrate. Fine-tuning Young's modulus of soft polymer substrates also allowed the control of particle deposition without modifying the used suspension.

12.1 Deformation of soft substrate due to interfacial forces

As described in many publications, the evaporation of a sessile drop on a non-heated substrate is a diffusion-limited process (Picknett and Bexon, 1977; Erbil

Droplet Wetting and Evaporation. DOI: http://dx.doi.org/10.1016/B978-0-12-800722-8.00012-6

et al., 2002; Butt et al., 2007). The solid–liquid interface plays hereby an important role, not only because much of the heat has to be transported from the supporting substrate to the evaporating drop but also because the contact line dynamic influences evaporation by modifying both the liquid–air interface area and the contact-line length, which directly affect the evaporation rate.

When an evaporating liquid drop is sitting on a substrate, interfacial forces between liquid, solid, and gas are balanced, and a certain tension is applied to the substrate. The horizontal component of this tension is balanced by the surface energies of the liquid–solid interface and the solid–gas interface. On soft substrates, the vertical component of the force must also be taken into account. In fact, the vertical component of the force deforms the substrate underneath the drop (Rusanov, 1975a,b).

For materials with a Young's modulus higher than some hundreds of megapascal, the deformation of the substrate is only relevant if the substrate is thin enough. To measure the deformation caused by a liquid drop, cantilevers can be used as sensors to monitor microdrop evaporation (Haschke et al., 2006).

The surface tension of the drop acts on the triple-phase-contact-line (TPCL), pulling up a ridge of material. The Laplace pressure acts underneath the drop, causing an indentation. The overall deformation looks like a crater. The geometry of such a deformation was first described by Rusanov (1975a,b), and this theoretical description was confirmed by experimentalists over the past few years (Pericet-Camara et al., 2008a; Lopes and Bonaccurso, 2012; Style and Dufresne, 2012; Style et al., 2013).

12.2 Influence of soft substrates on the evaporation of pure water

Technical applications like crop spraying (Xu et al., 2011), spray cooling (Sodtke and Stephan, 2007), heat exchange processes, combustion, measurement of diffusion coefficients in polymer films (Liu et al., 2011), and inkjet printing (Park and Moon, 2006) rely on the physics of the evaporation of sessile drops. Controlling the evaporation of sessile drops is therefore of crucial importance. For a reproducible and reliable handling of small quantities of liquids—for example, in microfluidic and lab-on-chip applications—evaporation plays a major role and must be well understood, predicted, and controlled. Moreover, drop evaporation on soft deformable substrates is a process that also occurs when sweat evaporates on human skin, which is essential for the thermoregulation of humans (Mcdonald et al., 2009).

To better understand the influence of the substrate's softness on the drop, evaporation experiments were performed on polydimethylsiloxane (PDMS)

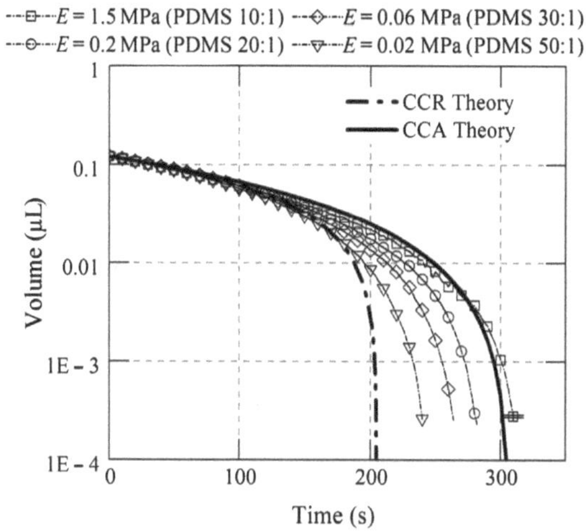

FIGURE 12.1 Volume change of sessile water drops during evaporation. The symbols represent the experimental data on different soft PDMS substrates. The solid and dashed lines represent the predicted volume change in the CCA mode and the CCR mode, respectively. Error bars are represented only for two data points (open squares) for clarity and are similar for all curves.

elastomer substrates with different mechanical properties (PDMS, Sylgard 184, Dow Corning, Wiesbaden, Germany).

Figure 12.1 shows the drop volume versus the time for drops with an initial volume $V_0 \approx 0.12$ μL on PDMS substrates with Young's moduli ranging from 0.02 to 1.5 MPa, and with a thickness of 60 μm. The solid curve represents the evaporation in the constant contact angle (CCA), and the dashed curve represents the constant contact radius (CCR) mode (Lopes and Bonaccurso, 2012).

In the initial phase of evaporation, the evaporation rate is similar for all substrates. This is consistent with the prediction that the evaporation rate is proportional to the contact radius of the drop, because all drops remain in the CCR mode with similar contact radii (Morse, 1910; Picknett and Bexon, 1977; Schonfeld et al., 2008). As evaporation goes on, the influence of the substrate becomes more evident. The evaporation rates start to differ, and differences between drop volumes increase. Drops on softer substrates (lower Young's modulus) evaporate faster than on harder substrates (higher Young's modulus). But what is the physical explanation for the effect of softness (Young's modulus) on the evaporation time of otherwise initially identical drops?

The left side of Figure 12.2 shows a plot of the normalized contact radii, a', as a function of the evaporation time. The radii are normalized to the respective initial contact radii, a_0. The softer the substrate, the longer the evaporation time in the CCR mode. Because the evaporation rate is proportional to a, evaporation

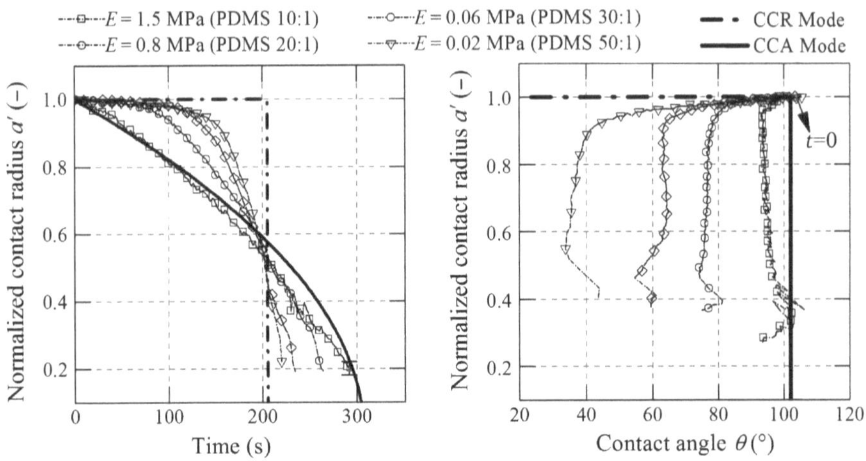

FIGURE 12.2 (Left) Normalized contact radii of sessile drops during evaporation. (Right) Geometry of a sessile drop during evaporation. The symbols show the experimental data on different soft PDMS substrates. The solid and dashed lines represent the radius of a drop evaporating in CCA and CCR mode, respectively. Error bars are shown for two data points only (open squares) for clarity and are similar for all curves.

times are shorter on softer substrates. On the hardest/softest substrate, the evolution of the contact radius was very similar to the expected contact radius evolution for the CCA/CCR evaporation theory.

The right side of Figure 12.2 shows the normalized contact radii, a', plotted as a function of the respective contact angles, θ. The expected CCR and CCA evaporation modes are thus indicated by the horizontal and vertical lines. The starting point of the evaporation was at the top right corner of the graph ($t = 0$), characterized by $\theta = 105°$ and $a' = 1$. All of the drops started evaporating in the CCR mode and switched to the CCA mode as soon as $\theta = \theta_r$. From then on, the contact radius of the drop decreased continuously.

Figure 12.2 shows the main difference between drop evaporation on substrates with different elasticities. The different receding contact angles lead to different transition times from CCR to CCA evaporation. The softest substrate showed the smallest θ_r leading to a longer duration of the CCR mode and to a shorter total evaporation time. Thus, it is clear that the evaporation time of a sessile drop was shortened by decreasing the Young's modulus of the substrate. To better understand what is driving the sessile drop to have a different receding contact angle θ_r, the deformation of the substrate was investigated for the different substrates by means of experiments and simulations.

Figure 12.3 shows the experimental and theoretical evaporation times of sessile water drops with an initial volume of 0.12 μL that depends on the receding contact

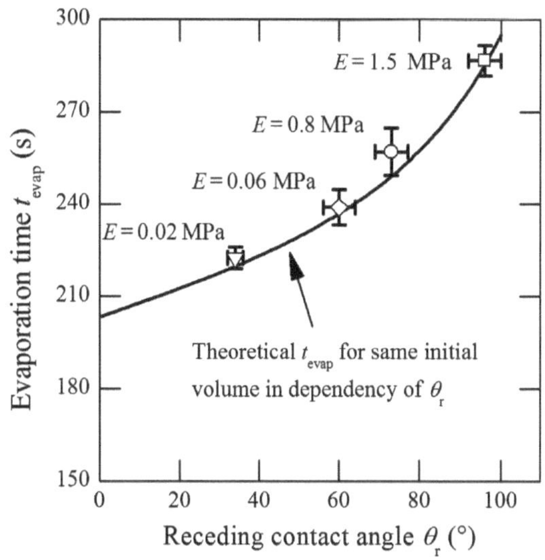

FIGURE 12.3 Evaporation time of sessile drops versus the receding contact angle. Evaporation time for $\theta_r = 0°$ was calculated using theoretical equations. *From Lopes and Bonaccurso (2012).*

angle. For the theoretical curve, it was assumed CCR evaporation from the start and CCA evaporation as soon as θ_r was reached, which roughly corresponded to the experiments. The bars show the scattering of the experimental data. The theoretical course of evaporation demonstrates the experimental evaporation time of drops, depending on θ_r. This confirmed that different evaporation times are related to different receding contact angles, which in turn are determined by the magnitude of substrate deformation, which in turn is controlled by its elasticity.

By producing substrates with given elasticity (Young's modulus), it was possible to control the contact line dynamics (by tuning the receding contact angle) and to empirically predict the evaporation time of drops on them. But how can we estimate the receding contact angle of a substrate for a given elasticity (Young's modulus)?

The left side of Figure 12.4 shows the measured receding contact angles versus the Young's moduli of the substrates. Previous works on the influence of solid micro-deformations on contact angle equilibrium (Shanahan, 1987) showed that the deformation height of the substrate at the drop rim, h_{rim}, is proportional to the inverse of the elastic modulus and therefore proportional to the deformation energy ($h_{max} \propto 1/E \propto \varepsilon_{defor}$). The extra energy stored in the deformation of the soft substrates acts as a barrier for the depinning of the TPCL. The right side of Figure 12.4 shows the dependence of the measured receding contact angle of the drop on h_{max}. An equation was derived to describe h_{max} in Lopes and Bonaccurso (2012),

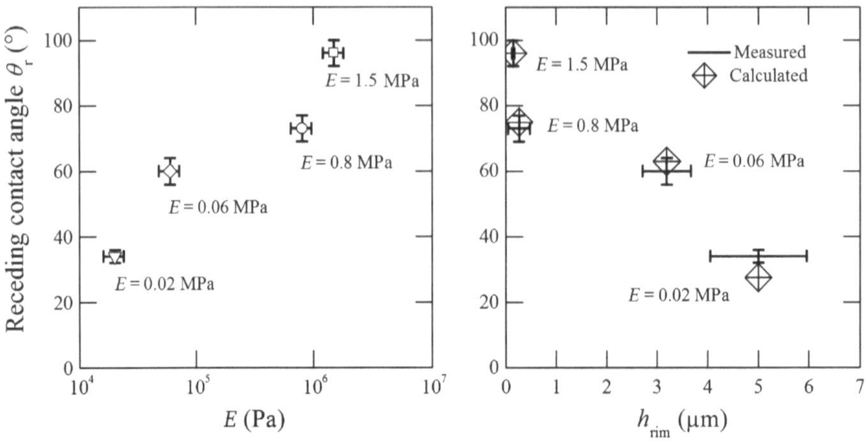

FIGURE 12.4 (Left) Receding contact angles versus Young's moduli of four different soft substrates. (Right) Measured (crosses) and calculated (diamonds) receding contact angles versus rim height h_{rim}.

allowing the control of the evaporation mode and therefore the evaporation time of sessile drops by tuning the elasticity (Young's modulus) of the substrate, which controls the deformation energy stored in the substrate, which modifies the receding contact angle of a liquid–solid system.

12.3 Using soft substrates to control the deposition pattern formed during suspension droplet evaporation

One characteristic particle-deposition-related effect common to nearly all coffee drinkers is the so-called coffee stain effect (Deegan et al., 1997). After the evaporation of a drop of coffee on a surface, a typical circular stain with dark borders remains. This ring is formed by minute coffee powder grains accumulating at the rim of the drop during an interplay of processes associated with the evaporation of water (Deegan et al., 1997): pinning of the contact line and evaporative cooling and convective flow inside the drop due to Marangoni effects. The flow inside the drop is directed toward the contact line because the evaporative flux is highest at this region (Hu and Larson, 2005; Ristenpart et al., 2007; Eggers and Pismen, 2010).

As described in the previous section, it is possible to control the evaporation mode and, therefore, the evaporation time of sessile drops by tuning the elasticity of the substrate. This section explains how this can be used to control the particle deposition patterns created during the sessile drop evaporation of water-silica suspensions.

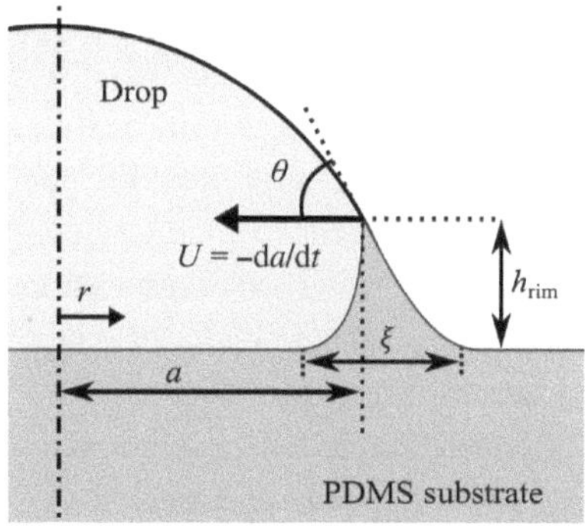

FIGURE 12.5 Scheme of the sessile drop sitting on a deformable substrate showing the contact radius a, contact angle θ, height of the substrate ridge at the drop rim h_{rim}, contact line velocity U, distance ξ, and the radial coordinate r. Deformation of substrate is not to scale.

It is possible to describe the receding velocity of the contact line during the evaporation of pure liquids by balancing the main forces acting on the contact line (Lopes and Bonaccurso, 2013):

$$U = \frac{\gamma}{\mu_f}\left(\cos\theta_r - \cos\theta_0 + \frac{C\gamma\sin\theta_r}{aE}(1-\Delta)\right) \tag{12.1}$$

Here, $U = -da/dt$ is the receding velocity of the contact line; θ_r and θ_0 are, respectively, the receding and the equilibrium contact angles (see Figure 12.5); C is the proportionality constant described in the previous section; γ is the surface tension of the liquid; a is the contact radius of the sessile drop; and E is the Young's modulus of the substrate. The dissipated part is denoted by a constant Δ, as was previously done in a work by Shanahan (1988), and μ_f is a friction parameter that can range between 0.05 and 0.1 Pa s for substrates with $\theta_0 \approx 100°$, liquids with dynamic viscosity from 1 to 100 mPa s, and spreading velocities U between 0.01 and 1 m s^{-1} (Carlson et al., 2012). Equation (12.1) can be used to predict the contact line velocities of evaporating water drops during the movement of the contact line during the CCA evaporation mode on deformable substrates. This equation is valid as long as a is considerably larger than the distance ξ, within which the largest amount of the deformation energy is stored.

Similar to the evaporation of pure water drops, the evaporation of drops of water-silica suspensions on soft substrate also is influenced by the substrate elasticity and can be controlled by the substrate softness (Lopes and Bonaccurso, 2013). But, unlike pure water drops, only two stages, instead of three, were observed during the evaporation of silica-suspension drops. The additional stage, the final stage of evaporation, which was nearly absent for pure water drops, was characterized by a renewed pinning of the contact line and almost CCR evaporation, while the drop turned into a thin film and finally disappeared completely. The deposited particles were likely responsible for the second contact line pinning (known as self-pinning), in accordance with previous publications (Nguyen et al., 2013; Weon and Je, 2013).

The receding contact angles θ_r of water-silica suspensions decreased with decreasing Young's modulus for pure water and for the water-silica suspension. Furthermore, the receding contact angles for the suspension were lower than for pure water. This lower receding contact angle for the water-silica suspension was an indicator of an extra pinning force applied on the contact line, which was already observed in previous work and occurred due to the accumulation of silica particles in this region (Deegan et al., 1997; Nguyen et al., 2013; Weon and Je, 2013): more silica particles accumulate at the contact line of the drop due to the longer pinning time on soft substrates and the resultant flow toward the contact line inside the drop.

The receding contact line velocity during the CCA evaporation stage of water-silica drops was nearly constant. This observation motivated the investigation of the contact line velocity. Figure 12.6 using Eq. (12.1), shows the plots of contact line receding velocities of pure water and water-silica suspension drops evaporating on substrates with different Young's moduli. Because the model did not account for particles, the velocities were only calculated for pure water. The following parameters were used: equilibrium contact angle $\theta_0 = 110°$; dissipation constant $\Delta = 0.1$, as in the work by Shanahan (1988); and $C = 0.3$. The friction parameter $\mu_f = 0.07$ Pa s

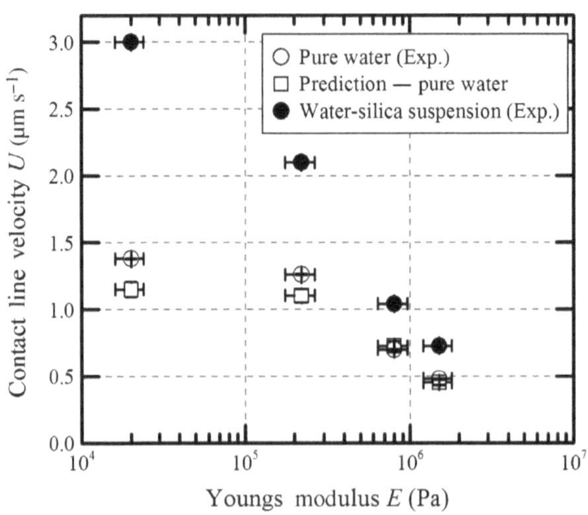

FIGURE 12.6 Measured contact line velocities of an evaporating drop of pure water (open circles) and water-silica suspension (filled circles) on substrates with different Young's moduli. Predicted velocity for pure water is calculated with Eq. (12.1).

delivered the best fit for the contact line receding velocities on harder substrates ($E = 1.5$ MPa and $E = 0.8$ MPa) and is similar to the values found in recent works (Carlson et al., 2012). Considering the capillary energy, the friction at the contact line, and the deformation energy, it is possible to model the decreasing contact line velocity of the drop rim with increasing Young's modulus of the substrate.

The velocity U for water-silica suspensions was also controlled by the Young's modulus of the substrate, just like the experiments with pure water. Furthermore, a relation between deposition pattern and contact line velocity was found (Lopes and Bonaccurso, 2013).

Figure 12.7 shows the deposit profile after evaporation of the suspension on substrates with different Young's moduli. There were hardly any particles in the outer region on the harder substrates ($E = 1.5$ MPa and $E = 0.8$ MPa), and particle deposition occurred mainly during the last stage of evaporation when the contact line was pinned, in agreement with previous works (Marín et al., 2011). As in previous works on coffee staining (Deegan et al., 1997), a crater-like deposit in the inner region with more particles at the rim than in the center of the stain was observed. The maximum height of the deposit decreased with decreasing Young's modulus of the substrate—leading to a decrease from 5.6 to 2.9 in the ratio between the maximum deposit height and the deposit height in the center of the stain, a parameter for characterizing the degree of coffee staining.

Videos acquired with the inverse microscope showed that the deposition in the inner region occurred in the last moments of drop evaporation ($t < 2$ s), leaving no time for readjustments in the distribution of the particles inside the drop. Thus, the stain pattern depended mainly on how particles were distributed just before the beginning of the third and final stage of evaporation and hardly changed at all during the last seconds of evaporation. This observation differs from other works on the evaporation of suspensions: when the CCR evaporation mode dominates, a large mobility of the particles is usually observed at the end of evaporation when the liquid flow toward the contact line rapidly increased (Deegan et al., 2000; Widjaja and Harris, 2008; Marín et al., 2011). But unlike what was observed previously, in the experiments with controllable evaporation, mode particles accumulated at the contact line as the contact line receded (during CCA evaporation mode).

Figure 12.8 shows this accumulation of particles during the second evaporation stage. The first image shows the drop at the beginning of the second stage of evaporation. Particles were distributed fairly homogeneously in the drop, and only a few particles accumulated close to the contact line due to the internal flow caused by the enhanced evaporation flux at the contact line, as described by Deegan et al. (1997). The subsequent images show how more and more particles were dragged by and accumulated at the drop rim due to the receding contact line. This is indicated also in the schematic representation on the right side of Figure 12.8. Note that the distribution of particles in the bulk of the drop

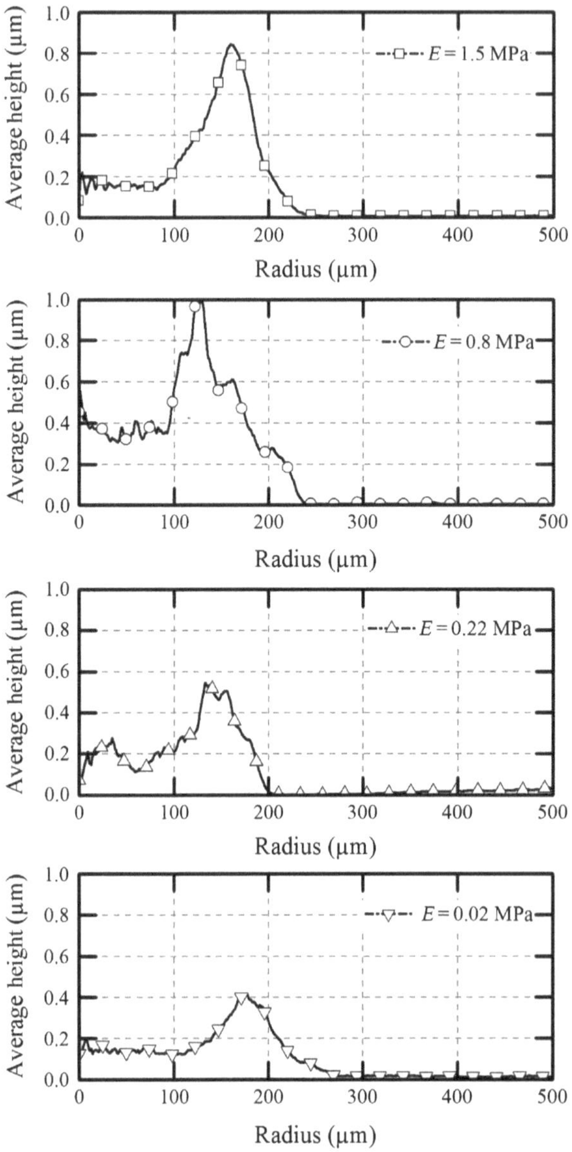

FIGURE 12.7 Average height profile of the silica deposits after the evaporation of water on substrates with different Young's moduli. Thick lines are mean values; thin lines represent the standard deviation subtracted from the mean value. The region between the curves is hashed for clarity of representation. Decreasing the Young's modulus reduced the coffee stain effect.

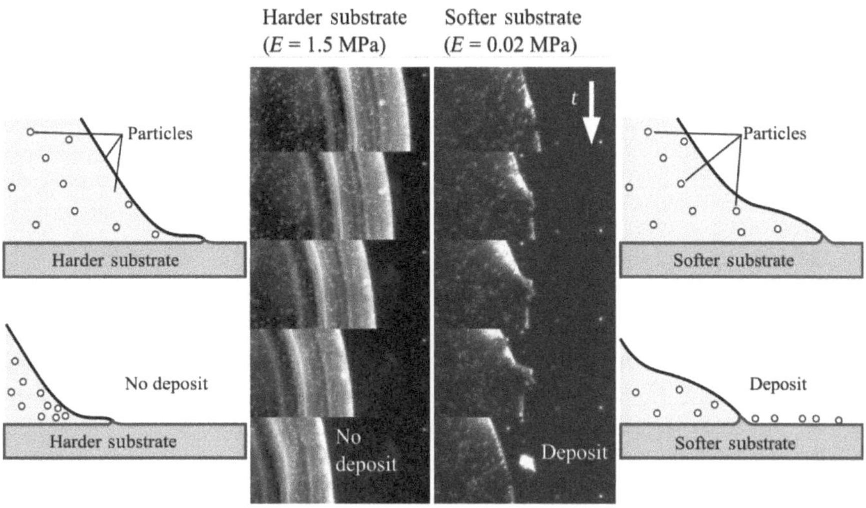

FIGURE 12.8 Sequence of images acquired with an inverse microscope showing the contact line of a water-silica suspension during the stage 2 evaporation (CCA mode) on harder ($E = 1.5$ MPa) and softer ($E = 0.02$ MPa) substrates, respectively, and schematic descriptions of the process. On the harder substrate the contact line dragged along the particles, leaving no deposit behind; a deposit of particles was left behind on the softer substrate.

remained homogeneous during this process. Therefore, the accumulation of particles due to the movement of the contact line was a main step in the formation of the stain profile during evaporation. To explain the different maximal deposit heights observed in the experiments, it is necessary to consider the outer region of the stain, which was formed during the second stage of evaporation.

Figure 12.8 presents a sequence of images of the receding contact line during the evaporation of the water-silica suspensions on harder ($E = 1.5$ MPa) and softer ($E = 0.02$ MPa) substrates. On harder substrates no deposits were left behind during the passing of the contact line. This was also observed in previous publications for the system silica particles and silicon wafer, when the attraction between particles and substrate was weakened by electrostatics (Reynolds et al., 2010). On the other hand, for softer substrates particles or agglomerates of particles were deposited during contact line retraction. Two parameters seemed responsible for the deposition of the particles on the softer substrate: the contact angle of the droplet and the contact line velocity.

The evaporative flux of water is highest at the rim of pinned droplets and is inversely proportional to the contact angle of the drop (Hu and Larson, 2005; Ristenpart et al., 2007; Eggers and Pismen, 2010). Thus, the smaller contact angles on softer substrates increase the evaporative flux at the contact line, increasing also the fluid flow toward the drop rim and facilitating the deposition of particles.

The contact line velocity also controls the deposition process, which we know from dip-coating processes. During dip-coating, a substrate is extracted with a constant speed from a liquid bath containing suspended particles. A thin liquid film is formed on the substrate due to the interplay of viscosity between the liquid and interfacial forces. The velocity of the substrate during dip-coating, which corresponds to the contact line velocity, plays a major role in the thickness of the liquid film and thus on the amount of particles deposited on the substrate. The higher the contact line velocity U, the more particles are deposited on the substrate (Berteloot et al., 2013). The Landau–Levich equation can predict the thickness of a liquid film when a substrate is withdrawn from a liquid bath: $h_\infty = 0.946\sqrt{\gamma/\rho g}\mathrm{Ca}^{2/3}$, with the capillary number $\mathrm{Ca} = \eta U/\gamma$.

The thickness h_∞ of the water film when a surface is withdrawn with velocities U in the range of experimental drop velocities can be calculated using the preceding equations. Here, h_∞ was smaller than the diameter of the silica particles (2 μm) up to a velocity $U \sim 1.5\ \mu\mathrm{m\ s}^{-1}$. Thus, the receding contact line would collect all particles and leave none behind. For higher withdraw velocities h_∞ increased beyond the silica particle diameter. Thus, some particles could be dragged inside the film, and the receding contact line would not be able to collect all of the particles. Consequently, some were left behind.

The deposition pattern resulting from the evaporation of drops of water-silica suspensions could be modified to some extent by the elasticity of the substrate. In fact, the evaporation modes of drops of a suspension can be controlled by tuning the elasticity of the substrate, similar to the case of water. Contact angle and contact line velocity, like the dip-coating process, controlled the thickness of the liquid film (Landau–Levich film) trailing behind the rim of the receding drop. Solid suspended particles smaller than the thickness of the Landau–Levich film became trapped within the film as the contact line receded and were more likely to be deposited during the CCA evaporation mode. Suspended particles larger than the thickness of the film moved together with the contact line and were accumulated until the contact line was pinned. Substrate elasticity can thus be used to control the deposition of particles also during the contact line movement (CCA evaporation mode; Lopes and Bonaccurso, 2013).

Glossary

CCA Constant contact angle
CCR Constant contact radius
PDMS Polydimethylsiloxane

Nomenclature

a Contact radius (m)
a^* Normalized contact radius (with respect to initial contact radius) (1)
Ca Capillary number (1)
d_p Particle diameter (m)
D Diffusion coefficient ($m^2\ s^{-1}$)
E Young's modulus (Pa)
g Gravitational acceleration ($m\ s^{-2}$)
G Shear modulus (Pa)
h Drop height (m)
h_{rim} Deformation height at rim (m)
h_∞ Landau–Levich film thickness (m)
K Curvature of interface ($1\ m^{-1}$)
L Distance between particles (m)
M Molar mass ($kg\ mol^{-1}$)
P^V Vapor pressure (Pa)
r Radial coordinate (m)
R Gas constant ($J(K \cdot mol)^{-1}$)
RH Relative vapor pressure (1)
S_i Displacement field (m)
T Temperature (K)
t Time (s)
t_{evap} Evaporation time (s)
U Contact line velocity ($m\ s^{-1}$)
V Volume (m^3)
$\dot{V}$ Volume evaporation rate ($m^3\ s^{-1}$)
$\dot{W}_{cap}$ Capillary work rate (W)
$\dot{W}_{CL}$ Contact line friction dissipation (W)
$\dot{W}_{def}$ Deformation work rate (W)
z Vertical coordinate (m)
Z_d Substrate displacement (m)

Greek Symbols

γ Surface tension ($N\ m^{-1}$)
γ_l Surface tension of the liquid–gas interface ($N\ m^{-1}$)
γ_S Surface energy of the solid–gas interface ($N\ m^{-1}$)
Δ Work rate dissipation constant (1)
$\Delta\theta$ Contact angle hysteresis (°)
ΔP_L Laplace pressure (Pa)
ΔP Pressure difference (Pa)
ε_{defor} Deformation energy ($J\ m^{-1}$)
η Dynamic viscosity ($Pa \cdot s$)
θ Contact angle (°)
θ_a Advancing contact angle (°)
θ_r Receding contact angle (°)

λ	Capillary length (1)
μ_f	Friction parameter (Pa s)
ν	Poisson's ratio (1)
ξ	Distance within the largest amount of deformation energy is stored (m)
ρ	Density (kg m^{-3})

References

Berteloot, G., Daerr, A., Lequeux, F., Limat, L., 2013. Dip coating with colloids and evaporation. Chem. Eng. Process. Process Intensif. 68, 69–73.

Butt, H.-J., Golovko, D.S., Bonaccurso, E., 2007. On the derivation of Young's equation for sessile drops: nonequilibrium effects due to evaporation. J. Phys. Chem. B 111, 5277–5283.

Carlson, A., Bellani, G., Amberg, G., 2012. Contact line dissipation in short-time dynamic wetting. Europhys. Lett. 97, 44004.

Deegan, R.D., Bakajin, O., Dupont, T.F., Huber, G., Nagel, S.R., Witten, T.A., 1997. Capillary flow as the cause of ring stains from dried liquid drops. Nature 389, 827–829.

Deegan, R.D., Bakajin, O., Dupont, T.F., Huber, G., Nagel, S.R., Witten, T.A., 2000. Contact line deposits in an evaporating drop. Phys. Rev. E 62, 756–765.

Eggers, J., Pismen, L.M., 2010. Nonlocal description of evaporating drops. Phys. Fluids 22, 112101.

Erbil, H.Y., Mchale, G., Newton, M.I., 2002. Drop evaporation on solid surfaces: constant contact angle mode. Langmuir 18, 2636–2641.

Haschke, T., Bonaccurso, E., Butt, H.-J., Lautenschlager, D., Schoenfeld, F., Wiechert, W., 2006. Sessile-drop-induced bending of atomic force microscope cantilevers: a model system for monitoring microdrop evaporation. J. Micromech. Microeng. 16, 2273–2280.

Hu, H., Larson, R.G., 2005. Analysis of the microfluid flow in an evaporating sessile droplet. Langmuir 21, 3963–3971.

Liu, C., Lopes, M.C., Pihan, S.A., Fell, D., Sokuler, M., Butt, H.-J., et al., 2011. Water diffusion in polymer nano-films measured with microcantilevers. Sens. Actuators B Chem. 160, 32–38.

Lopes, M.C., Bonaccurso, E., 2012. Evaporation control of sessile water drops by soft viscoelastic surfaces. Soft Matter 8, 3875–3881.

Lopes, M.C., Bonaccurso, E., 2013. Influence of substrate elasticity on particle deposition patterns from evaporating water-silica suspension droplets. Soft Matter 9, 7942–7950.

Marín, Á.G., Gelderblom, H., Lohse, D., Snoeijer, J.H., 2011. Order-to-disorder transition in ring-shaped colloidal stains. Phys. Rev. Lett. 107, 085502.

Mcdonald, R.E., Fleming, R.I., Beeley, J.G., Bovell, D.L., Lu, J.R., Zhao, X., et al., 2009. Latherin: a surfactant protein of horse sweat and saliva. PLoS One 4, e5726.

Morse, H.W., 1910. On evaporation from the surface of a solid sphere. Proc. Am. Acad. Arts Sci. 45, 7.

Nguyen, T.A.H., Hampton, M.A., Nguyen, A.V., 2013. Evaporation of nanoparticle droplets on smooth hydrophobic surfaces: the inner coffee ring deposits. J. Phys. Chem. C 117, 4707–4716.

Park, J., Moon, J., 2006. Control of colloidal particle deposit patterns within picoliter droplets ejected by ink-jet printing. Langmuir 22, 3506–3513.

Pericet-Camara, R., Best, A., Butt, H.-J., Bonaccurso, E., 2008a. Effect of capillary pressure and surface tension on the deformation of elastic surfaces by sessile liquid microdrops: an experimental investigation. Langmuir 24, 10565–10568.

Picknett, R.G., Bexon, R., 1977. Evaporation of sessile or pendant drops in still air. J. Colloid Interface Sci. 61, 336–350.

Reynolds, T.D., Kalpathy, S.K., Kumar, S., Francis, L.F., 2010. Dip coating of charged colloidal suspensions onto substrates with patterned wettability: coating regime maps. J. Colloid Interface Sci. 352, 202–210.

Ristenpart, W.D., Kim, P.G., Domingues, C., Wan, J., Stone, H.A., 2007. Influence of substrate conductivity on circulation reversal in evaporating drops. Phys. Rev. Lett. 99, 234502.

Rusanov, A.I., 1975a. Theory of the wetting of elastically deformed bodies. I. Deformation with a finite contact angle. Colloid J. USSR, 37.

Rusanov, A.I., 1975b. Theory of the wetting of elastically deformed bodies. II. Equilibrium conditions and work of deformation with a finite contact angle. Colloid J. USSR, 37.

Schonfeld, F., Graf, K., Hardt, S., Butt, H.J., 2008. Evaporation dynamics of sessile liquid drops in still air with constant contact radius. Int. J. Heat Mass Transfer 51, 3696–3699.

Shanahan, M.E.R., 1987. The influence of solid micro-deformation on contact angle equilibrium. J. Phys. D Appl. Phys. 20, 945–950.

Shanahan, M.E.R., 1988. The spreading dynamics of a liquid drop on a viscoelastic solid. J. Phys. D Appl. Phys. 21, 981–985.

Sodtke, C., Stephan, P., 2007. Spray cooling on micro structured surfaces. Int. J. Heat Mass Transfer 50, 4089–4097.

Style, R.W., Dufresne, E.R., 2012. Static wetting on deformable substrates, from liquids to soft solids. Soft Matter 8, 7177–7184.

Style, R.W., Boltyanskiy, R., Che, Y., Wettlaufer, J.S., Wilen, L.A., Dufresne, E.R., 2013. Universal deformation of soft substrates near a contact line and the direct measurement of solid surface stresses. Phys. Rev. Lett. 110, 066103.

Weon, B.M., Je, J.H., 2013. Self-pinning by colloids confined at a contact line. Phys. Rev. Lett. 110, 028303.

Widjaja, E., Harris, M.T., 2008. Particle deposition study during sessile drop evaporation. AIChE J. 54, 2250–2260.

Xu, L., Zhu, H., Ozkan, H.E., Bagley, W.E., Krause, C.R., 2011. Droplet evaporation and spread on waxy and hairy leaves associated with type and concentration of adjuvants. Pest Manag. Sci. 67, 842–851.

Chapter 13

SOLUBLE SUBSTRATE

Christophe Pirat, Jean Colombani and Alexandra Mailleur
Institute of Light and Matter, Université de Lyon, Université de Lyon 1 and CNRS, Villeurbanne, France

The studies of drop evaporation focus on the behavior of the liquid and on the influence of the environment. The substrate always plays a role on the evaporation, through heat exchange and via its wetting properties, but most of the time it is chosen as inert. Generally two model situations can be envisaged with regard to the evaporation of a sessile drop on an inert substrate. Either the contact line is free to move, so that the contact angle remains constant as the radius of the drop decreases, or the contact line is to be pinned, and then the contact angle decreases over time. In both situations the loss of mass obeys a power law with different exponents. In addition, mass loss due to evaporation, mainly at the contact line, is compensated by an outward flow. When colloidal particles are initially present in the liquid, the latter are deposited at the contact line, thus promoting the pinning.

Yet, many situations exist in natural or industrial configurations, where liquid and solid react during the evaporation of a sessile liquid drop. In the case of mineral solids, the reaction at stake is mainly dissolution, and it may lead to a substantial modification for the solid surface as well as for the evaporation dynamics.

Droplet Wetting and Evaporation. DOI: http://dx.doi.org/10.1016/B978-0-12-800722-8.00013-8

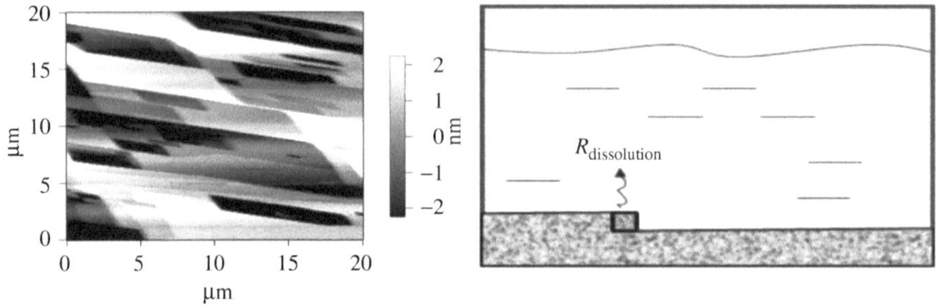

FIGURE 13.1 (Left) Atomic force microscopy picture of the cleavage surface of a gypsum ($CaSO_4$, $2H_2O$) single crystal in water. The dissolution of the solid has led to the digging of parallelepiped (due to the monoclinic lattice of gypsum) etch pits. These pits are enclosed by atomic steps. (Right) The solvation of surface ions leads to the migration of the atomic steps and to the enlargement and deepening of the pits.

This topic is quite new, and no coherent view exists at the moment. We present in this chapter the state-of-the-art of our knowledge on the evaporation of drops on a dissolving substrate.

13.1 Interplay between evaporation and dissolution

13.1.1 Dissolution

Dissolution is a heterogeneous reaction of the form solid + liquid → liquid. Its driving force is the difference in chemical potential between the solid and the initial liquid. When the two chemical potentials equalize, the dissolved species reaches the so-called solubility limit, or saturation concentration, of the solid in the liquid, and the reaction stops.

The microscopic mechanism of dissolution is the detachment and solvation of units (mostly ions) from the solid. This detachment proceeds through the migration of atomic steps, the digging of etch pits, and so on (Figure 13.1).

Many expressions of the dissolution flux have been observed, depending on the material. Its simplest shape is $J_{diss} = k\ (1 - c/s)$, where k is the dissolution rate constant, c is the concentration of dissolved matter, and s is its solubility in the liquid (Rickard and Sjöberg, 1983; Colombani and Bert, 2007). The value of the dissolution rate constant varies in many orders of magnitude from one solid to another. Available values span from ~10^{-14} mol m^{-2} s^{-1}—for instance, for kaolinite—to ~10^{-2} mol m^{-2} s^{-1}—for instance, for rock salt.

Whereas the studies on dissolution were originally limited to macroscopic investigations, the use since the early 2000s of recent far-field (vertical scanning interferometry) and near-field (atomic force microscopy) tools and the comparison with kinetic Monte Carlo molecular simulations have led to a better understanding of the microscopic mechanisms of dissolution. Questions such as "Why is the dissolution flux independent of the density of etch pits at the surface? Why is the dissolution flux dependent on the history of the surface? How do the atomic mechanisms combine to build a macroscopic matter flux?" have thereby begun to find answers (Lasaga and Luttge, 2001; Arvidson and Luttge, 2010).

13.1.2 Interaction Between Dissolution and Evaporation Kinetics

Interaction between dissolution and evaporation will take place if the kinetics of both phenomena are comparable. The evaporation flux of liquid water in normal conditions is $J_e \sim 10^{-7}\ m^3\ m^{-2}\ s^{-1}$. The overall dissolution flux is not necessarily the above-mentioned flux J_{diss}. Indeed, this flux characterizes the transfer of matter from solid to liquid at the surface. But if the diffusion flux J_{diff} of the dissolved species from the surface to the bulk liquid is small compared to the dissolution flux, these species will accumulate close to the surface, the concentration will increase, the driving force will decrease, and the reaction will slow down. The overall flux will then be a combination of J_{diss} and J_{diff} (Colombani, 2008). Two limiting cases are usually defined. For slow-dissolving solids (like oxides), dissolution is the slowest step and therefore drives the whole kinetics, called reaction-driven or surface-driven, kinetics. On the contrary, for fast-dissolving solids (like salts or sugars), diffusion is the slowest step, and it controls the kinetics, so-called diffusion-driven or transport-driven, kinetics.

For the study of drop evaporation, fast-dissolving solids are preferred to guarantee a significant release of matter in the liquid. In this case, diffusion is expected to drive the dynamics. The diffusion flux writes $J_{diff} \sim D\ \Delta c/h$, where D is the diffusion coefficient of the dissolved species, and Δc is the concentration difference inside the drop of height h. If we consider the exemplifying case of a water drop evaporating on NaCl, D is $10^{-9}\ m^2\ s^{-1}$, and Δc amounts at most to the solubility of NaCl in water—that is, 17%. For a drop of height $h \sim 1$ mm, this leads to a diffusion flux $J_{diff} \sim 2 \times 10^{-7}\ m^3\ m^{-2}\ s^{-1}$. Two conclusions can be drawn from this order of magnitude computation:

- The dissolution kinetics of all fast-dissolving solids ($\geq 10^{-9}\ m^3\ m^{-2}\ s^{-1}$) should be faster than the mass transport flux J_{diff} in the drop. Therefore, the overall dynamics of dissolution should be governed by diffusion and be therefore quite identical for all these solids, D varying little from one material to another.
- The evaporation and overall dissolution fluxes are comparable for fast-dissolving solids, and both mechanisms should therefore interfere during drop evaporation on a soluble substrate.

13.1.3 Coupling Mechanisms Between Dissolution and Evaporation

Dissolution is expected to have an influence on the behavior of the liquid and of the solid. Dissolution will first induce the buildup of vertical concentration gradients inside the drop. It is also likely to anchor the triple line. The divergence of the evaporation flux at the pinned triple line will then lead to outward flows inside the drop, as in the case of the evaporation of colloidal suspensions, therefore creating horizontal concentration gradients. These matter gradients may create hydrodynamic instabilities, leading either to gravitational convection (solutal Rayleigh–Bénard) or capillary convection (solutal Marangoni–Bénard convection). These flows will in turn modify the evaporation and dissolution kinetics.

In addition, these flows of matter inside the drop are expected to lead to a transfer of matter from the center to the periphery of the drop, thereby inducing the deposit of a ring at the end of the evaporation (like in the coffee stain situation; see Chapter 4). Because the dissolution proceeds via etch pitches deepening, the roughness of the substrate is also likely to be largely increased.

In the following, we provide a short overview of recent works that deal with dissolution and/or solute concentration effects during the evaporation of a sessile drop.

13.2 Evaporation in presence of solute concentration with or without dissolution

A study of the effect of saline Rayleigh convection on an evaporating sessile drop deposited on a nonsoluble substrate was recently carried out by Kang et al. (2013). The authors investigated the evaporation dynamics using NaCl aqueous drops on a hydrophobic smooth substrate and showed that the evaporation-induced density gradient affects the evaporation dynamics. Starting with a homogeneous concentration for the solute, concentration and temperature gradients progressively build up due to the fluid motion initiated by the evaporation at the drop surface. The resulting surface tension gradient will be the cause of a Marangoni flow, while another type of flow, due to the evaporation-induced density gradient (so-called evaporation-induced Rayleigh convective flow) and the resulting spatially nonuniform buoyancy force, will be generated as well. The internal flow in an almost hemispherical evaporating drop was visualized by a particle image velocimetry technique. It was shown that a stable toroidal vortex takes place, the flow being directed upward at the center, whatever the initial concentration in the range 0.01–10.00 wt% mixture (see Figure 13.2 (Top)).

Assuming that the problem is axially symmetric, a set of nondimensionalized governing equations was derived for the mass, momentum, and solute conservation

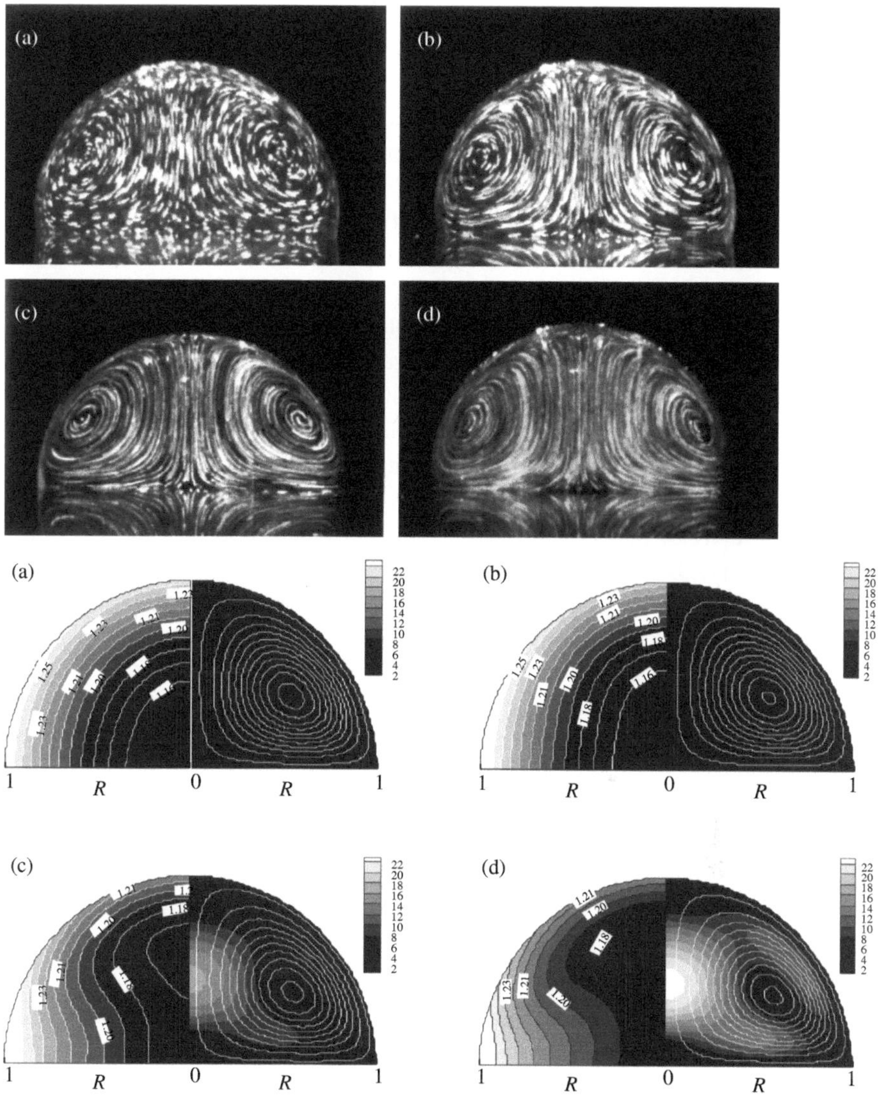

FIGURE 13.2 (Top) Flow visualization with fluorescent tracers for NaCl concentrations of (a) 0.01 wt%; (b) 0.1 wt%; (c) 1 wt%, with an exposure time of 20 s; and (d) 10 wt%, with an exposure time of 2 s. (Bottom) Effect of Rayleigh number on the concentration distribution and flow field in the pseudo-steady regime for $Pe = 0.197$: (a) $Ra = 5 \times 10^3$; (b) $Ra = 10^4$; (c) $Ra = 5 \times 10^4$; and (d) $Ra = 10^5$. The left half of the plane is for the concentration distribution, and the right half of the plane is for the velocity field. *Reproduced with permission from Kang et al. (2013). Copyright 2013, AIP Publishing LLC.*

equations, introducing the nondimensional Peclet (Pe), salinity Rayleigh (Ra), and Prandtl (Pr) numbers, respectively. $Pe = a_0u_0/D$, $Ra = g\beta c_0 a_0^3/\nu D$, and $Pr = \nu/D$, where $a_0 = 10^{-3}$ m stands for the initial base radius of the drop, $u_0 = 10^{-6}$ m s^{-1} is the constant contracting velocity of the drop surface, $g = 9.81$ m s^{-2} is the acceleration of gravity, $\beta = 7.10^{-3}$ wt%$^{-1}$ is the solutal expansion coefficient, $c_0 = 1$ wt% is the initial concentration, $\nu = 10^{-6}$ m s^{-2} is the kinematic viscosity, and $D = 10^{-9}$ m s^{-2} is the molecular diffusivity of the solute.

The main findings are as follows. Because almost no fluid motion was observed during the evaporation of a drop of pure water, it was assumed that the thermal contribution to the flow was negligible. Interestingly, it was observed that the boundary condition at the drop surface was no slip instead of the free surface condition. It is likely due to either the presence of a surfactant in the liquid or a contaminant in the air, even if the experiments were carefully conducted in a closed chamber at a temperature of 25°C and a relative humidity of 45%. This effect has been recently debated for other systems where the slip condition at a liquid–gas interface plays a crucial role (Bolognesi et al., 2014; Berkelaar et al., 2014; Ybert and Di Meglio, 1998).

Because the convection time is greater than the molecular diffusion time, the low flow field (of creeping-flow type) can be decoupled from the concentration field. It follows that the body force ($Ra(a/a_0)^2(c - c')/c_0$), where a is the base radius, c is the concentration, and c' is the concentration at the origin), proportional to the effective concentration, is rapidly time-invariant and that this pseudo-steady state is rapidly reached, even in the presence of convection, as checked by the authors experimentally and numerically (Figure 13.2 (Bottom)). Thus, in this configuration of $Pr >> 1$, the flow field can be considered by neglecting its time history, and it is only governed by the Rayleigh number, while the concentration field is driven by the Peclet number. As the body force determines the fluid motion, a scaling analysis shows that the flow velocity is proportional to (RaPe) in the low Rayleigh-number-limit, while it scales approximately as ($Ra^{1/2}$) in the high Rayleigh-number-limit, probably because of the mixing enhancement (see Figure 17 in Kang et al., 2013, for a comparison of computed and measured velocity versus evaporation time for various Rayleigh numbers).

It is well known that a polymeric coating could affect the wettability of a substrate. It is particularly the case for water drops deposited on substrates coated with a hydrophilic soluble polymer (i.e., a good solvent for the drop). Tay et al. (2010) have extensively studied the wetting dynamics and the drying of water drops deposited onto soluble-polymer-coated silicon substrates, using either a neutral polymer (PDMA) or two charged polymers—namely, a cationic poyelectrolyte (PDADMAC) or a zwitterionic polymer (PZ). In all cases, around a 200 nm-thick coating is deposited by spin coating. The authors conclude that both the spreading and the pinning dynamics depend on the type of polymer used. Basically, they observe that after an initially

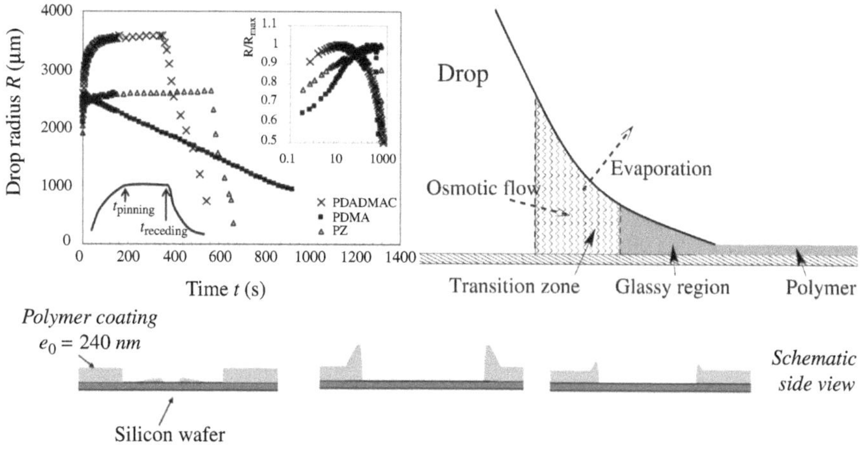

FIGURE 13.3 (Top Left) Radius evolution for a water drop with the tree polymers during spreading, pinning, and receding. (Top Right) Schematic drawing of the mechanisms at stake in the vicinity of the contact line during evaporation. (Bottom) After complete evaporation, top view and schematic height profile of the deposit from left to right PDMA, PDADMAC, and PZ polymers. *Reproduced in part from Tay et al. (2010) with permission of The Royal Society of Chemistry.*

spreading stage, for charged polymers, the drop undergoes pinning and then recedes at the very end of the evaporation, while it recedes immediately after the spreading for PDMA (with a constant contact angle; Figure 13.3 (Top Left)).

Since the polymers are not chemically grafted onto the substrate, they are essentially dissolved into the drop. A detailed characterization of the residual deposit shows that, for charged polymers, it forms a ring-like accumulation at the periphery of the drop, while it forms a crater at the center of the drop for the neutral polymer (Figure 13.3 (Bottom)).

The spreading dynamics is similar for the three configurations, but the difference in the evaporation dynamics between neutral and charged polymers can be explained as follows. Rapidly after the spreading, polymer molecules accumulate in the vicinity of the contact line, resulting in a glassy zone that will not contribute further to the evaporation. A concentration gradient builds up and makes a transition zone between the bulk and the glassy area, whose size will be essentially driven by the nature of the polymer. The dynamics of the contact line will depend on the relative strength of the evaporative flux (that will tend to make the contact line recede), and the osmotic pressure present in the transition zone (that will tend to promote a flux of water toward the edge; Figure 13.3 (Top Right)).

For the charged polymers, these two contributions equilibrate, which causes the pinning of the contact line in conjunction with the formation of a deposit of polymer. For the neutral polymer, the osmotic pressure is negligible, because it originates from the high concentration of electric charges. As a consequence, the contact line starts to recede immediately after spreading, leaving a continuous glassy deposit on the surface. It must be noted that when salt is added to charged polymers, as expected, the pinning disappears because the osmotic pressure weakens due to the induced screening of the charges in the transition zone. For a detailed study of the spreading dynamics on soluble substrates, the reader can refer to Dupas et al. (2013) and the references herein.

If we now focus on the evaporation of a drop deposited on an initially flat and smooth soluble substrate (i.e., not a soluble coating), it is expected that the solvent will quickly initiate a process of dissolution. The consequences are twofold. First, a concentration gradient will be established inside of the drop. On the other hand, the surface condition will be changed during evaporation.

Cordeiro and Pakula (2005) investigated numerically the behavior of evaporating nanometer-sized drops at nonsoluble and soluble surfaces. They performed in-lattice Monte Carlo simulations to investigate the evaporation dynamics, as well as the deposit formation. They used the *cooperative motion algorithm* method with periodic boundary conditions by considering liquid drops on solid in equilibrium with a vapor phase. The details about the method can be found in Cordeiro and Pakula (2005). Roughly, this method involves an ensemble of beads on a tri-dimensional lattice. There is no empty site. Three kinds of beads are considered in their approach: pure solvent particles, substrate particles, and vacuum. Each bead can replace one of its neighbors. At each time step, the probability of a jump for a bead is given by the boltzmann factor $p = \exp(-E/kT)$, the sum of all the displacements being zero. kT is the temperature at equilibrium, and E is the energy of interaction of the moving bead given by $E_i = n_{ij}e_{ij} + n_{ik}e_{ik}$ for the site i, with e the interaction energy between two beads and n the number of interactions. The jump is considered if and only if p is greater than a random number. To account for evaporation, if a solvent bead is surrounded by vacuum beads, then it is transformed in a vacuum bead. Gravity is not taken into account. When the energy interactions are carefully set via an appropriate density function, it is possible to observe microwell and ring formation for a swelling substrate.

The numerical results are obtained for a drop with an initial contact angle being roughly 80°, measured by fitting the shape of the generated drop as a spherical cap. It shows that if the evaporation takes place on a nonsoluble homogeneous and smooth substrate, as expected, the contact radius decreases and the contact angle remains constant (Figure 13.4 (Right a)). For a soluble substrate, first the system follows this regime, but a microwell is progressively dug, and a peripheral ring is formed underneath the drop that will cause a pinning of the contact line. Figure 13.4 (Left) shows a comparison of the evaporation dynamics

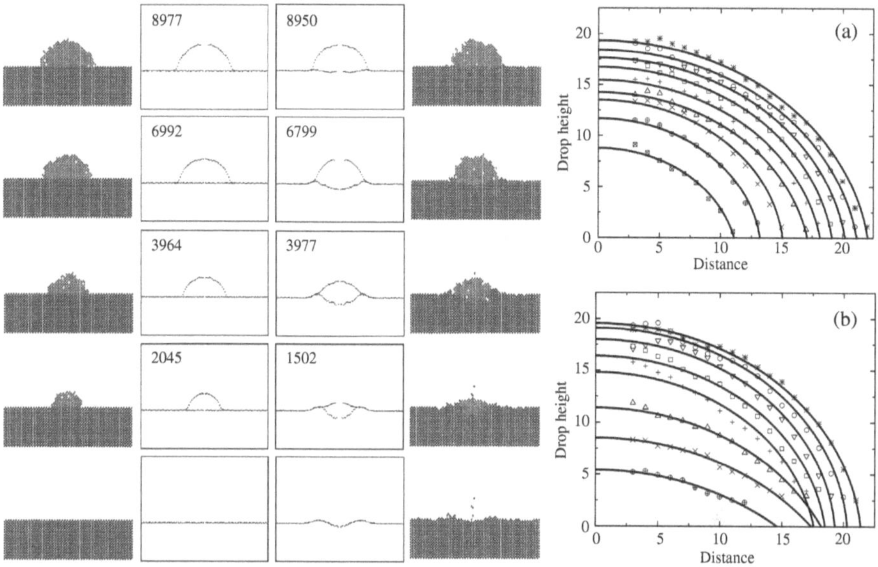

FIGURE 13.4 (Left) Snapshots of evaporating drops. Left: nonsoluble substrate, Right: soluble substrate. For each snapshot, the radial average is also shown, together with the number of solvent beads remaining within the drop. (Right) Evolution during the drop evaporation, and for each profile, the corresponding fit with a spherical cap. (a) Nonsoluble substrate, (b) soluble substrate. *Reprinted with permission from Cordeiro and Pakula (2005). Copyright (2005) American Chemical Society.*

for nonsoluble and soluble substrates. As a result, the contact radius becomes constant, and the contact angle starts to decrease (Figure 13.4 (Right b)).

This study supports that the evaporation/dissolution of a drop modifies the topography of the surface and in turn affects the evaporation dynamics, partly because of the induced pinning of the triple line. It must be noted that the authors claim that the results can be extended to micrometer-sized or even millimeter-sized drops.

It should be pointed out that dissolution-induced watermarks can unexpectedly occur during evaporation. Such an example is presented in Belmiloud et al. (2012). The dynamics of the drying of a sessile drop of ultrapure water on silicon substrates have been studied. The silicon surfaces were made by dHF-etching (0.5% HF for 5 min, so-called hereafter "hydrophobic"), while more hydrophilic surfaces were obtained by exposing these surfaces to the air for 1 day (so-called hereafter hydrophilic), allowing for the formation of a hydrophilic layer of oxidized Si. The experiments have been conducted in a well-controlled ultra-clean environment (clean room class 1, relative humidity of 40%, temperature of 22°C) with deoxygenated water. Sessile drop evaporation was carried out in both hydrophobic (just after dHF) and hydrophilic configurations. On a hydrophobic Si surface, a pinned phase

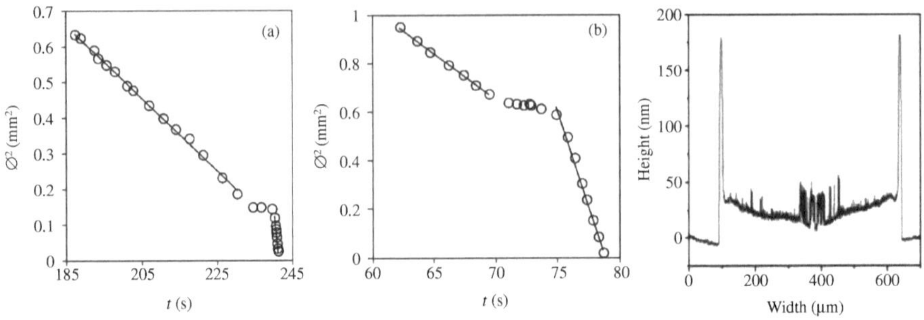

FIGURE 13.5 (Left) Evolution of the square diameter of a water drop deposited on a Si wafer (a) and a SiO_2 wafer (b) at the end of the evaporation—that is, before and after the formation of a watermark (plateau). (Right) Topology taken at the center across a ring-shaped watermark for a hydrophilic Si wafer. *Reprinted with permission from Belmiloud et al. (2012). Copyright 2012, The Electrochemical Society.*

was first observed (constant contact radius, decreasing contact angle starting from 70°), followed by a receding phase (contact angle of 32°, decreasing radius), as shown in Figure 2 of Belmiloud et al. (2012).

By comparing the evaporation on Si and SiO_2 surfaces in the receding phase, it is observed, as expected, that smaller evaporating drops dry slower with a higher receding contact angle. In both cases, the liquid-solid surface area decreases linearly with time. Focusing on the very last stage of evaporation, a similar behavior is observed for Si and SiO_2 wafers. Interestingly, it is shown that just before complete evaporation, the contact line jumps due to the depinning on a residual ring, or so-called watermark (Figure 13.5 (Left)).

After the depinning, the decrease in contact surface is faster than before the pinning begins, while the contact angle is measured almost constant around 2° for both Si and SiO_2 surfaces. A high-resolution profilometry (HRP) of the surfaces clearly shows a so-called coffee ring made of thin residues, while a cluster of bigger residues is visible at the center (Figure 13.5 (Right)).

The thickness of the ring is larger for SiO_2 than for Si. In this specific case, the deposits are known to be a result of chemical reactions. First, the ambient oxygen that diffuses into the water drop will create silica and silicic acids. The resulting $Si(OH)_4$ molecules will form hydrated polymeric chains due to their rising concentration throughout the evaporation. At the final stage, these silica-based residues of very low solubility in water will precipitate and will show dynamics similar to those observed with evaporating colloidal solutions.

Beyond this specific case, which shows how complex the cleaning process of wafers for microelectronics applications can be, it must be pointed out that it is common for the process of evaporation of a sessile drop to involve some sort of reaction with the substrate that might significantly affect both the evaporation dynamics and the substrate integrity.

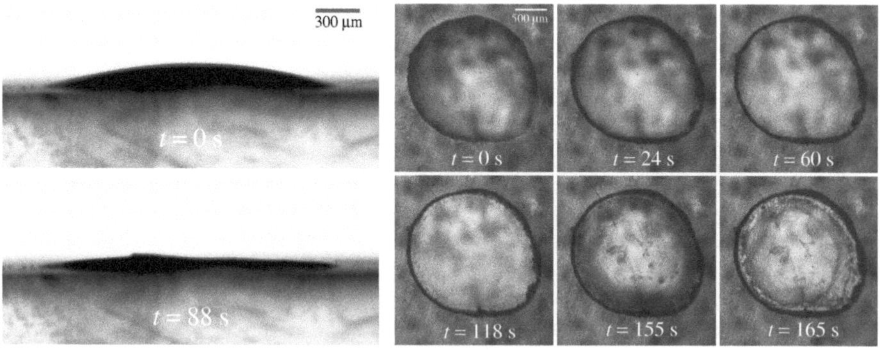

FIGURE 13.6 (Left) Side view of a 3 μL pure water drop deposited on a NaCl substrate at initial time and after complete evaporation ($t = 88$ s). The initial contact angle is about 15°. (Right) Top view of a 3 μL pure water drop on a NaCl substrate at different times. The deposit is clearly visible.

13.3 Water drop evaporation on a flat crystal salt substrate

The evaporation of a pure water drop deposited on NaCl soluble single crystals has been explored. After the crystals have been cleaved (10 × 10 mm, 2 mm in height), they are polished with sandpaper (grit size down to 3 μm). A water drop of a few microliters is gently deposited on the substrate. A contact angle measurement device is used to follow the evaporation dynamics. The drop evaporates at ambient pressure inside a transparent closed cell ($100 \times 100 \times 100$ mm^3) to avoid any perturbations from the ambient. The side view is recorded throughout the evaporation via a CCD camera fitted with a macro lens positioned laterally. To complete this observation, an upright reflected light microscope fitted with a high-resolution camera is used to observe surface modifications after evaporation.

These preliminary experiments showed unambiguously the strong influence of the substrate dissolution on the evaporation process for a water drop. After drop deposition, it is observed that the contact line is nearly circular for well-polished crystals substrates. First, it induces a pinning of the triple line during almost the whole evaporation (Figure 13.6 (Left)), instead of a receding triple-line typical of water drops on soft inert substrates. Second, the evaporation/dissolution interplay leads to an increase of the roughness of the solid surface. Finally, like in colloidal suspensions, the evaporation of a pure water drop on a soluble substrate results in the deposit of a ring at the periphery of the drop. Top view observations clearly reveal the accumulation of a ring-shaped deposit at the edge of the drop that seems to be homogeneous at the end of the evaporation (Figure 13.6 (Right)).

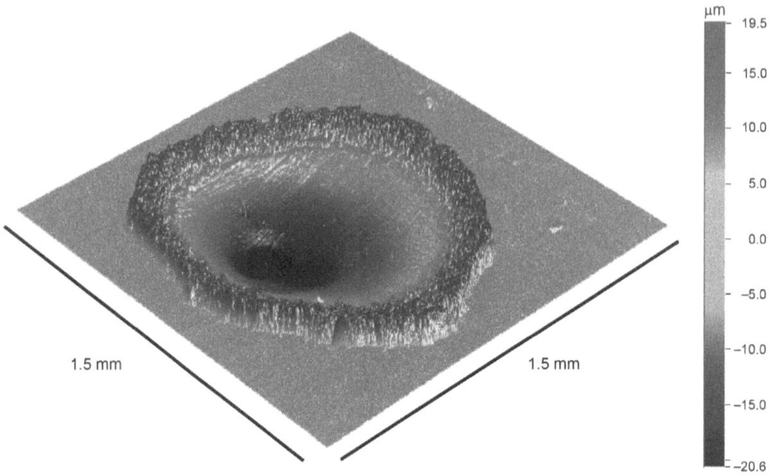

FIGURE 13.7 Three-dimensional profile obtained by profilometry after complete evaporation.

A typical 3D profilometry of the surface is shown in Figure 13.7. It reveals a digging at the center as deep as 15 µm, while the deposit reaches around 15 µm above the surface. These results are consistent with those obtained for the numerical study of the drop evaporation on the soluble substrate. On the contrary, no digging was measured for the evaporation with watermarks formation on a silicon wafer discussed above.

In the long term, this work demonstrates the influence of the dissolution on the kinetics of evaporation of a drop and the formation of the deposit formed after the evaporation process.

Glossary

Solvation also called dissolution, is the process of attraction and association of molecules of a solvent with molecules or ions of a solute.

Osmotic pressure minimum pressure which needs to be applied to a solution to prevent the inward flow of water across a semipermeable mebrane.

References

Arvidson, R.S., Luttge, A., 2010. Mineral dissolution kinetics as a function of distance from equilibrium—New experimental results. Chem. Geol. 269, 79–88.

Belmiloud, N., Tamaddon, A.H., Mertens, P.W., Struyf, H., Xu, X., 2012. Dynamics of the drying defects left by residual ultra-pure water droplets on silicon substrate. ECS J. Solid State Sci. Technol. 1 (1), 34–39.

Berkelaar, R.P., Dietrich, E., Kip, G.A., Kooij, E.S., Zandvliet, H.J., Lohse, D., 2014. Exposing nanobubble-like objects to a degassed environment. Soft Matter 10 (27), 4947–4955.

Bolognesi, G., Cottin-Bizonne, C., Pirat, C., 2014. Evidence of slippage breakdown for a superhydrophobic microchannel. Phys. Fluids 26, 082004.

Colombani, J., 2008. Measurement of the pure dissolution rate constant of a mineral in water. Geochim. Cosmochim. Acta 72, 5634–5640.

Colombani, J., Bert, J., 2007. Holographic interferometry study of the dissolution and diffusion of gypsum in water. Geochim. Cosmochim. Acta 71, 1913–1920.

Cordeiro, R.M., Pakula, T., 2005. Behavior of evaporating droplets at nonsoluble and soluble surfaces: modeling with molecular resolution. J. Phys. Chem. B 109, 4152–4161.

Dupas, J., Verneuil, E., Talini, L., Lequeux, F., Ramaioli, M., Forny, L., 2013. Diffusion and evaporation control the spreading of volatile droplets onto soluble films. Interfacial Phenom. Heat Transf. 1 (3), 231–243.

Kang, K.H., Lim, H.C., Lee, H.W., Lee, S.J., 2013. Evaporation-induced saline Rayleigh convection inside a colloidal droplet. Phys. Fluids 25, 042001.

Lasaga, A., Luttge, A., 2001. Variation in crystal dissolution rate based on a dissolution stepwave model. Science 291, 2400–2404.

Rickard, D., Sjöberg, E.L., 1983. Mixed kinetic control of calcite dissolution rates. Am. J. Sci. 283, 815–830.

Tay, A., Lequeux, F., Bendejacq, D., Monteux, C., 2010. Wetting properties of charged and uncharged polymeric coating—effect of the osmotic pressure at the contact line. Soft Matter 7 (10), 4715–4722.

Ybert, C., Di Meglio, J., 1998. Ascending air bubbles in protein solutions. Eur. Phys. J. B 4, 313–319.

Chapter 14

DROPLETS ON LIQUID INTERFACES

David Brutin
Mechanical Engineering Department, Aix-Marseille University, Marseille, France

While most of the situations covered here involve solid substrates, in this chapter, we examine liquid surfaces. This liquid/liquid system is extremely complex because the liquid substrate is never inert. Several new phenomena are involved here, such as solubilization, surfactant transfer, and classic ones like diffusion, convection, and evaporation. Some of the behaviors of droplet spreading and evaporation on liquid surfaces exhibit self-pulsation or self-propulsion. The source of energy is still a source of interest, but it seems to result from chemical-to-mechanical energy conversion through hydrodynamic instabilities. Strong surface tension gradients are the origin of the self-motion and deformation of small droplets floating on liquid surfaces. Pimienta and Antoine (2014) identified three mechanisms leading to spontaneous mobility. This chapter concentrates mostly on two of them.

14.1 Introduction

The movement of microscopic organisms through liquids is due predominantly to the viscous forces, with the effect of the inertial forces being negligible.

Droplet Wetting and Evaporation. DOI: http://dx.doi.org/10.1016/B978-0-12-800722-8.00014-X

The best-known problem that neglects all inertial forces is Stokes' solution for the slow, steady fluid motion past a sphere, in which the velocity field can be described in terms of singularities situated at the center of the sphere. The movement of microscopic organisms is determined by placing distributions of these singularities inside the surface of the organism and satisfying all boundary conditions. The motions that are considered involve organisms that propagate some kind of disturbance along filaments of circular cross-section with small radii.

The motion demonstrated by liquid droplets on surfaces is possible for two non-miscible liquids or with the use of surfactants. The objects are mostly considered nondeformable in practice, as shown later in the chapter. The liquid droplet's shape induces its motion through phenomena which do not exist when the droplet sits on a solid substrate. Solubilization and surfactant transfer will combine with spreading and evaporation to induce gradients of surface tension. The different situations encountered, such as self-motion and self-pulsation, are developed hereafter.

14.2 Self-motion of camphor grains

Since 1686, the "camphor grain dance" has been known. If a grain of camphor is placed on the surface of water, it starts moving around in jerks and continues its motion for many seconds. The mechanism of this dance is due to the nonuniformity of the grain's surface. One corner of it dissolves in water more rapidly than the other parts. Camphor lowers the water's surface tension; hence, the area in contact with the above weak corner has a lower surface tension than the rest of the grain. The higher surface tension of the opposite side pulls the grain in and it moves so that the weak corner forms the rear of the floating body. The dance stops when the dissolution stops; this mass transfer across the solid-liquid interface also stops when the liquid surface tension is below a critical value (Figure 14.1).

14.3 Self-motion, self-rotation, and self-pulsation

When the compound deposited on the surface is liquid, the droplets can deform and adapt their shape, minimizing the free energy. The solvent used as propellant is water-miscible and the droplets' lifetime is only a few minutes. Different situations are encountered depending on the surfactant concentration as presented in Figure 14.2 for the same oil droplet volume deposited. In this example, for a low CTAB concentration of $0.25\ \mathrm{mmol\ L^{-1}}$, small pulsating and moving droplets (1) are observed, which can lead to moving droplets when the tiny droplets' expulsion process is not axi-symmetric. Increasing the CTAB concentration to $0.5\ \mathrm{mmol\ L^{-1}}$ leads to pulsating droplets (3) with regular pulsation that is later

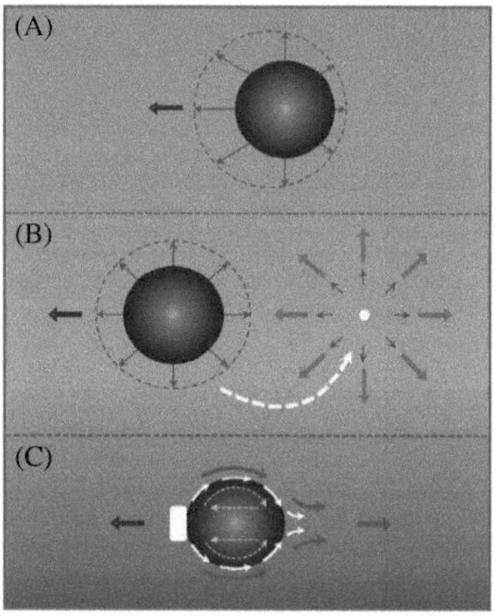

FIGURE 14.1 Top-view schemes of three identified mechanisms of surface tension driven motility: (A) Surface tension imbalance around the floating system; (B) distant surface liquid mixing pushing the object; (C) inhomogeneous water/oil surface tension creating an outside flow which drives the drop forward. The red arrows represent surface tension forces. The green and yellow arrows indicate the fluid flows in the liquid substrate and inside the drop, respectively, while white arrows show the trajectory of surfactant molecules expulsed from the floating object (Pimienta and Antoine, 2014). (For interpretation of the references to color in this figure legend, the reader is referred to the online version of this book.)

explained. When the concentration reaches 10 mmol L^{-1}, then, self-rotating droplets are observed with a number of arms evolving.

The situation of pulsating droplets (2) is analyzed by Antoine and Pimienta (2013). The authors analyzed and modeled the phenomenon of regular self-pulsations by a simple mass-spring mechanical model of the sessile drop of oil (dichloromethane). The authors decompose the process into several steps: (i) the rim movement is quasi-periodic with the period slightly increasing. (ii) One clearly observed the presence of rebounds for both the outward and inward rim movement. The lower part of the rim remains attached to the droplet and follows the movement of the upper part but with a delay. (iii) The pulsations are clearly not time-symmetric: the ascending spreading parts are slightly longer than the receding parts. This time asymmetry increases significantly over time. (iv) When the minimum of the drop radius is at its lowest, the following maximum is above the maximum average. This mechanical correlation seems to appear every two to three pulsation periods. (v) Then, the period of the pulsations is much larger than the natural oscillation period of the drop. The final stage of the droplet consists of the complete disappearance of the droplet by solubilization and evaporation.

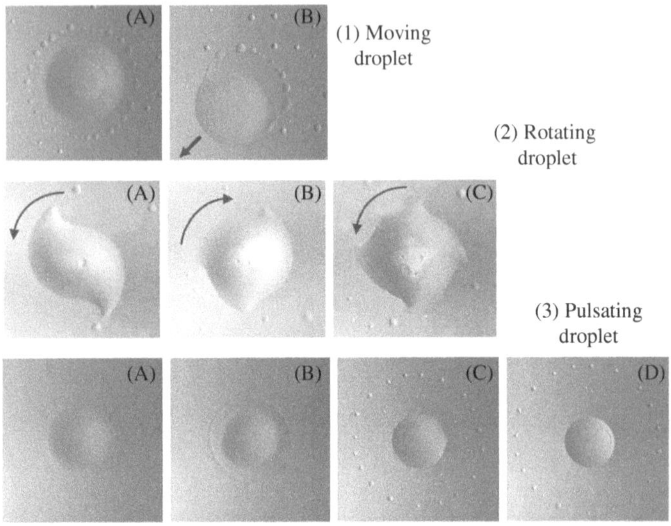

FIGURE 14.2 Top views of 25 μL dichloromethane (oil) droplet in different situations: (1) radial spreading and directional motion when placed on a 0.25 mmol L^{-1} CTAB aqueous solution. The directional motion is indicated by a red arrow. (2) Two, three, and four-armed rotors observed when placed on a 10 mmol L^{-1} CTAB aqueous solution. The direction of the rotating motion is indicated by a red arrow. (3) Four successive top views of the drop during the pulsating regime observed when placed on a 0.5 mmol L^{-1} CTAB aqueous solution (Pimienta and Antoine, 2014).

These stages are summarized in Figure 14.3, which presents the evolution of the droplet radius decrease experimentally measured and modeled by Antoine and Pimienta (2013). The authors based their model on the rim breakup and reformation of mass through a dewetting process. Furthermore, the model takes into account the fact that both solutal and thermal Marangoni convection are leading to these patterns. Indeed, evaporation is an obvious phenomenon occurring. Dichloromethane is highly volatile and evaporation has been shown to be responsible for one-third of the mass decrease while the other two-thirds are lost by dissolution. The heat required for evaporation results in a lower surface temperature. The evaporation is more intense for curved surfaces and thin liquid layers, but the cooling effect at the droplet surface strongly depends on the thickness of the drop and heat diffusion in the liquid layer. For the dichloromethane/water system, the temperature decrease is estimated to be a few degrees difference between that in the droplet center and that on its rim. The water surface is expected to be hotter than the droplet surface. Since surface tension decreases linearly with an increase in surface temperature at a rate of about 0.1 mN m^{-1} K^{-1}, then it is reasonable to think that evaporation can increase the drop surface tension by a few tenths of mN m^{-1}. The rigidity and elasticity of the surfactant-covered droplets have been also underlined as playing a nonnegligible role.

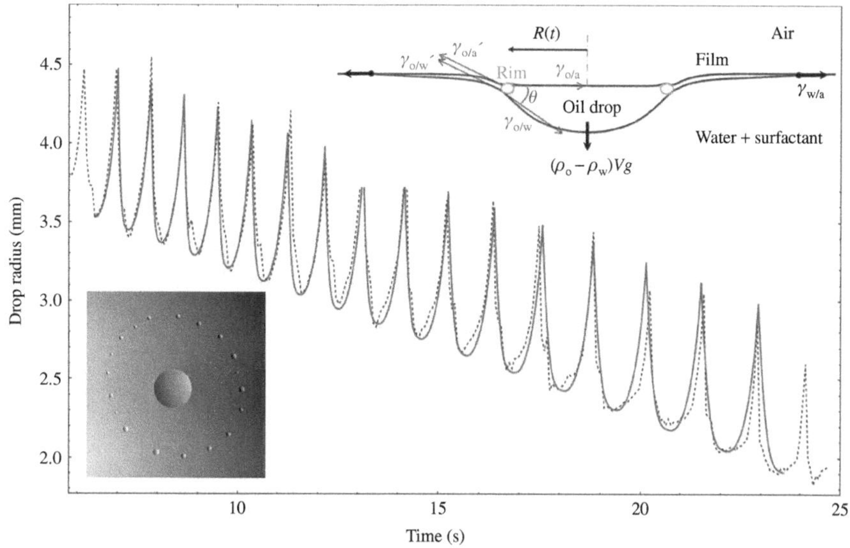

FIGURE 14.3 Drop radius evolution during a pulsating regime of a 25 μL dichloromethane drop placed on a 0.5 mmol L^{-1} CTAB aqueous solution. The dashed blue line corresponds to the experimental data whereas the solid red curve results from the mass-spring model detailed. Time modulation of the effective spreading parameter $S_{eff}(t)$ used in the mass-spring model was used (Antoine and Pimienta, 2013). (For interpretation of the references to color in this figure legend, the reader is referred to the online version of this book.)

Glossary

Surfactant A compound that lowers the surface tension.

References

Antoine, C., Pimienta, V., 2013. Mass-spring model of a self-pulsating drop. Langmuir 29, 14935–14949.

Pimienta, V., Antoine, C., 2014. Self-propulsion on liquid surfaces. Curr. Opin. Colloid Interface Sci. 19(4), 290–299. Available from: http://dx.doi.org/10.1016/j.cocis.2014.04.001.

Chapter 15

RADIATIVE HEAT TRANSFER IN DROPLETS

David Brutin, Florian Carle and Fabrice Rigollet
Mechanical Engineering Department, Aix-Marseille University, Marseille, France

Evaporating droplets are usually volatile fluids that are also semitransparent to infrared rays. Radiative heat transfer is disregarded while it contributes the heat transfer from all interfaces (both substrate and liquid interfaces). Infrared cameras make it possible to observe the radiative heat flux coming from the heated substrate and passing through the semitransparent fluid and then through the surrounding gas. Thus, the temperature provided by the camera is not the real fluid temperature but an averaged temperature of a column going from the substrate to the camera. In order to access a more precise column of fluid temperature, several assumptions are needed and developed in this chapter.

15.1 Theoretical aspects

A body emits thermal radiation whose wavelengths and intensities depend on its temperature. Our heated substrate can be considered as a black body—that is, it

Droplet Wetting and Evaporation. DOI: http://dx.doi.org/10.1016/B978-0-12-800722-8.00015-1

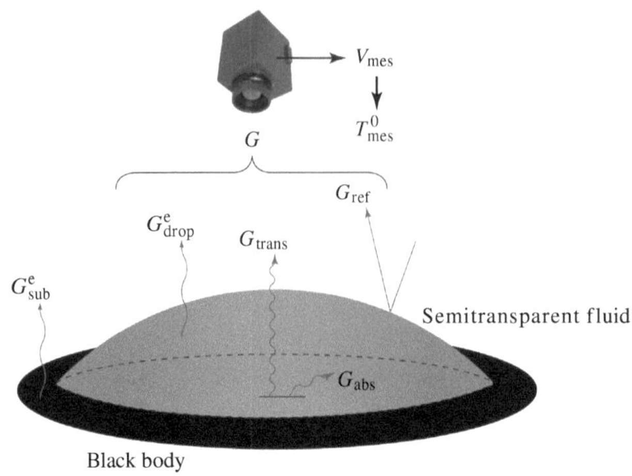

FIGURE 15.1 Emission, reflection, transmission, and absorption processes associated with semitransparent fluid (Carle, 2014).

emits thermal radiation G^e_{sub} following Stephan–Boltzmann's law $\varepsilon\sigma T^4$, where ε is emissivity and $\sigma = 5.67 \cdot 10^{-8}$ W m^{-2} K^{-4} is Boltzmann's constant. This radiation reaches the infrared camera captor matrix, which converts photons into electrons in a spectral band. Here in the calculations, the spectral band is of 3–5 μm. The measured current V_{mes} will then be converted into a temperature T^0_{mes} based on the camera calibration curve, creating a false color image of the thermal scene (Figure 15.1).

However, in the case of a real body, this conversion is more complex because the camera must convert radiosity—that is, the total radiation leaving the surface—and will provide the apparent temperature of a black body. One can consider, to a first approximation, that all objects behave as gray bodies; in other words, their properties do not depend on the wavelength, and Kirchhoff's law is applicable—that is, absorptivity α is equal to emissivity ε. The radiosity G is the total radiation leaving the thermal scene (Eq. (15.1)), composed of emitted radiation G_e—coming from the droplet at temperature T_d, reflected radiation G_{ref}—flux reflected by the droplet coming from the environment at the temperature T_a—and transmitted radiations G_{trans} from the substrate through the fluid:

$$G = G_e + G_{ref} + G_{trans} \tag{15.1}$$

$$V^0(T^0) = \sigma T^4_{cam} = E\sigma T^4_{droplet} + R\sigma T^4_{amb} + T\sigma T^4_{sub} \tag{15.2}$$

E, R, and T correspond to emission, reflexion, and transmission factors (Huetz-Aubert et al., 1995), respectively, and are functions of the emissivity ε, the reflectivity ρ, and the transmissivity τ. These factors can be simplified with assumptions and expressed as the following (Incropera et al., 2007):

$$E = (1 - \rho)(1 - \tau) \tag{15.3}$$

$$R = \rho \tag{15.4}$$

$$T = (1 - \rho)\tau \tag{15.5}$$

Therefore, by taking into account all of the previous statements, Eq. (15.2) can be rewritten to describe the real temperature; detailed calculations can be found in Sobac (2012):

$$T = \sqrt[4]{\frac{1}{(1 - \rho)\varepsilon} T_{\text{cam}}^4 - \rho T_{\text{amb}}^4 - \tau(1 - \rho) T_{\text{sub}}^4} \tag{15.6}$$

To obtain a relevant value of the temperature, knowing the fluid's infrared properties (emissivity, reflectivity, and transmittivity) is mandatory. These properties are well known and documented for opaque plane surfaces of usual materials, but there is a clear lack of knowledge in the case of semitransparent fluids. These types of fluids (liquids or gas) have their infrared properties modified by their thickness, angle, and temperature: $\varepsilon(e, \theta, T)$.

During the evaporation, fluids are evaporated from ambient temperature to a maximum of 60°C. It has been shown that ethanol absorptivity α (equal to ε for gray bodies) only decreases to 17%, for an increase of 1000°C (Hameury et al., 2007). Under these conditions, infrared properties can be, and will be in this study, considered constant for the experiments temperature ranges $\varepsilon(e, \theta)$.

The angle of emission plays a role in the properties (Siegel and Howell, 1993; Huetz-Aubert et al., 1995), as shown by Figure 15.2, where the directional emissivity $\varepsilon(\theta)$ is plotted as a function of the refractive index ratio n_2/n_1 (Table 15.1); these values were obtained from electromagnetic theory.

This theory is not perfect, with issues in the description of the influence of surface properties on the emissivity, but it is very useful when experimental data are lacking, such as in this case for liquids. Because all of the droplet studies in this case have a diameter below the capillary length and form spherical caps, the directional emissivity must be known. For the low refractive ratio of all of the fluids used (Tables 15.2 and 15.3; average value of 1.37), the directional emissivity starts to be influenced by the angle of emission for values above 40° and becomes strongly modified for values above 60°. The hemispherical total emissivity can be obtained by the integration (Eq. (15.7)):

$$\varepsilon(e, \theta) = \frac{1}{\pi} \int_S \varepsilon'(\theta) \cos\theta \, d\omega \tag{15.7}$$

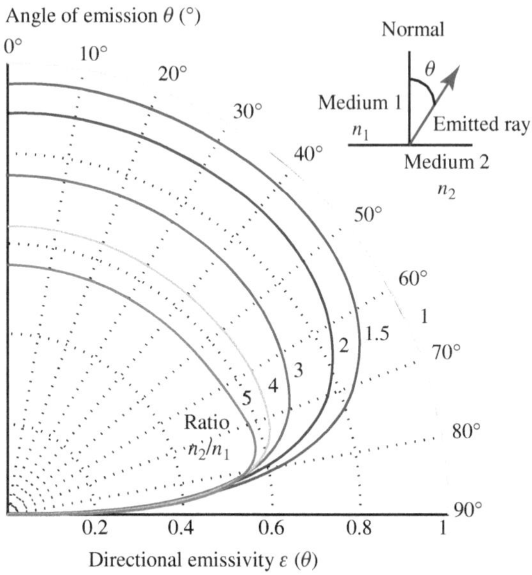

FIGURE 15.2 Directional emissivity predicted from electromagnetic theory as a function of the emission angle. *Data plotted from Siegel and Howell (1993).*

TABLE 15.1 Refraction Index Ratio n_2/n_1 for Alcohols and Alkanes

n	1	2	3	4	5	6	7	8	9	10
C_nH_{2n}+OH	1.326	1.359	1.384	1.397	1.409	1.418	1.423	1.429	–	–
C_nH_{2n+2}	–	–	–	–	1.358	1.375	1.387	1.398	1.405	1.411

TABLE 15.2 Alcohols Transmittivity for Minimum (0.1 mm) and Maximum (0.7 mm) Observed Fluid Thickness

n	Thickness	1	2	3	4	5	6	7	8
C_nH_{2n} + OH	0.1 mm	0.824	0.934	0.892	0.979	0.950	0.943	0.959	0.963
	0.7 mm	0.112	0.265	0.211	0.294	0.274	0.323	0.347	0.372

Each fluid has a similar exponential decrease in transmittivity with thickness.

TABLE 15.3 Alkanes Transmittivity for Minimum (0.1 mm) and Maximum (0.7 mm) Observed Fluid Thickness

n	Thickness	5	6	7	8	9	10
C_nH_{2n+2}	0.1 mm	0.978	0.975	0.969	0.966	0.962	0.959
	0.7 mm	0.833	0.818	0.769	0.770	0.750	0.746

Each fluid has a similar exponential decrease in transmittivity with thickness.

Because all of the droplets have a contact angle below 40°, the influence of the angle of visualization on the reflectivity has been neglected, and the droplets can be considered as several horizontal layers of semitransparent isothermal fluid $\varepsilon(e)$.

The last parameter that influences the infrared properties is the fluid thickness, with the transmittivity being particularly influenced. The transmittivity τ of all of the fluid was measured using a spectrophotometer (FTIR Nicolet Nexus 560) for thickness e ranging from 0.1 mm until the fluid appears opaque. The experimental test cell that must be used is composed of CaF_2 windows (exhibiting quasi-total transparent properties in the range of the spectrophotometer) and PTFE spacers to obtain the adequate optical fluid thickness. Prior to the measurement with the fluid, the spectrophotometer performs a background measurement of the radiative power that is transmitted through filtered atmospheric air inside the empty test cell. The measurements conducted in the range 3–5 μm, which matches the infrared camera sensitivity wavelength band. The spectrophotometer compares the monochromatic luminance $L_\lambda(0)$ reaching the first layer of fluid and the monochromatic luminance leaving the fluid $L_\lambda(e)$. As shown in Figure 15.3, all of the fluids have a similar transmittivity that follows an exponential decay with the optical fluid thickness $k_\lambda(T)e$ (Eq. (15.8)):

$$\varepsilon_\lambda(e,T) = 1 - \tau_\lambda(e) = 1 - \frac{L_\lambda(e)}{L_\lambda(0)} = 1 - e^{k_\lambda(T)e} \tag{15.8}$$

At the vicinity of the triple line, the transmittivity reaches a maximum, and the radiative flux reaching the infrared camera is directly provided by the heated substrate without any absorption from the fluid. However, when the thickness increases, the transmittivity decreases—that is, the fluid absorbs a part of the radiation. The camera is measuring the radiation coming from the substrate and several

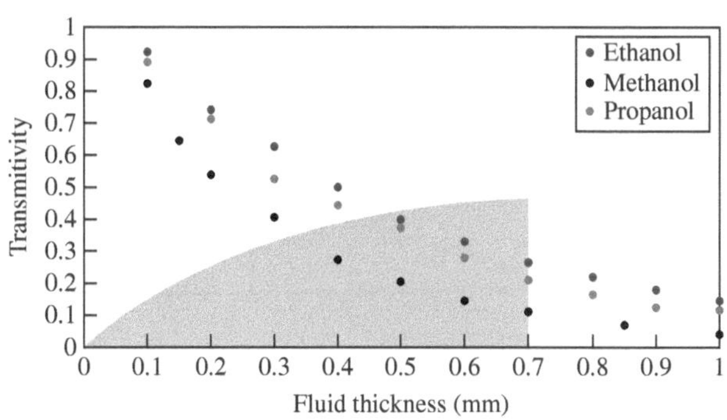

FIGURE 15.3 Transmittivity as a function of the fluid thickness in the wavelength range of 3–5 μm. A typical droplet profile is provided for comparison, with a maximum height of 0.7 mm (Carle, 2014).

layers of fluid, resulting in a temperature integrated on all of the visible layers. When the transmittivity becomes null, the fluid becomes opaque to radiation, and the camera only sees the surface of the droplet. The global emissivity of the fluid can be obtained from these values of transmittivity by integration of the monochromatic emissivity represented by the Plank luminance at the fluid temperature.

15.2 Application

In order to quantify the influence of the semitransparent properties in the final fluid temperature, we present in Figure 15.4 a radial temperature profile of an ethanol sessile droplet under evaporation extracted from an infrared camera. Then, using each pixel temperature and Eq. (15.6), a corrected temperature is calculated. The substrate temperature is constant at 50°C. One can see a clear difference in behavior between the two curves. The triple-line temperature is now close to the substrate temperature, which was not the case before. The apex droplet temperature is down shifted of 5.5°C. The radial temperature gradient is sharply increased (factor 2.6), especially near the triple line, where hydrothermal waves are observed. At this location, the temperature gradient before correction is 0.6°C mm^{-1} and after correction, it is 1.6°C mm^{-1}. Such a correction helps us to better estimate the surface tension gradient related to the temperature gradient and thus the intensity of the thermo-capillary convection.

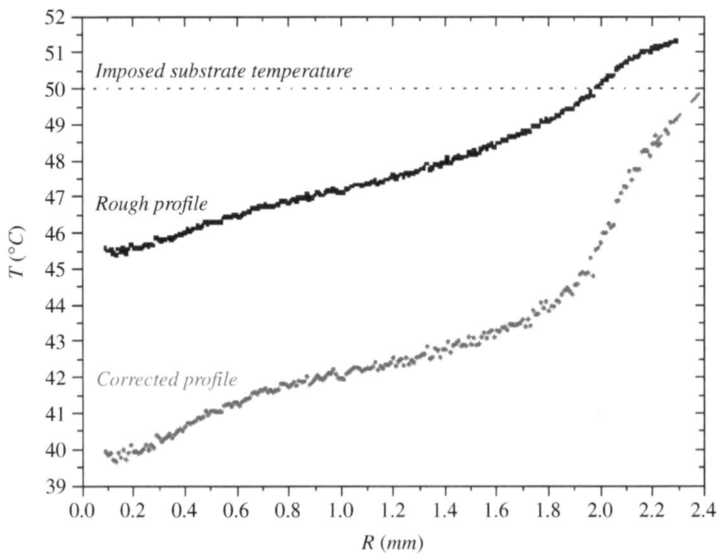

FIGURE 15.4 Droplet radial temperature profile—rough and corrected temperature ($R = 2.4$ mm, $T_s = 50$°C).

Glossary

Emmissivity effectiveness of a material in emitting energy as thermal radiation.

Reflectivity or reflectance is the fraction of incident electromagnetic radiation that is reflected at an interface.

Transmittivity fraction of electromagnetic radiation that passes through a sample.

References

Carle, F., 2014. Flow Motion in Sessile Droplets: Evaporation and Nanoparticles Assembly (Ph.D. thesis). Aix-Marseille University.

Hameury J., Scoarnec V., Filtz J.R., 2007. Mesure des propriétés radiatives infrarouges de liquides semi-transparents.

Huetz-Aubert, M., Klarsfeld, S., De Dianous, P., 1995. Rayonnement thermique des matériaux semi-transparents. Editions T.I., b8215.

Incropera, F.P, DeWitt, D.P., Bergman, T.L., Lavine, A.S., 2007. Fundamentals of Heat and Mass Transfer, sixth ed. John Wiley & Sons, Hoboken, NJ.

Siegel, R., Howell, J.R., 1993. Thermal Radiation Heat Transfer, third ed. CRC Press, Boca Raton, FL.

Sobac, B., 2012. Evaporation de Gouttes Sessiles: des Fluides Purs Aux Fluides Complexes (Ph.D. thesis). Aix-Marseille University.

Section III

FLOW INSTABILITIES

Chapter 16

FLOW STABILITY

Marc Medale, Sergey Semenov and David Brutin
Mechanical Engineering Department, Aix-Marseille University, Marseille, France

Patterns related to the flow structure in the sessile droplet and its surrounding air–vapor environment could reveal the leading mechanisms that drive the various instabilities observed and explain the dynamical regimes that take place in the course of evaporation. Therefore, this chapter presents some state-of-the-art methods and algorithms that can help us to better understand the various flow instabilities observed in evaporating sessile droplets. Although it is obviously a transient phenomenon, we have nevertheless chosen as first approach a frozen-time technique associated with a one-sided model to study the flow structure sensitivity of the evaporation rate in order to study its related bifurcation diagram. Then, a two-sided and unsteady model is used to investigate highly nonlinear regimes associated with high evaporation rates. The approaches we followed in this study are first presented, and the obtained results and their limits are then discussed. It turns out that the physically refined models can be used to reproduce and investigate most of the triggering mechanisms of the flow instabilities that take place during the evaporation process of a sessile droplet.

Droplet Wetting and Evaporation. DOI: http://dx.doi.org/10.1016/B978-0-12-800722-8.00016-3

16.1 Introduction

In this section the effects of fluids flows that develop during the evaporating process of a sessile droplet deposited on a substrate are investigated, because they play an important part in the related dynamics of the whole evaporation process. At least three main mechanisms drive the fluid flow in this process under earth conditions: (i) drainage in the droplet; (ii) buoyancy in the liquid and vapor phases; and (iii) thermo-capillary forces, which could induce fluid flow in the liquid, the vapor phase, or both of them. Owing to its spherical-cap shape, the evaporation rate in a sessile droplet deposited on a heated substrate is predominant in the vicinity of its contact line for contact angles lower than 90°. This induces drainage from the droplet center toward its periphery, resulting in a radial flow (Stefan flow) as long as the contact line is pinned to the substrate (Deegan et al., 1997, 2000; Hu and Larson, 2005a,b). Outside the droplet, vapor diffusion toward the surrounding air unavoidably induces gradients in vapor concentration, which under gravity conditions result in solutal buoyancy in the vapor phase. Moreover, because evaporation is by essence an endothermic process, thermal gradients also appear in both the liquid and vapor phases, so thermal buoyancy also takes place in both fluid phases. Finally, thermal gradients, and most likely solutal ones (for binary and multicomponent liquids), appear along the liquid–vapor interface. They induce surface-tension gradients that can drive the flow in both phases and become the prevailing driving mechanism in micro-gravity conditions.

Therefore, the dynamics of the evaporating sessile drop result from the competition between these three driving mechanisms, whose ratio continuously evolves in the course of the droplet evaporation process as its geometry changes accordingly. For droplet diameters up to a few millimeters, thermo-capillary force commonly dominates buoyancy ($Bd << 1$), and the resulting thermo-convective flow is at least one order of magnitude greater than the one resulting from drainage alone (Stefan flow) (Deegan, 2000). Moreover, the flow structure is also very different in these two cases: it is organized into torus roll(s) in the former case, whereas it is a purely radial flow in the latter (Hu and Larson, 2005a,b). Furthermore, at high evaporation rates, this thermo-convective torus fluid flow is prone to become unstable, and fully three-dimensional structures appear and can form hydrothermal waves (Sefiane et al., 2010; Brutin et al., 2011; Carle et al., 2012; Karapetsas et al., 2012). Figure 16.1 displays infrared snapshots taken from above at four evaporation times (every 10% of the total evaporation time, starting from the beginning on the left) for ethanol (upper row) and FC-72 (lower row) droplets deposited on a heated substrate.

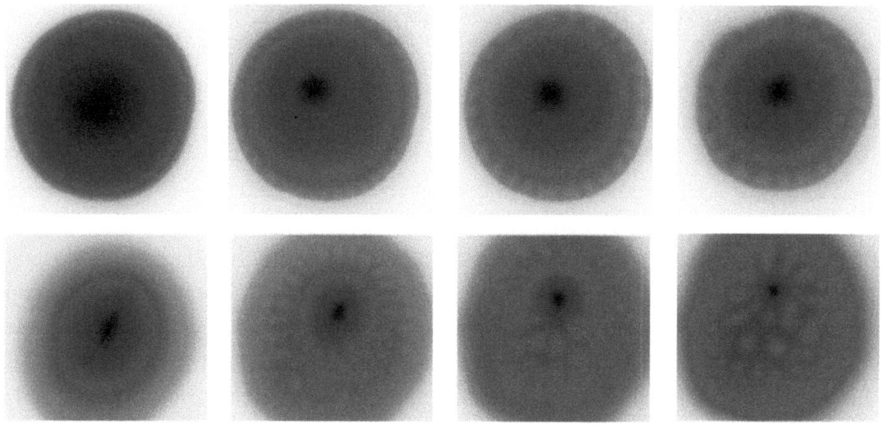

FIGURE 16.1 Infrared snapshots (top views) of two evaporating sessile droplets deposited on a heated substrate at four times during the evaporation process. Upper row: ethanol droplet (initial radius $R_0 = 2.2$ mm, substrate temperature $T_s = 46°C$); lower row: FC-72 droplet (initial radius $R_0 = 3.8$ mm, substrate temperature $T_s = 51°C$). *Drawn from Sobac and Brutin (2012).*

These snapshots reveal that complex dynamical flow structures develop in a very different ways, depending on the evaporating conditions. For the ethanol case (Figure 16.1, upper row), the fluid flow seems to develop first as a torus roll located close to the outer ring of the droplet, and this torus is then destabilized into cells, whose azimuthal wavelength decreases with time. For the FC-72 case (Figure 16.1, lower row), which evaporates at a much higher rate than ethanol because it is a much more volatile liquid in air, the outer torus roll seems to persist for a while, and meanwhile a chaotic cell pattern develops from the very early evaporation times and invades the whole central part of the liquid droplet.

Besides several very convincing sessile droplet experiments, which display quite complex instability routes, linear and nonlinear stability analyses of evaporating plane liquid layers have recently been undertaken (Doumenc et al., 2010; Sáenz et al., 2013). However, owing to the very different flow structure of the base state resulting from geometrically different configurations, these latter studies are unfortunately not representative of the stability problem of thermo-convective flows that take place during the evaporation process of a sessile droplet. Furthermore, we must take into account that both time-dependent evolution of the spherical-cap shape of the liquid–vapor interface and the dynamics of the contact line are still not obtainable from most off-the-shelf numerical models and consequently have not yet been solved, to the authors' best knowledge. Thus, alternate directions and simpler configurations can be considered as ways to contribute to a better understanding of the flow stability of a sessile droplet.

16.2 Relevance of frozen-time approaches

The evaporating process of a sessile droplet is obviously a transient phenomenon, and it usually displays three main stages from a purely geometrical point of view. In the first one, the contact line is pinned to the substrate at a constant droplet diameter, and evaporation proceeds with decreasing contact angle, up to a lower limit. When reaching it, the second evaporation stage takes place with a receding contact line at an approximately constant contact angle, resulting in a decreasing droplet diameter. Finally, the third stage occurs when the droplet looses its spherical-cap shape and evolves into a very thin film as it enters the microscopic scales.

Fortunately, it turns out from experimental observations (cf. Figure 16.1) that most of the flow instabilities arise in the first stage, where evaporation proceeds at constant droplet diameter and decreasing contact angle. It is also clear that in this first stage the rate of change of the contact angle with respect to time is much smaller than the rate of change of any thermal or fluid-flow perturbations, so a quasi-steady-state can be assumed throughout the first stage for stability analysis purposes. Moreover, for low evaporation rates, the evaporating process of a sessile droplet is often controlled by vapor diffusion in the surrounding air outside the droplet (Hu and Larson, 2002). In these cases the duration of this first evaporation stage is much longer than the vapor diffusion time, so here again a quasi-steady-state can be assumed in this stage, and a one-sided model can also be relevant (Doumenc et al., 2010). It is one of the simplest representative models, so it has been used in most of the present work, and it is introduced here.

The simplifying assumptions we have made to derive the physical model related to the one-sided frozen-time model are as follows (cf. Figure 16.2). The substrate and liquid drop are only considered; the droplet geometry is fixed (pinned contact line and constant contact angle). Heat transfers across the liquid–vapor interface are accounted in the energy balance throughout a global heat transfer coefficient that includes a linearized liquid–vapor phase change considering a uniform and constant vapor concentration gradient in the surrounding droplet environment

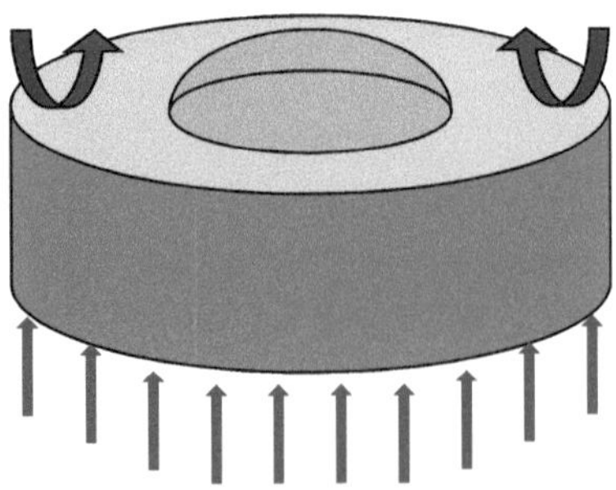

FIGURE 16.2 Sketch of the one-sided frozen-time evaporation model of a sessile drop deposited on a heated substrate.

(Doumenc et al., 2010). All thermo-physical properties are assumed constant, taken at room temperature, and the lower wall of the substrate is heated at constant heat flux.

From the preceding assumptions, the governing equations of the one-sided frozen-time evaporation model reduce in the solid substrate and liquid droplet to the following:

- Heat transfer in substrate:
  - Heat diffusion in solid: $\frac{\partial \theta_s}{\partial t} = \kappa_{sl} \Delta \theta_s$
  - Boundary conditions:
    - at its lower wall: $\frac{\partial \theta_s}{\partial n} = \varphi_{\text{input}}$;
    - along its lateral wall: $\frac{\partial \theta_s}{\partial n} = 0.$
- Heat transfer in liquid droplet:
  - Diffusion-advection equation in droplet: $\frac{\partial \theta_l}{\partial t} + \vec{V} \cdot \vec{\nabla} \theta_l = \Delta \theta_E$
  - Boundary condition at liquid–vapor interface: $\frac{\partial \theta_l}{\partial n} = -\text{Bi}(1 + \theta_l)$
- Heat flux continuity at substrate-droplet interface: $\frac{\partial \theta_l}{\partial n} = -\kappa_{sl} \frac{\partial \theta_s}{\partial n}$
- Liquid flow in droplet:
  - Incompressible Navier–Stokes equations:
    - Continuity: $\vec{\nabla} \cdot \vec{V} = 0$
    - Momentum: $\frac{1}{\text{Pr}} \left(\frac{\partial \vec{V}}{\partial t} + \vec{V} \cdot \nabla \vec{V} \right) = -\vec{\nabla} p + \Delta \vec{V} + \text{Ra} \theta \vec{e}_z$
  - Boundary conditions:
    - at rigid wall: $\vec{V} = \vec{0}$;
    - at liquid–vapor interface: $\vec{V} \cdot \vec{n} = 0$ and $\vec{t} \cdot \overline{\overline{\tau}} \cdot \vec{n} = -\text{Ma} \nabla_s(\theta_l)$

where θ_s and θ_l are dimensionless temperatures in the substrate and liquid droplet, respectively; $\vec{V}$ is the dimensionless liquid velocity vector; p is the dimensionless dynamic pressure; κ_{sl} is the solid to liquid ratio of thermal conductivity coefficients; Bi is the Biot number (as defined in Doumenc et al., 2010), Ma is the Marangoni number, Pr is the Prandtl number, Ra is the Rayleigh number; and $\vec{n}$ and $\vec{t}$ are the outward unit normal and tangent vectors, respectively.

The following section addresses the computational algorithms and methods we have implemented to study the fluid flow stability problem that arises in the droplet evaporation process.

16.3 Asymptotic Numerical Method of frozen-time model for sessile droplet evaporation

The numerical methods described in this section are used to determine the ranges of the control parameters that make the system either stable or unstable. Indeed, at critical values of these control parameters, the stability of the system changes, so qualitative and quantitative changes in the system behavior may occur. Unfortunately, these critical points coincide to singular solutions to the governing equations, so specialized algorithms should be used to accurately compute them along with the critical values of the control parameters (Keller, 1977). In the computational fluid mechanics community, the classical way to determine a bifurcation diagram usually consists of performing a sequence of the following two tasks for a series of control parameter values (Dijkstra et al., 2014): (i) compute the base state associated with a given value of the control parameters; and (ii) compute the linear stability of this base state, because the system becomes unstable when the growth rate is greater than zero (i.e., the real part of the rightmost eigenvalue of the linearized system), whereas it remains stable in the opposite case.

The present section is only concerned with the first of these two steps—that is, continuation or path-following algorithms, which are designed to map steady-state solutions in the parameter space by computing branches of solutions for a given range of control parameters, along with critical values and their corresponding singular solution, if any. Among continuation algorithms, first-order predictor-corrector ones (Euler predictor, Newton-Raphson based corrector) with pseudo-arc-length parameterization have been widely used for decades (Keller, 1977; Doedel et al., 1991; Seydel, 1994; Doedel, 1997). On the other hand, high-order predictors have also been introduced, one of which is the Asymptotic Numerical Method (ANM; Cochelin et al., 2007). It combines high-order Taylor series expansion, discretization technique, parameterization strategy, and it results in an efficient nonlinear solution method. ANM has been successfully applied to coupled fluid flow and heat transfer problems in previous works (Medale and Cochelin, 2009).

The computation of bifurcation points and emanating branches are other particularly important issues. They have been tackled in the ANM framework by computing either the zeros of Padé approximates or a scalar bifurcation indicator. Not only are these methods more expensive, but they sometimes lack robustness. On the other hand, power series analysis (Van Dyke, 1974, 1975) is definitely a much more powerful way to accurately detect and compute simple bifurcation points (Cochelin and Medale, 2013).

Let us now present the key points of the AMN algorithm along with its implementation related to the evaporation process of a sessile droplet in the framework of a one-sided model. First of all, let us recall that the governing equations can be cast in a more general form as follows:

$$R(U, \lambda) = L(U) + Q(U, U) - \lambda F = 0$$

where U is the generalized unknown vector, which gathers the discrete values of the nodal velocity, pressure, and temperature fields; λ is the generic scalar control parameter of the problem (the magnitude of applied heat flux in the present case); F is the discrete load vector; and $L(U)$ and $Q(U,U)$ are the discrete linear and quadratic operators, respectively, resulting from the Finite Element discretization of the governing equations (Medale and Cochelin, 2009). Starting from a known set (U_0, λ_0), the ANM consists in expanding the unknown (U, λ) set in a Taylor series with respect to an unknown path parameter s up to an arbitrary truncation order n:

$$\begin{cases} U(s) = U_0 + sU_1 + s_2U_2 + s_3U_3 + s_4U_4 + \cdots + s_nU_n \\ \lambda(s) = \lambda_0 + s\lambda_1 + s^2\lambda_2 + s^2\lambda_3 + s^2\lambda_4 + \cdots + s^n\lambda_n \end{cases}$$

Because the path parameter s has been introduced in the problem as a new unknown, one has to provide an extra equation to close the resulting augmented system. In our model this parameterization equation is chosen of bifurcation type (Cochelin, 1994)

$$s = (U - U_0)\cdot U_1 + (\lambda - \lambda_0)\lambda_1.$$

Inserting these high-order Taylor series expansions in the original set of equations, along with the parameterization equation, and then equating like powers of s transforms the original nonlinear problem into n sets of linear ones. Moreover, these n sets of linear algebraic systems inherit a very nice feature from a computational point of view: they both have the same operator—namely, the tangent matrix. Therefore, the solution of these n sets of linear algebraic systems gives access to an analytical representation of one part of the branch, whose length is related to the radius of convergence of the Taylor series (Cochelin, 1994). The ANM continuation algorithm proceeds step by step along a branch of solution, the starting point of the next step being the last one of the current step, until a steady bifurcation is encountered, if any. In this particular case, a singularity arises in both the governing equations and their associated algebraic system, so power series analysis should be performed to accurately and efficiently compute the singular solution associated to the bifurcation point (Cochelin and Medale, 2013).

16.4 Parametric study with the one-sided steady-state model

To contribute to a better understanding of the various instability mechanisms that take place in the evaporation process of a sessile droplet, we have first investigated the influence of the evaporation rate on the pattern selection for several dynamic Bond numbers (Bd = Ra/Ma)—that is, various ratios of the driving mechanisms. Keeping in mind that the frozen-time approach can only be

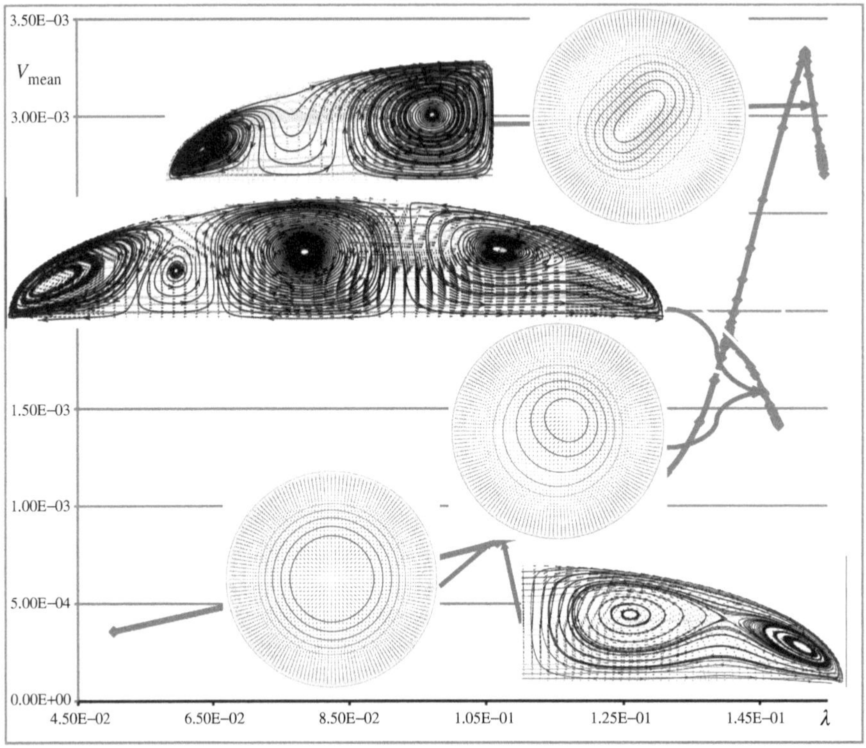

FIGURE 16.3 Bifurcation diagram of the one-sided frozen-time evaporation model of a sessile drop deposited on a heated substrate for Bd = Ra/Ma = 1.

representative of low evaporation rate cases in which the dynamics are low with respect to the time duration of the first evaporation stage, it is nevertheless noteworthy that symmetry breakings already occur in such simple cases. As an illustration, Figures 16.3 and 16.4 present the bifurcation diagrams for Bd = 1 and 1/1.75, respectively, and Pr = 6, plotting in dimensionless variables the mean droplet velocity versus the applied heat flux at the substrate lower wall. In these two figures, the insets represent top views of isotherms and velocity field over the droplet-free surface, along with isotherms, streamlines, and velocity vectors in radial or diametric cross sections. Starting the continuation from the origin (isothermal motionless droplet in saturated environment) in both cases, one first proceeds along the fundamental branch. In Figure 16.3 (Bd = 1), its slope increases slowly at first and then rises much more steeply. Then, it intersects two secondary branches, so two bifurcations occur in this diagram. The fundamental branch is characterized by the axisymmetric mode, which displays concentric circular isotherms in the top view, whereas the cross section reveals that the liquid flow structure is made up of two concentric torus rolls. It is noteworthy that these rolls

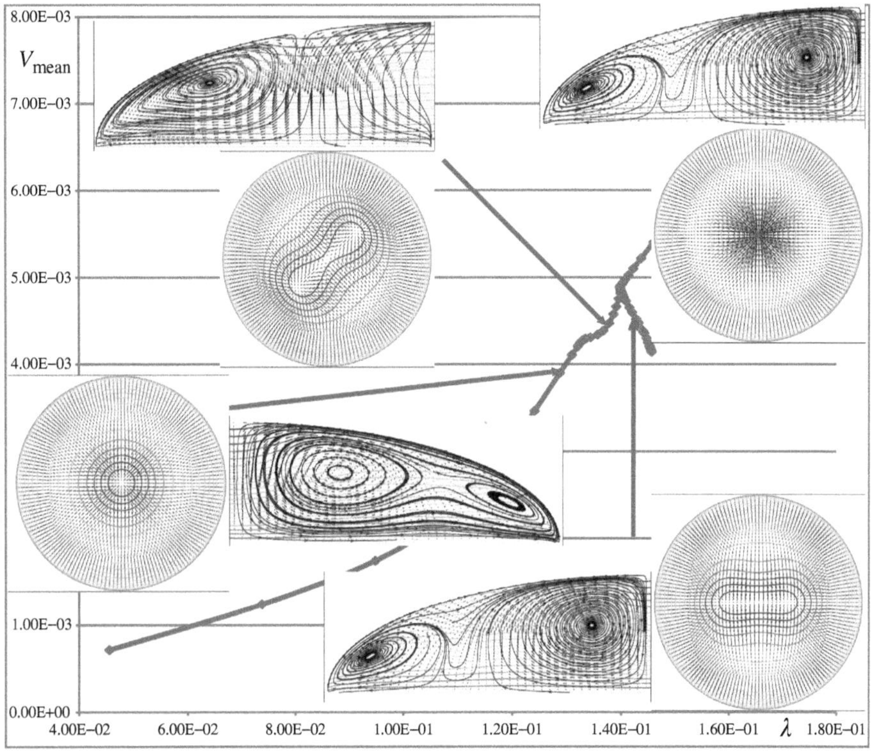

FIGURE 16.4 Bifurcation diagram of the one-sided frozen-time evaporation model of a sessile drop deposited on a heated substrate for Bd = Ra/Ma = 1/1.75.

rotate in the same direction (upward along the free surface) because they are predominantly oriented by Marangoni force for the considered values of the dynamic Bond number. Therefore, in this fluid flow configuration, the higher the fluid flow velocity, the higher the shear stress in between the corotating rolls, so instability is very likely to occur and indeed does as expected. Fully three-dimensional fluid flow structures appear along the two bifurcated branches, which exist for relatively high evaporation rates, as in experiments. The two concentric torus roll structures are replaced by a ring roll in which the fluid flow organizes itself into asymmetric donut-like structures to reduce shear stress.

Figure 16.4 presents a comparable behavior for Bd = 1/1.75, despite the fact that the flow structure along the bifurcated branches are a little bit different than for Bd = 1, owing to the stronger influence of Marangoni force with respect to buoyancy. In this case, after the first bifurcation of the fundamental branch, two cells appear inside the outer torus, which gather in the bifurcated branch.

Furthermore, it turns out from this parametric study that as Bd diminishes, cellular convection overcomes torus-like convection in the droplet core region.

In addition, the droplet geometry (wetting angle, its spherical-cap aspect ratio) determines the fluid flow pattern (number of structures). Finally, all the asymmetric fluid flow structures encountered on the bifurcated branches are unstable, so a transient solution should occur toward either a stable state or an unsteady regime.

The one-sided model performs a one-way coupling between the liquid and gas phases in which one assumes a linear behavior in the gas phase around an *a priori* prescribed base state of both heat and mass transfers along with the phase change itself. Therefore, the only nonlinearities that are accounted for arise in the liquid phase. Despite this quite restrictive assumption, the outcome of the one-sided parametric study is that three-dimensional instabilities can be triggered in the droplet. But one can wonder if the instabilities coming from the gas phase would not be more unstable than the ones coming from the liquid phase or if the actual interactions between the liquid and gas phases could not in turn be more destabilizing. We now address this question by considering a two-sided model, which is clearly more suitable to correctly account for fluid flow, heat, and mass transfers in the surrounding air and vapor phases. It enables us to investigate the highly nonlinear regimes, whereas the full two-way coupling is now accounted for. Moreover, since the dynamics of unstable regime exhibits much shorter time-scales than the total evaporation time, an unsteady approach is further required. Therefore, to address such computations with affordable computational resources and restrained elapsed times, we have considered an axisymmetric two-dimensional two-sided model presented in the following section.

16.5 The two-sided model to investigate highly nonlinear regimes

Here we present two-sided numerical models of evaporating ethanol drops on heated substrates. These models are based on real experimental studies (Carle et al., 2012; Sobac and Brutin, 2012) and take into account heat and mass transfers in both liquid and gaseous adjacent phases. The first model uses a steady-state formulation of the problem equations as evaporation occurs into ambient atmosphere and can be considered as a steady-state process. The second one uses unsteady approach as evaporation occurs into a closed chamber and only the first 20 s of the process are available experimentally (parabolic flight experiment). Two-sided models allow us to predict the vapor transfer from the droplet surface into ambient atmosphere or into a closed chamber and to determine the vapor flux distribution along the droplet surface, which results in physically correct (corresponding to actual experiments) interfacial conditions at the droplet surface.

Because in experiments the droplet evaporation proceeds into ambient atmosphere or into a chamber of a size much bigger than the droplet's size, the gaseous phase is usually modeled as a semi-infinite medium; the computational domain, however, cannot be infinite and must be bounded either by the chamber walls or by an artificial boundary. Therefore, for the problem of vapor diffusion into unbounded ambient atmosphere, the size of computational domain is chosen to be 100 times bigger than the size of a droplet, which results in an evaporation rate error (due to the presence of outer boundary) of less than 1% as compared to the rate of droplet evaporation into a semi-infinite medium. Besides the droplet and vapor phases, a heat conducting solid substrate is included in the models.

Thus, due to the presence of a gaseous phase, the computational domain in two-sided models is much bigger than in equivalent one-sided models. Therefore, one of the assumptions made is the axial symmetry of the model, which reduces the dimension of the problem down to two independent spatial coordinates (radial and vertical) and substantially speeds up the computational process as compared to a fully 3D model. The origin of the cylindrical system of coordinates is located in the plane of substrate-droplet interface, and the z-axis coincides with the axis of the droplet's symmetry.

The materials' parameters in the models are drawn from the experimental ones. The temperature difference between the heated substrate and the ambient environment is chosen to be 15°C, like in particular experiments. Because the temperature change occurs mostly in the air domain (the medium with the lowest heat conductivity in the system), its dynamic viscosity, heat conductivity, and vapor diffusivity were taken as functions of local temperature.

Because the flow velocities in the atmosphere and the droplet are much lower than the speed of sound, the air and liquid are modeled as incompressible Newtonian fluids (Navier–Stokes equations). The boundary conditions at the droplet surface for Navier–Stokes equations include the balance of normal (Laplace pressure) and tangential (thermocapillary forces) components of the full stress tensor, and the continuity of the tangential component of the velocity (no-slip condition) and discontinuity of the normal component of the velocity due to evaporation process (Stefan flow). No-slip boundary condition was applied on all solid surfaces. In case of evaporation into a semi-infinite medium, at the outer (artificial) boundary of the gaseous domain, the conditions of zero normal stress (open boundary), ambient temperature, and ambient vapor concentration (zero for ethanol vapor) were applied.

The vapor diffusion-convection equation with temperature-dependent diffusion coefficient was solved in the air domain with the boundary condition of saturated vapor concentration at the droplet surface, which in turn was also considered as a function of the local droplet's temperature (Clausius–Clapeyron equation). No penetration boundary condition was applied for this equation at solid walls.

The equation of conductive heat transfer was solved in the solid domain, conductive and convective heat transfer in fluid domains (liquid and air) with boundary conditions of constant heater temperature underneath the substrate, and constant ambient temperature at the outer boundaries. At the liquid–air interface a condition of the heat sink due to evaporative cooling (latent heat of vaporization) was applied as a function of the local evaporation rate.

The Boussinesq approximation was used in order to account for buoyancy in both liquid and air domains. These forces were calculated based on the dependence of ethanol and air density on temperature and of air density on vapor concentration.

Integration of the local evaporation rate along the droplet surface gives the total droplet's evaporation rate. Knowing this quantity, as well as assuming the pinning of the droplet's contact line and a spherical cap shape of the droplet, the velocity profile of the liquid–air interface was calculated and used in corresponding boundary conditions.

In order to discretize and solve numerically this complex system of strongly coupled equations of flow, diffusion, and heat transfer in three phases with temperature-dependent properties, a commercial software COMSOL Multiphysics (www.comsol.com) was used. It implements the Finite Element Method in the weak formulation of the physics equations, and all boundary conditions are implemented in the form of constraints with Lagrange multipliers. Second-order finite elements on a triangular mesh have been used for the spatial discretization, and a generalized alpha method has been used for the temporal discretization.

The ultimate aim of this modeling is to reproduce the experimentally observable hydrothermal waves and understand their nature and principles of their development in sessile droplets, whereas the current objective is to model numerically the real experimental situation, including all relevant physical processes, not only in the liquid droplet but also in the gaseous phase.

The results of 2D axisymmetric steady-state modeling of a sessile drop of ethanol evaporating into ambient atmosphere under terrestrial gravity conditions on top of a heated cylindrical substrate are presented in Figure 16.5. One can see two bifurcating branches of steady-state solutions. The bifurcation is related to different patterns of flow in the surrounding atmosphere but not in the drop. Both thermal and solutal gravitational vortices, rotating in different directions, are present in the air phase for both branches of stationary solutions. On the other hand, the flow pattern within the drop was the same (one thermo-capillary vortex; see the inset "drop" in Figure 16.5) for a studied range of substrate temperatures (up to 15°C above the ambient temperature), and did not reveal any flow bifurcations within the droplet. The model showed an agreement with the experimental evaporation rate within 10% error, which corresponds to experimental error bars.

Another 2D axisymmetric numerical model is based on the parabolic flight experiment: a sessile drop of ethanol on top of a heated substrate, evaporating into a closed chamber under microgravity conditions. Because evaporation occurs into a closed chamber, the problem was solved in its unsteady formulation. First, the preheating of the substrate and of the chamber was modeled without the drop

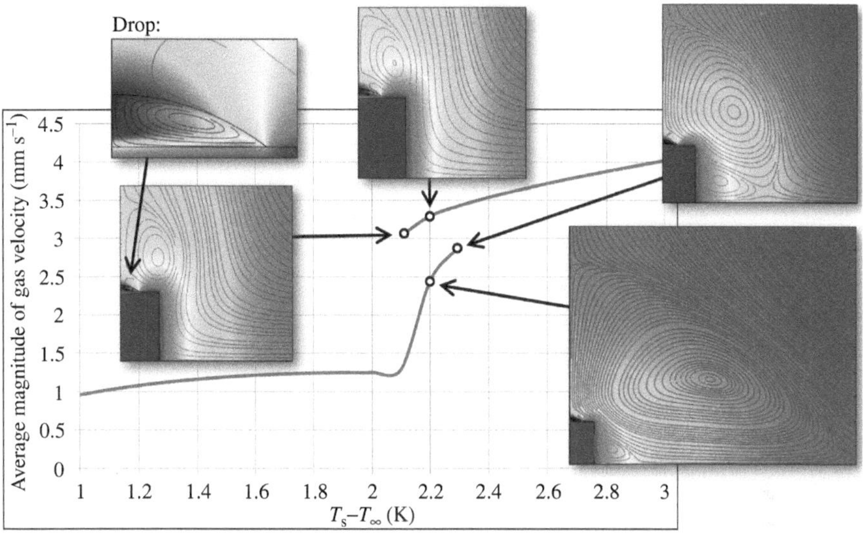

FIGURE 16.5 Axisymmetric steady-state numerical model of a sessile drop of ethanol evaporating into ambient atmosphere under terrestrial gravity conditions on top of a heated cylindrical substrate. Streamlines in the drop and atmosphere: flow field; color: temperature field; $T_S - T_\infty$ is the temperature difference between the substrate and ambient air.

evaporation, as has been done in the real experiment. Next, the drop evaporation starts instantly (the stage of drop injection and spreading is not modeled) and the time-dependent process of evaporation is computed under the condition of pinned triple line. The results show that after a short transitional stage at the beginning of evaporation, both numerical and experimental evaporation rates become more or less stable (but not yet steady, and still evolving in time) and agree well with each other. This unsteady model revealed thermo-convective instabilities within the droplet (Figure 16.6), which did not appear in the previous equivalent steady-state problem (Figure 16.5). From a qualitative point of view, very similar thermo-convective instabilities were obtained with a one-sided unsteady model (Karapetsas et al., 2012).

It is interesting to note that using absolutely the same physics as in our unsteady numerical model (parabolic flight experiment, microgravity), but replacing the closed chamber with an open atmosphere, and switching to a steady-state problem formulation results in a stationary solution with only one thermocapillary vortex within the droplet, as in the previous steady-state simulation. Even using the multicellular flow structure (Figure 16.6) as an initial condition for the steady-state problem results in the stationary solution with one thermocapillary vortex within the droplet.

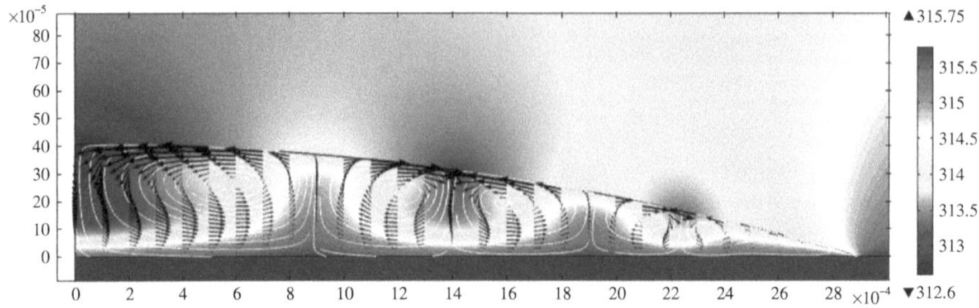

FIGURE 16.6 Two-dimensional axisymmetric unsteady numerical model of a sessile drop of ethanol evaporating into a closed chamber under microgravity conditions on top of a heated substrate. Streamlines and arrows in the drop: flow field; color: temperature field (in Kelvins).

Thus, comparing the steady-state and unsteady numerical models, one can conclude that these two approaches result in a big qualitative difference in obtained solutions. The unsteady solution of the problem demonstrates the appearance and development of multicellular thermo-convective instabilities within the droplet, while the steady-state approximation is not suitable for that purpose. Therefore, the unsteady approach is always preferable when thermo-convective instabilities are studied. Because the experimentally observed hydrothermal waves in sessile drops are three-dimensional and not axisymmetric, further developments of two-sided models and their extension into three dimensions is needed for a better understanding of the phenomenon and both qualitative and quantitative validations of the model against available experimental data.

16.6 Concluding remarks

This chapter examines the instabilities that can occur during the evaporation process of a sessile droplet deposited on a heated substrate. For that purpose, we consider progressively refined levels of approximation, either at the physical level or at the geometrical one. To start with, we first considered a frozen-time one-sided evaporation model that was developed and implemented in the framework of a continuation algorithm. Despite its very crude assumptions, this model shows the existence of unstable asymmetric modes at high evaporation rates that will obviously trigger azimuthal unsteadiness, which is very likely related to the hydrothermal waves that have been observed experimentally. However, in such cases the various nonlinearities in the gas phase could also trigger instabilities, so a one-sided model is no longer relevant, and we have replaced it with a two-sided model. This latter model is now able to fully account for the fluid flow, heat, and mass transfers in liquid and vapor phases to investigate the dynamical regimes associated with the unstable modes. Nevertheless, because the two-sided model should model the gas phase up to a large extent with respect to the droplet size, only axisymmetric computations were considered to manage the related studies with affordable elapsed times and computational resources. These axisymmetric computations reveal that subtle fluid flow, heat, and mass transfers exist in the gas phase surrounding the sessile droplet. Therefore, the next step on the way to a more comprehensive model is to address the two-sided model in fully three-dimensional computations in order to get a much better insight into this very intricate instability problem.

Glossary

Asymptotic Numerical Method High-order predictor-corrector continuation algorithm based on Taylor series expansion up to an arbitrarily high truncate order n that can be used to convert a highly nonlinear algebraic system into a set of n linear ones. The solution of this latter set can be used to obtain a piecewise analytical representation of the problem solution.

Bifurcation point Critical point where the solution of the governing equations is singular, the stability of the system changes, and thus qualitative and quantitative changes in the system behavior may occur.

Continuation algorithm Numerical procedure that can be used to follow the steady-state solution path from an initial known and regular solution.

Finite Element Method A numerical method that can be used to convert the weak integral form associated with the governing equations into an algebraic system to be solved on a computer.

Thermo-convective instability The flow has got unstable under thermal loads or their effects. In the present case, evaporation is an endothermic process therefore the environment should provide energy to the liquid to sustain it and thermo-convective instability could hence be involved.

Nomenclature

α	Liquid thermal diffusivity
β	Thermal expansion coefficient of liquid
Bd	Dynamic Bond number ($\mathrm{Bd} = \mathrm{Ra}/\mathrm{Ma} = \rho g \beta e^2/\gamma$)
Bi	Biot number ($\mathrm{Bi} = (h + L \cdot \partial Q_m/\partial T)e/\lambda$)
ΔT	Temperature difference across liquid droplet
d	Droplet diameter
e	Droplet height
γ	Liquid–air thermal surface tension coefficient
h	Convective heat transfer coefficient at liquid–air interface
Q_m	Mass flow rate of evaporation
λ	Liquid thermal conductivity
L	Liquid–vapor latent heat
Ma	Marangoni number ($\mathrm{Ma} = \gamma e^2 \Delta T/(\rho \nu \alpha d)$)

ν	Liquid dynamic viscosity
Pr	Prandtl number ($\mathrm{Pr} = \nu/\alpha$)
ρ	Liquid density
Ra	Rayleigh number ($\mathrm{Ra} = g\beta e^4 \Delta T/(\nu\alpha d)$)

References

Brutin, D., Sobac, B., Rigollet, F., Le Niliot, C., 2011. Infrared visualization of thermal motion inside a sessile drop deposited onto a heated surface. Exp. Therm. Fluid Sci. 35, 521–530.

Carle, F., Sobac, B., Brutin, D., 2012. Hydrothermal waves on ethanol droplets evaporating under terrestrial and reduced gravity levels. J. Fluid Mech. 712, 614–623.

Cochelin, B., 1994. A path following technique via an asymptotic numerical method. Comput. Struct. 53, 1181–1192.

Cochelin, B., Medale, M., 2013. Power series analysis as a major break-through to improve the efficiency of asymptotic numerical method in the vicinity of bifurcations. J. Comput. Phys. 236, 594–607.

Cochelin, B., Damil, N., Potier-Ferry, M., 2007. Méthode Asymptotique Numérique. Hermès–Lavoisier, Paris.

Deegan, R.D., 2000. Pattern formation in drying drops. Phys. Rev. E 61 (1), 475–485.

Deegan, R.D., Bakajin, O., Dupont, T.F., Huber, G., Nagel, S.R., Witten, T.A., 1997. Capillary flow as the cause of ring stains from dried liquid drops. Nature 389, 827–829.

Deegan, R.D., Bakajin, O., Dupont, T.F., Huber, G., Nagel, S.R., Witten, T.A., 2000. Contact line deposits in an evaporating drop. Phys. Rev. E 62 (1), 756–765.

Dijkstra, H.A., Wubs, F.W., Cliffe, A.K., Doedel, E., Dragomirescu, I.F., Eckhardt, B., et al., 2014. Numerical bifurcation methods and their application to fluid dynamics: analysis beyond simulation. Commun. Comput. Phys. 15, 1–45.

Doedel, E., 1997. Nonlinear numerics. J. Franklin Inst. 334B, 1049–1073.

Doedel, E., Keller, H.B., Kernevez, J.P., 1991. Numerical analysis and control of bifurcation problems. (i) Bifurcation in finite dimensions. Int. J. Bifurcat. Chaos 3, 493–520.

Doumenc, F., Boeck, T., Guerrier, B., Rossi, M., 2010. Transient Rayleigh-Bénard-Marangoni convection due to evaporation: a linear non-normal stability analysis. J. Fluid Mech. 648, 521–539.

Hu, H., Larson, R.G., 2002. Evaporation of a sessile droplet on a substrate. J. Phys. Chem. B 6 (106), 1334–1344.

Hu, H., Larson, R.G., 2005a. Analysis of the micro fluid flow in an evaporating sessile droplet. Langmuir 21 (9), 3963–3971.

Hu, H., Larson, R.G., 2005b. Analysis of the effects of Marangoni stresses on the micro flow in an evaporating sessile droplet. Langmuir 21 (9), 3972–3980.

Karapetsas, G., Matar, O.K., Valluri, P., Sefiane, K., 2012. Convective rolls and hydrothermal waves in evaporating sessile drops. Langmuir 28, 11433–11439.

Keller, H.B., 1977. Numerical Solution of Bifurcation and Nonlinear Eigenvalue Problems. Academic Press, New York, NY.

Medale, M., Cochelin, B., 2009. A parallel computer implementation of the asymptotic numerical method to study thermal convection instabilities. J. Comput. Phys. 228, 8249–8262.

Sáenz, P.J., Valluri, P., Sefiane, K., Karapetsas, G., Matar, O.K., 2013. Linear and nonlinear stability of hydrothermal waves in planar liquid layers driven by thermocapillarity. Phys. Fluids 25, 094101.

Sefiane, K., Steinchen, A., Moffat, R., 2010. On hydrothermal waves observed during evaporation of sessile droplets. Colloids Surf. A 365 (1–3), 95–108.

Seydel, R., 1994. Practical Bifurcation and Stability Analysis: from equilibrium to chaos (2nd edition). Interdisciplinary Applied Mathematics, Vol. 5. Springer-Verlag, Berlin, Heidelberg, New York, London, Paris, Tokyo, Hong Kong.

Sobac, B., Brutin, D., 2012. Thermocapillary instabilities in an evaporating drop deposited onto a heated substrate. Phys. Fluids 24, 032103.

Van Dyke, M., 1974. Analysis and improvement of perturbation series. J. Mech. Appl. Math. 27, 423–450.

Van Dyke, M., 1975. Computer extension of perturbation series in fluid mechanics. SIAM J. Appl. Math. 28, 720–734.

Chapter 17

THERMAL ORIGINS

Charles A. Ward and Aaron H. Persad
Thermodynamics and Kinetics Lab, Department of Mechanical and Industrial Engineering, University of Toronto, Toronto, ON, Canada

This chapter provides detailed examples of thermo-gravitational, thermo-capillary and thermo-solutal instabilities. When a fluid is confined and in normal Earth gravity, the fluid configuration is vertically anti-symmetric. If suddenly exposed to weightlessness, the fluid will transition to a symmetric configuration over a period of time. Thermo-capillary instabilities have been observed during sessile droplet evaporation in the absence of buoyancy-driven convection and radiation. When the droplet rests on a thermally insulating substrate, no thermo-capillary convection is measured. However, if the substrate is thermally conductive, a transition to thermo-capillary flow has been observed and the transition depends on the evaporation rate. When a droplet evaporates into multicomponent atmospheres, thermo-solutal instabilities arise if one of the atmosphere species has a heat of solution with the droplet component. The thermo-solutal instabilities appear as temperature and pressure pulsations in the liquid and in the vapor phases that are coupled by the evaporation flux.

Droplet Wetting and Evaporation. DOI: http://dx.doi.org/10.1016/B978-0-12-800722-8.00017-5

17.1 Thermo-gravitational instabilities: contact line in free fall

If a fluid partially fills a cylinder and is put into a state of complete free fall, thermodynamics can be used to predict the configuration the system would adopt if it were given enough time. Provided the amount of fluid is below a limiting value, two of the possible configurations are single- and double-interface. A study using the Japanese Dropshaft supported the predictions but also indicated that the system needed a longer period of time than it took the Dropshaft to reach equilibrium. The theory was extended to a near-free-fall condition in which g_e was small compared to normal gravity, g_0. The double-interface configuration was predicted to be the one adopted, but with the contact angle at the lower interface, θ_l, greater than that at the upper interface, θ_u. By measuring these angles, the value of g/g_0 could be determined. The effect of the random accelerations of the Space Shuttle orbiter (g-jitter) on the fluid configuration is shown to make the configuration of the fluid appear to be subjected to a higher acceleration than that indicated by the NASA accelerometers. Adsorption at the solid–liquid interface appears to be the mechanism producing the difference in θ_l and θ_u.

17.1.1 Introduction

The near-free-fall environment of an orbiting platform has many potential applications for improving our understanding of the processes that are important in science and engineering on Earth. But the interpretation of the measurements must take into account the random accelerations (g-jitter) to which the platform is subjected. The importance of these accelerations seems to depend on the phenomena being examined (Amin, 1988; Schneider and Straub, 1989; Iwan et al., 1991; Farooq and Homsy, 1994; van der Geld et al., 2012). We examine their importance in determining the contact angle at a water-(borosilicate) glass interface in the near-free-fall circumstance.

When a single-component, two-phase fluid is enclosed in a cylinder and is subjected to normal gravity, g_0, under equilibrium conditions, a common observation is that the fluid is at the bottom of the container. Such a configuration is called the single-interface configuration (Figure 17.1) and is completely antisymmetric. When the system is in complete free fall, the symmetry-breaking effect of gravity is removed, but there are other thermodynamic factors—for example, the surface tensions of the solid–liquid, solid–vapor, and liquid–vapor interfaces—that can keep the system from adopting a completely symmetric configuration.

In principle, complete free fall could be achieved by accelerating the system by an amount $a(t)$. The effective acceleration, g_e, would become

$$g_e \equiv [g_0 - a(t)]. \tag{17.1}$$

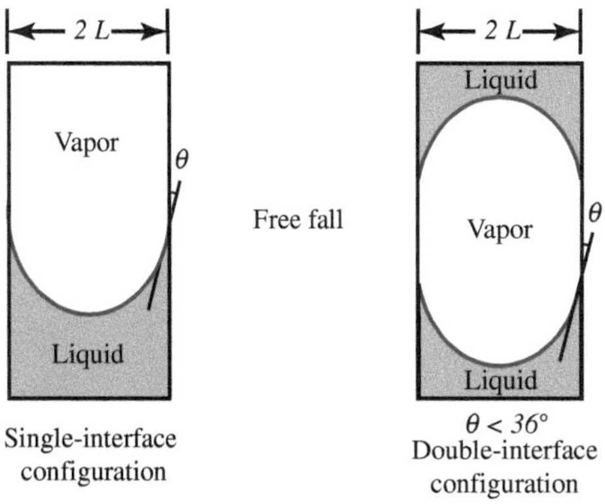

FIGURE 17.1 Possible configurations of a fluid held in a cylinder when in free fall (Sasges et al., 1996).

If there were no g-jitter, then g_e would be exactly zero, and the system would be in complete free fall.

We use thermodynamics to predict the configuration the system would adopt in this ideal circumstance, and compare the predicted fluid configuration with that observed in the near-free-fall environment of the Japanese Dropshaft (Sasges et al., 1996).

The thermodynamic theory is then extended so the effective acceleration is small enough that the system adopts the double-interface configuration (see Figure 17.1), but the upper and lower contact angles (θ_u and θ_l indicated in Figure 17.3) are not equal. We show that the difference in these angles is related to the value of g_e/g_0 (Ward et al., 2000).

17.1.2 Dependence of the Fluid Configuration on the Contact Angle at the Three-Phase Line

Consider a cylinder partially filled with liquid. Suppose it is initially subjected to normal gravity and that the ratio of the number of fluid moles in the cylinder, N, to the capacity of the cylinder, N_t is less than 0.52. Then suppose g_e is reduced to zero in the new environment. The two possible configurations the system could adopt are shown in Figure 17.1 (Sasges et al., 1996; Ward et al., 2000). We disregard any difference between the contact angles of the two configurations and ask which one is thermodynamically stable.

We assume that each of the systems shown in Figure 17.1 is maintained isothermal, is of constant volume, and has a fixed number of moles, denoted constant T, V, and N, respectively. Under these constraints, the Helmholtz function, $F(T, V, N)$, acts as the thermodynamic potential. The configuration that the fluid adopts in free fall is the one that minimizes $F(T, V, N)$.

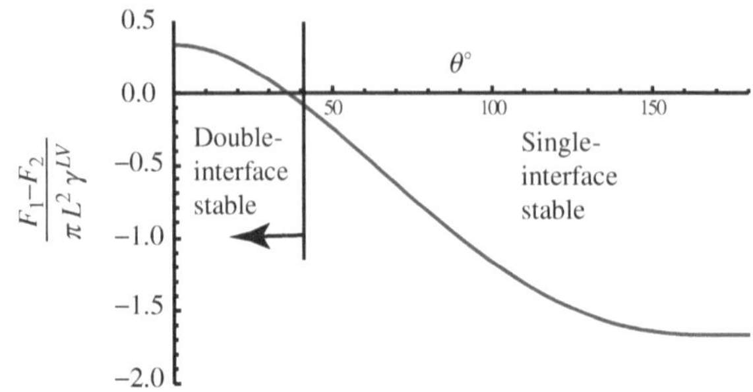

FIGURE 17.2 The difference between the Helmholtz function for the single-interface (F_1) and the double-interface (F_2) configurations is shown as a function of the contact angle, θ.

It has been previously shown that provided g_e is zero, the difference between the value of the Helmholtz function for the single-interface and the double-interface configurations, $F_1 - F_2$ can be expressed (Sasges et al., 1996)

$$\frac{F_1 - F_2}{\pi L^2 \gamma^{LV}} = \left[\cos\theta - \frac{2(1 + \sin\theta + \sin^2\theta)}{3(1 + \sin\theta)}\right] \tag{17.2}$$

where L is the radius of the cylinder, γ^{LV} is the surface tension of the fluid, and θ is the contact angle at the three-phase line.

A plot of this function is shown in Figure 17.2. Note that the relative magnitude of F_1 and F_2 depends only on the contact angle, θ. If θ is less than $\sim 36°$, the prediction is that F_1 is larger than F_2, meaning that the thermodynamically stable configuration in this contact angle range is F_2, the double-interface configuration. This is a vertically symmetric fluid configuration. However, if the contact angle is greater than $\sim 36°$, the thermodynamically stable configuration is the single-interface configuration, and it is not vertically symmetric.

17.1.3 Environment of the Japanese Dropshaft

The Japanese Dropshaft was used in the first attempt to examine the predictions indicated in Figure 17.2. That facility provided ~ 10 s of near free fall ($\sim 10^{-3} g_0$) and of that period 2.5 s of $1 \pm 2 \times 10^{-4} g_0$. A total of 9 (borosilicate) glass cylinders, ranging from ~ 20 to ~ 65 mm in diameter, that had different capacities, N_t, were prepared, partially filled with *water*—N/N_t from 0.2 to 0.52—and the contact angle measured. This angle was found to be $\sim 0°$ in each case.

Before being dropped, some cylinders were laid on their sides, but others were oriented vertically. Because the predicted equilibrium configuration of a fluid-cylinder system is independent of its initial configuration, comparing the final configurations with the initial configuration can be used to determine if there were enough time for a fluid-cylinder system to go to equilibrium. If there were enough time, there should be no dependence on the initial configurations; the final configuration should be a double-interface configuration in each case. A video recording was made of the cylinders during the period of near-free-fall. (Other liquids (hexadecane, propoanol, and butanol) were also considered, but because of space limitations, we discuss only the results with water.)

The experimental results indicated the final fluid configuration depended on the initial orientation—horizontal (i.e., laying on its side) or vertical—of the cylinders at the time of the drop. Those oriented horizontally were observed to spontaneously leave the single-interface and move toward the double-interface configuration, but none of the cylinders had reached the final equilibrium configuration. So these observations suggest that the 10 s of near free fall provided by the Dropshaft did not provide enough time for a system to reach the equilibrium configuration. However, they supported the hypothesis because the initially horizontal-oriented cylinders made the transition to a double-interface configuration, even though not completely in the time available.

17.1.4 Dependence of the Fluid Configuration on $G_E = G_0$

The thermodynamic theory was extended to allow g_e to be nonzero, but small enough for a system to adopt the double-interface configuration and to allow the the liquid–vapor interface to be approximated as spherical (Ward et al., 2000). This requires the Bond number, B, to be small compared to unity:

$$B \equiv \frac{g_e L^2 W}{4 v_f \gamma^{LV}} \tag{17.3}$$

where W is the molecular weight of the fluid, and v_f is the specific volume of the liquid at saturation. The extension of the theory then gave (Ward et al., 2000)

$$g_e = \left(\frac{4 v_f \gamma^{LV}}{W L \delta z} \right) (\cos \theta_u - \cos \theta_l) \tag{17.4}$$

where δz is the difference in the interface heights on the centerline of the cylinder:

$$\delta z \equiv (z_u - z_l) \tag{17.5}$$

Note that if θ_u were equal to θ_l, then the prediction would be that g_e was zero.

The theory also gave the equations that allowed θ_u and θ_l to be determined by measuring H_u, H_l, and h_u (Figure 17.3). Video images were taken when the space shuttle orbiter was on station. These measurements and Eq. (17.4) were used to

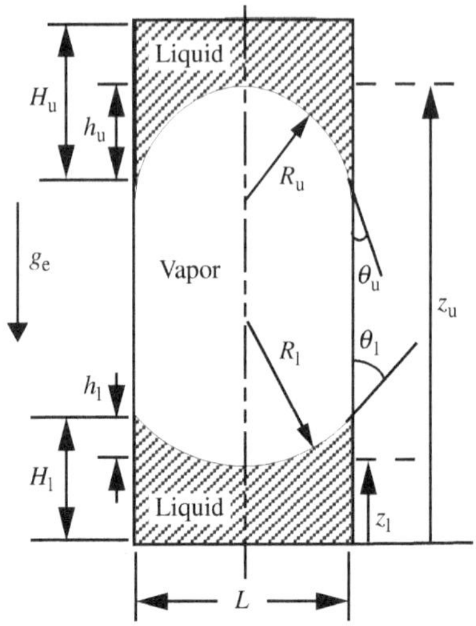

FIGURE 17.3 A (borosilicate) glass cylinder partially filled with water that is subject to an effective gravity field, g_e, of the space shuttle orbiter.

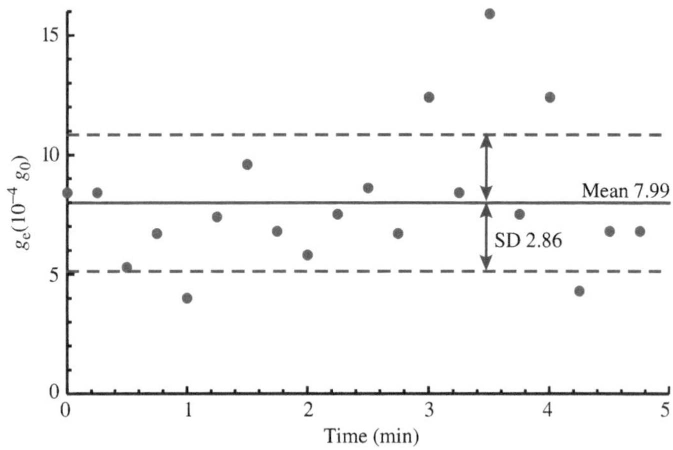

FIGURE 17.4 Predicted values of g_e made from measurements of the configurations of water when it is in the double-interface configuration (see Figures 17.1 and 17.3) and subject to the random accelerations (g-jitter) of the space shuttle orbiter.

calculate g_e at 15 s intervals. The results are shown in Figure 17.4. The mean value of (g_{ez}) directed downward toward the bottom of the cylinder that was determined from the contact angle measurements, Figure 17.4, was $8 \pm 3 \times 10^{-4}\ g_e/g_0$.

On the same orbiter and at the same time as our experiment, NASA operated both the OARE (Orbital Acceleration Research Experiment) and the SAMS (Space Acceleration Measurement Systems). These accelerometers indicated that the accelerations were $-7.4 \times 10^{-5} < g_e/g_0 < 7.5 \times 10^{-5}$ (Rogers et al., 1997).

Note that these accelerations are above and below zero. This is in contrast to the acceleration inferred from the fluid configurations: those from the fluid configuration were never negative.

17.1.5 Mechanism of Contact Angle Dependence on Pressure

After the space shuttle study had been completed, a study of the pressure dependence of the contact angle was undertaken (Wu et al., 2007). This subsequent study was conducted at normal gravity and showed that the contact angle depends strongly on the pressure in the liquid phase at the three-phase line. By using a small bore (borosilicate) glass tube of 1.12 mm in diameter, it was possible to put water into the double-interface configuration when the system was subject to normal gravity, g_0. It was shown experimentally that the pressure in the liquid phase had the same profile as though the phase were continuous, supporting a theory advanced earlier (Ward and Sasges, 1998).

With the tube oriented vertically, the pressure at the lower three-phase line of the double interface was larger than that at the upper three-phase line. The pressure difference could be increased by increasing the difference in height between of the two three-phase lines. The maximum difference between the three-phase line heights at which the system was brought into equilibrium was 28.86 mm. Under this condition, the contact angle at the lower three-phase line was $87.5° \pm 2.1°$, and at the upper three-phase line, it was $8.1° \pm 2.2°$. The difference in the pressure in the liquid phases at lower and upper three-phase lines in this experiment was ~230.1 Pa. When the system had reached the equilibrium configuration, there were no fluctuations in the contact angles at the upper or lower interface, as observed on the shuttle flight, Figure 17.4.

A method was subsequently developed that was based on the Gibbs dividing surface approximation (Gibbs, 1906) for measuring the amount adsorbed at a solid–liquid interface, n^{SL}, as a function of the cylinder radius, and θ (Ward et al., 2008; Ward and Wu, 2008; Wu et al., 2014). For water adsorbing on silica at 298 K, when the three-phase line was pinned at a diameter of 8.77 mm, the contact angles were shown experimentally to vary from 31.5° to 132.2°, while n^{SL} varied from -73.89 to -273.69 mol m^{-2}.

If the experiment flown on the shuttle could be approximated as being in equilibrium, adsorption at the solid–liquid interface could easily account for the difference in contact angles observed at the two three-phase lines. However, as indicated in Figure 17.4, g-jitter prevents the liquid–vapor system from reaching equilibrium, or, in other words, g-jitter causes the system to adopt a configuration corresponding to a larger value of g/g_0 than that indicated by the electronic accelerometers.

17.1.6 Conclusion

When water and its vapor are enclosed in a (borosilicate) glass cylinder and put in *complete* free-fall, provided the amount of fluid in the cylinder is less than a limiting value and the temperature remains unchanged, thermodynamics indicates the configuration adopted depends on the contact angle, θ: if θ is less than $\sim 36^\circ$, the prediction is that the fluid-cylinder system will spontaneously make a transition to the double-interface configuration shown in Figure 17.1, and if θ is greater, the system is predicted to remain in the familiar single-interface configuration.

A study conducted using the Japanese Dropshaft gave support to the predictions: cylinders initially lying on their sides were observed to make the transition to double-interface configuration during the 10 s of near free fall, but those that were initially oriented vertically did not. Because thermodynamic predictions are independent of the initial configuration, a possible interpretation is that the free-fall time of the Dropshaft was not sufficient for equilibrium to be established.

An experiment conducted on the space shuttle had much longer near-free-fall time, and a water-cylinder system was observed to adopt the double-interface configuration. The configuration corresponded to that predicted in a system in gravity with g/g_0 small compared to unity: the contact angle at the lower three-phase θ_l was greater than that at the upper θ_u. It could have been adsorption at the solid–liquid interface that was responsible for the observed difference between θ_l and θ_u.

When the measured values of these angles were used in the theory, the gravity was predicted to be that shown in Figure 17.4. The mean value of the predicted gravity is greater and in a different direction than that reported by the NASA accelerometers.

17.2 Thermo-capillary instabilities: evaporation of H_2O or D_2O

In the absence of buoyancy-driven convection, we compare the evaporation of H_2O and D_2O, both hydrogen-bonded liquids, from both thermally insulating and thermally conducting funnels. Water evaporates without Marangoni convection from a thermally insulating funnel, even though the Marangoni number greatly exceeds the Pearson–Nield prediction of 81 as the critical number. When the liquids evaporate from a thermally conducting funnel, the energy conservation principle cannot be satisfied unless energy transport by Marangoni convection is taken into account. The surface-thermal capacity of D_2O is found to be larger than that of H_2O, reflecting the relative dipole moments of these molecules.

17.2.1 Introduction

During water evaporation at temperatures below 4°C, there are three possibly active modes of energy transport: buoyancy-driven convection, thermal conduction, and Marangoni convection, but as a mode of energy transport, radiation is generally negligible. In energy analyses, the usual procedure has been to also neglect energy transport by Marangoni convection, but recent results indicate this mode of energy transport is the dominant mode during sessile droplet evaporation (Ghasemi and Ward, 2010, 2011).

Water maintained at the mouth of a funnel as it evaporates steadily can be used to determine experimentally the conditions at which Marangoni convection is initiated. The Pearson–Nield method has been the standard procedure for predicting these conditions (Pearson, 1958; Nield, 1964), but it is questionable whether their approach is applicable, because it does not actually consider evaporation or heating through the vapor. The Pearson–Nield model indicates that Marangoni convection starts when the Marangoni number exceeds 81, but water evaporation from the mouth of a polymethyl methacrylate (PMMA) funnel with Marangoni numbers greater than 27,000 has been reported in which there was no measurable Marangoni convection (Thompson et al., 2009; Das et al., 2010; MacDonald and Ward, 2011, 2012).

An expression for the energy transport by Marangoni convection has been proposed in terms of a transport coefficient, the surface-thermal capacity c_σ. The values of c_σ for water and for heavy water (D_2O) have been measured (Duan and Ward, 2005a,b). We review these measurements and find that if the energy transport by Marangoni convection is *not* taken into account during the evaporation experiments, the energy conservation principle cannot be satisfied.

Water and heavy water (D_2O) have a special property that can be used to simplify the analysis of their respective evaporations: below 4°C, water becomes progressively lighter as its temperature is reduced. Heavy water has the same type of property, but it reaches its maximum density at 11.2°C (Duan and Ward, 2005b). More important, these two waters have different values for their dipole moments: 6.19 and 6.24 ($\times 10^{-30}$ C m) for H_2O and D_2O, respectively (Design Institute for Physical Properties, 2011). Thus, the hydrogen bonding between D_2O molecules is stronger than that between H_2O molecules. As will be seen, this difference in the dipole moments is important in the values of c_σ, and this gives insight into the molecular origin of c_σ.

17.2.2 Energy Transport at a Liquid–Vapor Interface During Steady Evaporation

Provided a liquid has a maximum density at a particular temperature, the apparatus shown schematically in Figure 17.1 can be used to eliminate buoyancy-driven convection during the evaporation. For a spherical liquid–vapor interface of radius R_0 that is axisymmetric and has an interfacial temperature gradient of $\nabla T^{LV} \cdot i_\theta$, the energy flux along the interface by Marangoni transport may be expressed (Duan and Ward, 2005a) as

$$j_{\mathrm{st}}^{LV} = -\left(\frac{c_\sigma v_\theta^{LV}}{R_0}\right)(\nabla T^{LV} \cdot i_\theta), \tag{17.6}$$

where the flow speed along the interface is denoted as v_θ^{LV}, and the energy transport along the interface by Marangoni convection is

$$\dot{Q}_{\mathrm{st}} = 2\pi \int_0^{\theta_{\mathrm{m}}} j_{\mathrm{st}}^{LV} R_0^2 \sin\theta \, \mathrm{d}\theta. \tag{17.7}$$

As indicated in Figure 17.5, the angle θ_{m} defines the extent of the spherical interface. The thermal flux by conduction from the liquid to the interface and that from the vapor to the interface are, respectively,

$$j_{\mathrm{c}}^{L} = -\kappa^{L}(\nabla T^{L} \cdot i_{\mathrm{r}}), \tag{17.8}$$

$$j_{\mathrm{c}}^{V} = \kappa^{V}(\nabla T^{V} \cdot i_{\mathrm{r}}), \tag{17.9}$$

and the total heat conduction to the spherical, axisymmetric interface that we consider, $\dot{Q}_{\mathrm{c}}$, is given by

$$\dot{Q}_{\mathrm{c}} = 2\pi \int_0^{\theta_{\mathrm{m}}} (j_{\mathrm{c}}^{L} + j_{\mathrm{c}}^{V}) R_0^2 \sin\theta \, \mathrm{d}\theta. \tag{17.10}$$

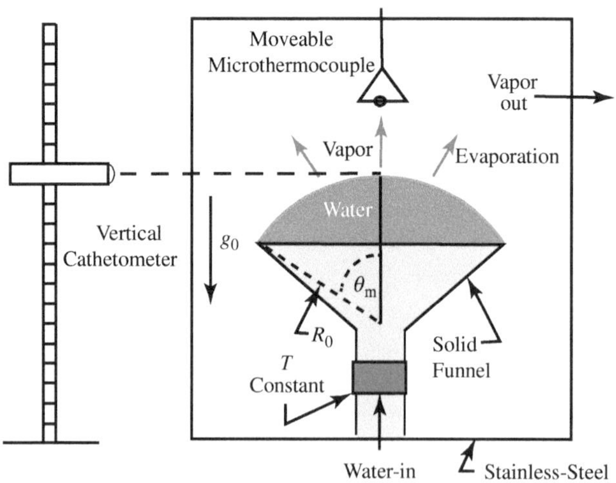

FIGURE 17.5 Schematic of the apparatus used to measure the surface-thermal capacity of H_2O and D_2O (Duan and Ward, 2005a).

The steady rate of energy consumption required to evaporate the liquid at a measured rate of J_{ev} may be expressed

$$\dot{E}_{ev} = 2\pi \int_0^{\theta_m} J_{ev}[h^V(T_I^V) - h^L(T_I^L)]R_0^2 \sin\theta \, d\theta, \tag{17.11}$$

where $h^V(T_I^V)$ and $h^L(T_I^L)$ are the enthalpies of the vapor and liquid phases evaluated on their respective sides of the interface. The difference between the enthalpies varies by less than 1% in the experiments we consider. Thus, $\dot{E}_{ev}$ can be simplified to

$$\dot{E}_{ev} = J_{ev}[h^V(T_I^V) - h^L(T_I^L)], \tag{17.12}$$

where J_{ev} is the evaporation rate.

If the interfacial flow is not turbulent (Ward and Duan, 2004), then imposing the conservation-of-energy principle gives

$$\dot{E} = \dot{Q}_c + \dot{Q}_{st}. \tag{17.13}$$

In the absence of turbulence, we suppose c_σ is constant (Duan and Ward, 2005a,b), and then from Eqs. (17.7), (17.10), and (17.12):

$$c_\sigma = \frac{\dot{Q}_c - \dot{E}}{2\pi \int_0^{\theta_m} v_\theta^{LV}(\nabla T^{LV} \cdot i_\theta) R_0 \sin\theta \, d\theta}. \tag{17.14}$$

This expression for c_σ can be applied once an expression for v_θ^{LV} is available.

17.2.3 Energy Transport by Marangoni Convection

The apparatus shown schematically in Figure 17.5 was used to establish the steady evaporation of the liquids H_2O and D_2O while each was maintained at the mouth of a stainless-steel funnel. H_2O was maintained at the mouth of a PMMA funnel while it evaporated steadily (Ward and Duan, 2004; Duan and Ward, 2005a,b; Thompson et al., 2009). The temperature of the water entering the throat of each funnel was brought to a value of slightly less than 4, while the temperature at the liquid–vapor interface was cooled below this value by evaporation. Thus, the lighter H_2O or D_2O was above the heavier, and no buoyancy-driven convection would be expected under these conditions in either case.

Liquid was poured in the funnel throat at a flow rate of J_{ev} that was accurately measured with a syringe pump. Under steady conditions, J_{ev} is equal to the total evaporation rate. The pressure in the vapor phase was controlled with a vacuum system and used to control the evaporation rate. The system was brought to steady state by adjusting both J_{ev} and the pressure in the vapor phase with the vacuum system. The liquid–vapor interface could be viewed from outside the chamber with a cathetometer. The evaporation was judged to be steady when the height of the liquid–vapor interface on the centerline of the funnel did not move by more than $\pm 10\ \mu m$ during the course of an experiment.

Once steady-state evaporation had been established, the temperature on the interface was measured as a function of θ and as a function of depth. Both were measured with a calibrated thermocouple ($\pm 0.03°C$) that had a bead diameter of 50 μm and was mounted on a positioning micrometer ($\pm 10\ \mu m$) that had 5 degrees of freedom. The measurements of the interfacial temperature fields indicated they were axisymmetric and, as shown in Figures 17.6 and 17.8, that there were interfacial temperature discontinuities at the liquid–vapor interface, as have been seen in previous experiments (Fang and Ward, 1999). These measurements indicate that the interfacial vapor temperature is greater than that of the interfacial liquid in all the cases considered.

17.2.3.1 Marangoni flow speed

For the Marangoni flow to be steady, we suppose that the fluid-shear stress at the interface, $\sigma_{r\theta}(R_0, \theta)$, is equal to the gradient in the surface tension along the interface:

$$\sigma_{r\theta}(R_0, \theta) = \nabla \gamma^{LV} \cdot i_\theta. \tag{17.15}$$

If the dependence of v_θ on r is disregarded, then a partial differential equation is obtained from Eq. (17.15) that may be transformed by a change of variables into a differential equation and approximately integrated to obtain (Ward and Duan, 2004)

$$v_\theta(R_0, \theta) = -\left(\frac{\gamma_T^{LV}}{\eta}\right)\left(\nabla T^{LV} \cdot i_\theta\right)\left[ln\left(1 - \frac{\delta_u}{R_0}\right)\right], \tag{17.16}$$

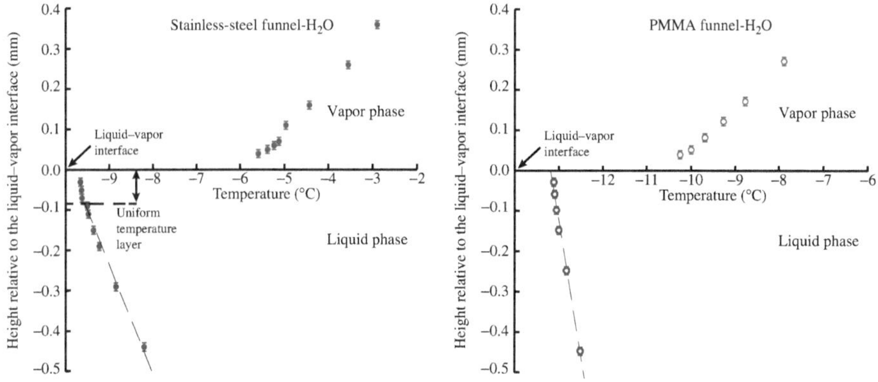

FIGURE 17.6 Temperature of water measured on its centerline as the water evaporated steadily into its vapor from a stainless-steel (left) or from a PMMA (right) funnel.

where η is the fluid viscosity, γ_{T}^{LV} is the temperature coefficient of the surface tension, and δ_{u} is the radial distance below the interface where the return flow parallel to the interface vanishes. The values of γ_{T}^{LV} and η are fluid properties, and their values are known for both of the liquids we consider.

The measured temperature as a function of depth for H_2O evaporating from the stainless-steel and from the PMMA funnels are shown in Figure 17.6. That of D_2O evaporating from the stainless-steel funnel is shown in Figure 17.7. Thus, for both liquids evaporating from the stainless-steel funnel, δ_{u} is greater than zero, and Eq. (17.16) indicates $v_{\theta}(R_0, \theta)$ is as well. But for water evaporating from the PMMA funnel, δ_{u} and $v_{\theta}(R_0, \theta)$ vanish.

For H_2O evaporating steadily while maintained at the mouth of the stainless-steel funnel, the pressure in the vapor phase was 300 Pa, and the average evaporation flux was 3380 ± 31 mg $(m^2\ s)^{-1}$. For the PMMA funnel, the vapor-phase pressure was 231 Pa, but the average evaporation flux was only 604 ± 30 mg $(m^2\ s)^{-1}$. Note that the average evaporation flux from the stainless-steel funnel was more than five times larger than that in the PMMA funnel, even though the pressure in the vapor phase above the stainless-steel funnel was *higher* than that above the PMMA funnel. These results suggest that Marangoni flow strongly enhances evaporation.

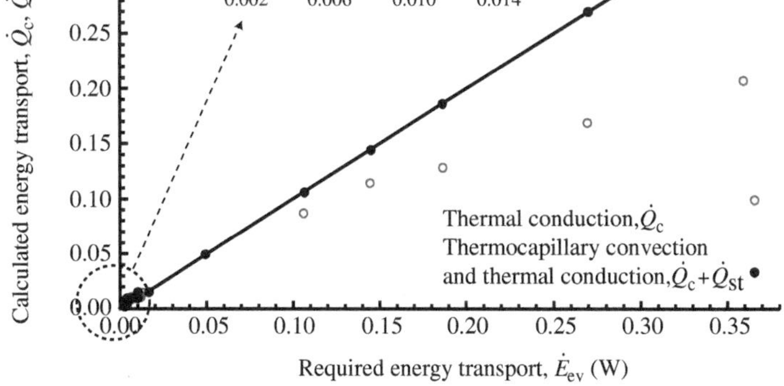

FIGURE 17.7 Comparison of the heat required to evaporate the liquid at the observed rate, $\dot{E}_{ev}$, with the heat by thermal conduction alone (open circles), and including Marangoni convection (solid circles).

17.2.3.2 Measurement of the surface-thermal capacity of H_2O and D_2O

The expression for the thermal energy required to evaporate a liquid, $\dot{E}$, at a measured evaporation rate, J_{ev}, is given in Eq. (17.13). The value of $\dot{E}$ was determined by the syringe pump reading and the specific enthalpies of the liquid and vapor. $\dot{E}$ has one value for each experiment. If conservation of energy is satisfied, then the energy transported to the interface by thermal conduction, Q_c, and by Marangoni convection, Q_{st}, must sum to $\dot{E}$.

We investigate first whether thermal conduction alone provides enough energy to evaporate H_2O at the measured evaporation rate. As seen in Figure 17.8, for the six experiments with average evaporation fluxes less than or equal to 63 mg $(m^2\ s)^{-1}$, the thermal conduction provides all the energy required to evaporate H_2O, and the data points lie on the 45 line, indicating that for these experiments:

$$\dot{E}_{ev} = \dot{Q}_c. \tag{17.17}$$

So for these six experiments, energy transport by Marangoni convection is negligible.

Next we determine the value of c_σ for H_2O from the six evaporation with average evaporation fluxes greater than 63 mg $(m^2\ s)^{-1}$. For these experiments, thermal conduction supplied a progressively smaller portion of the total. At the highest evaporation rate, thermal conduction supplied only 54% of the required energy transport rate, $\dot{E}_{ev}$. Marangoni convection supplies the rest. For these experiments, Eq. (17.14) was used to calculated c_σ for H_2O. The average value

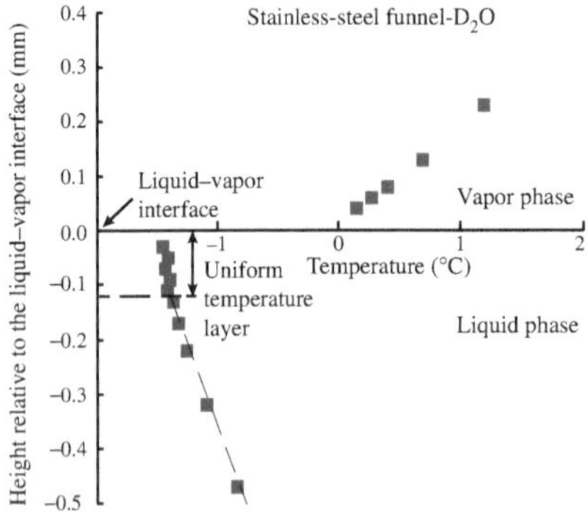

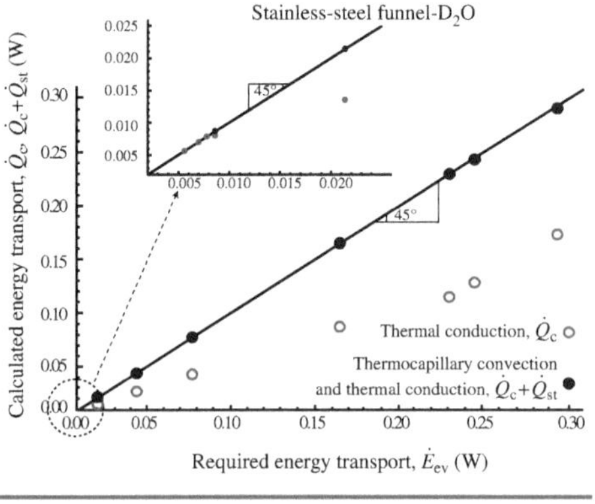

FIGURE 17.8 Left: Temperature measurements of heavy water on its centerline as the liquid evaporated steadily into its vapor from a stainless-steel funnel. Right: Comparison of the energy transport rate required to evaporate the D_2O at the observed rate, J_{ev}, with the heating by thermal conduction alone (open circles), and including Marangoni convection (solid circles) (right).

obtained was 30.6 ± (SDV)0.8 kJ $(m^2\ K)^{-1}$. The small value of the standard deviation (SDV), 2.6%, supports the assumption that c_σ can be approximated as constant in the experiments with different evaporation fluxes. In other words, c_σ appears to be independent of the evaporation flux.

Similar experiments were conducted with liquid D_2O evaporating from the stainless-steel funnel (Duan and Ward, 2005b). As seen in Figure 17.8, when the average evaporation flux was less than or equal to 89 ± 1 mg $(m^2\ s)^{-1}$, thermal conduction again provided the energy required to evaporate the liquid at the measured rate, and Eq. (17.17) was satisfied for these four experiments. But at the highest evaporation fluxes, thermal conduction provided only 59% of $\dot{E}_{ev}$. For the seven experiments with evaporation fluxes greater than 89 ± 1 mg $(m^2\ s)^{-1}$, Eq. (17.14) was used to determine the value of c_σ for D_2O. An average value of 32.5 ± 0.8 kJ $(m^2\ K)^{-1}$ was found. The small SDV value of these measurement, 2.4%, supports the hypothesis that c_σ is independent of the evaporation flux.

17.2.4 Discussion and Conclusion

A linear stability analysis that used statistical rate theory (SRT) to take the observed interfacial temperature discontinuities into account in the experiments has been reported (Ward and Fang, 1999; MacDonald and Ward, 2011, 2012). In the stability analysis, the initial state was assumed to be one in which the energy required to evaporate the liquid came through the vapor, and the perturbations were applied only in the liquid phase. If the funnel had *no thermal conductivity*, the governing equations indicated that the temperature perturbation changed the temperature in the liquid phase *uniformly*. Thus, it was not possible for the perturbation to generate a temperature gradient along the liquid–vapor interface—a necessity for Marangoni convection to exist—and a liquid evaporating from such a funnel was predicted to be stable, independent of the evaporation flux. Thus, the explanation for the greater stability of water evaporating from a PMMA funnel compared to that of water evaporating from a stainless-steel funnel appears to result from the type of perturbations that could be generated in an insulating funnel.

For both waters evaporating from a thermally conducting funnel, there was a small range of evaporation fluxes that preceded the onset of Marangoni convection. For H_2O this range extended up to evaporation fluxes of 63 ± 1, and for D_2O up to fluxes of 89 ± 1 mg $(m^2\ s)^{-1}$. The energy supplied for these small evaporation fluxes could be coming through the vapor phase (Thompson et al., 2009).

For evaporation from the insulating funnel, as seen in Figure 17.6, the energy supplied to the interface came by conduction through the liquid and through the vapor phase. Removing the energy supplied through the funnel strongly reduced the evaporation flux, as described earlier.

It has been previously hypothesized that the molecular origin of c_σ is the hydrogen bonding of water molecules (Duan and Ward, 2005a). This hypothesis is supported by the comparison of H_2O and D_2O liquids evaporating from the thermally conducting funnel. The dipole moment of D_2O is greater than that of H_2O. Thus, when a surface tension gradient pulls molecules toward a lower temperature, for a given temperature gradient, more D_2O molecules would be moved because of the temperature gradient. This gives a possible explanation for the higher value of c_σ for D_2O, 32.5, than for H_2O, 30.6 kJ $(m^2\ K)^{-1}$.

17.3 Thermo-solutal instabilities

Temperature and pressure pulsations are observed when a droplet evaporates into a multicomponent atmosphere that has a heat of solution with the droplet. The temperature pulsations in the liquid are correlated with the vapor-phase pressure pulsations. We show that it is evaporation that couples these variables and use statistical rate theory (SRT) to predict the temperature pulsations in the liquid phase using the vapor pressure pulsations and vice versa. The theoretical predictions of the pulsations agree with the measurements. No pulsations are observed when the system is single-component.

17.3.1 Introduction

Multicomponent droplet evaporation is often encountered in nature and in many industries. In multicomponent systems, different components can interact with one another to produce new effects that are not present in single-component systems. These effects must be considered as part of the evaporation system. For example, numerous investigations have used infrared (IR) thermography to study the evaporation of ethanol, methanol, or FC-72 droplets in room atmospheres (Brutin et al., 2010, 2011; Carle et al., 2012; Karapetsas et al., 2012; Sefiane et al., 2013). Those investigators observed temperature pulsations and claimed that the pulsations were hydrothermal waves (HTWs) that resulted from Marangoni instabilities when a droplet evaporates. Barthes et al. (2007) used shadowgraphs to observe temperature pulsations in FC-72. By comparing their results with degassed and nondegassed FC-72, they believed that the onset of Marangoni instabilities was due to either the presence of noncondensable gases in the liquid or the multicomponent nature of FC-72 itself. Only recently have doubts been raised about Marangoni convection being the mechanism of the pulsations (Karapetsas et al., 2012; Sefiane et al., 2013); Sefiane et al. (2013) suggested that adsorption of humidity from the room air could be another mechanism worth investigating.

Adsorption as the mechanism producing the temperature pulsations is supported by the work of Velarde et al. (2001). They used shadowgraphs to study the evaporation of toluene from cylindrical and annular pools. When the toluene evaporated in the presence of another vapor like pentane or heptane, they observed traveling wave trains with surface patterns similar to the hydrothermal waves observed with IR thermography.

Others have observed temperature and pressure pulsations when droplets evaporated into multicomponent atmospheres and a component from the atmosphere had a heat of solution with the droplet (Persad et al., 2013). But the pulsations were absent when the evaporating system was single-component. These results were repeatable: when ethanol, methanol, and water evaporated into their own vapors, no temperature or pressure pulsations were observed, but when they evaporated into multicomponent atmospheres, temperature and pressure pulsations were observed in the liquid and vapor phases (Persad et al., 2013).

The presence of humidity was found to trigger temperature pulsations when sessile droplets of alcohols evaporated into atmospheres that had relative humidities (RH) above threshold values; the threshold RH value for a methanol droplet is $\sim 70\%$, and for an ethanol droplet $\sim 45\%$ (Persad et al., 2013). Note that each of these liquids has a heat of solution with water (Bertrand et al., 1966; Freire et al., 2006).

The mechanism producing the temperature and pressure pulsations was shown to be the heat of solution from adsorption–absorption of a vapor species in a multicomponent atmosphere into a droplet that causes pulsations in the evaporation flux and results in temperature and pressure pulsations. The proposed mechanism was further examined by reversing the roles of the two components; for a stably stratified water droplet evaporating into a methanol-water atmosphere (also stably stratified), temperature and pressure pulsations were observed and were found to be coupled by the evaporation flux.

The adsorption–absorption mechanism was also investigated theoretically. The SRT expression for the evaporation flux, which has no fitting parameters (Ward and Fang, 1999), was used to calculate the pressure pulsations from inputting the measured temperature pulsations of the droplet, the average evaporation flux, and the temperatures across the liquid–vapor interface. The predicted pressure pulsations agreed closely with the measured water-vapor partial pressure pulsations. Similar agreement was found when the droplet temperature pulsations were predicted by using the measured water-vapor partial pressure pulsations as the input into SRT.

17.3.2 Experimental Apparatus

Persad et al. (2013) used a microthermocouple to examine the temperature in the liquid and vapor phases to overcome some of the limitations of IR thermography and shadowgraphs. One limitation is knowing precisely where the temperature is measured; IR thermography and shadowgraphs measure the average temperature over the optical path (Schöpf et al., 1996; Brutin et al., 2010; Sobac and Brutin, 2012), but microthermocouples measure the temperature of the bead, which can be precisely placed. Another limitation is that IR thermography and shadowgraphs do not give a direct reading of the temperature; it has to be computed from knowledge of the material emissivity and its dependence on temperature and droplet depth—and the optical wavelength makes it difficult to accurately determine the temperature (Brutin et al., 2010; Sobac and Brutin, 2012)—or from knowledge of the density, index of refraction, or concentration (Schöpf et al., 1996). In contrast, microthermocouples do not have to be calibrated for different fluids and can readily measure the temperature in the liquid and vapor phases. Microthermocouples can also be used in water studies even though water is IR opaque.

The experimental apparatus used by Persad et al. is shown schematically in Figure 17.9. A sessile droplet was formed on the substrate and evaporated by removing vapor from the chamber with a mechanical pump. The composition of the atmosphere was monitored with a residual gas analyzer, and the water vapor was measured with a relative humidity sensor. The droplet height was monitored

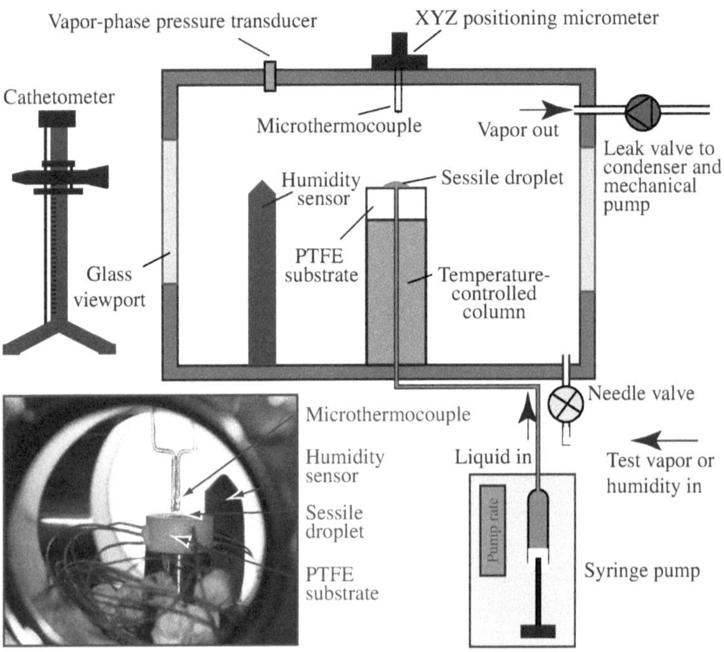

FIGURE 17.9 Schematic and picture of the experimental apparatus. *Reprinted (adapted) with permission from Persad et al. (2013). Copyright 2013 American Chemical Society.*

with a cathetometer to within $\pm 10\ \mu m$ through the chamber's glass viewports. The evaporation chamber was cleaned between experiments by connecting it to an ultra-high-vacuum system with a pressure of $<10^{-5}$ Pa for at least 24 h. For the steady evaporation studies, the evaporating droplet was replenished by injecting liquid at the droplet's base such that the droplet height remained steady within $\pm 10\ \mu m$ during evaporation for at least 30 min. The vapor-phase pressure was measured to an accuracy of ± 4 Pa. The temperature was measured with a type-k microthermocouple that was calibrated to an accuracy of ± 0.03 K. The microthermcouple could be moved in the *X*, *Y*, and *Z* axes with a precision of $\pm 10\ \mu m$ to measure the temperatures in the atmosphere and in the droplet.

17.3.3 Experimental Support of the Proposed Pulsation Mechanism

The noise in the thermocouple reading had a peak-to-peak amplitude (ppA) of 0.04°C (see noise profile in Figure 17.10) in air with a RH of 50–55%. With room air in the chamber, an ethanol droplet of approximately the same size as that used in the IR study (base diameter of 4 mm) was formed on the substrate and allowed to evaporate. The temperature in the bulk ethanol pulsated with a

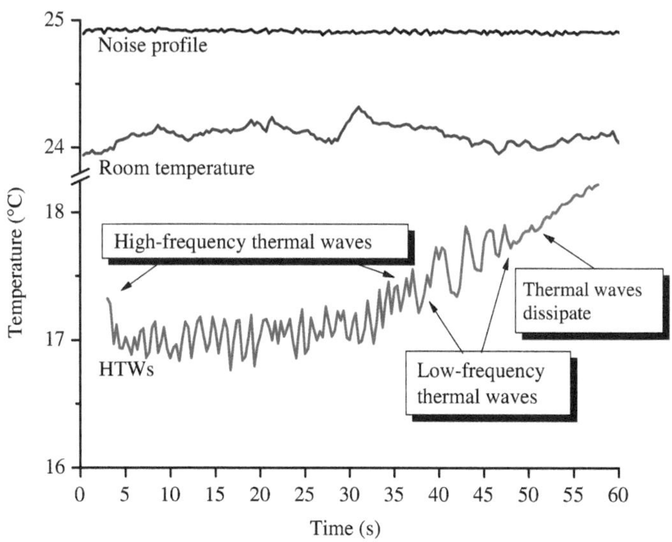

FIGURE 17.10 The noise profile of the microthermocouple in the chamber is shown. The room temperature is also shown with an RH of ~50%. An ethanol droplet evaporates from the surface of a PTFE substrate and the temperature in the bulk liquid shows three modes of temperature pulsations as described in the text. *Reprinted (adapted) with permission from Persad et al. (2013). Copyright 2013 American Chemical Society.*

ppA ten times greater than the thermocouple noise and with a frequency ~0.5 Hz, as indicated in Figure 17.10.

Water condensing on the ethanol droplet that was cooled by evaporation may have caused the temperature pulsations. This possibility was investigated and would have required that the temperature of the droplet surface be below 7°C. Because the measured temperature of the droplet was 17°C (see Figure 17.2), condensation on the droplet surface could not have been the source of the observed temperature pulsations.

Persad et al. (2013) noted that the temperature pulsations indicated in Figure 17.10 were similar to those reported for ethanol droplets evaporating into room air with IR thermography. In the initial stages of the evaporation, numerous higher-frequency temperature pulsations with a ppA of 0.4°C occurred spontaneously. These were followed by lower-frequency pulsations. Finally, as the droplet evaporated and became flatter, the pulsations dissipated (Brutin et al., 2011; Carle et al., 2012; Karapetsas et al., 2012; Sobac and Brutin, 2012). The agreement between the IR thermography and the microthermocouple measurements indicated that any intrusive effects of the microthermocouple were negligible.

When the experiments were repeated with a methanol droplet, even larger temperature pulsations were measured. This is consistent with the IR thermography results of others who reported that the temperature pulsations in an evaporating methanol droplet were larger than those in an ethanol droplet (Karapetsas et al., 2012). Note that the heat of solution of water and methanol is larger than that of water and ethanol (Bertrand et al., 1966).

In their next experiment, Persad et al. (2013) formed a methanol droplet on the substrate and precooled it to −15.1°C. The precooling was necessary to keep the lowest temperature at the droplet's base and thus prevent the onset of buoyancy-driven convection. The droplet's size was kept constant as it evaporated by adjusting the vapor-phase pressure and the injection rate of methanol into the droplet's base. When the microthermocouple was put in the methanol droplet, no pulsations were recorded.

The RH in the chamber was then increased at the time indicated in Figure 17.11. When the RH had reached ~70%, the temperature recorded by the microthermocouple that was in the droplet began to pulsate with a ppA that was three times greater than the noise level. As the RH continued to increase, so did the ppA of the temperature pulsations. But when the RH was reduced to below 70%, the temperature pulsation were no longer detectable. When the RH was again increased to above 70% the temperature pulsation reappeared, had a ppA of 0.60°C, and continued for over 30 min. Thus, an RH value greater than 70% appears necessary in order for the temperature in the methanol droplet to pulsate. A similar study with ethanol indicated its RH threshold value as ~45%.

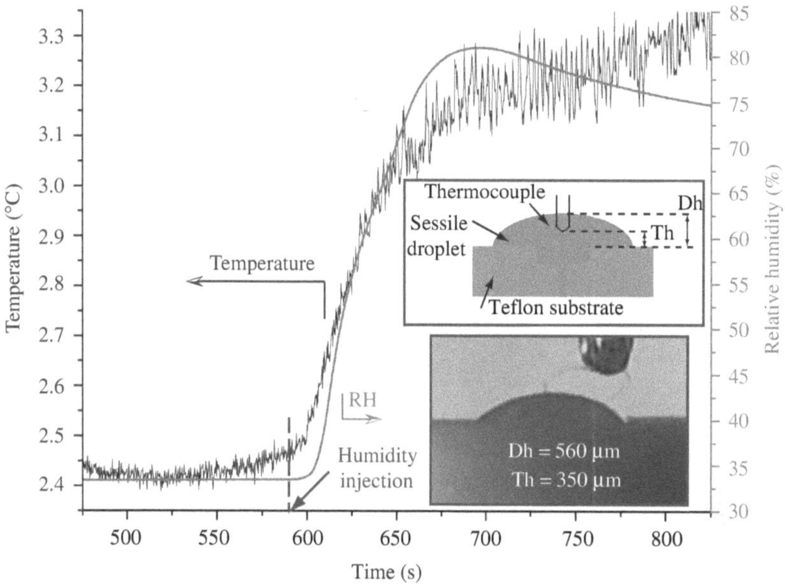

FIGURE 17.11 When a sessile methanol droplet evaporates into an atmosphere with an RH above a threshold value of ~70%, temperature pulsations appear in the bulk liquid. Insets show the profile of the droplet and the position of the microthermocouple bead. *Reprinted (adapted) with permission from Persad et al. (2013). Copyright 2013 American Chemical Society.*

Note that in the steady experiments of Persad et al. (2013), the pulsations could be maintained for long periods (more than 30 min) by maintaining the RH above the threshold value. However, in unsteady experiments, the pulsations would completely vanish within a minute or two (Brutin et al., 2011; Carle et al., 2012; Karapetsas et al., 2012; Sobac and Brutin, 2012; Sefiane et al., 2013).

The microthermocouple was then moved near the droplet's apex but still in the liquid phase, position 1 in Figure 17.12. The ppA of the temperature pulsations recorded at this position was unchanged. Similarly, when the microthermocouple was moved radially 200 μm to the left and to the right of the centerline along the droplet surface, positions 2 and 3 in Figure 17.12, the temperature pulsations still had a ppA of 0.60°C. Thus, temperature pulsations are not just a surface phenomena but are present throughout the droplet; Sefiane et al. (2013) also made similar observations.

With the methanol droplet still evaporating into the atmosphere with an RH value greater than 70%, the microthermocouple was moved into the vapor phase. The pulsations were measured 60, 110, and 340 μm above the droplet apex, as indicated in Figure 17.12, by positions 4, 5, and 6, respectively. Temperature

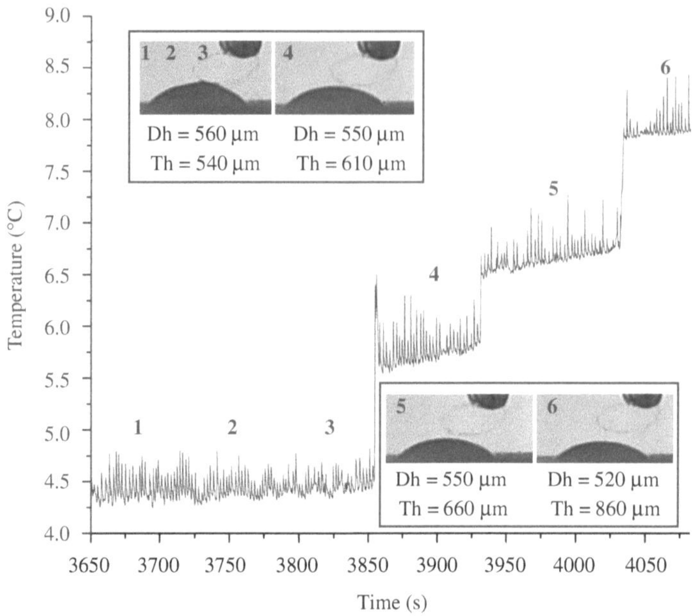

FIGURE 17.12 Temperature pulsations are present in the liquid and vapor phases when a sessile droplet of methanol evaporates into a humid atmosphere with an RH of 72.4%. The numbers above the graph correspond to different positions of microthermocouple bead, as shown in the insets. *Reprinted (adapted) with permission from Persad et al. (2013). Copyright 2013 American Chemical Society.*

pulsations were observed at each position in the vapor phase, and their ppA did not diminish with height above the interface but reached a maximum of 0.70°C.

After cleaning the chamber, Persad et al. (2013) reversed the roles of water and methanol; they formed a water droplet on the PTFE substrate and brought it into steady-state evaporation. In order to stably stratify the droplet, they kept the base of the steadily evaporating sessile water droplet at 4°C (temperature of maximum density). When the water droplet only evaporated into its own vapor and the microthermocouple was placed in the droplet, no temperature pulsations were observed. They also monitored the vapor-phase pressure with a pressure transducer and did not observe any pulsations. The absence of temperature and pressure pulsations under these conditions has been reported in water evaporation studies many times in the past (Fang and Ward, 1999; Duan et al., 2008; Ghasemi and Ward, 2010).

But after methanol was injected into the chamber to form a methanol atmosphere, the water droplet began to evaporate unsteadily and at a higher rate than

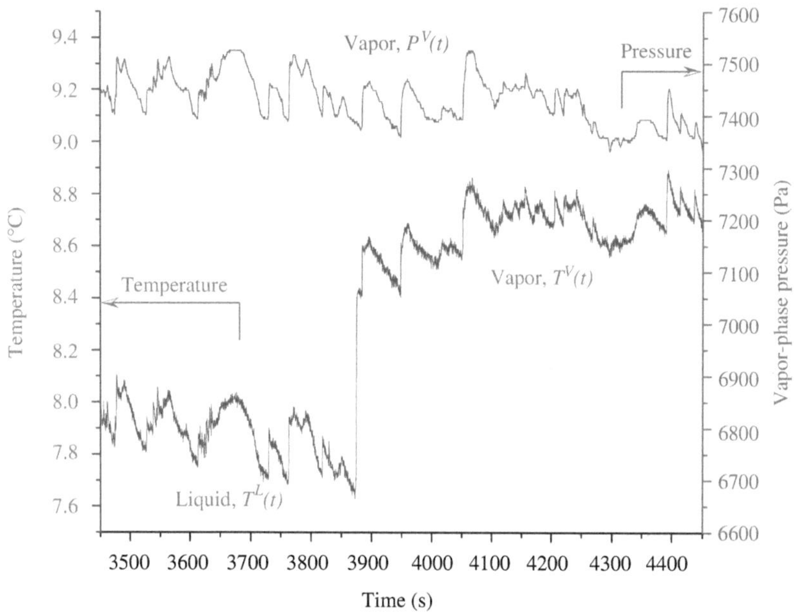

FIGURE 17.13 After methanol injection into the atmosphere, temperature pulsations appear in the bulk liquid and vapor phases. The vapor-phase pressure also pulsates and correlates strongly with the temperature pulsations in both the liquid and vapor phases. *Reprinted (adapted) with permission from Persad et al. (2013). Copyright 2013 American Chemical Society.*

when it evaporated into its own vapor. The droplet temperature also began to pulsate unsteadily. After 40 min, the temperature in the bulk liquid and vapor phases continued to pulsate as shown in Figure 17.13; the bulk-liquid temperature is denoted by $T^L(t)$, and the temperature in the bulk vapor is denoted by $T^V(t)$.

Pressure pulsations were also observed in the vapor phase and are denoted by $P^V(t)$ in Figure 17.13. The mole fraction of water in the atmosphere was measured with a quadrupole mass analyzer, and it was 0.143 ± 0.004. The partial pressure pulsations of the water in the atmosphere, P_w^V, could then be obtained by multiplying the vapor-phase pressures in Figure 17.13 by the water mole vapor fraction and are shown in Figure 17.14.

The temperature and pressure pulsations shown in Figure 17.13 were examined using the Pearson product-moment criterion for linear correlation (Asuero et al., 2006). The correlation coefficient between the bulk liquid temperature and

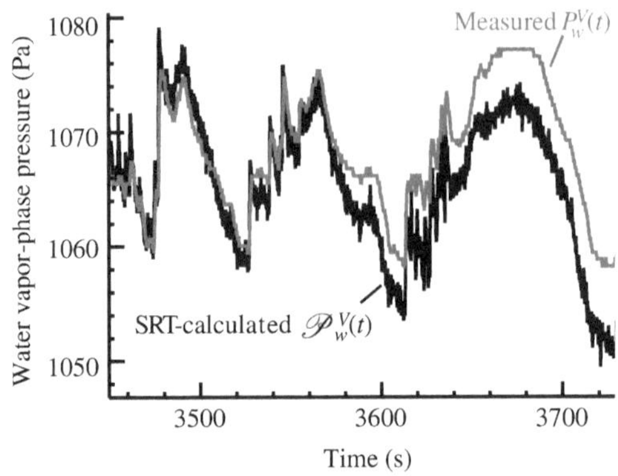

FIGURE 17.14 The measured partial pressure of water vapor, $P_w^V(t)$, as function of time is shown as a thin line. The thick line is the SRT calculation of the partial pressure of water vapor, $\mathscr{P}_w^V$, which corresponds to that in Figure 17.13. *Reprinted (adapted) with permission from Persad et al. (2013). Copyright 2013 American Chemical Society.*

vapor-phase pressure was 0.85, and it was 0.53 between the bulk vapor temperature and the vapor-phase pressure. These values indicated a linear dependence between each set of variables.

Persad et al. (2013) assumed that the methanol from the atmosphere adsorbed-absorbed in the water droplet. The droplet would remain stably stratified because methanol is less dense than water. Furthermore, Figure 17.12 indicates the vapor mixture would also be stably stratified (assuming that the atmospheres were ideal) because the vapor-phase temperature increased with height. Thus, they believed that it was not buoyancy-driven convection that caused the observed pulsations; others have expressed a similar opinion (Brutin et al., 2010).

In order to determine the average evaporation flux of the droplet in the methanol-water atmosphere and to calculate the concentration of methanol in the droplet, Persad et al. (2013) used image analysis techniques. From the droplet shape, they calculated the liquid–vapor surface tension and used that value to determine that the mole fraction of methanol in the droplet was $55 \pm 15\%$. Interestingly, the mole fraction of methanol in the vapor mixture measured with the quadrupole mass analyzer was $56 \pm 11\%$. Thus, the methanol mole fractions in the liquid and vapor phases were not measurably different. However, the water mole fraction in the droplet was $\sim 45\%$ but only $\sim 14.3\%$ in the vapor phase, and Persad et al. calculated that the average evaporation flux of water from the droplet was 475 mg $(m^2\ s)^{-1}$.

17.3.4 Theoretical Support of the Proposed Pulsation Mechanism

The full SRT expression for the net evaporation flux of a liquid, j_{ev} (Ward and Fang, 1999):

$$j_{ev} = 2P_s(T_I^L)\left(\sqrt{\frac{m}{2\pi k_B T_I^L}}\right)\sinh\left\{\ln\left[\left(\frac{T_I^V}{T_I^L}\right)^4 \frac{P_s(T_I^L)}{\mathscr{P}_w^V}\right] + 4\left(1-\frac{T_I^V}{T_I^L}\right)\right.$$
$$+\left(\frac{1}{T_I^V}-\frac{1}{T_I^L}\right)\sum_{l=1}^{3n-6}\left[\frac{\left(\frac{\hbar\omega_l}{k_B}\right)}{2}+\frac{\left(\frac{\hbar\omega_l}{k_B}\right)}{e^{\left(\frac{\hbar\omega_l}{k_B}\right)/T_I^V}-1}\right] - \ln\left[\prod_{l=1}^{3n-6}\frac{e^{-\left(\frac{\hbar\omega_l}{k_B}\right)/2T_I^L}}{1-e^{\left(\frac{\hbar\omega_l}{k_B}\right)/T_I^L}}\right]$$
$$\left.+\ln\left[\prod_{l=1}^{3n-6}\frac{e^{-\left(\frac{\hbar\omega_l}{k_B}\right)/2T_I^V}}{1-e^{\left(\frac{\hbar\omega_l}{k_B}\right)/T_I^V}}\right] + \frac{v_f(T_I^V)}{k_B T_I^V}\left(\mathscr{P}_w^V + 2\gamma^{LV}(T_I^L)(C_0)^{-1} - P_s(T_I^L)\right)\right\} \quad (17.18)$$

where k_B is the Boltzmann constant, $\hbar$ is the reduced Planck constant, n is the number of atoms in a nonlinear molecule, m is the molecular mass, ω_l is an internal molecular vibration frequency, P_s is the saturation vapor pressure, and v_f is the liquid specific volume at saturation. When SRT is applied to water evaporation, expressions for $P_s(T)$ and $v_f(T)$ for water can be found in the work by Duan et al. (2008). The ω_l values for water are given by McClelland (1973). The measured quantities used in the SRT expression are the liquid and vapor temperatures (T^L and T^V), the curvature of the liquid–vapor interface at the apex (C_0), and the vapor phase pressure (P^V). The SRT expression assumes that the system is in steady state and that the droplet's height is steady during evaporation. Also note that there are no fitting parameters in the SRT mass flux expression. When j_{ev} is known, the SRT expression can be used to calculate the droplet temperature, $\mathscr{T}^L$, or the water-vapor partial pressure, $\mathscr{P}_w^V$, and the results can be compared to their measured values, $T^L(t)$ or $P_w^V(t)$.

The SRT-calculated $\mathscr{P}_w^V$ values are shown in Figure 17.14. When compared with the mass-analyzer-measured $P_w^V(t)$ values, there was no measurable difference between the predictions and the measurements from 3450 to 3570 s. But after 3570 s, the $\mathscr{P}_w^V$ values decreased with time compared to the $P_w^V(t)$ values because the droplet's interface height decreased relative to the stationary microthermocouple bead. Similarly, SRT was applied to predict the temperature pulsations of the droplet, $\mathscr{T}^L$. The SRT-calculated $\mathscr{T}^L$ values and the measured $T^L(t)$

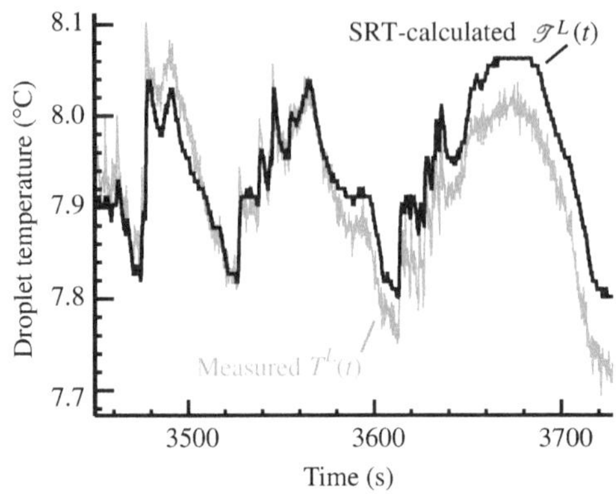

FIGURE 17.15 The measured droplet temperature, $T^L(t)$, as a function of time is shown as a thin line. The thick line is the SRT calculation of droplet temperature, $\mathscr{T}^L$. The abscissa corresponds to that in Figure 17.13. *Reprinted (adapted) with permission from Persad et al. (2013). Copyright 2013 American Chemical Society.*

are shown in Figure 17.15. Again, the small disagreement between the two values reflects the unsteady nature of the evaporation (Persad et al., 2013; Figure 17.12).

17.3.5 Comparison of the Proposed Mechanism with Others

Persad et al. noted that in the predictions of $\mathscr{P}_w^V$, the only pulsating input into SRT was the *liquid*-phase temperature. Similarly, in the predictions of $\mathscr{T}^L$, only the pulsating *vapor*-phase pressure was used as an input in SRT. Because it is a theory of evaporation kinetics (SRT) that allows the pulsations in one phase to be used to predict the pulsations in the other phase, it appears that multicomponent evaporation is the source of the temperature and pressure pulsations and of the coupling between the variables identified in Figure 17.13 (Persad et al., 2013).

Marangoni convection has been proposed by others as a mechanism that could produce the liquid-phase temperature pulsations. But investigations of that mechanism have ignored the temperature and pressure pulsations in the vapor phase (Barthes et al., 2007; Brutin et al., 2011; Karapetsas et al., 2012; Sobac and Brutin, 2012). It is also not clear that Marangoni convection alone can explain the coupling between the temperature and pressure pulsations.

Furthermore, Marangoni convection does not explain why the temperature and pressure pulsations only appear in a multicomponent system. Earlier experiments with sessile droplets of water evaporating into their own vapors have indicated that a temperature gradient spontaneously formed on the liquid–vapor interface from the three-phase line to the droplet apex, resulting in Marangoni flow (Duan, 2009; Ghasemi and Ward, 2010). But no temperature or pressure pulsations were reported in the evaporation system. Conservation of energy was satisfied in the experiments when energy transport by Marangoni convection was taken into account indicating that there was no energy consumed (or released) by a pulsating mechanism. Thus, Marangoni convection can be present during evaporation without producing temperature or pressure pulsations and supports the idea that a second component in the atmosphere that has a heat solution with the droplet is essential for the pulsations to appear.

Others have claimed that the pulsations may be physical (as opposed to only thermal) and could result from corrugations of the droplet surface. For example, surface corrugations appear on the solidification front during the solidification of binary alloys due to thermo-solutal instabilities (Shibata et al., 1978; Coriell et al., 1979). These instabilities can become oscillatory in nature when the concentrations in the binary alloys exceed critical values (Coriell et al., 1979). Sefiane et al. (2010) investigated the role of interfacial surface corrugations on the temperature pulsations of evaporating sessile droplets. They showed numerically that oscillatory deformations on the liquid–vapor surface qualitatively agreed with the spatial and temporal behavior of the temperature pulsations they observed with IR thermography. Their work indicated that the corrugated features were on the length scales of 50–100 μm. But Persad et al. (2013) did not observe any deformations of the liquid–vapor surface when the interface was monitored with a cathetometer and measured with an accuracy of ± 10 μm. As a result, it does not appear that surface corrugations are the source of the pulsations.

17.3.6 Summary and Conclusion

When evaporation systems are multicomponent, interactions between different species can lead to temperature and pressure pulsations in the droplet and in the atmosphere. The mechanism producing the pulsations is the heat release from the heat of solution that gives rise to a local increase in temperature and an increase in the local evaporation flux. The increased evaporation then locally cools the droplet, and the process repeats, creating temperature and pressure pulsations. The presence of humidity triggers temperature pulsations in evaporating alcohol sessile droplets when the humidity is above a threshold value. The pulsations do not result from water condensation on the alcohol droplet or from Marangoni or buoyancy-driven convection. For a stably stratified water droplet evaporating into a methanol-water atmosphere (also stably stratified), the measured temperature pulsations were used with a theory of evaporation kinetics (SRT) to predict the pulsations of the water partial pressure, and the predictions agreed closely with mass spectrometer measurements of the atmosphere. Similarly, SRT was used to predict the temperature pulsations in the droplet using the vapor-phase pressure, and good agreement was found with the measurements.

References

Amin, N., 1988. The effect of g-jitter on heat transfer. Proc. R. Soc. Lond. A 419, 151–172.

Asuero, A.G., Sayago, A., Gonzalez, A.G., 2006. The correlation coefficient: an overview. Crit. Rev. Anal. Chem. 36, 41–59.

Barthes, M., Reynard, C., Santini, R., Tadrist, L., 2007. Non-condensable gas influence on the Marangoni convection during a single vapour bubble growth in a subcooled liquid. Europhys. Lett. 77, 14001.

Bertrand, G.L., Millero, F.J., Wu, C., Hepler, L.G., 1966. Thermochemical investigations of the water-ethanol and water-methanol solvent systems. I. Heats of mixing, heats of solution, and heats of ionization of water. J. Phys. Chem. 70, 699–705.

Brutin, D., Zhu, Z.-Q., Rahli, O., Xie, J.-C., Liu, Q.-S., Tadrist, L., 2010. Evaporation of ethanol drops on a heated substrate under microgravity conditions. Microgravity Sci. Technol. 22, 387–395.

Brutin, D., Sobac, B., Rigollet, F., Niliot, C.L., 2011. Infrared visualization of thermal motion inside a sessile drop deposited onto a heated surface. Exp. Therm. Fluid Sci. 35, 521–530.

Carle, F., Sobac, B., Brutin, D., 2012. Hydrothermal waves in ethanol droplet evaporating under terrestrial and reduced gravity levels. J. Fluid Mech. 712, 614–623.

Coriell, S.R., Cordes, M.R., Boettinger, W.J., Sekerka, R.F., 1979. Convective and interfacial instabilities during unidirectional solidification of a binary alloy. J. Cryst. Growth 49, 13–28.

Das, K.S., MacDonald, B.D., Ward, C.A., 2010. Stability of evaporating water when heated through the vapor and the liquid phases. Phys. Rev. E 81 (3), 036318.

Design Institute for Physical Properties, 2011. DIPPR Project 801, Full version (Electronic Resource): Evaluated Standard Thermophysical Property Values. Design Institute for Physical Property Data/AIChE.

Duan, F., 2009. Local evaporation flux affected by thermocapillary convection transition at an evaporating droplet. J. Phys. D Appl. Phys. 42, 102004.

Duan, F., Ward, C.A., 2005a. Surface excess properties from energy transport measurements during water evaporation. Phys. Rev. E 72, 056302.

Duan, F., Ward, C.A., 2005b. Surface-thermal capacity of D_2O from measurements made during steady-state evaporation. Phys. Rev. E 72, 056304.

Duan, F., Thompson, I., Ward, C.A., 2008. Statistical-rate-theory determination of water properties below the triple point. J. Phys. Chem. B 112, 8605–8613.

Fang, G., Ward, C.A., 1999. Temperature measured close to the interface of an evaporating liquid. Phys. Rev. E 59, 417–428.

Farooq, A., Homsy, G.M., 1994. Streaming flows due to g-jitter-induced natural convection. J. Fluid Mech. 271, 351–378.

Freire, M.G., Gomes, L., Santos, L.M.N.B.F., Marrucho, I.M., Coutinho, J.A.P., 2006. Water solubility in linear fluoroalkanes used in blood substitute formulations. J. Phys. Chem. B 110, 22923–22929.

Ghasemi, H., Ward, C.A., 2010. Energy transport by thermocapillary convection during sessile-water-droplet evaporation. Phys. Rev. Lett. 105, 136102.

Ghasemi, H., Ward, C.A., 2011. Mechanism of sessile water droplet evaporation: Kapitza resistance at the solid-liquid interface. J. Phys. Chem. C 115, 21311–21319.

Gibbs, J.W., 1906. On the equilibrium of heterogeneous substances. In: Bumstead, H.A., Van Name, R.G. (Eds.), The Scientific Papers of J. Willard Gibbs, vol. 1. Dover, New York, NY.

Iwan, J., Alexander, D., Amiroudine, S., Ouazzani, J., Rosenberger, F., 1991. Analysis of the low gravity tolerance of Bridgman–Stockbarger crystal growth. J. Cryst. Growth 113, 21–38.

Karapetsas, G., Matar, O.K., Valluri, P., Sefiane, K., 2012. Convective rolls and hydrothermal waves in evaporating sessile drops. Langmuir 28, 11433–11439.

MacDonald, B.D., Ward, C.A., 2011. Onset of Marangoni convection for evaporating liquids with spherical interfaces and finite boundaries. Phys. Rev. E 84, 046319.

MacDonald, B.D., Ward, C.A., 2012. Onset of Marangoni convection for evaporating sessile droplets. J. Colloid Interface Sci. 383, 198–207.

McClelland, B.J., 1973. Statistical Thermodynamics. Chapman and Hall, London (Chapter 5).

Nield, D.A., 1964. Surface tension and buoyancy effects in cellular convection. J. Fluid Mech. 19 (3), 341–352.

Pearson, J.R.A., 1958. On convection cells induced by surface tension. J. Fluid Mech. 4, 489–500.

Persad, A.H., Sefiane, K., Ward, C.A., 2013. Source of temperature and pressure pulsations during sessile droplet evaporation into multicomponent atmospheres. Langmuir 29 (43), 13239–13250.

Rogers, M.J.B., Hrovat, K., McPherson, K., Moskowitz, M.E., Reckar, T., 1997. Acceleration Data Analysis and Presentation Techniques. Nasa report, NASA.

Sasges, M.R., Ward, C.A., Azuma, H., Yoshihara, S., 1996. Equilibrium fluid configurations in low gravity. J. Appl. Phys. 79, 8770–8782.

Schneider, S., Straub, J., 1989. Influence of the Prandtl number on laminar natural convection in a cylinder caused by g-jitter. J. Cryst. Growth 97, 235–242.

Schöpf, W., Patterson, J.C., Brooker, A.M.H., 1996. Evaluation of the shadowgraph method for the convective flow in a side-heated cavity. Exp. Fluids 21, 331–340.

Sefiane, K., Steinchen, A., Moffat, R., 2010. On hydrothermal waves observed during evaporation of sessile droplets. Colloids Surf. A Physicochem. Eng. Asp., 95–108.

Sefiane, K., Fukatani, Y., Takata, Y., Kim, J., 2013. Thermal patterns and hydrothermal waves (HTWs) in volatile drops. Langmuir 29, 9750–9760.

Shibata, K., Sato, T., Ohira, G., 1978. Morphological stabilities of planar solid-liquid interfaces during unidirectional solidification of dilute Al-Ti and Al-Cr alloys. J. Cryst. Growth 44, 419–434.

Sobac, B., Brutin, D., 2012. Thermocapillary instabilities in an evaporating drop deposited onto a heated substrate. Phys. Fluids 24, 032103.

Thompson, I., Duan, F., Ward, C.A., 2009. Marangoni numbers of greater than 27,000 during water evaporation without Marangoni convection. Phys. Rev. E 80, 056308.

van der Geld, C.W.M., Colin, C., Segers, Q.I.E., Pereira da Rosa, V.H., Yoshikawa, H.N., 2012. Forces on a boiling bubble in a developing boundary layer, in microgravity with g-jitter and in terrestrial conditions. Phys. Fluids 24, 082104.

Velarde, M.G., Nepomnyashchy, A.A., Hennenberg, M., 2001. Onset of oscillatory interfacial instability and wave motions in Bénard layers. Adv. Appl. Mech. 37, 167–238.

Ward, C.A., Fang, G., 1999. Expression for predicting liquid evaporation flux: statistical rate theory approach. Phys. Rev. E 59, 429–440.

Ward, C.A., Duan, F., 2004. Turbulent transition of thermocapillary flow induced by water evaporation. Phys. Rev. E 69, 056308.

Ward, C.A., Sasges, M.R., 1998. Effect of gravity on the contact angle: A theoretical investigation. J. Chem. Phys. 109, 3651–3660.

Ward, C.A., Wu, J., 2008. Effect of contact line curvature on solid-fluid surface tensions without line tension. Phys. Rev. Lett. 100, 256103.

Ward, C.A., Rahimi, P., Sasges, M.R., Stanga, D., 2000. Contact angle hysteresis generated by the residual gravitation field of the space shuttle. J. Chem. Phys. 112 (16), 7195–7202.

Ward, C.A., Wu, J., Keshavarz, A., 2008. Measurement of the adsorption at solid-liquid interfaces from the pressure dependence of contact angles. J. Phys. Chem. B 112, 71–80.

Wu, J., Farouk, T., Ward, C.A., 2007. Pressure dependence of the contact angle. J. Phys. Chem. B 111, 6189–6197.

Wu, C., Hadi Zandavi, S., Ward, C.A., 2014. Prediction of the wetting condition from the zeta adsorption isotherm. Phys. Chem. Chem. Phys. 16 (46), 25564–25572. Available from: http://dx.doi.10.1039/c4cp03585b.

Chapter 18

HYDROTHERMAL WAVES

Guiping Zhu, Tao Wei and Fei Duan
School of Mechanical & Aerospace Engineering, Nanyang Technological University, Singapore

Hydrothermal waves are reviewed for the evaporating droplets theoretically and experimentally. The experimental visualization of the thermal instability is from two techniques—laser refracted shadowgraphy and infrared thermography—for the sessile droplets on the substrate. The effects on hydrothermal waves in the view of the infrared thermography are further discussed on the basis of the fluids with various boiling temperatures, the substrates with various thermal conductivities, gravity, substrate temperature, heat flux, and so on. Analyses of number of waves or number of cells are also provided. In the theoretical study, we introduce the linear stability analysis in an unpinned sessile evaporating droplet for convective cells and hydrothermal wave configuration, and in a pinned sessile droplet for normal deformation of liquid−vapor interface.

18.1 Theoretical concepts

Hydrothermal waves (HTWs) are thermally induced traveling waves in the liquid. Smith and Davis (1983) first predicted the type of thermocapillary instability

Droplet Wetting and Evaporation. DOI: http://dx.doi.org/10.1016/B978-0-12-800722-8.00018-7

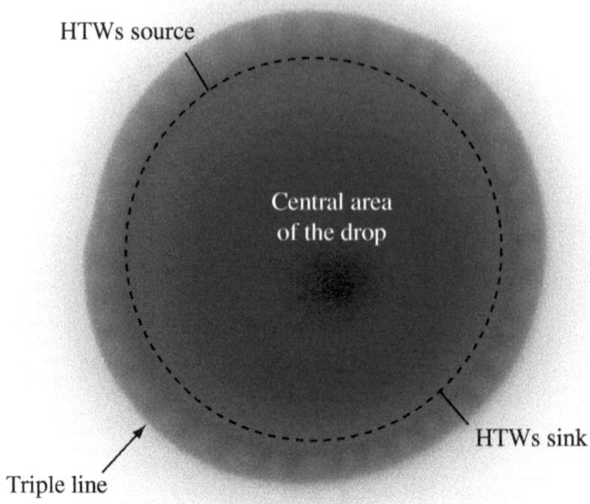

FIGURE 18.1 Infrared image of an evaporating drop of ethanol with the presence of HTWs (Sobac and Brutin, 2012).

when they studied the stability of a thin fluid layer that owns a free surface with an interfacial temperature gradient. They demonstrated that unstable fluid layers, when subjected to sufficiently large horizontal temperature gradients, generated steady longitudinal rolls and unsteady HTWs. The HTWs propagated in parallel and almost perpendicularly to the temperature gradient, or obliquely in two mirror-image directions. Riley and Neitzel (1998) managed to demonstrate the existence of HTWs in a layer of silicone oil experimentally. For the droplet, the self-excited HTWs were found by Sefiane et al. (2008) in the evaporation of droplet sitting on a heated substrate. The definition of HTWs in the droplet is summarized as that HTWs are the spontaneous development of thermal-convective instabilities induced by surface tension gradient along the free surface during the evaporation of droplets. Karapetsas et al. (2012) defined a special kind of HTWs characterized with the three-dimensional, spiral-like, oscillatory, traveling wave trains that are organized radially with a uniform azimuthal angle. Additionally, the dynamics of HTWs have been described by Sobac and Brutin (2012). HTWs, which consist of right and left propagating waves, circle in circumferential direction from source to sink near the three-phase contact line (triple line), where the evaporative flux and temperature are both larger than those of the central part in the early stage of evaporation of the droplet, as shown in Figure 18.1.

In most previous works, the theoretical development of propagating HTWs has been focused on the planar liquid layer with nondeformable free interface in the absence of evaporation as a constant temperature gradient is imposed along the layer (Smith and Davis, 1983; Smith, 1986; Davis, 1987). With the linear stability theory, Smith and Davis (1983) revealed that sufficiently large horizontal temperature

TABLE 18.1 Summary of Main Theoretical Investigations on HTWs

Authors	Dimension	Effects	Pr
Smith and Davis (1983)	3D (Cartesian coordinate)	Ma	10^{-3}–10^{2}
Laure and Roux (1989)	3D (Cartesian coordinate)	Ma, Gr	10^{-3}–1
Parmentier et al. (1993)	3D (Cartesian coordinate)	Ma, Ra	10^{-2}–10
Mercier and Normand (1996)	3D (Cartesian coordinate)	Ma, Ra	7
Priede and Gerbeth (1997)	2D (Cartesian coordinate)	Ma, Ra	10^{-3}–10^{3}
Sefiane et al. (2010)	3D (axisymmetric spherical coordinate)	Ma, E, T_w, k_w	>1
Sobac and Brutin (2012)	3D (axisymmetric cylindrical coordinate)	Ma, Bo, T_w, R_0	22.4
Karapetsas et al. (2012)	3D (axisymmetric cylindrical coordinate)	Ma, Re, E	1.7

Note: 3D, three dimensional; 2D, two dimensional; E, evaporation number; T_w, temperature of the substrate; k_w, thermal conductivity of the substrate; R_0, initial wetting radius; Ma, Marangoni number; Ra, Rayleigh number; Bo, static Bond number; and Re, Reynolds number.

gradients made the liquid layers unstable and that the generated propagating HTWs could obtain energy from the imposed horizontal temperature gradient for small Prandtl number (Pr) and from the vertical temperature gradient for large Pr, respectively. The instabilities moved in the same direction as the horizontal temperature gradients for the low Pr cases, while they went perpendicularly for the high Pr cases. The physics of the HTWs propagation was explained as the interaction between the thermal field and flow field dominated by the inertial (low Pr) or viscous (high Pr) effect (Smith, 1986). The model on the basis of Pr described the observation in the experiments discussed by Pearson (1958). For the sessile droplet on a substrate, motivated by the evaporating experiments reported by Sefiane et al. (2008), the presence of HTWs was mainly investigated by Karapetsas et al. (2012) by applying the quasi-steady-state approximation. In their model, the thermal gradient was not imposed externally but resulted from phase change, and the effect of vapor recoil was disregarded. The results of their energy analysis showed that the induced HTWs shared the same mechanisms as those presented by Smith (1986). The most representative theoretical works on HTWs are summarized and listed in Table 18.1.

18.2 Experimental observations

In the investigations on the flow and hydrodynamic instabilities in the liquids, HTWs are classified to relate to the onset of Marangoni flow or thermocapillary convection induced by the surface tension gradient. HTWs travel from the temperature gradients at the liquid surface. Early investigations of HTWs were found in thin fluid layers suffered from a fixed value of temperature gradient (Chun and Wuest, 1979; Schwabe and Scharmann, 1979; Kölker, 1980). The initial works can be traced back to the experiments of Bénard (1900) and Pearson (1958), in which crucial cells or convective patterns were discovered. Following the works of Smith and Davis (1983), the observation of HTWs was also extended by taking into account buoyancy forces. The experimental investigation was conducted by Kirdyashkin (1984) on thermocapillary-buoyancy-driven flow of ethyl alcohol in a slot with much smaller fluid depths. The experimental investigations were also presented by other groups with fixed Pr to obtain the effect of buoyancy and geometric factors (Villers and Platten, 1992; De Saedeleer et al., 1996). As the Marangoni number, Ma, increased, the flow started with a single steady convection cell and then transited to multiple steady cells and finally an oscillatory state. The presence of HTWs was experimentally demonstrated in differentially heated pools of ethanol (Schwabe et al., 1992) and silicon oil (Riley and Neitzel, 1998; Garnier and Chiffaudel, 2001). Schwabe et al. (1992) reported that the value of Ma for these transitions was 600 and 3000 for ethanol in slot and annular-pan geometries, respectively, with Pr at 17. In this thin annular geometry, two different kinds of HTWs were discussed and named short-and long-wavelength waves. The short-wavelength condition occurred in layers with liquid depths no higher than 1.4 mm. For a 1-mm-thick liquid layer, the wavelength and the dimensionless wave speed were agreeable with the experimental results by Riley and Neitzel (1998). In the annular liquid pool, HTWs took the form of spirals at relatively high liquid thicknesses. When the fluid depth was less than the capillary length, the HTWs propagated radially from the inner to the outer edge of the annular space. The experimental investigations on HTWs in a fluid layer have been widely explored. Different kinds of HTWs were summarized based on the Ph.D. works of Garnier for a 2D annular geometry and are shown in Figure 18.2 (Sobac and Brutin, 2012).

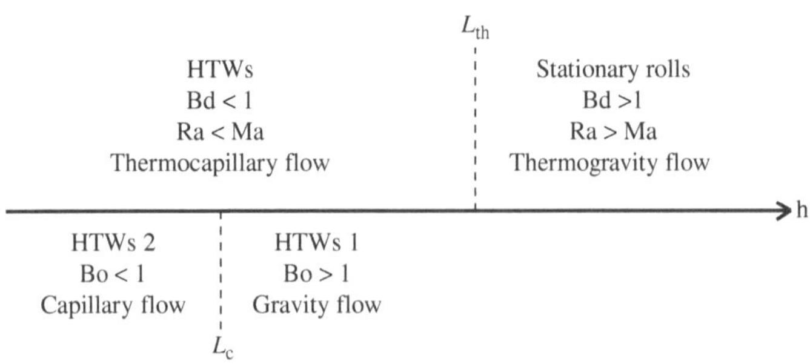

FIGURE 18.2 Different forms of thermocapillary and thermogravitational flow categorized in terms of liquid depth on the basis of the works of Garnier (Sobac and Brutin, 2012).

Figure 18.2 illustrates the various regimes of the instabilities, depending on the liquid height and physical properties of the fluids. The capillary length, L_c, and thermal length, L_{th}, are defined as $L_c = \sqrt{\sigma/(\rho g)}$ and $L_{th} = L_c\sqrt{\gamma/(\sigma\beta)}$, in which $\gamma = -(\partial\sigma/\partial T)$, σ is the surface tension, and β is the coefficient of thermal expansion. L_{th} separates the regimes of stationary roll due to thermogravity flow from that of HTWs induced by the thermocapillary flow as the liquid depth, h, increases. In the region of HTWs, dynamic Bond number, Bd, in terms of the ratio of Rayleigh number (Ra) to Marangoni number (Ma), is smaller than 1. L_c is applied to differentiate HTWs Type 1, which moves with a constant angle of propagation, because Bo is less than 1 from HTWs Type 2, which propagates with the changing angle due to the gradient temperature, because Bo is larger than 1.

For droplets, the experimental work of HTWs was extended from the investigations on thermocapillary convection. For example, Hegseth et al. (1996) conducted experiments for the pedant methanol droplet evaporation in a sealed chamber by using a particle tracking method. In the liquid droplet, the 2-μm polystyrene spheres tracer particles were illuminated by a laser light sheet. The images were taken by the reflected light from the particles and the droplet edge focused into a CCD camera. Sequences of images were used to calculate the velocity field to build the flow patterns in the droplet. The thermal instability can be observed from the flow tracking. In addition, the temperature and pressure pulsations, measured in an evaporative droplet by Persad et al. (2013), were found to be related to HTWs. However, the most direct observations for the thermal instabilities are from laser refracted shadowgraphy and infrared (IR) thermography.

18.2.1 Laser Refracted Shadowgraphy

For a sessile droplet, the laser shadowgraphy principle was first proposed by Allain et al. (1985) for measuring the contact angle of a droplet sitting on a nontransparent substrate. The method uses the droplet as a convex mirror for large, collimated, and incident laser beams. However, the application was narrow for limited low light reflectivity and incapability of flow visualization inside the droplet. For simultaneous study of dynamic contact-angle and internal flow, the concept of hybrid optical system involving the laser-refracted-shadowgraphy method was utilized for the experimental investigation of droplet spreading and contact angle on the effect of evaporation and thermocapillary convection (Chao and Zhang, 2001), as shown in Figure 18.3A. For a sessile droplet on a nontransparent metal substrate with contact angle at θ, there are both reflection and refraction when a parallel beam falls on it. The maximum angular aperture of the reflection cone is 4θ. The final refracted ray out of the droplet has a smaller angular aperture. Both the reflected and refracted shadowgraphy images were taken. Normally, the refracted-shadowgraphy images

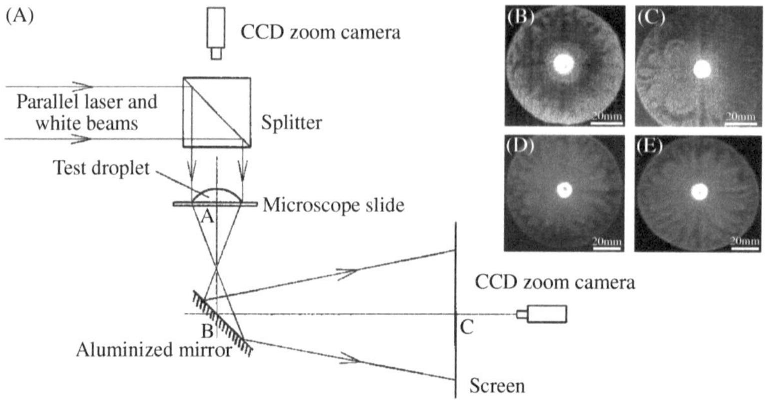

FIGURE 18.3 (A) Schematic image of the experimental optical system for laser refracted shadowgraphy. The refractive images of thermocapillary convection in the sessile droplets of *n*-pentane (B) and R-113 (C) (Chao and Zhang, 2001). With the modified experimental optical system, the instant refracted shadowgraphs of the R-113 droplet: (D) at 4.80 s and (E) at 5.81 s from the origin of spreading time (Zhang and Chao, 2002).

were much clearer due to the smaller ray angular aperture and higher reflectivity of the smooth metal surface. Chao and Zhang (2001) tested three volatile liquids, cyclohexane, *n*-pentane, and Freon-113 (R-113) in open air at the temperature of 23°C and the relative humidity of 44%. The patterns of thermal instability were observed instantaneously in the droplets of *n*-pentane in Figure 18.3B and R-113 in Figure 18.3C, even though they just stated that the shadows were induced by thermocapillary convection in evaporation. The patterns were found to be correlated to the evaporation rates of the fluids. The cyclohexane droplets did not exhibit the wave, while the R-113 droplet showed the pattern at a very early evaporation stage.

Zhang and Chao (2002) and Zhang and Yang (1982) applied the hybrid optical system for flow visualization of an R-113 droplet on a glass or even nontransparent metal substrate. The refracted shadowgraphy images were much clearer. Figure 18.3D and E demonstrates the instant refracted shadowgraphs of an R-113 droplet at 4.80 s and 5.81 s from the origin of spreading time, respectively. Both images show the strong shadow of thermal waves.

Even though the laser refracted shadowgraphy methods can demonstrate the patterns of HTWs, the quantitative works have not been conducted with the optical method. The IR thermography provides a more detailed experimental analysis for HTWs.

18.2.2 IR Thermography

Sefiane et al. (2008) designed the experiments for HTWs by using an IR camera. The setup consisted of a syringe pump for supplying volatile liquid, a horizontal substrate mount, temperature controller, and IR analysis equipment. The sessile droplets containing water, ethanol, methanol, and FC-72 were maintained on various solid substrates made of polytetrafluoroethylene (PTFE), Macor ceramics, titanium, and copper. The evaporation was under open conditions under controlled humidity and at the same temperature as the solid substrate. The IR camera was mounted directly to face vertically downward onto the evaporating drop in order to obtain images of the drop behavior. The investigation showed spontaneous HTWs and patterns in evaporating pinned droplets. The results were explained in terms of self-induced HTWs. The additional and detailed experimental results with the theoretical formulation are reported by Sefiane et al. (2010). Using the same nonintrusive testing method, Girard et al. (2010) studied the local temperature, contact line evolution, and evaporation flux of sessile water droplets on the heated copper substrate. Brutin et al. (2011) presented experimental results of HTWs with an IR camera coupled with a microscopic lens that had a spatial resolution of 10 μm. The evaporation of three different fluids—ethanol, methanol, and FC-72—was characterized as a function of time. The evolution of these instabilities was linked with the temperature difference between the heating substrate and the room temperature.

To avoid any perturbations from external flow, the experimental cell was further designed to cover the evaporation of a sessile drop deposited through a micropipet onto a heated substrate in the investigation of Sobac and Brutin (2012), and the schematic is shown in Figure 18.4A. The droplet was evaporated under the atmospheric pressure. The droplet was observed from both the side and the top. The substrate was a complex assembly of layers mainly composed of a heated aluminum cylinder. A heat flux meter was placed overhead to determine the heat flow absorbed by the drop on the substrate during evaporation. The black mast paintings and a smooth film of PTFE were coated upon the sensor. Sobac and Brutin (2012) simultaneously investigated the kinetics of evaporation, heat transfers, the triple-line dynamics, and thermal convection inside the pure ethanol droplet. We can see the simultaneous images for HTWs in the top view in the bottom inset image of Figure 18.5A and the side-view configuration of an ethanol droplet on the PTFE substrate in evaporation in the top inset image of Figure 18.5A. The experimental setup was then modified into a confined box by Carle et al. (2012) for the microgravity experiments in the parabolic flying aircraft.

In the experiments for ethanol, methanol, and FC-72 under open environmental conditions, Brutin et al. (2011) presented the complexity of droplet evaporation mechanisms, as shown in Figure 18.4B–D. For the methanol droplet in Figure 18.4B, many convection cells due to HTWs were observed near the droplet periphery. The cells had a relative thin corona in comparison with those of the ethanol droplet. The thermal flow motion in the droplet was faster than the other with ethanol and FC-72. For the ethanol droplet in Figure 18.4C, they observed the same convection cell structures, but there were fewer cells, and they were in a wider corona near the droplet perimeter. In the evaporation, the flow motion changed

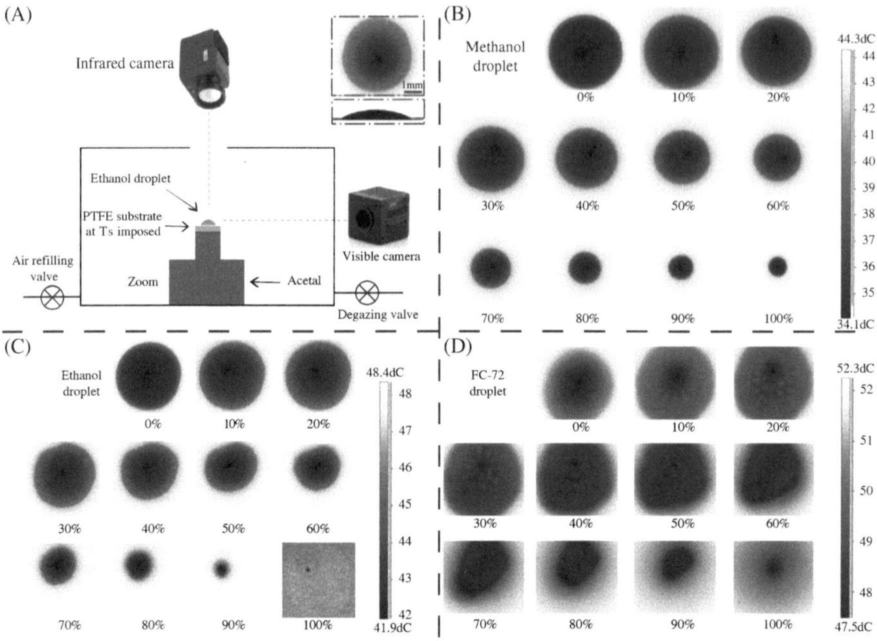

FIGURE 18.4 (A) Experimental setup of IR thermography for sessile droplet evaporation, the top view for IR thermography in top inset image and the side view for the droplet shape in bottom inset image (Sobac and Brutin, 2012). IR sequences at top view for methanol (B), ethanol (C), and FC-72 (D) droplets; 100% is total time of evaporation (Brutin et al., 2011).

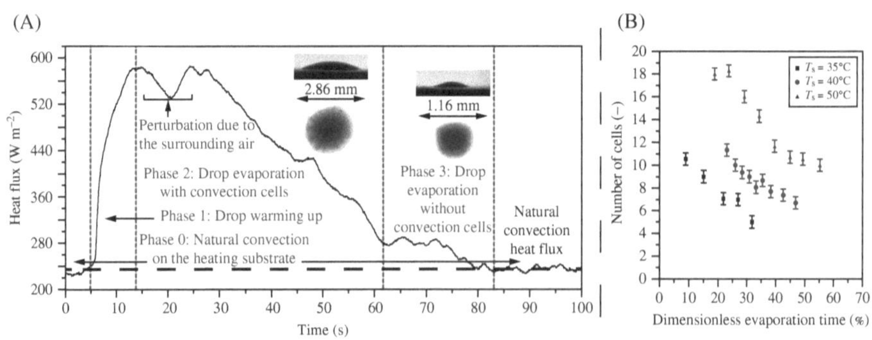

FIGURE 18.5 Different phases in droplet evaporation: (A) Heat flux of an evaporating ethanol sessile droplet at different times. (B) Number of thermo-convective cell evolutions for different substrate temperatures with the ethanol droplet (Brutin et al., 2011).

quickly and became large convection cells inside the droplet. The total time of evaporation was almost the same as that of methanol, because the heat of vaporization is comparable but was much longer than that of the FC-72 droplet, as shown in Figure 18.4D, because FC-72 has much lower latent than methanol and ethanol. During the evaporation of the FC-72 droplet, multiple internal convective cell structures were seen and dissipated very fast. It suggested that the evaporation dynamics of FC-72 was completely different from methanol and ethanol.

From the evaporation procedure of sessile droplets of methanol, ethanol, and FC-72, the IR camera and visual camera were used to capture the thermal and optical images. Corresponding to the measured heat flux, the process of droplet evaporation was classified into three phases in Figure 18.5A: Phase 1, the sessile droplet was warmed with a sharp increase of the heat flux. The first phase was thus only driven by the heat capacity of the fluid; Phase 2, the droplet was evaporated with the heat flux reaches a maximum value, and the HTWs were observed and the substrate surface was cooled due to the evaporation cooling; and Phase 3, liquid film was evaporated, and the convection cells in the droplet disappeared.

The influence of the substrate temperature on the number of convection cells is illustrated in Figure 18.5B in the case study of the ethanol droplet (Brutin et al., 2011). For three substrate temperatures at 35°C, 40°C, and 50°C in the experiments, the number of counted convection cells increased with an increase of surface temperature, while the number of convection cells decreased as a function of the time until it reached zero when the droplet became a layer.

The interfacial thermal patterns were recorded through IR thermography by Sefiane et al. (2008) for the methanol, ethanol, and FC-72 droplets, which have variety of volatilities and boiling points. Four different substrates (PTFE, Macor Ceramics, titanium, and copper) were adopted in the experiments with a variety of thermal conductivity. The substrates were additionally coated with a thin layer of a fluoropolymer Cytop to ensure homogeneous surface properties. The interfacial thermal activities of the sessile droplets can be seen in Figure 18.6A to demonstrate that the generated HTWs occurred spontaneously in the evaporation process of the sessile droplets. On the same substrate and evaporation conditions, the water droplet could not gain temperature gradient, as shown in Figure 18.6A. However, the thermal activities in the higher-volatility liquid were stronger. The HTWs were observed in both ethanol and methanol droplets in the form of a wave train of alternating warm and warm/cool regions. The IR visualization indicated that the temperature of the droplets was colder in the center and hotter close to the three-phase contact line. The HTWs propagated in both directions randomly. However, they stated that the thermally driven cells rather than HTWs were observed emerging near the FC-72 droplet center and drifting toward the edge. With these assumptions, the total number of waves for HTW cells of methanol and ethanol droplets were counted as a function of the dimensionless time,

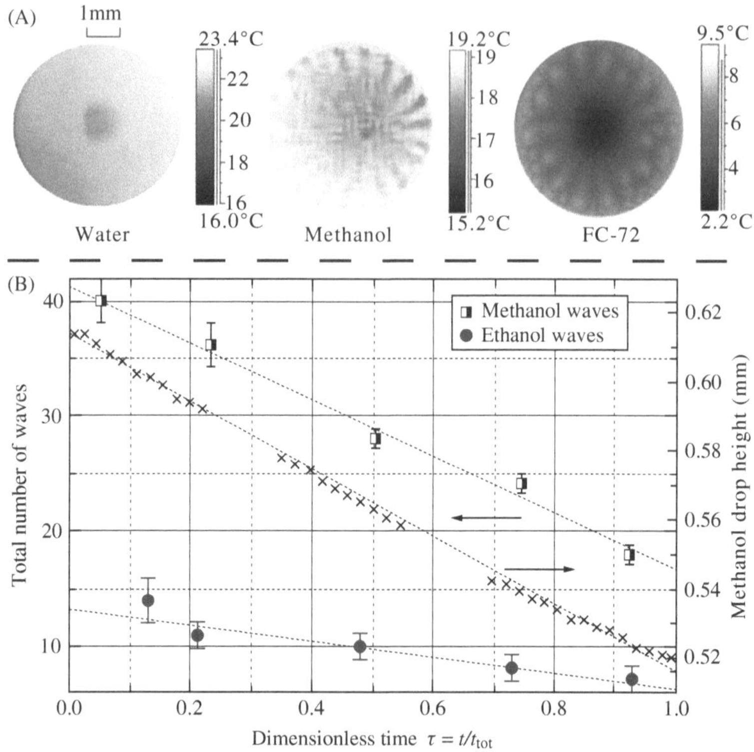

FIGURE 18.6 (A) Patterns in sessile droplets evaporating on a silicon substrate by IR thermography for water, methanol, and FC-72 droplets (Sefiane et al., 2008; Karapetsas et al., 2012). (B) Temporal evolution of the number of observed waves for methanol and ethanol droplets evaporating on a titanium substrate; the droplet height evolution of a methanol droplet is also shown (Sefiane et al., 2008).

t/t_{tot}, and illustrated in Figure 18.6B, in which t_{tot} is the duration of an experimental run. The number of cells of HTWs in the evaporating methanol droplet was larger than that of the ethanol droplet at the same instantaneous time, and the number of cells was observed to decrease approximately linearly with a decrease of the height of the droplet during the evaporation.

Sefiane et al. (2010) reported more detailed observations of HTWs of the ethanol droplets on the four different substrates of PTFE, Macor ceramics, titanium, and copper, which have different thermal conductivities. As seen in Figure 18.7A, during the droplet evaporation, the number of visible waves decreases with time for the particular substrate. The greater temperature fluctuations were observed at a later stage of the droplet evaporation; the frequency of the waves also decreased, indicating fewer waves.

The effect of various substrate thermal conductivities on the number of waves was compared between Macor ceramics and titanium. As shown in Figure 18.7B, the

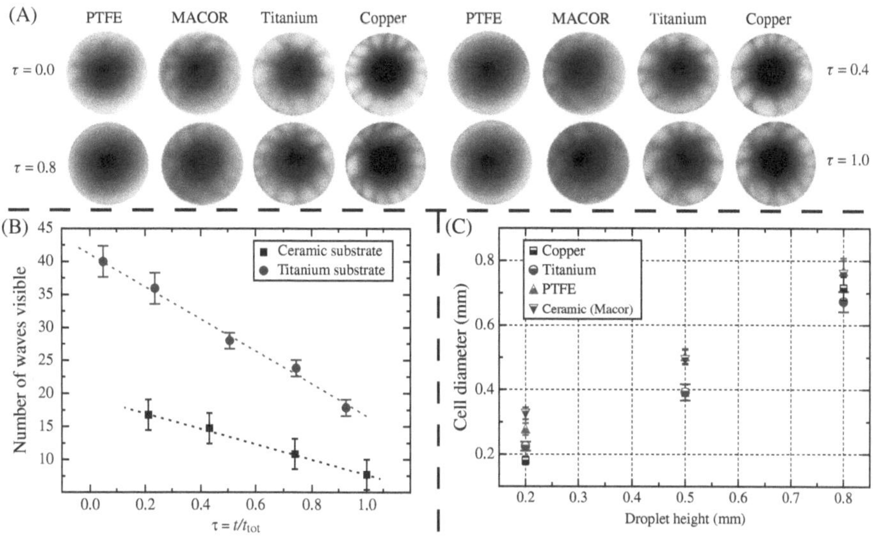

FIGURE 18.7 (A) Effect of surface property on HTW patterns for ethanol droplets. (B) Effect of substrate material on the visible cells. (C) Cell size variation in FC-72 droplet with different substrate thermal conductivity (Sefiane et al., 2008, 2010).

total number of visible waves (cells) against dimensionless time, t/t_{tot}, has a decreasing trend for both substrates, but at the relatively same time period, the number of waves of the evaporating droplets on the higher conductivity substrate, titanium, was much higher than that of ceramic substrate with less thermal conductivity.

The formed pattern of FC-72 on the various substrates was analyzed further by examining the correlation of the size of the cells on the local height of the droplet. The dependence was plotted in Figure 18.7C. It is seen that the size of the cells increased approximately linearly with droplet height in the experiments. The cell dimension, however, exhibited a rather weak dependence on the thermal properties of the substrate.

For the droplet evaporation involving HTWs, the experiment analysis from Sefiane et al. (2008) demonstrated the increasing tendency that the propagation angle varied with Pr. Additionally, the larger conductivity of the substrate gave rise to a larger evaporation rate as well as a higher propagation angle. Ristenpart et al. (2007) indicated that the direction of the Marangoni flow depends on the relative thermal conductivities between the substrate and the liquid. The experimental study from Sefiane et al. provided both the snapshots of wave evolution patterns for the working fluid droplets and the effect of substrate conductivity on the diameter of the wave cells for the FC-72 droplet (Sefiane et al., 2008, 2010). The direction of the flow reversed at a critical value of the contact angle, with the relative thermal conductivities ranging from 1.45 to 2. An asymptotic analysis indicated the reverse of the flow direction also depended on the ratio of the substrate thickness to the contact radius of the droplet (Xu et al., 2009). The verification of the analysis was corroborated by investigating water and isopropanol droplets evaporating on a glass substrate. The experimental investigation also indicated a not-constant-angle mode of propagation under the microgravity condition (Carle et al., 2012).

By analyzing experimental data at different times by Sefiane et al. (2008), the number of waves was found to depend on the droplet thickness, fluid property, and substrate thermal conductivity. The phenomena could be explained by the difference in experimental conditions, such as thermal resistance of the substrates and the dynamic of the triple contact line. A systematic analysis was carried out for estimating the number of HTWs (Sobac and Brutin, 2012). The geometrical and thermal parameters were studied for the effect on the evaporation mechanism by varying the initial wetting area and the substrate temperature. The number of HTWs exhibited a power law relationship with time evolution in 1 decade.

$$N_{\mathrm{w}} = a\left(\frac{d_0}{L_{\mathrm{c}}}\right)\left(\frac{\Delta T}{T_{\mathrm{s}}}\right)\left(\frac{t}{t_{\mathrm{F}}}\right)^b \tag{18.1}$$

where N_{w} is the number of HTWs, d_0 is the initial wetting diameter, L_{c} is the capillary length, T_{s} is the temperature of the substrate, ΔT is the temperature difference between the substrate and the atmosphere, and t/t_{F} is the normalized time with t_{F} indicating the time for evaporation. Here, a and b are constant with the value of $101.7 \pm 2.9\%$ and $-0.196 \pm 5.8\%$ in the study of Sobac and Brutin (2012).

A transition phase was added between droplet evaporation with HTWs and drop evaporation without HTWs by Sobac and Brutin (2012), as shown in Figure 18.8A. In the second phase of evaporation (constant contact-angle mode), the decrease of droplet volume and the recess of the contact line made the droplet unstable. The HTWs were difficult to track at this stage. In the first phase of evaporation (constant contact-area mode) of the droplet after warming up, N_{w} for HTWs was observed to develop spontaneously. With an increase of time, the value decreased.

At the quasi-stationary regime, the HTWs had the same behaviors that illustrated a motion around the edge of the droplet from the source point to the well, as shown in Figure 18.8B, in the experimental IR images of the sessile ethanol droplet. The waves also propagated on the edge of the droplet without preference in the left or right direction, which depended on the droplet height and substrate temperature in 300 s. The N_{w} scale was plotted as a function of dimensionless time in Figure 18.8C for the influence of both radius and surface temperature. Sobac and Brutin (2012) showed that the power law curve fitted the experimental data well for one specific case for ethanol droplet evaporation with initial radius at 2.34 mm and substrate temperature at 39.6°C. The number of waves evolves with a power law for all cases. We can see that the evolution of the number of

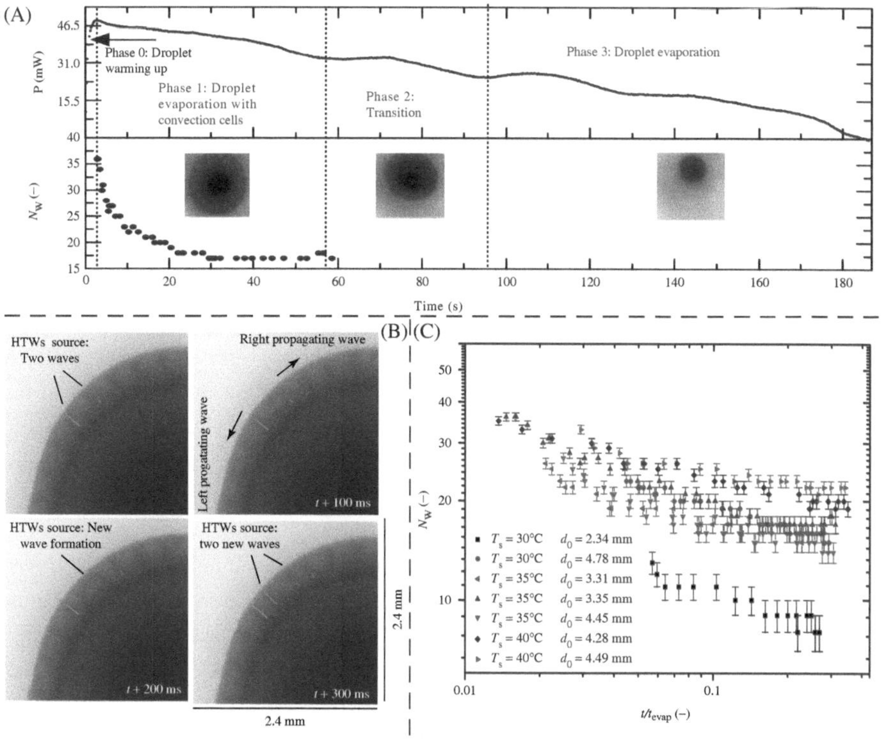

FIGURE 18.8 (A) Phase description and number of HTWs at different times. (B) Enlarged source location of the wave for visualization of the wave propagation with time evolution for Figure 18.1. (C) Number of HTWs, N_W, versus the normalized time for various initial wetting diameters and substrate temperatures in log–log scale (Sobac and Brutin 2012).

HTWs is influenced by two parameters directly related to the driving force of the thermocapillary instabilities: (i) at a fixed imposed substrate temperature, N_W increased with the drop size; and (ii) at a fixed drop size, N_W increased with the increase of the substrate temperature.

From Figure 18.2, the HTWs should be formed when the depth of the fluid is less than the capillary length. In this circumstance, investigation under a reduced gravity would offer more useful information. In the early stage, Smith and Davis (1983) applied a linear stability analysis for thermocapillary convection in a horizontal free-surface fluid layer with a horizontal temperature gradient. The results indicated that the convection became unstable in the form of HTWs with waves propagating upstream obliquely for a return flow pattern. The experimental verification of the linear stability analysis was then provided by Riley and Neitzel (1998). The experimental results showed that the angle of propagation, the frequency of oscillation, and the phase speed of the HTWs' instability decreased with the increasing fluid layer thickness. However, the wavelength of the waves stayed almost constant. The linear stability analysis was later utilized to investigate the effect of gravity on the stability of thermocapillary convection in a fluid layer (Chan and Chen, 2010). The properties of the HTWs were characterized by the Grashof number, Gr. The type of propagation of the HTWs was changed by taking gravity into consideration. Microgravity experiments are the best way to rule out the effect of gravity.

Carle et al. (2012) conducted the experiments on evaporating ethanol droplets in a parabolic flight airplane to reduce the gravity effect. The experiments in the microgravitational field were compared with that under the normal gravity in Figure 18.9A and B. Under normal gravity, the HTWs propagated radially around the apex at the location of higher evaporation. The HTWs were spaced by an almost constant angle along the axial symmetry of the three-phase line. The HTWs moved orthoradially around the droplet edge either from a source point where they were established to a well or all in the same direction but without a preferred direction (clockwise or counterclockwise). Under the microgravitational conditions, the temperature gradient was not as easily defined as it was under normal gravity, but the apex maintained a temperature below the temperature of the three-phase line. The HTWs had the same trend, but their movements were not as systematic as the observation under the normal gravity. The instabilities developed during the transitional phase shown in Figure 18.8A. In the quasi-stationary region, the HTWs propagated from a source point to a well. These instabilities also propagated on the droplet edge, while the angle of propagation was not as constant as it was under normal gravity. Carle et al. (2012) indicated the change was from the significant fluctuations in the level of microgravity. However, the evolution of the number of HTWs, N_w, was in the same magnitude under both the gravitational conditions, as shown in Figure 18.9B. Since the formation of HTWs was found to rely on the length of the apex-contact line interface of the droplet, the number of HTWs decreased as the droplet evaporated as a function of time at different substrate temperatures and levels of gravity.

Sefiane et al. (2013) invented a setup of multilayered substrate and mirror to experimentally investigate the existence of HTWs and their implication for energy transport for an evaporating droplet on a substrate. The experimental apparatus is illustrated in Figure 18.10A, which used one IR camera for both the top view and the bottom view of the FC-72 sessile droplet during evaporation. The HTWs were then demonstrated from both the droplet surface and the bottom for the entire droplet. Based on the careful calibration, the experimental measurements of temperature and heat flux at the liquid-wall interface were reported. Figure 18.10B and C shows the IR top and bottom view of a 2.5-μL FC-72 droplet on substrate at 26°C and 66°C and the correspondingly extracted temperature, HTWs and heat-flux distributions on the liquid–solid interface. A higher substrate temperature manifested

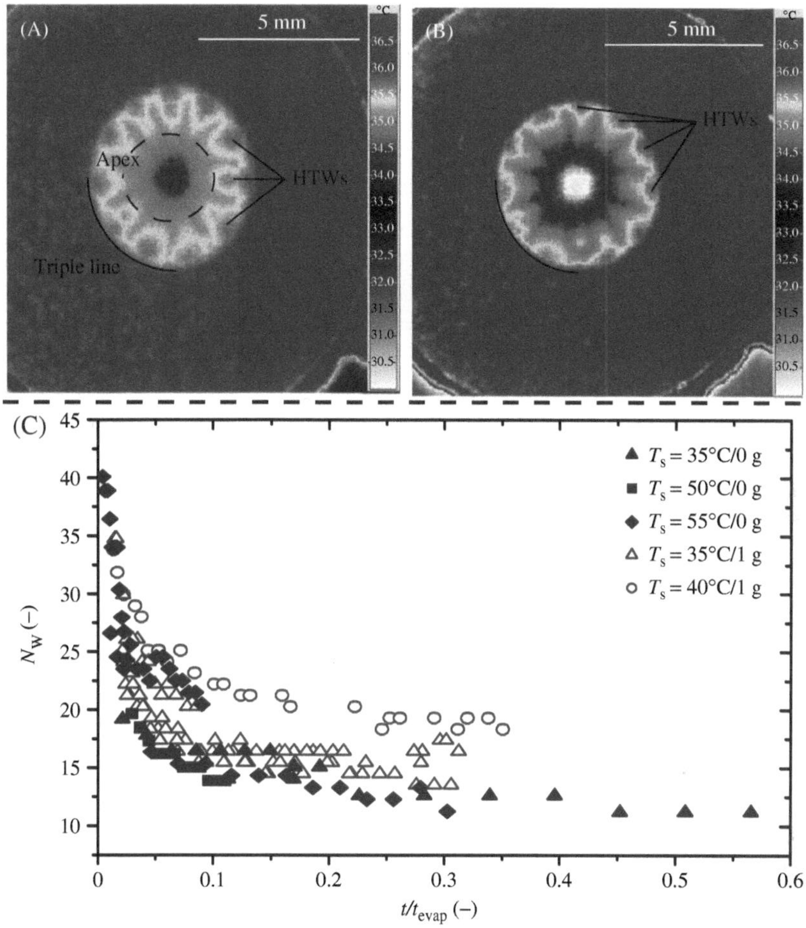

FIGURE 18.9 (A) Droplet IR image of ethanol droplets on a solid substrate with the presence of HTWs under normal gravity. (B) Droplet IR image with the presence of HTWs under the reduced gravity. (C) Variation with normalized time, which is time divided by the total evaporation period (Carle et al., 2012).

many features such as clear heat-sink effect near the droplet edge and clearly visible HTWs in the IR bottom view. On the bottom surface, the highest heat flux could be two orders of magnitude higher than the lowest heat flux. It was concluded that the energy transport during evaporation was affected by the presence of HTWs. For the heat flux distribution shown in Figure 18.10B and C, the unexpected high heat transfer in the center of the droplet was a result of the HTWs transferring hot liquid to the liquid–vapor interface. The setup of IR thermography also proved to be a good tool for measuring droplet heat transfer at solid–liquid interfaces. The high spatial resolution, along with the good thermal resolution,

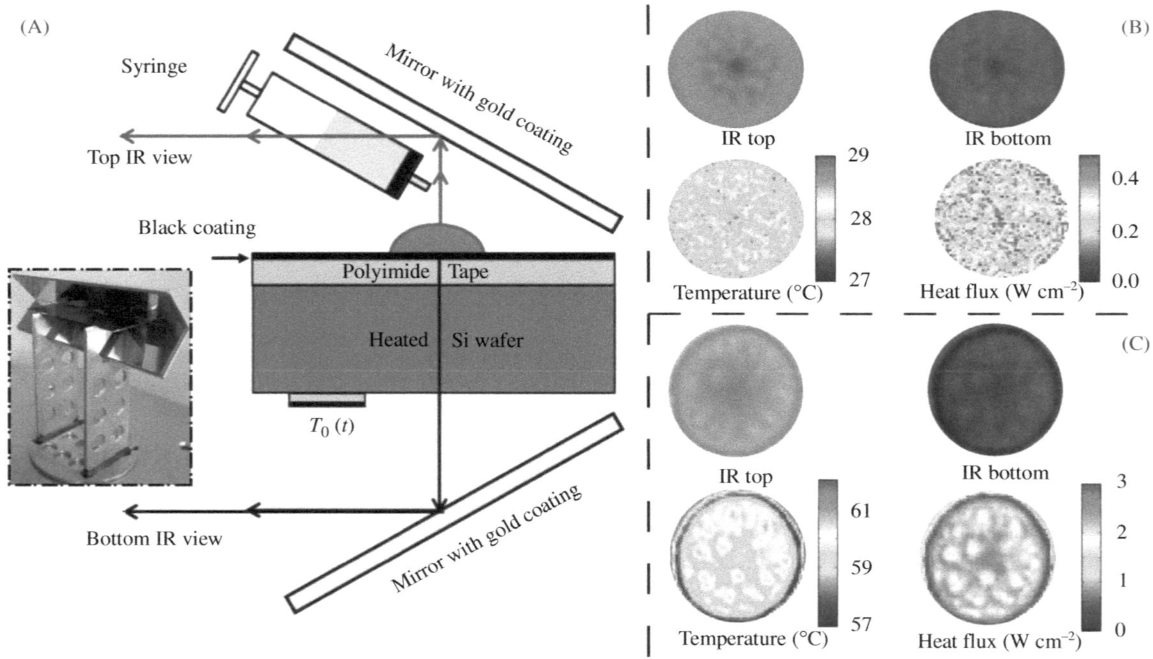

FIGURE 18.10 (A) Double IR thermography setup (Sefiane et al., 2013); the photo is in the inset. The IR top and bottom views for temperature the corresponding heat-flux distributions of a 2.5-μL FC-72 droplet on substrate at 26°C (B) and 66°C (C) (Sefiane et al., 2013).

enabled the detailed thermal information. In addition, the method IR provided the access to the observations of HTWs in droplets even when the droplets are very thin. The HTWs were shown to contribute to energy transport for evaporation in the study. The presence of HTWs suggested that the heat-flux distribution was nonuniform. It is indicated that those models with the constant flux should be improved to take account of HTWs and a nonuniform heat-flux distribution on the contact surface between droplets and substrate surfaces.

The IR thermographic experimental results raise the challenge in the study of HTWs and their application. Although the detailed theoretical analyses are still waiting for the development, some efforts have been made for pinned and unpinned droplets.

18.3 Mathematical model and simulation

18.3.1 Linear Stability Analysis of Receding Sessile Evaporating Droplets

18.3.1.1 Model

Karapetsas et al. (2012) considered the dynamics of an axisymmetric evaporating liquid droplet of a single-component, incompressible, and Newtonian fluid with constant density, which is bounded above by its vapor and below by a horizontal, smooth, uniformly heated substrate. The configuration is shown in Figure 18.11A,

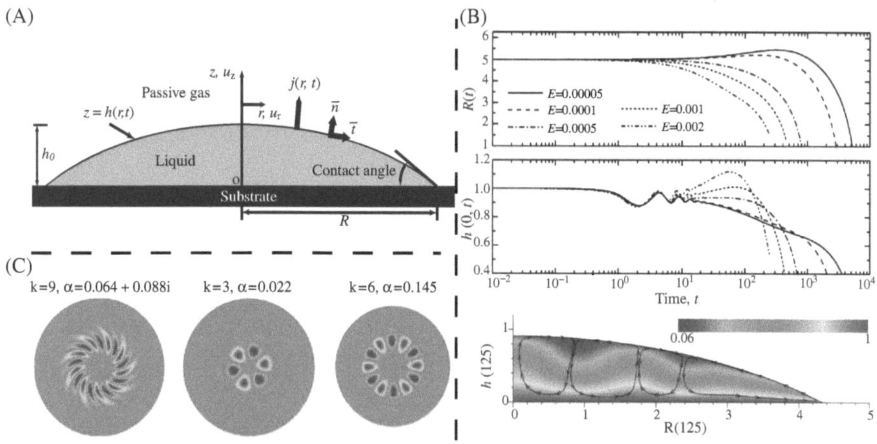

FIGURE 18.11 (A) Schematic of an evaporating sessile receding droplet on a uniformly heated substrate. (B) Time evolution of the position of the contact line of radius and height in the top and middle images; contour plots of temperature and streamlines for Re = 20, Pr = 7, and $E = 0.0005$ at $t_{qs} = 125$ in the bottom image. (C) Contour plots of the temperature perturbations of the most unstable modes for $E = 0.0005$ (Karapetsas et al., 2012).

a 3D formulation in cylindrical coordinate system, (r, φ, z), with the radial and polar angle directions embedded in the substrate and the z-axis directed upward, is employed for the model. The location of the liquid–vapor interface is defined by $z = h(r, t)$, whereas the liquid–solid and the solid–vapor interfaces are both located at $z = 0$. The vectors $\vec{n}(r,t)$ and $\vec{t}(r,t)$ are the outward normal and tangential unit vectors to the liquid–vapor interface, respectively,

$$\vec{n}(r,t) = \frac{(-\partial h/\partial r, 1)}{[1+(\partial h/\partial r)^2]^{1/2}} \tag{18.2}$$

$$\vec{t}(r,t) = \frac{(1, \partial h/\partial r)}{[1+(\partial h/\partial r)^2]^{1/2}} \tag{18.3}$$

The substrate temperature held at a constant T_w, which can be higher than the saturation temperature T_s at the given vapor pressure, while the constant far-field temperature is T_∞. Suppose that the surface tension of liquid–vapor interface is represented by the following linear function of temperature:

$$\sigma(T_I) = \sigma_0 - \gamma(T_I - T_\infty) \tag{18.4}$$

where T_I denotes the interfacial temperature and σ_0 is the surface tension at reference temperature T_∞. For commonly pure liquids, the constant $\gamma = -(\mathrm{d}\sigma/\mathrm{d}T)_{T=T\infty} > 0$. They assumed that the droplet had initially maximum thickness $h(r = 0, t = 0) = h_0$, initial position of the triple-line $r(z = 0, t = 0) = R_0$, and initially uniform temperature $T(r, z, t = 0) = T_w$. The droplet is considered to be very thin, and thus the aspect ratio $\varepsilon = h_0/R_0 << 1$. The assumption justifies the lubrication approximation. They adopted the one-sided model for the evaporating thin-films given by Burelbach et al. (1988), which effectively decouples the dynamics of the droplet from that of the vapor.

Karapetsas et al. (2012) scaled the problem by introducing a set of lubrication-type dimensionless variables. The temperature and velocity fields in the droplet, $T(r, z, t)$ and $\vec{v} = (u_r, u_z)$, are governed by the following dimensionless Navier–Stokes, continuity, and energy equations, where u_r and u_z correspond to the radial and vertical components of the velocity vector, respectively:

$$\nabla \cdot \vec{v} = 0 \tag{18.5}$$

$$\mathrm{Re}\left(\frac{\partial \vec{v}}{\partial t} + \vec{v} \cdot \nabla \vec{v}\right) = -\nabla p + \nabla^2 \vec{v} + St\,\vec{g} \tag{18.6}$$

$$\mathrm{RePr}\left(\frac{\partial T}{\partial t} + \vec{v} \cdot \nabla T\right) = \nabla^2 T \tag{18.7}$$

These governing equations are subject to the following dimensionless boundary conditions:

At $z = 0$:

$$u_z = 0 \tag{18.8}$$

$$u_r - \beta\frac{\partial u_r}{\partial z} = 0 \tag{18.9}$$

$$T = 1 \tag{18.10}$$

where β is the dimensionless slip coefficient for the Navier slip condition shown in Eq. (18.9). However, an alternative slip-condition, $u_r = \beta_{c1}(\cos\phi_s - \cos\phi)$, can be more suitable for a thin droplet with a moving contact line, in which β_{cl} is a new slip coefficient; ϕ_s and ϕ are the static and dynamic contact angles.

At $z = h(r, t)$:

$$\vec{n} \cdot [-p\delta + \nabla\vec{v} + (\nabla\vec{v})^T] = \varepsilon^{-1}\text{Ma}^{-1}(\nabla_s\sigma + 2\kappa\vec{n}\,\sigma) \tag{18.11}$$

where δ is the identity tensor, ∇_s is the surface gradient operator, and κ denotes the mean curvature of the interface.

$$EJ = \frac{\partial h}{\partial t} + \vec{v} \cdot \nabla h \tag{18.12}$$

Equation (18.12) is the kinematic boundary condition, in which E is the evaporation number, and J is the nondimensional mass flux. This model is closed by a constitutive equation with nonequilibrium parameter, K, from the Hertz-Knudsen law.

$$KJ = T_I \tag{18.13}$$

18.3.1.2 Simulation

In the study of Karapetsas et al. (2012), the disturbed variables were introduced by following the azimuthal perturbations or motion,

$$X(r, \varphi, z, t) = X_b(r, z, t_{qs}) + X_p(r, z)e^{ik\varphi + \alpha_X t} \tag{18.14}$$

where $X_b(r, z, t_{qs})$ represents the variable of basic state frozen at t_{qs} to which an infinitesimal disturbance represented by the second term of the right hand side is applied. Here, k is disturbance wave number in azimuthal direction, α_X is, the complex eigenvalue.

$$\alpha_X = \alpha_R + i\alpha_I \tag{18.15}$$

where α_R indicates the growth rate of the disturbance, and the phase speed is α_I/k.

Substituting these perturbed variables in Eq. (18.14) into the governing Eqs. (18.5–18.7) and linearizing in perturbation term, one can obtain a set of eigenvalue equations of the form

$$Px = \alpha_X Qx \tag{18.16}$$

where x denotes the eigenvector.

18.3.1.2.1 Base state

Based on quasi-steady-state approximation, Karapetsas et al. (2012) considered a "frozen solution" as the base state to which the linear stability analysis was conducted. The presence of inertia was indicated to be related to interfacial oscillations. The top and middle images of Figure 18.11B display the change of radius and oscillation of the liquid–gas interface for various values of the evaporation number, E, respectively. During the early stages of evaporation, the contact line remains pinned until a certain point, and the receding takes place earlier for larger E. Two interesting phenomena were reported: (i) the droplet may spread before it starts receding if the evaporation is not very intense, and (ii) oscillation occurs on the liquid–gas interface at the axis of symmetry whose amplitude and frequency are not sensitive to E. They attribute these phenomena to the presence of inertia. As shown in the bottom image of Figure 18.11B, there are multiple convective cells for a large value of Pr. The number and size of these cells decrease with time due to evaporation.

18.3.1.2.2 Hydrothermal waves

According to the evolution of eigenvalues of the most unstable mode with quasi-steady-state approximation linear stability analysis, Karapetsas et al. (2012) excluded those time intervals at early stages where the eigenvalues were affected significantly by the presence of interfacial oscillations due to the inertia. However, after the critical time, the base state oscillations decayed, and the disturbances growth rates would be much larger than the rate of change of the base state variables. Thus, quasi-steady-state approximation analysis became valid beyond this point.

The disturbed temperature contours of the most unstable mode in $r-\varphi$ plane are shown in Figure 18.11C, which demonstrates the growth rate and wavenumber decreased with the evaporation of droplet. Karapetsas et al. (2012) presented that the traveling waves (HTWs) in the azimuthal direction were implied by the modes with nonzero imaginary parts and relatively small base state growth rate at early times in first image of Figure 18.11C, which resembles the interfacial thermal patterns reported by Sefiane et al. (2008) (see the second image of Figure 18.6A). For the times beyond which the quasi-steady-state approximation is valid ($t > 20$), the most dangerous modes cannot predict imaginary part which indicates the presence of stationary rolls, as seen in the second and third images of Figure 18.11C.

Karapetsas et al. (2012) then performed an "energy budget" analysis to clarify further the energetics of the mechanism for the HTWs instability. The results indicated that the dissipative terms (viscous dissipation and thermal diffusion) provided a restoring effect and were stabilizing. The positive contributions to the rates of change of kinetic and thermal energy of disturbances merely resulted from Marangoni stresses and convection, respectively. Therefore, they suggested that HTW instabilities in the evaporating droplets were driven by Marangoni stresses and convection effect, which shares a similar mechanism as the HTWs in thin liquid layers (Smith and Davis, 1983; Smith, 1986; Davis, 1987).

18.3.2 Linear Stability Analysis of Pinned Sessile Evaporating Droplets

18.3.2.1 Model

Sefiane et al. (2010) considered the evaporation of the spherical cap droplet with a constant contact line radius on a horizontal solid substrate. They adopted a 3D, noninertial, spherical coordinates with a rotating angular velocity characterizing the velocity of each moving element of the evaporating interface, the droplet geometry was described with a moving spherical cone frame.

The governing equations describing the flow in liquid and vapor phases were given by

$$\frac{\partial \vec{\omega}}{\partial t} + (\vec{v} \cdot \nabla)\vec{\omega} = \left[(\vec{\omega} + 2\vec{\Omega})\cdot\nabla\right]\vec{v} + \nu\nabla^2\vec{\omega} \tag{18.17}$$

$$\frac{\partial \vec{v}}{\partial t} + (\vec{v} \cdot \nabla)\vec{v} = -\frac{1}{\rho}\nabla p - 2\vec{\Omega} \times \vec{v} + \nu\nabla^2\vec{v} \tag{18.18}$$

where $\vec{\omega}$ is the vorticity vector, and $\vec{\Omega}$ is the rate of rotation of the frame. The presence of rotational term is due to the rotation of fluids relative to the noninertial frame of reference. These governing equations are subject to the following boundary conditions:

At solid–liquid interface:

$$\vec{v} = \vec{0} \tag{18.19}$$

At liquid–vapor interface:

$$L\vec{j} \cdot \vec{e}_r = -k_{th}\nabla T \cdot \vec{e}_r \tag{18.20}$$

$$\vec{j} \cdot \vec{e}_r = \vec{v}_I(\rho - \rho_v) \cdot \vec{e}_r \tag{18.21}$$

where L is the latent heat of evaporation, $\vec{j}$ is the local mass flux, $\vec{e}_r$ is the unit vector in the radial direction, k_{th} is the thermal conductivity, ρ is the liquid density, $\vec{v}_I$ denotes the velocity of free surface, and ρ_v represents the vapor density. Then the system is closed by Clausius–Clapeyron equation

$$K(\theta) = Y_s'(\theta) = \exp\left[\frac{L}{R_u T_s}\left(1 - \frac{T_s}{T'(\theta)}\right)\right] \tag{18.22}$$

where local evaporation equilibrium constant $K(\theta)$ is a function of polar angle θ, $Y'_s(\theta)$ is the local perturbed saturation vapor mass fraction at the interface, R_u is the universal gas constant, and $T'(\theta)$ is the local perturbed temperature at the interface.

It is worth noting that a new Marangoni number was defined, which includes the change of surface geometry and characterizes the local variation of the mass flux at the liquid–vapor interface. Sefiane et al. (2010) assumed that the Clausius–Clapeyron equation governed the saturation concentration in a thin layer of vapor above the liquid–vapor interface. The far-field concentration in the gas phase was considered to be uniform. Furthermore, they justified the quasi-steady-state approximations for heat, momentum, and mass transports in vapor, as well as heat transfer in the substrate according to the order of magnitude analysis of various characteristic times. The large value of thermal conductivity of the substrate also justified the assumption of constant temperature on the surface.

18.3.2.2 Simulation

Sefiane et al. (2010) adopted similar perturbation equations appropriate for spherical coordinates. The notable differences between the two are that the latter includes the disturbed vapor mass fraction $Y'(r, \theta, \varphi, t)$ and expresses perturbed fields of velocity, temperature, and mass fraction according to the perturbed velocity of the free surface

$$\vec{v}_I = \vec{v}_I^* + \vec{v}_I' \tag{18.23}$$

where $\vec{v}_I^*$ and $\vec{v}_I'$ represent, respectively, the basic-state interface velocity and perturbation velocity. The small perturbation of interfacial velocity is represented by the normal mode of the form

$$\vec{v}_I' = \hat{\vec{v}}_I(r) Y_l^m(\theta, \varphi) e^{\alpha t} \tag{18.24}$$

where $\hat{\vec{v}}_I(r)$ is the magnitude of the perturbation of radial component of the interfacial velocity, and $Y_l^m(\theta, \varphi)$ denotes the Bessel function, and the other disturbed variables take the similar form.

Sefiane et al. (2010) performed the stability analysis of normal deformation of liquid–vapor interface by introducing an interfacial perturbation on φ- and θ-directions with three kinds of trial polynomial functions in order to reproduce and recall the patterns observed by IR thermography. They used an initial steady-state drop with a spherical-cap interface to simulate the patterns of the perturbed interface by parametric study. According to the applied trial functions, the patterns obtained on the droplet showed a decreasing wavelength when approaching the triple line. A deformation in the azimuthal direction might also be predicted

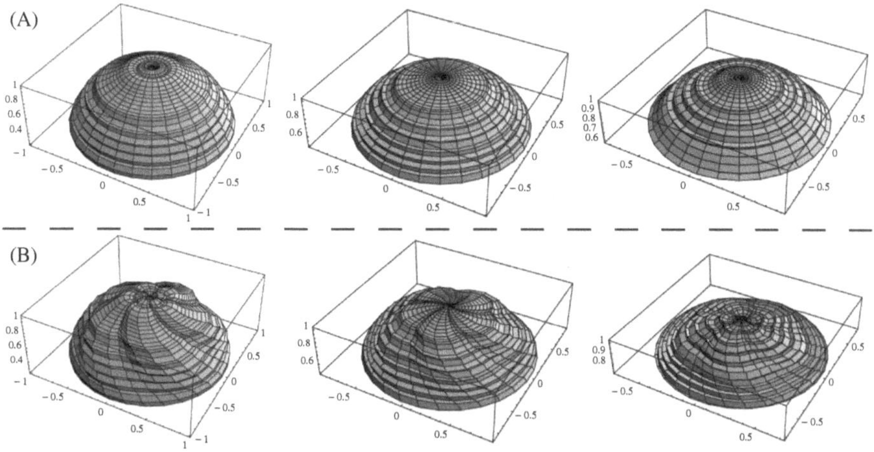

FIGURE 18.12 (A) Surface wavy perturbation obtained with the trial function 2. (B) Azimuthal motion obtained with the trial function 3 (Sefiane et al., 2010).

with the other trial functions. The resulting waves were of similar wavelength traveling from the apex to the contact line region. In a second stage, Sefiane et al. (2010) changed the perturbation function with the trial function 2 and reproduced that the generated waves again propagated from the apex to the contact line, with the wavelength decreasing as shown in Figure 18.12A. With an azimuthal perturbation, they showed the HTWs with the trial function 3, in which the wavelength depends on the position. Sefiane et al. (2010) found that the wavelength, oscillation frequencies, and wave amplitudes of the normal modes all varied during the evaporation process of the pinned droplets. As shown in Figure 18.12B, HTWs were found on the free surface, and the wave number decreased as evaporation with the third perturbation functions, which is very similar to the ones observed in their experiments. These qualitative results also resemble the simulation results obtained by Karapetsas et al. (2012).

The investigations of HTWs have been conducted experimentally and theoretically. Further developments based on new experimental techniques and new simulation methods should provide further understanding of HTWs fundamentally.

Glossary

Shadowgraphy optical method that reveals non-uniformities in transparent media like air, water, or glass.

References

Allain, C., Ausserre, D., Rondelez, F., 1985. A new method for contact-angle measurements of sessile drops. J. Colloid Interface Sci. 107 (1), 5–13.

Bénard, H., 1900. Les tourbillons cellulaires dans une nappe liquide. Rev. Gén. Sci. Pures Appl. 4, 254–266.

Brutin, D., Sobac, B., Rigollet, F., Le Niliot, C., 2011. Infrared visualization of thermal motion inside a sessile drop deposited onto a heated surface. Exp. Therm. Fluid Sci. 35 (3), 521–530.

Burelbach, J.P., Bankoff, S.G., Davis, S.H., 1988. Nonlinear stability of evaporating/condensing liquid films. J. Fluid Mech. 195, 463–494.

Carle, F., Sobac, B., Brutin, D., 2012. Hydrothermal waves on ethanol droplets evaporating under terrestrial and reduced gravity levels. J. Fluid Mech. 712, 614–623.

Chan, C.L., Chen, C., 2010. Effect of gravity on the stability of thermocapillary convection in a horizontal fluid layer. J. Fluid Mech. 647, 91–103.

Chao, D.F., Zhang, N., 2001. Effects of evaporation and thermocapillary convection on volatile liquid droplets. J. Thermophys. Heat Transf. 15 (4), 416–420.

Chun, C.H., Wuest, W., 1979. Experiments on the transition from the steady to the oscillatory Marangoni-convection of a floating zone under reduced gravity effect. Acta Astronaut. 6 (9), 1073–1082.

Davis, S.H., 1987. Thermocapillary instabilities. Annu. Rev. Fluid Mech. 19 (1), 403–435.

De Saedeleer, C., Garcimartin, A., Chavepeyer, G., Platten, J., Lebon, G., 1996. The instability of a liquid layer heated from the side when the upper surface is open to air. Phys. Fluids 8 (3), 670–676.

Garnier, N., Chiffaudel, A., 2001. Two dimensional hydrothermal waves in an extended cylindrical vessel. Eur. Phys. J. B Condens. Matter Complex Syst. 19 (1), 87–95.

Girard, F., Antoni, M., Sefiane, K., 2010. Infrared thermography investigation of an evaporating sessile water droplet on heated substrates. Langmuir 26 (7), 4576–4580.

Hegseth, J.J., Rashidnia, N., Chai, A., 1996. Natural convection in droplet evaporation. Phys. Rev. E 54 (2), 1640.

Karapetsas, G., Matar, O.K., Valluri, P., Sefiane, K., 2012. Convective rolls and hydrothermal waves in evaporating sessile drops. Langmuir 28 (31), 11433–11439.

Kirdyashkin, A.G., 1984. Thermogravitational and thermocapillary flows in a horizontal liquid layer under the conditions of a horizontal temperature gradient. Int. J. Heat Mass Transf. 27 (8), 1205–1218.

Kölker, H., 1980. The behaviour of nonrotational striations in silicon. J. Cryst. Growth 50 (4), 852–858.

Laure, P., Roux, B., 1989. Linear and non-linear analysis of the Hadley circulation. J. Cryst. Growth 97 (1), 226–234.

Mercier, J.F., Normand, C., 1996. Buoyant-thermocapillary instabilities of differentially heated liquid layers. Phys. Fluids 8 (6), 1433–1445.

Parmentier, P.M., Regnier, V.C., Lebon, G., 1993. Buoyant-thermocapillary instabilities in medium-Prandtl-number fluid layers subject to a horizontal temperature gradient. Int. J. Heat Mass Transf. 36 (9), 2417–2427.

Pearson, J., 1958. On convection cells induced by surface tension. J. Fluid Mech. 4 (05), 489–500.

Persad, A.H., Sefiane, K., Ward, C.A., 2013. Source of temperature and pressure pulsations during sessile droplet evaporation into multicomponent atmospheres. Langmuir 29 (43), 13239–13250.

Priede, J., Gerbeth, G., 1997. Convective, absolute, and global instabilities of thermocapillary-buoyancy convection in extended layers. Phys. Rev. E 56 (4), 4187–4199.
Riley, R., Neitzel, G., 1998. Instability of thermocapillary–buoyancy convection in shallow layers. Part 1. Characterization of steady and oscillatory instabilities. J. Fluid Mech. 359, 143–164.
Ristenpart, W., Kim, P., Domingues, C., Wan, J., Stone, H., 2007. Influence of substrate conductivity on circulation reversal in evaporating drops. Phys. Rev. Lett. 99 (23), 234502.
Schwabe, D., Scharmann, A., 1979. Some evidence for the existence and magnitude of a critical marangoni number for the onset of oscillatory flow in crystal growth melts. J. Cryst. Growth 46 (1), 125–131.
Schwabe, D., Möller, U., Schneider, J., Scharmann, A., 1992. Instabilities of shallow dynamic thermocapillary liquid layers. Phys. Fluids A Fluid Dyn. 4 (11), 2368–2381.
Sefiane, K., Moffat, J., Matar, O., Craster, R., 2008. Self-excited hydrothermal waves in evaporating sessile drops. Appl. Phys. Lett. 93 (7), 074103.
Sefiane, K., Steinchen, A., Moffat, R., 2010. On hydrothermal waves observed during evaporation of sessile droplets. Colloids Surf. A Physicochem. Eng. Asp. 365 (1), 95–108.
Sefiane, K., Fukatani, Y., Takata, Y., Kim, J., 2013. Thermal patterns and hydrothermal waves (HTWs) in volatile drops. Langmuir 29 (31), 9750–9760.
Smith, M.K., 1986. Instability mechanisms in dynamic thermocapillary liquid layers. Phys. Fluids 29 (10), 3182–3186.
Smith, M.K., Davis, S.H., 1983. Instabilities of dynamic thermocapillary liquid layers. Part 1. Convective instabilities. J. Fluid Mech. 132, 119–144.
Sobac, B., Brutin, D., 2012. Thermocapillary instabilities in an evaporating drop deposited onto a heated substrate. Phys. Fluids 24 (3), 032103.
Villers, D., Platten, J., 1992. Coupled buoyancy and Marangoni convection in acetone: experiments and comparison with numerical simulations. J. Fluid Mech. 234, 487–510.
Xu, X., Luo, J., Guo, D., 2009. Criterion for reversal of thermal marangoni flow in drying drops. Langmuir 26 (3), 1918–1922.
Zhang, N., Chao, D.F., 2002. A new laser shadowgraphy method for measurements of dynamic contact angle and simultaneous flow visualization in a sessile drop. Opt. Laser Technol. 34 (3), 243–248.
Zhang, N., Yang, W.-J., 1982. Natural convection in evaporating minute drops. J. Heat Transf. 104 (4), 656–662.

Section IV

COMPLEX FLUIDS DRYING

Chapter 19

DROPLETS OF COLLOIDS

David Brutin and Florian Carle
Mechanical Engineering Department, Aix-Marseille University, Marseille, France

19.1 Introduction

While nanofluids are widely studied in several scientific communities (physics, biology, chemistry, etc.), several questions remain regarding their heat transfer efficiency, stability, and toxicity to human health. One important application for nanofluids is inkjet printing (Chhasatia and Sun, 2011), though nanofluids are very promising for two-phase heat transfer as well (Barber et al., 2011).

Using metallic nanoparticles, inkjet printer technology is evolving to allow for the printing of metallic inks on soft substrates for photovoltaic and several others applications. These metallic inks are mostly made of silver, and consequently, they are very expensive (about $500 per milliliter). To create controlled patterns using inkjet-printing technology, a chain of droplets, whose volume has been calculated, must create lines of constant diameter following evaporation. The homogeneity of the line thickness is important because it will affect the electrical resistivity of the resulting pattern.

If the concentration of nanoparticles is too low, an o-ring pattern will appear, which means that the matter will not have been uniformly deposited on the

Droplet Wetting and Evaporation. DOI: http://dx.doi.org/10.1016/B978-0-12-800722-8.00019-9

substrate. If the concentration is too high, then the final thickness will also be too high, exceeding a certain thickness above which the electrical resistivity will no longer change; thus, nanoparticles, which are extremely expensive, will have been wasted. Thus, the search for the optimal concentration that avoids o-ring formation and consequently enables a uniform deposition by minimizing the amount of nanoparticles is of great interest. Several authors are currently looking at the driving parameters that lead to such pattern formation (Brutin, 2013; Askounis et al., 2014). However, because the evaporation of even a pure fluid droplet remains a complex topic, there is much research to be done to understand the evaporation of complex fluids (Conway et al., 1997; David et al., 2007; Kelly-Zion et al., 2011).

19.2 Colloidal system stability

Despite the wide range of applications, the major drawback is the fact that nanofluids tend to be unstable and to flocculate. The stability of nanofluids can be greatly improved by choosing the optimal synthesis method. The choice of the method will depend on not only the fluid/particle pair but also the nanofluid stability, which plays a primary role in determining the properties of the nanofluid. To be able to completely understand the following study on the formation of cracks, a brief review of the particle interaction theory and ion neutrality is provided here. The particles will be subject to molecular interactions, such as collision between the fluid and the particle atoms and Brownian motion, and also to unwanted interactions, such as van der Waals forces (short range electrostatic interaction) or other ionic interactions. These latter phenomena will have a disastrous effect on the stability of nanofluids because large packets of material will flocculate, thus quickly reducing the effectiveness of the nanofluid.

This phenomenon is well explained by the Derjaguin, Landau and Verwey Overbeek theory (DLVO theory (Verwey et al., 1999)). The DLVO theory suggests that the stability of a dispersion of colloids depends of the total potential energy V_T:

$$V_T = V_A + V_R + V_S,$$

where V_A and V_R are the attractive and repulsive contributions, respectively, and V_S is the potential energy due to the solvent. This last contribution plays a negligible role compared to the two other potentials for nanometer-scale separation of the particles. Basically, the suspension stability results from the van der Waals forces (minus the electrical double layer forces) between two particles brought close to each other by Brownian motion. The total potential energy is dependent upon the separation between the particles (Figure 19.1).

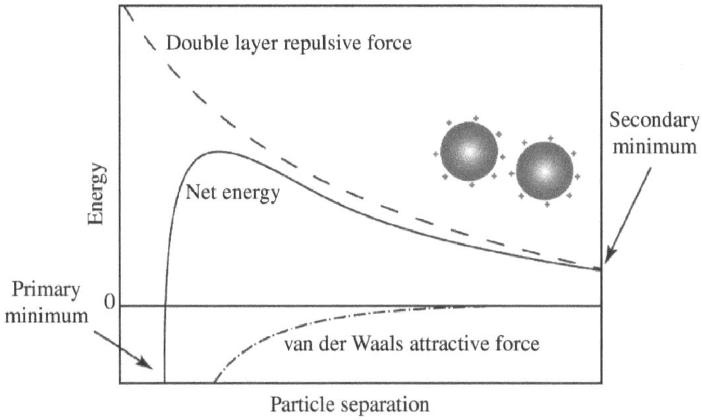

FIGURE 19.1 Potential energy as a function of the particle separation.

At very small interparticle distances, van der Waals forces are predominant over electrostatic repulsion forces. As a result, the particles are forced to stick together, and the dispersion will then flocculate; this contact is particularly strong and irreversible. The strength of this attractive force decreases inversely with the particle separation. Therefore, repulsive electrostatic forces become dominant, and the particles repulse each other. The net energy admits two different levels of energy minima; all of the particles will tend to stay in these two positions of energy minima. The primary minimum is located at the contact of the particle (flocculation of the nanofluid), and the second one is behind an energy potential barrier. This second minimum is where all of the particles should be to produce a stable nanofluid. Unfortunately, if the particles have sufficient energy to go through this barrier, the nanofluid will flocculate. To avoid flocculation, the repulsive forces must be as high as possible to have the potential barrier be as high as possible.

High repulsive forces can be achieved by using surfactants or functional groups to form either a steric barrier (e.g., polymers and plasticizers) or an ionic barrier (e.g., electrostatic stabilization from the electric double layer and from sol-gel precursors) to keep the particles separated and to avoid cluster formation. These additions greatly modify the suspensions but have yet to be investigated for their influence on the crack patterns. The strong effect of surface functional groups hereinafter referred as surface functional groupments on the final crack patterns of nanofluid deposits is described in this section.

Surface functional groups have never before been taken into account in studies on drying, crack patterns, or self-assembly of nanoparticles prior to Carle (2014). These functional groups are, however, widely used in several communities—for example, as biological tracers with fluorescent spheres, cancer treatment using their surface functional groups to bond a dedicated protein (Arima and Iwata,

2007), or even for inkjet printers, using metallic nanoparticles to create monoatomic layers on various hydrophilic or hydrophobic substrates (Chhasatia and Sun, 2011). For all of these applications, surface functional groups are crucial for nanoparticles to be compatible with a variety of conjugation strategies. Both the convective self-assembly of nanoparticles and the final crack patterns are affected by physicochemical interactions with surface functional groups. Interactions between surface functional groups (Frisbie et al., 1994) and solvents (Halicioglu and Jaffe, 2002) were investigated, but more research is required on the influence of the interaction between surface functional groups itself during the formation of close-packed particle assemblies and also on fracturing during drying.

19.3 Mono-dispersed colloidal systems

The evaporation dynamics of a nanofluid sessile droplet follow the same behavior as those of pure fluid sessile droplets in the first stages of evaporation. Figure 19.2 allows for the description, through the example of an aldehyde-sulfate nanofluid droplet, of colloid evaporation dynamics.

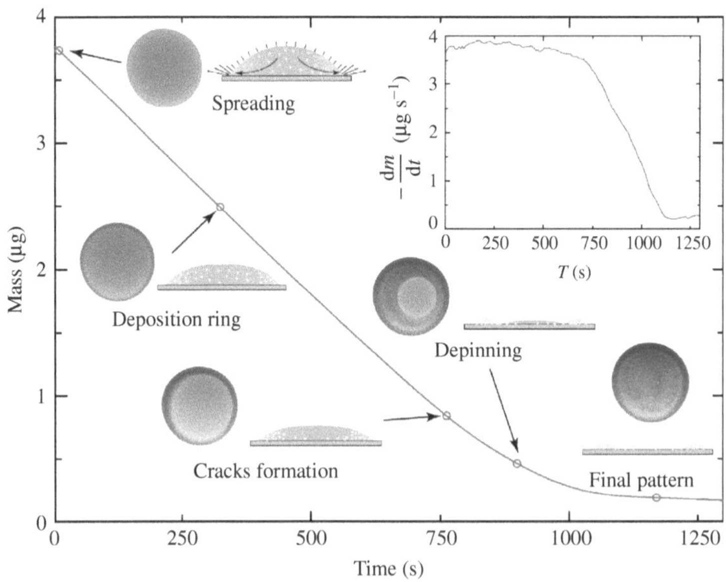

FIGURE 19.2 Evaporation of a 4 μL aldehyde-sulfate nanofluid droplet onto a lysine substrate at 24°C and 50% of humidity. Droplet top view pictures show the major evaporation events, along with a schematic representation of droplet interface and particle motion. The inset shows the droplet evaporation flux rate. *Figure from Carle (2014).*

In the very first instant of the droplet creation, as soon as the fluid touches the substrate, the droplet spreads to reach its maximum wetting diameter. The spreading influence of nanofluids was observed (Chengara et al., 2004; Nikolov et al., 2010), but it was not obvious in all of the experiments performed here. After a few minutes, a circular deposit is clearly visible at the edge of the drops. As we previously observed in Section 5.3, the sessile droplet evaporation profile exhibits a nonuniformity at the triple line; this nonuniformity leads to a radial flow toward the triple line to compensate for the evaporation loss at the contact line (Deegan et al., 1997). Therefore, nanoparticles are dragged outward with this flow in a fairly quick process and then form a deposit in the vicinity of the triple line. Here, the contact line is pinned during most of the evaporation process, whereas the contact angle and the height decrease linearly. As the solvent evaporates, the particles are forced to concentrate near the triple line in a close-packed array (Figure 19.3).

Figure 19.3 depicts remarkable cracks that have been found during SEM observations of the dried droplets. In addition to being dramatic nanoscenery, these pictures summarize the physics involved in the crack formation patterns. Picture (A), which was taken at the center of an aldehyde-sulfate droplet on glass, confirms the preferential paths of the cracks in the direction of the evaporation.

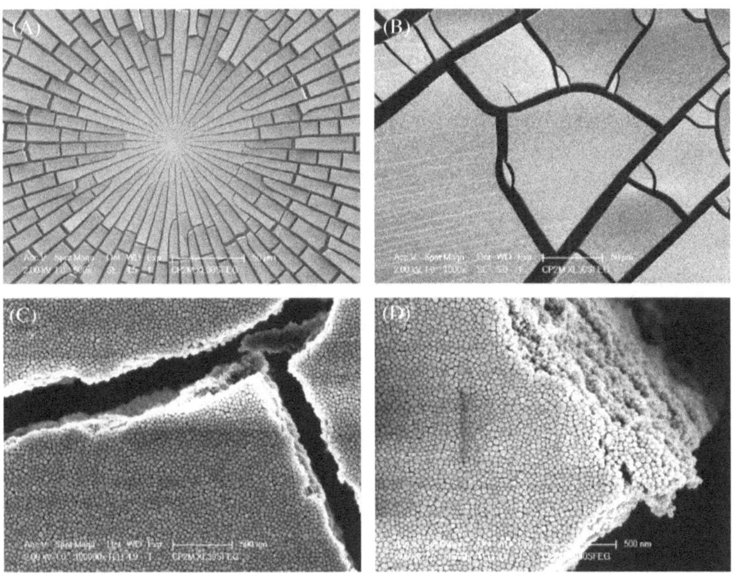

FIGURE 19.3 Remarkable SEM observations of particles deposits. (A) Aldehyde sulfate on glass, (B) carboxylate on lysine, (C) and (D) carboxylate on gold. The scale represents 50 μm for the top pictures and 500 nm for the bottom pictures. *Figures from Carle (2014).*

The depinning of the triple line and its receding toward the center induce the cracks to meet at the center of the droplet. Pictures (C) and (D) provide a glimpse of the forcefulness of the formation of the cracks. The compacted layers of nanoparticles are ripped apart by stress due to evaporation. The interesting aspect of these SEM observations is the ability to locally observe the sharpness of the cracks, enabling determination of the rigidity of the material created by the deposition, and the ability to observe the fragility of the deposits, with the broken pieces near the cracks observed in the image (B).

Coming back to drying, the pinned evaporation continues for approximately 60% of the total time of the evaporation until the contact line recedes when the critical contact angle is exceeded. The depinned contact line shrinks toward the center. The motion of the solvent induces a stress in all deposits, resulting in the nucleation of cracks. Once all of the solvent has evaporated, the dried particles completely cover the initial wetting area due to their relatively high concentration and form regular crack patterns. Experiments were performed several times for each configuration, and the crack patterns were found to be replicable.

Several conditions must be met for the droplets to form a deposition ring on the substrate: first, the solvent must not completely wet the substrate (initial contact angle larger than 0°); otherwise, the particle deposition will form a layer that will tend to crack, following the direction of evaporation. Then, the triple line must pin a major part of the evaporation to initiate a deposition position that remains the same to accumulate enough particles.

Stick slip droplets tend to form irregular concentric rings (Bhardwaj et al., 2009). This phenomenon, when forced to occur regularly, can be used to form patterns for industrial applications: homogeneous rings with a perfect spatial distribution by means of solid spheres in the droplets (Xu et al., 2007) and evaporation onto patterned hydrophobic substrates in the form of pillars (Fan and Stebe, 2004). As an interesting counterexample, the coffee ring effect can be suppressed by using electrowetting (see Section 5.3).

Other nanoparticles than polystyrene are used in the literature. Denkov et al. (1992) used latex particles. In the case of mixtures of more complex fluids, at low concentrations, Govor et al. (2004) evidenced the formation of nanoparticle rings following the evaporation of droplets for a liquid matrix of a binary mixture of nitrocellulose, amyl-acetate, and hexane. The authors used 6 nm nanoparticles of $CoPt_3$ at concentrations below 1%, leading to rings measuring 0.6–1.5 μm in diameter. They experimentally observed the phase separation that leads to the formation of a bilayer structure. The $CoPt_3$ nanoparticles located on the contact line assembled along the line. In this case, the accumulation at the drop edge was pronounced. Conway et al. (1997) first investigated the influence of particles on the evaporation dynamics of a drop. Using 200 and 750 nm polystyrene beads, the author observed the formation of an o-ring and a crater for these bead diameters and monitored the drop height and mass over time. The author provides a linear normalized model of the drop mass and normalized model of the drop height using a root square function.

Using 1 μm particles with a disparity going from 0.5 to 2 μm, Marin et al. (2011) demonstrated that the particles that settle first along the contact line of a coffee ring pattern have an ordered, crystalline structure. Toward the center of the drop, a transition to a disordered particle arrangement was observed. The authors propose a model that explains many results described in the literature where large-scale crystalline deposits were observed for particle sizes ranging from 6 to almost 50 nm. In this range of particle sizes, the critical velocity required for a disordered phase is unreachable within the droplet, which explains the absence of a disordered phase. The authors also state that apart from the order-to-disorder transition, there are also transitions within the ordered phase, from square to hexagonal packing and vice versa (Figure 19.1D). In an unconfined system, the most efficient packing lattice is the hexagonal compact. However, in the evaporating drop the particles are confined in a wedge formed by the glass slide on one side and the liquid–air interface with contact angle on the other side. Such a confinement indeed leads to a sequence of hexagonal and square packing structures that depend on the most efficient packing for the available space. Indeed, when a new layer is formed, the confinement by the wedge favors the square packing; see Figure 19.1. Away from the step edge, the available space increases and allows for the denser, hexagonal packing structure (Figure 19.4).

19.4 Bidispersed colloidal systems

O-ring formation during colloidal drop evaporation was first explained by Deegan (2000). By using sulfate-terminated polystyrene spheres of two sizes, 1 and 0.1 μm, at a maximum volume concentration of 2% dried on glass plates, the author observed the formation of different patterns, depending on the concentration and size of the spheres used. For spheres measuring 100 nm and concentrations ranging from 0.063% to 0.25%, Deegan observed the formation of multiple rings for initial droplets measuring 6 mm in diameter. In all cases, particles remained at the drop center and were not only located at the drop edge. In this study, we observed two different final patterns, depending on the concentration: an o-ring pattern and a nearly uniform deposition pattern on the substrate.

Chhasatia and Sun (2011) used bidispersed colloids made of micro- and nanoparticles and observed the interaction between these particles and the contact line. For 100 nm and 1.1 μm particles, they observed that, at the end of the evaporation process, the nanoparticles were located near the triple line, followed by the microparticles (Figure 19.5). For a hydrophobic substrate, the constant contact angle

FIGURE 19.4 Order-to-disorder transition in the particle stain left by an evaporating drop. (A) A 3 μL sessile water droplet evaporates from a glass substrate. (B) Ring-shaped stain of red particles (the coffee stain) left on the substrate after evaporation. (C) A close-up by an optical microscope of the bottom layer of the stain, taken from the white square in (B), shows that the outermost lines of the stain (left) have an ordered, crystalline structure. Toward the center of the drop (right), a transition to a disordered particle arrangement is observed. (D) A top view of the ring stain, taken from the red square in (C) with a scanning electron microscope (SEM), shows that the first lines of particles (left) are arranged in a hexagonal array, while the next lines (brighter in the image) are arranged as square, followed by again hexagonal array. (For interpretation of the references to color in this figure legend, the reader is referred to the online version of this book.) *Images from Marin et al. (2011).*

mode dominates the evaporation process, leaving little carrier liquid left to rearrange particles according to their sizes at the mixed mode. For a hydrophilic substrate of $0° < \theta < 45°$, particle separation is incomplete in the final deposition. Three separate regions exist from the outer edges to the center of the drop, including a region with only nanoparticles, a mixture of micro- and nanoparticles, and an inside region with only microparticles. The width of the middle region where micro- and nanoparticles overlap depends on the interactions of the surface tension and friction forces acting on the particles. For a hydrophilic substrate of $\theta = 0°$, nanoparticles accumulate at the contact line, forming several rows that prevent the contact line from receding. This enhanced pinning by nanoparticles permits radially outward evaporative flow toward the pinned contact line and pushes some of the microparticles to exceed through the liquid–vapor interface. Reduced surface tension acting on those exceeded microparticles prevents them from moving inward, whereas most other microparticles move radially inward to

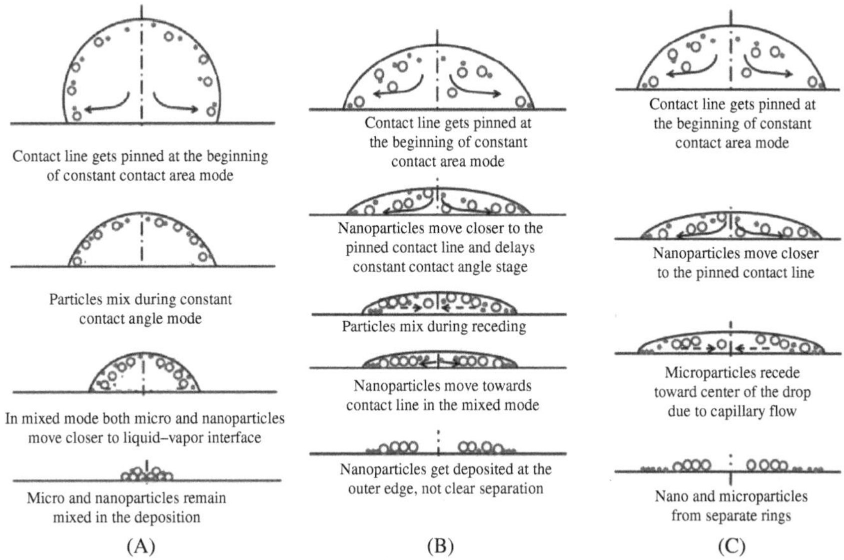

FIGURE 19.5 Schematic of the evaporation process of bidispersed particles (100 nm and 1.1 μm) on three substrates with contact angles of: (A) $\theta > 45°$ (mixed), (B) $0° < \theta < 45°$ (partial separation), and (C) $\theta = 0°$ (complete separation). *Sketches from Chhasatia and Sun (2011).*

the center of the drop. By modifying substrate wettability, the surface tension force acting on the particles is tuned. The control of particle separation according to their sizes can therefore be achieved.

The authors have also found that the increase in the size ratio of the bidispersed particles improves particle separation. Because the number of particles affects contact line pinning, the boundaries between mixed, partially separated, and completely separated regimes shift toward a higher contact angle value for a higher particle loading. By controlling the substrate wettability, particle size ratio, particle loading, and relative humidity, the deposition morphology of bidispersed particles can be further controlled.

During the drying of droplets containing latex particles measuring from 40 nm to 5 μm, Monteux and Lequeux (2011) observed a rearrangement that led to the formation of rings. The smallest nanoparticles were deposited on the outer side of the ring, while the largest particles were located on the inner side. The authors conclude that this deposition mechanism leads to a tight deposition but leaves a thin liquid film of pure water at the edge of the drop. The deposited particle ring size can be precisely controlled by changing the particle size. The segregation effect may consequently play an important role in polymer and biological fluids, which are composed of polydisperse particles.

Pattern formation has also been observed for biological fluids, such as whole blood, very recently and shows similar patterns to those of nanofluids (Brutin et al., 2011). The drying dynamics observed for this complex biological fluid show typical regimes that are treated in Section 4.5. Finally, pattern formation is also influenced by the wettability of a droplet on a substrate. The relationship between the spreading of a droplet and the evaporation dynamics has been previously discussed in Section 1.1. The triple line dynamics and the pinning of a drop play a major role in the drying process dealt with in Section 1.3.

19.5 Parameters influencing the patterns

Obviously the colloids concentration is the first parameter that influences the pattern. Brutin in 2013 evidenced a critical concentration to observe the classical o-ring formation. Above this concentration a more homogeneous deposition is observed. The author used polystyrene nanoparticles measuring 24 nm that were suspended in deionized water to study the effect of nanoparticle concentration and surrounding humidity on the pattern formation and drying dynamics of sessile droplets. Droplets were deposited on rough Nuflon 16X-coated substrates. The patterns formed were observed to have been influenced mainly by the concentration of nanoparticles but not at all by the drying time. Two patterns were observed: an o-ring formation for concentrations below 0.47% and a flower pattern above 1.15%. The effect of humidity was proportional to the drying rate, but no effect on pattern formation was observed. The experimentally measured evaporative mass flux was slightly higher compared to that predicted by a purely diffusive model, which may indicate a slight enhancement in heat transfer at the triple line due to the local improvement in the substrate heat capacity by nanoparticle deposition. For pure fluids, the effect of the substrates thermal properties was investigated, even at ambient temperature, because the latent heat of evaporation is mostly affected by element with the highest heat capacity, which is the substrate (Sobac and Brutin, 2012). One explanation provided by the author is that the diffusion coefficient of our water-based nanofluid, even at low concentration, in humid air is different from that of pure water in dry air. Experiments are currently underway to address this discrepancy. Another explanation is based on the experimental observation in the literature of an increase in the spreading of nanofluids leading to an increase in the drop perimeter and a simultaneous decrease in the contact angle. The consequence is an improvement in the evaporation rate. This assumption is also under investigation at this time (Figure 19.6).

Chhasatia et al. (2010) investigated the effect of relative humidity on the contact angle and nanoparticle deposition during the evaporation of a picoliter droplet of a nanofluid. They observed, for the pinned droplet mode, a spreading effect with the increase in relative humidity. The observed variations in the contact angle

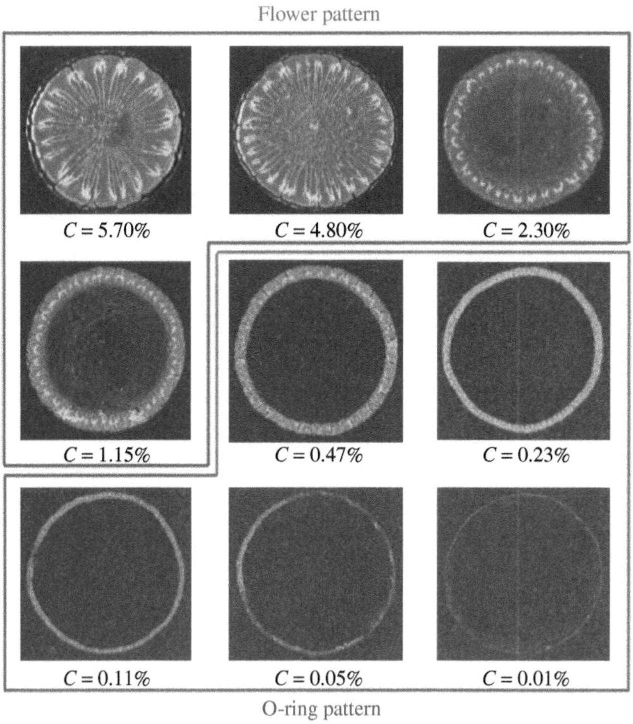

FIGURE 19.6 Influence of colloids concentrations ranging from 0.01% to 5.70% on the pattern formation (controlled glovebox humidity of 50% and room temperature of 25.0 ± 0.6°C). *Image from Brutin (2013).*

were linear but exhibited different slopes depending on the relative humidity. The authors conclude that due to the small volume of the droplet, the relative humidity becomes an important parameter driving the evaporation process and consequently the deposition. They also observed an increase in the initial spreading of the drop with increasing humidity.

Highly concentrated colloidal suspension exhibits spreading and evaporation instabilities, as observed by Hadj-Achour and Brutin (2014). The authors report a fingering instability that occurs during the spreading and evaporation of a silver metallic ink nanosuspension droplet concentrated at about 20% in mass so far above the low concentrations classically observed. The fingering instability has been widely studied for both Newtonian and non-Newtonian fluids. However, this report describes the first time that a fingering instability is observed for the spreading of a nanosuspension sessile droplet. In this study, the authors demonstrate that in certain cases, the contact line evolves through different spreading regimes according to De Coninck et al. (2001) with an enhancement in the evaporation rate due the formation of the fractal patterns.

19.6 Nanoparticles surface charges

Working with nanoparticles, the surface charges exposed at their vicinity cannot be neglected. Carle and Brutin, 2013 studied the influence of this charge on the pattern and the crack periodicity. Working with three substrates of different surface energies and the same nanoparticles but only with different surface charges, they evidenced strong differences in the final pattern. Figure 19.7, if observed vertically, shows the final crack patterns of each nanofluid on three substrates (surfaces are characterized by their increased surface energy) and if monitored horizontally, shows the effect of surface functional groups on cracks (characterized by their increased electrical charge in solution in milliequivalent per gram).

Two different sets of cracks are visible on the droplet dryouts: large radial cracks and smaller orthoradial ones with a much shorter characteristic length. In such an

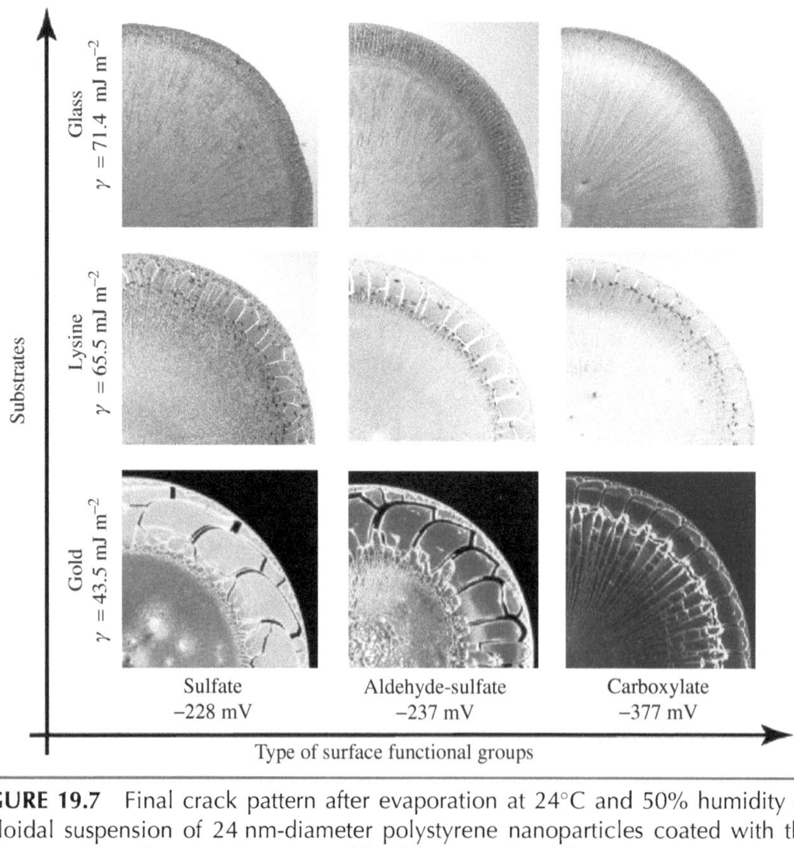

FIGURE 19.7 Final crack pattern after evaporation at 24°C and 50% humidity of a colloidal suspension of 24 nm-diameter polystyrene nanoparticles coated with three different surface functional groups with different surface free energies. *Image from Carle and Brutin (2013).*

isotropic material, cracks tend to form and grow in the direction in which the release of strain energy is the highest (Goehring et al., 2011). In this configuration, the stress in the deposit is due to capillary forces that develop as the solvent evaporates. At the early stages of the evaporation, particles are free to move in the solvent via Brownian motion, and the separation between the particles is subjected to the DLVO theory, where van der Walls forces are balanced by a ionic repulsive forces. As the solvent evaporates, the average distance between particles must be reduced to ensure that all of the particles remain inside the fluid; if a particle becomes exposed to air at the edge of the drop contact line, the surface energy will locally increase (Holmes et al., 2008) and will change the balance of forces. Particle reorganization in the external layer of the deposit can accommodate some of the strain, but the particles inside the lower layer tend to bind to the substrate and resist deformation, thereby preventing any possible relaxation (Chiu et al., 1993). This phenomenon leads to a stress increase, which is subsequently relaxed by plastic deformation and crack nucleation.

Large cracks are the first to form, and they do so in the direction of evaporation to release the stress induced by solvent evaporation and by particle adhesion to the substrate. The stress in a gel had previously been quantified for a confined system (Zarzycki, 1988); however, for droplets, the experimental determination of the stress in such an open configuration is almost impossible. As a result, numerical analysis should instead be used. The opening of the material allows the remaining solvent, hitherto confined under the deposit, to evaporate; such drainage increases the stress in large plates of particles. The stress is relieved with the formation of smaller orthoradial cracks between the large cracks. Due to the axisymmetrical aspect of the system, the stress generated by evaporation is identical at each point in the drop. In the vicinity of a crack, the stress is null and increases with distance from the crack. When the stress reaches a maximum and exceeds the fracture strength, a new crack nucleates. The same phenomenon occurs in the deposit in the vicinity of this new crack, which forms another crack, thus inducing crack formation with a regular periodicity.

These crack networks, which form a general pattern in the dryouts, are composed of a thick corona on the rim containing the majority of the particles and of a central area containing much fewer particles. Although the particle concentration is too high to form "coffee stain" patterns, the height of the corona is greater than that of the central area. The additional height explains the formation of three crack patterns. As previously explained for confined configurations, the stress increases in the material with distance, but it also increases with the thickness of the deposits. Figure 19.8 confirms this assumption for the droplet configuration and displays the ratio of length wave/deposit height as a function of the initial contact angle. This figure demonstrates that this ratio increases linearly with the contact angle (i.e., with decreasing surface energy). The three clusters of points represent droplets on the three substrates. On glass, the thin deposit and the high

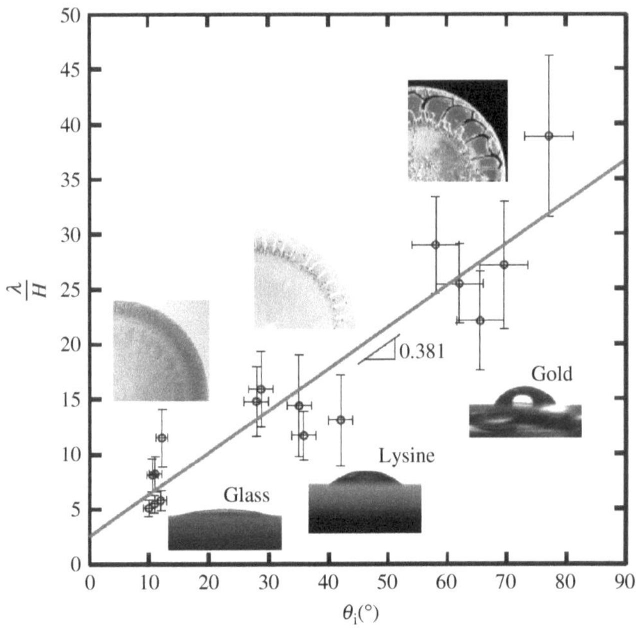

FIGURE 19.8 Crack spacing/deposit height ratio plotted as a function of the initial contact angle. Side and top views show both the behavior of aldehyde sulfate droplets on the substrates and the crack patterns formed. *Image from Carle and Brutin (2013).*

surface energy of the substrate induced a small wavelength. In this case, the material must relax stress more frequently. However, on gold, the low adhesion and thick deposit allow the material to better resist stress. Thus, crack formation is not required as often as on glass—that is, gold induces a larger wavelength.

References

Arima, Y., Iwata, H., 2007. Effect of wettability and surface functional groups on protein adsorption and cell adhesion using well-defined mixed self-assembled monolayers. Biomaterials 28 (20), 3074–3082.

Askounis, A., Sefiane, K., Koutsos, V., Shanahan, M.E.R., 2014. The effect of evaporation kinetics on nanoparticle structuring within contact line deposits of volatile drops. Colloids Surf. A, 441, 855–866. Available from: http://dx.doi.org/10.1016/j.colsurfa.2012.10.017.

Barber, J., Brutin, D., Tadrist, L., 2011. A review on boiling heat transfer enhancement with nanofluids. Nanoscale Res. Lett. 6, 1–16.

Bhardwaj, R., Fang, X., Attinger, D., 2009. Pattern formation during the evaporation of a colloidal nanoliter drop: a numerical and experimental study. New J. Phys. 11, 075020.

Brutin, D., 2013. Influence of relative humidity and nano-particle concentration on pattern formation and evaporation rate of pinned drying drops of nanofluids. Colloids Surf. A 429, 112–120.

Brutin, D., Sobac, B., Loquet, B., Sampol, J., 2011. Pattern formation in drying drops of blood. J. Fluid Mech. 667, 85–95.
Carle, F., 2014. Flow Motion in Sessile: Evaporation and Nanoparticles Assembly (Ph.D. dissertation). Aix-Marseille University.
Carle, F., Brutin, D., 2013. How surface functional groups influence fracturation in nanofluid droplet dry-outs. Langmuir 29, 9962–9966. Available from: http://dx.doi.org/10.1021/la401428v.
Chengara, A., Nikolov, A.D., Wasan, D.T., Trokhymchuk, A., Henderson, D., 2004. Spreading of nanofluids driven by the structural disjoining pressure gradient. J. Colloid Interface Sci. 280 (1), 192–201.
Chhasatia, V.H., Sun, Y., 2011. Interaction of bi-dispersed particles with contact line in an evaporating colloidal drop. Soft Matter 7, 10135–10143.
Chhasatia, V.H., Joshi, A.S., Sun, Y., 2010. Effect of relative humidity on contact angle and particle deposition morphology of an evaporating colloidal drop. Appl. Phys. Lett. 97, 231909.
Chiu, R.C., Garino, T.J., Cima, M.J., 1993. Drying of granular ceramic films: I, effect of processing variables on cracking behavior. J Am. Ceram. Soc. 76 (9), 2257–2264.
Conway, J., Korns, H., Fish, M.R., 1997. Evaporation kinematics of polystyrene bead suspensions. Langmuir 13, 426–431.
David, S., Sefiane, K., Tadrist, L., 2007. Experimental investigation of the effect of thermal properties of the substrate in the wetting and evaporation of sessile drops. Colloids Surf. A 298, 108–114.
De Coninck, J., De Ruijter, M.J., Vou, M., 2001. Dynamics of wetting. Colloid Interface Sci. 6, 49–53.
Deegan, R.D., 2000. Pattern formation in drying drops. Phys. Rev. E 61, 475–485.
Deegan, R.D., Bakajin, O., Dupont, T.F., Huber, G., Nagel, S.R., Witten, T.A., 1997. Capillary flow as the cause of ring stains from dried liquid drops. Nature 389, 827–829.
Denkov, N., Velev, O., Kralchevski, P., Ivanov, I., Yoshimura, H., Nagayama, K., 1992. Mechanism of formation of two-dimensional crystals from latex particles on substrates. Langmuir 8 (12), 3183–3190.
Fan, F., Stebe, K.J., 2004. Assembly of colloidal particles by evaporation on surfaces with patterned hydrophobicity. Langmuir 20 (8), 3062–3067.
Frisbie, C.D., Rozsnyai, L.F., Noy, A., Wrighton, M.S., Lieber, C.M., 1994. . Functional group imaging by chemical force microscopy. Science 265 (5181), 2071–2074.
Goehring, L., Clegg, W.J., Routh, A.F., 2011. Wavy cracks in drying colloidal films. Soft Matter 7 (18), 7984.
Govor, L.V., Reiter, G., Bauer, G.H., Parisi, J., 2004. Nanoparticles ring formation in evaporating micro-size droplets. Appl. Phys. Lett. 84 (23), 4774–4776.
Hadj-Achour, M., Brutin, D., 2014. Fractal pattern formation in nanosuspension sessile droplets via evaporation-spreading on a glass substrate. Colloids Interface Sci. Commun. 1, 43–46.
Halicioglu, T., Jaffe, R.L., 2002. Solvent effect on functional groups attached to edges of carbon nanotubes. Nano Lett. 2 (6), 573–575.
Holmes, D.M., Tegeler, F., Clegg, W.J., 2008. Stresses and strains in colloidal films during lateral drying. JECS 28 (7), 1381–1387.
Kelly-Zion, P.L., Pursell, C.J., Vaidya, S., Batra, J., 2011. Evaporation of sessile drops under combined diffusion and natural convection. Colloids Surf. A 381, 31–36.
Marin, A.G., Gelderblom, H., Lohse, D., Snoeijer, J.H., 2011. Order-to-disorder transition in ring-shaped colloidal stains. Phys. Rev. Lett., 085502.
Monteux, C., Lequeux, F., 2011. Packing and sorting colloids at the contact line of a drying drop. Langmuir 27, 2917–2922.
Nikolov, A., Kondiparty, K., Wasan, D., 2010. Nanoparticle self-structuring in a nanofluid film spreading on a solid surface. Langmuir 26 (11), 7665–7670.
Sobac, B., Brutin, D., 2012. Thermal effects of the substrate on water droplet evaporation. Phys. Rev. E 86, 021602.

Verwey, E.J.W., Overbeek, J.T.G., 1999. Theory of the Stability of Lyophobic Colloids. Courier Dover Publications, Mineola, NY, USA.

Wasan, D.T., Nikolov, A.D., 2003. Spreading of nanofluids on solids. Nature 423 (6936), 156–159.

Xu, J., Xia, J., Lin, Z., 2007. Evaporation-induced self-assembly of nanoparticles from a sphere-on flat geometry. Angew. Chem. Int. Ed. 46 (11), 1860–1863.

Zarzycki, J., 1988. Critical stress intensity factors of wet gels. Journal of Non-Crystalline Solids 100 (1-3), 359–363.

Chapter 20

DROPLETS OF IONIC SOLUTIONS

David Fairhurst

Department of Physics and Mathematics, Nottingham Trent University, Nottingham, UK

Of all the types of liquid droplets, those containing ionic solutions are probably the most abundant. Soluble salts are present all over Earth as well as on other planets, so they appear in any naturally occurring droplets. In fact, it takes care in the laboratory to purify liquids and keep such solutes out. Polymers, microparticles, and surfactant molecules are less common in nature, so they are less likely to be found in naturally occurring droplets. Despite, or perhaps because of, the ubiquity and variety of ionic solution droplets, there is no single focused body of scientific literature on such droplets. Arguably the most well-known and certainly one of the most visually arresting consequences of droplets is the rainbow, but droplets also give rise to other atmospheric phenomena, both on Earth and on other planets. The presence of dissolved compounds in the liquid influences the shape and size of these optical effects, measurements of which can even be used as analytic probes of the droplets' composition. In our atmosphere, aerosols containing dust or salt make up a large fraction of the atmosphere and influence the greenhouse effect, and consequently have an important role in determining global climate. On the ground, water droplets containing salt are responsible for

Droplet Wetting and Evaporation. DOI: http://dx.doi.org/10.1016/B978-0-12-800722-8.00020-5

damaging rocks, roads, and buildings, although the extent of the damage depends subtly on the dissolved species and its crystallization morphology. Droplets containing dissolved species are used in spray drying to create powders, in crop spraying to deliver nutrients or pesticides, and in electrospraying experiments, where they are ingeniously made to hover. Even when deposited on an inert substrate, molecular solution droplets exhibit interesting behaviors: evaporation-induced crystallization can be used to grow crystal polymorphs not usually stable during bulk crystallization, and drying droplets are known to "creep" across surfaces and even out of containers, escaping the constraints of the initial droplet.

This chapter provides an overview of these varied properties and applications of droplets of ionic solutions. As in other chapters, ionic solution droplets are considered in various forms: levitated, sessile, and pendant. However, unlike other chapters, here we also briefly consider droplets that are not evaporating, or at least for which evaporation does not play a critical role in the described effects; it is not necessary for rain droplets to evaporate to produce a rainbow. These "non-evaporating" examples are included for two reasons: first, we know that a constant volume raindrop is in a dynamic equilibrium between evaporation and condensation, and second (and more important), they are hopefully of interest!

20.1 Rainbows on Earth and other planets

Despite their familiarity, the childhood rhymes we are taught to remember the (six? seven?) colors of the rainbow and the mythical and religious stories surrounding their origins, when a bright rainbow appears in a stormy sky, we still feel the need to stop and marvel. However, rarely do we think about the optics involved in creating this personal arc of color. In particular, how does the refractive index of the raindrops alter the pattern? Figure 20.1 shows a split rainbow captured over the ocean, with the upper half created by freshwater droplets in the atmosphere (refractive index $n = 1.3330$) having a rainbow angle around 1° greater than the lower part, formed by sea spray ($n = 1.3383$). Similar shifts should also be observed for other atmospheric optic effects, including fogbows, halos (rings seen around 20° from the sun), and glories (rings seen when the sun shines past you onto a cloud, often visible from an airplane window) (Cowley, 2014). A fascinating example is the recent work of Markiewicz et al. (2014), who examined extraterrestrial glories measured by the *Venus Express* orbiter. The characteristic angles of the glory "are unambiguously and uniquely determined by the dominant particle size of the cloud layer" with "a very accurate estimate of the cloud particles size to be 1.2 μm with a very narrow size distribution." In addition, "the refractive index ... is higher than that of sulfuric acid previously proposed for the clouds' composition" (e.g., by Hansen and Arking, 1971) and suggests that either ferric chloride or elemental

FIGURE 20.1 The angle a rainbow makes depends on the refractive index of the liquid. Here, the salty spray of the Pacific Ocean creates a lower rainbow around 1° smaller than the freshwater rainbow above. *Photograph taken by J. Dijkema in 1981 in the Pacific Ocean, 500 km SE of Japan © G.P. Können, permission obtained. See more at: http://www.osa-opn.org/home/gallery/after_images/10-oct-10_00_00_00/.*

sulfur is responsible for the higher than anticipated refractive index. So, by analyzing the optical effects of droplets on Venus, planetary scientists are able to improve their knowledge of the Venusian atmosphere.

20.2 Atmospheric aerosols

Besides producing interesting optical effects, small droplets of aqueous solutions in the atmosphere (often referred to as atmospheric aerosols) have important consequences for radiation transfer, atmospheric chemistry, and, thus, climate (Andreae and Crutzen, 1997). For example, the overall reflectance of Earth to incoming radiation (albedo) is subtly dependent on the chemical composition of atmospheric aerosols, which can either increase the albedo and offset warming due to greenhouse gases or, for other chemicals, result in lower levels of reflected radiation and contribute to an increase in mean temperature (Satheesh and Moorthy, 2004). The composition of atmospheric aerosols ranges from the more abundant natural aerosols, with predominantly sea salt in droplets larger than 0.5 μm or natural sulfates in smaller droplets (Fitzgerald, 1991), to the anthropogenic aerosols containing soot and organic molecules. The aerosols can also act as nucleation sites for clouds and raindrops that themselves will evaporate while falling (Kinzer and Gunn, 1950) following a modified version of the D^2 equation, due to the moving air stream and contributing to the measured exponential distribution of raindrop sizes (Marshall and Palmer, 1948). The presence of dissolved impurities in cloud aerosols also has the effect of stabilizing them against both evaporation and coalescence, and this explains the persistence of sea mists or smog: a phenomenon known as the Köhler effect (1936), after the scientist who discovered that "the concentration of chlorine in raindrops is the same as in the droplets in fogs and clouds, which are millions of times smaller. From this it follows that raindrops never arise from condensation of the droplets in fogs and clouds." This effect can counteract the Kelvin effect that predicts that small droplets are more volatile due to lower vapor pressure (Laplace pressure across a more tightly curved interface (Thompson, 1871)). Much more recent work by Ghosal et al. (2005) using X-ray photoelectron spectroscopy has shown that concentrations of atmospheric Zimpurities (potassium bromide and potassium iodide) have enhanced concentrations of the halide ion at the surface of aerosol droplets, most noticeably for the larger iodide anion. These measurements are in quantitative agreement with molecular dynamics simulations and have implications for the dynamics of atmospheric chemistry, because the surface concentration determines reaction rates.

20.3 Spray drying and levitated droplets

One of the very first studies of the evaporation behavior of droplets of ionic solutions was by Ranz and Marshall (1954), in which they undertook a range of experiments to investigate the evaporation of water drops containing dissolved and suspended materials in both stationary and moving air flows. They used two configurations for their droplets: suspended on a thin thermocouple with changing droplet diameter and from the end of a microburette, enabling the addition of liquid to maintain the droplet's volume and monitor the evaporation rate. One of the key assumptions they made in their analysis was that the surface concentration is always at saturation, irrespective of the concentration in the bulk of the droplet. In their words, "The drop evaporates at any average concentration as though it were saturated throughout.... This simplification is based on the fact that diffusion coefficients in water are on the order of 10^{-5} cm^2 s^{-1}, so small that solids will concentrate in the surface by evaporation much faster than they can diffuse toward the center of the drop." Although Charlesworth and Marshall (1960) later showed this approximation to be incorrect by demonstrating that evaporation rates and surface temperatures for calcium chloride droplets varied with initial concentrations, it remains a useful simplification in many cases. They also identify two regimes during droplet evaporation. The first follows the behavior of pure liquid droplets with a steadily decreasing surface area and a surface temperature drop determined by considering the temperature vapor–pressure relationship and the balance of heat transferred via conduction, convection, and radiation to the latent heat of evaporation (equation 13 in Ranz and Marshall, 1954). In the second case, solid phases start to appear on the surface, reducing the evaporation rate and requiring the heat of crystallization to be included in the energy balance. This additional heat results in a smaller surface temperature drop, a larger vapor pressure, and a reduction in the evaporation rate. Experiments performed using NH_4NO_3 and NaCl confirmed their predictions.

Charlesworth and Marshall (1960) undertook extensive experimental work to investigate the drying behavior of 16 different liquids over a range of drying conditions, including droplet volume, concentration, air flow, and temperature. Droplet temperatures and images were recorded during drying, but a visual method could not be used to determine droplet mass due to the formation of solid structures around the liquid. Instead, the authors suspended single droplets on a 400-μm glass filament, the calibrated deflection of which they used to determine the mass of the droplets during the entire evaporation process. From their observations, they created a generalized description of the various evaporation pathways, reproduced in Figure 20.2.

The first stage of drying was similar for all experiments during which the surface is at least partially liquid, because a complete solid crust had not yet formed. The second stage was after a complete solid shell or crust had formed.

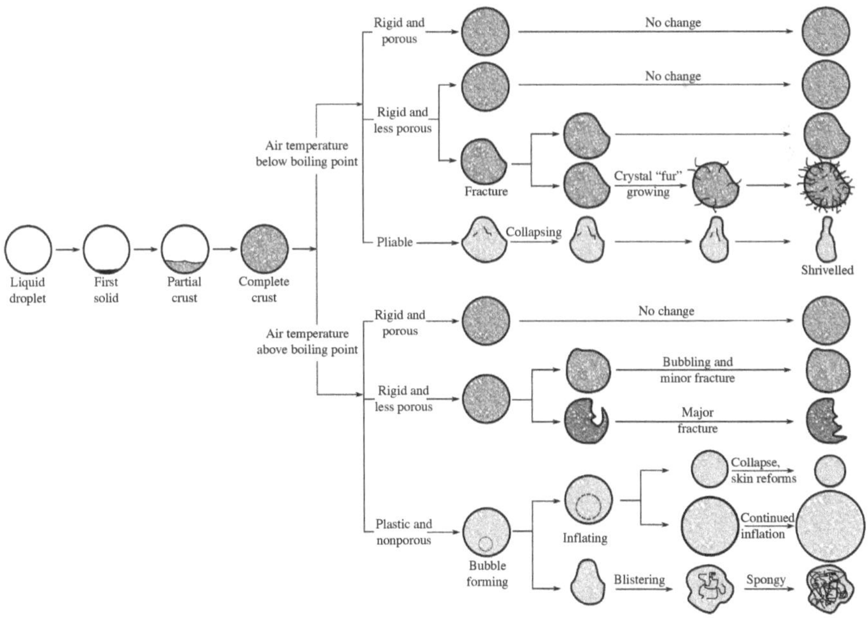

FIGURE 20.2 Generalized appearance changes in suspended drying droplets. *Image reproduced from Charlesworth and Marshall (1960), copyright © 1960 American Institute of Chemical Engineers.*

The concentration at the droplet edge can be calculated by assuming transport within the droplet is by diffusion only (with solute diffusivity D) and that the evaporation rate does not depend on concentration and is characterized by an evaporation constant $K = -R\frac{dR}{dt} = -\frac{\bar{v}}{4\pi R}\frac{dm}{dt}$, where R is the droplet radius, $\bar{v}$ is the partial specific volume and m is the droplet mass. Another approximation avoids the complication of treating the moving boundary by adding a layer of solute at the edge to compensate for the solvent evaporation, thereby keeping the radius constant, leading to

$$c(R,t) = c_0 \exp\left(\frac{\phi^2 Dt}{R_0^2 - 2Kt}\right)$$

where R_0 is the initial droplet radius, c_0 and c are the initial and saturation droplet mass fractions, and ϕ is the positive real root of $\tanh y = \frac{y}{\frac{K}{D}+1}$ and is well approximated by $\phi \approx 1 + \frac{K}{D}$ for $\frac{K}{D} > 1$. The characteristic time for a solid to appear at the droplet surface is then given by

$$t_c = \frac{R_0^2 \ln\left(\frac{c_s}{c_0}\right)}{\phi^2 D + 2K \ln\left(\frac{c_s}{c_0}\right)}$$

From their experimental investigations, the authors propose an empirical simplification, which they found worked satisfactorily for all solutes over most experimental conditions and surprisingly does not feature the solute diffusivity:

$$t_c = \frac{R_0^2 c_0}{2K}\left[1 - \frac{c_0}{c_s}\right]$$

Although there is some scatter in the data obtained for seven different salts, the time at which crystals first appear can be estimated by $t_c/2$ and the complete shell forms close to t_c.

The following various possible sequences of events occur in the second stage (after the full crust has formed). Table 20.1 maps these to the various salts investigated. Sequences 1–3 apply for air temperatures below the boiling point of the liquid.

1. The shell is porous and the liquid retreats into it, as detected by "twinkling" within the crust.
2. The shell is rigid and less porous, so it creates stresses in the solid, which can lead to cracking and exposing liquid to the air again. Later on, bubbles can form within the shell.
3. The shell is pliable and impervious, leading to dimpling or wrinkling with subsequent volume loss. In some cases the skin stiffens on drying, which can then lead to rupture and implosion.

Sequences 4–6 are for air temperatures considerably above the liquid boiling point, where the crust slows down the evaporation, allowing the droplet to heat

TABLE 20.1 Summary of Materials and Results from the Comprehensive Work by Charlesworth and Marshall (1960)

Dissolved Material	Below Boiling	Above Boiling	Comments
Sodium sulfate	2		Thin-walled hollow sphere; crust 5% droplet diameter; outside smooth, inner surface rough.
Potassium sulfate	1		Open crust structure that does not reduce evaporation rate noticeably; incomplete thin crust with craters, composed of large crystals (Figure 20.3A).
Copper sulfate	2	5a	Impermeable crust almost stops evaporation; white, shiny, smooth.
Ammonium nitrate	2		Lower concentration droplets evaporated more quickly; overall concentration when crust forms independent of initial concentration; large temperature drop due to large heat of crystallization.
Calcium chloride			Used to demonstrate that evaporation rate does depend on concentration, contrary to assumption of Marshall and Ranz.
Sodium acetate			
Coffee extract		6c	Pliable skin; shrivels with further evaporation; becomes noticeable asymmetric; above boiling repeated inflating of skin forming spongy structure.
Ammonium sulfate	2	5b, 6a	
Ammonium chloride	1		
Sodium chloride	1	5a	
Potassium nitrate		6b	
Lithium hypochlorite		6a	
Sucrose		6a	
Blue dye			
Whole fresh milk		6b	
Polyvinyl acetate		6b	

up; reduced mass transfer leads to increased heat transfer. In these cases the water boils, creating pressure within the shell, which can have various consequences:

4. The shell is wide open, and liquid is forced out, where it quickly evaporates with minimal pressure buildup.
5. The shell is rigid and impermeable, so the pressure causes fracture.

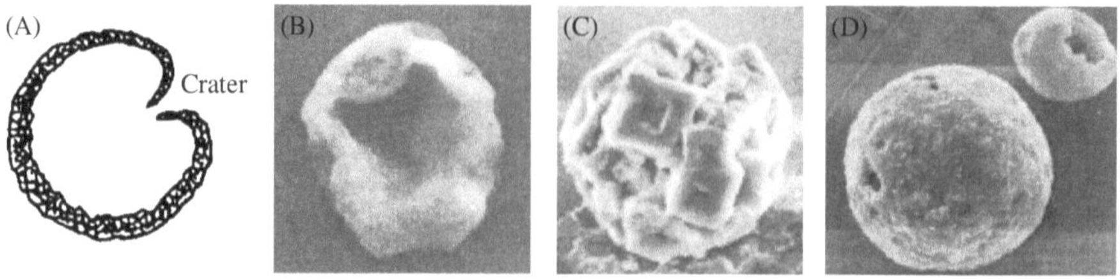

FIGURE 20.3 (A) Schematic of a crater forming in a hollow crust from an evaporated droplet of potassium sulfate (Charlesworth and Marshall, 1960). (B)–(D) Scanning electron microscope pictures of hollow particles formed from dried droplets of sea salt. *Reprinted from Cheng et al. (1988), with permission from Elsevier.* (B) Naturally occurring sea salt aerosol, diameter 10 μm. (C) Laboratory-formed particle in freefall at 60% humidity, diameter 26 μm. (D) Laboratory-formed particle in freefall at less than 30% humidity, diameter 26 μm.

6. The shell is plastic and impermeable, and it inflates due to the pressure. In some cases the stretched skin then becomes sufficiently porous to allow liquid to pass through (6b); otherwise, the skin ruptures and then collapses (6a and 6c).

Surprisingly, the final state of the dried droplet is relatively insensitive to which sequence the liquid droplet has followed. Charlesworth and Marshall (1960) state that in "virtually all cases the final particle consisted of a hollow, thin, nearly spherical crust. The outer surface was usually smooth, whereas the inner surface was rough and uneven. Occasionally, the inner core contained an open network of large crystals." The authors also note that the weight-loss curves, obtained using the deflection of the thin glass filament, are very similar between ionic salt solutions and pure water: "Generally the coincidence of the weight-loss curves for droplets of water and of solutions persisted past the formation of the solid phase and almost to the completion of the crust." This observation is reinforced in later work by Ferron and Soderholm (1990), who also found that the ratio of evaporation times only became greater than two for "particles containing a high salt concentration with a relative humidity ... near to [that] ... of the saturated salt solution." Further particular details of some of the materials investigated are presented in Table 20.1. Interestingly, droplets of ammonia nitrate show an effect similar to the famous Mpemba effect in which hot water freezes faster than cold (Mpemba and Osborne, 1969): 2 μL droplets with lower concentrations of salt were observed to evaporate more quickly. For example, a droplet with an initial concentration of 50% reached 90% after 510 s, compared to a 25% droplet reaching 90% after 420 s and a 12.5% droplet after only 370 s. This effect may well be caused by the highly exothermic crystallization of ammonia nitrate. Finally, Charlesworth and Marshall commented that they saw no evidence for any radial circulatory flow of fluid within the droplets, only motion parallel to the surface, either due to rotation of the entire droplet or due to surface tension–induced flows along the surface.

Since this early work, there have been many further publications on the behavior of such droplets, and the review article by Mezhericher et al. (2010) provides a thorough overview of the strengths and weaknesses of many of the proposed theoretical descriptions. These approaches can be divided into two broad categories. The first and most common approach considers species diffusion of both solute and solvent within a droplet with diffusivity values taken from the experiment. Additionally, the droplet temperature is assumed to be uniform, which can be justified for small Biot numbers (ratio of heat transfer inside and outside the droplet). The biggest difficulty in these calculations arises due to the moving boundary condition as the droplet radius decreases. An alternative and more recent strategy uses a reactive engineering approach, the predictions of which are in good agreement with the experiment but are restricted because it requires an empirical relation between droplet concentration and vapor pressure, which is not known for a wide range of relevant materials.

Although many of the works cited in this section were motivated by a desire to understand and improve the processes used in the chemical engineering industry, such as spray drying and freeze drying, the findings do have implications beyond these discipline boundaries. Cheng et al. (1988) found similar hollow particles to those seen by Charlesworth and Marshall from sea salt aerosols both in the marine environment and in freefall under laboratory conditions, as shown in Figure 20.3B–D.

20.4 Droplets in electric fields

Continuing from the mechanical methods for suspending and weighing evaporating droplets discussed in the previous sections, we now turn to drops in electrical fields, in particular electrically levitated droplets. In this noncontact approach, the electric field is adjusted to keep a liquid droplet or dried particle at the same position as water evaporates or condenses. Because the charge on the droplet is constant, its mass can be determined accurately once the electric field strength is calibrated, and because the mass of dissolved solute (assumed to be nonvolatile) is fixed, the concentration of the droplet is known.

Tang and Munkelwitz (1983) investigated the evaporation of aqueous droplets of NaCl and $(NH_4)_2SO_4$ using this method and later continued their work for the atmospherically important aerosols $(NH_4)_2SO_4$, NH_4HSO_4, $(NH_4)_3H(SO_4)_2$, Na_2SO_4, $NaHSO_4$, and $NaNO_3$, along with simultaneous light scattering to determine droplet or particle size (Tang and Munkelwitz, 1994). Figure 20.4 shows typical behavior for a particle of $(NH_4)_2SO_4$. During condensation, the particle is hygroscopic (absorbs water from the atmosphere) and eventually deliquesces around 80% concentration (dissolves in the absorbed liquid). On evaporation, the

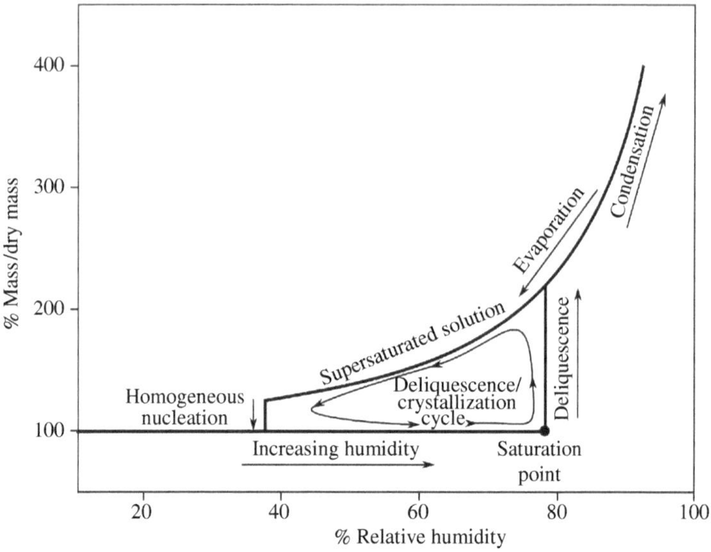

FIGURE 20.4 Schematic of condensation and evaporation curves for 12 μm diameter solid $(NH_4)_2SO_4$ particle showing deliquescence above the saturation point at 80% humidity and a supersaturated liquid droplet on the descending branch down to around 37%. The deliquesence/crystallization cycle used by Desarnaud and Shahidzadeh-Bonn (2011) is also marked. *Reprinted from Tang and Munkelwitz (1994), with permission from Wiley.*

liquid droplet shows significant supersaturation because it is stable well below the saturation point, down to a concentration of around 37%. Two interesting observations from this work are that $NaHSO_4$ is so hygroscopic that the liquid droplets persisted in a vacuum for several days without crystallizing and that $NaNO_3$ droplets, which usually crystallize between 20% and 30%, "can persist in the liquid form to a molality as high as 300, at which value one solvent molecule is shared by five or six solute molecules!" (Tang and Munkelwitz 1994).

In his PhD thesis, Cohen (1987) provides a detailed study of the properties of electrically levitated droplets of 11 different single-electrolyte solutions (NaCl, NaBr, KCl, KBr, NH_4Cl, Na_2SO_4, $(NH_4)_2SO_4$, $CaCl_2$, $MnCl_2$, $MnSO_4$, $FeCl_3$) and three mixed-electrolyte solutions (NaCl and KCl, NaCl and KBr, NaCl and $(NH_4)_2SO_4$). These investigations permit accurate measurements of equilibrium liquid activity by relating the droplet concentration to the ambient relative humidity. In addition, electrically levitated droplets are ideal for studying crystal nucleation because there are no container walls on which crystals are typically heterogeneously nucleated, allowing much higher values of the supersaturation to be accessed from which homogeneous nucleation will occur.

The effect of dissolved salts can also make a noticeable difference to electrowetting droplets (see Chapter 5 for a more in-depth discussion of droplet electrowetting). For example, Lee et al. (2013) found that the contact angle of a droplet of weak electrolyte solution (pyroboric acid, $H_2B_4O_7$) will be lower than that of pure water at the same applied voltage and will therefore saturate at a lower voltage, too. However, although the contact angle of droplets of a strong electrolyte (sodium chloride, NaCl) will be even lower, they do not reach the same saturation angle, so they may be less useful in electrowetting devices.

20.5 Dryout patterns, crystal nucleation, and morphology from sessile droplets

So far, we have considered only pendant or levitated ionic droplets, but in the following sections, we address the behavior of sessile droplets. The addition of the solid substrate changes droplet behavior in several ways that have been discussed previously (e.g., contact line pinning affecting evaporation dynamics, substrate conductivity affecting heat transfer and temperature gradients in the droplet, and radial capillary flow affecting deposition patterns). These effects are still relevant for ionic solutions, with the additional specific complications of the interactions between the dissolved species and substrate and/or triple line, which may affect surface adsorption, act as a nucleation site for crystallization, and influence the morphology of the crystals that form. To explain why the substrate may affect crystal nucleation, both Sear (2007), using a simple numerical model, and Djikaev and Ruckenstein (2008), with a theoretical approach, demonstrated that crystal nucleation rates are orders of magnitude greater in the vicinity of a gas–liquid–solid triple line, arising either from an impurity or from a preexisting crystal at the liquid–air interface or at the contact line where the droplet meets the substrate. In fact, the line tension (arising from the triple line) is a more important factor in many cases than the difference in surface tensions. For evaporation with pinned contact lines, there is another reason why crystals would nucleate at the triple line: the solute concentration there will be enhanced through the coffee stain effect. Several works investigate the consequences of these enhanced nucleation rates in droplets of ionic solutions.

Takhistov and Chang (2002) published a range of experimental observations on various evaporating water droplets containing dissolved $CuSO_4$ and NaCl, along with various nonelectrolyte solutions of albumin, citric acid, urea, sugar, and gold colloids. They varied the wettability of the glass substrate from 3° (on bare glass) to 45° (on glass washed in ethyl alcohol) to 105° (using a thin layer of castor oil). They observed that "the larger-scale patterns were quite similar for all solutions, but the smaller ones could differ." In general, on hydrophilic surfaces, they saw large crystals deposited in the center of the droplet with high initial concentrations and smaller ring-like deposits for lower concentrations: crystal deposition at the edge enhanced contact line pinning. On hydrophobic surfaces, they nearly always saw central deposits because the contact line was not pinned. However, they noted an interesting exception for a small intermediate range of initial concentration values that showed ring-like deposits.

The details of the deposits depended sensitively and subtly on the particular liquid. For example, the Mullins–Sekerka instability (1963) leads to crystal protrusions growing more quickly than the rest of the crystal, which produces crystals with a fractal or dendritic appearance, like snowflakes. Also, the effect of solute concentration on surface tension changes sign between surfactant solutions and particle suspensions (for which surface tension decreases with concentration) and ionic solutions for which there is a noticeable increase in surface tension as concentration is increased. Dutcher et al. (2010) proposed a semi-empirical model that fits the experimental surface tension data for a large range of ionic salts. Their explanation is that most dissolved salts are less likely to be found near the liquid surface, and then, for entropic reasons, the interface is energetically less favorable. In turn, this effect leads to a reversal of the direction of any solutal Marangoni flows that will alter the detail of the final deposits. They claim that the Marangoni effect can also explain the observed weak dependence of the evaporation rate of NaCl on concentration (a threefold decrease for a four-order-of-magnitude increase in concentration), but this could also be because of a smaller droplet radius due to higher concentration and thus higher surface tension.

Many of the deposits also contained concentric ring-like deposits, due to the slip-stick motion of the interface. On surfaces titled by an angle α to the horizontal, the slip-stick behavior is enhanced by gravitational film draining, and the spacing of the rings decreases with increasing tilt. A robust observation is that the slip-stick deposits for different angles can be collapsed by rescaling the vertical spacing y between rings to $\frac{y}{\sin\alpha}$. Finally, they also observed a dimple or circular trough in the liquid surface close to the growing crystal during evaporation in salt solutions, although this proved difficult to image conclusively. They proposed that this was caused by a drainage flow toward the growing crystal at the periphery, which locally reduced the liquid surface faster than could be replenished from the remainder of the droplet.

Although not as wide-ranging as the previous work, a later investigation by Shahidzadeh-Bonn et al. (2008) into the crystallization behavior of two salts (Na_2SO_4 and NaCl) presents careful experimentation and insightful analysis that should have applications beyond their chosen systems. They were motivated to understand why only some salt solutions cause significant damage to porous rocks when they evaporate: "What is clear is that some salts preferably crystallize on the exterior wall of the stones, a phenomenon called efflorescence, whereas others prefer to crystallize within the porous medium, called subfluorescence. The latter is far more damaging than the former." Evaporation was allowed to occur from sessile droplets and from square capillaries, with both hydrophilic surfaces (by washing glass slides with sulfochromic acid, resulting in a critical surface tension $\gamma_c \approx 71\ \mathrm{mN\ m^{-1}}$) and hydrophobic surfaces (prepared with fluoroalkyl silane solution with a $\gamma_c \approx 15\ \mathrm{mN\ m^{-1}}$). NaCl is known to form cubic crystals and is not particularly damaging to rocks. Na_2SO_4 has two crystalline forms: mirabilite is the thermodynamically stable hydrated crystal, and thenardite is anhydrous. Particular care was taken to prepare the Na_2SO_4 solutions, which can be sensitive to preparation methods.

Their findings are summarized in Table 20.2 and Figure 20.5. They deduce that "sodium chloride crystals are found to form preferentially in contact with a nonpolar area (air or hydrophobic solid)" and that "not only the transport properties in the liquid but also the interfacial properties of the different crystalline phases govern the crystallization dynamics." In summary, "we find that depending on the type of salt used and the experimental conditions, salt crystals form either at the interface of the substrate with the salt solution or at the solution–air interface." Later work by Kulichikhin et al. (2011) compared the drying patterns

TABLE 20.2 Summary of Evaporation and Crystallization Behavior for Salt Droplets at Different Initial Relative Concentration Values and on Surfaces with Different Wettabilities, by Shahidzadeh-Bonn et al. (2008)

Salt	c_0/c_s	Behavior on Hydrophilic Surface	Behavior on Hydrophobic Surface
Sodium sulfate	1.12	• Mirabilite forms at air–liquid interface • Crystals lower γ below γ_c • The droplet spreads • Flat circular crystal deposit	• Mirabilite forms at air–liquid interface • γ does not drop below γ_c • Droplet does not spread • Crust of hydrated crystals • "Ping-pong ball" deposit
Sodium sulfate	1.00	• Concentration increase → γ increase → θ increase • At $c_0/c_s = 1.8$ thenardite nucleates at contact line • Dendritic crystals form in precursor film • Precursor film increases evaporation rate	• Very similar to hydrophilic behavior at same concentration • No precursor film • No dendritic crystals beyond droplet
Sodium chloride	1.00	• Crystals form at air–liquid interface • Crystals lowering γ • (Slight) droplet spreading • Ring stain deposit	• Crystals grow on substrate at contact line • Cauliflower-like growth on top of crystals

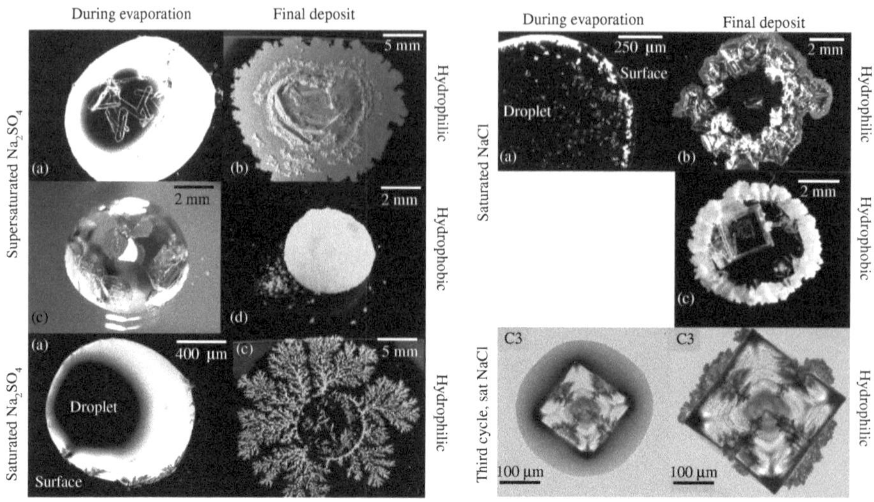

FIGURE 20.5 Images of aqueous droplets of sodium sulfate and sodium chloride during and after drying, on hydrophobic and hydrophilic substrates, with initial concentrations either saturated or supersaturated. *Reprinted (adapted) with permission from Shahidzadeh-Bonn et al. 2008. Copyright (2008) American Chemical Society.* The bottom row shows a droplet during and after three cycles of evaporation and deliquescence forming a large single crystal. *Reprinted with permission from Desarnaud and Shahidzadeh-Bonn (2011).*

from microliter droplets of NaCl solutions placed on inert polymer surfaces with surfaces coated with the biopolymer mucin and confirmed that the interaction between the solute and the substrate was indeed important for determining the crystal morphology that occurs.

Metastable polymorphs of other materials, which are difficult to obtain through traditional methods, have also been nucleated at the contact line of evaporating droplets—for example, by Capes and Cameron (2006) for the drug paracetamol and by Poornachary et al. (2013) for mannitol (sugar alcohol) and glycine.

In addition to aiding nucleation of metastable crystal polymorphs, droplet evaporation can also be used to create large single crystals. Desarnaud and Shahidzadeh-Bonn (2011) repeatedly cycled droplets around the deliquescence (condensation)/crystallization loop shown in Figure 20.4 by changing the environment from dry to humid. Evaporating saturated (6.1 M) droplets of NaCl on a hydrophilic surface, they also found that crystals formed on the air–liquid interface, close to the contact line. They then increased the humidity to redissolve the crystals. The second time around the cycle, crystals nucleated later but grew more quickly, toward the center of the droplet, because there was no room at the edge to accommodate the larger crystals. After the third cycle, a single crystal was formed, as shown in the bottom row of Figure 20.5. With hydrophobic surfaces, only two cycles were required until a single crystal formed, but when droplets of water were added rather than vapor from the atmosphere (deliquescence), more cycles were needed. They reasoned that impurities ejected from the bulk during each crystallization do not redissolve when the concentration decreases. Without impurities, the nucleation rate is lower, so eventually it only occurs at one point in the droplet, but the crystal growth rate is faster due to higher supersaturation and the absence of impurities.

Rodríguez-Ruiz et al. (2013) showed that very small (150 pL) droplets under a much larger volume of oil (to avoid saturation) behave the same as those left to evaporate in air. They observed a pinned contact line with decreasing contact angles for a droplet of water, NaCl, Na_2CO_3, and $CaCl_2$.

20.6 Creeping salt droplets

Many evaporating salt solutions show the peculiar behavior known as creeping: crystals grow *beyond* the extent of the initial liquid, as depicted in Figure 20.6 and as just described by Shahidzadeh-Bonn et al. (2008) when dendritic crystals were seen to form in the precursor film surrounding a droplet. This phenomenon was first reported by Washburn (1927): "It is not an uncommon experience, in almost every chemical laboratory, to see the glass walls of beakers or crystallizing dishes, above the surfaces of evaporating solutions, lined or coated with a solid deposit of the material which had been in solution. This deposit is usually present on the walls not only throughout the space through which the solution has been lowered by evaporation but often several centimeters above the original level of

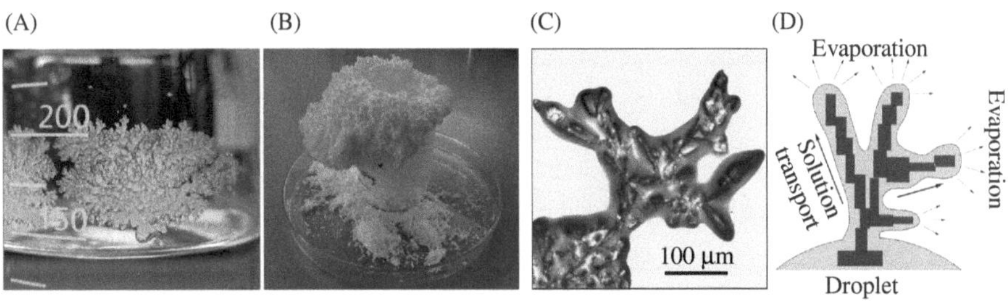

FIGURE 20.6 (A) Creeping from a saturated solution of NaCl up the walls of a glass beaker. (B) Further creeping (efflorescence) from NaCl with $Fe(CN)_6^{4-}$ additive. (C) *In situ* microscopy image from a droplet of $K_2Cr_2O_7$ creeping via seaweed pattern crystals. (D) Schematic diagram showing fluid flow and evaporation during creeping of $K_2Cr_2O_7$. *Reprinted (adapted) with permission from van Enchevort and Los (2013). Copyright (2013) American Chemical Society.*

the solution. In fact, the deposit often creeps to the top of the containing vessel and down on the outside, sometimes even spreading out on the table top around the dish ... in much the same manner that it would have done if the bottom of the dish had been cracked." He placed several aqueous salt solutions in glass crystallizing dishes, listed in Table 20.3, as well as many other solute/solvent combinations. Although not a rigorous study, he did correctly identify two important ideas: capillary flow between crystal and glass, and the importance of the "wettability" of the solid on the glass. Clearly, a solid cannot "wet" another solid, but the term is used loosely to describe the interfacial energy between the two solids. Around the same time, Druce (1927) performed similar studies on a range of aqueous salts. The first ten salts listed in Table 20.3 are his descending order of "tendency to creep." A more thorough scientific study was later performed by Hazlehurst et al. (1936), who challenged Washburn's proposed mechanism: "It is difficult to picture how the capillaries are formed originally, for the protrusion of the crystals beyond the limits of the solution seems much like raising oneself by ones bootstraps ... the prime factor ... of creeping is the ability of the *solid solute to wet preferentially the solid surface over which the creeping takes place*." They used glass rods dipped into solutions and also used droplets for which they observed that the deposits "were found over an area far larger than that of the original drop," and started when crystals within the droplet were transported to the contact line and then pushed out beyond the edge of the liquid. Du and Stone (1996) investigated crystalline deposits from evaporating saturated ammonium chloride solutions, in particular when the deposits grow *away* from the walls of the container into what they described as "salt trees," and they observed that these patterns differ from other branched objects as they grow away from the source of material, due to the capillary flow of liquid.

Recently, van Enchevort and Los (2013) undertook the definitive work on a wide range of creeping salt droplets, which are listed in Table 20.3, along with relevant comments based on their careful microscopic observations. Their key conclusions are:

1. Crystal patterns fall into two categories: "seaweed" patterns, which are branched but with no pronounced orientational order (as shown in Figure 20.6), and compact crystals.
2. Creeping is driven by solvent evaporation from the liquid in front of the growing crystal. Consequently, evaporation is the rate-limiting step.
3. Fresh solution reaches the growing front by flowing along either the top or the edge of the crystal ("top supplied creeping," TSC) or through the gap between crystal and substrate ("bottom supplied creeping," BSC).
4. The criteria for TSC or BSC growth can be formulated in terms of the interfacial energies of vapor, liquid, crystal, and substrate and correctly predicts the observations. However, geometry is also important, because small crystals

TABLE 20.3 Summary of Various Salts Investigated for Creeping Behavior from Evaporating Droplets

Salt	Creeping Property	Description from van Enchevort and Los (2013)	References
NH_4Cl	BSC	Dense, compact patterns	Druce (1927), van Enchevort and Los (2013)
$ZnSO_4$		No creeping, overgrows droplet surface	Druce (1927), van Enchevort and Los (2013)
$K_2Cr_2O_7$	TSC, some BSC	Most well developed, branched aggregate patterns	Druce (1927), Washburn (1927), van Enchevort and Los (2013)
$CuSO_4.5H_2O$	TSC	Block-shaped single crystals, no branching	Druce (1927), Washburn (1927), Takhistov and Chang (2002), van Enchevort and Los (2013)
Na_2SO_4		Depends on substrate wettability—see text	Druce (1927), Shahidzadeh-Bonn et al. (2008)
$FeSO_4$			Druce (1927)
$BaCl_2.2H_2O$		No creeping, overgrows droplet surface	Druce (1927), Washburn (1927), van Enchevort and Los (2013)
KNO_3			Druce (1927), Washburn (1927)
$MnSO_4$			Druce (1927)
NaCl		See Takhistov and Chang (2002) discussion earlier	Druce (1927), Washburn (1927), Shahidzadeh-Bonn et al. (2008), Takhistov and Chang (2002)
$(NH_4)_2Cr_2O_7$	TSC	Similar to $K_2Cr_2O_7$ except needle-like crystals, followed by recrystallization	van Enchevort and Los (2013)
$NaNO_3$			Washburn (1927)
KCl	TSC	Block-shaped single crystals, escapes from beaker	Huang and Huang (1976), van Enchevort and Los (2013)
KH_2PO_4	TSC	Two polymorphs	van Enchevort and Los (2013)
CdI_2	Anticreeping	Crystal pushes droplet inward	van Enchevort and Los (2013)
$KAl(SO_4)_2.12H_2O$	TSC	Block-shaped single crystals, no branching	van Enchevort and Los (2013)
$NH_4H_2PO_4$	TSC	Very little creeping, thick blocky crystals	van Enchevort and Los (2013)

The first ten salts are listed in decreasing "tendency to creep," as described by Druce (1927), which is not in complete agreement with the descriptions of van Enchevort and Los (2013).

grow more quickly and rough surfaces favor BSC, whereas smooth surfaces seem to promote TSC.

5. Some salts show no creeping behavior, and others show anticreeping, in which the deposited crystal pushes the remaining liquid toward the center of the original droplet. This behavior is reminiscent of the contraction stage seen in pillar formation in evaporating polymer droplets by Baldwin and Fairhurst (2014).

As van Enchevort and Los (2013) conclude, "crystal growth by creeping, a quite common phenomenon in nature and the laboratory, is a process of fascinating complexity," which is not currently fully understood.

Glossary

Creeping The phenomenon in which crystals grow from a solution beyond the initial extent of the liquid, spreading up the walls of the container or across the substrate.

Efflorescence (from the French "to flower out") Occurs when hydrated salts lose their water of hydration from the crystal structure to the atmosphere through evaporation, leading to deposition of the anhydrous salt on the surface of the material.

Subfluorescence Similar to efflorescence, but the anhydrous salt is deposited within the medium rather than on the surface, and it can lead to structural damage.

References

Andreae, M.O., Crutzen, P.J., 1997. Atmospheric aerosols: biogeochemical sources and role in atmospheric chemistry. Science 276, 1052–1058.

Baldwin, K.A., Fairhurst, D.J., 2014. The effects of molecular weight, evaporation rate and polymer concentration on pillar formation in drying poly (ethylene oxide) droplets. Colloids Surf. A Physicochem. Eng. Asp. 441, 867–871.

Capes, J.S., Cameron, R.E., 2006. Contact line crystallization to obtain metastable polymorphs. Cryst. Growth Des. 7 (1), 108–112.

Charlesworth, D.H., Marshall, W.R., 1960. Evaporation from drops containing dissolved solids. AIChE J. 6 (1), 9–23.

Cheng, R.J., Blanchard, D.C., Cipriano, R.J., 1988. The formation of hollow sea-salt particles from the evaporation of drops of sea water. Atmos. Res. 22 (1), 15–25.

Cohen, M.D., 1987. Studies of Concentrated Electrolyte Solutions Using the Electrodynamic Balance (Ph.D. thesis). California Institute of Technology.

Cowley, L., 2014. Atmospheric Optics. Available from: <http://www.atoptics.co.uk/> (accessed 07.04.14).

Desarnaud, J., Shahidzadeh-Bonn, N., 2011. Salt crystal purification by deliquesence/crystallization cycling. Europhys. Lett. 95, 48002.

Djikaev, Y.S., Ruckenstein, E., 2008. Thermodynamics of heterogeneous crystal nucleation in contact and immersion modes. J. Phys. Chem. A 112, 11677–11687.

Druce, J.G.F., 1927. The creeping of solutions. J. Pharm., 119, 333.

Du, R., Stone, H., 1996. Evaporatively controlled growth of salt trees. Phys. Rev. E 53 (2), 1994–1997.

Dutcher, C.S., Wexler, A.S., Clegg, S.L., 2010. Surface tensions of inorganic multicomponent aqueous electrolyte solutions and melts. J. Phys. Chem. A 114, 12212–12230.

Ferron, G.A., Soderholm, S.C., 1990. Estimation of the times for evaporation of pure water droplets and for stabilization of salt solution particles. J. Aerosol Sci. 21 (3), 415–429.

Fitzgerald, J.W., 1991. Marine aerosols: a review. Atmos. Environ. 25A (3/4), 533–545.

Ghosal, S., Hemminger, J.C., Bluhm, H., Mun, B.S., Hebenstreit, E.L.D., Ketteler, G., et al., 2005. Electron spectroscopy of aqueous solution interfaces reveals surface enhancement of halides. Science 307 (5709), 563–566.

Hansen, J.E., Arking, A., 1971. Clouds of Venus: evidence for their nature. Science 171 (3972), 669–672.

Hazlehurst Jr., T.H., Martin, H.C., Brewer, L., 1936. The creeping of saturated salt solutions. J. Phys. Chem. 40, 439–452.

Huang, B.-J., Huang, J.-C., 1976. Creeping-film phenomenon of potassium chloride solution. Nature 261, 36–38.

Kinzer, G.D., Gunn, R., 1950. The evaporation, temperature and thermal relaxation of freely falling waterdrops. J. Meteorol. 8 (2), 71–83.

Köhler, H., 1936. The nucleus in and the growth of hygroscopic droplets. Trans. Faraday Soc. 32, 1152–1161.

Kulichikhin, V.G., Elenskii, A.A., Kharlov, A.E., Shabanov, M.P., Yampol'skaya, G.P., 2011. Peculiarities of the surface crystallization of sodium chloride on mucin films. Colloid J. 74 (2), 207–214.

Lee, C.-P., Fang, B.-Y., Wei, Z.-H., 2013. Influence of electrolytes on contact angles of droplets under electric field. Analyst 138, 2372–2377.

Markiewicz, W.J., Petrova, E., Shalygina, O., Almeida, M., Titov, D.V., Limaye, S.S., et al., 2014. Glory on Venus cloud tops and the unknown UV absorber. Icarus 234, 200–203.

Marshall, J.S., Palmer, W.M.c.K., 1948. The distribution of raindrops with size. J. Meteorol. 5, 165–166.

Mezhericher, M., Levy, A., Borde, I., 2010. Theoretical models of single droplet drying kinetics: a review. Drying Technol. 28, 278–293.

Mpemba, E.B., Osborne, D.G., 1969. Cool? Phys. Educ. 4, 172–175.

Mullins, W., Sekerka, R.F., 1963. Morphological stability of a particle growing by diffusion or heat flow. J. Appl. Phys. 34, 323–328.

Poornachary, S.K., Parambil, J.V., Chow, P.S., Tan, R.B.H., Heng, J.Y.Y., 2013. Nucleation of elusive crystal polymorphs at the solution-substrate contact line. Cryst. Growth Des. 13, 1180–1186.

Ranz, W.E., Marshall Jr., W.R., 1954. Evaporation from drops. Chem. Eng. Prog. 48 (3), 141–146.

Rodríguez-Ruiz, I., Hammadi, Z., Grossier, R., Gómez-Morales, J., Veesler, S., 2013. Monitoring picoliter sessile microdroplet dynamics shows that size does not matter. Langmuir 29, 12628–12632.

Satheesh, S.K., Moorthy, K.K., 2004. Radiative effects on natural aerosols: a review. Atmos. Environ. 39, 2089–2110.

Sear, R., 2007. Nucleation at contact lines where fluid-fluid interfaces meet solid surfaces. J. Phys. Condens. Matter 19, 466106.

Shahidzadeh-Bonn, N., Rafaï, S., Bonn, D., Wegdam, G., 2008. Salt crystallization during evaporation: impact of interfacial properties. Langmuir 24, 8599–8605.

Takhistov, P., Chang, H.-C., 2002. Complex stain morphology. Ind. Eng. Chem. Res. 41, 6256–6269.

Tang, I.N., Munkelwitz, H.R., 1983. An investigation of solute nucleation in levitated solution droplets. J. Colloid Interface Sci. 98 (2), 430–438.

Tang, I.N., Munkelwitz, H.R., 1994. Water activities, densities, and refractive indices of aqueous sulfates and sodium nitrate droplets of atmospheric importance. J. Geophys. Res. 99 (D9), 18801–18808.

Thompson, W., 1871. On the equilibrium of vapour at a curved surface of liquid. Philos. Mag. 4, 42 (282), 448–452.

Van Enchevort, W.J.P., Los, J.H., 2013. On the creeping of saturated salt solutions. Cryst. Growth Des. 13, 1838–1848.

Washburn, R., 1927. The creeping of solutions. J. Phys. Chem. 31 (8), 1246–1248.

Chapter 21

DROPLETS WITH SURFACTANTS

Sergey Semenov[1], Victor Starov[2] and Ramon G. Rubio[3]
[1]*Mechanical Engineering Department, Aix-Marseille University, Marseille, France,* [2]*Loughborough University, Department of Chemical Engineering, Loughborough, UK,* [3]*Complutense University of Madrid, Department of Physical Chemistry I, Madrid, Spain*

In this chapter, the general problem of evaporation of sessile drops of pure liquids and surfactant solutions is studied from both the experimental and the theoretical points of view. The kinetics of evaporation of aqueous solutions of the surfactant SILWET L77 on highly hydrophobic substrates has been studied experimentally for a wide range of surfactant concentrations, temperatures, and relative air humidity. The influence of the surfactant on the diffusion limited evaporation of a droplet has been analyzed. Five distinct stages of the process of droplet's spreading/evaporation were observed, three of which are modeled theoretically.

Numerical tools have been used for the simulation of a nonisothermal problem of evaporating sessile drops of pure liquids. The results of these simulations have been used for the development of a mathematical model of evaporating sessile drops. The model uses only one fitting parameter—the average temperature of the droplet surface—which is usually not accessible experimentally. Afterward the theory developed for pure liquids is successfully applied for surfactant solutions. The difficulty and necessity of predicting advancing and receding contact angles for further development of the theory are discussed.

Droplet Wetting and Evaporation. DOI: http://dx.doi.org/10.1016/B978-0-12-800722-8.00021-7

21.1 Introduction

21.1.1 Industrial Applications of Evaporating Fluids

The spreading and evaporation of sessile droplets of different (simple and complex) liquids have a broad variety of industrial applications, including spray drying and production of fine powders; spray cooling; fuel preparation; air humidifying; heat exchangers; drying in evaporation chambers of air conditioning systems; fire extinguishing; fuel spray auto ignition (diesel); solid surface templates from evaporation of nanofluid drops (coffee ring effect); spraying of herbicides and pesticides on hydrophobic leaves (Ivanova and Starov, 2011); painting, coating, and inkjet printing; printing microelectromechanical systems and pixel arrays, microlens manufacturing, and stretching of single DNA molecules; biological tissue engineering (Campbell and Weiss, 2007); surface modification; and biomedical applications, such as blood analysis (Sobac and Brutin, 2014) and surfactant replacement therapy.

In general, industrial interest is always focused on some particular aspects of the preceding applications, such as development of new technologies and extraction of their profitable potential, optimization of processes, economy of resources (e.g., ink, surfactant, herbicide, pesticide, other acting substance), and enhancement of some effects (e.g., increase of the droplet spreading area, speed of spreading, quality of coatings, and so on).

Because of such a wide range of actual and potential industrial applications, drops evaporation has been under investigation for many years. Different research studies (theoretical, experimental, and numerical) have been focused on different aspects of the phenomenon and on different experimental conditions: constant pressure and temperature, elevated pressure, fast compression, still gas atmosphere, turbulent reacting flows, and strong and weak pinning substrates. Researchers considered the influence of different physical processes: heat transfer (radiation, conduction, and convection), mass diffusion in bi- and multicomponent fluids, droplets interaction in sprays (see Chapter 20), Marangoni (thermo- and soluto-capillary) convection in droplets, surfactant adsorption, and action of surface forces in the vicinity of the apparent three-phase contact line. However, a complete understanding and our ability to predict experimentally observed phenomena are still lacking, especially for complex fluids (surfactant solutions, nanosuspensions, biological fluids).

21.1.2 Surfactants

The word "surfactant" stands for a surface active agent, a substance that efficiently lowers the surface tension of a liquid or interfacial tension between two liquids or a liquid and a solid. It is a class of mostly organic compounds whose molecules are amphiphilic, which means that they contain both hydrophilic (water-soluble component) and hydrophobic (water-insoluble component) groups. A key characteristic of a surfactant in solution is its surface activity—that is, the amount adsorbed at the interface relative to the solvent. It can be expressed as a derivative of the interfacial tension, γ, with respect to the surfactant bulk concentration, c, as c tends to zero.

Depending on the geometry of the molecule, both hydrophilic and hydrophobic groups can be schematically represented with a "head" or a "tail," as shown in Figure 21.1A. Further classification is based on the nature of a polar hydrophilic part of the molecule (usually "head"). It can be one of four types: anionic (bears a negatively charged group), cationic (positively charged group), zwitterionic (contains both anionic and cationic groups), or nonionic (bears no charge, but is polar). The hydrophobic part of the molecule is usually a hydrocarbon or other carbon-based chain. A significant exception is the trisiloxane surfactants, whose hydrophobic part is a trisiloxane group. The orientation of surfactant molecules on the surface of an aqueous solution is such that a hydrophobic tail is sticking out from the solution into the gaseous phase, while the hydrophilic head is submerged into the solution and surrounded by the water phase (Figure 21.1B). In this way the molecule occupies a stable position at the minimum of potential energy of its interaction with two phases. The same principle of potential energy minimization governs the processes of molecular aggregation.

At a low surfactant concentration, its molecules exist in a solution in the form of individual monomers (Figure 21.1A). They diffuse across the bulk and adsorb on interfaces, creating molecular monolayers. At increased surfactant concentrations, different molecular aggregates form in addition to monomers: spherical micelles (Figure 21.2A), bilayers (Figure 21.2C), and spherical vesicles

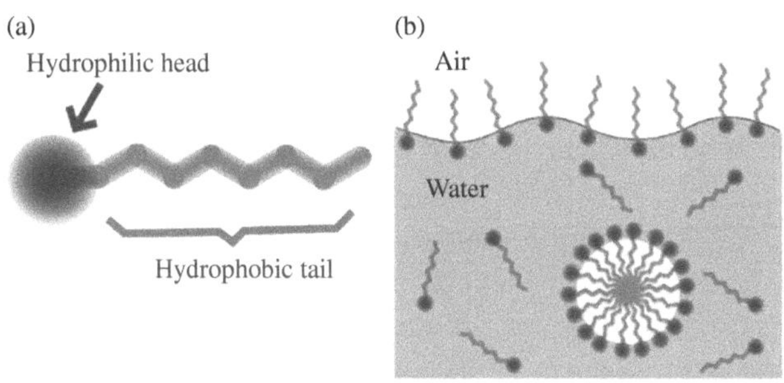

FIGURE 21.1 (A) Schematic representation of a surfactant molecule. (B) Preferred orientation of surfactant molecules on the surface of an aqueous solution and in a micelle.

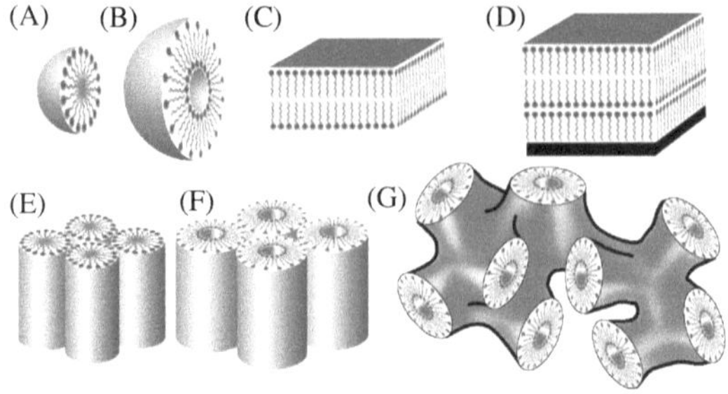

FIGURE 21.2 Surfactant aggregates and solution phases: (A) spherical micelle; (B) spherical vesicle; (C) bilayer aggregate; (D) multilayered lamella; (E) cylindrical micelle; (F) cylindrical vesicle; (G) sponge-like L_3 phase.

(Figure 21.2B). These aggregates are formed when surfactant concentration exceeds some critical value, which is called a critical micelle concentration (CMC) or a critical aggregation concentration (CAC). Adsorption of additional layers of surfactant molecules on interfaces and the formation of multilayered lamellar structures (Figure 21.2D) and cylindrical aggregates (Figure 21.2E and F) are also possible. Further increase of the surfactant concentration can lead to a formation of liquid crystalline phases in the bulk of solution—for example, dense packing of micelles (not shown), bicontinuous cubic phase (not shown), and sponge-like L_3 phase (Figure 21.2G).

The characteristics of surfactant aggregates (size and shape) are determined by several parameters of the solution, such as concentration of surfactants and salts, pH of the solution, and temperature. Based on studied ranges of solution parameters, multidimensional phase diagrams of surfactant aggregates can be built.

21.2 Spreading and evaporation of fluid droplets

21.2.1 Divergence of Evaporation Flux

The problems of spreading and evaporation of droplets are linked to a number of factors. For example, the conventional diffusion model of evaporation (Picknett and Bexon, 1977) of a sessile droplet with an explicit three-phase contact line introduces an unphysical behavior at the droplet edge: the divergence of local evaporation flux (Hu and Larson, 2005).

Other physical models can be used in order to deal with these problem. For example, the kinetic model of evaporation assumes that the evaporation rate is limited by the rate of molecules transfer across the liquid–gas interface (e.g., evaporation into vacuum) and not by the vapor diffusion process. Rate of evaporation in the kinetic model can be calculated by using the Hertz-Knudsen-Langmuir equation (based on the kinetic theory of gases) and its modifications containing mass accommodation coefficients (or evaporation/condensation coefficients; Sazhin et al., 2007; Semenov et al., 2013a, p. 143). The kinetic model is free of the problem of divergence of evaporation flux, but in most real applications of drying drops, the evaporation regime is far from the pure kinetic model, and diffusion limited evaporation dominates.

Thus, an obvious and logical way out of the situation is to apply a hybrid diffusive/kinetic model of evaporation, in which the evaporation rate is limited by both transport phenomena simultaneously: by vapor diffusion and by the rate of molecules transfer across the interface. Virtually it means that the evaporation rate is limited by the slowest of these two processes. This model is more complex but more consistent with the real physics of the phenomenon in question, especially when the physics of the contact line is considered in more detail.

21.2.2 Singularity of Viscous Stress

Another difficulty arises in hydrodynamics when the no-slip boundary condition at the solid–liquid interface is applied for problems with a moving three-phase contact line. A moving contact line is incompatible with the no-slip boundary condition. In this case a well-known problem of singularity of viscous stress appears. There are a variety of approaches to deal with this. For example, an application of a generalized Navier (instead of no-slip) boundary condition introduces a slip velocity at the solid surface and makes it possible to prevent stress singularity and to model the motion of the explicit contact line. The value of the slip velocity locally depends on applied shear viscous stress and uncompensated Young stress in the vicinity of a contact line. This approach is well discussed in a review paper by Qian et al. (2006).

Another approach is to introduce an adsorbed liquid film that is formed on a solid surface in contact with a vapor phase in front of a moving contact line (precursor film). In this way the explicit three-phase contact line is eliminated and replaced with a smooth transition of the liquid bulk into the adsorbed liquid layer. The absence of the explicit contact line means the absence of the singularities associated with it. Usually the adsorption of thin liquid layers goes in line with the notion of disjoining/conjoining pressure, which characterizes the action of surface forces in thin liquid films. The isotherm of disjoining/conjoining pressure determines the equilibrium thickness of the adsorbed liquid layer for a specified vapor pressure above the solid–gas interface. This approach was introduced by Derjaguin (1940) and has been applied in many studies afterward (Starov et al., 2007; Ajaev et al., 2010). Savva and Kalliadasis (2011) compared the slip and precursor film models under the limit of quasi-static spreading of a two-dimensional droplet and showed the equivalence relations for these two models.

These improved models resolve the singularity problems, but at the same time they substantially complicate the solution and introduce new unknown parameters or coefficients that are difficult or impossible to measure unambiguously (evaporation coefficient, slip length, disjoining pressure isotherm).

21.2.3 Contact Angle Prediction

The most complex problem in the field is the difficulty of predicting the static and dynamic contact angle and its hysteresis. Even for pure liquids, it depends on a number of physical phenomena and hardly measurable parameters—for example, roughness of the contact surface, reorientation of surface molecules under contact with wetting liquid, chemical inhomogeneity of the solid surface, molecular kinetics and action of surface forces in the vicinity of the contact line. A significant static contact angle hysteresis has been reported also for pure liquids, even on sufficiently smooth and chemically homogeneous substrates (Extrand, 2002) and for definitely molecularly smooth surfaces—free liquid films (Platikanov et al., 2003). Suciu and Goto (2010) discussed the molecular nature of contact angle hysteresis and its correlation with the hysteresis of adhesion between two surfaces.

A number of researchers (Starov et al., 2007; Diaz et al., 2010, p. 574; Boinovich and Emelyanenko, 2011) demonstrated that the contact angle hysteresis for a pure liquid on a smooth homogeneous solid substrate can be explained by a theoretical model based on the concept of disjoining pressure and an adsorbed liquid film on a solid surface.

Different possible mechanisms responsible for the hysteresis of the contact angle and their combinations are yet to be understood. Without that understanding, the prediction of the contact angle is a very complicated scientific challenge.

On top of these challenges, additional ones appear when surfactant molecules are present in the systems under consideration. The addition of surfactants dramatically changes the energy of interfaces (liquid–gas, liquid–solid, solid–gas), reducing interfacial tensions and thus affecting the values of advancing and receding contact angles and the kinetics of droplets' spreading and evaporation. It is easy to contaminate interfaces with surfactants and difficult to clean them. For this reason surfactants can unintentionally be present in some experiments in the form of contaminants. It is important to remember this when you are interpreting experimental results.

If surfactants are intentionally added into a liquid for the purpose of study, then processes of surfactant adsorption on different surfaces of the experimental system make it difficult to control the instant concentration (and not the initial one) of the surfactant during the experiment. In order to precisely model such variations of the surfactant concentration, it would require performing extensive computations, which are not always possible due to the absence of all required input data and parameters of the experimental system (such as kinetic rates of adsorption on different surfaces). In light of this, both theoretical predictions and experimental measurements of contact angles for both pure substances and surfactant solutions can be problematic.

21.2.4 Superspreading

Surfactants increase the spreading radius and thus indirectly enhance the evaporation rate, because it is proportional to the droplet radius. Also, surfactants can strongly affect the rate of spreading and the uniformity of this process. In some cases the superspreading behavior is observed experimentally: a very rapid spreading on highly hydrophobic surfaces (Venzmer, 2011; Nikolov and Wasan, 2014). Those surfactants that produce the superspreading effect are called superspreaders (usually siloxane-based surfactants). Trisiloxane surfactants are characterized by a critical wetting concentration (CWC; Ivanova et al., 2010). CWC is the concentration at which an abrupt transition from a partial wetting (finite measurable contact angle) to a complete wetting (vanishing contact angle) is observed for moderately hydrophobic surfaces. Thus, for concentrations above the CWC, a superspreading phenomenon is observed.

The nature of superspreading phenomenon has been under study for many years. Different candidate mechanisms have been proposed to explain superspreading (Venzmer, 2011). Most of the suggested hypotheses are controversial, and there is no a general agreement in the scientific society regarding the problem of superspreading. There are still a lot of unanswered questions. For more information on superspreading, the reader is referred to Chapter 6.

Thus, the roles of surfactants in the spreading and evaporation processes and corresponding underlying mechanisms are not clear to the end and have to be understood in order to be able to optimize their applications. Modeling the problems of evaporating surfactant solutions is a very complex task and therefore often requires simplifications and different methodologies (theoretical and experimental approaches, numerical and molecular dynamics simulations).

21.3 Sessile droplets of surfactant solutions

The presence of surfactant in a liquid results in the adsorption of surfactant molecules on all the interfaces of the system (liquid–gas and solid–liquid). It causes the lowering of interfacial tensions and a change in the equilibrium contact angle, until the CMC/CAC is reached (the adsorption at the solid–liquid interface may saturate below the CMC!). If surfactant concentration in the bulk is above CMC/CAC, then surfactant molecules form aggregates in the solution bulk rather than continue adsorbing at interfaces. It means that for concentrations above CMC/CAC, the surface tension of the solution no longer changes with the bulk concentration, and the contact angle remains constant.

However, the problem is more complex in the case of an evaporating sessile droplet, because the surfactant concentration in the droplet bulk constantly changes, not only due to the unsteady process of adsorption but also due to solvent evaporation as well. Thus, the bulk concentration is slightly decreased by adsorption. The evaporation of a solvent works in an opposite direction: it decreases the droplet volume and thus increases the surfactant bulk concentration. The kinetics of adsorption at both interfaces compete with the evaporation one, usually being slower (Ritacco et al., 2010a, p. 199, 2010b, p. 204).

These two competing processes, together with the phenomena of contact angle hysteresis and contact line pinning, can result in a very complex evolution of the contact angle and the contact line position in the course of droplet evaporation.

Even for pure liquids, it is difficult to predict the advancing and receding contact angles. Therefore, instead of predicting it, for the purpose of simplified modeling, we use the experimentally measured advancing and receding contact angles and their dependence on surfactant concentrations. In this respect, the values of these contact angles are accepted as arbitrary parameters of the following model.

21.3.1 Influence of Surfactants on Droplet Evaporation

How does the surfactant influence the process of evaporation of a sessile droplet? One way is the change in the saturated vapor pressure just above the surface of the surfactant solution. Surfactant molecules form a monolayer on the liquid–vapor interface, so some area of that interface will be covered by heavy nonevaporating surfactant molecules. This could possibly affect the evaporation rate of the solvent from such a surface. However, the influence of the surfactant monolayer at the liquid–vapor interface on the droplet evaporation rate was not detected in our experiments (Agogo et al., 2012; Semenov et al., 2013b, p. 10028), which means that this effect is weak enough to be less than experimental error allows. Based on this experimental evidence, one can conclude that the direct influence of the surfactant monolayer on the process of diffusive evaporation is negligible. It means that in this case the "bottleneck" that limits the evaporation rate is not the surfactant monolayer on the surface of the solution but the surrounding atmosphere through which vapor diffuses into the ambient environment.

Another, and the most effective, way the surfactant influences the droplet evaporation rate is indirectly through the change in the droplet geometry (lowering the contact angle and increasing the droplet spreading area). It is well known that in the diffusive regime of evaporation, the droplet evaporation rate is directly proportional to the linear size of the droplet (the contact line radius or the perimeter of a sessile droplet). Such a relation is a consequence of the diffusive nature of the vapor transport and has nothing to do with a distribution of the local evaporation flux along the droplet's surface, as was demonstrated earlier (Hu and Larson, 2002; Semenov et al., 2010, 2014). Therefore, if the surfactant is present in a sessile droplet under diffusion limited evaporation, then it lowers the contact angle and increases the contact line radius, thus increasing the evaporation rate.

The change of the droplet geometry affects not only the diffusive processes in surrounding air but also the process of heat transfer in the droplet bulk and in the substrate. The latent heat of vaporization requires a consumption of heat due to evaporation, which causes the cooling of the droplet surface. At the same time, heat transfer through the droplet and the substrate replenishes this evaporative heat loss. The balance between these two processes determines the temperature of the droplet surface and consequently the pressure of saturated vapor above the droplet surface that in turn, along with ambient air humidity, controls the rate of vapor diffusion.

Yet another possible mechanism affecting the heat transfer in the droplet is the thermo-capillary Marangoni convection inside the droplet. The important outcome of Marangoni convection is the convective heat flux in the droplet bulk, which can facilitate the heat transfer from the substrate to the droplet surface and consequently raise the average temperature of the droplet surface and increase the evaporation rate. In this respect, the presence of a surfactant at the droplet's surface creates soluto-capillary stresses, which can effectively suppress thermo-capillary convection in the droplet, reducing the rate of heat transfer from the substrate to the droplet surface, and thus reducing the evaporation rate. However, this effect of the suppression of Marangoni convection is weak and can be masked by other, more pronounced accompanying phenomena, such as the droplet's geometry change.

To summarize, we can conclude that *within the studied range of surfactant concentrations* (up to 12.5 CAC of SILWET L77) (Agogo et al., 2012; Semenov et al., 2013b, p. 10028), the process of a diffusion limited evaporation of a sessile droplet is fully determined by the drop's geometry and by the temperature distribution along its surface. Thus, if we know the temperature distribution and the contact angle at each moment of the evaporation process, then we can predict the evaporation rate and the time evolution of the droplet's shape based on a solution of the vapor diffusion equation that is not influenced by the surfactant concentration. This conclusion allows us to continue our research by studying pure liquids in the first place and then generalizing or extending the obtained results to the case of nonzero concentrations of surfactants.

21.3.2 Nonisothermal Numerical Model of Evaporation of Sessile Drops of Pure Liquids

The numerical model developed by Semenov et al. (2010, 2013a, p. 143) aimed at the prediction of the evaporation rate of a sessile droplet of pure water in a nonisothermal regime of evaporation but without forced heating or cooling of the substrate. The model took into account the vapor diffusion into ambient atmosphere, Stefan flow in the air generated by vapor production, evaporative cooling, thermal Marangoni convection in the drop, heat conduction in the substrate, and heat conduction and convection in the air and the droplet. Figure 21.3 shows the relative impact of different thermal phenomena (latent heat of vaporization, thermal Marangoni convection, convective heat flux in the air due to Stefan flow) on the normalized total evaporation rate of a droplet of pure water on top of a highly heat-conductive substrate (copper) as a function of the contact line radius, L (Semenov et al., 2013a, p. 143). Note: Figure 21.3 is based on a hybrid diffusive-kinetic model of evaporation, and droplet sizes $L > 1\ \mu m$ correspond to a pure diffusive regime of evaporation (Semenov et al., 2013a, p. 143). The effect of evaporative cooling (see squares in Figure 21.3) reduces the evaporation rate

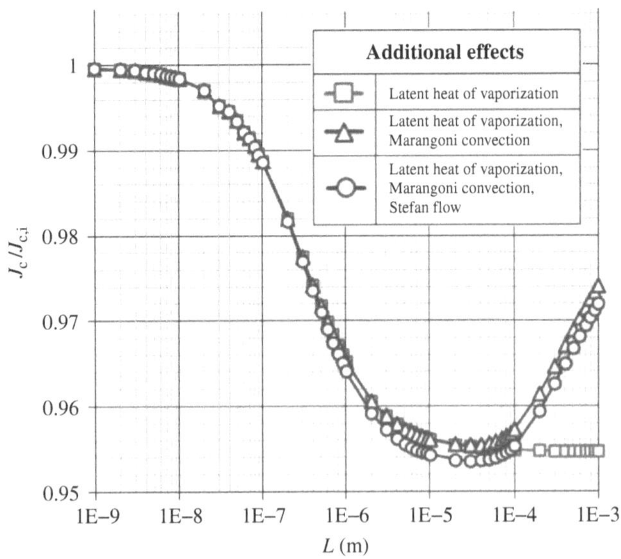

FIGURE 21.3 Influence of different thermal phenomena on the droplet's evaporation rate. J_c and $J_{c,i}$ are the total molar flux of the droplet evaporation in the nonisothermal model and the isothermal one, respectively. *Redrawn from Semenov et al. (2013a, p. 143).*

compared to the isothermal model. For the droplets over 0.1 mm in size, Marangoni convection (triangles in Figure 21.3) slightly increases the evaporation rate due to enhancement of the heat transfer from the substrate to the droplet surface. The effect of convective heat transfer due to Stefan flow (circles in Figure 21.3) is negligible for the whole range of sizes of the droplets. Our numerical model shows that the preceding reduction in the evaporation rate due to thermal effects becomes more pronounced for lower heat conductivities of the substrate (Figure 21.4A).

All of these thermal effects eventually lead to a reduction in temperature of the droplet surface. In order to understand how this temperature reduction affects the droplet evaporation in comparison with the isothermal model, we extracted from our numerical model an additional characteristic of the problem: the average temperature of the droplet surface: $T_{av} = \frac{1}{S}\int_S T \, dS$, where S is the area of the droplet surface, and T is its local temperature. In *isothermal* models the temperature is the same everywhere in the system and equal to the ambient one, T_{amb}. Therefore, the saturated vapor concentration at the droplet surface, c_{sat} (mol m^{-3}), is a function of T_{amb}. Thus, according to the *isothermal* diffusive model, the theoretical evaporation rate, J_i (i stands for "isothermal"), of the sessile droplet is

$$J_i = J_{\pi/2}(L, T_{amb})F(\theta) = 2\pi D(c_{sat}(T_{amb}) - c_{amb})LF(\theta),$$
$$\text{or shortly } J_i/J_{\pi/2}(L, T_{amb}) = F(\theta), \tag{21.1}$$

where $J_{\pi/2}(L, T_{amb})$ is the droplet's evaporation rate for the contact angle $\theta = \pi/2$, D is the diffusion coefficient of vapor in air, L is the contact line radius, and c_{amb}

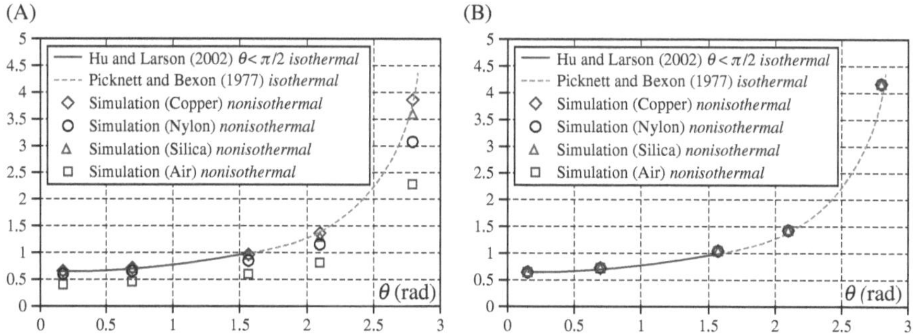

FIGURE 21.4 The droplet's evaporation rate, J, as a function of contact angle, θ, for different heat conductivities of the substrate. J is normalized with (A) $J_{\pi/2}(L, T_{amb})$, as in Eq. (21.1); and (B) $J_{\pi/2}(L, T_{av})$, as in Eq. (21.2). Theoretical *isothermal* curves ($T_{av} = T_{amb}$, $J = J_i$) correspond to a function $F(\theta)$. For *nonisothermal* simulation markers, T_{av} and J are directly extracted from our nonisothermal numerical model. *Redrawn from Semenov et al. (2010).*

is the ambient vapor concentration. $F(\theta)$ is the function of the contact angle determined analytically by Picknett and Bexon (1977) and equivalent to another one, $F_P(\theta) = 2F(\theta)$, derived by Popov (2005):

$$F(\theta) = \begin{cases} (0.6366 \cdot \theta + 0.09591 \cdot \theta^2 - 0.06144 \cdot \theta^3)/(\sin\theta), & 0 \le \theta < \pi/18 \\ (0.00008957 + 0.6333 \cdot \theta + 0.1160 \cdot \theta^2 & \pi/18 \le \theta \le \pi, \\ -0.08878 \cdot \theta^3 + 0.01033 \cdot \theta^4)/(\sin\theta), & \end{cases}$$

$$F_P(\theta) = \frac{\sin\theta}{1+\cos\theta} + 4\int_0^\infty \frac{1+\cosh(2\theta\tau)}{\sinh(2\pi\tau)} \tanh[(\pi-\theta)\tau]d\tau.$$

Note: $F(\pi/2) = 1$. Using T_{av}, we modified Eq. (21.1) by making c_{sat} a function of T_{av} (below m stands for "modified"):

$$J_m = J_{\pi/2}(L, T_{av})F(\theta) = 2\pi D(c_{sat}(T_{av}) - c_{amb})LF(\theta), \\ \text{or shortly } J_m/J_{\pi/2}(L, T_{av}) = F(\theta). \tag{21.2}$$

The results are presented in Figure 21.4. As one can see from Figure 21.4A, markers (evaporation rates, J, extracted from our nonisothermal numerical simulations) do not coincide with the isothermal curves. This means that J_i and J have different values (see Eq. (21.1)). However, modified normalization (Eq. (21.2); Figure 21.4B) makes all simulation markers appear on top of the isothermal curves. This means that J_m and J have very close values. Thus, the empirically modified formula, Eq. (21.2), gives an excellent (within numerical accuracy) prediction of the nonisothermal evaporation rate, J, and we can write

$$J = J_m = J_{\pi/2}(L, T_{av})F(\theta), \quad \text{where } J_{\pi/2}(L, T_{av}) = 2\pi D(c_{sat}(T_{av}) - c_{amb})L. \tag{21.3}$$

Note: T_{av} equals to T_{amb} for the *isothermal* model; therefore, Eq. (21.1) is the isothermal limit of Eq. (21.3).

We conclude that the average temperature, T_{av}, of the droplet surface well determines the diffusion limited evaporation of a sessile droplet. The nonuniformity of the temperature along the droplet surface does not play a significant role in the prediction of the total evaporation rate, J.

21.3.3 Mathematical Model of Nonisothermal Evaporation of Sessile Drops of Pure Liquids

Let us consider the temperature of a droplet from the experimental point of view. In most of the published experimental results the temperature of the droplet surface is not reported because measuring it is a complicated task. However, there are experiments where the temperature inside of a sessile droplet was measured directly with the help of a tiny thermocouple (David et al., 2007).

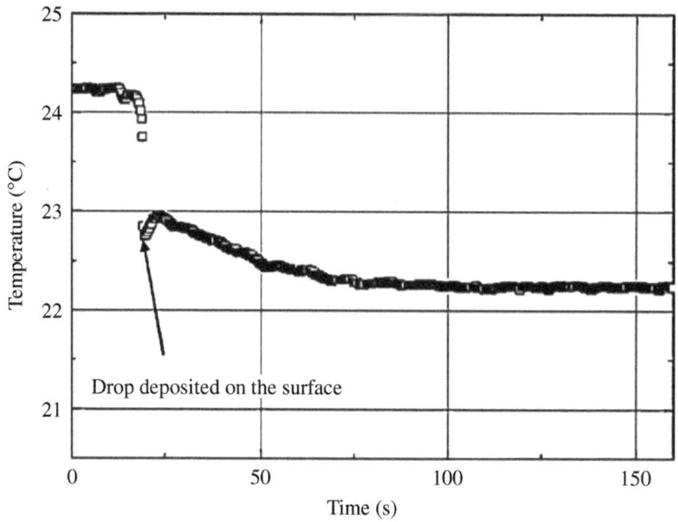

FIGURE 21.5 Temperature measurements inside of an evaporating sessile droplet. *Redrawn from David et al. (2007).*

Figure 21.5 shows an example of such measurements. One can clearly see the moment of the droplet deposition, when the temperature suddenly drops. During the whole evaporation time, the temperature inside of the droplet remains below the ambient one and almost constant. This observation allows us to make a hypothesis that the average temperature, T_{av}, of the droplet surface also remains nearly constant during the whole process of the droplet evaporation. This hypothesis was adopted in the following theory.

Semenov et al. (2011, 2012, 2014) developed an analytical description of time evolution of the droplet shape, taking into account the following assumptions: (i) the evaporation is a diffusion limited quasi-steady process; (ii) the advancing, θ_{ad}, and receding, θ_r, contact angles are taken as arbitrary parameters of the model and determined experimentally; (iii) the nonisothermal evaporation rate is described by the Eq. (21.3); (iv) the average temperature of the droplet surface, T_{av}, remains constant during the whole process of evaporation, $T_{av} = \text{const}$; (v) T_{av} is adopted as a fitting parameter, because it is not available from most experimental data; and (vi) the droplet shape is a spherical cap, because surface forces dominate over the body forces.

Based on the preceding six assumptions, the volume of the droplet can be expressed in the following form:

$$V = f(\theta)L^3, \quad \text{where } f(\theta) = \frac{\pi}{3}\,\frac{(1-\cos\theta)^2(2+\cos\theta)}{\sin^3\theta}. \tag{21.4}$$

Noting that $\rho dV/dt = -JM$ (where M is the molar mass of evaporating substance, and ρ is the mass density of the droplet) and using Eqs (21.3) and (21.4), we get

$$\frac{d(f(\theta)L^3)}{dt} = -2\pi\frac{DM}{\rho}(c_{sat}(T_{av}) - c_{amb})LF(\theta), \text{ or simply } \frac{d(f(\theta)L^3)}{dt} = -\beta LF(\theta) \tag{21.5}$$

where $\beta = 2\pi\frac{MD}{\rho}(c_{sat}(T_{av}) - c_{amb})$ is a new fitting parameter because it depends only on one unknown quantity T_{av}, which was previously adopted as a fitting parameter. Thus, Eq. (21.5) contains only two unknown time-dependent variables: L and θ. If the time dependence of one of them is known, then the time dependence of another one can be obtained by solving this equation.

Figure 21.6 schematically represents possible stages of the spreading/evaporation process of a sessile droplet with pinning/depinning (A) and for surfactant solutions (B). We will concentrate our attention on the three longest stages of evaporation: (I), (S), and (II). The short initial spreading stage and pinning/depinning

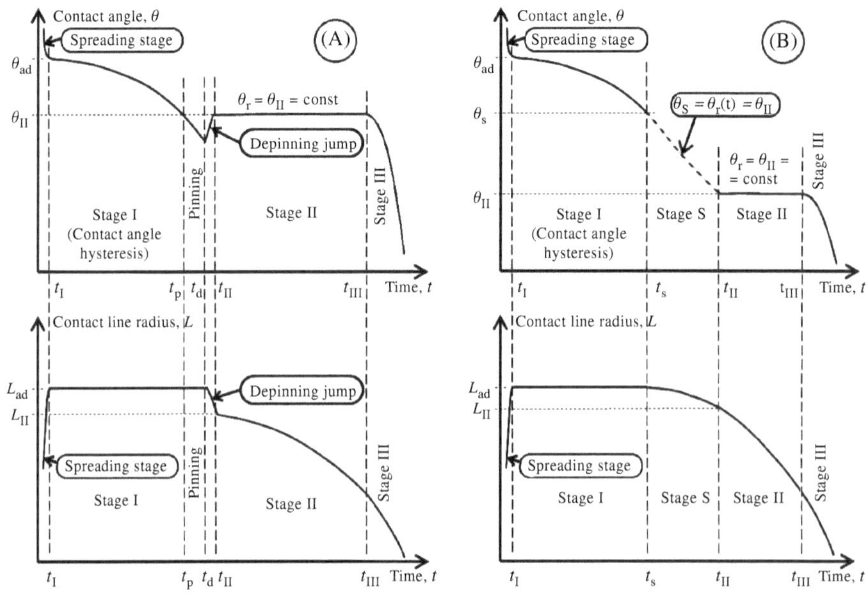

FIGURE 21.6 Schematic representation of spreading/evaporation stages for sessile droplets: (A) with contact line pinning followed by a depinning jump; (B) with intermediate stage (S), when receding contact angle, θ_r, is time-dependent and fully determined by the *surfactant* adsorption on interfaces. Times t_I, t_S, t_{II}, and t_{III} correspond to the beginnings of stages (I), (S), (II), and (III), respectively. Times t_p and t_d correspond to the beginning of pinning and depinning, respectively. $L_{ad} = L(t_I)$, $L_{II} = L(t_{II})$, $\theta_{ad} = \theta(t_I)$, $\theta_S = \theta_r(t_S)$, $\theta_{II} = \theta_r(t_{II})$.

are not considered here, and neither is the last evaporation stage (III), where both contact angle, θ, and contact line radius, L, quickly decrease down to zero value.

For pure liquids on solid, smooth, and homogeneous substrates, the two simplest regimes of evaporation can be described by Eq. (21.5): (I) pinned contact line, $L = \text{const} = L_{ad} = L(t_I)$, contact angle reduces from its advancing, θ_{ad}, down to its receding, θ_r, value (contact angle hysteresis); or (II) constant receding contact angle, $\theta = \text{const} = \theta_r(t_{II}) = \theta_{II}$, and reducing contact line radius, L (Figure 21.6A). These two particular solutions are presented here in a nondimensional form (for more details, see Semenov et al., 2014):

$$\text{(I)}\ L = \text{const} = L_{ad} \Rightarrow K(\theta) = \tau_I, \tag{21.6}$$

where $L_{ad} = L(t_I)$ is the maximum value of L at time t_I when the stage of droplet spreading is finished, and the static advancing contact angle, $\theta_{ad} = \theta(t_I)$, is reached; $K(\theta) = \int_{\theta}^{\pi/2} \frac{f'(\theta)}{F(\theta)} d\theta$, $f'(\theta)$ is the first derivative of $f(\theta)$; $\tau_I = \frac{t - t_I}{L_{ad}^2}\beta + K(\theta_{ad})$ is a nondimensional time for stage (I), and t is a real time.

$$\text{(II)}\ \theta = \text{const} = \theta_{II} = \theta_r(t_{II}) \Rightarrow \ell_{II} = \sqrt{1 - \tau_{II}}, \tag{21.7}$$

where $\tau_{II} = \frac{2F(\theta_{II})}{3f(\theta_{II})} \frac{(t - t_{II})}{L_{II}^2}\beta$ is a nondimensional time for stage (II), $\ell_{II} = \frac{L}{L_{II}}$, $L_{II} = L(t_{II})$ is the value of L at the moment t_{II} when the contact line starts to recede with a constant receding contact angle (for complex liquids, L_{II} can be different from L_{ad}—e.g., due to depinning jump (Figure 21.6A) or due to the presence of an intermediate stage between stages (I) and (II) (Figure 21.6B)). The theory for stages (I) and (II) is validated against experimental data from the literature for pure water on different substrates. The results, presented in Figure 21.7, demonstrate a good agreement with experiments.

21.3.4 Application of the Theory Developed for Pure Liquids to Surfactant Solutions

We performed an experimental study on the kinetics of the evaporation of aqueous solutions of the surfactant SILWET L77 (known as a superspreader) on highly hydrophobic substrates (Agogo et al., 2012; Semenov et al., 2013b, p. 10028, 2014). A pH 7.0 buffer was used in this study as a solvent to prevent hydrolysis of the SILWET L77. Silicon wafers covered by amorphous Teflon were used as hydrophobic substrates. The macroscopic contact angle of either pure water or a buffer solution was (118 ± 2°) on those substrates. The volume of deposited drops did not exceed 4 μL in order to avoid the influence of Earth's gravity on the shape of the drop, which was assumed to be a spherical cap. Five independent measurements were performed for each experimental point reported, and the average was used. The studied range of surfactant concentrations was

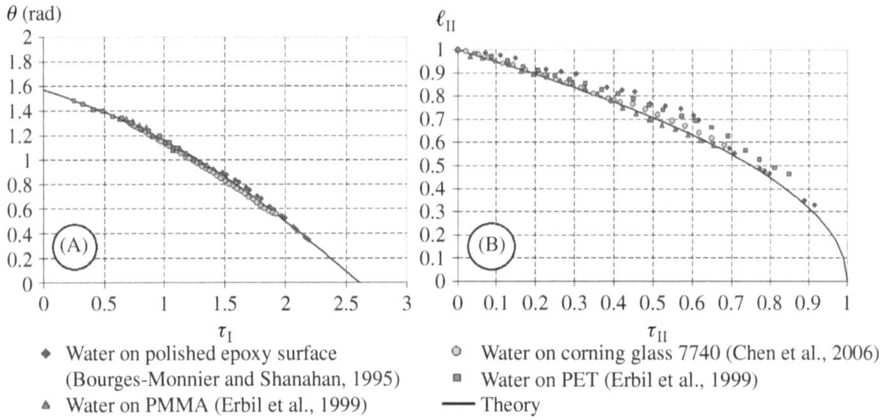

FIGURE 21.7 Validation of mathematical model of nonisothermal evaporation of sessile droplets of water against experimental data available from literature: (A) regime with $L = \text{const}$, (B) regime with $\theta_r = \text{const}$. *Redrawn from Semenov et al. (2012).*

from 0 (pure water) to 12.5 CAC (for SILWET L77 CAC $= 0.1\ \text{g L}^{-1}$; Wagner et al., 1997). The experiments were performed under controlled conditions (different ambient temperatures and air humidity).

The process of evaporation of surfactant solutions is different from the one for pure liquids in the same way that surfactant adsorption on all interfaces (liquid–air, solid–liquid, solid–air) determines the values of advancing and receding contact angles. Bigger value of initial concentration, C, of the surfactant in the solution will result in a smaller value of advancing contact angle until a critical wetting concentration (CWC $= 0.40\ \text{mmol L}^{-1} = 0.25\ \text{g L}^{-1} = 2.5$ CAC; Ivanova et al., 2010) is reached (Figure 21.8).

Immediately after the droplet deposition, its volume starts to decrease due to evaporation, resulting in an increase in the bulk concentration of the surfactant. After the spreading stage is finished, the increasing bulk concentration raises the amount of adsorbed surfactant molecules on the liquid–air and solid–liquid interfaces, which reduces corresponding interfacial tensions and thus reduces the value of the receding contact angle, θ_r. This process makes the receding contact angle a function of time: $\theta_r = \theta_r(t)$.

As soon as the contact angle $\theta(t)$ during stage I reaches its receding value $\theta_r(t)$, the contact line starts to recede, and evaporation proceeds to stage (S) (S stands for "surfactant"), as shown in Figure 21.6B. During that stage, the contact angle equals its time-dependent receding value, $\theta(t) = \theta_r(t)$, which is completely determined by the surfactant adsorption, and continues to reduce from $\theta_S = \theta_r(t_S)$ down to its minimum value $\theta_{II} = \theta_r(t_{II})$. After θ_{II} is reached, the receding contact angle remains constant, and evaporation proceeds to stage (II). Reaching the

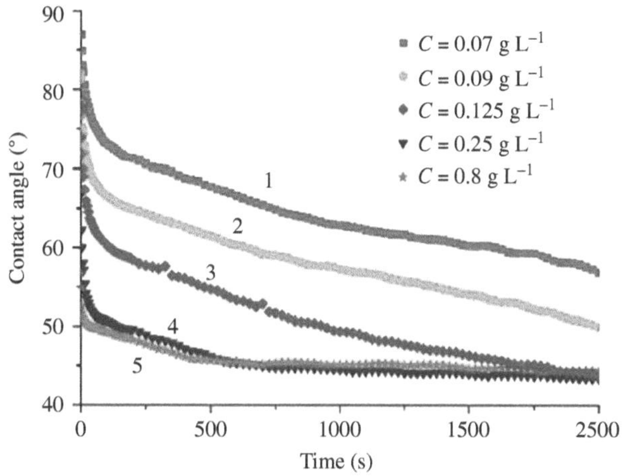

FIGURE 21.8 Time dependence of the contact angle for different concentrations of SILWET L77 surfactant at 24°C and 50% relative humidity: (i) $C = 0.07\ \mathrm{g\,L^{-1}}$; (ii) $C = 0.09\ \mathrm{g\,L^{-1}}$; (iii) $C = 0.125\ \mathrm{g\,L^{-1}}$; (iv) $C = 0.25\ \mathrm{g\,L^{-1}}$ (CWC); (v) $C = 0.8\ \mathrm{g\,L^{-1}}$. The final third stage is not shown. *Redrawn from Semenov et al. (2013b, p. 10028).*

minimum value of the receding contact angle, θ_{II}, indicates the stopping of adsorption of the surfactant molecules on interfaces, because the formation of surfactant aggregates in the bulk of the solution becomes energetically more favorable than the adsorption.

If the initial surfactant concentration, C, is high enough (supposedly above CWC), then the time-dependent value of the receding contact angle reaches its minimum before the end of stage (I). That is, by the end of stage (I), when the contact angle $\theta(t)$ reaches its receding value $\theta_r(t)$, the receding contact angle is already at its minimum: $\theta_r(t) = \theta_{\mathrm{II}}$. In this case, the intermediate stage (S) is absent, and stage (II) with a constant receding contact angle follows stage (I), as in the case of pure liquids, but with a lower value of the receding contact angle.

In order to describe the evolution of the droplet shape during stage (S), we need to know the time dependence of the receding contact angle, because $\theta(t) = \theta_r(t)$ for stage (S). Because its theoretical prediction is complicated, we use the experimentally recorded value of $\theta_r(t)$. As soon as $\theta_r(t)$ is known, it can be used in the preceding model, Eq. (21.5), for a prediction of the evolution of contact line radius, $L(t)$:

$$\text{(S)} \quad \frac{\mathrm{d}L(t)}{\mathrm{d}t} = -\frac{1}{3L(t)f(\theta_r(t))}\left(\beta F(\theta_r(t)) + L^2(t)\frac{\mathrm{d}f(\theta_r(t))}{\mathrm{d}t}\right). \tag{21.8}$$

Equation (21.8) is easy to solve numerically. If we assume constancy of the receding contact angle, $\theta_r(t) = \text{const}$, then Eq. (21.8) can be solved analytically, and the obtained solution becomes identical to a dimensional form of the Eq. (21.7), which describes stage (II) of the evaporation process. Thus, Eq. (21.8) in general can describe both stages (S) and (II) simultaneously, because it would be the one stage with time-dependent receding contact angle, $\theta_r(t)$. Therefore, in the case of the presence of stage (S), we can merge it with stage (II) and apply Eq. (21.8) for both stages in the following nondimensional form:

$$\text{(S and II)} \quad \frac{d\ell_S}{d\tau_{S,II}} = -\frac{1}{f_S}\left(\frac{F_S}{2\ell_S} + \frac{\ell_S}{3}\cdot\frac{df_S}{d\tau_{S,II}}\right), \qquad (21.9)$$

where $\ell_S = \frac{L}{L_S}$, $L_S = L(t_S)$, $f_S = \frac{f(\theta_r(t))}{f(\theta_S)}$, $F_S = \frac{F(\theta_r(t))}{F(\theta_S)}$, $\theta_S = \theta(t_S)$, and $\tau_{S,II} = \frac{2F(\theta_S)}{3f(\theta_S)}\frac{(t-t_S)}{L_S^2}\beta$ is a nondimensional time for stages (S) and (II).

Under the limit $t_S \to t_{II}$ (that is, in the absence of stage S), we get $L_S \to L_{II}$, $\ell_S \to \ell_{II}$, $\theta_S \to \theta_{II}$, $\theta_r(t) \to \text{const} = \theta_{II}$, thus $f_S \to 1$, $F_S \to 1$, $\frac{df_S}{d\tau_{S,II}} \to 0$, and $\tau_{S,II} \to \tau_{II}$, and Eq. (21.9) transforms into the following:

$$\frac{d\ell_{II}}{d\tau_{II}} = -\frac{1}{2\ell_{II}}, \qquad (21.10)$$

which is absolutely identical to the equation of stage (II), the solution to which is given by Eq. (21.7). Thus, we conclude that Eq. (21.9) for both stages (S) and (II) is a generalization of Eq. (21.7) for stage (II).

Validation against experimental data shows that Eq. (21.9) well describes the evolution of contact line radius, $L(t)$, for surfactant solutions, as shown in Figure 21.9, which confirms the validity of the derived mathematical model and justifies the preceding assumptions. The only fitting parameter of the theory, β, was found for each experimental run separately, and then its average value and random error were calculated (with 90% confidence level): $\beta = (1.40 \pm 0.09)\cdot 10^{-3}\ \text{mm}^2\ \text{s}^{-1}$. This average value of β, as well as experimentally determined advancing, θ_{ad}, and time-dependent receding, $\theta_r(t)$, contact angles were used to build the theoretical curves in Figure (21.9).

21.3.5 Future Development of the Theoretical Model

Future development of the theory of evaporating drops of surfactant solutions requires a prediction of advancing and receding contact angles as well as their time-dependences, which remains a problem of open research. In order to attain the solution of this extremely complex problem, it is necessary to perform a direct computer simulation of surfactant solutions. Such a simulation has to take into account the molecular kinetics of surfactants: processes of surfactant adsorption/desorption on interfaces and at the contact line, processes of association/dissociation of surfactant

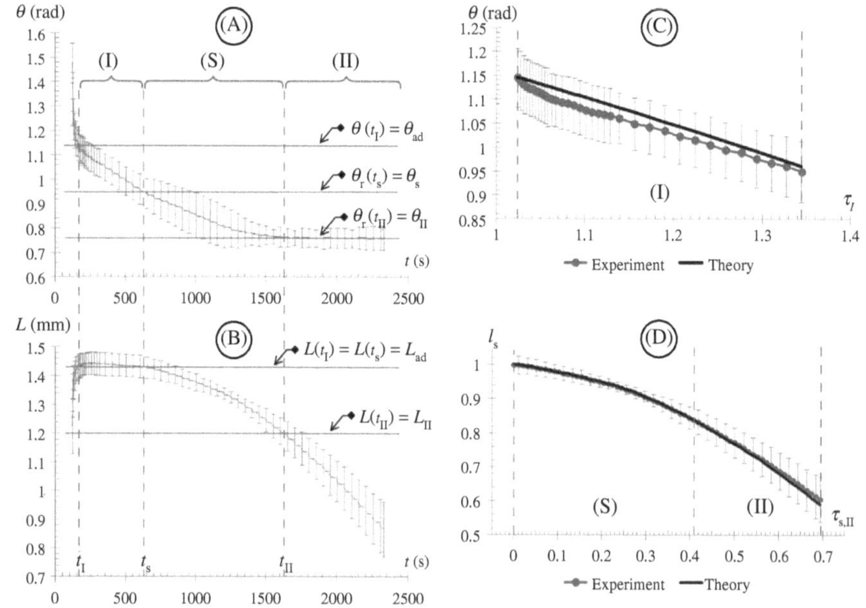

FIGURE 21.9 Experimental measurements of the contact angle (A) and the contact line radius (B) of an evaporating sessile droplet of aqueous surfactant (SILWET L77) solution with the initial concentration $C = 1.25$ CAC, at 30% and 24°C of relative air humidity and ambient temperature, respectively. (C) Nondimensional representation of stage (I) and its validation against the theory, Eq. (21.6). (D) Nondimensional representation of stages (S) and (II) and their validation against the theory, Eq. (21.9).

aggregates in the bulk, and the influence of all these processes on the interfacial tensions and the dynamics of the contact angle.

The progress in this direction is hindered by a large number of unknown parameters in the problem statement, which are not easy to access or control experimentally. Such parameters include, for example, the rate constants of all the kinetic reactions, molecular structure and a maximum surface excess of the adsorbed surfactant, and critical concentrations for the formation of different aggregates.

Although the search of quantitative characteristics of the model is essential, another vital issue is a qualitative validation of the theory and a search of the exact physical mechanisms that are responsible for the observed phenomena.

21.4 Conclusions

This study focuses on the problem of evaporating sessile drops of surfactant solutions. It gives a short overview of industrial applications of evaporating sessile droplets, describes the general challenges in the field of research, and presents the numerical and theoretical models of evaporating drops of pure liquids and surfactant solutions validated against experimental data.

We did experimental study of the kinetics of evaporation of aqueous solutions of the surfactant SILWET L77 on highly hydrophobic substrates. According to the experimental evidence, the direct influence of the surfactant monolayer at the droplet surface on the process of diffusive evaporation is negligible. Within a studied range of surfactant concentrations (up to 12.5 CAC of SILWET L77), the process of a diffusion limited evaporation of a sessile droplet is fully determined by the drop's geometry and by the temperature distribution along its surface. Thus, if we know the temperature distribution and the contact angle at each moment of the evaporation process, we can predict the evaporation rate and the time evolution of the droplet's shape based on a solution of the vapor diffusion equation.

Using numerical tools, the nonisothermal problem of evaporating drops of pure liquids was solved. It was found that the average temperature of the droplet surface well determines the diffusion limited evaporation of a sessile droplet. The nonuniformity of the temperature along the droplet surface does not play a significant role in the prediction of the total evaporation rate from the droplet surface.

In our experimental work, we observed the following five distinct stages of the spreading/evaporation process for sessile droplets of surfactant solutions. Short spreading stage: contact line radius quickly increases, while contact angle decreases until it reaches its static advancing value. Stage (I): constant contact line radius (pinned contact line) and reducing contact angle (contact angle hysteresis). Stage (S): recession of the contact line, with changing receding contact angle that is fully determined by the surfactant adsorption on interfaces. Stage (II): recession of the contact line with constant receding contact angle (maximum surfactant adsorption is reached). Short final stage (III): both contact line radius and contact angle quickly reduce until the drop disappears.

A mathematical model of the phenomenon in question has been developed based on the results of our computer simulations and on a number of assumptions. The theory uses only one fitting parameter—the average temperature of the droplet surface—because it is not available in many experiments. The values of advancing and receding contact angles are not predicted by the theory and therefore taken directly from the experiments. This theory well predicts the evaporation rate and the evolution of the droplet shape for the three particular stages of evaporation: (I), (S), and (II).

Future development of the theoretical model needs a prediction of advancing and receding contact angles. This is an extremely complex task because it requires a direct computer simulation of surfactant solutions and a prediction of the surfactant adsorption. In addition to the preceding results, the role of surface forces in the formation of contact angles is yet to be understood.

Nomenclature

Latin

c $[\mathrm{mol\ m^{-3}}]$ molar concentration of vapor
C $[\mathrm{g\ L^{-1}}]$ initial surfactant concentration
D $[\mathrm{m^2\ s^{-1}}]$ vapor diffusion coefficient
f function of the contact angle θ in the formula for the droplet volume
F function of the contact angle θ in the total flux of evaporation
J $[\mathrm{mol\ s^{-1}}]$ total evaporation flux from the droplet surface
K additional functions of the contact angle θ
ℓ nondimensional three-phase contact line radius
L [m] three-phase contact line radius
M $[\mathrm{kg\ mol^{-1}}]$ molar mass of the liquid (and vapor)
S $[\mathrm{m^2}]$ droplet/air interface or its area
t [s] time
T [K] temperature
V $[\mathrm{m^3}]$ droplet volume

Greek

β $[\mathrm{m^2\ s^{-1}}]$ proportionality coefficient in the expressions for the droplet evaporation rate
θ [rad] contact angle
ρ $[\mathrm{kg\ m^{-3}}]$ mass density
τ nondimensional time, or integration variable

Subscripts

I, S, II, III	stages, or beginnings of stages (I), (S), (II), (III)
$\pi/2$	corresponding to the contact angle $\theta = \pi/2$
ad	advancing (contact angle)
amb	ambient
av	average over the droplet's surface
c	molar
d	depinning
i	isothermal
m	modified
p	pinning
P	Popov's (function F)
r	receding (contact angle)
sat	saturated

References

Agogo, H., Semenov, S., Ortega, F., Rubio, R.G., Starov, V.M., Velarde, M.G., 2012. Spreading and evaporation of surfactant solution droplets. Prog. Colloid Polym. Sci. 139, 1–6.

Ajaev, V.S., Gambaryan-Roisman, T., Stephan, P., 2010. Static and dynamic contact angles of evaporating liquids on heated surfaces. J. Colloid Interface Sci. 342, 550–558.

Boinovich, L., Emelyanenko, A., 2011. Wetting and surface forces. Adv. Colloid Interface 165, 60–69.

Bourges-Monnier, C., Shanahan, M.E.R., 1995. Influence of evaporation on contact angle. Langmuir 11, 2820–2829.

Campbell, P.G., Weiss, L.E., 2007. Tissue engineering with the aid of inkjet printers. Expert Opin. Biol. Ther. 7, 1123–1127.

Chen, C.-T., Tseng, F.-G., Chieng, C.-C., 2006. Evaporation evolution of volatile liquid droplets in nanoliter wells. Sens. Actuators A 130–131, 12–19.

David, S., Sefiane, K., Tadrist, L., 2007. Experimental investigation of the effect of thermal properties of the substrate in the wetting and evaporation of sessile drops. Colloid Surf. A 298, 108–114.

Derjaguin, B.V., 1940. Theory of capillary condensation and related capillary phenomena taking into account a disjoining action of polymolecular liquid films (in Russian). Zh. Fiz. Khim. USSR 14, 137.

Diaz, M.E., Fuentes, J., Cerro, R.L., Savage, M.D., 2010. Hysteresis during contact angles measurement. J. Colloid Interface Sci. 343, 574–583.

Erbil, H.Y., McHale, G., Rowan, S.M., Newton, M.I., 1999. Determination of the receding contact angle of sessile drops on polymer surfaces by evaporation. Langmuir 15, 7378–7385.

Extrand, C.W., 2002. Water contact angles and hysteresis of polyamide surfaces. J. Colloid Interface Sci. 248, 136–142.

Hu, H., Larson, R.G., 2002. Evaporation of a sessile droplet on a substrate. J. Phys. Chem. B 106, 1334–1344.

Hu, H., Larson, R.G., 2005. Analysis of the microfluid flow in an evaporating sessile droplet. Langmuir 21, 3963–3971.

Ivanova, N., Starov, V., Rubio, R., Ritacco, H., Hilal, N., Johnson, D., 2010. Critical wetting concentrations of trisiloxane surfactants. Colloid Surf. A 354, 143–148.

Ivanova, N.A., Starov, V.M., 2011. Wetting of low free energy surfaces by aqueous surfactant solutions. Curr. Opin. Colloid Interface Sci. 16, 285–291.

Nikolov, A., Wasan, D. 2014. Current opinion in superspreading mechanisms. Adv. Colloid Interface Sci. Available from: http://dx.doi.org/10.1016/j.cis.2014.03.006.

Picknett, R.G., Bexon, R., 1977. The evaporation of sessile or pendant drops in still air. J. Colloid Interface Sci. 61 (2), 336–350.

Platikanov, D., Nedyalkov, M., Petkova, V., 2003. Phospholipid black foam films: dynamic contact angles and gas permeability of DMPC bilayer films. Adv. Colloid Interface 100–102, 185–203.

Popov, Yu. O., 2005. Evaporative deposition patterns: spatial dimensions of the deposit. Phys. Rev. E 71, 036313.

Qian, T., Wang, X.-P., Sheng, P., 2006. Molecular Hydrodynamics of the moving contact line in two-phase immiscible flows. Commun. Comput. Phys. 1 (1), 1–52.

Ritacco, H.A., Fainerman, V.B., Ortega, F., Rubio, R.G., Ivanova, N., Starov, V.M., 2010a. Equilibrium and dynamic surface properties of trisiloxane aqueous solutions. Part 2. Theory and comparison with experiment. Colloid Surf. A 365, 204–209.

Ritacco, H.A., Ortega, F., Rubio, R.G., Ivanova, N., Starov, V.M., 2010b. Equilibrium and dynamic surface properties of trisiloxane aqueous solutions. Part 1. Experimental results. Colloid Surf. A 365, 199–203.

Savva, N., Kalliadasis, S., 2011. Dynamics of moving contact lines: a comparison between slip and precursor film models. EPL 94, 64004.

Sazhin, S.S., Shishkova, I.N., Kryukov, A.P., Levashov, V. Yu, Heikal, M.R., 2007. Evaporation of droplets into a background gas: kinetic modelling. Int. J. Heat Mass Transfer 50, 2675–2691.

Semenov, S., Starov, V.M., Rubio, R.G., Velarde, M.G., 2010. Instantaneous distribution of fluxes in the course of evaporation of sessile liquid droplets: computer simulations. Colloid Surf. A 372, 127–134.

Semenov, S., Starov, V.M., Rubio, R.G., Agogo, H., Velarde, M.G., 2011. Evaporation of sessile water droplets: universal behaviour in presence of contact angle hysteresis. Colloid Surf. A 391, 135–144.

Semenov, S., Starov, V.M., Rubio, R.G., Agogo, H., Velarde, M.G., 2012. Evaporation of sessile water droplets in presence of contact angle hysteresis. Math. Model. Nat. Phenom. 7 (4), 82–98.

Semenov, S., Starov, V.M., Rubio, R.G., 2013a. Evaporation of pinned sessile microdroplets of water on a highly heat-conductive substrate: computer simulations. Eur. Phys. J. Spec. Top. 219, 143–154.

Semenov, S., Trybala, A., Agogo, H., Kovalchuk, N., Ortega, F., Rubio, R.G., et al., 2013b. Evaporation of droplets of surfactant solutions. Langmuir 29, 10028–10036.

Semenov, S., Trybala, A., Rubio, R.G., Kovalchuk, N., Starov, V., Velarde, M.G., 2014. Simultaneous spreading and evaporation: recent developments. Adv. Colloid Interface 206, 382–398.

Sobac, B., Brutin, D., 2014. Desiccation of a sessile drop of blood: cracks, folds formation and delamination. Colloid Surf. A 448, 34–44.

Starov, V.M., Velarde, M.G., Radke, C., 2007. Wetting and Spreading Dynamics. CRC Press (Surfactant science series, 138), Boca Raton, FL, USA.

Suciu, C.V., Goto, H., 2010. Interconnection between the contact angle (adhesion) hysteresis, slippage on a nanorough solid wall and phononic friction. Tribol. Int. 43, 1091–1098.

Venzmer, J., 2011. Superspreading—20 years of physicochemical research. Curr. Opin. Colloid Interface 16, 335–343.

Wagner, R., Richter, L., Weißmuller, J., Reiners, J., Klein, K.D., Schaefer, D., et al., 1997. Silicon-modified carbohydrate surfactants: IV. The impact of substructures on the wetting behavior of siloxanyl-modified carbohydrate surfactants on low-energy surfaces. Appl. Organomet. Chem. 11, 617–632.

Chapter 22

DROPLETS OF POLYMERS

Ludovic Pauchard
Université Paris-Sud, CNRS. Laboratoire FAST, Bat 502, Campus Universitaire, Orsay, France

Drying of a sessile drop of polymer solutions results in intriguing complex shapes. Due to the nonuniform evaporative flux from the surface of the sessile drop, the evaporation-driven flow inside the drop may distribute the polymers in an irregular and undesirable manner. A modification of the rheological properties of the system and the occurrence of instabilities take place. As a result a nonuniform deposit is left. Understanding the broad variety of final morphologies is highly relevant to coating processes or biointerface mechanics for instance. They represent also a particularly spectacular example in the growing field of surface instabilities. Coupling the flow of a liquid phase and the mechanical instabilities of elastic material governs the final morphology of the sessile drop. Indeed, under solvent loss, polymer accumulates, resulting in the formation of an elastic skin. Depending on the experimental conditions (drying conditions, contact angle, and polymer concentration), different modes of deformation can take place: axisymmetric or secondary instabilities that break the axisymmetry can be observed.

Droplet Wetting and Evaporation. DOI: http://dx.doi.org/10.1016/B978-0-12-800722-8.00022-9

22.1 Introduction

The drying process of complex fluid drops is ubiquitous. This process is an interesting nonequilibrium phenomenon observed in our daily life, such as the ring-like spots that remain on dishes after we wash them, the enhanced edges in watercolor paintings, or the skin that forms on top of hot milk. Also, this an important problem in many industrial applications in the fabrication of functional material via self-assembly or in the printing industry, especially the ring-like deposit that must be avoided in order to obtain uniform dried ink drops (De Gans and Schubert, 2004). During evaporation the drop shape is mainly governed by two effects: the evolution of the rheological properties and the occurrence of instabilities. Indeed, different types of instabilities can take place due to the temperature gradient (the solvent evaporation cools the drop surface) and/or the solute concentration gradients; the resulting Rayleigh-Bénard or Bénard-Marangoni instabilities then modify the shape evolution (Mitov and Kumacheva, 1998; De Gennes, 2001; Nguyen and Stebe, 2002; Bassou and Rharbi, 2009). In this way the control of the shape of the solute deposit is a significant and complex problem because hydrodynamics, contact line dynamics, and mass and heat transfers are mutually coupled. Moreover the final deposit can be strongly affected by mechanical instabilities resulting from skin formation at the free surface (Deegan et al., 1997; Pauchard and Allain, 2003a).

While evaporating, a pure solvent sessile droplet recedes with a constant contact angle, whose value is determined by surface tension equilibrium, and retains a spherical cap shape. However, this occurs only in the ideal case of ultra-pure solvents and perfectly cleaned surfaces. The presence of nonvolatile compounds in the droplet brings considerable changes in the desiccation processes. Indeed the evaporation-driven flow inside the sessile drop induces the transport of various components and can display very intriguing final complex shapes. The resulting morphologies depend on several physical parameters, including contact angles, drying kinetics, and concentrations in nonvolatile solutes. In general, the drop three-phase line is pinned due to a rapid deposition and adhesion of the nonvolatile species on the substrate at the drop edge. This pinning is responsible for an outward flow, which brings the nonvolatile solutes toward the three-phase line. The outward flow results from the combined action of the increased evaporation rate at the drop edge and the contact line pinning caused by the solutes deposition there (Deegan et al., 1997). An analysis of the outward flow was reported by Hu and Larson (2002) and detailed of the shape of the final deposit by Routh and Russel (1998).

If the amount of nonvolatile material is very small, different stages take place during the sessile drop drying. First, a pinning of the contact line is caused by the irregularities of the substrate; the contact angle decreases until a critical angle.

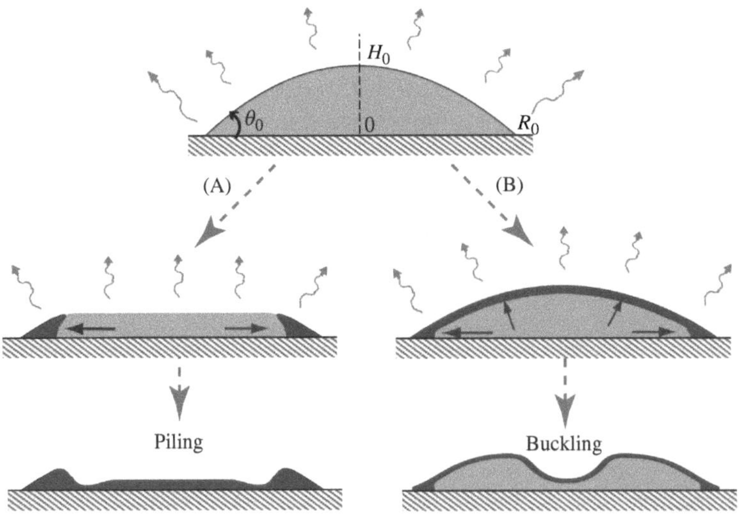

FIGURE 22.1 Sketch of a drying sessile drop: (A) coffee ring formation due to the singular evaporative flux at the drop edge, and (B) concentration gradient at the evaporation surface.

Then the drop recedes, keeping its contact angle constant. Finally, the contact line becomes pinned again; this latter self-pinning takes place as soon as the polymer concentration becomes so high near the edge that the solution loses mobility there, as a piling of the solute. As a result, the drop edge is higher than the center of the drop, resulting in a dimple (Figure 22.1A). Increasing the polymer concentration, the second pinning—that is, self-pinning—occurs earlier, thus after a lower shrinkage of the drop contact base; in addition, the dimple becomes less pronounced as the polymer concentration increases, as reported by Kajiya et al. (2006).

When the amount of nonvolatile solutes is large (typical polymer volume fraction, $\phi_p > 10^{-1}$), the contact line remains pinned at its initial position. The drop then evaporates with a constant surface of contact with the substrate. This situation is more frequently encountered in practical applications. Consequently large concentration gradients of nonvolatile solutes form at the air/solution interface. This results in a modification of the rheological properties of the medium which undergoes for instance a sol-gel transition or a glass transition at the free surface of the drying drop and consequently to the formation of a skin (Figure 22.1B).

We focus next on drying drops of concentrated polymer solutions that result in a rapid and strong pinning of the three-phase line under conditions where the systems are stable in terms of hydrodynamical instability and unstable in terms of mechanical instability.

22.2 Drying process of complex fluids

The lifetime t_D of a sessile drop gives the time needed for the drop to dry. Time t_D describes the rate of volume variation and therefore depends on drop size, contact angle, and the relative humidity $RH = n_{w\infty}/n_{wsat}$, where $n_{w\infty}$ and n_{wsat} are the water concentrations in air at infinity and at saturation, respectively. Therefore $t_D = V_0/S_0E^{-1}$ is defined from the initial volume V_0, vapor/drop surface area S_0 of the sessile drop, and evaporation rate E. As long as the transfer of water in air is limited by diffusion, the evaporation rate is characterized by the mean solvent flux per surface unit as: $E = -(1/S)(dV/dt)$, where V is the volume, and S the vapor/drop surface area, is given by (Pauchard and Allain, 2003b):

$$E = A(\theta_0)\frac{D_w}{R_0}\frac{n_{wsatp} - n_{w\infty}}{n_l} \tag{22.1}$$

where D_w is the diffusion coefficient of water into air (typically $D_w = 25 \times 10^{-6}\ m^2\ s^{-1}$ at room temperature), R_0 is the initial contact base radius of the drop, n_{wsat} is the water concentration in air at equilibrium with the drop fluid, n_l is the number of water moles per unit volume in liquid water, and $A(\theta_0)$ is a numerical factor related to the shape of the isoconcentration curves of water in air, which varies only slightly with θ_0—typically from 1.3 to 1 when θ_0 increases from 20° to 90°). For a polymer solution n_{wsatp} is smaller than n_{wsat} (for pure water) and can be expressed using the Flory–Huggins equation: $n_{wsatp}/n_{wsat} = (1 - \phi_p) \times \exp[\phi_p + \chi\phi_p^2]$ (De Gennes, 1979); the χ-parameter characterizes the affinity between the solvent and the polymer ($\chi \sim 0.5$, for aqueous Dextran solutions (Vink, 1971)). Then the characteristic drying time of a sessile drop is found to vary as:

$$t_D = g(RH, \theta_0)R_0^2/D_w \tag{22.2}$$

where $g(RH,\theta_0)$ is a known function. Therefore, t_D is given by a power law with an exponent 2 on R_0 in good agreement with experimental results (Pauchard and Allain, 2003a).

For low contact angles, the apex height H_0 of the drop is such that $H_0 << R_0$, and the velocity of the fluid is dominated by a radial component, v, only. The Darcy law, describes how the resulting gradient of the capillary pressure ∇p drives the flow of a polymeric solution in a layer of thickness h (Ozawa et al., 2005):

$$v = \frac{2h^2}{3\eta(\phi_p)}\nabla p \tag{22.3}$$

The viscosity η depends on the polymer concentration ϕ_p. At low concentrations, the viscosity increases slowly. Above an overlap threshold, ϕ^*_p, as is typical in polymer solutions, the viscosity increases significantly. Because the polymer concentration of the solutions considered in the following is larger than ϕ^*_p, the

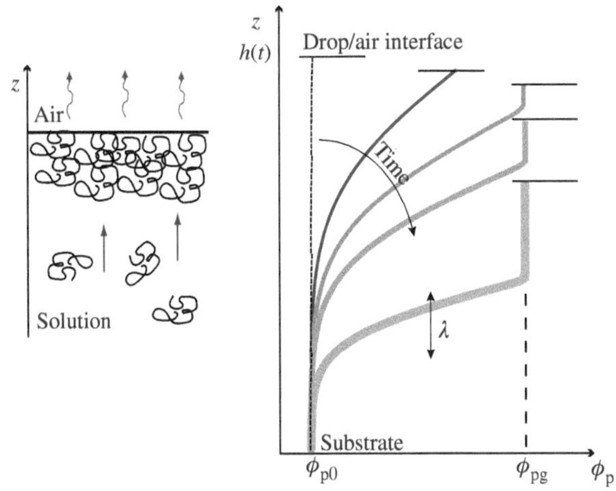

FIGURE 22.2 (Left) Accumulation of polymer chains at the drop surface. (Right) Polymer volume fraction, ϕ, profiles at various times during the drying process (layer of thickness *h(t)* on a flat substrate whose normal lies in *z*). The initial polymer concentration is ϕ_0; when ϕ_p reaches a certain value ϕ_{pg}, a skin phase is formed at the drop surface. λ is the diffusion length characterizing the concentration gradient.

polymer chains are entangled, resulting in a semidilute regime. For high viscosity, the flow near the contact line results in a piling process there.

For high contact angles, the concentration near the free surface increases during the drying process and forms a polymer-rich region (Figure 22.2). The diffusion process can be described by the following one-dimensional equation along the z-axis normal to the plane of the substrate ($0 < z < h(t)$):

$$\frac{\partial \phi_p}{\partial t} = \frac{\partial}{\partial z}\left[D_m \frac{\partial \phi_p}{\partial z}\right] \tag{22.4}$$

where D_m is a the polymer/solvent mutual diffusion coefficient, which in general depends on ϕ_p. A no-flux boundary condition is applied at the substrate ($z = 0$), while at the top surface ($z = H_0 - Et$):

$$D_m \frac{\partial \phi_p}{\partial z} = E\phi_p \tag{22.5}$$

The resulting polymer volume fraction, ϕ_p, profiles are plotted at various times during the drying process in Figure 22.2.

Following Eq. (22.5), the order of magnitude of the concentration gradient at the drop surface is: $(\phi_{ps} - \phi_{p0})/\sqrt{D_m t}$, where ϕ_{ps} and ϕ_{p0} are the values of the polymer volume fraction at the drop surface and in the drop core, respectively. When ϕ_{ps} reaches a certain value ϕ_{pg}, characteristic of a gelled or glassy phase, an elastic phase is formed as a skin layer.

22.3 Skin formation

How a skin phase is formed at the free surface of a polymer solution is an important question in controlling the drying process and the drop shape. The way the polymer accumulates at the free surface is described by the importance of the evaporation to diffusion in the drying drop. A dimensionless parameter that describes both processes in a layer of initial thickness h_0 is a Péclet number defined as:

$$\mathrm{Pe} = Eh_0/D, \tag{22.6}$$

where D is the diffusion coefficient of the nonvolatile solutes in the liquid layer. When diffusion dominates ($\mathrm{Pe} << 1$), the solutes distribution is predicted to remain uniform through the layer as it dries. However, when evaporation is the dominant mechanism ($\mathrm{Pe} >> 1$), accumulation of solutes at the air-liquid interface takes place. The concentration profile is expected to develop in a typical length scale $\lambda = D/E$ (typical values are $\lambda \sim 8 \times 10^{-12}/5 \times 10^{-8} \sim 100\ \mu\mathrm{m}$ under ambient drying conditions (Boulogne et al., 2013)). Under these conditions, the transport of polymers to the drop-air interface strongly suggests the formation of a skin during the drying process. This results in the formation of a consolidation front that grows in thickness while the free surface descends toward the substrate due to the solvent removal (Figure 22.2).

Equations (22.4) and (22.5) give the relations between the evaporation rate and the volume fraction of polymer in the skin phase. Coupling the latter and the Péclet number (Eq. (22.6)) results in a criterion to characterize the formation of well-structured skin layers as reported by Okuzono et al. (2006):

$$\mathrm{Pe} > \frac{\phi_{\mathrm{pg}} - \phi_{\mathrm{p0}}}{(1 - \phi_{\mathrm{p0}})\phi_{\mathrm{p0}}} \tag{22.7}$$

The region described by Eq. (22.7) in the parameter space (ϕ_{p0}, Pe) is shown in Figure 22.3. This result suggests that the skin formation can be controlled by Pe and ϕ_{p0} for fixed ϕ_{pg}. Therefore, the skin phase appears when the evaporation rate is high or when the initial polymer concentration is high.

When its structural strength is sufficient, the skin withstands internal stress and can be regarded as an elastic shell. Assuming that the planar stress in the skin layer is isotropic, the stress, σ, is related to the pressure difference ΔP in the liquid inside the drop relative to the atmospheric pressure as: $\sigma h\kappa = \Delta P$, where κ is the mean curvature of the surface. This is an analogous equation of the Laplace formula for the pressure difference, ΔP, between two media caused by surface tension at the surface of separation (Landau and Lifshitz, 1986). The stress is also related to the change of the surface area ΔA of the skin layer as: $\sigma = Y\Delta A/A_0$,

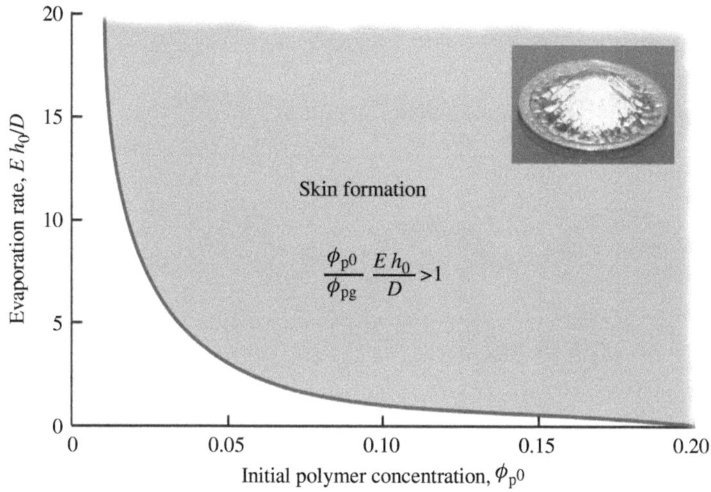

FIGURE 22.3 Diagram of skin formation following the evaporation rate and the initial concentration of the solution (Arai and Doi, 2013). Above the solid line, well-structured skin layers are formed.

where A_0 is the surface area of the skin in the stress-free state and Y is the Young modulus of the skin material (Arai and Doi, 2013).

By appropriate selection of physical quantities, constrained shrinkage of the skin can result in various surface instabilities.

22.4 Surface instabilities

The drying of drops of concentrated polymer solution involves three processes: three-phase line pinning, polymer accumulation, and rheological transformations. First, the three-phase line is pinned at the rim of the drop due to a rapid deposition of the polymer solute on the substrate at the drop edge. Next, the polymer accumulates near the evaporation surface, leading to large concentration nonuniformities. Finally, rheological properties also change near the surface due to this local increase of polymer concentration. Moreover, depending on the initial contact angle (at fixed initial polymer concentration, $\phi_{p0} > \phi^*_p$, and room drying conditions), different behaviors are observed.

For low contact angles ($\theta_0 < 30°$), the drop progressively flattens with solvent evaporation; the drop shape is not strongly modified. In the final stage, a dried film with a depressed central zone is formed for the lowest concentrations (sketch in Figures 22.1A and 22.4A, $\theta_0 = 20°$).

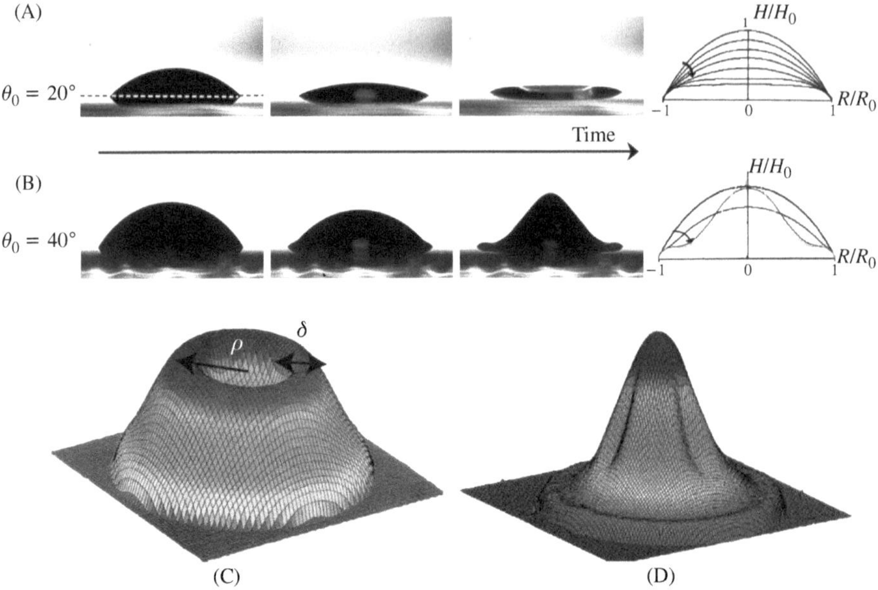

FIGURE 22.4 Side views showing stable (A) and unstable (B) evolutions of polymer solution sessile drops on glass slides (drop base diameter is 4 mm; just after deposition the dashed line corresponds to the drop/substrate contact; the contact angle is controlled by using substrates of different surface energies). (A) For low contact angles ($\theta_0 < 20°$), the drop progressively flattens; at the end of the drying process, the drop forms a film with a dimple in the periphery and a depression at the center. (B) For high contact angles ($\theta_0 > 30°$), the apex height reaches a peak leading to the typical "Mexican hat" shape at the end of the drying process. In both cases superposition of dimensionless profiles of sessile drops are shown during desiccation. (C) and (D) Three-dimensional map of drops at the final stage of the drying process (measurements use a mechanical profilometer): for high initial contact angles ($\theta_0 = 70°$) and room drying conditions (RH ~ 50%), a circular fold forms and results in a dip at the center; for low initial contact angles ($\theta_0 = 30°$) and low relative humidity (RH ~ 30%), the drop axisymmetry is broken by the formation of radial wrinkles. The initial volume of drops is $<10\ mm^3$. (A): Pauchard and Allain, 2003b; (C): *Reprinted with permission from Gorand et al., 2004. Copyright 2015 American Chemical Society.*

For larger contact angles (typically $30° < \theta_0 < 60°$), the apex height first decreases and then quickly increases, resulting in a distorted drop shape. At time t_B—for example, the onset of the instability—the apex height variation differs from the regular decrease and the profiles get distorted (Figure 22.4B; $\theta_0 = 40°$). The surface area remains constant, while the volume keeps decreasing, only slower. An elastic skin builds up at the drop surface. The skin does not however prevent evaporation, so the enclosed volume decreases while the contact line remains pinned. Surface instabilities take place to allow the decrease of the

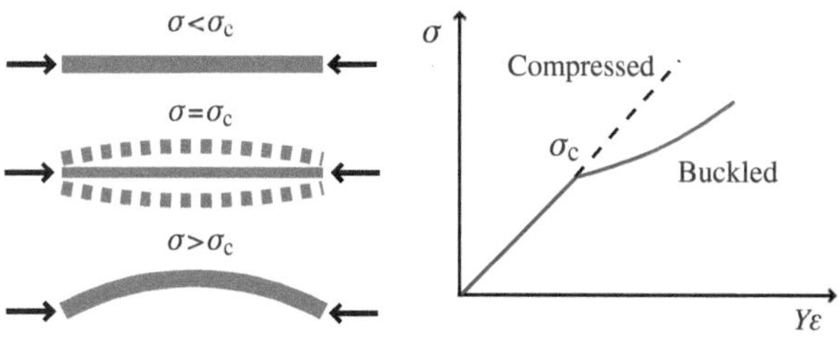

FIGURE 22.5 Response of a beam to an applied compressive stress—the critical stress, σ_c. In the state of unstable equilibrium, the beam becomes unstable, resulting in a lateral deflection.

enclosed volume in spite of constraints due to the constant contact base and the skin rigidity. When the internal pressure becomes less than a critical value, the drop cannot keep the spherical cap shape, and it buckles. At the final stage, the drop has the shape of a "Mexican hat" (Pauchard and Allain, 2003a,b).

Indeed, these observations can be explained by assuming a close relation with buckling processes for thin solids sheets. Also, compression (in-plane displacement) of the surface that flattens costs more and more energy, and it becomes energetically more favorable to convert a part of the stretching energy into bending energy (out-of-plane displacement-curvature): a buckling instability takes place at a critical stress σ_c (Figure 22.5).

For still larger contact angles ($60° < \theta_0$), the apex height first decreases, following a same trend as in the two previous cases, and then going faster. The final drop shape displays, in the latter case, a circular fold that can only be seen on profilometry measurements (Figure 22.4C). In the 3D map, the skin thickness, τ, can be estimated from the curvature of the circular fold, ρ, by minimizing the bending energy with respect to the change of length, δ, leading to $\tau \sim \rho\delta^2$ (Pauchard and Rica, 1998; Gorand et al., 2004). The corresponding thickness value is 400 μm in rough agreement with optical observations.

Axisymmetric deformation mode and stability properties of elastic caps were investigated by nonlinear quasi-static simulation and scaling analysis by Head (2006).

Moreover, the diagram in Figure 22.6 reports the drop evolution as a function of the drying conditions and initial contact angle. Two domains can be clearly identified: at large relative humidity RH and small initial contact angle θ_0, no instability develops while at small RH and large θ_0, instability develops.

The limit between the two domains can be related to the following timescales:

- the drying timescale t_D,
- the timescale t_S corresponding to the formation of a glassy skin: when the polymer volume fraction at the drop surface, ϕ_{ps}, becomes equal to ϕ_{pg},

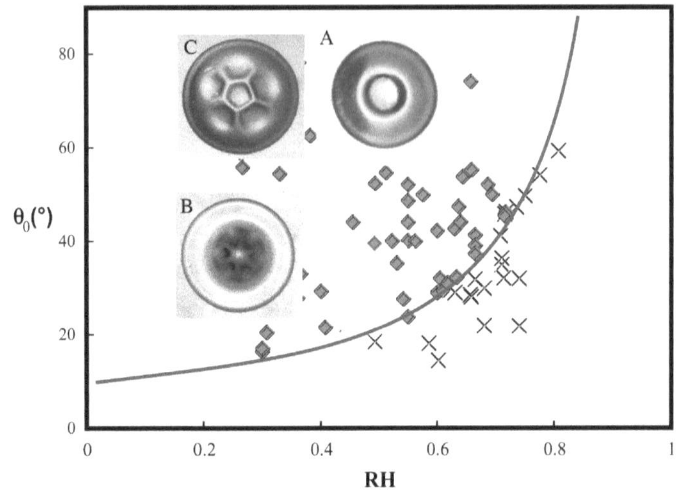

FIGURE 22.6 Diagram showing the different behaviors displayed by sessile drops of polymer solutions following the initial contact angle, θ_0, and the relative humidity, RH. The solid line displays theoretical predictions. Below this line drops continuously flatten (crosses). Above this line, high distortions related to the buckling instabilities of a solid skin at the free surface of the drop (diamonds) form. Inset: Top views at end of the desiccation ($R_0 = 2$ mm). In particular, at RH ~ 50%, the drop keeps its axisymmetry (A); at low RH, a peak first forms at the drop center, and later a complex pattern progressively builds up involving a cascade of buckling (B,C). The initial volume of drops is $<10\ \text{mm}^3$. (Pauchard and Allain, 2003a: http://dx.doi.org/10.1103/PhysRevE.68.052801).

characteristic of a glassy or gelled phase. Following Eq. (22.5), and assuming that the evaporation rate is constant and determined by water diffusion in air, we find:

$$t_S = \frac{D_m(\phi_{pg} - \phi_{p0})^2}{\phi_{pg} E^2} \propto \frac{(\phi_{pg} - \phi_{p0})^2}{\phi_{pg}(1 - \text{RH})^2} R_0^2 \tag{22.8}$$

The buckling instability occurs as soon as the rigid skin forms at the drop surface—that is, $t_B \equiv t_S$ (Pauchard and Allain, 2003a,b). Consequently, comparing the two characteristic times, t_S and t_D, allows the determination of the condition for buckling instability occurrence. Indeed, if $t_D < t_S$, the drop is dried before a glassy skin forms. Under this condition, the drop progressively flattens under solvent evaporation, resulting in a flat film with a depressed zone in its center. However, if $t_D > t_S$, buckling instability takes place during the drying process. Thus, assuming that the limit of the stability domain is given by $t_S = t_D$ leads to the following condition:

$$A(\theta_0)B(\theta_0) = \left(\frac{D_m\left(\phi_{pg} - \phi_{p0}\right)^2}{D_w} \frac{n_l}{n_{wsat}} \right) \frac{1}{1 - \text{RH}} \tag{22.9}$$

The geometrical factor $B(\theta_0) = V_0/(R_0S_0)$ increases with increasing θ_0. In the parameter space (θ_0, RH) (Figure 22.6), the solid line is calculated from Eq. (22.9): a fairly good agreement is observed with the experimental results.

Moreover, the domain where the instability takes place reveals various patterns, in which the drop axisymmetry is or is not broken. For an intermediate relative humidity (RH ~ 50%), only the primary instability takes place and the drop shape remains symmetrical. Drops display a peak, or a trough builds up on the drop axis. However, for small relative humidity (RH ~ 30%), secondary instabilities occur. For a low contact angle, the drop axisymmetry breaks and radial wrinkles form (Figures 22.4D and 22.6). The number of radial wrinkles increases as the skin gets thinner. For a large contact angle, a complex pattern progressively builds up involving a cascade of buckling (Figure 22.6). These various types of shapes are closely related to different modes of buckling instabilities.

In spite of their different structures and physicochemical properties, the same types of shape instabilities may be observed in drops of colloidal suspensions (Pauchard et al., 1999; Pauchard and Couder, 2004). This observation of a mechanical buckling instability in a physicochemical process may help to solve a variety of problems in such fields as coating processes or biointerface mechanics.

Glossary

Buckling When a thin rod under an axial compression is unable to support the load, a sudden deflection occurs. This failure is a called buckling process.

Nomenclature

RH Relative humidity
D_w Diffusion coefficient of water into air ($m^2\ s^{-1}$)
D_m Polymer/solvent mutual diffusion coefficient ($m^2\ s^{-1}$)
E Evaporation rate ($m\ s^{-1}$)
t_D Drying time (s)
t_B Buckling time (s)
H Drop apex height (m)
R Drop base radius (m)
h Flat layer thickness (m)
τ Solid skin thickness (m)
Y Young modulus (Pa)
ϕ_p Polymer volume fraction
ϕ_{pg} Polymer volume fraction at the gelation point

References

Arai, S., Doi, M., 2013. Anomalous drying dynamics of a polymer solution on a substrate. Eur. Phys. J. E 36, 63–69.

Bassou, N., Rharbi, Y., 2009. Role of Bénard-Marangoni instabilities during solvent evaporation in polymer surface corrugations. Langmuir 25, 624.

Boulogne, F., Giorgiutti-Dauphiné, F., Pauchard, L., 2013. The buckling and invagination process during consolidation of colloidal droplets. Soft Matter 9, 750–757.

De Gans, B.J., Schubert, U.S., 2004. Inkjet printing of well-defined polymer dots and arrays. Langmuir 20 (18), 7789–7793.

De Gennes, P.G., 1979. Scaling Concepts in Polymer Physics. Cornell University Press, Ithaca, NY.

De Gennes, P.G., 2001. Solvent evaporation of spin cast films: "crust" effects. Eur. Phys. J. E 6, 421.

Deegan, R.D., Bakajin, O., Dupont, T.F., Huber, G., Nagel, S.R., Witten, T.A., 1997. Capillary flow as the cause of ring stains form dried liquid. Nature 389, 827–829.

Gorand, Y., Pauchard, L., Calligari, G., Hulin, J.P., Allain, C., 2004. Mechanical instability induced by the desiccation of sessile drops. Langmuir 20, 5138.

Head, D.A., 2006. Modeling the elastic deformation of polymer crusts formed by sessile droplet evaporation. Phys. Rev. E 74, 021601.

Hu, H., Larson, R.G., 2002. Evaporation of a sessile droplet on a substrate. J. Phys. Chem. 106, 1334.

Kajiya, T., Nishitani, E., Yamaue, T., Doi, M., 2006. Piling to buckling transition in the drying process of polymer solution drop on substrate having a large contact angle. Phys. Rev. E 73, 011601.

Landau, L.D., Lifshitz, E.M., 1970. Theory of Elasticity (Course of Theoretical Physics), vol. 7. Pergamon Press.

Mitov, Z., Kumacheva, E., 1998. Convection-induced patterns in phase-separating polymeric fluids. Phys. Rev. Lett. 81, 3427.

Nguyen, V.X., Stebe, K.J., 2002. Patterning of small particles by a surfactant-enhanced Marangoni-Bénard instability. Phys. Rev. Lett. 88, 164501.

Okuzono, T., Ozawa, K., Doi, M., 2006. Simple model of skin formation caused by solvent evaporation in polymer solutions. Phys. Rev. Lett. 97, 136103.

Ozawa, K., Nishitani, E., Doi, M., 2005. Modeling of the drying process of liquid droplet to form thin film. Jpn. J. Appl. Phys. 4, 4229–4234.

Pauchard, L., Rica, S., 1998. Contact and compression of elastic spherical shells: the physics of a "ping–pong" ball. Philos. Mag. B 78, 225.

Pauchard, L., Allain, C., 2003a. Stable and unstable surface evolution during the drying of a polymer solution drop. Phys. Rev. E 68, 052801.

Pauchard, L., Allain, C., 2003b. Buckling instability induced by polymer solution drying. Europhys. Lett. 62, 897–903.

Pauchard, L., Couder, Y., 2004. Invagination during the collapse of an inhomogeneous spheroidal shell. Europhys. Lett. 66 (5), 667.

Pauchard, L., Parisse, F., Allain, C., 1999. Influence of salt content on crack patterns formed through colloidal suspension desiccation. Phys. Rev. E 59, 3737.

Routh, A.F., Russel, W.B., 1998. Horizontal drying fronts during solvent evaporation from latex films. AIChE J. 44, 2088.

Vink, H., 1971. Precision measurements of osmotic pressure in concentrated polymer solutions. Eur. Polym. J. 7, 1411.

Chapter 23

DROPLETS OF BIOLOGICAL FLUIDS

David Brutin and Wassim Bou Zeid
Mechanical Engineering Department, Aix-Marseille University, Marseille, France

Evaporation of sessile drops of biological fluid has been studied as a potential interest for medical applications. Most of the studies were performed on biological fluids and blood serum but a few have been performed on whole human blood. This chapter summarizes the parameters that influence the evaporation, gelation, cracking, and delamination processes of a drop of whole human blood. It also gives a literature review on the evaporation of biological fluids. Even for this complex fluid, the initial rate of evaporation is well explained using a purely diffusive model of a pure fluid.

23.1 Blood serum

Biological fluids are very difficult to model because they contain different components (proteins, electrolytes) that interact with one another and can affect the physical mechanisms inside drying droplets. This makes it difficult to understand the physical mechanisms leading to pattern formation. The final complex patterns in dried droplets of blood serum were investigated to differentiate healthy people

Droplet Wetting and Evaporation. DOI: http://dx.doi.org/10.1016/B978-0-12-800722-8.00023-0

from patients with diseases (see Table 23.1). The drying technique was used to store specimens of serum in locations at different temperatures (Zhuang et al., 1982). The method was used as a biological marker of detecting hepatitis B in blood serum dried on filter paper. Also, Yakhno et al. (2003) investigated the patterns left from drops of blood plasma from women who had given birth both prematurely and full term. The patterns left from serum drops of women with normal childbirth are identical, whereas with premature childbirth, a thick crystallized outer ring containing large crystals is observed.

23.2 Whole blood drying

Blood is a complex colloidal suspension that behaves like a non-Newtonian fluid. Human blood is composed of different cells that form about 45% of whole blood volume and 25% of the total blood mass (Thieret, 2008). They are classified as red blood cells (erythrocytes), later called RBCs; white blood cells (leukocytes), later called WBCs; and platelets (thrombocytes). The cells' proportions and characteristics are detailed in Table 23.2. The other 55% is blood plasma, which is composed of water (over 90%), ions, electrolytes (salts), plasma proteins (most of which are by weight 7%), and other substances transported by blood. The main physical properties of whole human blood, such as density, viscosity, and surface tension, are given in the literature Brutin et al. (2011; Table 23.3).

The different stages of blood drop evaporation are shown in Figure 1 of Brutin et al. (2011). The initial drop diameter presented is 5.9 mm. The

TABLE 23.1 Literature Concerning the Evaporation of Drops of Blood Serum

References	Substrate Type	Observations
Zhuang et al. (1982)	Filter paper	Markers of hepatitis B
Shabalin and Shatokhina (2007)	Glass	Pattern differences evidenced for different diseases
Yakhno (2010)	Glass	Different protein structures of patients with diseases
Buzoverya et al. (2012a,b)	Glass	Different types of cracks, depending on pathologies

TABLE 23.2 Distribution of Bio-Colloids Inside the Blood Drop

Bio-Colloids	Quantity (mm^{-3})	Typical Size (μm)	Percentage (%)
Red blood cells	4.9×10^6	8	97
White blood cells	4.3×10^3	15	2
Platelets	2.5×10^5	3	1

TABLE 23.3 Literature Concerning the Evaporation of Drops of Whole Human Blood

References	Substrate Type	Observations
Brutin et al. (2011)	Glass	Cracks pattern formation in drying drops of blood
Brutin et al. (2011)	Glass	Definition of drying stage of a drop of blood
Brutin et al. (2012)	Glass, aluminum, gold	Drying dynamics and pattern formation
Bou-Zeid et al. (2013)	Glass	Drying rate is related to crack angular deviation
Bou-Zeid and Brutin (2014)	Glass	Humidity influence on spreading and cracks

evaporation/desiccation process can be divided into the following five stages. During the evaporation process, the drop is always pinned on the substrate. The total evaporation time for that case is 36 min, and the durations of the stages are presented as a percentage of the total time of evaporation.

- Stage 1 (from 0% to 20%): the RBCs move out from the center of the drop to the desiccation line that recedes. A red deposit is observed at the edge of the drop. The measured receding speed of the desiccation line is 0.7 μms^{-1}.
- Stage 2 (from 20% to 50%): crystallization appears at the edge of the drop and propagates inward. At the same time, the desiccation process continues from the edge toward the inside of the drop. A dark red torus is observed, which demonstrates a different fluid composition. This torus contains a high concentration of RBCs. These cells are now clearly observed moving from the center to the edge of the drop.
- Stage 3 (from 50% to 70%): the dark red torus desiccates rapidly, and at the same time, the color of the central part of the drop changes to an increasingly lighter red. The drop is almost totally desiccated, and the first crack appears around the drop between the future corona and central part of the drop.
- Stage 4 (from 70% to 85%): the central part of the drop desiccates, producing much smaller plaques. Desiccation of the corona finishes, and circular drying spots are observed around the corona.
- Stage 5 (from 85% to 100%): the large plaques of the corona move slightly as soon as they are totally desiccated. This concludes the desiccation of the drop, and no further changes are observed.

All patterns observed with dried drops of blood present similar characteristics: the central part of the drop, the wide mobile plaques of the corona with wide white cracks, and a fine periphery. Whereas the central part of the drop and the fine periphery stick onto the glass plate due to the absence of RBCs, the corona is where mobile deposits form, and these do not adhere to the substrate. The explanation of this wetting phenomenon lies in the proteins that exist on the RBCs (glycoproteins). The function of these proteins is to avoid wettability with a wall and, in particular, the internal wall of organs: veins.

At the beginning of drop evaporation, the fluid is homogeneous in colloids; this is why a thin periphery can dry on the glass plate. However, because Marangoni convection is occurring, the RBCs (and other heavy colloids) accumulate at the drop triple line, which is receding. The RBCs accumulate by Marangoni convection to form a solid deposit that will be called hereafter "the corona." The remaining fluid, which consists mainly of serum without heavy colloids, leads to typical small patterns adhering to the glass substrate after its evaporation.

One of the parameters that influences the drying rate, and thus the final pattern, is room humidity. The influence of the relative humidity (RH) on the pattern at the end of the drying phase has been investigated by considering the same volume of drops of blood evaporating in different RH levels. Final drop images are presented in Figure 23.1, which shows morphological and structural evolutions of a drying drop of whole blood for different values of RH. For each RH, the bottom images are taken at the final evaporation time t_F of the drying phase. One clearly observes a structural change and, consequently, the morphological evolutions of the final drying pattern. When the evaporation of the drop of blood is complete, the final pattern is characterized by the presence of three distinguished regions: a central part composed of a sticking deposit and a network of small cracks, a

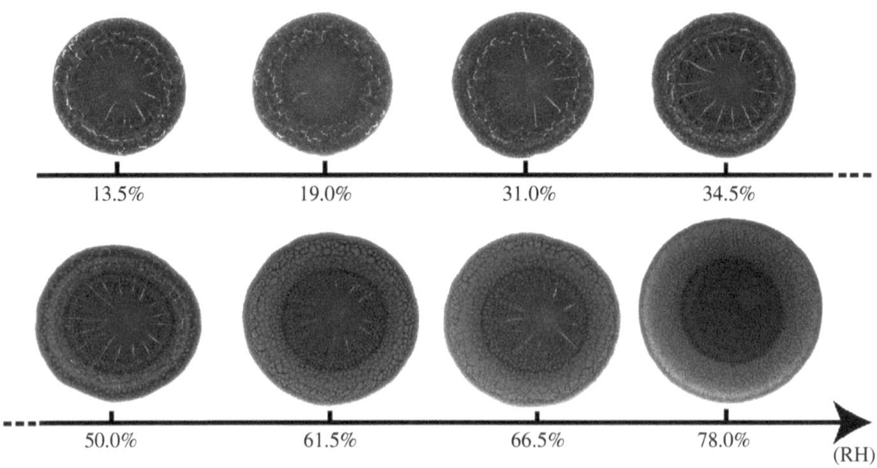

FIGURE 23.1 Top images (same scale) of deposit left behind after the complete evaporation of a sessile drop of whole blood. All experiments are performed for the same drop volume ($V = 14.2$ μL) and for various ranges of RH (microscope ultraclean glass substrate, room temperature: 23.8°C, room pressure: 1005 hPa).

corona composed of wide mobile plaques with white cracks organized radially, and a fine periphery completely adhering to the substrate. It can be observed that the width of mobile plaques in the corona and the fine periphery region become larger as the RH increases. Moreover, the mobile plaques in the corona are more displaced toward the center of the drop, leaving a wider deposit made of plaques with dark cracks. Also, Figure 23.1 shows that mobile plaques depend on the drying rate. At low drying rates (high RH), more completely adhering mobile plaques are observed than at high drying rates (lower RH). Indeed, low adhering areas are highlighted by the circulated red light region that is found in each of the mobile plaques in the corona area.

The final drying pattern and crack nucleation vary with the kinetics of the evaporation rate as presented in Figure 23.2. Under our experimental conditions, the transfer of water in the air is limited by diffusion and is controlled by the RH in the surrounding air. The drying process of a sessile drop of blood is characterized by an evolution of the solution into a gel saturated with solvent. When the gel is formed, the new porous matrix formed by the aggregation of particles continues to dry by evaporation of the solvent that causes gel to consolidate. As the liquid progressively recedes into the porous medium, it forms at first menisci at the air/solvent interface due to capillary tension between the particles and then

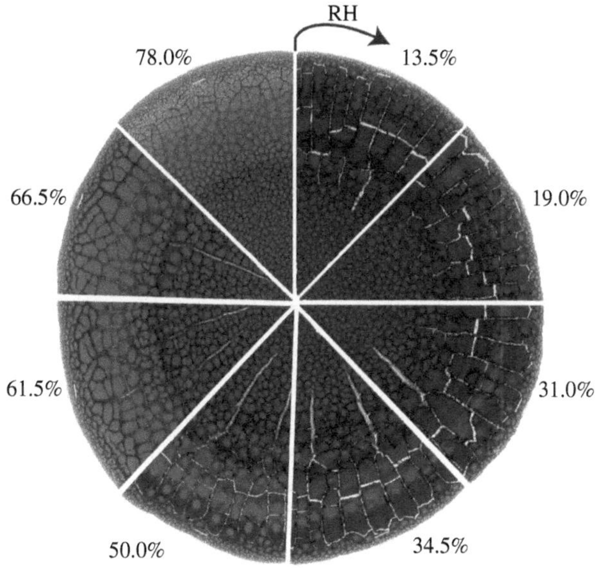

FIGURE 23.2 Final deposition morphology for various RH values (varying between 13.5% and 78.0%). Area between dashes represents the fine periphery of the corona region. The whole human blood was collected in samples containing liquid anticoagulant (Bou-Zeid and Brutin, 2013).

causes liquid bridges between the particles. During solvent evaporation, curvature of the solvent–air menisci is responsible for a capillary pressure in the liquid phase. This depression induces shrinkage of the porous matrix that is constrained by the adhesion of the deposit to the glass substrate and the evaporation of solvent. As tensile stresses builds up, the internal stresses become too great, and fractures appear so as to release mechanical energy. Assuming that gelation is due to particle accumulation, we attribute these differences in pattern formation to the competition between the drying process and the adhesion of the gel on the substrate.

Final dried images presented in Figure 23.2 show that the final adhering area of mobile plaques is strongly dependent on the RH values. By changing the drying rate, the mechanical properties of the drying gel are modified, such as the adhesion energy of the gel to the glass substrate (according to the Griffith theory). Indeed, the surface area of mobile plaques of drops of blood dried at RH from 13.5% to 50.0% becomes progressively larger with a lower adhering region. This adhering region shrinks until the formation of a circular adhering region that leads to a delamination process. For images of RH above 50%, mobile plaques are smaller with a higher adhering region. This observation is due to the buckling process that is rapidly overcome by the adhesion of the gel to the substrate. In desiccated colloidal gel dried at RH = 70% and RH = 46%, L. Pauchard experimentally showed that complete adhesion or deadhesion is a function of the cell surface area (Pauchard, 2006). Moreover, the large fine periphery of the stick deposit onto the glass substrate is due to the receding of the gel front, causing the absence of RBCs. This result is due to a change in the internal flow that transports RBCs from the center to the periphery of the drop of blood.

The influence of RH on the final drying pattern and on the geometrical parameters was investigated by considering the same small volume of drops of blood evaporating at different RH values (Bou-Zeid and Brutin, 2013). Figure 23.3 shows that a linear increase in the contact angle θ_{eq} leads to a decrease in the final wetting diameter D_f of the blood drop as RH decreases. The higher equilibrium contact angle at high evaporative rates than at low evaporative rates can be explained by a dependence of the dynamic contact angle on the contact line velocity at the moment when the drop touches the glass substrate (at the initial spreading) during the spreading process. Therefore, a large wetting diameter can be attributed to a low equilibrium contact angle. The evaporation of the blood drop can be described by the purely diffusive model by considering the variations of the contact angle and the wetting radius as a function of RH:

$$-\frac{\mathrm{d}m}{\mathrm{d}t} = \pi R_{RH} D_{diff} C_v (1 - RH) f_{RH}$$

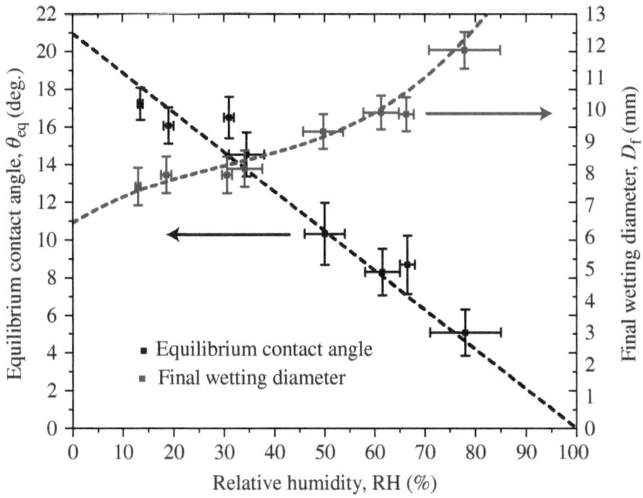

FIGURE 23.3 Variation of the equilibrium contact angle θ_{eq} and the final wetting diameter of the blood drop D_f as a function of RH values. Ambient conditions ($T_a = 24.5 \pm 0.5°C$, $P_a = 1005.2 \pm 0.6$ hPa) and RH varying between 13.5% and 78.0%. The dashed lines are the linear fit for the equilibrium contact angle (θ_{eq}) and the cubic fit for the final wetting diameter (D_f) passing by the measurement points. This purely diffusive model shows a good correlation with our experimental results, which decreases with increasing RH (Figure 23.4).

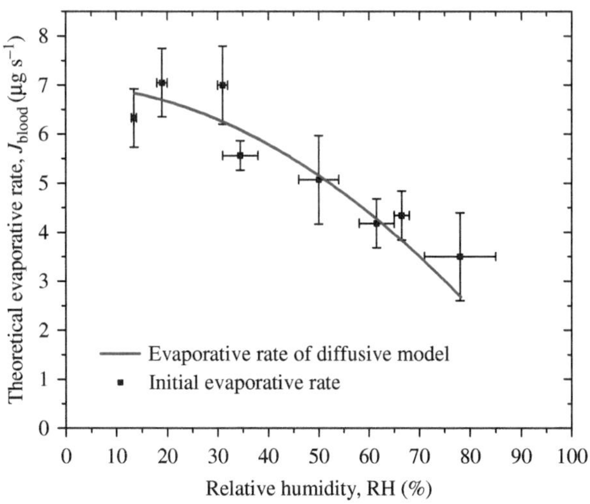

FIGURE 23.4 Variation of the initial rate of evaporation as a function of the RH: experimental results (points) and theoretical model (red curve). (For interpretation of the references to color in this figure legend, the reader is referred to the online version of this book.)

with

$$R_{RH} = \left(\frac{3V}{\pi} \frac{\sin^3 \theta_{RH}}{(2+\cos \theta_{RH})(1-\cos \theta_{RH})} \right)^{\frac{1}{3}}$$

where

$$\theta_{RH} = 21(1 - RH)$$

and

$$f_{RH} = (3.59 \times 10^{-2})RH^2 - (7.18 \times 10^{-2})RH + 1.33$$

In the corona region and at low drying rates (high RH), disordered plaque arrangements exhibit a larger plaques area compared to high drying rates (low RH). By changing the drying rate, the mechanical constraint field inside the drying gel of the blood drop is modified. This mechanical constraint modifies the drying stresses and, consequently, the crack's formation and dynamic. Also, this is due to the variation of geometrical parameters with RH. Sobac and Brutin (2014) used both an approximated function of Allain and Limat model (1995) and the assumption of Annarelli et al. (2001) to calculate the mean crack spacing in a drying drop of blood:

$$\Lambda = 0.522 - 0.716 \log(k)$$

with

$$k = J_0 \times 2h_{max} \times \left(\frac{1}{D_m C_\infty} \right)$$

where J_0 is the initial evaporative flux, Λ is the dimensionless crack spacing (λ/e), D_m is the product of the capillary characteristic velocity (γ/μ) and the hydraulic radius of the pore gel r ($D_m = 3 \times 10^8$ m^2 s^{-1}), and C_∞ is the solvent volume fraction approximately equal to 55% (volume fraction of solvent in blood).

23.3 Whole blood drop spreading

We now examine the spreading/evaporation dynamics of a blood drop as it touches and subsequently wets a clean glass substrate until its complete evaporation. A pendant drop of blood that touches a microscope glass substrate spreads quasi-instantaneously until it reaches a final wetting radius and an equilibrium contact angle of less than 20° (Bou-Zeid and Brutin, 2014). At the beginning of the evaporation process, the blood drop is homogeneous in bio-colloids. As the drop of blood spreads on the glass substrate, the triple line moves radially outward with a contact line velocity U_{CL}, and then it becomes pinned to the substrate (evaporation continue with a constant contact area mode). In the case of biological fluids, proteins adsorb on the glass substrate near the contact line, which anchors the triple line. In hydrophilic situations, an internal outward flow develops inside the blood droplet

(Brutin et al., 2011). Subsequently, the micrometer-sized colloids (RBCs) accumulate near the contact line, reinforcing the contact line pinning and thus forming a layer of biological deposit. For a small drop ($V = 14.2\ \mu L$) with a contact base radius of less than or equal to the capillary length of water, the drop of blood retains its spherical cap shape during the entire spreading/evaporation process. Thus, the instantaneous configuration of the drop can be described by the time-dependent wetting radius, $r(t)$, where $r(t)$ is the wetting radius of the spreading droplet. However, the blood droplets spread faster at high evaporative rates than at low evaporative rates (see Figure 23.5). Consequently, the total time of spreading, t_s, increases with increasing RH values. All curves exhibit the same qualitative behavior as characterized by the two distinct spreading regimes.

In the experiments, the evaporation occurs in a quasi-steady manner because the diffusion time is much smaller than the time of evaporation $C_v(1 - \mathrm{RH})/\rho$, which is of the order of 10^5. In the diffusive-limited case (hydrophilic situation), this evaporative mass flux, J, along the blood drop surface that is proportional to the vapor concentration difference between the drop surface and the ambient air is not uniform and tends to diverge near the contact line (Hu and Larson, 2002).

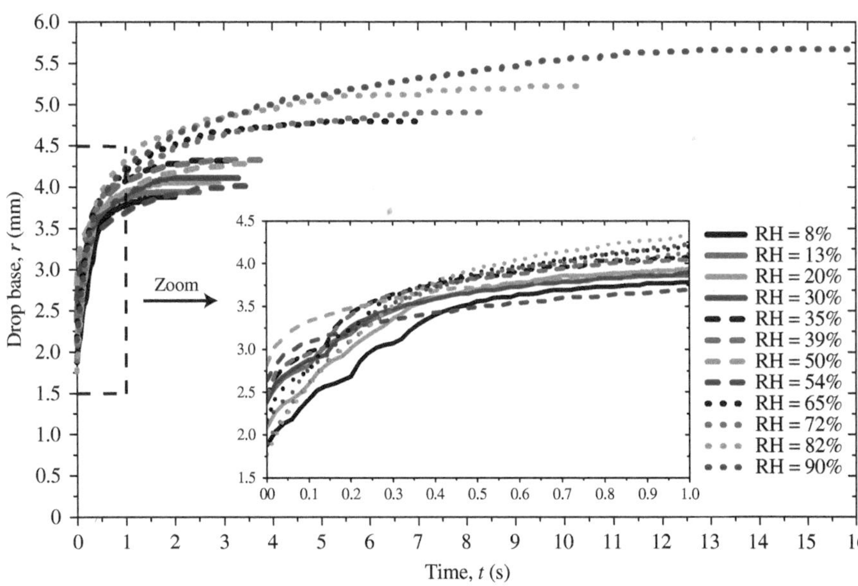

FIGURE 23.5 Evolution of the radius r as a function of time t for drops of blood with viscosity $\mu = 10.0$ mPa s and initial deposited radius (average deposit radius: $R_0 = 2.3 \pm 0.3$ mm) for different humidity levels ranging from 8% to 90%. All experiments were performed on the same drop volume $V = 14.2\ \mu L \pm 1.0\%$ (average mass $m = 13.5$ mg $\pm$ 9.5%). Atmospheric conditions ($T_a = 25.5 \pm 0.5°C$, $P_a = 1005.2 \pm 0.5$ hPa). The inset graph represents a magnification of the spreading time from 0 to 1 s.

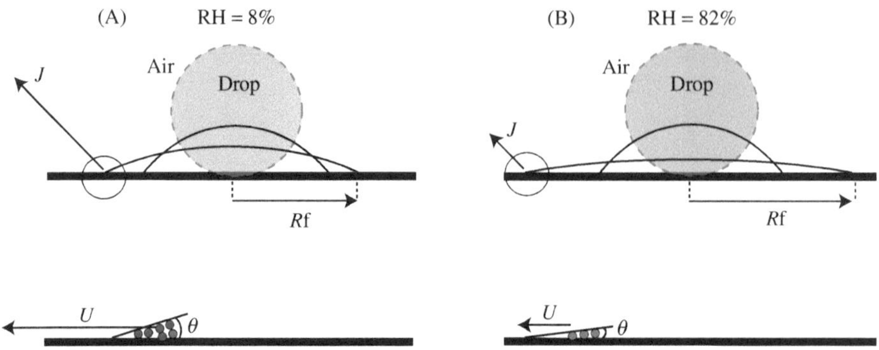

FIGURE 23.6 Schematic representation of spreading drops for the case of (A) RH = 8% and (B) RH = 82%. Initially, the blood drop has a spherical shape, and at equilibrium the drop has the shape of a spherical cap. The arrows indicate the velocity and evaporative flux of water vapor from the drop surface to the surroundings in the vicinity of the triple line. The dashed square marks the area close to the triple line, where the drop geometry can be approximated by a wedge.

For small contact angles, this diverging evaporative flux leads to a diverging velocity field ($u(\rho, \theta) = (1/\rho_l)J(\rho)$) because of the singular corner geometry of the droplet close to the contact line (Gelderblom et al., 2012). The nonuniform evaporation mass flux leads to temperature gradients over the drop surface (self-cooling), which leads to differences in surface tension and drives a Marangoni flow inside the droplet (Hu and Larson, 2006). This internal flow transports the micrometer-sized colloids and liquid toward the rim to replenish the liquid near the contact line that evaporates significantly faster. The intensity of the rate of evaporation scales with this internal flow motion inside the drop of blood. Indeed, an increased rate of evaporation is observed in the vicinity of the triple line at high evaporative rates (Figure 23.6A) compared to low evaporative rates (Figure 23.6B). This increase in the evaporation rate enhances the transportation mechanism of colloids toward the edge of the blood drop and thus increases the flow velocity near the contact line. Consequently, the mean velocity near the contact line decreases with increasing RH.

In the diffusion-limited evaporation case and for low contact angles ($\theta_{eq} < 20°$), the spreading/evaporation process can be divided into two regimes: a fast first regime and a slow second regime. At the first regime, the spreading process is controlled by viscous forces and surface tension forces, because all of the experimental radius curves collapse on one another (Figure 23.7). The curves of the spreading radius were scaled by dividing the spreading time by viscous-capillary characteristic time, $t_D = 4\pi\mu/\gamma R_f$. The radius evolution can be expressed by a power law of the form $r^*(t) = k(t/t_D)^n$, with a preexponential factor $k = 2.19 \times 10^{-2}$ and a wetting exponent $n = 0.65 \pm 0.11$, which was obtained by fitting all the experimental

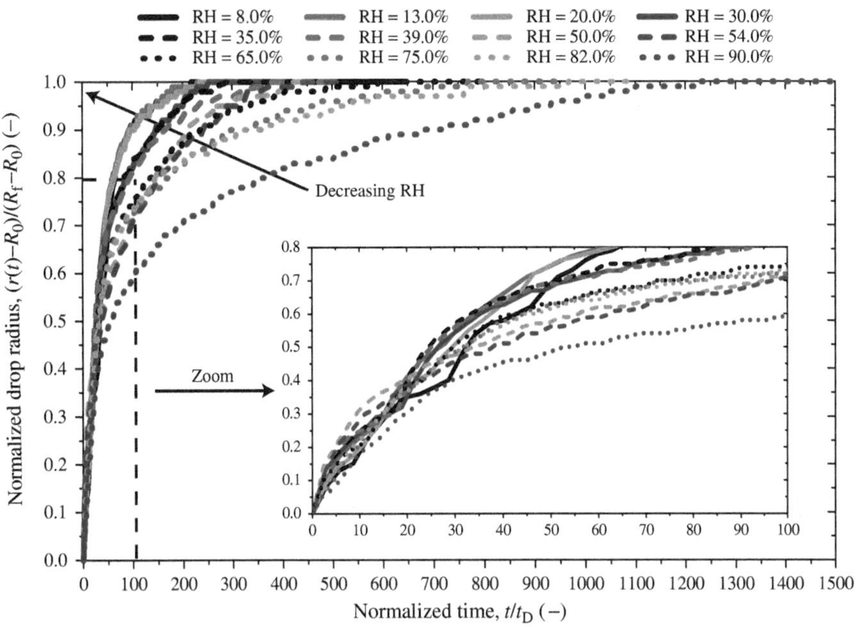

FIGURE 23.7 Normalized drop radius, $r^*(t) = (r(t) - R_0)/(R_f - R_0)$, as a function of normalized time, t/t_D, for RH varying from 8% to 90%. The inset graph presents a magnification of t/t_D from 0 to 100.

curves. Indeed, the evolution of the wetted drop radius depends on the dominant physical forces that are driving and resisting the spreading process. From Table 23.4, it was concluded that $F_{\text{capillary}} > F_{\text{gravity}} > F_{\text{viscosity}} > F_{\text{inertia}}$. Thus, the effects of gravity are negligible, and capillary forces dominate the flow motion inside the drop of blood. In this regime, capillary forces rapidly drive the drop spreading because of the curved surface at the contact point (between the drop and the glass substrate), whereas the viscous forces of the fluid resist deformation. The value of the wetting exponent ($n = 0.65 \pm 0.11$) can be explained by the balance of the viscous stress ($\sim \mu(\mathrm{d}r/\mathrm{d}t)/L$) with the capillary pressure ($\gamma R/r^2$). For a viscosity equal to or larger than 10 mPa s, Legendre and Maglio (2013) observed a wetting exponent of 2/3 when the viscous-capillary characteristic time $t\mu = \mu R_0/\gamma$ was used. This shows that bio-colloids inside the blood drop have no significant effect at the early stages of spreading.

At late times, the spreading process is controlled by the competition between viscous forces (shear-thinning) and by the difference between the saturated pressure at the drop interface and the air pressure, because all of the experimental radius curves collapse on one another (Figure 23.8). The curves of the spreading radius were scaled by dividing the spreading time by the evaporative time, t_F. When the

TABLE 23.4 Estimation of Characteristic Dimensionless Numbers for the Two Regimes of Blood Drop Spreading (RH = 65%)

Number	Formula	Signification	Regime I	Regime II	Ratio I/II
R (mm)	–	Mean radius	3.18	4.54	0.7
h (mm)	$\frac{\overline{R}(1-\cos\theta)}{\sin\theta}$	Mean height	1.05	0.41	2.5
U_{CL} (mm s^{-1})	–	Mean velocity	1.92	0.11	17.4
Re (–)	$\frac{\rho\overline{U}_{CL}\overline{h}}{\mu}$	Inertia/viscous	0.209	0.004	44.4
Ca (–)	$\frac{\mu\overline{U}_{CL}}{\gamma}$	Viscous/capillary	2.75×10^{-4}	1.58×10^{-5}	17.0
Bo (–)	$\frac{\rho g\overline{h}^2}{\gamma}$	Gravity/capillary	0.160	0.024	6.5
We (–)	$\frac{\rho\overline{U}_{CL}^2\overline{h}}{\gamma}$	Inertia/capillary	5.74×10^{-5}	7.41×10^{-8}	774
Fr (–)	$\frac{\overline{U}_{CL}^2}{g\overline{h}}$	Inertia/gravity	3.59×10^{-4}	3.00×10^{-6}	120
St (–)	$\frac{\mu\overline{U}_{CL}}{\rho g\overline{h}^2}$	Viscous/gravity	1.72×10^{-3}	6.38×10^{-4}	2.7

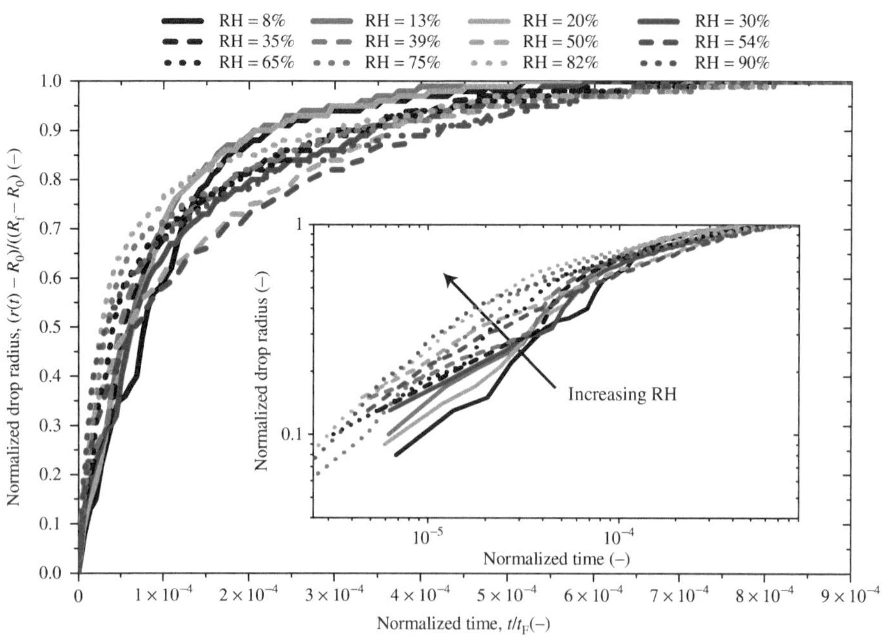

FIGURE 23.8 Normalized drop radius, $(r(t) - R_0)/(R_f - R_0)$, as a function of normalized time, t/t_F, for different humidities.

evaporation process is limited by the diffusion of water into the air (without heating), a purely diffusive model was used to predict the total evaporation time, t_F:

$$t_F = \frac{4RTm_0}{\pi \Delta PDD_{diff}}$$

For the second regime of spreading, the radius evolution can be expressed by a power law of the form $r^*(t) = k(t/t_D)^n$, with a wetting exponent $n = 0.19 \pm 0.03$. From the literature, it is known that the rate of spreading of shear-thinning fluid is lower than that of Newtonian fluid ($m = 1$) because of the shearing forces near the moving contact line (Carre and Eustache, 2000):

$$\cos\theta_{eq} - \cos\theta_D \left(\frac{U_{CL}}{\theta(U_{CL})}\right)^m$$

Indeed, the shear-thinning rheology decreases the curvature of the liquid–vapor interface near a moving contact line. Thus, a weaker dependence of the capillary number Ca with the dynamic contact angle, θ_D, appeared compared to a Newtonian fluid ($\theta_D \sim Ca^{1/3}$). Contrary to these observations for shear-thinning fluids, the wetting exponent is found to be 0.19 (higher than that of Tanner's law, $n = 0.10$) because blood is a complex colloidal suspension and different physical mechanisms are involved inside the blood droplet. Table 23.4 shows that the second regime of spreading is governed by viscous dissipation near the contact line. However, the calculation of the Bond number ($Bo < 1$) revealed that the shape of the drops is a hemispherical cap and that gravitational effects can be neglected. The effect of gravitational forces can lead to acceleration in the spreading ($r \propto t^{1/8}$), and a very small contribution can be added to the internal flow (Cazabat and Cohen Stuart, 1986). For low contact angles, the rate of evaporation from the free surface is not uniform and diverges near the contact line. A small temperature gradient develops at the vapor–liquid interface because the surface of the droplet is cooled by evaporation. Thus, a surface tension gradient is generated that can induce a Marangoni flow (Hu and Larson, 2005). Consequently, an evaporatively driven Marangoni flow could enhance the spreading rate of the second regime. Another possible mechanism that could be responsible for spreading enhancement is the disjoining pressure, $\Pi(e)$. This produces a pressure gradient that competes with the capillary pressure and viscous resistance to flow (Semenov et al., 2014). This demonstrates that the accumulation of bioparticles near the moving contact line plays an important role in the spreading dynamics because the velocity of the contact line is increasing rather than decreasing.

At the final stage of the evaporation process, complex patterns composed of cracks are formed on the glass substrate. In the experiments, the transfer of water in the air is limited by diffusion and therefore controlled by RH. The evaporation of whole blood drops on glass was investigated in previous studies (Brutin et al., 2011; Sobac and Brutin, 2011). The authors described the structural evolution of the blood drop leading to the final pattern observed and analyzed the competition between gelation and evaporation (Sobac and Brutin, 2011). Several physical mechanisms were involved, including Marangoni flow, crack formation, gelation, and adhesion. Also, the authors described the different regimes of evaporation kinetics (Brutin et al., 2011): a first regime driven by convection, diffusion, and gelation; a transition phase occurring with the complete gelation of the system; and a regime that is only driven by diffusion (solvent moves by permeation into the porous matrix). For low contact angles ($\theta < 90°$) and for pinned contact lines, Hu and Larson (2002) obtained the evaporation rate of a droplet for any contact angle:

$$\dot{J}_{\mathrm{HL}} = \pi R D_{\mathrm{diff}} C_{\mathrm{v}}(1 - \mathrm{RH}) f(\theta)$$

where $f(\theta)$ is a function of the contact angle ($f(\theta) = 0.27\theta^2 + 1.30$), D_{diff} is the diffusivity of the vapor in air saturation, C_{v} is the saturation concentration value, and $C_{\mathrm{v}}(1 - \mathrm{RH})$ is the difference of the vapor concentration between the interface assumed at saturation and the far-field from the drop surface. For contact angles less than 40°, the function of the contact angle $f(\theta)$ is almost flat. Consequently, the influence of the contact angle is negligible, and the evaporation rate from the droplet surface remains almost constant over time.

The drying process of a sessile drop of blood is characterized by an evolution of the solution into a gel saturated with solvent. When the gel is formed inside the drop, the porous matrix formed by the aggregation of particles continues to dry by the evaporation of the solvent, which causes the gel to consolidate. Indeed, the system is defined by the two interfaces shown in Figure 23.9: the solid–solvent

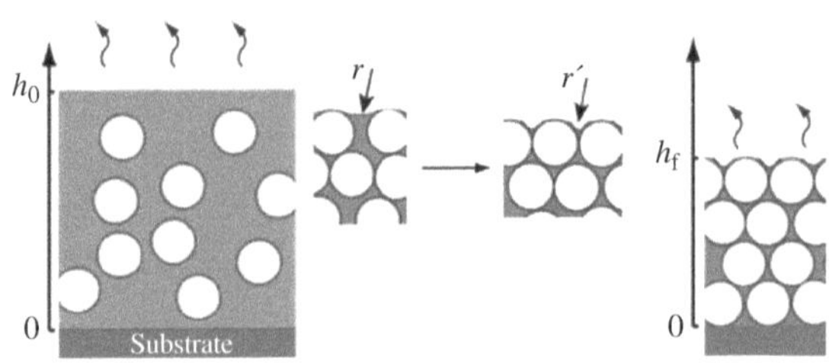

FIGURE 23.9 Formation of a porous matrix during the evaporation process (Lazarus and Pauchard, 2011). (Left) Particles are dispersed in the solvent. (Right) Concentrated suspension of particles (porous matrix) in contact with one another, surrounded by solvent-filled interstices. The system is bounded by two interfaces: the solid–gel interface and the gel–air interface. A curvature of the solvent–air menisci occurs at the evaporation surface and increases during the evaporation process. The competition between adhesion and evaporation leads to crack formation.

interface and the solvent–air interface. At this latter interface, menisci form and are responsible for a capillary pressure, P_{cap}, in the liquid phase:

$$P_{cap} = \frac{\alpha \gamma_{(s,a)} \cos(\theta)}{r_p}$$

where $\gamma_{(s,a)}$ is the solvent–air surface tension, r_p is the pore radius, and α is a geometrical constant with a value of approximately 10 (Brinker and Scherer, 1990). As evaporation continues, the curvature of the menisci increases, which leads to a depression in the solvent filling the interstices. However, this depression induces gradual shrinkage of the porous matrix that is constrained by the adhesion of the gel to the glass substrate and the evaporation of the solvent. As tensile stresses build up, the internal stresses become important, and cracks appear so as to release mechanical energy. However, the dynamics of the cracks depend on the evolution of mechanical stresses and on the thickness of the gel layer or deposit. Below a critical thickness, the layer can be free of cracks at the end of the evaporation process (Bohn et al., 2005).

During the evaporation of the blood drop, a gelled foot forms at the contact line and propagates radially inward, while the central area of the blood drop remains liquid. As evaporation continues, this gelled foot leads to the development of ortho-radial constraints in the gel and thus to the formation of radial cracks. The first cracks nucleate near the contact after the beginning of the evaporation process. Recently, Sobac and Brutin (2011) performed experiments on drying drops of blood and observed that these cracks nucleate at a critical particle mass concentration of 29.6%, depending on the initial mass of the drop. After complete evaporation of the blood drop, the final drying pattern is characterized by the presence of four distinguished regions (Figure 23.10):

- A central region ($r/R_{max} < 0.50$) composed of a sticking deposit and a network of small-scale disordered cracks.
- A transition area ($0.50 < r/R_{max} < 0.76$) located between the central area and the corona region. This area is composed of a network of small and large radial cracks.
- A corona region ($0.76 < r/R_{max} < 0.95$) composed of polygonal cells or wide mobile plaques delimited by white cracks.
- A fine periphery ($r/R_{max} > 0.95$), which is completely adhering to the solid substrate.

23.4 Other fluids

Other studies have focused on understanding the formation of patterns left by protein solutions. For instance, a regularly radial crack is often observed. The

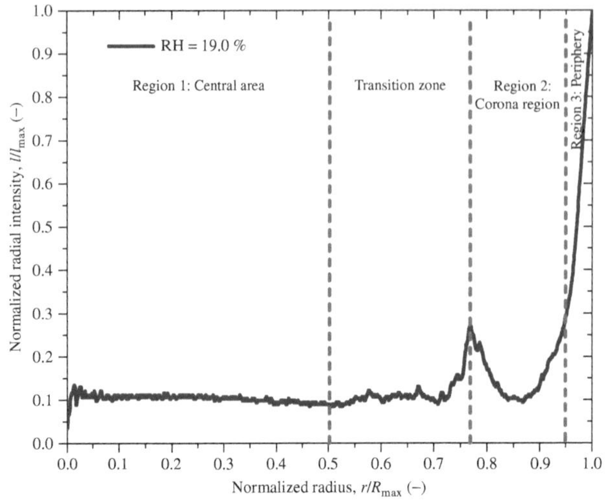

FIGURE 23.10 Variation of the cumulated ortho-radial light intensity (I/I_{max}) as a function of the dimensionless radial position (r/R_{max}) at 19.0% RH.

evaporating, gelling, and cracking dynamics of a deposited drop of bovine serum albumin–salt solution were investigated (Annarelli et al., 2001). At the initial stage of drop evaporation, the authors observed a regularly spaced cracks at the edge of the gelled foot. This was shown to be due to competition between the adhesion of the biological deposit on the solid substrate and the drop evaporation-induced ortho-radial stresses. Also, Yakhno et al. (2008) performed experiments with drops of protein-salt-water. They showed that the patterns observed in dried drops are a result of the protein's phase transition (gelation). An attachment of the dried drop to the glass plate due to the adsorption of protein and salt crystallization was observed in the dried drop. Buzoverya et al. (2012a,b) dried drops of human albumin serum (HAS) on a glass slide and found different structures of patterns left by a dried protein solution (HAS), depending on the salt and solvent concentrations.

Glossary

Delamination repeated cyclic stresses, impact, and so on can cause layers to separate, forming a mica-like structure of separate layers, with significant loss of mechanical toughness.

Biocolloids any colloid derived from any living organism.

Gelation the formation of a solid by cooling; freezing.

References

Annarelli, C.C., Fornazero, J., Bert, J., Colombani, J., 2001. Crack patterns in drying protein solution drops. Eur. Phys. J. E 5, 599–603.

Allain, C., Limat, L., 1995. Regular patterns of cracks formed by directional drying of a colloidal suspension. Phys. Rev. Lett. 74, 6672–6677.

Bou-Zeid, W., Brutin, D., 2013. Influence of relative humidity on spreading, pattern formation and adhesion of a drying drop of whole blood. Colloids and Surfaces A: Physicochemical and Engineering Aspects 430, 1–7.

Bou-Zeid, W., Brutin, D., 2014. Effect of relative humidity on the spreading dynamics of sessile drops of blood. Colloids Surf. A 456, 273–285.

Bou-Zeid, W., Vicente, J., Brutin, D., 2013. Influence of evaporation rate on crack's formation of a drying drop of whole blood. Colloids and Surfaces A: Physicochemical and Engineering Aspects 432, 139–146.

Bohn, S., Pauchard, L., Couder, Y., 2005. Hierarchical crack pattern as formed by successive domain divisions. Phys. Rev. E 71, 046214.

Brinker, C.J., Scherer, G.W., 1990. Sol-Gel Science: The Physics and Chemistry of Sol-Gel Processing. Academic Press, Boston, MA.

Brutin, D., Sobac, B., Loquet, B., Sampol, J., 2011. Pattern formation in drying drops of blood. J. Fluid Mech. 667, 85–95.

Brutin, D., Sobac, B., Nicloux, C., 2012. Influence of Substrate Nature on the Evaporation of a Sessile Drop of Blood. J. Heat Transfer 134, 061101.

Buzoverya, M.E., Shcherbak, Y.P., Shishpora, I.V., Potekhinab, Y.P., 2012a. Microstructural analysis of biological fluids. Tech. Phys. 82, 123–128.

Buzoverya, M.E., Shcherbak, Y.P., Shishpora, I.V., 2012b. Experimental investigation of the serum albumin fascia microstructure. Tech. Phys. 57, 1270–1276.

Carre, A., Eustache, F., 2000. Spreading kinetics of shear-thinning fluids in wetting and dewetting modes. Heat Mass Transfer. 16, 2936–2940.

Cazabat, A.M., Cohen Stuart, M.A., 1986. Dynamique de l'étalement de gouttes sur une surface. Loi de tanner et effet de la gravité. C. R. Acad. Sci. Paris II 301, 1337–1339.

Gelderblom, H., Bloemen, O., Snoeijer, J.H., 2012. Stokes flow near the contact line of an evaporating drop. J. Fluid Mech. 12, 1–16.

Hu, H., Larson, R.G., 2002. Evaporation of a sessile droplet on a substrate. J. Phys. Chem. B 106, 1334–1344.

Hu, H., Larson, R.G., 2005. Analysis of the effects of Marangoni stresses on the microflow in an evaporating sessile droplet. Langmuir 21, 3972–3980.

Hu, H., Larson, R.G., 2006. Marangoni effect reverses coffee-ring depositions. J. Phys. Chem. B 110, 7090–7094.

Lazarus, V., Pauchard, L., 2011. From craquelures to spiral crack patterns: influence of layer thickness on the crack patterns induced by desiccation. Soft Matter 7, 2552–2559.

Legendre, D., Maglio, M., 2013. Numerical simulation of spreading drops. Colloids Surf. A 432, 29–37.

Pauchard, L., 2006. Patterns caused by buckle-driven delamination in desiccated colloidal gels. Europhys. Lett. 74, 188–194.

Semenov, S., Trybala, A., Rubio, R.G., Kovalchuk, N., Starov, V., Velarde, M.G., 2014. Simultaneous spreading and evaporation: recent developments. Adv. Colloid Interface Sci. 206, 382–398.

Shabalin, V.N., Shatokhina, S.N., 2007. Diagnostic markers in the structures of human biological liquids. Singapore Med. J. 48, 440.

Sobac, B., Brutin, D., 2011. Structural and evaporative evolutions in desiccating sessile drops of blood. Phys. Rev. E 84, 011603.

Sobac, B., Brutin, D., 2014. Desiccation of a sessile drop of blood: cracks, folds formation and delamination. Colloids Surf. A 448, 34–44.

Tarasevich, Y.Y., Irina, V.V., Isakova, O.P., 2011. Desiccating colloidal sessile drop: dynamics of shape and concentration. Colloid Polym. Sci. 289, 1015–1023.

Thieret, M., 2008. Biology and Mechanics of Blood Flows, Part II: Mechanics and Medical Aspects. Springer, New York, NY.

Yakhno, T.A., 2008. Salt-induced protein phase transitions in drying drop. J. Colloid Interface Sci. 318, 225–230.

Yakhno, T.A., 2010. Protein phase instability developed in plasma of sick patients: clinical observations and model experiments. Nat. Sci. 2, 220–227.

Yakhno, T.A., Sedova, O.A., Sanin, A.G., Pelyushenko, A.S., 2003. On the existence of regular structures in liquid human blood serum (plasma) and phase transitions in the course of its drying. Tech. Phys. 48, 399–403.

Zhuang, H., Coulepis, A.G., Locarnini, S.A., Gust, I.D., 1982. Detection of markers of hepatitis B infection in serum dried on to filter paper: an application to field studies. Bull. World Health Organ. 60, 783–787.

Chapter 24

COMPLEX FLUIDS DROPLETS IN LEIDENFROST STATE

Stéphane Dorbolo[1], Laurent Maquet[1], Benjamin Sobac[2,4], Alexey Rednikov[2], Pierre Colinet[2] and Florian Moreau[3]

[1]Université de Liège—GRASP Laboratory, Liège, Belgium, [2]Université Libre de Bruxelles—TIPs Fluid Physics, Brussels, Belgium, France, [3]PPrime ENSMA, Poitiers, France, [4]Mechanical Engineering Department, Aix-Marseille University, Marseille, France

A droplet may levitate on its own vapor when dropped on a hot surface. This non-wetting situation is a great opportunity to study the evaporation of complex fluids. In this chapter, we present two case studies in order to evidence general statements about the behavior of complex fluids in Leidenfrost droplets. The first case concerns water and a surfactant mixture. While the droplet is evaporating, it is covered by a thin shell of sodium dodecyl sulfate. Eventually, the droplet explodes. The second case concerns microparticles (glass or basalt). During evaporation, the droplet is encapsulated in a monolayer of grains.

24.1 Introduction

The evaporation rate of a liquid droplet deposited on a surface increase in plate temperature. This statement is only true until it reaches a certain temperature, called the Leidenfrost temperature, after the scientist who first reported the phenomenon (Quéré, 2013). Above this temperature, the evaporation rate is so fast that when a droplet is released over the plate, the droplet levitates over its own

Droplet Wetting and Evaporation. DOI: http://dx.doi.org/10.1016/B978-0-12-800722-8.00024-2

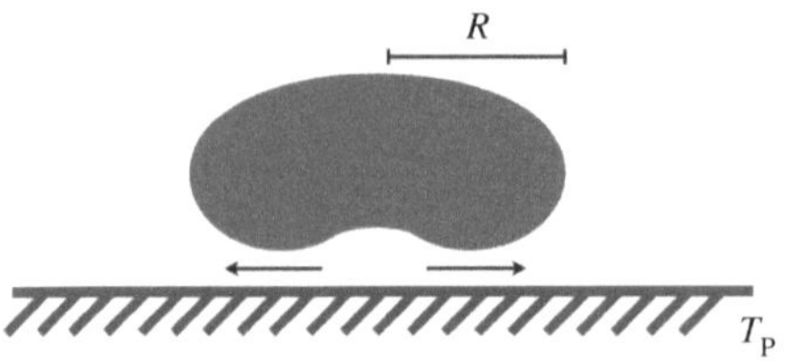

FIGURE 24.1 Sketch of a droplet in Leidenfrost state. See Chapter 7 for an extended discussion about the shape of the droplet.

vapor (Figure 24.1). Because the vapor under the droplet acts as a thermal insulator, the evaporation rate drops drastically. This particular transition has been intensively revisited in the last decade.

The Leidenfrost transition temperature has been shown to depend not only on the liquid nature but also on the substrate properties. The bulk thermal conductivity of the substrate has to be large enough to transmit enough energy from the plate to the droplet. The surface roughness is also crucial. Indeed, the asperity size and distribution over the surface determine the critical thickness of the vapor layer and consequently the Leidenfrost temperature (Kim et al., 2011). In short, to lower the Leidenfrost temperature, the substrate has to be a good thermal conductor and as smooth and flat as possible.

As far as applications are concerned, the Leidenfrost effect may be undesired during cooling processes (heating systems, cooling of a nuclear reactor; Kim et al., 2011). On the other hand, this effect can be desired as the droplet is in a perfect nonwetting situation. This latter point of view is the one adopted in this chapter.

In this nonwetting situation, the droplet has no contact with any solid, and therefore it is very sensitive to any tilting of the plate because the friction is strongly reduced by the presence of the vapor film. The shape of the droplet is nevertheless different than in a non-wetting situation (Sobac et al., 2014). However, by engraving the surface, it has recently been shown that a regular pattern makes it possible to control the movement of the droplet (Dupeux et al., 2011; Soto, 2014). The trick is to manipulate the vapor flow below the droplet to induce its motion by viscous entrainment. The manipulation of the drops then becomes possible without any contact.

The droplet can also be thought of as a chemical reactor without borders. This very promising feature has recently been exploited by (Elbahri et al., 2007). Moreover, self-assembly can be triggered by a Leidenfrost droplet and be used in nanofabrication processes (Elbahri et al., 2007; Abdelaziz et al., 2013). Finally, the droplet is also able to transport small objects. Taking advantage of the nearly zero friction with the substrate, the droplet can transport small electronic compounds (Hashmi et al., 2012).

According to works by Tsapis et al. (2005), when nanobeads are inserted in the droplet, they move toward the interface to form a shell. An interesting question is whether this process generalizes to other scales. In this chapter, we focus

FIGURE 24.2 Picture of a droplet in Leidenfrost state. This drop is about 10 mm of diameter.

on what happens to the Leidenfrost effect when a material is added to water in two case studies. The main phase is water. In the first case, we add a surfactant (SDS: sodium dodecyl sulfate), and in the second case, we add microparticles. We therefore have a solution and a two-phase system, respectively. We will show that when a droplet of these mixtures is set in its Leidenfrost state, in both cases, the secondary phase (surfactant or microparticles) moves toward the interface, forming a layer that wraps around the droplet. Eventually, spectacular phenomena like explosions may occur. A model is proposed that takes into account the main physical ingredients.

24.2 Experimental setup

The experimental setup is common to both case studies. A polished aluminum plate was placed on a heating system. The plate was slightly curved so the Leidenfrost droplet would remain near the center of the plate. A thermocouple was screwed at the surface of the plate. A PID device controlled the heating using the thermocouple information as feedback. Once the set temperature point was reached, the droplet of the concerned mixture was released over the plate. Figure 24.2 presents a typical picture of a drop in the Leidenfrost state.

The experimental data were obtained by using either high-speed imaging or a high-resolution camera. The high-speed camera was an IDT-Motion Y4, which can take up to 30,000 frames per second. The high-resolution camera was a Canon 400D.

24.3 Water + surfactant

The surfactant used is SDS. The critical micellar concentration (CMC) is 2.37 g L^{-1}. The mixtures were prepared with several concentrations. Therefore, for a given concentration, the volume of the droplet determines the total amount

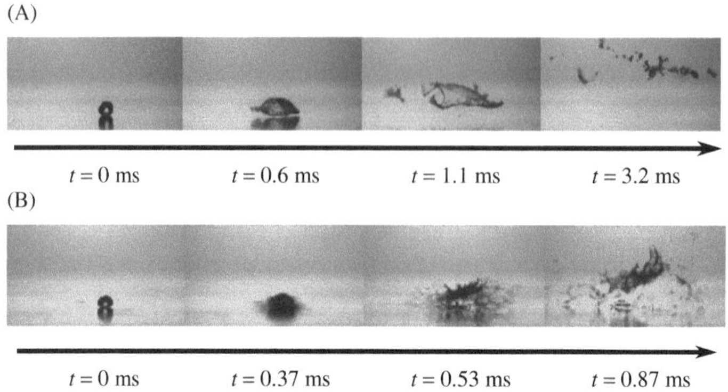

FIGURE 24.3 Typical droplet explosions. (Top) Initial concentration 0.1 CMC and initial volume 1000 μL. The droplets on the left measure approximately 2 mm in diameter. (Bottom) Initial concentration 1 CMC and initial volume 100 μL. Note that $t = 0$ ms corresponds to the beginning of the explosion.

of SDS. During the evaporation, the concentration keeps on increasing. Let us first examine some typical experiments (see two examples in Figure 24.3). Two droplets (one of 1 mL and another of 100 μL), both containing a solution of SDS at a concentration 0.1 CMC, are released on the plate heated at 300°C. Figure 24.3 presents two series of four snapshots of these droplets during the evaporation.

To begin with, the droplet is in the Leidenfrost state and it evaporates. By image analysis of pictures taken from the side, the volume is computed as a function of time. After a while, the droplet appears to be covered by a thin layer of a transparent material. Finally, the droplet suddenly explodes! Several features have to be discussed here. First, what is the process that is responsible for the generation of the shell? Second, what is the nature of this thin shell? Third, why does the droplet explode and what are the conditions to observe this unexpected phenomenon?

The initial amount of SDS m_{i} was tuned in two ways: by changing the initial volume and by changing the initial concentration. In Figure 24.4, the volume evolution of the droplets is represented as a function of time when the plate is at 300°C. The different symbols (see Figure 24.4 caption) represent different initial volumes V_0 (10, 100, and 1000 μL) and different initial concentrations ρ_0 (0.1, 1, and 10 times the CMC). The plain symbols represent the moment of the explosion. The horizontal dashed lines clearly show that the explosions occur at the same volume when the initial amount of SDS is the same. In other words, the explosion occurs at a defined critical mean concentration of SDS $\rho^* = \rho_{\mathrm{f}}$, which is the final concentration. This concentration is found to be about 100 CMC.

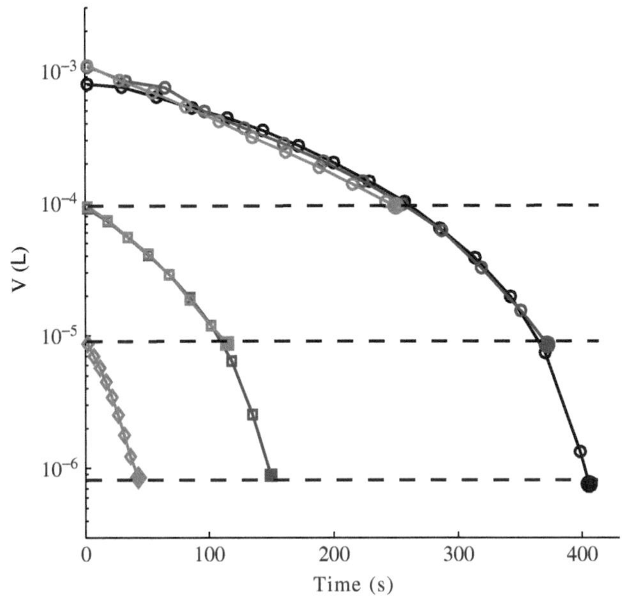

FIGURE 24.4 Volume versus time for water/SDS droplets of several initial sizes and initial SDS concentrations. Initial volume: 1000 μL, ○; 100 μL, □; 10 μL, ◇. Initial concentration: red, 10 CMC; green, 1 CMC; blue, 0.1 CMC. Filled symbols are used for the final volume before the explosion. (For interpretation of the references to color in this figure legend, the reader is referred to the online version of this book.)

Starting with the same amount of SDS—that is, 100 μL droplet of 10 CMC of SDS—we also varied the temperature of the plate. Figure 24.5 presents the evolution of the final mean concentration as a function of the temperature of the plate. The final mean concentration is found to decrease with the temperature of the plate. Moreover, no explosion is observed below 250°C.

The thin transparent shell was analyzed. Actually, the droplet remains mobile during the whole evaporation process until the explosion. By gently blowing on the droplet, it is possible to capture the droplet on a plate. In so doing, we found that the thin shell completely encapsulates the droplet. Interestingly, even if the droplet is encapsulated, it is still mobile and we conclude that the droplet remains in a levitating state. Several shells were collected and tested at a nuclear magnetic resonance device. The shell is shown to be made of SDS. The "solid" shell results from a phase transition of the SDS at high concentration—indeed, a phase transition at the surface.

The migration of SDS at the interface can be modeled as follows.

First we assume that the droplet is small and spherical. Initially, the concentration of SDS is uniform in the droplet. A heat conduction-limited model (Biance et al., 2003) gives the temporal evolution of the radius R as $\dot{R} \sim -\frac{\chi}{R}$ with

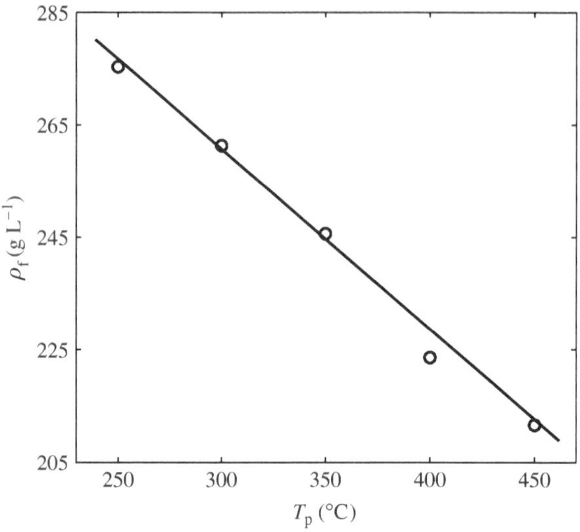

FIGURE 24.5 Final mean concentration versus the temperature of the plate, for a water/SDS droplet with an initial volume 100 μL and an initial concentration 10 CMC.

$\chi = \lambda_v \Theta / \rho_l \mathcal{L}$ where λ_v is vapor thermal conductivity, Θ is the plate superheat (difference between the temperature of the plate and the boiling temperature), ρ_l is the liquid density, and $\mathcal{L}$ is the latent heat of evaporation. The spatial distribution of surfactant does not remain uniform during the evaporation process. There is an accumulation of SDS at the surface of the droplet in a small layer of thickness e. The temporal evolution of the thickness of this layer scales as $e(t) \sim D/|\dot{R}| \sim \epsilon R(t)$, where D is the diffusion coefficient of micelles, and $\varepsilon = D/\chi$. With the assumption that the SDS mean concentration before explosion, ρ_f, is much larger than the initial concentration, ρ_i, expressing the SDS mass conservation leads to a direct link between the final mean concentration and the surface concentration ρ_Σ: $\rho_f \sim \varepsilon \rho_\Sigma$. Here, the key assumption is that when a defined value of the surface concentration is reached, the explosion is triggered. Then, this model shows that the final mean concentration does not depend on the initial volume and initial mean concentration, as experimentally found. Furthermore the final mean concentration decreases as the temperature of the plate increases (as $\varepsilon \propto \theta^{-1}$) but not as fast as experimentally observed (this is probably due to the simplicity of our model; Moreau et al., 2014).

24.4 Water + microparticle

Microparticles can also be used to wrap a droplet. To obtain such *liquid marbles*, one can use hydrophobic grains (Aussilous and Quéré, 2001). Such droplets have

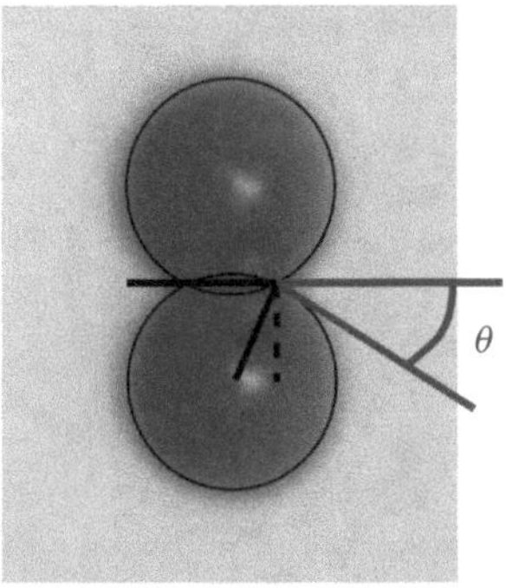

FIGURE 24.6 Basalt bead floating under an air–water interface. The upper part results from the reflection on the interface. The fit of circles on the bead and on its image makes it possible to extract the contact angle θ. The bead is about 100 μm in diameter.

FIGURE 24.7 A Leidenfrost drop made of water + fluorescein in which basalt beads have been poured. As the droplet evaporates, the beads cover a larger fraction of the surface and form a monolayer. The drop on the left is about 30 mm in diameter.

been tested in the Leidenfrost state (Aberle et al., 2011). The grains fully covered the droplet thermally insulating the droplet from the plate. In the present case, however, we considered glass beads and basalt beads which are both hydrophilic materials (Maquet et al., 2014). To check this, beads were gently put on the surface of the water and were observed to float. That means that buoyancy and surface tension surpass the weight. Figure 24.6 shows a picture of a basalt bead (106–125 μm) trapped at an air–water interface. By image analysis, the contact angle is found to be 30°.

The experimental procedure is as follows. A large droplet of water (much larger than the total volume of the grains in order to allow the grains to freely move) is released on the heated plate. Then a given quantity of grains is put in the droplet using a funnel. The grains cross the droplet interface and gather at the bottom of the Leidenfrost droplet. As the water evaporates, the grains start forming a stripe and cover the droplet more and more, starting from the bottom. Figure 24.7 shows three pictures taken at different moments of the evaporation process, showing the "invasion" of the whole droplet surface by a layer of grains. Eventually, the droplet is completely wrapped by the grains, similarly to a liquid marble. Later the covering layer buckles and shrinks until the water completely evaporates. It is crucial to note that this phenomenon cannot be observed when the water is replaced by

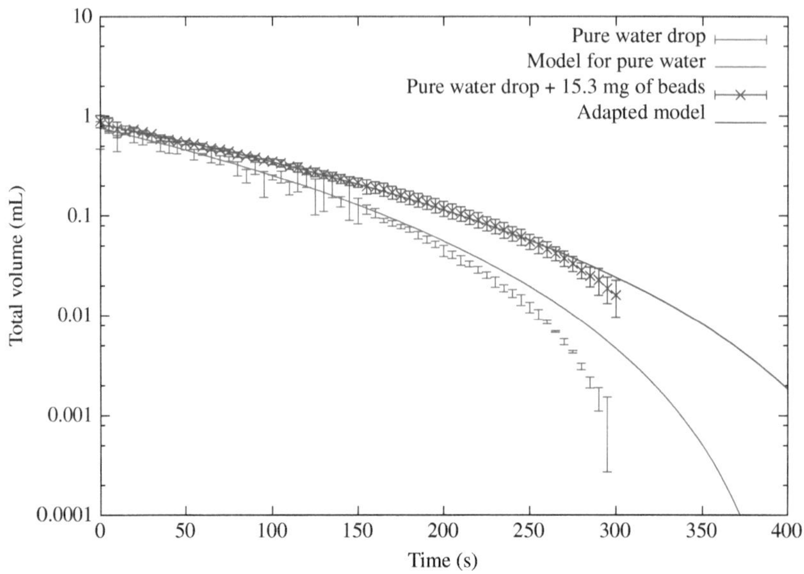

FIGURE 24.8 Evolution of the volume of the drop with time for pure water drops (red) and for drops with glass beads in them. The lines correspond to fit of the experimental data with the developed model (cf. text). (For interpretation of the references to color in this figure legend, the reader is referred to the online version of this book.)

ethanol. In this latter case, the grains cannot float on the interface as the surface tension of ethanol is three times smaller than for the water. Therefore, they cannot be trapped at the interface and they cannot form a monolayer. Consequently, the questions now are: first, does the presence of the grains affect the evaporation rate, and second, how do the grains self-organize in the droplet?

In Figure 24.8, the volume of the drop as a function of time is represented for different loads of grains and for different sizes of grains. The red symbols correspond to the evaporation of a drop of water without any grains. Remarkably, whatever the number of grains present in the drop, the evaporation rate is decreased by the presence of the grains. This result may seem surprising *a priori*. Indeed, it is known that the evaporation rate increases with the density of the liquid (Biance et al., 2003). In the present case, the grains increase the average density of the drop. This fact is amplified as the water evaporates—in other words, the ratio of volume of beads to volume of water keeps on increasing. The reason for the slowing of the evaporation should be found in the decrease of the free surface at the bottom of the droplet imposed by the grains. The bottom surface size is critical because the evaporation mainly occurs at the base of the droplet (see also Chapter 7).

The grains of mean radius R_b seem to self-organize into a monolayer that surrounds the droplet. The efficiency of the trapping of the grains at the interface when the coverage becomes complete has been estimated knowing the number of grains and knowing (as we measured) that the grains occupied 80% of the surface (2D Random Close Packing fraction of polydisperse disks, ϕ_0 (Berryman, 1983)). The efficiency of the trapping is then given by the ratio between the surface of the drop, S_d, and the surface that a number of grains N_b can cover, S_b.

$$E = \frac{S_d}{S_b} = \frac{S_d \, \phi_0}{N_b \, \pi \, R_b{}^2}$$

This quantity was computed and reported as a function of the total mass of grains in the drop. This efficiency is always found to be between 0.7 and 1. It can thus be claimed that grains self-organize as a monolayer.

As the monolayer first forms at the bottom of the drop, the vapor flow is modified by the presence of the beads. Indeed, the free surface has to be corrected lowered by the presence of the grains as they emerge from the surface. The surface occupied by the emerged part of the grains is related to the contact angle of the water on the grains, θ. The fraction of the surface intercepted by the emerged grains is given by

$$\phi_{occ} = \phi_0 \sin^2 \theta$$

Consequently, the area below the drop from which the vapor can flow is reduced by a constant factor given by $1 - \phi_{occ}$. The scaling of the evaporation rate proposed by Biance et al. (2003) can tentatively be corrected as follows:

$$\frac{dV_d}{dt} \simeq -BV_d^{3/4}\left(1 + \frac{V_b}{V_d}\,\frac{\rho_b - \rho_l}{\rho_l}\right)^{1/4}$$

with

$$B = \left(\frac{\lambda_v \Delta T}{Ll_c \rho l}\right)^{3/4} \left(\frac{\pi g \rho_v l_c}{6\eta_v}\right)^{1/4} (1 - \phi_{occ})^{3/4}$$

where the subscripts d, b, l, and v refer to the drop, the beads, the liquid, and the vapor, respectively. V is the volume, ρ is the density, λ is the thermal conductivity, η is the viscosity, and ΔT is the difference in temperature between the substrate and the drop (which is at the boiling temperature of the liquid). The corrections are (i) the second factor of the second member of the first equation, which takes into account the density variation of the droplet during the evaporation, and (ii) the factor $(1 - \phi_{occ})^{3/4}$, which takes into account the decrease of the free surface at the bottom of the drop by the grains. The solid blue curve in Figure 24.8 shows a good agreement with the measured volume decrease of the droplet containing grains.

24.5 Summary and perspectives

Leidenfrost drops are a wonderful tool to study complex liquid evaporation without contact line. Through both case studies, we show that interesting features can emerge, such as phase transition, explosion, and self-organization. To summarize, the evaporation induces an increase in the mean concentration of the added material. The concentration is not homogeneous, and, roughly speaking, all the added material becomes concentrated at the surface of the droplet. This often leads to encapsulation of a drop by the added material. As the concentration becomes very high, phase transition may occur. This was the case with SDS: SDS transforms into a gel-like phase that encapsulates the droplet. In the case of grains, the physical properties of the grains are relevant. Indeed, the encapsulation by a self-organized layer of grains is only possible if the grains are able to be trapped at the interface. The results of the evaporation can be of interest regarding both application and fundamental science. In the case of a surfactant, the liquid can be encapsulated by the surfactant shell if the droplet is pushed out of the hot plate before the explosion. In the case of the grains, the droplet after being completely covered, exhibits a nontrivial shape that is governed by capillary force and by geometric constraint of the grains. As a result, buckling is observed before complex reorganization of the grains. Finally, the capillary forces gather all the particles in a small cohesive packet. which can be interesting for cleaning a surface.

Glossary

CMC Critical Micelle Concentration. Surfactant molecules have 'janus' geometry. One end is hydrophobic, while the other is hydrophilic (at least, one end is more phobic than the other one). As a consequence, the molecule natural position in the water is at the surface in order to satisfy both ends which decreases the surface tension. On the other hand, in the bulk of the water, the surfactant molecule hydrophobic parts gather. The aggregation of the surfactant molecule is a micelle. The number of place at the surface being limited, above a given concentration, the CMC, the surfactant molecule does not decrease the surface tension anymore. Consequently, the surfactant molecules remain in the bulk: freely or forming micelles.

Super-hydrophobic 2D Random Close Packing: When identical spherical grains are placed on a surface, they form a packing. The densest packing is an hexagonal pattern. The surface fraction is given by the ratio between the total cross surface occupied by the spheres and the total available surface. The ideal packing is an hexagonal pattern for which the surface fraction is $\pi/2\sqrt{3} \approx 0.90$. For the square patter, the surface fraction is $\pi/4 \approx 0.78$. Finally, Random Close Packing corresponds to the average surface fraction occupied by the sphere when they are packed randomly.

References

Abdelaziz, R., Disci-Zayed, D., Hedayati, M.K., Pöhls, J.H., Zillohu, A.U., Erkartal, B., et al., 2013. Green chemistry and nanofabrication in a levitated Leidenfrost drop. Nat. Commun. 4, 2400.

Aberle, C., Lewis, M., Yu, G., Lei, N., Xu, J., 2011. Liquid marbles as thermally robust droplets: coating-assisted Leidenfrost-like effect. Soft Matter 7, 11314–11318.

Aussilous, P., Quéré, D., 2001. Liquid marbles. Nature 411, 924–927.

Berryman, J.G., 1983. Random close packing of hard spheres and disks. Phys. Rev. A 27 (2), 1053–1061.

Biance, A.-L., Clanet, C., Quéré, D., 2003. Leidenfrost drops. Phys. Fluids 15 (6), 1632–1637.

Dupeux, G., Merrer, M. Le, Lagubeau, G., Clanet, C., Hardt, S., Quéré, D., 2011. Viscous mechanism for Leidenfrost propulsion on a ratchet. Europhys. Lett. 96 (5), 58001.

Elbahri, M., Paretkar, D., Hirmas, K., Jebril, S., Adelung, R., 2007. Anti-lotus effect for nanostructuring at the Leidenfrost temperature. Adv. Mater. 19 (9), 1262–1266.

Hashmi, A., Xu, Y., Coder, B., Osborne, P.A., Spafford, J., Michael, G.E., et al., 2012. Leidenfrost levitation: beyond droplets. Sci. Rep. 2, 797.

Kim, H., Truong, B., Buongiorno, J., Hu, L., 2011. On the effect of surface roughness height, wettability, and nanoporosity on Leidenfrost phenomena. Appl. Phys. Lett. 98, 083121.

Maquet, L., Colinet, P., Dorbolo, S., 2014. Organization of microbeads in Leidenfrost drops. Soft Matter 10, 4061–4066.

Moreau, F., Colinet, P., Dorbolo, S., 2013. Leidenfrost explosions. Phys. Fluids 25 (9), 091111.

Quéré, D., 2013. Leidenfrost dynamics. Annu. Rev. Fluid Mech. 45, 197–215.

Soto, D., 2014. Non-wetting drops: from impacts to self-propulsion, PhD thesis, Université Pierre et Marie Curie.

Sobac, B., Rednikov, A., Dorbolo, S., Colinet, P., 2014. Leidenfrost effect: Accurate drop shape modeling and refined scaling laws. Physical Review E 90, 053011.

Tsapis, N., Dufresne, E.R., Sinha, S.S., Riera, C.S., Hutchinson, J.W., Mahadevan, L., et al., 2005. Onset of buckling in drying droplets of colloidal suspensions. Phys. Rev. Lett. 94, 018302.

Section V

EXTERNAL FORCES INFLUENCE

Chapter 25

GRAVITY

David Brutin and Florian Carle

Mechanical Engineering Department, Aix-Marseille University, Marseille, France

The formation of droplets happens constantly on Earth, whether it is a natural phenomenon or the result of human activity. However, creating a droplet under reduced gravity conditions is different. How can the droplet be deposited onto the substrate when the fluid floats in the test cell and when the scientist floats around his experiment? This chapter discusses how to bypass the disadvantages of weightlessness and use the advantages of microgravity; it also present a few studies performed under hypergravity using a centrifuge.

25.1 Access to reduced gravity conditions

Nature is governed by four fundamental interactions: strong nuclear interactions, weak nuclear interactions, electromagnetism, and gravitational forces. This last interaction, despite being the weakest of all, makes all physical bodies attract one another in proportion to their mass. The first consequence of this interaction is the weight P that can be directly felt from Earth. Here, we

Droplet Wetting and Evaporation. DOI: http://dx.doi.org/10.1016/B978-0-12-800722-8.00025-4

disregard all interactions from other astronomical objects—the sun, the moon—and objects on Earth:

$$P = m \cdot g = G_C \frac{m_{earth} \cdot m}{R_{earth}^2} \tag{25.1}$$

$G_C = 6.67 \cdot 10^{-11}\ \mathrm{N\,m^2\,kg^{-2}}$ is Cavendish's constant, m is the object/Earth mass, and R_{earth} is Earth's radius (approximately 6370 km). Microgravity, also misleadingly called zero-gravity, is a state in which a body (object or human) is no longer under the influence of contact forces, only body forces. The notion of weightlessness comes from the fact that mass is not a parameter that influences the fall. As shown by Galileo in his legendary experiment from the top of the Leaning Tower of Pisa around 1604, two objects with different masses will fall at the same speed if the air friction is similar for the two bodies. This experiment has been reproduced by David Scott in 1971 on the moon surface with a feather and a hammer during the Apollo 15 mission in the vacuum of space, which lacks air friction. After a fall of 1.2 s, the two objects touched the floor at the exact same time, confirming Galileo's theory that bodies fall with a uniform acceleration without any influence of mass. The more obvious way to experience a microgravity state is to go to the International Space Station (ISS) as it orbits around Earth. However, microgravity is often misunderstood. This state does not mean that the object has no weight. At the orbit apogee, the 450,000 kg payload of the ISS is 426 km from Earth's surface. From Eq. (25.1), one can derive Eq. (25.2), where

$$P_{ISS} = G_C \frac{m_{earth} \cdot m_{ISS}}{(R_{earth} + D_{ISS})^2} \tag{25.2}$$

Therefore, the ISS weight is only reduced by 6.2%; it still weighs 421,700 kg at its farthest point from Earth. Earth's gravity is constantly pulling objects toward the surface, causing them to fall. If an object is thrown into the gravity field, its trajectory will be curved, and the object will hit the surface rather quickly. However, if the initial speed is great enough to fall into a trajectory that matches Earth's curvature, the fall is perpetual, and the microgravity state can theoretically last forever. The ISS speed is approximately 7.66 km $\mathrm{s^{-1}}$, or 27,600 km $\mathrm{h^{-1}}$, and a fall of only 4.9 m for each 8000 m traversed horizontally can be perpetual. All of the objects and the crew inside the station fall at the same speed, and they therefore appear to be floating relative to one another in the ISS frame of reference.

Although critics of microgravity science argue that the experiments are extremely costly, microgravity experiments are a unique way to study phenomena that are usually hidden under normal gravity conditions. Microgravity environments are obviously particularly suited for space applications such as astronaut training, antenna deployment, or ISS experiment preparation. It is absolutely essential to test and validate all of the technical solutions before launching the experiments on satellites or on the space station. This work is done by all space agencies (ESA, NASA, CNES, JAXA, DLR, CNSA) that finance research activities onboard the ISS, rockets, scientific satellites, and so on. But space science starts on Earth with ground-based research to prepare space experiments.

All of this preparation is a great way to conduct high-quality fundamental research that cannot be done on Earth. Biomedical experiments are performed on future astronauts and human guinea pigs to learn how microgravity affects concentration and reaction time, blood flow, vision, displacement, and orientation, because top and bottom notions are not relevant under reduced gravity levels. These experiments allow us to have a better understanding of the human body and its mechanisms of adaptation to new conditions. Most of the time, the results are used to develop techniques and medications for the general public (McGuckin et al., 2005).

The microgravity state in droplets is also interesting because of the flow motion and instabilities that appear during evaporation. Hydrodynamic instabilities correspond to bifurcations that develop when a controlled parameter such as temperature gradient or geometry deviates from the threshold value. Several parameters can induce instabilities, so the flow motion inside the droplet is the result of several instabilities happening at the same time. Microgravity is useful for separating several phenomena and observing only one type of instability, such as the hydrothermal waves that appear due to surface tension gradients but are usually hidden by thermogravitational flow. All of the technology and knowledge obtained through microgravity experiments are, in the long run, reused for applications on Earth, such as satellite navigation systems, communication and data transmission, and optics and ophthalmology. In addition to all of these direct applications, the skills and expertise developed during space conquest are reused for applications on Earth, such as the space pen, nonstick coating in frying pans, and UV protection on sunglasses.

Launching an experiment in the ISS is extremely costly and time consuming. It takes more than 3 years to design an experiment and bring it onboard. Under the same constraints, scientific satellites provide a microgravity environment that is as good as the space station for periods ranging from 10 to 15 days. However, the chances of getting access to satellite facilities are slim. Fortunately, there are other ways to access microgravity on Earth (Figure 25.1).

Halfway to space, sounding rockets are regularly launched from uninhabited areas, and they create 3–12 min of microgravity in low orbits (below 200 km) before reentering the atmosphere and landing with a parachute for recovery. The small portions of orbits that sounding rockets follow are strongly elliptical due to the initial speed (orbit conjugal diameter can be less than 80 km). The payload in which the experiments run is in constant telemetric link with the researchers, allowing them to monitor the experiments and to collect data so that it is not lost in the case of a crash during the return to Earth. The only way to create a microgravity environment on Earth is to use a drop tower. The experiment is set up in a drag shield and dropped from the top of the tower. The tower height h is directly linked to the duration t of the microgravity phase ($t_{\mu g} = \sqrt{2h/g}$) to the 20 m Brisbane drop tower gives about 2 s, whereas the 146 m ZARM vacuum drop

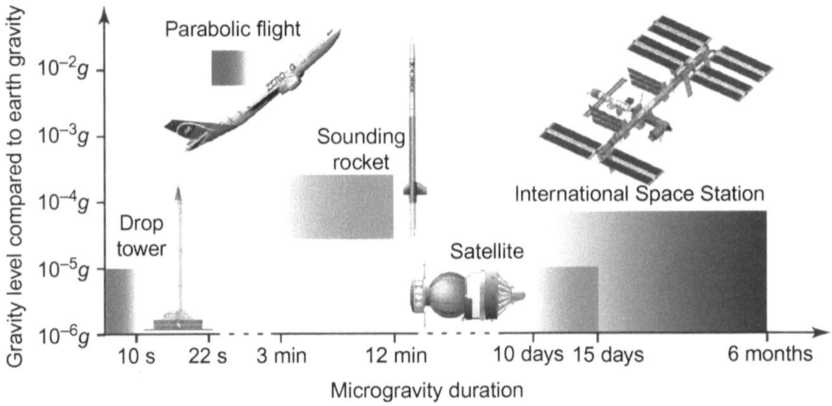

FIGURE 25.1 Access to reduced gravity levels through various platforms. *Figure inspired from ESA website (Carle, 2014).*

tower in Bremen allows up to 5.3 s in free fall and 9.3 s with a catapult phase before the drop. These facilities, among others, have the lowest gravity levels ($10^{-5}g$) and the shortest conception time from design to drop in a few months. However, the phenomena that are studied must have very short transition phases in order to benefit from the microgravity phase. The decelerations during landing (involving an air bag), fine polystyrene pellets, or electromagnetic brakes are strong, and the experiment needs to resist up to $30g$ for catapulted experiments.

25.2 Droplets under reduced gravity conditions

25.2.1 Parabolic Flights

Among the tools just mentioned to access reduced gravity conditions, parabolic flights are the most convenient in terms of project duration, cost, and research manipulation. Several papers have been written about spherical tiny droplets' evaporation of volatile liquids for combustion experiments. A few papers deal with sessile droplets' wetting and evaporating. In 2009, Brutin et al. performed experiments on sessile drop creation in microgravity. They focused on the droplets' contact angle and droplet interface in relation to gravity levels. They observed that under terrestrial gravity and hypergravity, the contact angle increased to reach a constant value from small droplets to big drops; they also evidenced in microgravity a different behavior with almost a constant contact angle that was very sensitive to the gravity level, even for small drops with a diameter smaller than the capillary length. On the drop creation, the authors confirmed the feasibility of creating posed drops under microgravity conditions even with low surface tension fluids such as FC-72 and HFE-7100 (below 10 mN m^{-1}). Two different contacts angles were evidenced considering water droplet creation: if the drop is created under normal gravity, then put under microgravity, or is created during the microgravity phase. The contact angle can be 10° lower compared to the same drop created in microgravity. The authors explained these differences. Among these explanations was the presence of gas trapped in the substrate roughness below the drop and physicochemical interactions. Also, a possible explanation for the change in drop contact angle can be found in the drop pressure change through the gravitational pressure.

The droplet interface equation is given by a pressure balance (Eq. 25.3). At the hydrostatic equilibrium, the overpressure inside the drop is given by the balance of the Laplace pressure and the hydrostatic pressure:

$$P_0 - \rho g z = \sigma\left(\frac{\mathrm{d}^2 z}{\mathrm{d}x^2} + \frac{1}{x}\frac{\mathrm{d}z}{\mathrm{d}x}\right) \tag{25.3}$$

where $L_c = \sqrt{\frac{\sigma}{\rho g}}$ is the capillary length and P_0 is the pressure at the top of the drop. The equation to obtain the interface profile is solved in polar coordinate. Two parameters are used to solve numerically Eq. (25.3): the pressure at the top of the drop (P_0) and the drop contact angle (θ). These two parameters have to be modified to verify the two other physical parameters: the drop base diameter and the drop height. Consequently, a drop contact angle is obtained. Based on the images obtained for three gravity levels during the parabolic flight, we extract the experimental drop interfaces and plot them on Figure 25.2.

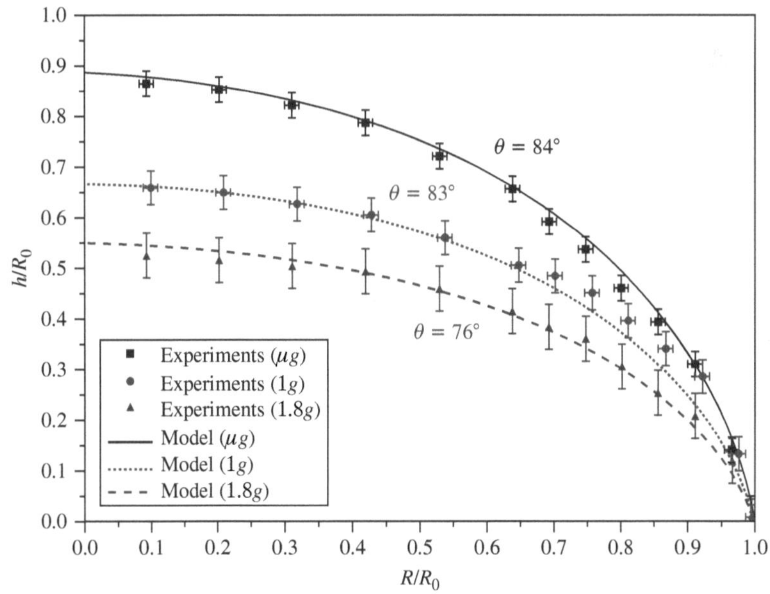

FIGURE 25.2 Normalized droplet height variation for three levels of gravity compared to the Laplace-hydrostatic model (water, Teflon, $R_0 = 4$ mm).

The contact angles obtained for these 8-mm diameter droplets are 85.3° in microgravity, 93.4° under normal gravity, and 89.5° under hypergravity. On the same figure, we plot the drop interface modeled using the Laplace-hydrostatic equation. We successfully predict all three drop interfaces within the error margin, and we obtain for the contact angle 84.0° in microgravity, 83.0° under normal gravity, and 76.0° under hypergravity. This increasing variation in the contact angle with decreasing gravity level is in accordance with the Laplace-hydrostatic equilibrium. We obtain a good agreement for the contact angle only in microgravity. When gravity exists, a gap of more than 10° exists. This is attributed to the difficult repeatability of contact angle experiments with gravity fluctuation and plane vibrations.

After the experiments on wetting, the same team studied the evaporation stage. Brutin et al. (2010) and Carle et al. (2012) presented results obtained during a parabolic flight campaign observing ethanol sessile drop evaporation under microgravity conditions. Droplets were created using a syringe pump by injection through a substrate. The droplets were recorded using both a video camera and an infrared camera. The experimental setup enables the simultaneous visualization in the infrared and the visible wavelengths and also allows access to the heat-flux density transferred to the droplet using a heat-flux meter. The authors evidenced the different stages observed in reduced gravity when the droplets evaporate. The first phase (a) of the experiment is a phase of hypergravity at 1.8g. The substrate is dry at the set temperature. The flux measured by the heat-flux meter during this phase corresponds to the loss by natural convection from a horizontal flat plate at 1.8g (approximately 200 W m^{-2}). A microgravity phase follows this hypergravity phase. At $t = 0$ s, a droplet is created by the injection of the fluid, and a transition regime is observed. The substrate temperature decreases rapidly by approximately 2°C, and the heat flux increases to 1500 W m^{-2}. These changes are caused by the warming of the droplet coupled with the spreading of the liquid on the substrate. This transitional phase lasts between 5 and 8 s. After the quasi-stationary regime is reached (phase c), we observe that these two variables (temperature and heat-flux) remain constant. The limiting phenomenon of evaporation, for a conductive substrate, is the vapor diffusion into the atmosphere (David et al., 2007). The evolution of the evaporation mass flow rate is consistent with the theoretical model of quasi-steady diffusion-driven evaporation implemented with the temperature variation.

During the evaporation, thermo-capillary instabilities develop. Figure 25.3 shows the hydrothermal waves (HTWs) on two droplets with similar diameters ($R_{1g} = 2.81$ mm and $R_{\mu g} = 2.77$ mm) that are evaporating on equally heated substrates ($T_S = 35$°C) under two levels of gravity. In both cases, the droplets of ethanol are hemispherical, and the initial contact angles (θ_i) are less than 40°. Utilizing the work of Garnier et al. (2006), the two cases can be compared using

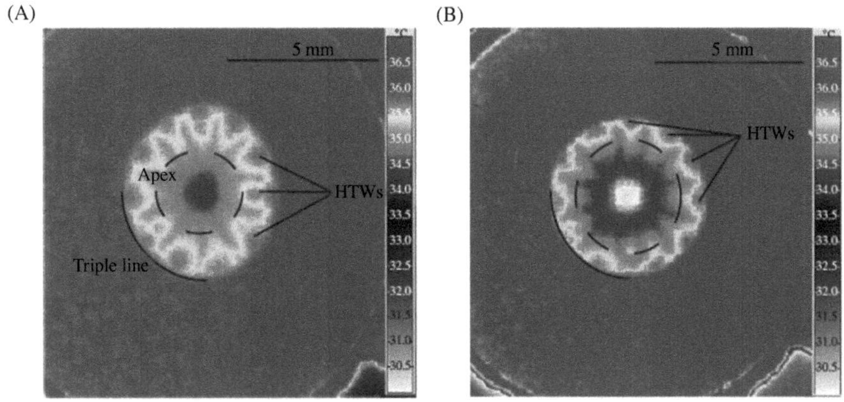

FIGURE 25.3 Infrared visualization of ethanol droplets under normal gravity (A) and reduced gravity (B) (normal gravity ($T_S = 35°C$; $P = 1013$: 25 mbar; $R_{1g} = 2.81$ mm), and reduced gravity ($T_S = 35°C$; $P = 835$ mbar; $R_{\mu g} = 2.77$ mm)).

characteristic dimensionless numbers. The Rayleigh number, representing the buoyancy forces, is compared to the Marangoni number, representing the thermo-capillary forces. In both cases shown, buoyancy forces are quite small, allowing the thermo-capillarity effect to dominate. In this configuration, the instabilities that develop are of type hydrothermal waves 2 (HTW_2). The use of a micro-gravity environment induces a 113-fold decrease in the static Bond number and a 74-fold decrease in the dynamic Bond number. Under normal gravity, the thermo-convective effects are dominated by the thermo-capillary ones but are not negligible ($Bd_{1g} = Ra = Ma \approx 0.08$), whereas under microgravity, only the thermo-capillary effects exist ($Bd_{\mu g} = 0.001$). Under normal gravity, a temperature gradient develops during the evaporation from the apex of the droplet and the contact line, resulting in a gradient of surface tension. This gradient then generates the thermo-capillary instabilities. These HTWs propagate radially around the apex, where most of the evaporation occurs. The HTWs are spaced by an almost constant angle along the axial symmetry of the triple line. Two different spatial dynamics were observed. The HTWs were found to run orthoradially around the edge of the droplet, either from a source point where they are established to a well where they collapse or all in the same direction but with no preferred direction (clockwise or counterclockwise), depending on the temperature of the substrate and the height of the droplet (Sobac and Brutin, 2011). Under microgravity, the temperature gradient is not as well defined as it is under normal gravity, but the apex maintains a temperature below the temperature of the triple line. In this configuration, HTWs have the same trend, but their movements are not as systematic as they are under normal gravity. The instabilities develop during the transitional phase. During the quasi-stationary regime, the HTWs propagate from a source point to a well. These instabilities also propagate on the edge of the droplet, but the angle of propagation is not as constant as it is under normal gravity. This change in propagation could be caused by the vibrations of the aircraft, which cause significant fluctuations in the level of microgravity. Despite this lack of stability, the evolution of the number of HTWs is similar under both gravitational conditions.

25.2.2 Drop Tower

Several drop towers exist over the world (including Germany, Australia, Japan, and Spain). The one used here is in Brisbane, Australia, at the Queensland University of Technology (QUT). The QUT drop tower facility enables testing for 2 s in reduced gravity. The experimental rack is contained inside a drag shield, and both are raised to the top of the tower and suspended by a single thin metal wire. The facility incorporates a simple wire cutting release mechanism to initiate the drop of the experimental rack independently from the drag shield at the same instance. The drag shield and experiment are caught by an airbag mechanism, where release vents allow for high-pressure air release during impact in order to reduce peak deceleration levels. A network of bungee cords and ropes are used to provide horizontal stability to the drag shield during deceleration. The experimental rack experiences typical levels of reduced gravity around $10^{-5}g$.

Diana et al. (2012) observed that the droplet shape is adequately described by the Young–Laplace equation in using the drop tower facility, whereby a variation occurred in the diameter and height upon the surface energy to obtain the lowest free energy shape (sphere-like). The dimensional variations showed that an increase in the Bond number correlated well with the change in the wetted diameter and the height of the droplet (Figure 25.4). The triple-line perimeters of the drops were observed to be lower in reduced gravity than in normal gravity, and differences in contact angle were observed between these two different gravity environments and do not seem well described by theory. The Young–Laplace equation, therefore, can be used to determine the contact angle in reduced gravity

FIGURE 25.4 Digital image of sessile drops of water in normal gravity exhibiting various diameters (different Bond numbers). (A) corresponds to a Bond number of less than 1 (diameter less than capillary length), (B) corresponds to a Bond number of about 1 that is equal to the capillary length), and (C) corresponds to a Bond number greater than 1 (diameter greater than the capillary length). *Images from Diana et al. (2012).*

for small droplets, but it was not optimal for describing the contact angle for larger drops (above the drops' associated capillary length).

Because hysteresis is the result of the drop pinning due to roughness, one could consider whether the hysteresis is affected by the microgravity conditions. The way in which the roughness affects the contact angle also could be gravity-dependent. In fact, weightlessness could promote the droplet liftoff (Wenzel versus Cassie–Baxter behavior) in the case where one assumes some air pockets are trapped at the rough solid–liquid surface.

25.3 Increased gravity conditions

Increased gravity conditions are investigated using centrifuge. Experimental results on the dynamics of sessile water drops are presented by Kabov and Zaitsev (2013) and Kabov et al. (2014). They performed experiments under normal gravity ($1g$), microgravity (μg), and hypergravity (up to $20g$). The microgravity experiments were conducted during the parabolic flights of the European Space Agency (ESA). The hypergravity experiments were carried out on the ESA large diameter centrifuge. The objective was to study the effect of the gravity on (i) the shape of a static sessile drop and (ii) the dynamic advancing contact angle in a growing sessile drop (Figure 25.5).

Different smooth and rough surfaces were used, with different contact angles and different contact angle hysteresis. Water was used for the working liquid. The main variable parameters were gravity (μg–$20g$); drop volume (1 μL–5 mL); liquid flow rate (0.06–16 mL min^{-1}); and contact angle (30°–130°). The drop shape was visualized from the top with the help of the Phase Schlieren System and from the side with the help of the shadow technique with resolution of 6 μm pix^{-1}. The spreading of a sessile liquid drop under the influence of gravity was observed experimentally on surfaces with relatively low contact angle hysteresis (<15°)

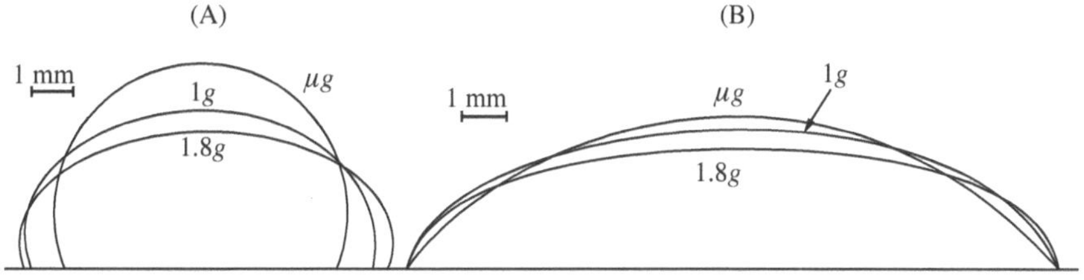

FIGURE 25.5 Parabolic flight experiments, static drop gravity effect on drop shape. (A) 0.146 mL water drop on Teflon surface (contact angle hysteresis is 12.6°). (B) 0.313 mL water drop on polyvinylacetate (contact angle hysteresis is 34.5°). *From Kabov et al. (2014).*

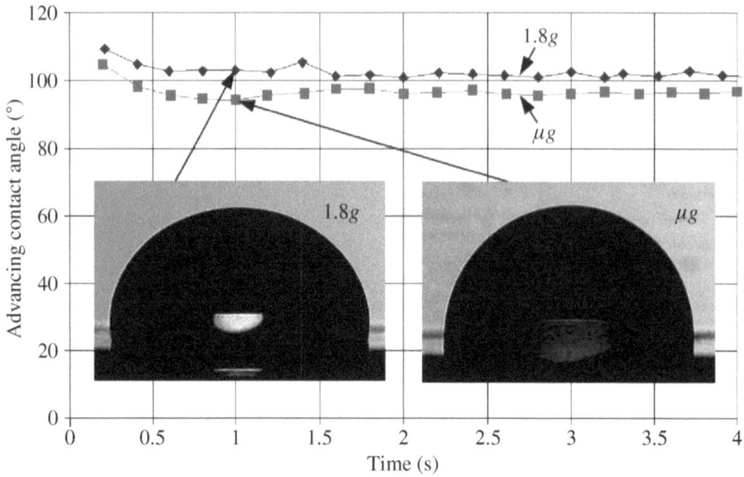

FIGURE 25.6 Parabolic flights experiments, dynamic drop. Dynamic advancing contact angle versus time in growing water drop on copper surface under different gravity levels. Flow rate is 1 mL min^{-1}. Shown are 16 μL drops. *From Kabov et al. (2014).*

(Figure 25.6). In this case good agreement is obtained between the experiment data and modeling (Picknett and Bexon, 1977) with assumption of constant contact angle.

For surfaces with relatively high contact angle hysteresis (>30°), the contact line is pinned while the contact angle adjusts for different gravity levels (Figure 25.6). The dynamic advancing contact angle was found to increase with the gravity.

References

Bartashevich, M.V., Kuznetsov, V.V., Kabov, O.A., 2010. Gravity effect on the axisymmetric drop spreading. Microgravity Sci. Technol. 22 (1), 107–114.

Brutin, D., Zhu, Z.Q., Rahli, O., Xie, J.-C., Liu, Q.-S., Tadrist, L., 2009. Sessile drop in microgravity: creation, contact angle and interface. Microgravity Sci. Technol. 21 (Suppl. 1), 67–76.

Brutin, D., Zhu, Z.Q., Rahli, O., Xie, J.-C., Liu, Q.-S., Tadrist, L., 2010. Evaporation of ethanol drops on a heated substrate under microgravity conditions. Microgravity Sci. Technol. 22, 387–395.

Carle F., 2014. Flow Motion in Sessile Droplets: Evaporation and Nanoparticles Assembly (Ph.D. thesis). Aix-Marseille University.

Carle, F., Sobac, B., Brutin, D., 2012. Hydrothermal waves on ethanol droplets evaporating under terrestrial and reduced gravity levels. J. Fluid Mech. 712, 614–623.

David, S., Sefiane, K., Tadrist, L., 2007. Experimental investigation of the effect of thermal properties of the substrate in the wetting and evaporation of sessile drops. Colloids and Surfaces A 298, 108–114.

Diana, A., Castillo, M., Brutin, D., Steinberg, T., 2012. Sessile drop wettability in normal and reduced gravity. Microgravity Sci. Technol. 24, 195–202.

Garnier, N., Chiffaudel, A., Daviaud, F., 2006. Hydrothermal waves in a disk of fluid. In: Mutabazi, I., Wesfreid, J.E., Guyon, E. (Eds.), Dynamics of spatio-temporal cellular structures: Henri Bénard centenary review, vol. 207. Springer, pp. 147–161.

Kabov, O.A., Zaitsev, D.V., 2013. The effect of wetting hysteresis on drop spreading under gravity. Doklady Phys. 58 (7), 292–295.

Kabov, O., Zaitsev, D., Gatapova, E., Semenov, A., Bykovskaya, E., Karnauhova, E., et al. (2014). Drop spreading and evaporation on a heated substrate under variable gravity conditions. In: Proceedings of the 15th International Heat Transfer Conference, IHTC-15, August 10–15, 2014, Kyoto, Japan. Paper # IHTC15-9504.

McGuckin, C.P., Forraz, N., Baradez, M.-O., Navran, S., Zhao, J., Urban, R., Tilton, R., Denner, L., 2005. Production of stem cells with embryonic characteristics from human umbilical cord blood. Cell Prolif. 38 (4), 245–255.

Picknett, R.G., Bexon, R., 1977. The evaporation of sessile or pendant drops in still air. Journal Colloid Interface Science 61, 336–350.

Sobac, B., Brutin, D., 2011. Triple-line behavior and wettability controlled by nano-coated substrates: Influence on sessile drop evaporation. Langmuir 27, 14999–15007.

Chapter 26

OSCILLATING/ VIBRATING SURFACES

Edward Bormashenko[1] and Didier Laux[2]
[1]*Department of Physics, Ariel University, Ariel, Israel,* [2]*France Institut of electronics and Systems Department, University Montpellier II, Montpellier, France*

In this chapter, vibrations of free and sessile droplets and low and high frequencies are discussed. Modes appearing under vertical and lateral vibrations of sessile droplets are examined. Boundary conditions taking place at the triple line are crucial for constituting modes. Both axisymmetric and nonaxisymmetric modes are addressed. Use of the vibration of droplets for the measurement of surface tension of liquids is covered. Vibrated droplets supply valuable information about wetting regimes that take place on flat and rough surfaces and may be successfully used for the study of wetting transitions observed on rough surfaces. Vibrated sessile droplets are useful for the development of microfluidic devices. The mechanisms of displacement of vibrated sessile droplets on ratchet and gradient surfaces are covered. Oscillation of droplets induced by electric and magnetic fields is discussed.

Droplet Wetting and Evaporation. DOI: http://dx.doi.org/10.1016/B978-0-12-800722-8.00026-6

26.1 Low frequencies

26.1.1 Oscillating Free Droplets

The vibrations of free liquid drops have been investigated for more than a century by Rayleigh (1879, 1894), and later by Lamb (1932), who established a general expression for the resonance mode frequencies ω_n:

$$\omega_n^2 = \frac{n(n-1)(n+2)\gamma}{\rho R^3} \tag{26.1}$$

where ρ and R are the density and radius of a droplet, respectively, and γ is the surface tension. Equation (26.1) was verified and corrected in a series of experimental and theoretical works (Cummings and Blackburn, 1991). It was also generalized for the situation in which a droplet is deformed and exposed to external fields (Suryanarayana and Bayazitoglu, 1991). Equation (26.1) was effectively exploited for the dynamic measurement of surface tension in the situations where the establishment of the surface tension by other methods is problematic, such as occurs with liquid metals and molten silicon (Fujii et al., 2000; Matsumoto et al., 2002; Sauerland et al., 1993). Remarkably, vibrated jets were first used for the dynamic measurement of surface tension by Bohr (1909).

26.1.2 Oscillating Sessile Droplets

The treatment of physical problems related to oscillating sessile droplets is much more complicated than that of free droplets, for several reasons. First of all, various regimes of oscillations are possible, due to the different possible regimes of movement of the triple (three-phase) line. The triple line may be pinned by the substrate or may be mobile (depinned; Noblin et al., 2004, 2009). The mobility (or immobility) of the triple line gives rise to a broad diversity of modes observed for oscillating sessile droplets. It should be stressed that various experimental situations are possible for the same liquid/solid pair. At low amplitude, the contact line remains pinned, and the drop presents eigen modes at different resonance frequencies.

The modes taking place for vertically vibrated sessile droplets in the situation where the triple line is pinned are depicted in Figure 26.1. The modes of the surface waves, shown in Figure 26.1, obey

$$q_n = \frac{2\pi}{\lambda_n} = \frac{\pi(n-1/2)}{L} \tag{26.2}$$

where $\lambda_n/2$ is half the pseudo wavelength, equal to the mean distance between the wave nodes (Figure 26.1); q_n is the corresponding wave vector; L is the half-perimeter

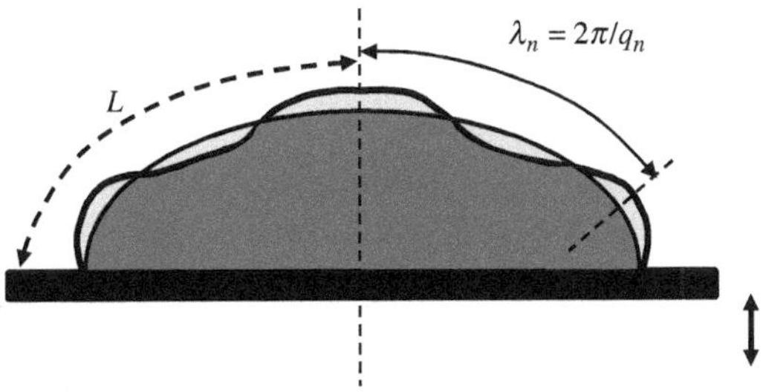

FIGURE 26.1 Vertical vibration of a sessile droplet. The triple line is pinned. The pseudo wavelength for mode $n = 3$.

of the maximal meridian cross section of the drop; and n is an integer or half-integer. To find the frequency of these modes, Noblin et al. (2004) used the dispersion relation for 1D capillary-gravity waves on a liquid bath (Noblin et al., 2004; Landau and Lifshitz, 1987). This is expressed by

$$\omega^2 = \left(gq + \frac{\gamma}{\rho}q^3\right)\tanh(qb) \tag{26.3}$$

where the mean depth b of the droplet is $b = V/[\pi(R\sin\theta)^2]$, with V being the drop volume.

At a higher amplitude of vibrations, the contact line moves; it remains circular, but its radius oscillates at the excitation frequency (Noblin et al., 2004, 2009). The transition between these two regimes arises when the variations of contact angles exceed the value allowed by contact angle hysteresis (Noblin et al., 2009). The transition from the pinned to depinned regime of oscillations was studied by Noblin et al. (2009).

The mathematical treatment of the oscillations of sessile droplets with a depinned triple lines is much more complicated. However, for a hemispherical sessile drop, as it has been shown recently by Lyubimov and coworkers, eigen frequencies depend on the boundary conditions at the triple line, and for a freely sliding contact line, they coincide with the natural frequencies of a spherical drop (Lyubimov et al., 2006). It was demonstrated that an increase of the wetting parameter reduces the oscillation frequency and that the boundary condition on the contact line is responsible for oscillation damping (Lyubimov et al., 2006). The effects of nonlinearity of oscillations due to the depinning of the triple line have also been considered (Perlin et al., 2004; Lyubimov et al., 2006).

The majority of papers devoted to vertical oscillations of droplets are devoted to axisymmetrically shaped droplets (Noblin et al., 2004, 2009). However, depinning of the triple line under a parametric excitation can induce a transition of shape and can break the drop's initial axial symmetry (Brunet and Snoeijer, 2011). In that case, a standing wave appears at the drop interface and induces a periodic motion, with a frequency that equals half the excitation frequency, leading to the formation of so-called star droplets (Brunet and Snoeijer, 2011).

We now examine horizontal (lateral) vibrations of sessile droplets. Lyubimov et al. performed a theoretical study of the nonaxisymmetric, spherical, harmonic oscillations of a hemispherical drop on a flat surface with a contact line constrained by the Hocking condition. Taking the nonaxisymmetric motions into account led to some fairly interesting effects: the degeneracy of the oscillation frequencies in the azimuthal number was eliminated, and the appearance of a "bending" oscillation mode became possible (Lyubimov et al., 2004). The physics of nonaxisymmetric modes observed in sessile droplets were also extensively discussed by Milne et al. (2014).

Horizontal vibrations of sessile droplets were studied by Celestini and Kofman (2006). For a fixed triple line, Eq. (26.4), predicting the first eigen mode, was proposed:

$$\omega_1 = \sqrt{\frac{6\gamma h(\theta)}{\rho(1-\cos\theta)(2+\cos\theta)}}R^{-3/2} \tag{26.4}$$

where θ is the contact angle, and $h(\theta)$ is the dimensionless, geometrical factor calculated by Celestini and Kofman (2006).

Dong et al. performed the numerical analysis of the lateral and vertical vibrations of a sessile drop on a hydrophobic surface with the computational fluid dynamic modeling. From simulations with a drop on a vibrating vertical surface, it was found that gravity enhances the first resonant mode and weakens the second mode, even though the positions of the resonance peaks do not differ substantially from those observed with the horizontal vibrations (Dong et al., 2006).

26.1.3 Characterization of Solid Substrates with Vibrated Droplets

The main macroscopic parameter describing the wetting of solid substrates is the equilibrium (or Young) contact angle. However, the experimental establishment of the Young contact angle turns out to be a nontrivial, challenging task, due to the phenomenon of the contact angle hysteresis (Bormashenko, 2013). The contact angle hysteresis appears even on ideal—that is, atomically flat—chemically homogeneous, isotropic, insoluble, nonreactive, and nondeformed solid surfaces (Marmur, 2009). Actually, the contact angles measured for droplets deposited even on ideal surfaces are enclosed between advancing (θ_A) and receding (θ_R) contact angles, which are equilibrium but metastable contact angles (Marmur, 2009; Bormashenko, 2013). Della Volpe et al. (2002) proposed using the vibrated solid substrates for experimental establishment of the Young contact angle. The authors transferred mechanical energy to the three-phase system in a controlled

FIGURE 26.2 Wetting transitions observed under vibration of a 15 μL water drop deposited on a micrometrically rough polydimethylsiloxane surface. (Left) The initial Cassie state. (Right) The Cassie impregnating state induced by vibrations.

way through a simple loudspeaker added to a standard Wilhelmy microbalance (Della Volpe et al., 2002). The authors stated that the true equilibrium contact angle was thus established experimentally. Della Volpe et al. also tested the validity of different semiempirical equations proposed for the theoretical evaluation of the equilibrium contact angle from advancing and receding contact angles. One of the most common empirical equations suggests that the cosine of the equilibrium (Young) contact angles equals the mean of cosines of the advancing and receding contact angles. The mean of cosines of these angles appeared to be in agreement with the value established experimentally by Della Volpe et al. in many cases; the presence of significant roughness, and mainly of heterogeneity, strongly reduced the validity of these approximations (Della Volpe et al., 2002).

Vibrated sessile droplets also have been applied to the characterization of rough surfaces. Meiron et al. (2004) vibrated the surface, took top-view pictures of sessile drops, monitored the drop roundness, and calculated the contact angle from the drop diameter and weight. The reliability of this method has been demonstrated by the fact that the Young ("ideal" as it called by Meiron et al.) contact angles of all surfaces, as calculated from the Wenzel equation using the measured apparent contact angles, were practically identical (Meiron et al., 2004).

The vibration of sessile droplets has also been successfully applied to the study of Cassie-Wenzel wetting transitions occurring on rough surfaces. The abrupt change in the apparent contact angle occurring on a rough surface is called the wetting transition. This change may be promoted by the vibration of sessile droplets (Bormashenko et al., 2007a,b). A typical wetting transition occurring under vibration of sessile droplets is depicted in Figure 26.2. Monitoring of wetting transitions in vibrated sessile droplets allowed the establishment of the pressure p, and the depinning force F corresponding to the onset of transition, calculated according to Eqs (26.5a) and (26.6). The pressure in the vibrated drop is given by

$$p = p_{\mathrm{i}} + \frac{2\gamma}{R} + \rho g l \tag{26.5a}$$

where

$$p_{\mathrm{i}} = \frac{\rho V A \omega^2}{\pi R^2 \sin^2 \theta^*} \tag{26.5b}$$

is the pressure increase caused by oscillation; R, l, and V are the radius, height, and volume of a droplet, respectively; $\theta^*$is the apparent contact angle; ρ is the density of a liquid; and A and ω are the amplitude and frequency of vibration, respectively. The depinning force acting on the unit length of the triple line was calculated as

$$F = \frac{pR}{4\sin\theta^*}(2\theta^* - \sin 2\theta^*) \tag{26.6}$$

with p given by Eqs (26.5a) and (26.5b). Experiments carried out with sessile droplets vibrated on honeycomb polymer surfaces led to the conclusion that wetting transitions are more likely to be a 1D than a 2D affair—that is, the transition occurs when a threshold force acting on the triple line F is exceeded (Bormashenko et al., 2007a,b). This was true for both horizontal and vertical vibrations of drops deposited on artificial and natural rough surfaces (Bormashenko et al., 2007a,b). The resonance Cassie-Wenzel transition was observed under the horizontal vibration of sessile droplets. The frequencies corresponding to the minimal amplitudes of vibration coincided with the mode frequencies of surface capillary-gravity waves inherent in the vibrated drop, given by Eq. (26.3) (Bormashenko et al., 2007b).

It was also established with vibrated sessile droplets that the lowest energy state of a droplet deposited on a rough surface corresponds to a Cassie impregnating wetting regime, when the droplet finds itself on a wet substrate, viewed as a patchwork of solid and liquid (Bico et al., 2002; Bormashenko et al., 2008). The vibration of rough surfaces allowed identification of pathways of wetting transitions: Cassie air trapping–Cassie impregnating and Cassie air trapping–Wenzel state pathways are possible (Bormashenko et al., 2008).

On the other hand, it was demonstrated that mechanical vibration can be used to overcome the energy barrier for transition from the sticky Wenzel state to the nonsticking Cassie state (Boreyko and Chen, 2009). It was also shown that the threshold for the dewetting transition follows a scaling law comparing the kinetic energy imparted to the drop with the work of adhesion (Boreyko and Chen, 2009). Thus, we conclude that the vibration of sessile droplets supplies useful information about the wetting properties of flat and rough surfaces.

26.1.4 Transport of Droplets with Vibration

The vibration of droplets was used successfully for the development of microfluidic devices, allowing movement of discrete liquid drops on a surface. When a liquid drop is subjected to an asymmetric lateral vibration on a nonwettable surface, a net inertial force acting on the drop causes it to move (Daniel et al., 2005). The direction and velocity of the drop motion are related to the shape, frequency, and amplitude of vibration, as well as the natural harmonics of the drop oscillation (Daniel et al., 2005). The transport of microgels encapsulated in droplets using vibrated unidirectional nanoratchets was demonstrated recently (Sekeroglu et al., 2011).

Chaudhury et al. performed a series of experimental and theoretical studies of the dynamics of droplets deposited on inclined surfaces attached to a mechanical oscillator. In the first of these studies, the drop was deposited on a silicon wafer silanized with decyltrichlorosilane (Chaudhury and Mettu, 2008). The substrate was vibrated in parallel to the support with bands of Gaussian white noise of different powers. The drops drifted downward on the inclined support, accompanied with random forward and backward movements. For a hysteresis-free surface, the drift velocity should only be the product of the component of the gravitational acceleration and the Langevin relaxation time, independent of the power of the noise. However, in the presence of hysteresis, as is the case here, the drift velocity depended strongly on the power of the noise. This result illustrated the role of hysteresis in the drifted motion of drops on a surface subjected to vibration (Chaudhury and Mettu, 2008).

A study was then conducted on vibrated droplets placed on an inclined superhydrophobic pillared surface (Chaudhury and Goohpattader, 2013). In this case, a droplet exhibited the critical speeding dynamics—namely, it moved very slowly at first, but then sped up rapidly after a critical velocity had been reached. During the mobile phase, some of the natural vibration modes of the drops were self-excited on a pillared surface but not on a smooth hydrophobic surface (Chaudhury and Goohpattader, 2013).

Daniel et al. demonstrated also that the vibration of the substrate may supply velocities as high as $5-10\ \mathrm{mm\ s^{-1}}$ to droplets deposited on substrates possessing the gradient of wettability (Daniel and Chaudhury, 2002). The same idea may be exploited for displacement of droplets deposited on substrates exposed to the thermal gradient (Mettu and Chaudhury, 2008).

26.1.5 Oscillation of Droplets Induced by Electric and Magnetic Fields

An interest in electrically induced oscillations of droplets arose from electrical spraying of liquids from capillaries. For both dielectric and conducting droplets, an electric field has two effects on the potential and kinetic energies of the drop. The first is the introduction of an electrical contribution to its potential energy (Cheng, 1985). The second effect is geometrical. Deforming the undistorted drop to a spheroid changes the kinetic and surface energies of a droplet (Cheng, 1985). The classification of modes that are possible in an electric field was performed by Rosenkilde (1969). The transverse-shear, toroidal, and pulsation modes were distinguished (Rosenkilde, 1969).

The modes for conducting droplets were calculated by Sample and Raghupathym (1970). Oscillations of magnetically levitated aspherical droplets were studied by Cummings and Blackburn (1991). They demonstrated that the expected single frequency of the fundamental mode is split into either three—when there is an axis of rotational symmetry—or five unequally spaced bands. The frequencies, on average, are higher than those of an unconstrained droplet (Cummings and Blackburn, 1991). The numerical model addressing droplet oscillations in high-gradient static magnetic fields was reported (Bojarevics and Pericleous, 2009).

26.2 High frequencies

Classically, shear moduli (G' and G'') assessment is ensured with rheometers in ranges of frequencies not exceeding a thousand of hertz. Using the time temperature superposition principle, it is possible to extend this range and to obtain high-frequency band master curves that are very useful because they are directly related to the microstructure of the polymer or the fluid studied. In the case of a rheological study of a droplet during evaporation, these approaches become inapplicable because the sample has to be placed between two plates (plane/plane configuration) or two cylinders (Couette geometry), so the evaporation or wetting is not possible or modified by the rheological experiment. Consequently, only nondestructive methods that can directly work on the droplet and its substrate are suitable. Hence, by their nondestructive nature, ultrasonic methods using a vibrating substrate with very tiny amplitudes are very interesting. In the literature, a specific and well-known ultrasonic method (QCM: Quartz Crystal Micro balance) has already been largely applied to monitor fluid evaporation and wetting. Since the first works of Sauerbrey (1959) and Kanazawa concerning Newtonian fluids (Kanazawa and Gordon, 1985), the literature concerning the subject is extremely wide, but one can refer, for instance, to the works of Zhuang or Yakhno concerning droplets investigated with the QCM method (Zhuang et al., 2008; Yakhno et al., 2005). In particular, Yakhno et al. (2007, 2010) focused on various liquids using a specific uncoated QCM device and the acoustic-mechanical impedance of the droplet. In the QCM approach, the droplet is deposited on a quartz crystal excited in shear vibration mode. When evaporation occurs, the loading impedance of the quartz is modified because the mass and viscosity of the droplet evolve. The resonance frequency and quality factor of the quartz are shifted and can be studied versus time to give the evaporation kinetic. Singular points on this evolution therefore give a parameter linked to rheological parameters of the droplet. However, this method presents some drawbacks: first, it is difficult to change the substrate on which the evaporation occurs, and second, the quartz resonance modification is related to many parameters that are evolving simultaneously, including mass, viscosity, contact angle, droplet area on the QCM, and so on. Then, it is sometimes tricky to easily and directly measure G', G'', and viscosity. This point is underscored by Zhuang, and he conducted much work on rigorous QCM application for investigating droplets wetting. More information can be found in his PhD thesis (Zhuang, 2011). The ultrasonic method proposed in this chapter can overcome some disadvantages previously mentioned because the estimation of G' and G'' is independent of precise mass knowledge, and the substrate can be modified very easily. Even if the ultrasonic principles are identical in QCM and ultrasonic shear reflectometry, the sensor and the experimental approach are totally different. After a brief presentation of some theoretical aspects concerning the reflection of a plane mechanical shear wave on the interface (elastic solid/viscoelastic material), we present various correlations between the ultrasound parameters measured by this method and the conventional parameters (desiccation and gelation times) classically measured, thanks to (mass vs. time) graphs assessment during blood droplet evaporation (Sobac and Brutin, 2011).

26.2.1 Theoretical Aspects: Shear Ultrasonic Reflectometry

The method is essentially based on the measurement of the complex reflection coefficient for an interface between an elastic solid and a viscoelastic material. Due to viscoelasticity, when the ultrasonic wave is reflected on such an interface, its amplitude decreases, and the wave undergoes a phase shift that is generally extremely small (typically a few nanoseconds for operating frequencies in the megahertz range). Experimentally, it is possible to work in reflection mode with one transducer acting as emitter and receiver or in transmission mode with two sensors in oblique incidence (O'Neil, 1948). For practical reasons we have used the reflection configuration (Gasparoux et al., 2008). In this configuration, an ultrasonic transducer, made of a piezoelectric crystal and a delay line (DL) made of silica or glass, for instance, is used. Because the ultrasonic attenuation is small in glassy materials, multiple reflections are possible, leading to the existence of many ultrasonic echoes in the DL. Then, the complex reflection coefficient at the interface (DL/viscoelastic material), $R^* = r_o e^{j\Phi}$, is calculated with Eqs (26.7) and (26.8), where A_i is the amplitude of the echo $n°i$, in the case of the interface (DL)/air; B_i is the amplitude for the interface (DL)/material; Δt_i is the time shift between these two echoes; and f is the operating frequency:

$$r_o = \left(\frac{B_i}{A_i}\right)^{1/i} \tag{26.7}$$

$$\Phi = -2\pi f\left(\frac{\Delta t_i}{i}\right) \tag{26.8}$$

If P represents the slope on the curve $\ln(B_i/A_i) = f(i)$, then r_o is given by e^P. Because Δt_i is proportional to i, the slope γ of the graph $\Delta t_i = f(i)$ leads to $\Phi = 2\pi f\gamma$. Experimentally, after these linear fittings, G' and G'' are evaluated with Eqs (26.9) and (26.10), where ρ is the mass density of the studied material, ρ_{DL} is the mass density of the DL, and V_{DL} is the shear ultrasonic velocity in the DL (Gasparoux et al., 2008):

$$G' = \left(\rho_{DL} V_{DL}\right)^2 \frac{(r_0^2 - 1)^2 - 4r_0^2 \sin^2 \Phi}{\rho(2r_0 \cos \Phi + 1 + r_0^2)^2} \tag{26.9}$$

$$G'' = -4\left(\rho_{DL} V_{DL}\right)^2 \frac{(r_0^2 - 1)r_0 \sin \Phi}{\rho(2r_0 \cos \Phi + 1 + r_0^2)^2} \tag{26.10}$$

If the mass density of the material is not known, one can work with the ratio (G''/G'), also called $\tan(\delta)$. This quantity is generally very interesting from a rheological point of view because behavior transitions (e.g., the beginning or end of a rubbery plateau for entangled polymers, alpha transition) are characterized by $\tan(\delta) = 1$. Such a parameter can be directly deduced from ultrasonic data without any knowledge of other physical parameters of the material. Such a fact is very interesting for the investigation of a system that is wetting and for which density is not easily accessible versus time (Figure 26.3).

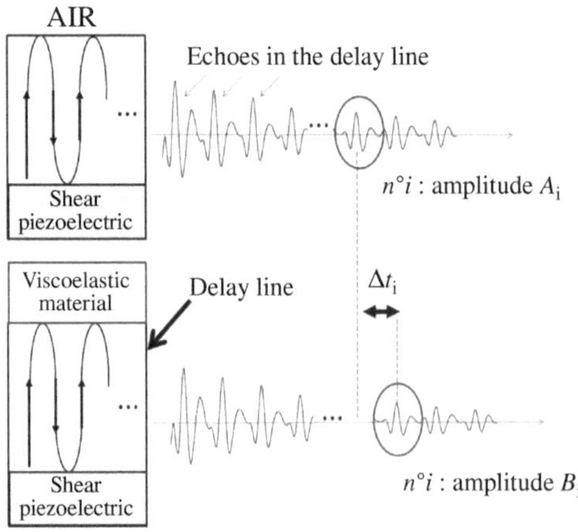

FIGURE 26.3 Shear ultrasonic reflectometry method.

26.2.2 Experimental Approach Dedicated to Droplets Evaporation Monitoring

Experimentally, we used a shear wave transducer with an incorporated DL and excited with a pulse generator (Olympus 5073 PR). For this study, it was a 5 MHz (Olympus/Panametrics) sensor with a 7 μs DL in silica. As presented in Cereser Camara et al. (2010), an additional DL (1 mm glass plate for blood wetting) was added on the sensor (Figure 26.4). Such an approach has several advantages: the high-quality ultrasonic sensor ensures plane waves emission in far field or in Fresnel zone, and having an additional disposable DL is very useful for studying aggressive materials or evaluating DL nature effects on wetting or evaporation. At last, the good choice of the DL material can be important because, in difficult experimental situations, it largely increases Δt_i and leads to a better estimation of G' and G''. In order to measure simultaneously mass and ultrasonic data, the ultrasonic sensor and its additional DL were mounted on a balance (Ohaus Explorer E10640). After droplet blood sampling directly on a finger, the droplet was deposited on the ultrasonic sensor, and the measurement was taken. At the end of evaporation, we obtained $m(t)$, $r_o(t)$, and $\gamma(t)$. A schematic representation of the experiment in given in Figure 26.4.

26.2.3 Results

Concerning mass versus time analysis, we followed the method described by D. Brutin and colleagues in Sobac and Brutin (2011). From this analysis, for each experiment, two specific times ($\tau_{D(mass)}$, desiccation time, and $\tau_{G(mass)} \approx 1.25\ \tau_D$

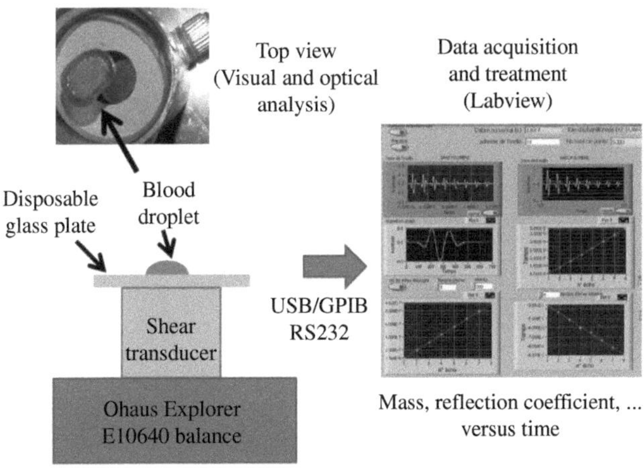

FIGURE 26.4 Experimental method for mass and ultrasonic data measurements during blood droplet evaporation and wetting.

(mass), gelation time) were measured. For all the experiments conducted, the evolution of r_o(t) was very characteristic and obtained with good accuracy. At the beginning of the experiment, when the mass decreases very rapidly, r_o does not evolve because blood behaves like a liquid with very small viscosity (typically a few millipascal second for high shear rates considered in our experiments). Because shear waves do not propagate in liquids, the entire ultrasonic signal is reflected at the interface (DL/blood). Then, when the evaporation is going on, the sol-gel transition occurs, and the blood droplet appears like a viscoelastic material with higher G' and G''. For the ultrasonic signal, it is like a harder material, and the waves can go from the DL into the droplet under study. One has to notice that for such high ultrasonic frequencies, the penetration depth is quite small and does not exceed a few micrometers. At the end of the process, when cracks or delaminations appear under or in the droplet, if hygrometry is not too high (Bou Zeid and Brutin, 2013), ultrasonic signals cannot propagate properly in the desiccated droplet, so r_o increases again. The comparison between mass versus time and r_o versus time) reveals $\tau_{D(US)}$ and $\tau_{G(US)}$ on the curve (r_o versus time). The subscript US stands for ultrasonic. Figure 26.5 shows the evolution of m and r_o versus time. Specific times $\tau_{D(mass)}$ and $\tau_{G(mass)}$, evaluated from the curve $m(t)$, are also reported. $\tau_{D(US)}$ is obtained when r_o begins to decrease significantly and $\tau_{G(US)}$ is clearly identified as the minimum of r_o(t). This clear determination of $\tau_{G(US)}$ is very important because, on mass, this evaluation is more complex.

In order to check the correlation between desiccation and gelation times obtained by the two methods, experiments have been performed with various

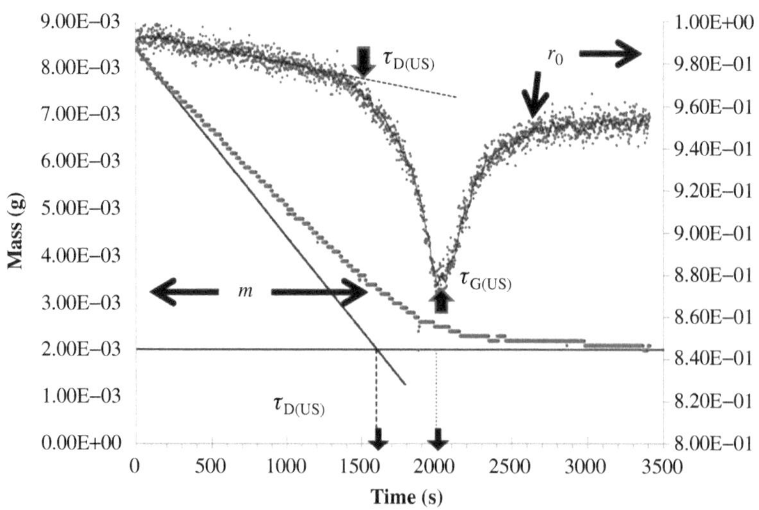

FIGURE 26.5 $m(t)$ analysis for $\tau_{D(mass)}$, $\tau_{G(mass)}$ assessment, and r_o(t) analysis for $\tau_{D(US)}$, $\tau_{G(US)}$ evaluation.

droplets volumes (8–50 μL). Hygrometry ranged between 26% and 46%, and temperature between 19°C and 28°C. All these variations participated to obtain a large range of desiccation and gelation times. We covered a large range of evaporation times, up to 4000 s. In Figures 26.6–26.8, we show $\tau_{D(mass)}$ versus $\tau_{D(US)}$, $\tau_{G(mass)}$ versus $\tau_{G(US)}$, and $\tau_{G(US)}$ versus $\tau_{D(US)}$, respectively.

In these figures, it is clear that $\tau_{D(mass)}=\tau_{D(US)}$ and that $\tau_{G(mass)}=\tau_{G(US)}$. Furthermore, we obtained $(\tau_{G(US)}/\tau_{D(US)}) \sim 1.26$, which is in line with the value given by Sobac and Brutin (2011). In order to estimate G' and G'' from ultrasonic data and Eqs (26.9) and (26.10), one has to know the density of blood during evaporation. Mass density evaluation is quite complex because it depends on evaporated volume versus time. Because whole blood density is around 1060 kg m^{-3} and because it is made of 50% water and 50% colloids, this leads to a density of colloids of 1120 kg m^{-3}. So at the beginning, density should be closed to 1060 kg m^{-3}, and when evaporation is totally finished, density is 1120 kg m^{-3}. In our experiments, this small variation of density of 5% has been neglected. We have taken 1060 kg m^{-3} during all the measurements. Regarding Eqs (26.9) and (26.10), it creates the same error on G' and G'' and reaches the end of the process (all water evaporated) at a maximum value of 5%. But also regarding Eqs (26.9) and (26.10), it will have no influence on tan(δ) evaluation. In Figure 26.9, we present the typical evolutions of G' and G'' during the experiment.

At the beginning, only G'' exists, and the system is like a viscous fluid. Then, an elastic part appears and grows until the specific time for which $G' = G''$. This time is clearly identified, and, regarding various curves of G' and G'' versus t

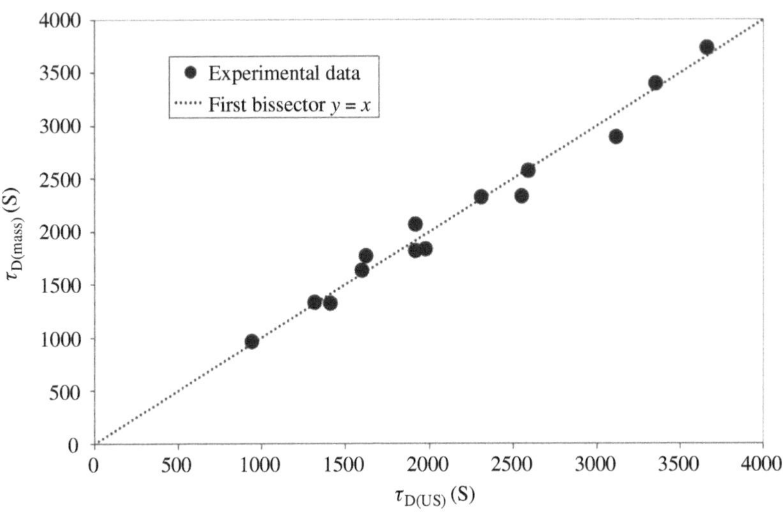

FIGURE 26.6 Blood droplet evaporation: $\tau_{D(mass)}$ versus $\tau_{D(US)}$.

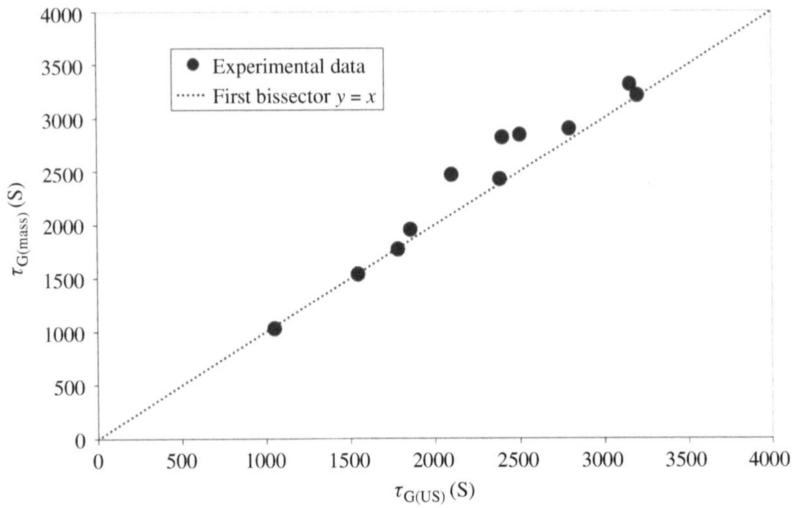

FIGURE 26.7 Blood droplet evaporation: $\tau_{G(mass)}$ versus $\tau_{G(US)}$.

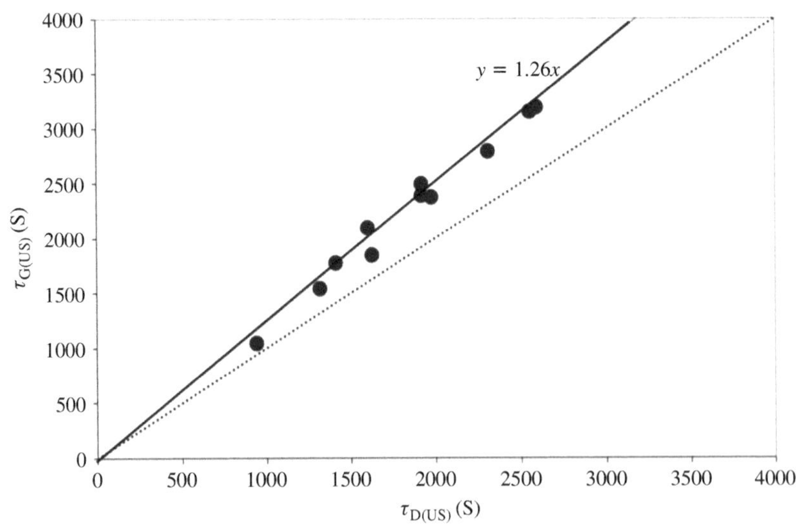

FIGURE 26.8 Blood droplet evaporation: $\tau_{G(US)}$ versus $\tau_{D(US)}$.

(not presented here), it seems to be between τ_D and τ_G previously evaluated. Such an element has to be confirmed, and studies are going on in this direction. But as the time for which $\tan(\delta) = 1$ has a clear rheological meaning, it is believed that the analysis of G' and G'' versus time during evaporation will lead to a better

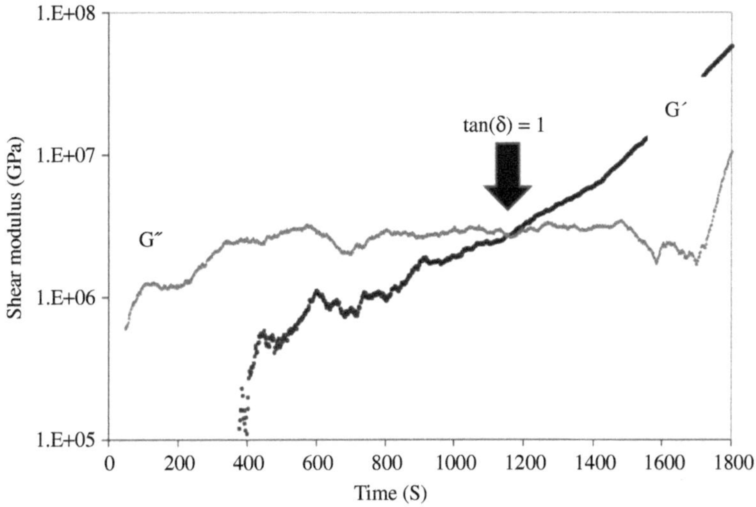

FIGURE 26.9 Typical evolutions of G' and G'' measured at 5 MHz during blood droplet evaporation.

(more precise) evaluation of gelation time and perhaps to a better understanding of the physical and rheological phenomena involved.

26.2.4 Conclusions

For whole blood droplet evaporation, we have shown that ultrasonic shear reflectometry is an interesting tool for "traditional" relevant times τ_D and τ_G estimation. Because it is a nondestructive approach that can be easily implemented on other experiments (thermical infrared analysis, optical studies), we recommend using it to obtain relevant rheological data during blood or more generally complex fluid droplets desiccation, wetting, or evaporation. In order to obtain a key parameter of (Sol/Gel) transition, the estimation of the time for which $\tan(\delta) = 1$ seems to be a better criterion than τ_D and τ_G usually measured. Studies are going on in this direction in order to investigate with this approach all the parameters that could influence wetting: hygrometry, substrate nature, temperature, droplet composition, and so on.

Glossary

Elastic modulus number that measures an object or substance's resistance to being deformed elastically when a force is applied to it.

Ultrasound an oscillating sound pressure wave with a frequency greater than the upper limit of the human hearing range (20 kHz).

References

Bico, J., Thiele, U., Quéré, D., 2002. Wetting of textured surfaces. Colloids Surf. A 206, 41–46.

Bohr, N., 1909. Determination of the surface-tension of water by the method of jet vibration. Philos. Trans. R. Soc. Lond. A 209, 281–317.

Bojarevics, V., Pericleous, K., 2009. Droplet oscillations in high gradient static magnetic field. Microgravity Sci. Technol. 21, 119–122.

Boreyko, J.B., Chen, Ch-H., 2009. Restoring superhydrophobicity of lotus leaves with vibration-induced dewetting. Phys. Rev. Lett. 103, 174502.

Bormashenko, E., 2013. Wetting of Real Surfaces. De Gruyter, Berlin.

Bormashenko, E., Pogreb, R., Whyman, G., Erlich, M., 2007a. Cassie-Wenzel wetting transition in vibrating drops deposited on rough surfaces: is dynamic Cassie-Wenzel wetting transition a 2D or 1D affair? Langmuir 23, 6501–6503.

Bormashenko, E., Pogreb, R., Whyman, G., Erlich, M., 2007b. Resonance Cassie-Wenzel wetting transition for horizontally vibrated drops deposited on a rough surface. Langmuir 23, 12217–12222.

Bormashenko, E., Pogreb, R., Stein, T., Whyman, G., Erlich, M., Musin, A., et al., 2008. Characterization of rough surfaces with vibrated drops. Phys. Chem. Chem. Phys. 27, 4056–4406.

Bou Zeid, W., Brutin, D., 2013. Influence of relative humidity on spreading, pattern formation and adhesion of a drying drop of whole blood. Colloids Surf. A 430, 1–7.

Brunet, P., Snoeijer, J.H., 2011. Star-drops formed by periodic excitation and on an air cushion—a short review. Eur. Phys. J. Spec. Top. 192, 207–226.

Celestini, F., Kofman, R., 2006. Vibration of submillimeter-size supported droplets. Phys. Rev. E 73, 041602-1–041602-6.

Cereser Camara, V., Laux, D., Arnould, O., 2010. Enhanced multiple ultrasonic shear reflection method for the determination of high frequency viscoelastic properties. Ultrasonics 50, 710–715.

Chaudhury, M.K., Goohpattader, P.S., 2013. Activated drops: self-excited oscillation, critical speeding and noisy transport. Eur. Phys. J. E 36, 15.

Chaudhury, M.K., Mettu, S., 2008. Brownian motion of a drop with hysteresis dissipation. Langmuir 24, 6128–6132.

Cheng, K.J., 1985. Capillary oscillations of a drop in an electric field. Phys. Lett. A 112 (8), 392–396.

Cummings, D.L., Blackburn, D.A., 1991. Oscillations of magnetically levitated aspherical droplets. J. Fluid Mech. 224, 395–416.

Daniel, S., Chaudhury, M.K., 2002. Rectified motion of liquid drops on gradient surfaces induced by vibration. Langmuir 18, 3404–3407.

Daniel, S., Chaudhury, M.K., de Gennes, P.-G., 2005. Vibration-actuated drop motion on surfaces for batch microfluidic processes. Langmuir 21, 4240–4248.

Della Volpe, C., Maniglio, D., Morra, M., Siboni, S., 2002. The determination of a "stable-equilibrium" contact angle on heterogeneous and rough surfaces. Colloids Surf. A 206, 47–67.

Dong, L., Chaudhury, A., Chaudhury, M.K., 2006. Lateral vibration of a water drop and its motion on a vibrating surface. Eur. Phys. J. E 21, 231–242.

Fujii, H., Matsumoto, L., Nogi, K., 2000. Analysis of surface oscillation of droplet under microgravity for the determination of its surface tension. Acta Mater. 48 (11), 2933–2939.

Gasparoux, J., Laux, D., Ferrandis, J.Y., Attal, J., Tordjeman, P., 2008. Large frequency bands viscoelastic properties of honey. J. Non Newtonian Fluid Mech. 153, 46–52.

Kanazawa, K.K., Gordon, J.G., 1985. The oscillation frequency of a quartz resonator in contact with liquid. Anal. Chim. Acta 175, 99–105.

Lamb, H., 1932. Hydrodynamics. Cambridge University Press, Cambridge, UK.

Landau, L., Lifshitz, E.M., 1987. Fluid Mechanics, second ed. Butterworth-Heinemann, Oxford.

Lyubimov, D.V., Lyubimova, T.P., Shklyaev, S.V., 2004. Non-axisymmetric oscillations of a hemispherical drop. Fluid Dyn. 39 (6), 851–862.

Lyubimov, D.V., Lyubimova, T.P., Shklyaev, S.V., 2006. Behavior of a drop on an oscillating solid plate. Phys. Fluids 18, 012101-1–012101-11.

Marmur, A., 2009. A guide to the equilibrium contact angles maze. In: Mittal, K.L. (Ed.), Contact Angle Wettability and Adhesion, vol. 6. VSP, Leiden, pp. 3–18.

Matsumoto, T., Nakano, T., Fujii, H., Kamai, M., Nogi, K., 2002. Precise measurement of liquid viscosity and surface tension with an improved oscillating drop method. Phys. Rev. E 65, 031201.

Meiron, T.S., Marmur, A., Saguy, I.S., 2004. Contact angle measurement on rough surfaces. J. Colloid Interface Sci. 274, 637–644.

Mettu, S., Chaudhury, M.K., 2008. Motion of drops on a surface induced by thermal gradient and vibration. Langmuir 24 (19), 10833–10837.

Milne, A.J.B., Defez, B., Cabrerizo-Vílchez, M., Amirfazli, A., 2014. Understanding (sessile/constrained) bubble and drop oscillations. Adv. Colloid Interface Sci. 203, 22–36.

Noblin, X., Buguin, A., Brochard-Wyart, F., 2004. Vibrated sessile drops: transition between pinned and mobile contact line oscillations. Eur. Phys. J. E 14, 395–404.

Noblin, X., Buguin, A., Brochard-Wyart, F., 2009. Vibrations of sessile drops. Eur. Phys. J. Spec. Top. 166, 7–10.

O'Neil, H.T., 1948. Reflection and refraction of plane shear waves in viscoelastic media. Phys. Rev. 75 (6), 928–935.

Perlin, M., Schultz, W.W., Liu, Z., 2004. High Reynolds number oscillating contact lines. Wave Motion 40, 41–56.

Rayleigh, L., 1879. On the capillary phenomena of jets. Proc. R. Soc. Lond. 29, 71.

Rayleigh, L., 1894. The Theory of Sound. Macmillan, London.

Rosenkilde, C.E., 1969. A dielectric fluid drop in an electric field. Proc. R. Soc. Lond. A Math. Phys. Sci. 312 (1511), 473–494.

Sample, S.B., Raghupathym, B., 1970. Quiescent distortion and resonant oscillations of a liquid drop in an electric field. Int. J. Eng. Sci. 8 (1), 97–109.

Sauerbrey, G., 1959. Use of quartz vibrator for weighting thin films on a microbalance. Z. Phys. 155, 206–210.

Sauerland, S., Lohöfer, G., Egry, I., 1993. Surface tension measurements on levitated liquid metal drops. J. Non Cryst. Solids 156–158 (2), 833–836.

Sekeroglu, K., Gurkan, U.A., Demirci, U., Demirel, M.C., 2011. Transport of a soft cargo on a nanoscale ratchet. Appl. Phys. Lett. 99, 063703.

Sobac, B., Brutin, D., 2011. Structural and evaporative evolutions in desiccating sessile drops of blood. Phys. Rev. E 84, 011603-1–011603-5.

Suryanarayana, P.V.R., Bayazitoglu, Y., 1991. Effect of static deformation and external forces on the oscillation of levitated droplets. Phys. Fluids A 3 (5), 967–977.

Yakhno, T., Yakhno, V., Sanin, A., Sanina, O., Pelyushenko, A., 2005. Dynamics of phase transitions in drying drops as an information parameter of liquid structure. Nonlinear Dyn. 39, 369–374.

Yakhno, T., Sanin, A., Pelyushenko, A., Kazakov, V., Shaposhnikova, O., Chernov, A., et al., 2007. Uncoated quartz resonator as a universal biosensor. Biosens. Bioelectron. 22, 2127–2131.

Yakhno, T., Sanin, A., Kazakov, V., Sanina, O., Vacca, C., Falcione, F., et al., 2010. In: Vernon, S. (Ed.), Uncoated Quartz Resonator as a Universal Biosensor, in Intelligent and Biosensors. INTECH, Somerset, Croatia, pp. 345–362.

Zhuang, H., 2011. Performance of the Quartz Crystal Microbalance in Liquid Environment. Thesis of National University of Singapore.

Zhuang, H., Lu, P., Lim, S.P., Lee, H.P., 2008. Study of the evaporation of colloidal suspension droplets with the quartz crystal microbalance. Langmuir 24, 8373–8378.

Chapter 27

ELECTRIC FORCES

David Brutin[1], Valérie Vancauwenberghe[1] and Paolo Di Marco[2]
[1]*Mechanical Engineering Department, Aix-Marseille University, Marseille, France,*
[2]*DESTEC, Engineering Thermodynamics and Heat Transfer, University of Pisa, Pisa, Italy*

The enhancement of heat and mass transfer using a static electric field is an interesting process for industrial applications due to its low energy consumption and potentially high level of evaporation rate enhancement. However, to date, this phenomenon is still not understood in the context of the evaporation of sessile drops. This chapter encompasses the existing literature about the effect of an electrical field established by two plate electrodes on sessile drops. First, the investigations about the influence of the electric field on the contact angle and the drop shape are discussed. Next, the studies about the enhancement of evaporation under an electrical field are summarized. Finally, we conclude with a discussion of possible future investigations.

27.1 Theoretical basis

Over the past few decades, the evaporation of sessile drops has been the subject of a great deal of interest due to its increasing use in heat and mass transfer applications and industrial processes. Most of these investigations have been focused

Droplet Wetting and Evaporation. DOI: http://dx.doi.org/10.1016/B978-0-12-800722-8.00027-8

on the complexity of the physical dynamics, such as the influence of the substrate surface on the drop wettability, the evaporation flux at the interface and the triple line, and the influence of the substrate temperature on the evaporation rate (Picknett and Bexon, 1977; Sobac and Brutin, 2011; Semenov et al., 2011; David et al., 2006, 2007; Dunn et al., 2008).

Under normal gravity, different physical mechanisms drive the evaporation of small drops: heat transfer by convection, heat transfer induced by the substrate conduction, diffusion of the vapor around the drop in the gaseous phase, and the molecular interactions between the drop and the substrate that change the wettability of the drop. From the existing literature, it is generally assumed that drop evaporation is mainly controlled by the diffusion of molecules from the liquid–gas interface to the surrounding gas (Brutin et al., 2012; Sobac and Brutin, 2012). This quasi-steady mechanism is expressed by the Laplace equation. For small contact angle ($\theta < 90°$), the evaporation flux J is given by Fick's law $J = -D\nabla c$, where D is the coefficient of diffusion and c corresponds to the concentration (Deegan et al., 2000; Hu and Larson, 2002). However, the rate of mass loss has been expressed for arbitrary contact angles by Popov (2005):

$$-\frac{\mathrm{d}m}{\mathrm{d}t} = \pi R D (c_\mathrm{s} - c_\infty) f(\theta) \tag{27.1}$$

where m is the mass of the drop, R is the wetting radius, t is the time, c_s is the saturate concentration of vapor at the drop interface, c_∞ is the concentration of vapor in the air, and $f(\theta)$ is a function of the contact angle. The effect of a static electric field is of great interest in many applications, including mass and heat transfer (Bologa et al., 1985; Eckert et al., 1996), particle manipulation (Liu et al., 2008), coating (Mugele and Baret, 2005), and drying (Ahmedou and Rouaud, 2009). Indeed, the generation of an electric field is a cost-effective solution due to its low power consumption.

The enhancement of the heat transfer in the evaporation process by the application of an electric field can be useful on Earth to improve the evaporation rate, as well as in space, under weightlessness, to replace the absence of the convective effect. Thus, the objective of research in this area is to overcome the diffusion limitation on the evaporation rate by applying an external electric field. Moreover, in the scope of the nanoparticle deposition by evaporation, which is limited on Earth due to the presence of capillary convection and gravity (the coffee ring effect), the use of an electric field would provide a tool to control the pattern formation. The applications are various on Earth, as well as in space, regarding the possibilities of heat transfer enhancement, such as cooling using drop evaporation in space or the uniform deposition of nanoparticles that is required for almost all industrial applications (e.g., coating, printing, DNA mapping). When a drop is exposed to an electric field, its equilibrium shape is established by the balance between the surface tension and the external force corresponding to gravity and the electric field. In the case of a sessile drop, the surface tension tends to make the drop spherical, whereas gravity tends to flatten it. Typically, the external field tends to elongate the drop along the field direction.

The augmented Young–Laplace equation describes the mechanical equilibrium between the surface tension, the electric field, and gravity (Bateni et al., 2006; Borzaadi and Bailey, 1978; Adamiak, 2001):

$$\gamma\left(\frac{1}{R_1} - \frac{1}{R_2}\right) = \Delta P_0 + (\Delta\rho)gz + \Delta P_e \tag{27.2}$$

where γ is the surface tension, R_1 and R_2 are the principal radii of curvature, ΔP_0 corresponds to the difference of pressure across the drop interface, $\Delta\rho$ is the density difference across the interface, g is the gravitational acceleration, z is the vertical distance from the reference, and ΔP_e is the electrical pressure corresponding to the difference in the normal component of the Maxwell stress tensor across the interface.

The presence of an electric field affects the wettability of liquids according to Eq. (27.2). The change in the contact angle, called the electrowetting effect, is caused by a potential applied to the interface of the three-phase system ϕ_{sl}. This phenomenon can be derived from the Lippmann equation using the approximation of a parallel capacitor for the solid–liquid interface and assuming the differential capacitance of the interface $C_{sl} = \varepsilon_0/\varepsilon_d$ (ε corresponding to the electric permittivity of the dielectric layer beneath the liquid drop and d corresponding to its thickness) (Mugele and Baret, 2005; Digilov, 2000; Kang, 2002):

$$cos(\theta_e) = cos(\theta_0) + \frac{C_{sl}(\phi_{sl} - \phi_{sl}^0)^2}{2\gamma_{sl}} \tag{27.3}$$

Here, θ_e is the contact angle under the electric field, θ_0 is the contact angle without the field, and ϕ_{sl} is the interfacial tension of the solid–liquid phase at the zero potential ϕ_{sl}^0. An electric field can be established in two ways, as shown in Figure 27.1.

In the first case, the electric field is directly created inside the liquid drops. One electrode is placed in contact with the drop surface. The applied potential between the two electrodes is several hundred volts. Many theoretical, experimental, and numerical studies based on this configuration have been performed for different applications (Mugele and Baret, 2005; Quiliet and Berge, 2001; Shamai et al., 2007) such as in microfluidics (actuation and drop manipulation; Muguele et al., 2010; Lee et al., 2002; Paik et al., 2003; Cho et al., 2003; Pollack et al., 2000), optics (Berge and Peseux, 2000; Lee and Yang, 2008), microelectromechanics (Jones, 2009; Chakraborty et al., 2011), and electrical engineering

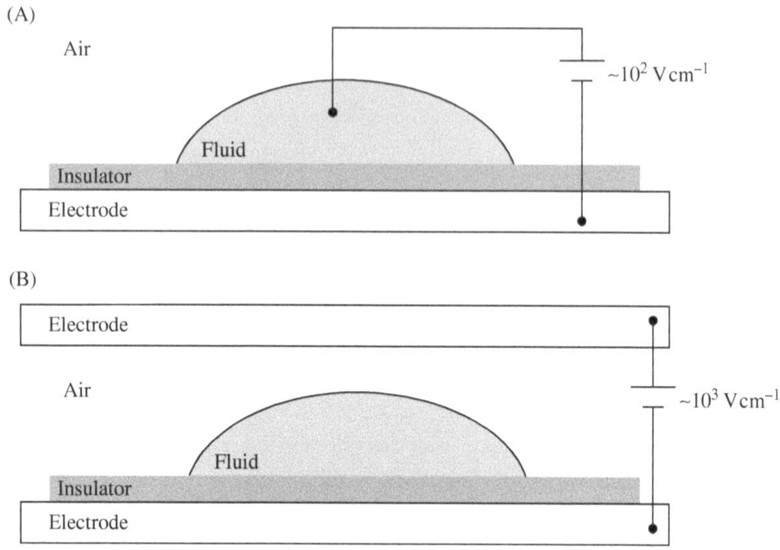

FIGURE 27.1 Schematic of the electrode configuration. (A) The electric field is established inside the drop. The upper electrode is in contact with the drop volume. (B) The electric field is established outside the drop.

(Drygiannakis et al., 2009; Quinn et al., 2005; Moon, 2002; Sedev, 2011). However, because the electrodes must be in contact with the liquid, such applications require a precise adjustment of the experimental setup and may not be convenient for large processes or those involving volume changes. In the other case, the electric field can be established between two plate electrodes without contact with the drop. This configuration is less popular than the first due to the order of magnitude of the applied potential, which is in the range of several kV (electric field $\sim 1{-}10\ \mathrm{kV\ cm^{-1}}$), and it constitutes a small part of the literature. However, this setup may be the most appropriate for evaporation applications.

27.2 Changes induced by an electrical field

When a dielectric fluid is placed in a uniform electric field E, a volumic electric force acts thereon. It is expressed by Eq. (27.4) (Panofsky and Philips, 1962):

$$f_e = \rho_e E - \frac{\varepsilon_0}{2} E^2 \nabla \varepsilon_r + \frac{\varepsilon_0}{2} \nabla \left(E^2 \frac{\mathrm{d}\varepsilon_r}{\mathrm{d}\rho} \rho \right) \tag{27.4}$$

The first term of this sum corresponds to the Coulomb's force that is connected to the free charge density of the fluid ρ_e. This force generally predominates over the other forces in the presence of the free charge in the fluid.

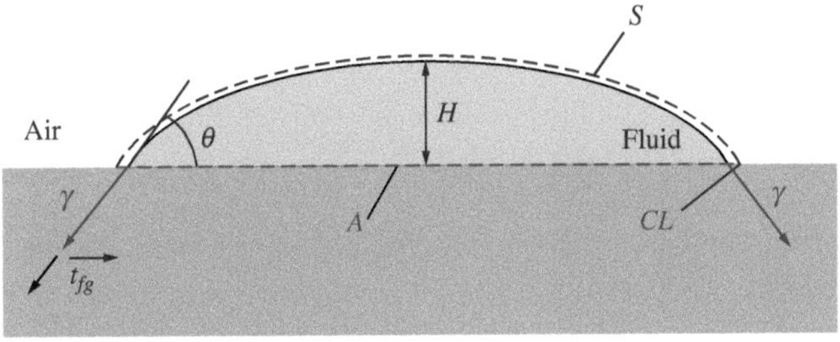

FIGURE 27.2 Schematic of the sessile drop.

The second term represents the dielectrophoretic force that results from the nonuniformity of the relative dielectric permittivity ε_r under the electric field. The last term corresponds to the electrostriction force that acts on the fluid when the electric field is nonuniform. The dielectrophoretic and electrostriction are independent of the field polarity (Landau and Lifsitz, 1984).

The volumic electric force in and around a fluid interface can be reformulated as the divergence of the Maxwell stress tensor T_e (Panofsky and Phillips, 1962; Harris and Basaran, 1995; Basaran and Scriven, 1990):

$$f_e = \mathrm{div}\overline{\overline{T}}_e \tag{27.5}$$

where its components are

$$t_{ik} = \varepsilon_0 \varepsilon_r E_i E_k - \frac{1}{2}\varepsilon_0 E^2 \left[\varepsilon_r - \frac{d\varepsilon_r}{d\rho}\rho\right]\delta_{ik} \tag{27.6}$$

Thus, the resulting electric force F_e acting on the surface S of a drop is given by Eq. (27.7). This force can be evaluated when the electric field at the drop interface is known (Di Marco et al., 2013):

$$F_e = \iint_S \overline{\overline{T}}_e \cdot n \, dS \tag{27.7}$$

where n is the normal outward unit vector.

When a drop sitting on a flat surface (Figure 27.2) is subjected to an electric field, the equilibrium can be calculated for a stationary case by considering the momentum balance given by Eq. (27.8):

$$\int_{CL} \gamma_{gl} t_{\vec{gl}} \, dL + \int_V \rho_l \vec{g} \, dV + \int_S \overline{\overline{T}}_{eg} \cdot \vec{n} \, dS + \int_A \overline{\overline{T}}_{el} \cdot \vec{n} \, dA - \int_S p_g \cdot \vec{n} \, dS - \int_A p_l \cdot \vec{n} \, dA = 0 \tag{27.8}$$

where A is the drop base area, CL is the contour of the drop base, S is the area of the gas–liquid interface, V is the drop volume, and the subscripts l and g refer to the liquid drop and the surrounded gas, considered here to be air, respectively. The integrals represent, respectively, the adhesion force, the drop weight, the electric force at the gas–liquid interface, the electric force at the drop base, and the force caused by the pressure at the gas–liquid interface p_g and the base of the drop.

The momentum balance can be simplified by considering that the value of the electric stress tensor is zero at the droplet base, which is grounded:

$$\frac{\pi D^2}{4}(p_f - \rho_g g H) = V(\rho_l - \rho_g)g + \pi D \gamma \, sin(\theta) \qquad (27.9)$$

where D is the drop base diameter, H is the drop height, and V is the drop volume. Equation (27.9) corresponds to

$$F_p = F_w + F_\gamma + F_e \qquad (27.10)$$

where the corresponding forces are, respectively, the internal overpressure, the weight, the adhesion, and the electric force (Marco et al., 2013). The influence of an electric field on the contact angle and wetting radius of a drop must be known to be able to predict the evaporation rate given by Eq. (27.1). This chapter compiles the studies based on the action of an external electric field (such as the configuration in Figure 27.1B) with regard to three factors: contact angle, shape, and evaporation rate.

27.3 Wetting contact angle

Only a few authors have studied the effect on the contact angle and the interface when an electric field is established outside a liquid drop (shown in Figure 27.1B). As depicted in Eq. (27.1), the evaporation rate of a sessile drop is a function of θ, which changes under an electric field. To predict this angle change, the Lippmann equation is proposed and has been fully implemented by many experimenters, who have correlated empirical results with a small variation in the contact angle, especially in electrowetting, as represented in Figure 27.1A. In fact, the Lippmann equation considers the effect of the electric field on the drop wettability as a consequence of interface polarization such as the classical capillary effect. This relationship supposes that the surface tension of the solid–vapor and liquid–vapor phases, γ_{sv} and γ_{lv}, is not affected by the electric field. This expression may not be considered for systems with a dielectric multilayer, nonaqueous liquids, and different electrolyte concentrations (Peykov et al., 2000). Digilov et al. proposed that the change in the contact angle may be due to the reduction of the line tension caused by the charge redistribution along the contact line of the three phases (Digilov, 2000; Bateni et al., 2005a,b).

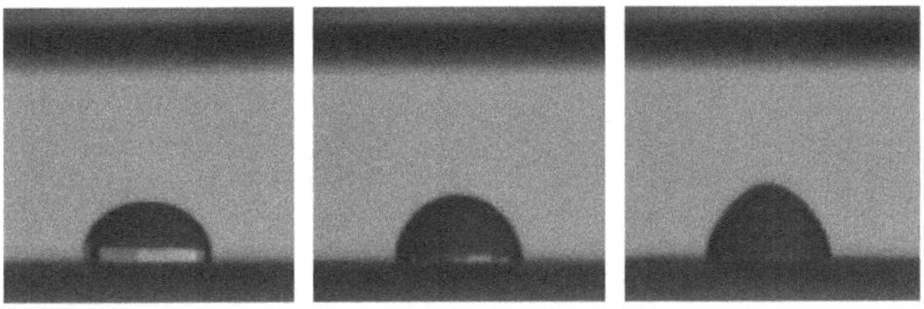

FIGURE 27.3 Water droplet of 80 μL deformed by the electric field. The substrate was an aluminum surface sandblasted and subjected to a hydrophobic treatment. The electrode spacing was 10 mm [50].

The experimental studies of the literature focused on the change in the contact angle for dielectric liquids under an external electric field varying from 1 to 25 kV cm^{-1}. The overall mean of these results suggested an increase or decrease in the contact angle with increasing magnitude of the electric field. In their works on the liftoff of a conducting drop, Roux et al. (Roux and Achard, 2009; Gliere et al., 2013) observed a decrease in the contact angle of a water drop placed on the lower plate electrode of a condenser surrounded by mineral oil with a high relative permittivity. Their investigations focused on the development of a theoretical model predicting the electric field threshold. They highlighted the effect of the contact angle hysteresis that induces an elongation of the droplet under the electric field without any movement along the contact line, decreasing the contact angle.

Roero (2006) studied the shape and contact angle of sessile water drops on various substrates subjected to an electric field to explore the generation of sound emission from wet high-voltage transmission lines. In his experiments, three substrates that received different treatments (untreated, hydrophobic, and hydrophilic) were used, and the variation of sessile water drops of various volumes and their instabilities under an increasing electric field were measured. The contact angle decreased with increasing electric field. Figure 27.3 illustrates his observation.

Bateni et al. (2004; 2005a,b) were interested in the changes in interfacial tension of a droplet under an external field. Based on experimental results (Figure 27.4), they developed a new methodology, called ADSA-EF (Axisymmetric Drop Shape Analysis—Electric Field), to study the drop shape and surface tension under an external electric field (more about this in Section 27.4). An increase in the surface tension of different conducting liquids with the increase in the electric field was observed, and the analysis of the regression suggested that this increase is proportional to the square root of the electric field strength. Moreover, they highlighted some influencing factors, such as the liquid polarity and the size of the liquid molecules, and the lack of effect of the direction of the electric field. They also suggested that the observed increase in the contact angles is due to a bulk effect rather than an interfacial effect

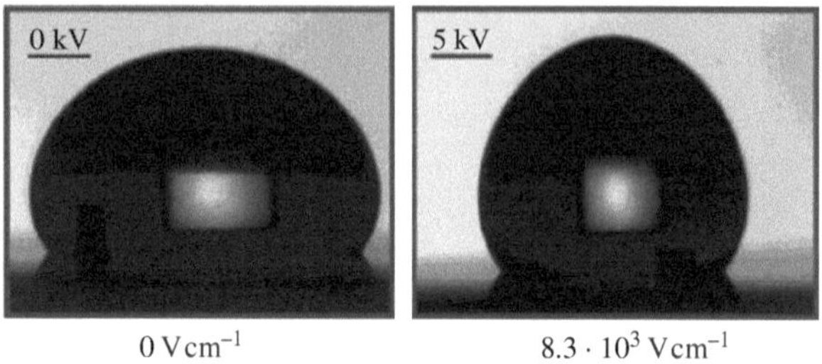

FIGURE 27.4 Deformation of water drops under an electric field. The substrate was Teflon silicon wafers, and the electrode spacing was 6 mm. The two drops did not have the same volume (Bateni et al., 2004).

(Bateni et al., 2005a,b). Further investigation is required to interpret the existing results, including the influence of the liquid and the substrate properties (conductivity, permittivity, wettability, surface treatment, etc.). Moreover, the effect of the electric field on the contact line has not been considered. However, the contact line can be pinned to the surface (Bateni et al., 2006) or less. This may change contact angle evolution when the bubble is elongated by the electric field and may explain differences in experiments.

27.4 Drop shape

The effect of an electrostatic field on the shape of a drop is a well-known phenomenon, elongating the drop along the direction of the field, as shown in Figure 27.6. In 1882, Lord Rayleigh pioneered an investigation of the interaction of charges and electric fields on liquids drops. He theoretically derived the critical amount of charges required to destabilize an isolated, surface-charged spherical drop. His work inspired the beginning of electrohydrodynamics (Melcher and Taylor, 1969).

Over the past three decades, the shape and stability of sessile drops under electric fields have been analyzed in many works. Rosenkilde (1969) examined the equilibrium and stability of an incompressible dielectric drop subjected to a uniform electric field. Assuming the drop to be ellipsoidal, he proposed a relationship between the equilibrium deformation, the strength of the electric field, the drop dimensions, and the surface tension. The cause of the instability for high fluid permittivities has also been presented (Rosenkilde, 1969).

Miksis (1981) proposed a numerical model to compute the shape of an axisymmetric drop under a DC electric field, producing good agreement with experiments. His results show a critical value of the dielectric constant under which the drops elongated and retained their originally spheroidal shape. Above this critical value, the drop developed a conical shape.

In their numerical works, Basaran and Wohlhuter (Wohlhuter and Basaran, 1992; Basaran and Wohlhuter, 1992) calculated the shape and stability of axisymmetric pendant and sessile drops or bubbles considering that the contact line is fixed or prescribed. Their first calculations addressed the assumption that conducting drops are linearly polarizable materials (Wohlhuter and Basaran, 1992). The equilibrium shape and stability of the drops are determined by simultaneously solving the Laplace and Young–Laplace equations. A study considering the nonlinear polarization of the drops was performed by substituting the Maxwell equation with the Laplace equation (Basaran and Wohlhuter, 1992). Their computations revealed the importance of varying the plate condenser spacing, which affects the electric field strength and the drop size.

Reznik et al. (2004) distinguished three scenarios of droplet shape evolution under strong electric fields determined by the values of the electric Bond number and based on the numerical solution of the Stokes equations. Under a weak electric field, the droplets are subjected to Maxwell stresses and exhibit steady-state shapes. Under strong electric fields and for low static contact angles, the droplets acquire a conical shape and jetting sets in from the droplet tip. The final scenario corresponds to the detachment of an almost whole droplet (i.e., a dripping phenomenon) that results from a strong electric field and a contact angle exceeding a critical value. The numerical model was in good agreement with the experimental results for polymer droplets (Reznik et al., 2004).

Bateni et al. (2004; 2005a,b) based their methodology ADSA-EF on an assumption of an axisymmetric drop shape. ADSA-EF can be used to measure the surface tension of a drop under an electric field with and without the gravity effect. From the Young–Laplace equation described by Eq. (27.2), they considered the electrical pressure ΔP_e, which depends on the permittivity of the fluids and the strength and direction of the electric field, as follows:

$$\Delta P_\mathrm{e} = \frac{1}{2}\left[\varepsilon_0 E_\mathrm{n}^{(g)2} - \varepsilon_0(\varepsilon_\mathrm{r} - 1)E_\mathrm{t}^2\right] \tag{27.11}$$

where E_t and E_n are, respectively, the tangential and the normal components of the electric field at the surface of the drop, and the superscripts l and g refer to the drop liquid and the surrounded fluid seen as air, respectively. When the drop is a conducting fluid, there is no electric field inside the drop, and the Young–Laplace equation is simplified as

$$\gamma\left(\frac{1}{R_1} - \frac{1}{R_2}\right) = \Delta P_0 + (\Delta\rho)gz + \frac{1}{2}\varepsilon_0 E_\mathrm{n}^{(g)2} \tag{27.12}$$

From Eq. (27.12) and the electrostatic field module, they numerically predicted the drop shape with an iterative scheme, as shown in Figure 27.5.

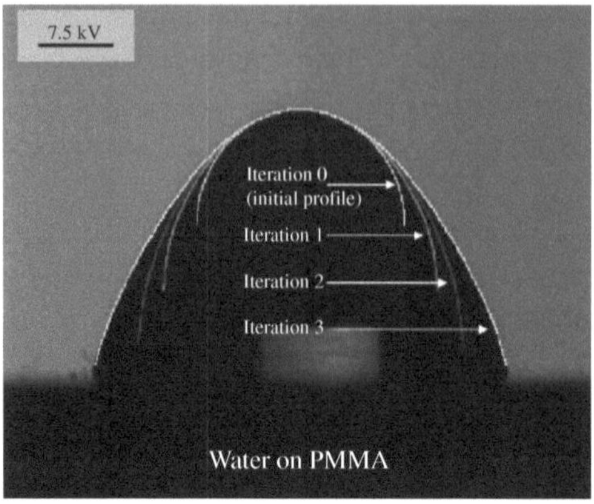

FIGURE 27.5 Prediction of the drop shape by iterative calculations (Bateni et al., 2004).

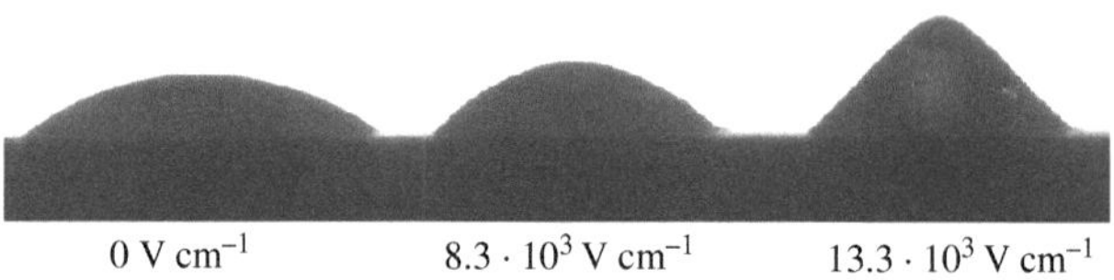

FIGURE 27.6 Shape of an ethanol droplet for different electric field strengths (Di Marco et al., 2013).

Di Marco et al. (2013) evaluated the electric force acting on a sessile ethanol drop interface and compared the experimental results with a theoretical model based on the equilibrium action of a sessile drop, taking into account the electric field and gravity. The experimental setup is composed of a flat aluminum plate equipped with an orifice for droplet injection. The plate is electrically grounded and coated with a Teflon layer. A washer-shaped stainless-steel electrode, 4 mm $\times$ 15 mm, is laid parallel to the plate at a distance of 6 mm and connected to the positive pole of a high-voltage supply, which can provide up to 8 kV DC. The resulting electric field in the cavity has an average strength of 13.3×10^3 V cm^{-1}. The shape of the obtained droplets is shown in Figure 27.6. The drops subjected to an electric field are elongated in the electric field direction due to the alteration of the interface curvature by the local electric stress, as reported previously. The contact angle and the curvature at the drop apex increase with increasing electric field strength.

The profile of the droplets and their geometric parameters (contact angle, drop dimensions, and volume) are treated by a dedicated image processing software that can identify and interpolate the droplet profile. These measurements make it possible to evaluate the resulting forces acting on the droplet due to weight, adhesion, and internal pressure. In the absence of an electric field, the experimental values of the force acting on the drop agree well with the theoretical predictions and this validated the method. Therefore, in the presence of electric field, the resulting electric force on the droplet can be evaluated by difference and compared with theoretical models.

Glossary

Permittivity measure of the resistance that is encountered when forming an electric field in a medium.

References

Adamiak, K., 2001. Interaction of two dielectric or conducting droplets aligned in the uniform electric field. J. Electrostat. 51–52, 578–584.

Ahmedou, S.A.O., Rouaud, O., 2009. Assessment of the electrohydrodynamic drying process. Food Bioprocess Technol. 2, 240–247.

Basaran, O.A., Scriven, L.E., 1990. Axisymmetric shapes and stability of pendant and sessile drops in an electric field. J. Colloid Interface Sci. 140, 10–30.

Basaran, O.A., Wohlhuter, F.K., 1992. Effect of nonlinear polarization on shapes and stability of pendant and sessile drops in an electric (magnetic) field. J. Fluid Mech. 244, 1–16.

Bateni, A., Susnar, S.S., Amirfazli, A., Neumann, A.W., 2004. Development of a new methodology to study drop shape and surface tension in electric fields. Langmuir 20, 7589–7597.

Bateni, A., Laughton, S., Tavana, H., Susnar, S.S., Amirfazli, A., Neumann, A.W., 2005a. Effect of electric fields on contact angle and surface tension of drops. J. Colloid Interface Sci. 283, 215–222.

Bateni, A., Ababneh, A., Elliott, J.A.W., Neumann, A.W., Amirfazli, A., 2005b. Effect of gravity and electric field on shape and surface tension of drops. Adv. Space Res. 36, 64–69.

Bateni, A., Amirfazli, A., Neumann, A.W., 2006. Effect of an electric field on the surface tension conducting drops. Colloids Surf. A Physiochem. Eng. Asp. 289, 25–38.

Berge, B., Peseux, J., 2000. Variable focal lens controlled by an external voltage: an application of electrowetting. Eur. Phys. J. E 3, 159–163.

Bologa, M.K., Pushkov, V.V., Berkov, A.B., 1985. Electric field induced heat transfer enhancement in a gas–solid suspension heat exchanger. Int. J. Heat Mass Transf. 28, 1245–1255.

Borzabadi, E., Bailey, A.G., 1978. The profiles of axially symmetric electrified pendant drops. J. Electrostat. 5, 369–380.

Brutin, D., Sobac, B., Rigollet, F., Le Nilliot, C., 2012. Infrared visualization of thermal motion inside a sessile drop deposited onto a heated surface. Exp. Therm. Fluid Sci. 35, 521–530.

Chakraborty, D., Sai Sudha, G., Chakraborty, S., DasGupta, S., 2011. Effect of submicron particles on electrowetting on dielectrics (EWOD) of sessile droplets. J. Colloid Interface Sci. 363, 640–645.

Cho, S.K., Moon, H., Kim, C.J., 2003. Creating, transporting, cutting, and merging liquid droplets by electrowetting-based actuation for digital microfluidic circuits. J. Microelectromech. Syst. 12, 70–80.

David, S., Sefiane, K., Shanahan, M.E.R., 2006. On the dynamics of wetting of water–methanol volatile sessile drops on smooth substrates. Defect Diffus. Forum 258–260, 469–473.

David, S., Sefiane, K., Tadrist, L., 2007. Experimental investigation of the effect of thermal properties of the substrate in the wetting and evaporation of sessile drops. Colloids Surf. A Physiochem. Eng. Asp. 298, 108–114.

Deegan, R.D., Bakajin, O., Dupont, T.F., Hubert, G., Nagel, S.R., Witten, T.A., 2000. Contact line deposits in an evaporating drop. Phys. Rev. E 62, 756–765.

Di Marco P., Pedretti F., Saccone G., 2013. Effect of an external electric field on the shape of a dielectric sessile drop Transfer, Fluid Mechanics, and Thermodynamics. In: Eighth World Conference on Experimental Heat Transfer, June 16–20, 2013, Lisbon, Portugal.

Digilov, R., 2000. Charge-induced modification of contact: the secondary electrocapillary effect. Langmuir 16, 6719–6723.

Drygiannakis, A.I., Papathanasiou, A.G., Boudouvis, A.G., 2009. Manipulation equilibrium shape transitions of microdroplets in electrowetting—a computational analysis. Microelectr. Eng. 86, 1365–1367.

Dunn, G.J., Wilson, S.K., Duffy, B.R., David, S., Sefiane, K., 2008. A mathematical model for the evaporation of a thin sessile liquid droplet: comparison between experiment and theory. Colloids Surf. A Physiochem. Eng. Asp. 323, 50–55.

Eckert, E.R.G., Goldstein, R.J., Ibele, W.E., Patankar, S.V., Simon, T.W., Strykowski, P.J., et al., 1996. Heat transfer—a review of 1993 literature. Int. J. Heat Mass Transf. 39, 885–963.

Gliere, A., Roux, J.M., Achard, J.L., 2013. Lift-off of a Conducting Sessile Drop in an Electric Field Microfluid Nanofluid. Published online Available from: http://dx.doi.org/10.1007/s10404-013-1144-6 (February).

Harris, M.T., Basaran, O.A., 1995. Equilibrium shapes and stability of nonconducting pendant drops surrounded by a conducting fluid in an electric field. J. Colloid Interface Sci. 170, 308–319.

Hu, H., Larson, R.G., 2002. Evaporation of a sessile droplet on a substrate. J. Phys. Chem. B 106, 1334–1344.

Jones, T.B., 2009. More about the electromechanics of electrowetting. Mech. Res. Commun. 36, 2–9.

Kang, K.H., 2002. How electrostatic fields change contact angle in electrowetting. Langmuir 18, 10318–10322.

Landau, D., Lifsitz, E.M., 1984. Electrodynamics of Continuous Media. Pergamon, New York, NY.

Lee, J., Moon, H., Fowler, J., Schoellhammer, T., Kim, C.-J., 2002. Electrowetting and electrowetting-on-dielectric for microscale liquid handling. Sens. Actuat. A 95, 259–268.

Lee, S.-L., Yang, C.-F., 2008. Numerical simulation for meniscus shape and optical performance of a MEMS-based liquid micro-lens. Opt. Express 16.

Liu, Y., Oh, K., Bai, J.G., Chang, C.-L., Yeo, W., Chung, J.-H., et al., 2008. Manipulation of nanoparticles and biomolecules by electric field and surface tension. Comput. Methods Appl. Mech. Eng. 197, 2156–2172.

Melcher, J.R., Taylor, G.I., 1969. Electrohydrodynamics: a review of the role of interfacial shear stresses. Ann. Rev. Fluid Mech. 1, 111.

Miksis, M.J., 1981. Shape of a drop in an electric field. Phys. Fluids 24, 1967.

Moon, H., 2002. Low voltage electrowetting-on-dielectric. J. Appl. Phys. 92, 4080–4087.

Mugele, F., Baret, J.-C., 2005. Electrowetting: from basics to applications. J. Phys. Condens. Matter 17, R705–R774.

Mugele, F., Duits, M., van den Ende, D., 2010. Electrowetting: a versatile tool for drop manipulation, generation, and characterization. Adv. Colloid Interface Sci. 161, 115–123.

Paik, P., Pamula, V.K., Pollack, M.G., Fair, R.B., 2003. Electrowetting-based droplet mixers for microfluidic systems. Lab Chip 3, 28–33.

Panofsky, W.K., Phillips, M., 1962. Classical Electricity and Magnetism. Addison-Wesley, Reading.

Peykov, V., Quinn, A., Ralston, J., 2000. Electrowetting: a model for contact-angle saturation. Colloid Polym. Sci. 278, 789–793.

Picknett, R.G., Bexon, R., 1977. The evaporation of sessile or pendant drops in still air. J. Colloid Interface Sci. 61, 336–350.

Pollack, M.G., Fair, R.B., Shenderov, A.D., 2000. Electrowetting-based actuation of liquid droplets for microfluidic applications. Appl. Phys. Lett. 77, 1725–1726.

Popov, Y.O., 2005. Evaporative deposition patterns: spatial dimensions of the deposit. Phys. Rev. E 71, 036313.

Quilliet, C., Berge, B., 2001. Electrowetting: a recent outbreak. Curr. Opin. Colloid Interface Sci. 6, 34–39.

Quinn, A., Sedev, R., Ralston, J., 2005. Contact angle saturation in electrowetting. J. Phys. Chem. B 109, 6268–6275.

Rayleigh, L., 1882. On the equilibrium of liquid conducting masses charged with electricity. Philos. Mag. 14, 184–195.

Reznik, S.N., Yarin, A.L., Theron, A., Zussman, E., 2004. Transient and steady shapes of droplets attached to a surface in a strong electric field. J. Fluid Mech. 516, 349–377.

Roero, C., 2006. Contact angle measurements of sessile drops deformed by a DC electric field. Cont. Angle Wettab. Adhes. 4, 165–176.

Rosenkilde, C.E., 1969. A dielectric fluid drop in an electric field. Proc. R. Soc. Lond. A 3112, 473.

Roux, J.M., Achard, J.L., 2009. Forces and charges on a slightly deformed droplet in the DC field of a plate condenser. J. Electrostat. 67, 789–798.

Sedev, R., 2011. Electrowetting: electrocapillarity, saturation, and dynamics. Eur. Phys. J. Spec. Top. 197, 307–319.

Semenov, S., Starov, V.M., Rubio, R.G., Agogo, H., Velarde, M.G., 2011. Evaporation of sessile water droplets: universal behavior in presence of contact angle hysteresis. Colloids Surf. A Physiochem. Eng. Asp. 391, 135–144.

Shamai, R., Andelman, D., Berge, B., Hayes, R., 2007. Water, electricity, and between... On electrowetting and its applications. Soft Matter 4, 38–45.

Sobac, B., Brutin, D., 2011. Triple-line behavior and wettability controlled by nanocoated substrates: influence on sessile drop evaporation. Langmuir 27, 14999–15007.

Sobac, B., Brutin, D., 2012. Thermal effects of the substrate on water droplet evaporation. Phys. Rev. E 86, 021602.

Takano, K., Tanasawa, I., Nishio S. Enhancement of a droplet from a hot surface by electric field. In: Proceedings of the Ninth International Heat Transfer Conference, vol. 4, pp. 69–74,90.

Wohlhuter, F.K., Basaran, O.A., 1992. Shapes and stability of pendant and sessile dielectric drops in an electric field. J. Fluid Mech. 235, 481–510.

Index

Note: Page numbers followed by "*f*" and "*t*" refer to figures and tables, respectively.

D

Kolibri 360 GmbH
Ferdinand-Jühlke-Straße 7
99095 Erfurt, Deutschland
produktsicherheit@kolibri360.de